COMPUTERS IN EARTH AND ENVIRONMENTAL SCIENCES

COMPUTERS IN EARTH AND ENVIRONMENTAL SCIENCES

Artificial Intelligence and Advanced Technologies in Hazards and Risk Management

Edited by

HAMID REZA POURGHASEMI

Associate Professor, Watershed Management Engineering, College of Agriculture, Shiraz University, Shiraz, Iran

ELSEVIER

Elsevier
Radarweg 29, PO Box 211, 1000 AE Amsterdam, Netherlands
The Boulevard, Langford Lane, Kidlington, Oxford OX5 1GB, United Kingdom
50 Hampshire Street, 5th Floor, Cambridge, MA 02139, United States

Notices
Knowledge and best practice in this field are constantly changing. As new research and experience broaden our
understanding, changes in research methods, professional practices, or medical treatment may become
necessary.

Practitioners and researchers must always rely on their own experience and knowledge in evaluating and using
any information, methods, compounds, or experiments described herein. In using such information or methods
they should be mindful of their own safety and the safety of others, including parties for whom they have a
professional responsibility.

To the fullest extent of the law, neither the Publisher nor the authors, contributors, or editors, assume any liability
for any injury and/or damage to persons or property as a matter of products liability, negligence or otherwise, or
from any use or operation of any methods, products, instructions, or ideas contained in the material herein.

Library of Congress Cataloging-in-Publication Data
A catalog record for this book is available from the Library of Congress

British Library Cataloguing-in-Publication Data
A catalogue record for this book is available from the British Library

ISBN: 978-0-323-89861-4

For information on all Elsevier publications
visit our website at https://www.elsevier.com/books-and-journals

Publisher: Candice Janco
Acquisitions Editor: Peter J. Llewellyn
Editorial Project Manager: Michelle Fisher
Production Project Manager: Vijayaraj Purushothaman
Cover Designer: Miles Hitchen

Typeset by STRAIVE, India

Working together
to grow libraries in
developing countries

www.elsevier.com • www.bookaid.org

Dedication

This book is dedicated to my wife (Kimia) for all her help and my son (Yara).

Contents

Contributors

Sohaib K.M. Abujayyab Department of Geography, Karabuk University, Karabuk, Turkey

Maryam Aghaei School of Physical Geography, University of Mohaghegh Ardabili, Ardabil, Iran

Adeel Ahmad Department of Geography, University of the Punjab, Lahore, Pakistan

Fatemeh Ahmadloo Poplar and Fast-growing Trees Research Division, Research Institute of Forests and Rangelands, Agricultural Research, Education and Extension Organization (AREEO), Tehran, Iran

Stavros G. Alexandris Department of Natural Resources Development and Agricultural Engineering, Agricultural University of Athens, Athens, Greece

Shakir Ali Department of Geology, University of Delhi, Delhi, India

Sajjad Ali Mahmoudi Sarab Khuzestan Agricultural and Natural Resources Research and Education Center, Agricultural Research, Education and Extension Organization (AREEO), Tehran, Khuzestan, Iran

Amir Alizadeh Spatial Sciences Innovators Consulting Engineering Company, Tehran, Iran

Hadi Alizadeh Spatial Sciences Innovators Consulting Engineering Company, Tehran, Iran

Meisam Amani Wood Environment & Infrastructure Solutions, Ottawa, ON, Canada

Ouafi Ameur-Zaimeche Laboratory of Underground Reservoirs: Petroleum, Gas and Aquifers, University of Kasdi Merbah Ouargla, Ouargla, Algeria

Abdulfattah Ahmad Amin Department of Road Construction, Erbil Technology College, Erbil Polytechnic University, Erbil, Kurdistan Region, Iraq

Atiyeh Amindin Department of Natural Resources and Environmental Engineering, College of Agriculture, Shiraz University, Shiraz, Iran

Mahdis Amiri Department of Watershed and Arid Zone Management, Gorgan University of Agricultural Sciences and Natural Resources, Gorgan, Iran

Charaf-Eddine Aouam Faculty of Hydrocarbons, Renewable Energies and Earth Sciences and the Universe, University of Kasdi Merbah Ouargla, Ouargla, Algeria

Ebrahim Asgari Department of Watershed Management Engineering, Faculty of Natural Resources, Yazd University, Yazd Province, Iran

Sedigheh Babaei Department of Natural Resources and Environmental Engineering, College of Agriculture, Shiraz University, Shiraz, Iran

Yasser Baleghi Department of Electrical and Computer Engineering, Babol Noshirvani University of Technology, Babol, Iran

Abdollah Bameri Natural Resources and Watershed Management Organization of Sistan and Baluchestan Province, Zahedan, Iran

Anil Bhardwaj Department of Soil and Water Engineering, Punjab Agricultural University, Ludhiana, India

Mojgan Bordbar Department of GIS/RS, Faculty of Natural Resources and Environment, Science and Research Branch, Islamic Azad University, Tehran, Iran

Ali Boustan Civil Engineering Department, Islamic Azad University, Kerman Branch, Kerman, Iran

Trupti Chandrasekhar Department of Earth Sciences, IIT Bombay, Mumbai, India

Songchao Chen INRAE, Unité InfoSol, Orléans, France

Asheer Chhetri Civil Engineering Department, College of Science and Technology, Rinchending, Bhutan

Christian Conoscenti Department of Earth and Marine Sciences (DISTEM), University of Palermo, Palermo, Italy

Manju Sara Dahal Civil Engineering Department, College of Science and Technology, Rinchending, Bhutan

Ali Danandeh Mehr Department of Civil Engineering, Antalya Bilim University, Antalya, Turkey

Saha Dauji Nuclear Recycle Board, Bhabha Atomic Research Centre; Homi Bhabha National Institute, Mumbai, India

Diego García de Jalón Department of Systems and Natural Resources, ETSI of Mountains, Forest and the Natural Environment, Polytechnic University of Madrid, Madrid, Spain

José A.M. Demattê Department of Soil Science, Luiz de Queiroz College of Agriculture, University of São Paulo, São Paulo, Brazil

Payam Ebrahimi Forests, Rangelands, and Watershed Research Department, Sistan Agricultural and Natural Resources Research and Education Center AREEO, Zabol, Iran

Mohammad Reza Ekhtesasi Faculty of Natural Resources and Desert Studies, Yazd University, Yazd, Iran

Ahmed Elbeltagi College of Environmental and Resource Sciences, Zhejiang University, Hangzhou, China; Agricultural Engineering Department, Faculty of Agriculture, Mansoura University, Mansoura, Egypt

Sayed Naeim Emami Soil Conservation and Watershed Management Research Department, Chaharmahal and Bakhtiari Agricultural and Natural Resources Research and Education Center, AREEO, Shahrekord, Iran

Saeedeh Eskandari   Forest Research Division, Research Institute of Forests and Rangelands, Agricultural Research, Education and Extension Organization (AREEO), Tehran, Iran

Seyed Rashid Fallah Shamsi   Department of Natural Resources and Environmental Engineering, College of Agriculture, Shiraz University, Shiraz, Iran

Zakariya Farajzadeh   Department of Agricultural Economics, College of Agriculture, Shiraz University, Shiraz, Iran

Hassan Fathizad   Department of Arid and Desert Regions Management, College of Natural Resources and Desert, Yazd University, Yazd, Iran

Khosro Fazelpoor   Department of Systems and Natural Resources, ETSI of Mountains, Forest and the Natural Environment, Polytechnic University of Madrid, Madrid, Spain

Shilan Felegari   Department of Soil Science, Faculty of Agriculture, University of Zanjan, Zanjan, Iran

G.P. Ganapathy   VIT University, Vellore, Tamil Nadu, India

E. Gayathiri   Guru Nanak College (Autonomous), Chennai, Tamil Nadu, India

Hemant Ghalley   Civil Engineering Department, College of Science and Technology, Rinchending, Bhutan

Gholamabbas Ghanbarian   Department of Natural Resources and Environmental Engineering, College of Agriculture, Shiraz University, Shiraz, Iran

Arsalan Ghorbanian   Department of Photogrammetry and Remote Sensing, Faculty of Geodesy and Geomatics Engineering, K. N. Toosi University of Technology, Tehran, Iran

Hammad Gilani   Department of Space Science, Institute of Space Technology, Islamabad, Pakistan

R. Gobinath   SR University, Warangal, Telangana, India

Ahmad Golchin   Department of Soil Science, Faculty of Agriculture, University of Zanjan, Zanjan, Iran

Abbas Goli Jirandeh   Spatial Sciences Innovators Consulting Engineering Company, Tehran, Iran

Mohsen Golian   Tehran Science and Research Branch, Islamic Azad University, Tehran, Iran

Farshad Haghighian   Research Division of Natural Resources, Chaharmahal and Bakhtiari Agricultural and Natural Resources Research and Education Center, AREEO, Shahrekord, Iran

Salim Heddam   Faculty of Science, Agronomy Department, Hydraulics Division, Laboratory of Research in Biodiversity Interaction Ecosystem and Biotechnology, University 20 Août 1955, Skikda, Algeria

Anis Heidari   Department of Reclamation of Arid and Mountainous Regions, Faculty of Natural Resources, University of Tehran, Karaj, Iran

Bahram Heidari   Department of Plant Production and Genetics, School of Agriculture, Shiraz University, Shiraz, Iran

Mohsen Hosseinalizadeh   Department of Arid Zone Management, Gorgan University of Agricultural Sciences and Natural Resources, Gorgan, Iran

Hooshyar Hossini   Department of Environmental Health Engineering, Faculty of Health, Kermanshah University of Medical Sciences, Kermanshah, Iran

J. Jayanthi   Guru Nanak College (Autonomous), Chennai, Tamil Nadu, India

Mohammad Kakooei   Department of Electrical and Computer Engineering, Babol Noshirvani University of Technology, Babol, Iran

Kleomenis Kalogeropoulos   Department of Geography, Harokopio University of Athens, Athens, Greece

Sahar Karami   Quantitative Plant Ecology and Biodiversity Research Lab, Department of Biology, Faculty of Science, Ferdowsi University of Mashhad, Mashhad, Iran

Christos A. Karavitis   Department of Natural Resources Development and Agricultural Engineering, Agricultural University of Athens, Athens, Greece

Alireza Karimi   Department of Soil Science, Faculty of Agriculture, Ferdowsi University of Mashhad, Mashhad, Iran

Hazhir Karimi   Department of Environmental Science, Faculty of Science, University of Zakho, Duhok, Kurdistan Region, Iraq

Mahdi Karimi   Spatial Sciences Innovators Consulting Engineering Company, Tehran, Iran

Narges Kariminejad   Department of Arid Zone Management, Gorgan University of Agricultural Sciences and Natural Resources, Gorgan, Iran

Abdullah Kaviani Rad   Department of Soil Science, School of Agriculture, Shiraz University, Shiraz, Iran

Rabah Kechiched   Laboratory of Underground Reservoirs: Petroleum, Gas and Aquifers, University of Kasdi Merbah Ouargla, Ouargla, Algeria

Tirumalesh Keesari   Isotope and Radiation Application Division, Bhabha Atomic Research Centre; Homi Bhabha National Institute, Mumbai, India

Farhad Khormali   Department of Soil Science, Gorgan University of Agricultural Sciences and Natural Resources, Gorgan, Iran

Mojdeh Mohammadi Khoshoui   Watershed Management Engineering, Faculty of Natural Resources and Desert Studies, Yazd University, Yazd, Iran

Mahboobeh Kiani-Harchegani   Department of Watershed Management Engineering, Faculty of Natural Resources, Yazd University, Yazd Province, Iran

Sungwon Kim   Department of Railroad Construction and Safety Engineering, Dongyang University, Yeongju, Republic of Korea

Ozgur Kisi   Department of Civil Engineering, School of Technology, Ilia State University, Tbilisi, Georgia

Aiding Kornejady   Spatial Sciences Innovators Consulting Engineering Company, Tehran, Iran

Elham Kouchaki   Water and Hydraulic Structure Department, Faculty of Civil Engineering, Islamic Azad University, Estahban Branch, Estahban, Iran

Petr Kubíček   Department of Geography, Masaryk University, Brno, Czech Republic

Luigi Lombardo   Faculty of Geo-Information Science and Earth Observation (ITC), University of Twente, Enschede, Netherlands

Sandhya Makkar   Lal Bahadur Shastri Institute of Management, Delhi, India

Sedigheh Maleki   Department of Soil Science, Gorgan University of Agricultural Sciences and Natural Resources; Department of Soil Science, Faculty of Agriculture, Ferdowsi University of Mashhad, Mashhad, Iran

Anurag Malik   Punjab Agricultural University, Regional Research Station, Bathinda, Punjab, India

Iman Mallakpour   Department of Civil and Environmental Engineering, University of California, Irvine, CA, United States

Vanesa Martínez-Fernández   National Museum of Natural Sciences, CSIC, Madrid, Spain

Masoud Masoudi   Department of Natural Resources and Environmental Engineering, College of Agriculture, Shiraz University, Shiraz, Iran

Ahmad Reza Mehrabian   Department of Plant Sciences and Biotechnology, Faculty of Life Sciences and Biotechnology, Shahid Beheshti University, Tehran, Iran

Majid Mohammady   Faculty of Natural Resources, Semnan University, Semnan, Iran

Marzieh Mokarram   Department of Range and Watershed Management, College of Agriculture and Natural Resources of Darab, Shiraz University, Shiraz, Iran

Ali Akbar Moosavi   Department of Soil Science, College of Agriculture, Shiraz University, Shiraz, Iran

Kamran Moravej   Department of Soil Science, Faculty of Agriculture, University of Zanjan, Zanjan, Iran

Hamidreza Mosaffa   Department of Water Engineering, Shiraz University, Shiraz, Iran

Hossein Mostafavi   Department of Biodiversity and Ecosystems Management, Research Institute of Environmental Sciences, Shahid Beheshti University, Tehran, Iran

Raoof Mostafazadeh   School of Natural Resources Management; School of Rangeland and Watershed Management, University of Mohaghegh Ardabili, Ardabil, Iran

Hasan Mozaffari   Department of Soil Science, College of Agriculture, Shiraz University, Shiraz, Iran

Mojtaba Naghdyzadegan Jahromi   Department of Water Engineering, School of Agriculture, Shiraz University, Shiraz, Iran

Zeynab Najafi   Department of Natural Resources and Environmental Engineering, College of Agriculture, Shiraz University, Shiraz, Iran

Farzaneh Khajoei Nasab   Department of Plant Sciences and Biotechnology, Faculty of Life Sciences and Biotechnology, Shahid Beheshti University, Tehran, Iran

Andrea Nascetti   Geoinformatics Division, KTH Royal Institute of Technology, Stockholm, Sweden

Hassan Khavarian Nehzak   School of Physical Geography, University of Mohaghegh Ardabili, Ardabil, Iran

Mohammad Nekooeimehr   Soil Conservation and Watershed Management Research Department, Chaharmahal and Bakhtiari Agricultural and Natural Resources Research and Education Center, AREEO, Shahrekord, Iran

Majid Niazkar   Department of Civil and Environmental Engineering, Shiraz University, Shiraz, Iran

Rigzin Norbu   Civil Engineering Department, College of Science and Technology, Royal University of Bhutan, Rinchending, Bhutan

Panagiotis D. Oikonomou   Vermont EPSCoR; Gund Institute for Environment, University of Vermont, Burlington, VT, United States

Yaser Ostovari   Chair of Soil Science, Research Department of Ecology and Ecosystem Management, TUM-School of Life Sciences Weihenstephan, Technical University of Munich, Freising, Germany

Sina Paryani   Department of GIS/RS, Faculty of Natural Resources and Environment, Science and Research Branch, Islamic Azad University, Tehran, Iran

Sangey Pasang   Department of Geography, Masaryk University, Brno, Czech Republic; Civil Engineering Department, College of Science and Technology, Royal University of Bhutan, Rinchending, Bhutan

Moujhuri Patra   Civil Engineering Department, College of Science and Technology, Rinchending, Bhutan

Raúl Roberto Poppiel   Department of Soil Science, Luiz de Queiroz College of Agriculture, University of São Paulo, São Paulo, Brazil

Hamid Reza Pourghasemi   Department of Natural Resources and Environmental Engineering, College of Agriculture, Shiraz University, Shiraz, Iran

Mehdi Pourhashemi   Forest Research Division, Research Institute of Forests and Rangelands, Agricultural Research, Education and Extension Organization (AREEO), Tehran, Iran

Soheila Pouyan   Department of Natural Resources and Environmental Engineering, College of Agriculture, Shiraz University, Shiraz, Iran

Vishnu Prasad   Department of Soil and Water Engineering, Punjab Agricultural University, Ludhiana, India

Hamidreza Rabiei-Dastjerdi   School of Computer Science and CeADAR, University College Dublin (UCD), Dublin, Belfield, Dublin 4, Ireland

M.G. Ragunathan   Guru Nanak College (Autonomous), Chennai, Tamil Nadu, India

Soroor Rahmanian   Quantitative Plant Ecology and Biodiversity Research Lab, Department of Biology, Faculty of Science, Ferdowsi University of Mashhad, Mashhad, Iran

Ahmad Rastegarnia   Department of Geology, Faculty of Science, Ferdowsi University of Mashhad, Mashhad, Iran

Jesús Rodrigo-Comino   Departamento de Análisis Geográfico Regional y Geografía Física, Facultad de Filosofía y Letras, Campus Universitario de Cartuja, University of Granada, Granada, Spain

Mojtaba Sadeghi   Department of Civil and Environmental Engineering, University of California, Irvine, CA, United States

Shahriar Sadeghi   Department of Geology, Faculty of Science, Imam Khomeini International University, Qazvin, Iran

Majid Sadeghinia   Department of Nature Engineering, Faculty of Agriculture and Natural Resources, Ardakan University, Ardakan, Iran

Ashwini Arun Salunkhe   SR University, Warangal, Telangana; Dr. D. Y. Patil Institute of Technology, Pimpri, Pune, India

Arnold R. Salvacion   Department of Community and Environmental Resource Planning, College of Human Ecology, University of the Philippines Los Baños, Los Baños, Laguna, Philippines

Mahmood Samadi   Spatial Sciences Innovators Consulting Engineering Company, Tehran, Iran

Alireza Sarvarinezhad   Natural Resources and Watershed Management Organization of Sistan and Baluchestan Province, Zahedan, Iran

Erhan Şener   Remote Sensing Center, Suleyman Demirel University, Isparta, Turkey

Şehnaz Şener   Department of Geological Engineering, Suleyman Demirel University, Isparta, Turkey

Munawar Shah   Space Education and GNSS Lab, National Center of GIS and Space Application, Institute of Space Technology, Islamabad, Pakistan

Alireza Sharifi   Department of Surveying engineering, Faculty of civil engineering, Shahid Rajaee Teacher Training University, Tehran, Iran

Ebrahim Sharifi Teshnizi   Department of Geology, Faculty of Science, Ferdowsi University of Mashhad, Mashhad, Iran

Ifrah Shaukat   Geography of Environmental Resources and Human Security, United Nations University, Bonn, Germany

Shashank Shekhar   Department of Geology, University of Delhi, Delhi, India

Safdar Ali Shirazi   Department of Geography, University of the Punjab, Lahore, Pakistan

Esmaeil Silakhori   Spatial Sciences Innovators Consulting Engineering Company, Tehran, Iran

Sukhdeep Singh   Department of Soil and Water Engineering, Punjab Agricultural University, Ludhiana, India

U. Sinthuja   Hindusthan College of Arts and Science; Sri Ramakrishna College of Arts and Science, Coimbatore, Tamil Nadu, India

Nikolaos Stathopoulos   Institute for Space Applications and Remote Sensing, National Observatory of Athens, BEYOND Centre of EO Research & Satellite Remote Sensing, Athens, Greece

Ruhollah Taghizadeh-Mehrjardi   Department of Geosciences, Soil Science and Geomorphology; CRC 1070 Resource Cultures, University of Tübingen, Tübingen, Germany; Faculty of Agriculture and Natural Resources, Ardakan University, Ardakan, Iran

Ali Talebi   Department of Watershed Management Engineering, Faculty of Natural Resources, Yazd University, Yazd Province, Iran

Enes Taşoğlu   Department of Geography, Karabuk University, Karabuk, Turkey

Mahboobeh Tayebi   Department of Soil Science, Luiz de Queiroz College of Agriculture, University of São Paulo, São Paulo, Brazil

S. Thavamani   Sri Ramakrishna College of Arts and Science, Coimbatore, Tamil Nadu, India

John P. Tiefenbacher   Department of Geography, Texas State University, San Marcos, TX, United States

Suren Timsina   Civil Engineering Department, College of Science and Technology, Royal University of Bhutan, Rinchending, Bhutan

Demetrios E. Tsesmelis   Laboratory of Technology and Policy of Energy and Environment, School of Science and Technology, Hellenic Open University, Patra; Department of Natural Resources Development and Agricultural Engineering, Agricultural University of Athens, Athens, Greece

Simge Varol   Department of Geological Engineering, Suleyman Demirel University, Isparta, Turkey

Constantina G. Vasilakou   Department of Natural Resources Development and Agricultural Engineering, Agricultural University of Athens, Athens, Greece

Tshering Wangchuk   Civil Engineering Department, College of Science and Technology, Royal University of Bhutan, Rinchending, Bhutan

Saleh Yousefi   Soil Conservation and Watershed Management Research Department, Chaharmahal and Bakhtiari Agricultural and Natural Resources Research and Education Center, AREEO, Shahrekord, Iran

Abrar Yousuf   Department of Soil and Water Engineering, Punjab Agricultural University, Ludhiana, India

Sara Zakeri-Anaraki   Desert Region Management, College of Agriculture, Shiraz University, Shiraz, Iran

Mohammad Zakwan   Civil Engineering Department, IIT Roorkee, Roorkee; Civil Engineering Department, MANUU, Hyderabad, India

Mehdi Zarei   Department of Soil Science, School of Agriculture, Shiraz University, Shiraz; Department of Agriculture and Natural Resources, Higher Education Center of Eghlid, Eghlid, Iran

Efthimios Zervas   Laboratory of Technology and Policy of Energy and Environment, School of Science and Technology, Hellenic Open University, Patra, Greece

Mohammad Zounemat-Kermani   Department of Water Engineering, Shahid Bahonar University of Kerman, Kerman, Iran

Acknowledgments

This book was supported by the Iran National Science Foundation (INSF) under Grant No. 99011055. Thanks to INSF.

Also, the editor Dr. Hamid Reza Pourghasemi gratefully thanks the College of Agriculture, Shiraz University.

1

Predicting dissolved oxygen concentration in river using new advanced machines learning: Long-short term memory (LSTM) deep learning

Salim Heddam[a], Sungwon Kim[b], Ali Danandeh Mehr[c], Mohammad Zounemat-Kermani[d], Anurag Malik[e], Ahmed Elbeltagi[f,g], and Ozgur Kisi[h]

[a]Faculty of Science, Agronomy Department, Hydraulics Division, Laboratory of Research in Biodiversity Interaction Ecosystem and Biotechnology, University 20 Août 1955, Skikda, Algeria [b]Department of Railroad Construction and Safety Engineering, Dongyang University, Yeongju, Republic of Korea [c]Department of Civil Engineering, Antalya Bilim University, Antalya, Turkey [d]Department of Water Engineering, Shahid Bahonar University of Kerman, Kerman, Iran [e]Punjab Agricultural University, Regional Research Station, Bathinda, Punjab, India [f]College of Environmental and Resource Sciences, Zhejiang University, Hangzhou, China [g]Agricultural Engineering Department, Faculty of Agriculture, Mansoura University, Mansoura, Egypt [h]Department of Civil Engineering, School of Technology, Ilia State University, Tbilisi, Georgia

1 Introduction

Given the major contribution of the physical and chemical variables in providing a direct or indirect indication of freshwater quality, their continuous control monitoring has become a majority-owned by water managers worldwide.[1-3] By correlating and aggregating several water quality variables, a large number of robust and accurate water quality indices (WQI) have been proposed worldwide.[4] However, all reported WQI values in the literature were mainly based on and included the most important variables, i.e., that is dissolved oxygen concentration (DO), biochemical oxygen demand (BOD), chemical oxygen demand (COD), and river water temperature.[5,6] The most commonly used water quality variable is DO concentration, although it has long been linked to water temperature (T_w), and its concentration in water is a key indicator of the physicochemical properties of a water body.[7] However, the contribution of DO as a basis for calculating WQI models has been discussed and analyzed in many studies.[6,8] Additionally, previous studies have shown that DO is strongly related to hydrological and weather variables (i.e., air temperature). Clearly, DO has been subject to significant debate, and reliable and robust estimates of DO concentration remain challenging issues.[9] Over the years, machine learning algorithms to estimate DO have been applied successfully, where the DO is linked to a suite of other variables.

Heddam[10] proposed an extremely randomized tree (ERT) for predicting DO concentration measured at 15 min, using data collected from two stations located in Charles and Mystic Rivers, USA. For comparison, results obtained using the random forest (RF), multilayer perceptron neural network (MLPNN), and the MLR models were compared to those obtained using the proposed ERT model. All models were developed using water temperature (T_w), water pH, specific conductance (SC), and phycocyanin pigment concentration (PC). From the obtained results, the RF model was the most accurate and exhibited very high accuracy with correlation coefficient (R) and Nash-Sutcliffe index (NSE) approximately superior to 0.99 and 0.98, respectively. Zhu et al.[11] introduced a new and innovative modeling strategy called transfer learning, which is mainly based on the bidirectional long short-term memory (BiLSTM) neural network for predicting the DO concentration in an aquatic system. The model was developed using data collected from two lakes in China namely, Lake Taihu and a Lake near Yixing, and water T_w, water pH, SC, turbidity (TU), chlorophyll concentration (CHL-a), and blue-green algae (BGA) were used as input variables. The BiLSTM was calibrated using data from the first station and tested using data of the second station, which is called transfer learning, and high accuracy was obtained with R^2 and NSE of 0.983 and 0.979. Zhu and Heddam[12] compared the MLPNN and extreme learning machine (ELM) models in predicting DO concentration in four rivers in China, namely Yipin, Huaxi, Wubu, and Tributary of the Huaxi River, and reported that the two approaches worked relatively equally with slight superiority to the ELM model. Antanasijević et al.[13] advocated a new philosophy for modeling DO concentration, which is adapted for a large number of neighboring stations. They proposed the use of a self-organizing network similarity index (LSI) integrated with Ward neural networks (WNNs) for in-depth multilevel splitting. High accuracies were achieved and an R^2 value of 0.917 was obtained. Nacar et al.[14] used three models to predict DO concentrations using data collected at two USGS stations located in the Broad River, USA. The employed models were the standard multiple linear regression (MLR), multivariate adaptive regression splines (MARS), and TreeNet gradient boosting machine. By including three water quality variables (i.e., T_w, pH, and SC), excellent results were obtained with NSE values higher than 0.980. A Hybrid model combining the teaching-learning optimization (TLBO) algorithm and regression analysis (MLR) was proposed by Nacar et al.[15] and applied for modeling the DO concentration in the Eastern Black Sea Basin stream, Turkey. The Results obtained using the MLR-TLBO were compared to those obtained using the MARS models, highlighting the superiority of the MARS model (NSE $\approx$ 0.940).

Ren et al.[16] compared deep belief networks (DBN), bagging, AdaBoost, decision tree (DT) and convolutional neural network (CNN) to predict DO concentrations measured in aquaculture ponds. The accuracy of the DBN was further improved using variational mode decomposition (VMD). The proposed machine learning models were developed using pH, T_w, TU, ammonia nitrogen, and water level. From the obtained results, it was demonstrated that the best algorithm for DO was the VMD-DBN, which has an NSE of 0.933 superior to the values achieved using bagging (NSE $\approx$ 0.901), AdaBoost (NSE $\approx$ 0.9262), DT (NSE $\approx$ 0.918) and CNN (NSE $\approx$ 0.881). More recently, Cao et al.[17] focused on the application of the gated recurrent unit (GRU) neural network combined with K-means clustering to predict DO in pond cultures. The proposed model was developed using a large number of inputs that contain both water quality and meteorological variables. The K-means-GRU yielded the best accuracies, with RMSE ranging between 0.353 and 1.256. A method based on the ELM approach has recently been introduced by Kuang et al.,[18] who suggested using a hybrid model combining the K-means clustering algorithm (K-means), improved genetic algorithm (IGA), and ELM. The proposed hybrid KIG-ELM was applied to predict DO in an aquaculture pond in Wuxi City, China, and it was demonstrated that the KIG-ELM yielded high accuracies with an NSE value of 0.929 compared to the values of 0.910, 0.877, 0.825, 0.836, and 0.791 obtained by K-means-GA-ELM, K-means-ELM, K-means-LSSVM, IGA-ELM and ELM, respectively. Recent advances in the application of nature-inspired algorithms as learning frameworks for a variety of machine learning models have demonstrated that it is possible to develop a highly robust model for predicting DO concentrations in rivers. Fitting into this framework, Fadaee et al.[19] compared adaptive neuro-fuzzy inference systems (ANFIS) optimized using biogeography-based optimization (BBO-ANFIS); ANFIS optimized using particle swarm optimization (PSO-ANFIS), and ANFIS optimized using butterfly optimization algorithm (BOA), for predicting DO concentrations at the Rock Creek River in Washington, USA. The authors used data measured at 15 min and reported that the performance of the ANFIS model was significantly improved using the nature-inspired algorithm during all seasons, exhibiting an NSE value of approximately 0.98, in the validation phase.

Bayesian model averaging (BMA) was used by Kisi et al.[20] for modeling the hourly DO concentrations using Tw, pH, and SC as input variables. The results obtained using BMA were compared to those obtained using ANFIS, MLPNN, MLR, ELM, and classification and regression tree (CART) models. The BMA yielded high accuracies with NSE values of 0.985 compared to 0.956, 0.941, 0.951, and 0.948 achieved using the ELM, MLPNN, ANFIS, and MLR, while the CART model was slightly superior to all models with an NSE value of 0.986. More recently, Huan et al.[21] proposed the application of the gradient boosting decision tree (GBDT) to select the most significant water quality variables influencing the variation of DO in an aquaculture pond. The results obtained showed that the error metrics used for model evaluation were significantly decreased and the GBDT-LSTM model yielded high accuracies with root mean square error (RMSE) and mean absolute error (MAE) of 0.197 mg/L and 0.299 mg/L, respectively, which are less than the values obtained using the ELM, MLPNN, and PSO-LSSVM models. Li et al.[22] showed that SC was the most significant water quality variable used as input for the SVR model developed for predicting the DO concentrations in the Pearl River located in China. By applying the maximal information coefficient (MIC) as an input selection method, the accuracy of the SVR model was significantly improved by 28.65%, 22.16%, and 56.27% in terms of RMSE, R^2, and NSE values respectively. Zhou[23] compared the LSTM and the hybrid transfer learning long-short term memory (TL-LSTM) model in forecasting DO for up to 10 h in advance. The obtained results illustrate that the TL-LSTM worked significantly more than the LSTM.

Therefore, the overall objective of this study was to propose machine learning models for predicting hourly dissolved oxygen concentrations using different kinds of input variables, that is, water quality (T_w), air quality (T_a), and hydrological variable (Q). This objective was achieved by analyzing two different scenarios: (i) modeling using only T_w, T_a, and Q as input variables, and (ii) by adding the periodicity to the previous variables. The specific objectives were: (1) application of a suite of models belonging to different paradigms, namely, deep learning, Gaussian process, support vector machines, genetic programming, group method of data handling, and the standard multiple regression model, and (2) evaluation and comparison of the accuracies of the proposed models in response to changes in the number and quality of input variables. The second objective was achieved by comparing six different input combinations with different input variables. The most important innovation of the present chapter is that the need to combine water quality with hydrological variable (Q) to obtain accurate and reliable results, was highlighted and justified using advanced machine learning models, and a reliable method can be easily obtained for DO estimation.

2 Materials and methods

2.1 Study area and data

The modeling approaches proposed in the present study were developed using data collected from two United States Geological Survey (USGS) stations (Fig. 1). The two stations were: (i) USGS 15015595 Unuk River below blue river Near Wrangell Alaska, USA (Latitude 56°14′26″, Longitude 130°52′49″NAD27, Hydrologic Unit Code 19010105), (ii) USGS15024800 Stikine River, City and Borough of Wrangell, Alaska, USA (Latitude 56°42′29″, Longitude 132°07′49″NAD27, Hydrologic Unit Code 19010207). Four water quality variables measured at hourly time steps were selected: dissolved oxygen concentration (DO: mg/L), river water temperature (Tw: °C), air temperature (Ta: °C), and river discharge (Q: Ft^3/s). For the USGS 15015595 station, data were measured from 01 January 2020, to November 11, 2020, by removing the incomplete patterns; in total, 5117 data were retained, of which 70% were randomly selected for training (3582 data) and 30% for validation (1535 data). In addition, for the second station, USGS15024800, data were measured during the period from January 01, 2020, to November 11, 2020, by removing the incomplete patterns in total, 7289 data were retained, among which 70% were randomly selected for training (5103 data) and 30% for validation (2186 data). The statistical description of the dataset is presented in Table 1.

2.2 Performance assessment of the models

The performance of the developed models was evaluated using four statistical indices: root mean square error (RMSE), mean absolute error (MAE), correlation coefficient (R), and Nash-Sutcliffe efficiency (NSE), calculated as follows:

$$RMSE = \sqrt{\frac{1}{N}\sum_{i=1}^{N}\left[(DO_{obs,i}) - (DO_{est,i})_i\right]^2}, (0 \leq RMSE < +\infty) \tag{1}$$

$$MAE = \frac{1}{N}\sum_{i=1}^{N}|DO_{obs,i} - DO_{est,i}|, (0 \leq MAE < +\infty) \tag{2}$$

$$R = \left[\frac{\frac{1}{N}\sum_{i=1}^{N}\left(DO_{obs,i} - \overline{DO_{obs}}\right)\left(DO_{est,i} - \overline{DO_{est}}\right)}{\sqrt{\frac{1}{N}\sum_{i=1}^{n}\left(DO_{obs,i} - \overline{DO_{obs}}\right)^2}\sqrt{\frac{1}{N}\sum_{i-1}^{n}\left(DO_{est,i} - \overline{DO_{est}}\right)^2}}\right], (-1 < R \leq +1) \tag{3}$$

$$NSE = 1 - \left[\frac{\sum_{i=1}^{N}[DO_{obs} - DO_{est}]^2}{\sum_{i=1}^{N}[DO_{obs,i} - \overline{DO_{obs}}]^2}\right], (-\infty < NSE \leq 1) \tag{4}$$

where, DO_{obs} and DO_{est} specify the observed and estimated daily river water temperature for the ith observation, and N is the number of data points, $\overline{DO_{obs}}$ and $\overline{DO_{est}}$ are the mean measured and mean estimated DO, respectively.

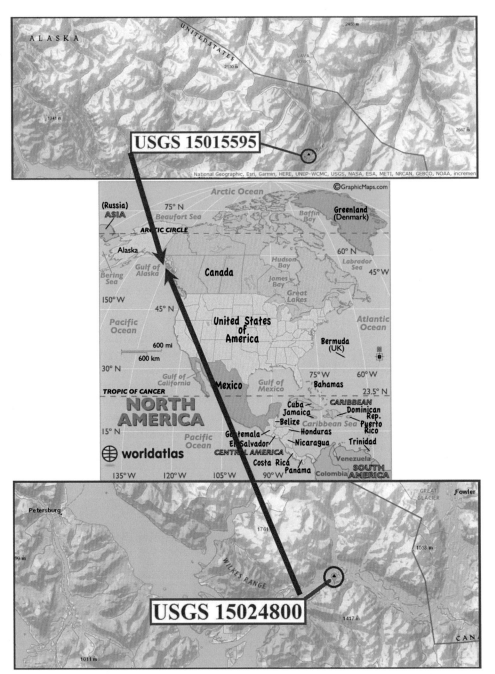

FIG. 1   Study area map showing the location of two stations.

3 Methodology

3.1 Gaussian process regression (GPR)

Over the last two decades, non-parametric models, including Gaussian process regression (GPR) have gained popularity in the field of soft computing field.[24] The GPR model was invented by Rasmussen and Williams.[25] The GPR model utilizes the Bayesian approach to solve multidimensional optimization problems.[25–31] By exploiting a hypothetical quantity, the complex associations among inputs outputs are handled. Distribution over functions produced by the GP (Gaussian process), a set of random variables $[f(z): z \in Z]$ with a mean $(\mu(z))$ and covariance function $(\text{cov}(z, z')$ described the GP follow a joint multivariate Gaussian distribution, which is written as[25]:

$$f(z) \sim GP[\mu(z), \text{cov}(z, z')], \tag{5}$$

TABLE 1 Summary statistics of water and air temperatures for the two stations.

Variables	Subset	Unit	X_{mean}	X_{max}	X_{min}	S_x	C_v	R
USGS 15015595								
T_w	Training	°C	4.320	10.800	0.000	2.498	1.729	−0.798
	Validation	°C	4.339	10.700	0.000	2.541	1.707	−0.816
	All data	°C	4.326	10.800	0.000	2.509	1.724	−0.803
T_a	Training	°C	5.441	33.900	−13.000	6.206	0.877	−0.626
	Validation	°C	5.584	33.700	−12.000	6.294	0.887	−0.655
	All data	°C	5.483	33.900	−13.000	6.216	0.882	−0.634
Q	Training	Ft^3/s	5866.975	50,100.000	555.000	6112.059	0.960	−0.436
	Validation	Ft^3/s	6134.450	43,600.000	555.000	6534.477	0.939	−0.445
	All data	Ft^3/s	5947.212	50,100.000	555.000	6212.957	0.957	−0.438
DO	Training	mg/L	13.040	14.500	10.500	0.711	18.339	1.000
	Validation	mg/L	13.053	14.500	10.500	0.688	18.980	1.000
	All data	mg/L	13.044	14.500	10.500	0.703	18.551	1.000
USGS 15024800								
T_w	Training	°C	4.373	10.000	−0.100	3.350	1.305	−0.680
	Validation	°C	4.373	10.000	−0.100	3.350	1.305	−0.680
	All data	°C	4.325	10.100	−0.100	3.319	1.303	−0.675
T_a	Training	°C	5.303	26.800	−20.400	6.888	0.770	−0.616
	Validation	°C	5.470	24.300	−20.200	6.817	0.802	−0.634
	All data	°C	5.353	26.800	−20.400	6.857	0.781	−0.622
Q	Training	Ft^3/s	69,191.250	209,000.000	5300.000	59,152.041	1.170	−0.596
	Validation	Ft^3/s	69,186.002	209,000.000	5300.000	59,186.823	1.169	−0.591
	All data	Ft^3/s	69,189.676	209,000.000	5300.000	59,143.184	1.170	−0.594
DO	Training	mg/L	12.655	14.700	11.500	0.617	20.512	1.000
	Validation	mg/L	12.624	14.700	11.600	0.621	20.320	1.000
	All data	mg/L	12.646	14.700	11.500	0.618	20.466	1.000

Abbreviations: X_{mean}, mean; X_{max}, maximum; X_{min}, minimum; S_x, standard deviation; C_v, coefficient of variation; R, coefficient of correlation with DO; T_w, river water temperature; T_a, air temperature; Q, river discharge; DO, dissolved oxygen.

Here, the mean and covariance functions are defined as[32]:

$$\mu(z) = E[f(z)] \tag{6}$$
$$\mathrm{cov}(z, z') = E[(f(z) - \mu(z))(f(z') - \mu(z'))] \tag{7}$$

where, $f(z)$ describes the signal function at z input variable (Fig. 2). The covariance function, also known as the kernel of GP, is used for all the training and testing datasets. The squared-exponential kernel is most commonly utilized in GPR models and is written as[25]:

$$k(z, z') = \sigma_f^2 \exp\left(-\sum_{p=1}^{h} \frac{\left(z_p - z_p'\right)^2}{2Z_p^2}\right), \tag{8}$$

where, h specifies the dimension of the input z, z_p states the pth dimension of the input variables, Z_p denotes the typical length scale equivalent to the pth dimension of input z and σ_f^2 denotes the noise-free signal variance. Readers can refer to Rasmussen and Williams[25] and Schulz et al.[33] for an exhaustive knowledge of GPR.

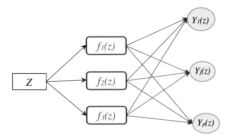

FIG. 2　Structure of Gaussian process regression model.

3.2 Genetic programming (GP)

In computer programming, genetic programming (GP) is an automatic problem-solving technique in which Darwin's principle of "survival of the fittest" is used to find the best solution among randomly generated potential solutions aka genes. In the classic genetic programming (GP) developed by Koza,[34] each gene is represented by a tree structure with a root node, inner nodes, and terminal nodes called leaves (see Fig. 3). Fig. 3 illustrates a genome including a root node (multiplication), inner nodes of addition and *Sin* function, and terminal nodes of X_1, X_2, and a random number C_1. Each node in a GP tree can adopt a function or terminal variables, such as C_1, X_1, and X_2, as shown in Fig. 3.

To solve regression problems using GP, the algorithm commences with the formation of an initial population of genes. Then, three evolutionary operators, *Reproduction*, *Crossover*, and *Mutation*, act on the initial population of genes to improve their fitness. The new population that shows the highest fitness survives to the next generation of the population. The evolutionary process was repeated until the individual met the objective goals. Details of the modeling processes in different GP variants have been reported by Hrnjica and Danandeh Mehr.[35] Here, a brief overview of the evolutionary process of monolithic GP is presented to ensure the integrity of this chapter. Fig. 4 shows the crossover operator between two parents producing two offspring. New individuals have the same materials as their parents, but in different combinations with them. Various studies have shown that the offspring fit the training set better than their parents.[36,37] The mutation is another evolutionary operator in which the genetic material of a parent gene is replaced with new materials at the mutation node. As illustrated in Fig. 5, the subtree $x_1 \times \sin x_2$ in the parent gave his place to the randomly created $\log x_2$ subtree in the offspring.

Reproduction is the way to survive the best gene at each iteration without any morph. A user-defined percentage of individuals with higher fitness is directly moved to the next generation. All the above-mentioned operations are performed on a set of training data and the model performance must be controlled using unseen testing data sets. To prepare an educated model, a set of suitable functions, input variables, rate of evolutionary operations, and the maximum depth of GP trees must also be considered in the modeling process.[37] To avoid overfitting, a lower number of functions and short trees are recommended.[35] A variety of GP codes are available to run the GP (see the list in[39]). A GPLAB toolbox for MATLAB (version 4.04)[40] was used in the present study.

3.3 Group method of data handling (GMDH)

The GMDH methodology is a subcategory of self-organization (inductive) machine learning methods, which are developed according to the black box concepts of artificial neural networks (ANNs). In other words, a GMDH model can be constructed similar to a forward ANN using several neurons in several layers. However, in the GMDH, each layer consists of various allocated hidden neurons that are based on a quadratic polynomial. Accordingly, new neurons are generated in the next hidden layer.[41] The connection between the input $(x_i, x_j, ..., x_n)$ and output (y) variables can be formed by the use of the Volterra-Kolmogorov-Gabor (VKG) series as follows:

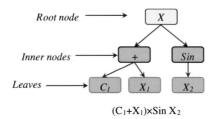

$(C_1 + X_1) \times \mathrm{Sin}\ X_2$

FIG. 3　Example of a gene (genome) and its associated function.

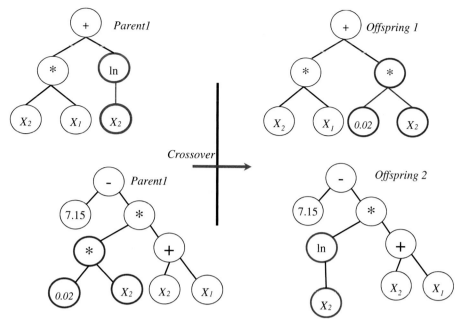

FIG. 4 An example of crossover operation acting on two parents and producing two offspring.[38]

$$y = c_0 + \sum_{i=1}^{n} c_i x_i + \sum_{j1}^{n} \sum_{k=1}^{n} c_{ij} x_i x_j + \sum_{i=1}^{n} \sum_{j=1}^{n} \sum_{k=1}^{n} c_{ijk} x_i x_j x_k + ..., \quad n = 0, 1, ..., 2^l \tag{9}$$

where, l denotes the number of layers, and C_0 and $c_{ij...k}$ represent the polynomial coefficients. According to the literature,[42,43] the second-order polynomial (Eq. 10) is the basic and fundamental structure of the GMDH model and can be used as an appropriate transfer function for neurons. Hence, a trained GMDH network provides the modeled/predicted output values (y^m) based on a second-order function (G_i) as follows:

$$y^m = G(x_i, x_j) = c_0 + c_1 x_i + c_2 x_j + c_3 x_i^2 + c_4 x_j^2 + c_5 x_i x_j \tag{10}$$

Eq. (10) defines that all the neurons of the hidden layers take a similar quadratic form, and accordingly, the architecture of the network is constructed. The main goal of tuning the polynomial coefficient is to find a function that can provide an acceptable approximation of the output parameter. With N observed data, the aim of training the network is to minimize the difference between the observed/actual (y_i^0) and the modeled/predicted (y_i^m) values using the mean square error, that is:

$$MSE = \frac{1}{N} \sum_{i=1}^{N} (y_i^0 - y_i^m)^2 \rightarrow minimum \tag{11}$$

To achieve this goal, the bias and weighting coefficients ($C_0, C_1, C_2, C_3, C_4, C_5$) in Eq. (10) can be calculated using regression techniques. During the training procedure, some quadratics, as in Eq. (10), were utilized to construct a structure of the heuristic solution to the given problem. Thus, an optimal fit in the entire set of the input-output datasets can be achieved by minimizing the network's error formula:

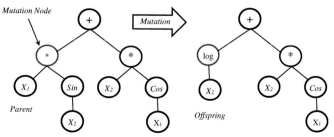

FIG. 5 Mutation operation acts on a genetic programming (GP) chromosome.[38]

$$E = \frac{1}{N} \sum_{i=1}^{N} (y_i - G_i())^2 \tag{12}$$

By introducing Y and C vectors as $Y = [y_i]^T$, $C = \{C_0, C_1, C_2, C_3, C_4, C_5\}$, the regression analysis for obtaining the solution leads to the least-square method as follow:

$$C = (A^T A)^{-1} A^T Y \tag{13}$$

Subsequently, vector A represents the following matrix:

$$A = \begin{bmatrix} 1 & x_1^1 & x_2^1 & x_1^1 x_2^1 & (x_1^1)^2 & (x_2^1)^2 \\ 1 & x_1^2 & x_2^2 & x_1^2 x_2^2 & (x_1^2)^2 & (x_2^2)^2 \\ \vdots & & & & & \\ 1 & x_1^N & x_2^N & x_1^N x_2^N & (x_1^N)^2 & (x_2^N)^2 \end{bmatrix} \tag{14}$$

Detailed explanations regarding the solution of Eq. (14) can be found in Anastasakis and Mort[44] and Azimi et al.[45]

3.4 Support vector regression (SVR)

The SVR is one of the most commonly used machine learning methods which were first introduced by Smola.[46] It is a type of SVM that was developed by Vapnik.[47] The SVR method generally has better generalization ability than the neural networks because it utilizes structural risk minimization. In the SVR modeling process, first, the support vectors are selected and their weights are calculated.

SVR can be defined by the following relation[48,49]:

$$f(x) = w^T \varphi(x) + b, \tag{15}$$

where f is a function of the relationship between the dependent and independent variables, and w, φ and b are the weight vector, mapping or transfer function, and bias term, respectively. The regression problem can be expressed as:

$$\text{minimize} \quad \frac{1}{2} \|w\|^2 + c \sum_{i=1}^{n} (\xi + \xi_k^*),$$

$$\text{subject to} \quad \begin{cases} y_k - w^T \varphi(x_k) - b \leq \varepsilon + \xi_k \\ w^T \varphi(x_k) + b - y_k \leq \varepsilon + \xi_k^* \\ \xi_k, \xi_k^* \geq 0 \end{cases} \quad k = 1, 2, \dots, N \tag{16}$$

where ξ and ξ^* indicate slack variables, and c and ε are the penalty parameter and boundary value, respectively. The solution of nonlinear regression by considering the Lagrangian multipliers ($\propto$, $\propto^*$) can be defined as:

$$f(x) = \sum_{k=1}^{n} (\propto_k - \propto_k^*) K(x, x_k) + b \tag{17}$$

where k and n are the kernel function and support vector quantity, respectively. In the present study, we applied a radial basis kernel, which can be defined as

$$K(x_k, x_l) = \exp\left(-\frac{\|x_k - x_l\|}{2\sigma^2}\right), \tag{18}$$

where σ is the standard deviation indicating the Gaussian noise level.[50] Fig. 6 illustrates a schematic representation of the SVR model.

3.5 Long short-term memory (LSTM)

Long short-term memory (LSTM) networks consider a type of Recurrent Neural network (RNN) and a development over the usual RNN layout, as stated by Huang et al.[51] There is only one threshold function as the difference between the LSTM and RNN in the standard structure of RNN structure. The RNN considers a form of an artificial neural network (ANN), where a recurrent structure is created by links between hidden neuron nodes. The original RNN

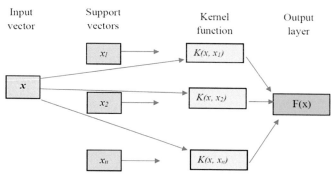

FIG. 6 Schematic representation of SVR model.

structure, namely a vanilla artificial recurrent neural network (ARNN), contains recurrent neural network chains, where each ARNN module comprises a hyperbolic tangent activation function layer sole structure.[52] As the building unit of the RNN layer, which is commonly called the LSTM network, the LSTM unit is used. LSTM helps an RNN for a long time to remember their inputs. The LSTM network solves the issue of RNN generating a gradient explosion when long time-series prediction is processed because the memory unit is connected to the LSTM network neurons to detect the capability effectiveness of the data used, which is useful for the processing and forecasting of significant events in long-term periods and time series delays. When a message reaches the LSTM network, it can be determined if, according to the rules, it is helpful. The data conforming to the algorithm rules are left, while the incompatible information will be neglected through the forgetting gate.

Eq. (1) and Fig. 7 indicate that the current state h_{t-1} and input X_t are affected by the secret state h_t, which is modified within the RNN. The sequence data can be processed efficiently by the RNN structure because past information can influence the future as follows:

$$h_t = f_w(h_{t-1}, x_t) \tag{19}$$

A standard LSTM layout has one cell and three gates divided into input, forget, and output. Over non-regular time intervals, the cell recalls values and the three gates control the datasets into and out of the cell. Each gate has the following specific function. First, the forget gate estimates the size of past data that needs to be derelict; to evaluate the output value between 0 and 1, it receives h_{t-1} of the unseen state and inputs data X_t by applying the activation function of the sigmoid. Second, the input gate estimates the volume of information required for a recall. This gate also uses the same function of the forget gate with h_{t-1} and input x_t and implements the matrix operation of the Hadamard product beside the hyperbolic tangent to the input variables. Third, the output gate considers an interior gate that determines the output summation of the existing state. The active *tan h* function is utilized to identify the information of the cell state that has been modified in advance, and the ht. value represents the existing outcome and works as ht-1 in the following sequence functions. This arrangement minimizes information loss and facilitates data distribution in all the

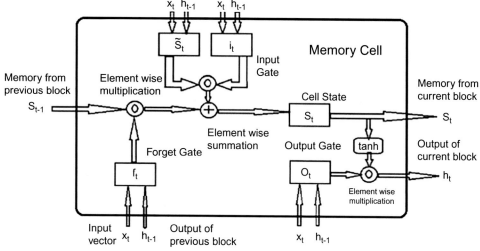

FIG. 7 Structure of the LSTM neural network.

series. The LSTM structural feature allows the problem of long-term dependence to be solved and has been applied to process sequential data as documented by.[51–54]

3.6 Multiple linear regression (MLR)

Multiple linear regression (MLR) is a simple, and well-known method used to build a linear model between one or more independent variables (predictors) and one dependent variable (predictand). In the present study, the predictand variable is the dissolved oxygen (DO) and the predictors are the water temperature (T_w), air temperature (T_a), and the river discharge (Q), and the equation of the MLR model is:

$$Y = f(x_i) \Rightarrow DO = \lambda_0 + \lambda_1 Q + \lambda_2 T_a + \lambda_3 T_w \tag{20}$$

λ_i are the models' parameters.

4 Results

In the present chapter, the proposed models were applied and compared according to six input combinations (Table 2) and the obtained results are discussed for each station separately.

4.1 USGS 15015595 station

The calculated correctness of the hired models to predict the dissolved oxygen (DO) concentration values during training and validation levels are presented in Table 3 at the USGS 15015595 station. Dark numbers represent the values of outstanding correctness during training and validation levels at the USGS 15015595 station. It can be observed in Table 3 that the models with the 1-input combination (i.e., GMDH1, LSTM1, GPR1, SVR1, MLR1, and GP1) were proven to be the advanced models with the lowest (i.e., RMSE and MAE) and the highest (i.e., R and NSE) statistical indicators during training and validation levels. In addition, a simple comparison of statistical indicators among the advanced models demonstrated that the GPR1 model ($R = 0.986$, NSE $= 0.969$, RMSE $= 0.120$ mg/L, and MAE $= 0.081$ mg/L for validation level) provided the best accuracy for predicting the DO concentration values during training and validation levels at the USGS 15015595 station. The scatter schemes between the measured and computed DO concentration values hiring the advanced models (i.e., GMDH1, LSTM1, GPR1, SVR1, MLR1, and GP1) for USGS 15015595 station are shown in Figs. 2A–F associating the best fit (Blue) line, optimized equation, and the coefficient of determination (R^2), respectively. It can be explored from the scatter schemes that the computed values matching the GPR1 model truly traced the measured values during the validation level. Fig. 8 shows the daily time series of measured and computed dissolved oxygen (DO) concentration values hiring the advanced models (i.e., GMDH1, LSTM1, GPR1, SVR1, MLR1, and GP1) at the validation level at the USGS 15015595 station. It can be seen from Fig. 9 that the measured values of DO concentration illustrate the fluctuation behaviors of the daily time series. The GPR1 and SVR1 models traced the measured values of DO concentration, while the GP1 model could not track the measured values of DO concentration and provided the worst achievement in predicting the DO concentration values during validation at the USGS 15015595 station.

TABLE 2 The input combinations of different models.

MLR	GPR	SVR	GMDH	LSTM	Input combination	Output
MLR1	GPR1	SVR1	GMDH1	LSTM1	HH, DD, MM, T_w, T_a, Q	DO
MLR2	GPR2	SVR2	GMDH2	LSTM2	T_w, T_a, Q	DO
MLR3	GPR3	SVR3	GMDH3	LSTM3	T_w, T_a	DO
MLR4	GPR4	SVR4	GMDH4	LSTM4	T_w, Q	DO
MLR5	GPR5	SVR5	GMDH5	LSTM5	HH, DD, MM, T_w, T_a	DO
MLR6	GPR6	SVR6	GMDH6	LSTM6	HH, DD, MM, T_w, Q	DO

HH, hour of the day (1–24); *DD*, day of the month (1–31); *MM*, month of the year (1–12); T_w, water temperature; T_a, air temperature; Q, discharge; *DO*, dissolved oxygen.

TABLE 3 Performances of different models at USGS 15015595 station.

Models	Training				Validation			
	R	NSE	RMSE (mg/L)	MAE (mg/L)	R	NSE	RMSE (mg/L)	MAE (mg/L)
GMDH1	**0.922**	**0.849**	**0.275**	**0.182**	**0.921**	**0.848**	**0.268**	**0.175**
GMDH2	0.854	0.730	0.369	0.240	0.864	0.745	0.347	0.222
GMDH3	0.843	0.711	0.381	0.255	0.853	0.726	0.360	0.239
GMDH4	0.835	0.697	0.391	0.255	0.849	0.720	0.364	0.235
GMDH5	0.914	0.836	0.287	0.185	0.907	0.822	0.290	0.184
GMDH6	0.899	0.808	0.311	0.205	0.896	0.803	0.305	0.200
LSTM1	**0.964**	**0.930**	**0.188**	**0.121**	**0.939**	**0.881**	**0.238**	**0.154**
LSTM2	0.914	0.835	0.288	0.190	0.895	0.801	0.307	0.196
LSTM3	0.861	0.741	0.361	0.233	0.858	0.734	0.354	0.227
LSTM4	0.910	0.828	0.294	0.194	0.871	0.758	0.338	0.209
LSTM5	0.954	0.910	0.213	0.129	0.910	0.828	0.285	0.173
LSTM6	0.952	0.906	0.218	0.149	0.928	0.860	0.257	0.170
GPR1	**0.990**	**0.979**	**0.102**	**0.061**	**0.986**	**0.969**	**0.120**	**0.081**
GPR2	0.972	0.944	0.168	0.115	0.960	0.920	0.195	0.134
GPR3	0.905	0.820	0.301	0.195	0.906	0.820	0.292	0.189
GPR4	0.953	0.908	0.215	0.148	0.944	0.889	0.229	0.156
GPR5	0.989	0.978	0.105	0.066	0.985	0.968	0.122	0.083
GPR6	0.986	0.972	0.118	0.074	0.982	0.962	0.133	0.089
SVR1	**0.987**	**0.975**	**0.113**	**0.059**	**0.984**	**0.963**	**0.132**	**0.082**
SVR2	0.936	0.873	0.253	0.155	0.936	0.869	0.249	0.155
SVR3	0.885	0.781	0.332	0.202	0.897	0.800	0.308	0.190
SVR4	0.897	0.796	0.321	0.197	0.902	0.802	0.306	0.190
SVR5	0.985	0.969	0.124	0.069	0.983	0.961	0.135	0.087
SVR6	0.984	0.968	0.126	0.070	0.980	0.956	0.145	0.089
MLR1	**0.838**	**0.703**	**0.387**	**0.249**	**0.850**	**0.722**	**0.363**	**0.232**
MLR2	0.836	0.699	0.389	0.253	0.848	0.718	0.365	0.236
MLR3	0.823	0.677	0.403	0.254	0.836	0.698	0.378	0.239
MLR4	0.819	0.670	0.408	0.255	0.832	0.692	0.382	0.239
MLR5	0.827	0.683	0.399	0.248	0.841	0.707	0.372	0.230
MLR6	0.820	0.672	0.406	0.256	0.833	0.692	0.381	0.240
GP1	**0.825**	**0.680**	**0.401**	**0.239**	**0.837**	**0.700**	**0.377**	**0.221**

For GP models, only the results using the best input combination are provided (GP1), which corresponds to MM, T_w, and T_a. Numbers given in bold correspond to the best models.

Fig. 12A highlights the box schemes for the spatial diffusion of computed DO concentration values based on the advanced models (i.e., GMDH1, LSTM1, GPR1, SVR1, MLR1, and GP1) during the validation level at the USGS 15015595 station. The compressed enumeration (e.g., median, 1st and 3rd quartiles) of spatial diffusion hiring the GPR1 and SVR1 models followed the measured DO concentration values similarly, while the GP1 and MLR1 models could not track the measured DO concentration values clearly during validation at the USGS 15015595 station. In addition, the full ranges (e.g., maximum and minimum) of the GPR1 and SVR1 models clearly trailed those of measured DO concentration. The violin scheme,[55] which illustrates the probability diffusion of measured and computed DO

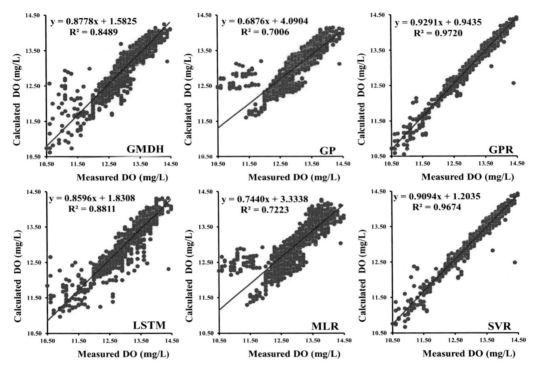

FIG. 8 Scatter schemes of measured versus computed dissolved oxygen (DO) concentration employing the advanced models at USGS 15015595 station (Validation level).

concentration values, is arranged as a box diagram with the regulation of the kernel density plot. The violin scheme (Fig. 13A) clarified that the GPR1 and SVR1 models accomplished diverse enumeration (e.g., minimum, 25th percentile, medium, and 75th percentile fields) of measured DO concentration very closely compared with the MLR1 and GP1 models at the validation level at the USGS 15015595 station. A polar scheme provided by Taylor[56] was employed to acquire visible support for model accomplishment considering three enumerations.[56–58] Fig. 14A shows the Taylor scheme between the measured and computed DO concentration values based on the advanced models (i.e., GMDH1, LSTM1, GPR1, SVR1, MLR1, and GP1) during the validation level at the USGS 15015595 station. The Taylor scheme indicated that the GPR1 model yielded the smallest RMSE value compared with the other advanced models. Although the GPR1 model could not achieve the measured values completely depending on the normalized standard deviation and correlation coefficient, the node corresponding to the GPR1 model was adjacent to the measured value compared with the other advanced models. In addition, the node corresponding to the GP1 model demonstrated the worst performances in predicting the DO concentration values based on the Taylor scheme's assistance.

4.2 USGS 15024800 station

The estimated correctness of the engaged models to predict the dissolved oxygen (DO) concentration values during training and validation levels are presented in Table 4 at the USGS 15024800 station. Dark numbers indicate the values of superior correctness during training and validation levels at the USGS 15024800 station. It can be seen from Table 4 that the 1-input combination-based models (i.e., GMDH1, LSTM1, GPR1, SVR1, MLR1, and GP1) were verified as advanced models during training and validation levels. In addition, the simple contrast of statistical indicators among the advanced models explained that the GPR1 model ($R = 0.991$, NSE $= 0.981$, RMSE $= 0.085$ mg/L, and MAE $= 0.062$ mg/L for validation level) generated the best accuracy for computing the DO concentration values during training and validation levels at the USGS 15024800 station. The scatter schemes between the measured and computed DO concentration values for the advanced models (i.e., GMDH1, LSTM1, GPR1, SVR1, MLR1, and GP1) for the USGS 15024800 station are underlined in Fig. 10, enclosing the best fit (blue) line, optimized equation, and the coefficient of determination (R^2), respectively. It can be observed from the scatter schemes that the computed values equivalent to the GPR1 model detected the measured values confidently during the validation level.

Fig. 11 illustrates the daily time series of measured and computed DO concentration values using the advanced models (i.e., GMDH1, LSTM1, GPR1, SVR1, MLR1, and GP1) during the validation level at the USGS 15024800 station.

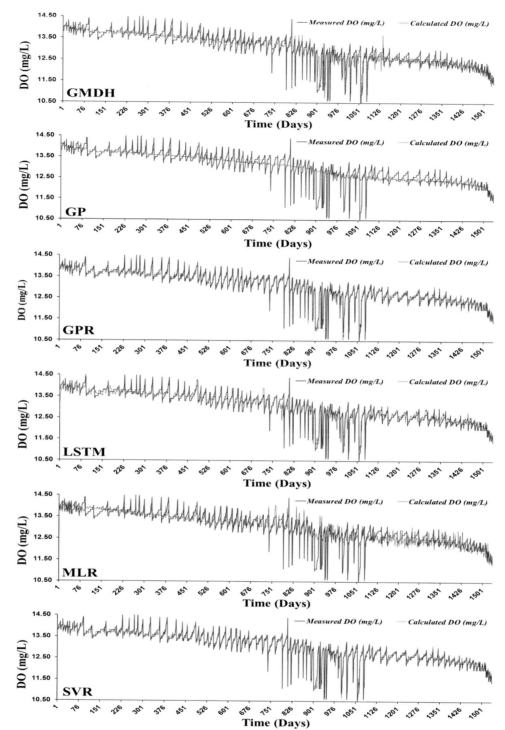

FIG. 9 Daily time series of measured versus computed dissolved oxygen (DO) concentration at USGS 15015595 station (Validation level).

It can be suggested from Fig. 11 that the measured values of DO concentration supply the fluctuation behaviors of the daily time series. The GPR1 and SVR1 models followed the measured values of DO concentration, whereas the MLR1 model could not pursue the measured values of DO concentration and support the worst accomplishment to predict the DO concentration values during validation at the USGS 15024800 station. Fig. 12B shows the box schemes for the spatial distribution of the computed DO concentration values based on the advanced models (i.e., GMDH1, LSTM1, GPR1, SVR1, MLR1, and GP1) during the validation level at the USGS 15024800 station. The compact list (e.g., median, 1st and 3rd quartiles) of spatial distribution engaging the GPR1 and SVR1 models trailed the measured DO

TABLE 4 Performances of different models at USGS 15024800 station.

Models	Training				Validation			
	R	NSE	RMSE (mg/L)	MAE (mg/L)	R	NSE	RMSE (mg/L)	MAE (mg/L)
GMDH1	**0.943**	**0.889**	**0.205**	**0.158**	**0.941**	**0.886**	**0.210**	**0.162**
GMDH2	0.828	0.685	0.346	0.258	0.834	0.695	0.343	0.257
GMDH3	0.768	0.590	0.395	0.273	0.764	0.583	0.401	0.281
GMDH4	0.792	0.627	0.376	0.283	0.796	0.633	0.376	0.283
GMDH5	0.932	0.868	0.224	0.167	0.928	0.860	0.232	0.172
GMDH6	0.937	0.877	0.216	0.167	0.935	0.873	0.221	0.171
LSTM1	**0.963**	**0.927**	**0.167**	**0.114**	**0.950**	**0.903**	**0.193**	**0.139**
LSTM2	0.846	0.713	0.330	0.245	0.802	0.643	0.371	0.281
LSTM3	0.823	0.675	0.351	0.251	0.799	0.636	0.375	0.268
LSTM4	0.821	0.658	0.360	0.264	0.821	0.658	0.363	0.272
LSTM5	0.942	0.887	0.207	0.149	0.918	0.841	0.248	0.180
LSTM6	0.928	0.861	0.230	0.157	0.916	0.839	0.250	0.175
GPR1	**0.995**	**0.990**	**0.061**	**0.044**	**0.991**	**0.981**	**0.085**	**0.062**
GPR2	0.964	0.929	0.164	0.111	0.954	0.909	0.187	0.127
GPR3	0.823	0.677	0.350	0.241	0.806	0.649	0.368	0.250
GPR4	0.926	0.857	0.233	0.165	0.920	0.846	0.244	0.173
GPR5	0.995	0.991	0.060	0.043	0.990	0.980	0.089	0.064
GPR6	0.988	0.976	0.096	0.070	0.984	0.968	0.111	0.081
SVR1	**0.989**	**0.978**	**0.092**	**0.053**	**0.985**	**0.968**	**0.110**	**0.072**
SVR2	0.909	0.827	0.257	0.173	0.902	0.814	0.268	0.179
SVR3	0.799	0.638	0.371	0.237	0.785	0.615	0.385	0.247
SVR4	0.892	0.794	0.280	0.204	0.890	0.790	0.284	0.205
SVR5	0.984	0.967	0.111	0.064	0.979	0.957	0.129	0.081
SVR6	0.983	0.966	0.114	0.069	0.979	0.957	0.129	0.082
MLR1	**0.810**	**0.656**	**0.362**	**0.291**	**0.812**	**0.657**	**0.363**	**0.290**
MLR2	0.676	0.456	0.454	0.356	0.684	0.466	0.454	0.356
MLR3	0.676	0.456	0.454	0.356	0.684	0.467	0.454	0.356
MLR4	0.673	0.453	0.456	0.358	0.680	0.462	0.456	0.358
MLR5	0.809	0.655	0.362	0.292	0.811	0.656	0.364	0.291
MLR6	0.799	0.638	0.371	0.298	0.803	0.643	0.371	0.295
GP1	**0.842**	**0.702**	**0.337**	**0.264**	**0.842**	**0.702**	**0.339**	**0.263**

For GP models, only the results using the best input combination are provided (GP1), which corresponds to MM, T_w, and T_a. Numbers given in bold correspond to the best models.

concentration values, whereas the GP1 and MLR1 models could not duplicate the measured DO concentration values during validation at the USGS 15024800 station. In addition, the entire extent (e.g., maximum and minimum) of GPR1 and SVR1 models copied those of measured DO concentration definitely.

The violin scheme (Fig. 13B) explained that the LSTM1, GPR1, and SVR1 models carried out the different lists (e.g., minimum, 25th percentile, medium, and 75th percentile fields) of measured DO concentration very strictly compared to GMDH1, MLR1, and GP1 models at the validation level at the USGS 15024800 station. Fig. 14B represents the Taylor scheme between the measured and computed DO concentration values based on the advanced models (i.e., GMDH1,

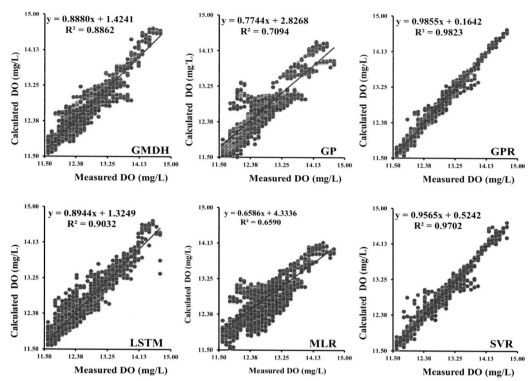

FIG. 10 Scatter schemes of measured versus computed dissolved oxygen (DO) concentration employing the advanced models at USGS 15024800 station (Validation level).

LSTM1, GPR1, SVR1, MLR1, and GP1) during the validation level at the USGS 15024800 station. The Taylor scheme showed that the GPR1 model provided the smallest RMSE value compared to the other advanced models. Although the GPR1 model could not attain the measured values absolutely depending on the normalized standard deviation and correlation coefficient, the point corresponding to the GPR1 model was contiguous to the measured values compared to the other advanced models. In addition, the point corresponding to the MLR1 model yielded the worst performances in predicting the DO concentration values based on the Taylor scheme's help.

5 Discussion

The underlying research showed that different machine learning (i.e., GMDH, GPR, SVR, MLR, and GP) and deep learning (i.e., LSTM) models replicated the nonlinear and nonstationary conduct of dissolved oxygen (DO) concentration values at two stations in the USA. Recognizing the advanced models (i.e., GMDH1, LSTM1, GPR1, SVR1, MLR1, and GP1), the best model was determined as the GPR1 model at two stations to predict the DO concentration values during training and validation levels, respectively. Considering the RMSE values of the GPR1 model at the two stations, the GPR1 model enhanced the accuracy of the GMDH1, LSTM1, SVR1, MLR1, and GP1 models by 55.22%, 49.58%, 9.09%, 66.94%, and 68.17% at the USGS 15015595 station. In addition, the GPR1 model upgraded the accuracy of GMDH1, LSTM1, SVR1, MLR1, and GP1 models by 59.52%, 55.96%, 22.73%, 76.58%, and 74.93% at the USGS 15024800 station. Because the prior information for literature (e.g., articles and technical reports) has demonstrated that the accomplishment correctness of machine learning and deep learning models can be governed predominantly by the data characteristics and model's internal structure,[59–61] the unused models in the addressed research can, therefore, present the best accomplishment correctness of DO concentration values based on the identified data series. To prevent the phenomena of complex concepts for the best model, different statistical approaches, such as the null hypothesis and the extension of data available, can be petitioned for model selection.[62]

Considering the previous articles for predicting DO concentration values using machine learning and deep learning models, Li et al.[63] employed the recurrent neural networks (RNN), long short-term memory (LSTM), and gated recurrent unit (GRU) models to predict the DO concentration values in fishery ponds. The GRU model predicted the DO concentration values accurately compared with the RNN and LSTM models. Wang et al.[64] developed a marine deep jointly informed neural network model to predict marine dissolved oxygen concentration values. The M-DJINN model

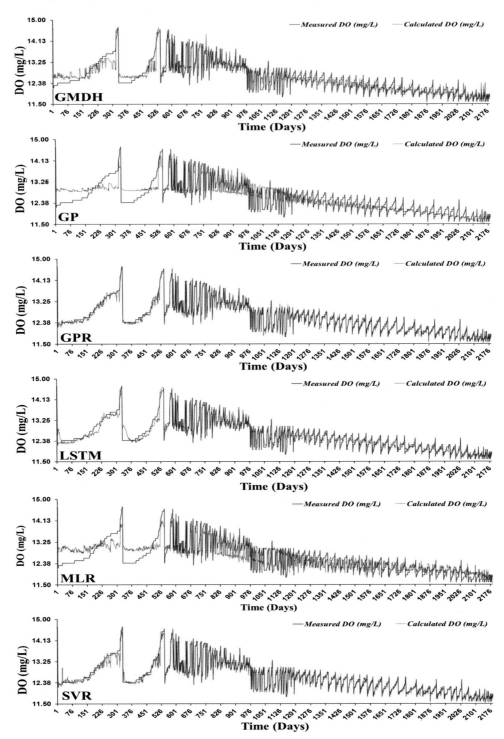

FIG. 11   Daily time series of measured versus computed dissolved oxygen (DO) concentration at USGS 15024800 station (Validation level).

improved the deep jointly informed neural network model based on different statistical indicators. Ji et al.[65] provided a support vector machine (SVM), multiple linear regression (MLR), back-propagation neural network (BPNN), and general regression neural network (GRNN) models to predict the DO concentration values in the Wen-Rui Tang River, China. The SVM model was found to be the best model for predicting DO concentration values compared to the MLT, BPNN, and GRNN models. Therefore, the addressed research was a distinguishing approach to compute DO concentration values in the United States. To augment the model accomplishment of addressed research, access to diverse

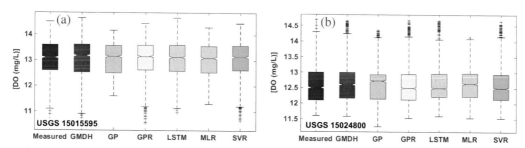

FIG. 12 Box schemes of measured versus computed dissolved oxygen (DO) concentration at both stations (Validation level).

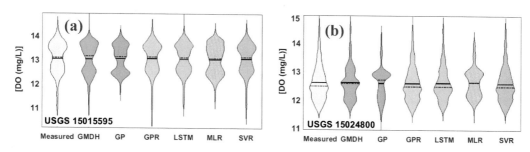

FIG. 13 Violin schemes showing the distributions of measured versus computed dissolved oxygen (DO) concentration at both stations (Validation level).

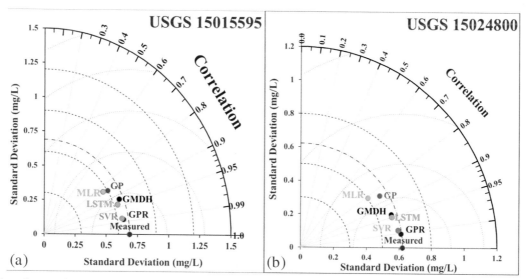

FIG. 14 Taylor schemes of measured versus computed dissolved oxygen (DO) concentration at both stations (Validation level).

conjunctions such as evolutionary algorithms and preprocessing methods can be considered to enhance the computed correctness of DO concentration values in rivers.

Finally, at the end of the present chapter, we can highlight some advantages and disadvantages of the proposed approaches for modeling DO concentration. Although, the present study successfully modeled DO base on fewer input variables, the high correlation between DO and river water T_w is certainly one of the reasons for both its successes and its failures, and the accuracies of proposed models should be verified without the inclusion of the T_w. Distinct from previous studies related to DO modeling, our modeling strategy is based on data at 15-min, which help in obtaining robust models as the modeled variable varied slightly form one pattern to another, contrary to the data measured daily or monthly. Thus, with other datasets measured at a large time interval, we can certainly affirm that the proposed models are reliable. In addition, given that the proposed models were tested using standard training-validation data splitting, an expanded study with different splitting ratios should be a good alternative for demonstrating the advantages and disadvantages of each proposed model.

6 Conclusions

The present study proposed the application of a new machine learning model belonging to the category of deep learning paradigm: long short-term memory (LSTM). LSTM was proposed as a competitive model to predict dissolved oxygen concentration in two rivers in the USA. The LSTM was compared to a standalone machine learning model, namely, the GMDH, SVR, GPR, GP, and MLR models. Based on data collected from in situ measurements, the proposed models were developed and applied at hourly time steps using three different variables: climatic (T_a), water quality (T_w), and hydrological characteristics (Q), which constitute one of the novelties of our investigation. The models were first calibrated using the training dataset and later successfully applied to a new validation dataset. Several input combinations were examined and the contribution of the periodicity (hours, day, and month number) in the improvement of the model performance was evaluated. In addition, among the three variables, T_w was included in the six combinations. Based on the results obtained in this study, several conclusions can be drawn. First, this study demonstrates that dissolved oxygen can be correlated and modeled with different kinds of predictors, not only using water quality variables as broadly reported in the literature, and highly complex models with highly nonlinear structures can be obtained, exhibiting excellent accuracies with a correlation coefficient (R) of approximately 0.99. This advancement was further improved by the inclusion of periodicity and can help explore the multitude of variables that may be available simultaneously. Second, the results obtained in the present study indicate that water T_w, T_a, and Q combined with the periodicity have great potential for modeling dissolved oxygen. However, this statement cannot be generalized, and a genetic programming model that worked relatively equally with the MLR model failed to provide a robust model and their accuracies did not exceed 0.840 and 0.700 in terms of R and NSE values. Third, the achieved results go against, to some extent, what had always been supposed about the superiority of deep learning models, for which the LSTM model was ranked third place less than SVR and GPR. Overall, the main contributions of this investigation lie in enlarging the application of the machine learning models and exploring different types of variables (i.e., water quality, air, and hydrological variables).

References

1. Chen K, Chen H, Zhou C, et al. Comparative analysis of surface water quality prediction performance and identification of key water parameters using different machine learning models based on big data. *Water Res.* 2020;171:115454. https://doi.org/10.1016/j.watres.2019.115454.
2. Heddam S, Ptak M, Zhu S. Modelling of daily Lake surface water temperature from air temperature: extremely randomized trees (ERT) versus Air2Water, MARS, M5Tree, RF and MLPNN. *J Hydrol.* 2020;588. https://doi.org/10.1016/j.jhydrol.2020.125130, 125130.
3. Tung TM, Yaseen ZM. A survey on river water quality modelling using artificial intelligence models: 2000-2020. *J Hydrol.* 2020;585:124670. https://doi.org/10.1016/j.jhydrol.2020.124670.
4. Abba SI, Hadi SJ, Sammen SS, et al. Evolutionary computational intelligence algorithm coupled with self-tuning predictive model for water quality index determination. *J Hydrol.* 2020;587. https://doi.org/10.1016/j.jhydrol.2020.124974, 124974.
5. Abba SI, Pham QB, Saini G, et al. Implementation of data intelligence models coupled with ensemble machine learning for prediction of water quality index. *Environ Sci Pollut Res.* 2020;27(33):41524-41539. https://doi.org/10.1007/s11356-020-09689-x.
6. Bui DT, Khosravi K, Tiefenbacher J, Nguyen H, Kazakis N. Improving prediction of water quality indices using novel hybrid machine-learning algorithms. *Sci Total Environ.* 2020;721. https://doi.org/10.1016/j.scitotenv.2020.137612, 137612.
7. Salih SQ, Alakili I, Beyaztas U, Shahid S, Yaseen ZM. Prediction of dissolved oxygen, biochemical oxygen demand, and chemical oxygen demand using hydrometeorological variables: case study of Selangor River, Malaysia. *Environ Dev Sustain.* 2020;23:8027-8046. https://doi.org/10.1007/s10668-020-00927-3.
8. Nayak JG, Patil LG, Patki VK. Development of water quality index for Godavari River (India) based on fuzzy inference system. *Groundw Sustain Dev.* 2020;10:100350. https://doi.org/10.1016/j.gsd.2020.100350.
9. Rajaee T, Khani S, Ravansalar M. Artificial intelligence-based single and hybrid models for prediction of water quality in rivers: a review. *Chemometr Intel Lab Syst.* 2020;200. https://doi.org/10.1016/j.chemolab.2020.103978, 103978.
10. Heddam S. Intelligent data analytics approaches for predicting dissolved oxygen concentration in river: extremely randomized tree versus random forest, MLPNN and MLR. In: Deo R, Samui P, Kisi O, Yaseen Z, eds. *Intelligent data analytics for decision-support systems in hazard mitigation.* Singapore: Springer Transactions in Civil and Environmental Engineering. Springer; 2021. https://doi.org/10.1007/978-981-15-5772-9_5.
11. Zhu N, Ji X, Tan J, Jiang Y, Guo Y. Prediction of dissolved oxygen concentration in aquatic systems based on transfer learning. *Comput Electron Agric.* 2021;180:105888. https://doi.org/10.1016/j.compag.2020.105888.
12. Zhu S, Heddam S. Prediction of dissolved oxygen in urban rivers at the three gorges reservoir, China: extreme learning machines (ELM) versus artificial neural network (ANN). *Water Qual Res J Canada.* 2020;55(1):106-118. https://doi.org/10.2166/wqrj.2019.053.
13. Antanasijević D, Pocajt V, Perić-Grujić A, Ristić M. Multilevel split of high-dimensional water quality data using artificial neural networks for the prediction of dissolved oxygen in the Danube River. *Neural Comput Appl.* 2020;32:3957-3966. https://doi.org/10.1007/s00521-019-04079-y.
14. Nacar S, Bayram A, Baki OT, Kankal M, Aras E. Spatial forecasting of dissolved oxygen concentration in the eastern Black Sea Basin, Turkey. *Water.* 2020;12(4):1041. https://doi.org/10.3390/w12041041.
15. Nacar S, Mete B, Bayram A. Estimation of daily dissolved oxygen concentration for river water quality using conventional regression analysis, multivariate adaptive regression splines, and TreeNet techniques. *Environ Monit Assess.* 2020;192(12):1-21. https://doi.org/10.1007/s10661-020-08649-9.

16. Ren Q, Wang X, Li W, Wei Y, An D. Research of dissolved oxygen prediction in recirculating aquaculture systems based on deep belief network. *Aquac Eng.* 2020;90:102085. https://doi.org/10.1016/j.aquaeng.2020.102085.

17. Cao X, Liu Y, Wang J, Liu C, Duan Q. Prediction of dissolved oxygen in pond culture water based on K-means clustering and gated recurrent unit neural network. *Aquacult Eng.* 2020;91:102122. https://doi.org/10.1016/j.aquaeng.2020.102122.

18. Kuang L, Shi P, Hua C, Chen B, Zhu H. An enhanced extreme learning machine for dissolved oxygen prediction in wireless sensor networks. *IEEE Access.* 2020;8:198730–198739. https://doi.org/10.1109/ACCESS.2020.3033455.

19. Fadaee M, Mahdavi-Meymand A, Zounemat-Kermani M. Seasonal short-term prediction of dissolved oxygen in rivers via nature-inspired algorithms. *CLEAN-Soil Air Water.* 2020;48(2). https://doi.org/10.1002/clen.201900300, 1900300.

20. Kisi O, Alizamir M, Gorgij AD. Dissolved oxygen prediction using a new ensemble method. *Environ Sci Pollut Res.* 2020;27:1–15. https://doi.org/10.1007/s11356-019-07574-w.

21. Huan J, Li H, Li M, Chen B. Prediction of dissolved oxygen in aquaculture based on gradient boosting decision tree and long short-term memory network: a study of Chang Zhou fishery demonstration base, China. *Comput Electron Agric.* 2020;175:105530. https://doi.org/10.1016/j.compag.2020.105530.

22. Li W, Fang H, Qin G, et al. Concentration estimation of dissolved oxygen in Pearl River Basin using input variable selection and machine learning techniques. *Sci Total Environ.* 2020;731:139099. https://doi.org/10.1016/j.scitotenv.2020.139099.

23. Zhou Y. Real-time probabilistic forecasting of river water quality under data missing situation: deep learning plus post-processing techniques. *J Hydrol.* 2020;589. https://doi.org/10.1016/j.jhydrol.2020.125164, 125164.

24. Yuan J, Wang K, Yu T, Fang M. Reliable multi-objective optimization of high-speed WEDM process based on Gaussian process regression. *Int J Mach Tool Manuf.* 2008;48(1):47–60. https://doi.org/10.1016/j.ijmachtools.2007.07.011.

25. Rasmussen CE, Williams CKI. *Gaussian Processes for Machine Learning.* MIT Press Cambridge; 2006.

26. Ceylan Z. Assessment of agricultural energy consumption of Turkey by MLR and Bayesian optimized SVR and GPR models. *J Forecast.* 2020;39:944–956. https://doi.org/10.1002/for.2673.

27. Inyurt S, Kashani MH, Sekertekin A. Ionospheric TEC forecasting using Gaussian process regression (GPR) and multiple linear regression (MLR) in Turkey. *Astrophys Space Sci.* 2020;365(6):1–17. https://doi.org/10.1007/s10509-020-03817-2.

28. Kopsiaftis G, Protopapadakis E, Voulodimos A, Doulamis N, Mantoglou A. Gaussian process regression tuned by Bayesian optimization for seawater intrusion prediction. *Comput Intell Neurosci.* 2019;2019. https://doi.org/10.1155/2019/2859429, 2859429.

29. Sihag P, Esmaeilbeiki F, Singh B, Pandhiani SM. Model-based soil temperature estimation using climatic parameters: the case of Azerbaijan Province, Iran. *Geol Ecol Landsc.* 2020;4(3):203–215. https://doi.org/10.1080/24749508.2019.1610841.

30. Sihag P, Tiwari NK, Ranjan S. Support vector regression-based modeling of cumulative infiltration of sandy soil. *ISH J Hydraul Eng.* 2020;26(1):44–50. https://doi.org/10.1080/09715010.2018.1439776.

31. Yaseen ZM, Sihag P, Yusuf B, Al-Janabi AMS. Modelling infiltration rates in permeable stormwater channels using soft computing techniques. *Irrig Drain.* 2020;70:117–130. https://doi.org/10.1002/ird.2530.

32. Abdessalem AB, Dervilis N, Wagg DJ, Worden K. Automatic kernel selection for Gaussian processes regression with approximate Bayesian computation and sequential Monte Carlo. *Front Built Environ.* 2017;3:52. https://doi.org/10.3389/fbuil.2017.00052.

33. Schulz E, Speekenbrink M, Krause A. A tutorial on Gaussian process regression: modelling, exploring, and exploiting functions. *J Math Psychol.* 2018;85:1–16. https://doi.org/10.1016/j.jmp.2018.03.001.

34. Koza JR. *Genetic programming: on the programming of computers by means of natural selection.* Vol. 1. MIT Press; 1992.

35. Hrnjica B, Danandeh Mehr A. *Optimized Genetic Programming Applications: Emerging Research and Opportunities: Emerging Research and Opportunities.* IGI Global; 2019.

36. Hrnjica B, Danandeh Mehr A, Behrem Š, Ağıralioğlu N. Genetic programming for turbidity prediction: hourly and monthly scenarios. *Pamukkale Üniversitesi Mühendislik Bilimleri Dergisi.* 2019;25(8):992–997.

37. Tür R. Maximum wave height hindcasting using ensemble linear-nonlinear models. *Theor Appl Climatol.* 2020;141:1151–1163. https://doi.org/10.1007/s00704-020-03272-7.

38. Danandeh Mehr A. An ensemble genetic programming model for seasonal precipitation forecasting. *SN Appl Sci.* 2020;2(11):1–14. https://doi.org/10.1007/s42452-020-03625-x.

39. Danandeh Mehr A. An improved gene expression programming model for streamflow forecasting in intermittent streams. *J Hydrol.* 2018;563:669–678. https://doi.org/10.1016/j.jhydrol.2018.06.049.

40. Silva S, Almeida J. GPLAB-a genetic programming toolbox for MatLab. In: *Proceedings of the Nordic MATLAB conference*; 2003:273–278.

41. Ivakhnenko AG, Ivakhnenko GA. The review of problems solvable by algorithms of the group method of data handling (GMDH). *Pattern Recogn Image Anal C/C Of Raspoznavaniye Obrazov I Analiz Izobrazhenii.* 1995;5:527–535.

42. Mahdavi-Meymand A, Zounemat-Kermani M. A new integrated model of the group method of data handling and the firefly algorithm (GMDH-FA): application to aeration modelling on spillways. *Artif Intell Rev.* 2020;53(4):2549–2569.

43. Najafzadeh M, Barani GA, Azamathulla HM. GMDH to predict scour depth around a pier in cohesive soils. *Appl Ocean Res.* 2013;40:35–41.

44. Anastasakis L, Mort N. *The Development of Self-Organization Techniques in Modelling: A Review of the Group Method of Data Handling (GMDH).* Research Report-University of Sheffield Department of Automatic Control and Systems Engineering; 2001.

45. Azimi H, Bonakdari H, Ebtehaj I, Gharabaghi B, Khoshbin F. Evolutionary design of generalized group method of data handling-type neural network for estimating the hydraulic jump roller length. *Acta Mech.* 2018;229(3):1197–1214.

46. Smola A. *Regression estimation with support vector learning machines.* Master's Thesis, Technische Universit at Munchen; 1996.

47. Vapnik VN. *The nature of statistical learning theory.* New York, USA: Springer-Verlag; 1995.

48. Shiri J, Kisi O, Yoon H, Lee K-K, Nazemi AH. Predicting groundwater level fluctuations with meteorological effect implications – a comparative study among soft computing techniques. *Comput Geosci.* 2013;56:32–44.

49. Tikhamarine Y, Malik A, Pandey K, Sammen SS, Souag-Gamane D, Kisi O. Monthly evapotranspiration estimation using optimal climatic parameters: efficacy of hybrid support vector regression integrated with whale optimization algorithm. *Environ Monit Assess.* 2020;192(11):696.

50. Cimen M, Kisi O. Comparison of two different data driven techniques in modeling surface water level fluctuations of Lake Van, Turkey. *J Hydrology.* 2009;378(3–4):253–262.

51. Huang L, Cai T, Zhu Y, Zhu Y, Wang W. LSTM-based forecasting for urban construction waste generation. *Sustainability*. 2020;12:1–12. https://doi.org/10.3390/su12208555.

52. Choi E, Cho S, Kim DK. Power demand forecasting using long short-term memory (LSTM) deep-learning model for monitoring energy. *Sustainability*. 2020;12:1109. https://doi.org/10.3390/su12031109.

53. Choi Y, Lee J, Kong J. Performance degradation model for concrete deck of bridge using pseudo-LSTM. *Sustainability*. 2020;12:3848. https://doi.org/10.3390/su12093848.

54. Zhang Q, Gao T, Liu X, Zheng Y. Public environment emotion prediction model using LSTM network. *Sustain*. 2020;12. https://doi.org/10.3390/su12041665, 1665.

55. Hintze JL, Nelson RD. Violin plots: a box plot-density trace synergism. *Am Statist*. 1998;52(2):181–184.

56. Taylor KE. Summarizing multiple aspects of model performance in a single diagram. *J Geophys Res Atmos*. 2001;106(D7):7183–7192.

57. Kim S, Alizamir M, Zounemat-Kermani M, Kisi O, Singh VP. Assessing the biochemical oxygen demand using neural networks and ensemble tree approaches in South Korea. *J Environ Manage*. 2020;270:110834.

58. Zounemat-Kermani M, Seo Y, Kim S, et al. Can decomposition approaches always enhance soft computing models? Predicting the dissolved oxygen concentration in the St. Johns River, Florida. *Appl Sci*. 2019;9(12), 2534.

59. Kim S, Alizamir M, Kim NW, Kisi O. Bayesian model averaging: a unique model enhancing forecasting accuracy for daily streamflow based on different antecedent time series. *Sustainability*. 2020;12(22):9720.

60. Shen C. A transdisciplinary review of deep learning research and its relevance for water resources scientists. *Water Resour Res*. 2018;54(11):8558–8593.

61. Zheng F, Maier HR, Wu W, Dandy GC, Gupta HV, Zhang T. On lack of robustness in hydrological model development due to absence of guidelines for selecting calibration and evaluation data: demonstration for data-driven models. *Water Resour Res*. 2018;54(2):1013–1030.

62. Salas JD, Smith RA, Tabios GQ, Heo JH. *Statistical computer techniques in water resources and environmental engineering*. Course notes, Department of Civil Engineering, Colorado State University, Fort Collins, Colorado; 2002.

63. Li W, Wu H, Zhu N, Jiang Y, Tan J, Guo Y. Prediction of dissolved oxygen in a fishery pond based on gated recurrent unit (GRU). *Inf Process Agric*. 2020;8. https://doi.org/10.1016/j.inpa.2020.02.002.

64. Wang L, Jiang Y, Qi H. Marine dissolved oxygen prediction with tree tuned deep neural network. *IEEE Access*. 2020;8:182431–182440.

65. Ji X, Shang X, Dahlgren RA, Zhang M. Prediction of dissolved oxygen concentration in hypoxic river systems using support vector machine: a case study of Wen-Rui Tang River, China. *Environ Sci Pollut Res*. 2017;24(19):16062–16076.

2

Fractal analysis of valley sections in geological formations of arid areas

Mojdeh Mohammadi Khoshoui[a] *and Mohammad Reza Ekhtesasi*[b]

[a]Watershed Management Engineering, Faculty of Natural Resources and Desert Studies, Yazd University, Yazd, Iran
[b]Faculty of Natural Resources and Desert Studies, Yazd University, Yazd, Iran

1 Introduction

Valleys in most arid and semiarid areas in Iran have a concave profile, as the rigid top layers have been destroyed over time, and only the lower soft layers beneath the rigid layer have remained. The evolution of valleys is of great importance in quantitative geomorphological studies, allowing quantification of data and analysis of variations. Quantitative and numerical measurements enable geomorphologists to evaluate and compare various landforms realistically and tangibly. In other words, the results of quantitative studies on valley changes allow the reconstruction of past processes that occurred in these landforms to predict future evolutions using the numerical results of such analyses. The pioneers of valley evolution analysis studied Narmada Valley evolutions in India regarding deposition in the geological periods, namely Oligocene–Miocene and Eocene.[1] In another study, the evolution of a valley in Scotland was studied using fluvial terraces.[2] The evolution of valleys has also been investigated in terms of tectonics, neotectonics, and colluviums.[3–6] Omhori is one of the most prominent pioneers in studying valley evolution by mathematical functions that explain the longitudinal profile and geomorphological evolution of valleys using collected data and the results of mathematical functions.[7] Thereafter, the geomorphological evolution of the longitudinal profile and rivers and their stream beds were investigated using mathematical functions.[8] The erosion pattern and glacial flow in the cross-sectional profile of U-shaped glacial valleys can be modeled using the finite element method (FEM).[9] To distinguish various types of valley sections, a new method based on multistage curvature was proposed in a study conducted in three regions in the western US, and a simple morphometric parameter called the minimum curvature was developed. The results indicated the good performance of this parameter in distinguishing glacial and nonglacial valleys.[10] Various geomorphic units can be analyzed by studying the longitudinal profiles of rivers using mathematical functions and curve-fitting functions.[11]

As a new field, fractal geometry[12, 13] has attracted the interest of scholars in the past two decades to obtain indicators for explaining the characteristics of drainage networks, valleys, and hydrological and geomorphological properties of catchments by studying the relationships of various phenomena. Mathematics is a powerful tool in the hands of naturalists for explaining processes and complexities in nature to quantify them in the form of mathematical relations. In the meantime, the evaluation of glacial and fluvial landforms in the Himalayas by multifractal analysis[14, 15] showed a more complicated structure of glacial landforms than fluvial ones.[16] Numerous studies have been conducted on fractal geometry worldwide, leading to new visions for understanding various phenomena. The great interest in fractal geometry can be related to the fact that fractal geometry can explain the formation and evolution of most geometric landforms, including. Considering the direct relationship between the fractal dimension and irregularity, the irregularity (disorder) of various variables increases with increasing fractal dimension.[27–30] The regularity or irregularity governing processes influencing the valley system can be identified by calculating the fractal dimension for better regional management. Due to the vulnerable ecosystems of arid areas, a correct understanding of the erosional and geomorphological status of valleys is considered a useful tool for managing such areas. There is no comprehensive information regarding the paleoclimate of Iran[25, 26] especially in Central Iran (Arid areas). Most studies are qualitative researches and these have focused mainly on the western and northwestern areas of Iran. Landforms in Central Iran and the highlands, especially in mountain valleys, cannot be formed under current weather conditions. Different climatic cycles create specific processes, which in turn create different landforms. The existence of geoforms such as very large cirques at moderate altitudes, wide valleys, and very large flagstones and rubbles downstream of hillsides at low altitudes on the floor of valleys is suggestive of a humid cold climate system and the dominance of mountain glaciers and glacial processes in the region. Most scholars believe that these landforms were formed during the Quaternary glaciation. Studies indicate the existence of at least two glaciation phases of Gunz and Würm in the Khezrabad Basin and probably across Central Iran.[24] The profile of valleys can be analyzed to understand paleoclimatic conditions, especially in the arid areas of Iran. there are no quantitative studies on fractal geometry in various Iranian geological formations in arid areas. Most studies on Iranian valleys have focused on the application of classic methods, dimensionless hypsometric curves,[8, 17–20] and mathematical functions,[7, 11, 17, 21, 22] and there has been less interest in the quantitative analysis of valleys to identify the nature of processes involved in their morphology and classify valleys formed by different processes. In addition to understanding the intensity of stream deposition, geomorphological ambiguities, and issues regarding factors involved in the formation and evolution of valleys can be identified by studying the evolution of valleys.[23] Given the limited quantitative studies on Iranian valleys by fractal geometry, this study investigates the morphology of valley sections in geological formations of the Yazd–Ardakan catchment using multifractal analysis and mathematical functions. The difference in the fractal properties of ground relief can be related to various geomorphological processes over time. Despite multiple monofractal methods for studying geomorphological fractal features, recent studies[9, 10] have shown that multifractal approaches are more appropriate for this purpose, as fractal parameters may differ depending on location.

2 Materials and methods

2.1 Study area

The Yazd–Ardakan catchment extends from the northern latitude of $31° 41'$ to $32° 38'$, and the eastern longitude of $53° 24'$ to $54° 57'$ in the Central Iranian Plateau (Fig. 1). This is one of the most important catchments in Yazd Province in terms of natural resources. The catchment with a general southwestern slope is discharged to the Siah-Kuh hole. Because the study area is located below the subtropical high-pressure center, the precipitation rate in the region is very low, along with large temperature fluctuations. Winds dominantly blow from the northwest, and the annual rainfall varies from less than 65 mm around the Siah-Kuh Desert to more than 250 mm at the Shirkuh highlands.[30] The region is mainly formed from Tertiary rocks, especially Neogene formations covered by conglomerates and Quaternary fluvial flooding deposits. The remains of the oldest geological units pertained to Precambrian and recent units, namely fluvial terraces, and dune sand deposits, are also observed in the region.[33] Kahar is the oldest formation in the region formed in the Precambrian period in the southwest of Yazd at the Khezrabad highlands. The Granite to granodiorite of Shirkuh formed in the Jurassic period covers a large area of mountainous regions of the catchment. The Limestone facies of the Cretaceous period in the informal Taft formation cover a large area of the northern hillsides of Shirkuh from 1600 to 4000 m. After Shirkuh Granite, this facies covers a large area of the western and southeastern mountainous regions of the catchment.

The climate mainly acts as a system in a historical period and creates a specific geomorphic system. Evidence of glacial geomorphologies, such as typical U-shaped valleys with a width of more than 200 m, cirques, moraines, and erratic boulders were observed in field surveys in the west of the study area (Fig. 2). These geoforms indicate the dominance of the glacial climate system during the Holocene period.[24] Accordingly, multifractal and classic methods were used to study the morphology of longitudinal and cross-sections of valleys to evaluate the performance of these methods in explaining climactic changes and the dominance of various climate systems. A total of nine sheets of topographic maps of four blocks, namely Naeen, Ardakan, Abadeh, and Yazd, were used in this study.[34] The geological map of the Yazd–Ardakan catchment[32] (Fig. 3) was also prepared, and three formations, Kahar, Granite, and Taft, were selected among the geological formations in this catchment. The reasons behind the selection of these three formations were spatial extension and the lack of dispersion, allowing random homogenous sampling over a large area. Sampling in each formation was carried out using $1 \times 1\,km^2$ plots to achieve a complete view of the drainage network and the valley with all stream orders given the extent of geological formations in the study area. Moreover, the use of larger plots may eliminate the geological formations. From the plots of these three formations, the longitudinal and cross-sections of 45 valleys and geomorphometric indicators such as the ratio of valley floor width to valley height (V_f)[35] and the ratio of valley depth to valley top width (F_r),[31] and hypsometric integral (HI)[36] were randomly calculated. In this section, fractal properties and mathematical functions are used to examine the morphology of the valleys (Fig. 4).

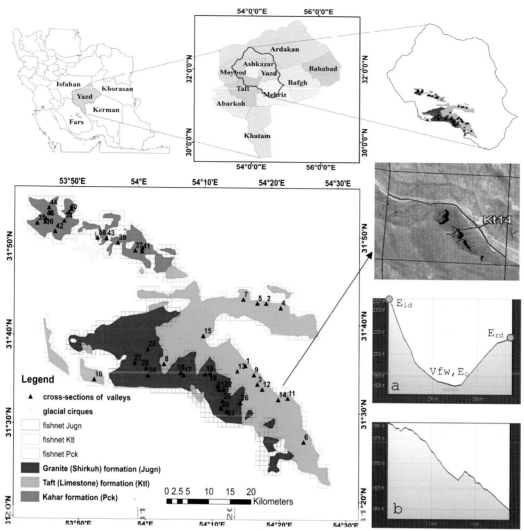

FIG. 1 The geographical location of the study area and valleys in three geological formations of Kahar, Granite, and Taft, (A) the cross-section of valley 14 in Taft formation (V_{fw}, valley floor width; E_{ld} and E_{rd}, respective elevations of the left and right valley divides; E_{sc}, the elevation of the valley floor for calculating the ratio of valley floor width to valley height, V_f), (B) the longitudinal profile of valley 14 in Taft formation.

FIG. 2 Geomorphologic evidence in the west of the study area, (A) glacial cirques, (B) moraines and erratic boulders.

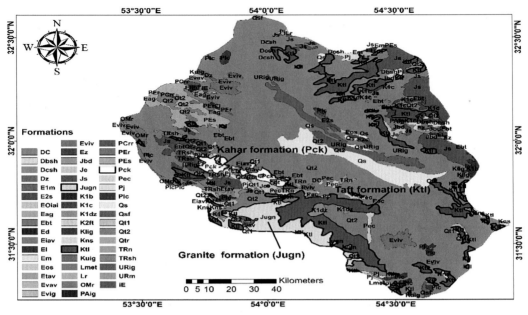

FIG. 3   The geological map of Yazd–Ardakan catchment.

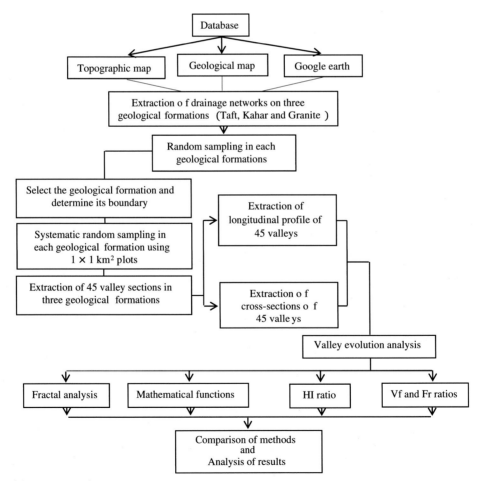

FIG. 4   Flowchart of the current study.

2.2 Hypsometric integral

Hypsometric curves were plotted by plotting the relative height (altitude) versus the relative area of the catchment. The Hypsometric integral is an indicator that depends on the catchment area, defined as the area under the curve.[36, 37] The hypsometric integral was calculated using Eq. (1):

$$HI = \frac{H_{avg} - H_{min}}{H_{max} - H_{min}} \tag{1}$$

where H_{avg} is average height, H_{min} minimum height, and H_{max} is the maximum height, all in terms of m. The hypsometric integral (HI) represents the regional distribution of the height of reliefs and the evolution of the landform drainage basin during the erosional cycle.[8] An HI greater than 0.5, with a convex curve, shows the young topography of the landform and highlands with deep valleys. An HI between 0.4 and 0.5 and the sigmoid curve show the maturity stage of reliefs, and an HI less than 0.4 represents a concave curve and the aging stage of the drainage basin.[38, 39]

2.3 The ratio of valley floor width to valley height (V_f)

A V_f greater than 2 indicates a U-shaped profile and a slight neotectonic uplift. Therefore, the river erodes the entire stream width, widening its bed. A V_f less than 1 is indicative of deep valleys with rivers that are usually associated with neotectonic uplifting. This ratio was calculated using Eqs. 2, where V_{fw} is the valley floor width, E_{ld} and E_{rd} are the left- and right-hand side hillside altitudes of the valley floor and E_{sc} is the valley floor height.[35]

$$V_f = 2V_{fw}/(E_{ld} - E_{sc}) + (E_{rd} - E_{sc}) \tag{2}$$

2.4 The ratio of valley depth to valley top width (F_r)

The graph relation or the ratio of valley depth to valley top width (F_r) is among the simply developed correlations in which D represents the valley depth and WI is the valley top width.[31]

$$F_r = D/Wl \tag{3}$$

2.5 The stream length gradient index

The index SL (stream length) was determined to investigate the effect of environmental variables on the curvature of the longitudinal stream profile. The index SL was calculated using Eq. (4), where ΔH and ΔL, respectively, show the height difference and horizontal distance difference at a certain section, and L is the stream length from the central point of the measured section to the stream origin.[37]

$$SL = (\Delta H/\Delta L).L \tag{4}$$

2.6 Fractal geometry (monofractal and multifractal analyses)

A fractal is a geometry that can be divided into several parts, each of which is a modified copy of the whole geometry in terms of size. The most widely used method for calculating the fractal dimension is the box-counting method, as it can be used for faster calculation of the fractal dimension of all elements, dimensions, and shapes than other methods. Furthermore, its algorithm can extract highly reliable results.[40] In this method, a grid of boxes is placed on the intended image. The grid size is specified by ε as the criterion for the scale. The number of box covering each part of the curve was then calculated (N_ε). This is repeated for boxes on different sides until ε approaches zero then is plotted in log-log graphic to obtain fractal dimension.[13]

$$D = \lim \frac{\log(N_\varepsilon)}{\log(1/\varepsilon)} \quad \varepsilon \to 0 \tag{5}$$

Then, log(Nε) is plotted against log(1/ε). The slope of the straight line representing the information well was considered the fractal dimension. Fig. 5 shows the procedure for calculating the fractal dimension of the drainage network on the geological formation by changing the scale of the box. In this diagram, the x- and y-axes show the network size and number of points in each square, respectively. This curve was then fitted to the curve obtained from fractal theory to

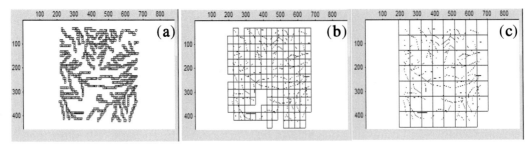

FIG. 5 Calculation of the fractal dimension of the drainage network on Kahar formation by changing the scale for a box size of (A) 8, (B) 16, and (C) 32.

achieve the correlation of the diagrams. In this way, the correlation between the number of points and the network size is obtained.[12, 13]

$$N = \varepsilon D + c \tag{6}$$

where D represents the fractal dimension or dimension correlation, and c is a constant. The fractal dimension can be calculated by box-counting or other methods with the help of Fractalyse.

An object or process is monofractal when a certain pattern is repeated, whereas when multiple different patterns govern a phenomenon, it is called a multifractal phenomenon. When only a fractal indicator, for example, its fractal dimension, is sufficient to build an object or process, it will be a monofractal object or process. Otherwise, it is a multifractal object or process. In other words, the intended object has different fractal dimensions in different parts. In multifractal analysis, instead of calculating the fractal dimension, a spectrum or function of fractal dimensions is obtained.[41] spectra D_q and $f(\alpha_q)$ were used in this study. The spectrum D_q, also known as the generalized correlation dimension, is calculated using Eq. 7:

$$D_q = \frac{1}{q-1} \lim \frac{\log \left(\sum_{i=1}^{N} \{P_i(r)\}^q \right)}{\log r} \quad r \to 0 \tag{7}$$

where, $P_i(r)$ is the probability of finding a point or mass in the ith sphere ($i = 1, 2, ..., N$) with a radius of r, and $\sum^{N} \{P_i(r)\}^q$ is called the partition function. The variable q, the moment degree of P_i, varies from $-$ to $+$. D_q is the same for all qs in monofractal structures, whereas D_q gradually decreases with increasing q in multifractal phenomena. Box-counting, fixed-radius, and fixed-mass methods were used to obtain multifractal dimensions. The box-counting method was used in this study. In this method, the phenomenon under study is covered by a network consisting of N boxes of size r, and $P_i(r)$ is obtained for the ith box. The technical spectrum $f(\alpha q)$ is also used for the multifractal analysis of phenomena, which is calculated from Eqs. (8)–(10).[14, 16, 41]

$$f(\alpha(q)) - \tau(q) \tag{8}$$
$$\tau(q) = (q-1)D(q) \tag{9}$$
$$\alpha(q) = \frac{d\tau(q)}{dq} \tag{10}$$

If the technical intensity, $\alpha(q)$, is plotted as a function of the technical spectrum $f(\alpha q)$, the multifractal characteristics are demonstrated. A second-order function is fitted to this diagram to obtain the quantitative features of this spectrum, where A, B, and C are constant. Among these variables, the asymmetry factor B is of special importance.[4]

$$f(\alpha(q) - A(\alpha(q) - \alpha_0)^2 + B(\alpha(q) - \alpha_0) + C \tag{11}$$

3 Results

Evaluating geomorphological indicators reveals the erosional stage and activity or inactivity of erosional processes in the study area. The results obtained from the hypsometric integral index showed that plots of all three geological formations have an average numerical value of 0.5, indicating the mature topography and the concave-convex curve of the study area (Table 1). Therefore, most geomorphic processes in the study area occurred in a balanced way, leading to

TABLE 1 The hypsometric integral (HI) values and the values of valley floor width to valley height (V_f) in three geological formations (H_{max}, maximum elevation; H_{min}, minimum elevation; H_{mean}, mean elevation; SL, stream length-gradient index; D, valley depth; F_r, valley depth to valley top width; WI, valley top width; V_f, valley floor width to valley top width).

Formation	Plot = Valley number	Longitude	Latitude	H_{max}	H_{min}	H_{mean}	HI	SL	V_{fw}	E_{ld}	E_{rd}	E_{sc}	D	WI	V_f	F_r
Taft	1	239810	3499049	2118	1851	1985	0.50	107	21	2039	2026	1988	50.6	330	0.47	0.15
	2	245365	3511881	2304	1663	1964	0.47	281	19	1842	1828	1801	41.8	215	0.55	0.19
	3	242423	3495018	2355	1944	2152	0.51	163	22	2226	2250	2197	53.4	171	0.53	0.31
	4	248756	3510758	2077	1651	1864	0.50	197	17	1695	1699	1684	15.8	138	1.25	0.11
	5	243299	3512023	2063	1638	1830	0.45	64	18	1686	1688	1669	18.8	219	1.01	0.09
	6	252694	3482406	2660	2070	2410	0.46	175	36	2500	2481	2444	37.0	217	0.77	0.17
	7	240037	3513010	2100	1680	1916	0.50	186	37	1896	1873	1823	50.0	212	0.60	0.24
	8	220559	3500395	3340	2880	3100	0.50	278	23	3073	3079	3062	11.0	81	1.64	0.14
	9	241711	3497030	2226	1892	2059	0.50	69	15	1984	1992	1956	36.0	172.8	0.46	0.21
	10	773231	3497342	3200	2840	3049	0.56	163	36	2879	2879	2868	11.0	217	3.27	0.05
	11	249420	3491742	2038	1827	1933	0.50	58	11	1849	1854	1844	10.5	81	1.36	0.13
	12	243506	3493926	2448	2041	2246	0.50	179	20	2190	2213	2141	72.1	200	0.33	0.36
	13	237805	3497999	2470	2046	2256	0.50	131	15	2291	2305	2268	14.6	146	0.49	0.10
	14	247159	3491569	2143	1869	2006	0.50	91	25	1987	2002	1952	50.2	156.5	0.59	0.32
	15	230159	3505816	2626	2192	2409	0.50	144	15	2224	2229	2213	16.0	197	1.14	0.08
Granit	16	224426	3498452	2926	2595	2760	0.50	152	19	2687	2694	2672	22.2	224.1	1.01	0.10
	17	225090	3497864	3177	2730	2951	0.49	153	26	2770	2790	2749	40.7	210.4	0.86	0.19
	18	231436	3497919	2644	2266	2453	0.50	44	20	2335	2341	2321	19.8	199.1	1.18	0.10
	19	230264	3497813	2530	2276	2403	0.50	73	28	2444	2447	2440	7.8	128.2	4.63	0.06
	20	233082	3495121	2750	2277	2511	0.49	77	25	2317	2316	2278	39.9	280	0.63	0.14
	21	233391	3494388	2659	2291	2474	0.50	51	20	2326	2327	2304	22.7	144	0.91	0.16
	22	234429	3494373	2515	2240	2377	0.50	65	17	2265	2264	2253	12.3	159	1.47	0.08
	23	234921	3489777	2779	2351	2564	0.50	201	32	2438	2432	2408	30.0	280	1.19	0.11
	24	233609	3490501	2910	2480	2691	0.49	109	28	2517	2518	2482	36.0	423.6	0.79	0.08
	25	234116	3491950	2689	2335	2511	0.50	97	38	2497	2517	2469	48.6	227.7	0.99	0.21
	26	238211	3491392	2883	2248	2559	0.49	88	23	2461	2461	2445	15.7	139	1.49	0.11
	27	782875	3500893	2561	2375	2468	0.50	90	19	2450	24,504	2447	3.3	83.7	5.85	0.04
	28	216882	3503713	2670	2328	2498	0.50	33	35	2383	2386	2378	8.7	141.1	4.86	0.05

Continued

TABLE 1 The hypsometric integral (HI) values and the values of valley floor width to valley height (V_f) in three geological formations (H_{max}, maximum elevation; H_{min}, minimum elevation; H_{mean}, mean elevation; SL, stream length-gradient index; WI, valley top width; F_r, valley depth to valley top width)—cont'd

Formation	Plot = Valley number	Longitude	Latitude	H_{max}	H_{min}	H_{mean}	HI	SL	V_{fw}	E_{ld}	E_{rd}	E_{sc}	D	WI	V_f	F_r
	29	216335	3499772	2804	2496	2650	0.50	65	71	2508	2508	2499	8.7	197	8.45	0.04
	30	216663	3498179	2728	2517	2622	0.50	58	30	2559	2568	2546	22.2	270.2	1.69	0.08
Kahar	31	216651	3524927	1921	1745	1832	0.49	43	45	1826	1829	1815	14.1	330	3.46	0.04
	32	762682	3532243	2037	1940	1989	0.50	31	25	1988	1990	1978	11.9	170	2.24	0.07
	33	767619	3533114	1827	1735	1780	0.49	26	62	1760	1760	1754	6.0	221	10.33	0.03
	34	765995	3530660	2048	1970	2009	0.50	31	50	1982	1978	1976	6.4	163	11.24	0.04
	35	759575	3530150	2368	2186	2276	0.49	53	40	2253	2256	2249	6.5	137	7.71	0.05
	36	761479	3530423	2364	2132	2247	0.49	55	36	2176	2171	2167	8.7	119	5.93	0.07
	37	782844	3524463	2176	1931	2053	0.50	65	55	1975	1981	1958	22.9	296.5	2.77	0.08
	38	773898	3527099	2177	1970	2074	0.50	74	37	2053	2060	2047	6.0	185	3.78	0.03
	39	778770	3526096	2377	1937	2154	0.49	82	40	2147	2165	2097	49.5	283	0.68	0.17
	40	767042	3532460	2084	1830	1957	0.50	91	30	1884	1902	1863	20.3	186	3.81	0.05
	41	216532	3524293	2026	1805	1913	0.49	83	61	1878	1878	1862	16.0	292	3.15	0.11
	42	763826	3528426	2355	2131	2243	0.50	87	41	2013	2005	1992	7.7	261	2.48	0.03
	43	776009	3526994	2309	1936	2116	0.48	90	20	1686	1675	1665	20.4	170	1.34	0.12
	44	762374	3533391	2057	1934	1996	0.50	39	133	1820	1820	1806	14.0	351	9.5	0.04
	45	761629	3531103	2270	2021	2139	0.47	48	74	1918	1910	1840	70.0	539	1	0.13

a transition from maturity to aging. Some plots show a numerical value greater than 0.5, which can be related to the detachment of a surface with a moderate deposition age (Table 1).

A total of 45 valleys were evaluated in the study area, and V_f was calculated (Table 1). This index varies in the study area depending on the type of sediment and their resistance to erosional factors. The index V_f indicates the youth or aging of the erosional cycle stages. The lowest V_f is observed for Valley 12 in Taft formation, and the highest V_f of 11.24, is seen in Valley 34 in the Kahar Formation. The average V_f values for the geological formations of Kahar, granite, and Taft are 4.63, 2.40, and 0.97, respectively. A comparison of the measured values of valleys in these geological formations indicates lateral stream erosion in the Kahar Formation and active downward stream excavation in the Taft formation. Similar results were observed for the numerical values of F_r (Table 1). In general, a lower F_r represents a U-shaped valley, but a higher F_r is indicative of a V-shaped valley. The highest F_r is observed for Valley 12 in Taft formation, and the mean F_r values for the Kahar, Granite and Taft formations are 0.07, 0.11, and 0.18, indicating a heterogeneous lithological composition and structural status of the valley.

The curvatures of the longitudinal and cross-sections of the valleys reflect very different geomorphologic features and tectonic and climatic events. Fig. 6 quantitatively describes the curvature of the longitudinal profile of valleys in three geological formations, Kahar, granite, and Taft, using different mathematical functions. The difference in mathematical functions shows the curvature of the longitudinal profile and the difference in the evolution of the valleys. The low-curvature profiles and the equilibrium stage between the deposition and excavation stages were fitted to the linear function. In contrast, valleys with greater longitudinal curvature, concave, and concave-convex profiles are fitted to polynomial functions. The curvature of the longitudinal profile indicates an extensive evolution of the valleys.

The curvature of the longitudinal profile in Taft formation is lower with a concave-convex curvature owing to the high strength, heterogeneity, and youth of this formation. The curvature of the longitudinal profiles of the valleys was higher in the Kahar and Granite formations. The index SL was used for further investigation of the effect of the three geological formations of Kahar, Taft, and Granite on the longitudinal profile of the valleys. This index is related to river power and is also sensitive to rock strength. According to the results, SL was higher in regions where the stream bed passed Taft formation (Table 1). The lowest SL of 26 was observed for the valley with the lowest depth in Plot 33 of the Kahar Formation, and the highest SL was observed in the deep valleys of Taft formation. Accordingly, SL is highly sensitive to lithology, and its value decreases from Taft to Kahar formation.

The capability of fractal geometry to explain the complexities of valley cross-sections were evaluated. Accordingly, the cross-sections of the 45 valleys in the study area were examined by multifractal analysis. Of these, six valleys in Plots 38, 39, 40, 41, 42, and 43 in the Kahar Formation, and three valleys in Plots 16, 23, and 25 in granite formation showed a multifractal nature. Fig. 7 displays the multifractal analysis of valley cross-sections in Plots 38, 39, 40, 41, 42, and 43 in the Kahar Formation. The spectrum D_q, also known as the generalized correlation dimension, gradually decreases with increasing q in these plots. This is characteristic of the spectrum of multifractal phenomena in comparison with that of monofractal structures. Another spectrum, $f(\alpha q)$, is also considered in the analysis of the properties of multifractal phenomena. The most important feature of multifractal phenomena should be explored in this spectrum. The apex of this inverse parabola is at $\alpha = 0$ and extends from $\alpha_{min} = -10$ to $\alpha_{max} = +10$. Similar results were observed for valleys in Plots 38, 39, 40, 41, 42, and 43 in the Kahar Formation. The cross-sectional profile of these valleys is also a multifractal parabola. The multifractal nature of a phenomenon suggests a complicated structure of this phenomenon. Moreover, the span of spectrum $f(\alpha q)$ indicates the multifractal degree and intensity of a phenomenon. Valley 39 has a higher multifractal degree than valleys in Plots 38, 40, 41, 42, and 43. Other valleys in the Kahar Formation, however, lack a multifractal nature, so that D_q increases with increasing q. Only three valleys in Plots 16, 23, and 25 in granite formation have a parabolic $f(\alpha q)$ spectrum (Fig. 8). Therefore, these sections have a multifractal nature, but the rest lack such a nature as D_q increases with increasing q. The multifractal nature is observed in none of the valley sections in the Taft formation, where D_q increases with increasing q (Fig. 9).

4 Discussion

The results obtained from the hypsometric integral index showed that most geomorphic processes in the study area occurred in a balanced way, leading to a transition from maturity to aging. A hypsometric analysis is an important and powerful tool for evaluating the reliefs, tectonics, lithology, climate, and erosional stages of different regions.[42, 43] The index V_f varies in the study area depending on the type of sediment and their resistance to erosional factors.

The high V_f in the Kahar Formation is suggestive of valley widening, leveling, and stability. The valley in these regions was laterally extended, leading to a U-shaped valley. The convex side slopes of the older valleys are lower with a larger valley span. The high V_f in the Kahar Formation can be related to high erodibility and age, leading to the formation of U-shaped valleys in comparison with Taft formation. Taft formation has high strength owing

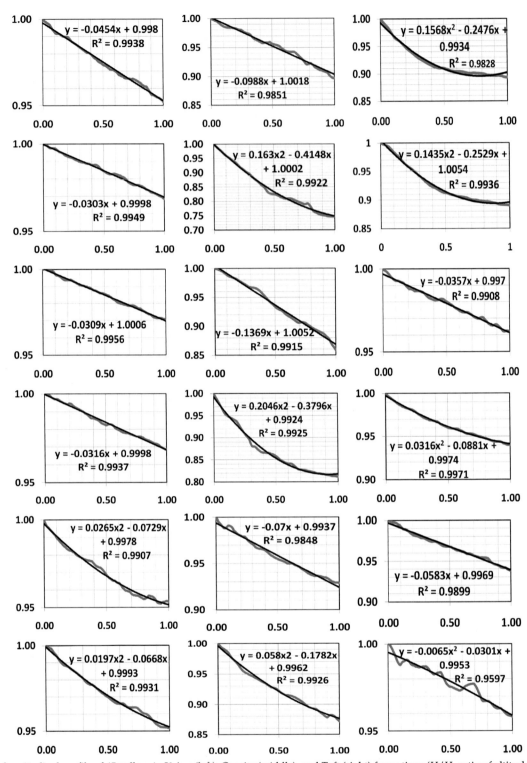

FIG. 6   The longitudinal profile of 45 valleys in Kahar (left), Granite (middle), and Taft (right) formations (H/H_0 ratio of altitude, where H is the stream altitude at the point of measurement, H_0 is the stream altitude from the river mouth at the headwaters; L/L_0 ratio of distance, where L is the stream distance from the river mouth at the point of measurement, L_0 is the stream distance from the river mouth at the headwaters).

(Continued)

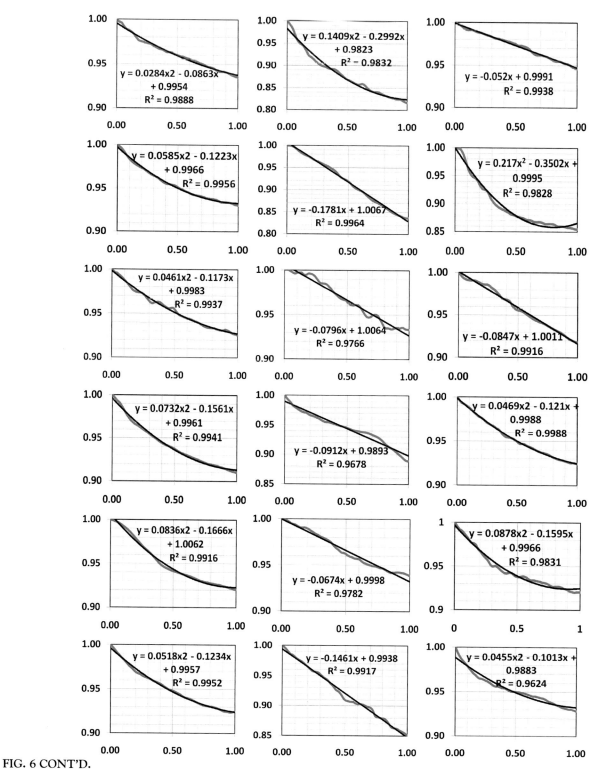

FIG. 6 CONT'D.

(Continued)

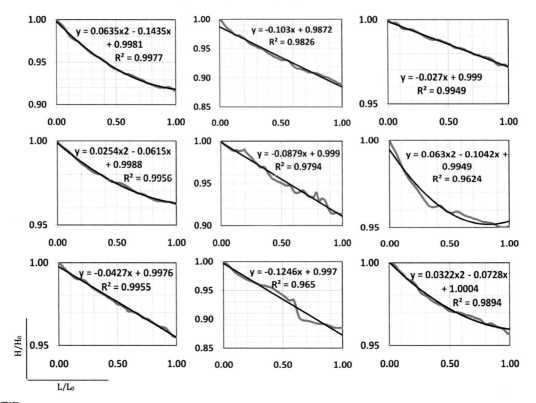

FIG. 6 CONT'D.

to lithological reasons and bulk structure. On the other hand, the results showed that slopes over 40% cover approximately 60% of Taft formation. In contrast, slopes less than 40% cover 50% of granite formation and all plots in the Kahar Formation. A higher land slope leads to higher shear force and water erosion. Considering the strength and lithology of the Taft and granite formations, streams create deep V-shaped valleys in these formations. Similar results were observed for the numerical values of F_r (Table 1). According to calculated V_f and F_r, the U-shaped valleys change to V-shaped valleys from Kahar formation to Taft formation.

The curvatures of the longitudinal and cross-sections of the valleys reflect very different geomorphologic features and tectonic and climatic events. The curvature of the longitudinal profile in Taft formation is lower with a concave-convex curvature owing to the high strength, heterogeneity, and youth of this formation. The curvature of the longitudinal profiles of the valleys was higher in the Kahar and Granite formations. The concave longitudinal profile of the valleys is seen in these two formations clearly, especially in the Granite formation. Owing to the sandstone stratification and shales in the Kahar Formation, the sandstone veins in this formation break the concave profile, preventing the formation of a concave longitudinal profile. Granite formation has a higher strength than Kahar formation but has a lower strength than Taft formation. Given the regional climatic conditions and the presence of granite formation under arid and cold conditions, this formation is prone to weathering, and severe weathering is observed in the form of granular rocks and the production of fine-grained sedimentary materials. However, the uniformity of granite formation causes the formation of a concave stream longitudinal profile in comparison with the Kahar formation. The results of the hypsometric integral (HI) also confirmed the mature topography and a concave-convex curve in the study area. Polynomial functions better approximate high-curvature profiles. Polynomial methods can model the longitudinal section of valleys more efficiently. The results of this study confirm those reported in other studies on the analysis of valley profiles using mathematical functions. Evaluations indicated that the difference in the function type demonstrates the evolution of valleys so that low-curvature and high-curvature valleys were fitted to linear and polynomial functions, respectively. However, none of the valley profiles in the study area were fitted to exponential and power-law functions, indicating the transition of valleys from maturity to aging stages.[11, 14] The results also showed the successful application of geomorphologic indicators of the hypsometric integral, V_f, F_r, and SL, for studying valley profiles. Furthermore, valleys in the Kahar and Granite formations show a different longitudinal evolution than those in the Taft formation, suggesting different erosional, lithological, and climatic systems in these three geological formations. Using these indicators, the evolution of valleys can be studied, and various aggradation and degradation

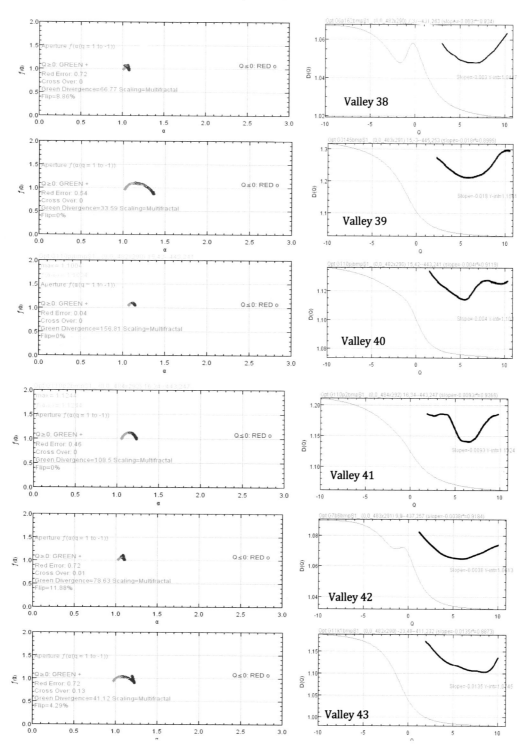

FIG. 7 The multifractal analysis of cross-sections of six valleys in Kahar formation (these sections have a multifractal nature).

processes across the valley can be explained.[11, 19, 44, 45] Monofractal and multifractal analyses of valley cross-sections in three geological formations, Taft, Kahar, and Granite, showed a transition from a monofractal to a multifractal nature. Meanwhile, none of the valley sections in the Taft formation showed a multifractal nature. The multifractal approach seems more appropriate in some valley sections in the Granite and Kahar formations, as fractal parameters may differ depending on location. Hence, it can be concluded that valley cross-sections in the same formation may follow different mathematical equations. Classic methods and mathematical functions analyze the different behaviors of the Kahar

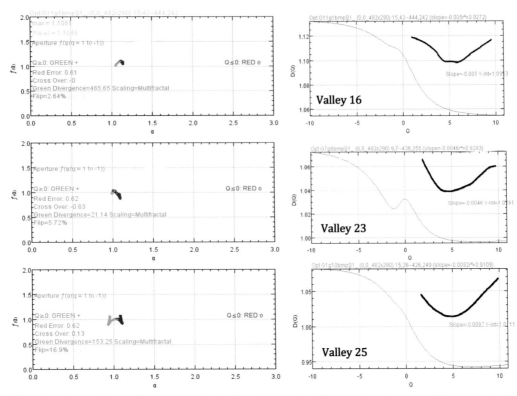

FIG. 8 The multifractal analysis of three valley cross-sections in Granite formation (these sections have a multifractal nature).

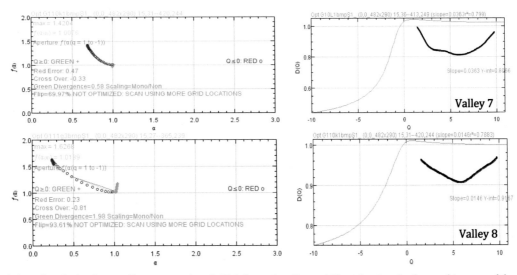

FIG. 9 The multifractal analysis of two valley cross-sections in Taft formation (the multifractal nature is observed in none of the valley sections).

and granite formations in comparison with Taft formation caused by the heterogeneous texture and lithology of these two formations. Taft formation with a homogenous texture lacks diverse stratification. However, the batholith body in granite formation consists of granite and granodiorite, and their different sensitivities to erosion have caused multifractal behavior in this formation. The Kahar Formation is composed of alternating shales and sandstones. Various lithological layers are observed in the Kahar formation, leading to this complicated behavior such that erodibility increases from Taft to Kahar formation. According to the multifractal analysis, this complicated behavior is attributed to sudden changes along the valley, causing climate change and the dominance of different climate systems.

The use of geomorphological indicators and mathematical functions helped analyze the complexities of valley cross-sections, so that the longitudinal profile of valleys in the Kahar and Granite formations were well evaluated in comparison with Taft formation. The erosional stage was also estimated during the transition from maturity to aging. As mentioned, the geoforms in the study area indicate the presence of glacial valleys in the region. In fact, classic methods and mathematical functions are not able to evaluate the morphology of valleys alone. In contrast, the multifractal analysis demonstrates the nonlinear nature of valley profiles in the Kahar and Granite formations. Given the need for only the valley profile, ease of analysis, accuracy, and, more importantly, is based on the nonlinear nature of valley profiles, multifractal analysis is an efficient method for evaluating the morphology and evolution of valleys. The results of this study are consistent with those obtained in glacial and fluvial landforms in the Himalayas, indicating the effectiveness of the multifractal analysis in evaluating complicated glacial landforms.[32]

5 Conclusion

Evaluating the longitudinal and cross-sections of valleys allows the study of the evolution and mechanism of various aggradation and degradation processes across valleys. The evolution of valleys caused by long-term variations of the stream bed may be followed normally or created because of sudden changes in the region or along the valley, leading to specific geomorphological shapes. The quantitative description of the longitudinal and cross-sections of valleys in the three geological formations indicated that the U-shaped valleys in the Kahar Formation changed to V-shaped valleys in Taft formation. The different curvatures of valley profiles in the study area can be related to the erosional and depositional fronts of streams flowing in these valleys. A relatively straight longitudinal profile with successive concave and convex curvatures indicates the dominance of erosion and deposition, respectively in convex and concave curvatures. The qualitative analysis of valleys is considered a useful tool for classifying valleys formed by different processes. According to the results, in monofractal valleys with a uniform and homogenous lithology, less erosion and sedimentation were observed, and the formation showed a higher strength. Owing to the different strengths of rock layers in multifractal valleys, erosion and sedimentation increase in this type of valley. Consequently, most erosion control plans should focus on multifractal hillsides in these regions. According to the multifractal analysis, this complicated behavior is attributed to sudden changes along the valley, causing climate change and the dominance of different climate systems. The results indicate the ability of fractal geometry to measure the geometrical complexities of this phenomenon. Mono-fractal and multifractal analyses of valley cross-sections in three geological formations, Taft, Kahar, and Granite, showed a transition from a monofractal to a multifractal nature. Meanwhile, none of the valley sections in the Taft formation showed a multifractal nature. The multifractal approach seems more appropriate in some valley sections of the Granite and Kahar formations. Therefore, the disorder (irregularity) of various variables increases with an increase in the fractal dimension. While calculating the fractal dimension, the order or disorder (regularity or irregularity) governing processes influencing the valley system can be used to better manage the region. Due to the vulnerable ecosystems of arid areas, a good understanding of the erosional and geomorphological status of valleys is an efficient and useful tool for managing such areas.

References

1. Sant D, Karanth R. Drainage evolution of the lower Narmada valley, western India. *Geomorphology.* 1993;8(2–3):221–244.
2. Tipping R. Fluvial chronology and valley floor evolution of the upper Bowmont valley, Borders region. *Earth Surf Proc Land.* 1994;19(7):641–657.
3. Nash DJ, Shaw PA, Thomas DS. Duricrust development and valley evolution: process–landform links in the Kalahari. *Earth Surf Proc Land.* 1994;19(4):299–317.
4. Stokes M, Mather AE. Tectonic origin and evolution of a transverse drainage: the Rı´o Almanzora, Betic cordillera, Southeast Spain. *Geomorphology.* 2003;50(1–3):59–81.
5. Zelilidis A. Drainage evolution in a rifted basin, Corinth graben, Greece. *Geomorphology.* 2000;35(1–2):69–85.
6. Wallerstein N, Thorne C. Influence of large woody debris on morphological evolution of incised, sand-bed channels. *Geomorphology.* 2004;57(1–2):53–73.
7. Ohmori H. Morphological characteristics of longitudinal profiles of rivers in the South Island, New Zealand. *Bull Dep Geog.* 1996;28:1–23. University of Tokyo.
8. Radoane M, Rādoane N, Dumitriu D. Geomorphological evolution of longitudinal river profiles in the Carpathians. *Geomorphology.* 2003;50(4):293–306.
9. Harbor JM. Numerical modeling of the development of U-shaped valleys by glacial erosion. *Geol Soc Am Bull.* 1992;104(10):1364–1375.
10. Prasicek G, Otto J-C, Montgomery DR, Schrott L. Multi-scale curvature for automated identification of glaciated mountain landscapes. *Geomorphology.* 2014;209:53–65.
11. Peckham SD. Longitudinal elevation profiles of rivers: Curve fitting with functions predicted by theory. In: Jasiewicz J, Zb Z, Mitasova H, Hengl T, eds. *Geomorphometry for Geosciences.* Poland: Int Soc Geomorphometry; 2015:137–140.
12. Mandelbrot B. How long is the coast of Britain? Statistical self-similarity and fractional dimension. *Science.* 1967;156(3775):636–638.
13. Mandelbrot BB. *The Fractal Geometry of nature/Revised and Enlarged Edition.* whf; 1983.

14. Biswas A, Zeleke T, Si BC. Multifractal detrended fluctuation analysis in examining scaling properties of the spatial patterns of soil water storage. *Nonlinear Process Geophys.* 2012;19(2).
15. Dutta S, Ghosh D, Chatterjee S. Multifractal detrended fluctuation analysis of pseudorapidity and azimuthal distribution of pions emitted in high energy nuclear collisions. *Int J Mod Phys A.* 2014;29(18):1450084.
16. Dutta S. Decoding the morphological differences between himalayan glacial and fluvial landscapes using multifractal analysis. *Sci Rep.* 2017;7 (1):1–8.
17. Bayati Khatibi M, Heidarzadegan P. Determination of the stages of geomorphological evolution of mountain valleys with classical and mathematical methods (case study: eleven basins and main valley of Sahand Mountain). *Geogr Dev Iranian J.* 2005;3(5):85–110 [in Persion].
18. Larue J-P. Longitudinal profiles and knickzones: the example of the rivers of the Cher basin in the northern French massif central. *Proc Geol Assoc.* 2011;122(1):125–142.
19. Saber R, Isik V, Caglayan A. Tectonic geomorphology of the Aras drainage basin (NW Iran): implications for the recent activity of the Aras fault zone. *Geol J.* 2020;55(7):5022–5048.
20. Khorshid Dost AM, Rezaei Moghadam MH, Khaleghi S. Analysis of river evolution and Erosion in Sangharcha basin using longitudinal and cross profiles and digital elevation model. *J Geogr Plann.* 2011;15(34):45–65 [in Persion].
21. Rajabi M, Bayatih KM. Investigation of glacial valley landforms (case study: glacier valleys of Sahand mountain). *Geogr Res.* 2008;64:105–121 [in Persion].
22. Tubau X, Lastras G, Canals M, Micallef A, Amblas D. Significance of the fine drainage pattern for submarine canyon evolution: the Foix canyon system, Northwestern Mediterranean Sea. *Geomorphology.* 2013;184:20–37.
23. Li Y, Liu G, Cui Z. Glacial valley cross-profile morphology, Tian Shan Mountains, China. *Geomorphology.* 2001;38(1–2):153–166.
24. Sharifi Paichoon M, Parnoon F. The Study and analysis of geometrical variations of longitudinal and cross profiles of Qaresou River during 1954–2014. *Geogr Dev Iranian J.* 2017;15(46):43–60 [in Persion].
25. Farahbkhsh Z. Hydrogeomorphology of Khezrabad Basin with Emphasis on Glacier Features. M.Sc. Thesis: Yazd University, Department of Geograph [in Persion].
26. Azizi G, Maleki S, Karimi M, Shahbazi R, Rostami H. Holocene vegetation and climate changes in Iran. *Quaternery J Iran.* 2018;3(11):205–229 [in Persion].
27. Karam A. Chaos theory, fractal & non-linear systems in geomorphology. *J Phys Geogr.* 2010;3(8):67–82 [in Persion].
28. Malekinezhad H, Talebi A, Ilderomi AR, Hosseini SZ, Sepehri M. Flood hazard mapping using fractal dimension of drainage network in Hamadan City, Iran. *J Environ Eng Sci.* 2017;12(4):86–92.
29. Mohammadi Khoshoui M, Ekhtesasi MR. Comparison of fractal dimension and geomorphologic characteristics in the management of Aqda Basin. *Environ Eros Res J.* 2019;9(1):62–84 [in Persion].
30. Mohammadi Khoshoui M, Ekhtesasi MR, Talebi A, Hosseini SZ. The application of fractal dimension and morphometric properties of drainage networks in the analysis of formation sensibility in arid areas (case study, Yazd-Ardakan Basin). *Desert Ecos Eng J.* 2019;8(24):1–18 [in Persion].
31. Graf WL. The geomorphology of the glacial valley cross section. *Arct Antarct Alp Res.* 1970;2(4):303–312.
32. GSI Geological survey and mineral explorations of Iran. 2014. Geological maps 1:100000 [in Persion].
33. Taghizadeh-Mehrjardi R, Minasny B, Sarmadian F, Malone B. Digital mapping of soil salinity in Ardakan region, Central Iran. *Geoderma.* 2014;213:15–28.
34. NCC National Cartographic Center. *Topographic Maps 1:25000;* 2014 [in Persion].
35. Bull WB, WB B, LD M. Tectonic geomorphology north and south of the Garlock Fault, California. In: *Proceedings of the 8th Annual Geomorphology Symposium on Geomorphology in Arid Regions.* New York: Binghamton University, United States; 1977:115–138.
36. Pike RJ, Wilson SE. Elevation-relief ratio, hypsometric integral, and geomorphic area-altitude analysis. *Geol Soc Am Bull.* 1971;82(4):1079–1084.
37. Hack JT. Stream-profile analysis and stream-gradient index. *J Res us Geol Surv.* 1973;1(4):421–429.
38. Pérez-Peña JV, Azor A, Azañón JM, Keller EA. Active tectonics in the Sierra Nevada (Betic cordillera, SE Spain): insights from geomorphic indexes and drainage pattern analysis. *Geomorphology.* 2010;119(1–2):74–87.
39. El Hamdouni R, Irigaray C, Fernández T, Chacón J, Keller E. Assessment of relative active tectonics, southwest border of the Sierra Nevada (southern Spain). *Geomorphology.* 2008;96(1–2):150–173.
40. Kusák M. Methods of fractal geometry used in the study of complex geomorphic networks. *Acta Univ Carol Geogr.* 2014;49(2):99–110.
41. Subhakar D, Chandrasekhar E. Reservoir characterization using multifractal detrended fluctuation analysis of geophysical well-log data. *Physica A.* 2016;445:57–65.
42. Huang X, Niemann JD. Modelling the potential impacts of groundwater hydrology on long-term drainage basin evolution. *Earth Surf Proc Land.* 2006;31(14):1802–1823.
43. Guarnieri P, Pirrotta C. The response of drainage basins to the late quaternary tectonics in the Sicilian side of the Messina Strait (NE Sicily). *Geomorphology.* 2008;95(3–4):260–273.
44. Pedrera A, Pérez-Peña JV, Galindo-Zaldívar J, Azañón JM, Azor A. Testing the sensitivity of geomorphic indices in areas of low-rate active folding (eastern Betic cordillera, Spain). *Geomorphology.* 2009;105(3–4):218–231.
45. Ayaz S, Dhali MK. Longitudinal profiles and geomorphic indices analysis on tectonic evidence of fluvial form, process and landform deformation of eastern Himalayan Rivers, India. *Geol Ecol Landscapes.* 2020;4(1):11–22.

3

A data-driven approach for estimating contaminants in natural water

Saha Dauji[a,c] *and Tirumalesh Keesari*[b,c]

[a]Nuclear Recycle Board, Bhabha Atomic Research Centre, Mumbai, India [b]Isotope and Radiation Application Division, Bhabha Atomic Research Centre, Mumbai, India [c]Homi Bhabha National Institute, Mumbai, India

1 Introduction

Water quality is an important factor for the quality of life and is affected by diverse forms of pollution caused by both anthropogenic and geogenic factors. Although water samples are routinely measured for some quality parameters, the evaluation and prediction of all water quality parameters is a daunting task. Among the inorganic pollutants, high chloride concentrations in groundwater have been a serious issue affecting the freshwater resources of many countries, and this problem worsens due to rising sea levels.[1–9] Widespread upward trends in chloride loads in groundwater have been reported in many studies in several parts of the world due to factors such as animal wastes and agrochemicals in agriculture, mine spillage, human excreta, landfills, industrial effluents, and de-icing in cold regions.[10]

High Cl^- in groundwater has also been reported in India, which is attributed to waterlogging due to poor soil drainage, excess irrigation return flows, evapotranspiration, and seawater intrusion. The arid and semi-arid regions of India, especially the northwestern parts, have shown very high saline water in shallow zones mainly due to evaporation.[11, 12] On the other hand, the coastal regions are also impacted by high Cl^- water due to seawater intrusion resulting from the reversal of the hydraulic gradient when freshwater aquifers are exploited indiscriminately. This scenario is commonly observed in high population density regions such as urban centers in Tamil Nadu, Orissa, Pondicherry, West Bengal, and Kerala.[11] Some parts of India became water-logged due to over-use of canal water and improper agricultural practices, which impair both the soil and water quality by inducing high salt content.[13, 14]

A secondary maximum contaminant level (SMCL) for Cl^- in drinking water was set at 250 mg/L by both the World Health Organization (WHO) and the US Environmental Protection Agency (USEPA).[15] However, this guideline limit relates only to aesthetics, that is, water tastes salty above 250 mg/L; therefore, this does not form an enforceable guideline. For aquatic life, the recommended acute criterion concentration for Cl^- is 860 mg/L.[16] Studies have reported that excessive intake of drinking water containing sodium chloride (>2.5 g/L) causes hypertension,[17] but it is believed that the concentration of sodium ions mainly causes this effect. Human health studies conducted so far have not observed chloride toxicity except in special cases, such as congestive heart failure.[18] The intake of large quantities of chloride did not cause any significant physiological changes in individuals as long as freshwater is also consumed intermittently. Other effects of high Cl^- in water include galvanic and pitting corrosion in pipes.[19] Excess Cl^- in water used in concrete could lead to corrosion issues of reinforcement[10, 20]; therefore, the Indian standard limits the maximum chloride content of water for mixing concrete.[21]

37

1.1 Estimation of chloride in water

The chloride concentration in water is mainly measured in laboratories, as on-site methods are neither very common nor very precise. Several analytical methods are available for the estimation of chloride in water.[22] The commonly used methods include titration or ion-selective electrode or chromatography, which are chemical-intensive, require good quality reagents and clean labs to arrive at accurate results. This makes Cl^- measurements on a routine basis a demanding task. Therefore, methods to estimate Cl^- from routinely measured in-situ parameters, such as total dissolved solids (TDS) or electrical conductivity (EC), would be very helpful since their estimation is relatively simple, fast, eco-friendly, and cost-effective. TDS and EC values are both reported to be strongly correlated to Cl^- concentrations in water[23] and thus can be potential proxy candidates for Cl^-.

Correlation expressions ($R^2 = 0.97$ to 0.98) developed between EC and dissolved salts in river water were obtained and applied to a very large area in the Murray River and its tributaries.[24] In another study,[23] measured Cl^- concentrations in streamflow not being available for the entire study period/area, the available data (450 surface water analyses data) was used to develop an exponential relationship ($R^2 = 0.95$) between Cl^- concentration and EC, which was thereafter used for estimation of unknown (not measured) values of Cl^-. In a later study, Bresciani et al.[25] arrived at an empirical equation ($R^2 = 0.9996$) between EC and Cl^-. The limitations of these studies are: (i) performance evaluation was carried out with the same data as was used for developing the model and (ii) the suitability of the chosen empirical expression and the choice of performance metric/s, which has mostly been confined to the coefficient of determination.

1.2 Data-driven tools applied for water quality

The development of data-driven techniques, such as neural networks, evolutionary algorithms, and tree-based algorithms, has helped to extract useful information from various available datasets and has found numerous applications in many engineering and natural science fields, including groundwater quality. Based on several hydrochemical parameters and rainfall data, groundwater was classified according to its quality in the Ardebil region, Iran, using the decision tree method.[26] The source of groundwater salinity was identified by Haselbeck et al.[27] using a self-organizing map, a different machine learning application. Engle and Brunner[28] proposed an approach for parsimonious variable selection, in which water from oil and gas wells in the northern U.S. Gulf Coast Basin was characterized based on an emergent self-organizing map analysis. Groundwater salinity mapping was reported by Mosavi et al.[29] using machine learning approaches, including boosted regression tree and random forest algorithms of the tree-based method. Norouzi and Moghaddam[30] assessed the groundwater quality in the Miandoab plain aquifer in Iran and employed a random forest algorithm.

Chloride has been an important ion in hydrology because of its application in understanding the processes affecting groundwater chemistry and the sources and processes impacting the overall groundwater quality.[31, 32] Considering the utility of Cl^- ions in hydrological investigations and the limitations associated with its measurement, it would be pertinent to use proxies that can provide information on Cl^- concentration in water and facilitate easy and rapid measurement with good accuracy.

1.3 Research significance

The innovations in this study are as follows:

- Use of a new proxy (TDS) for estimation of Cl^- in groundwater;
- Application of a new tool (DT) for the estimation of Cl^- in groundwater from TDS
- The seasonal influence (premonsoon/postmonsoon/all-season) and degree of saturation of soil (water-depleted or water-logged) on the relationship between the target variable (Cl^-) and the proxy (TDS) were considered for achieving more accurate estimates.

2 Data and methods

2.1 Data

2.1.1 Study area

In the present study, the northwest (NW) part of India was chosen as the study area, where groundwater salinization is a major issue. Most of the research studies have reported that water samples collected from the shallow aquifers are saline and unfit for drinking or irrigation at many investigated sites.[33] The districts selected for this study are

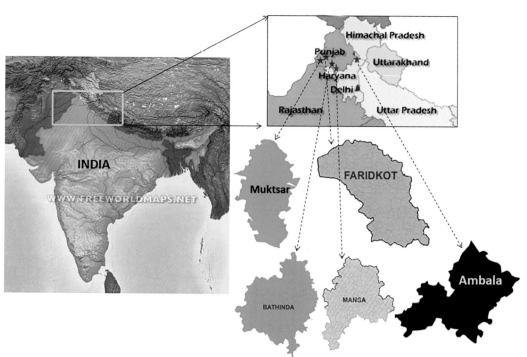

FIG. 1 Location of selected study sites in the northwest part of India. *Modified from www.mapsofindia.com.*

five in number and belong to the NW India: Muktsar (29.91°N to 30.67°N; 74.25°E to75.32°E), Faridkot (29.90°N to 34.90°N; 74.25°E to 75.42°E), Bathinda (29.56°N to 30.57°N; 74.63°E to 75.77°E), and Mansa (29.53°N to 30.20°N; 75.17°E to 75.77°E) in the state of Punjab; and Ambala (30.09°N to 30.55°N; 76.58°E to 77.20°E), in the adjacent state of Haryana (Fig. 1). The data consisted of the TDS and chloride concentrations measured in groundwater (wells) at different locations in the districts and were collected from the literature, as listed in Table 1. Precipitation during the monsoon majorly affects the water table of these districts; hence, the depths of the water table before and after the monsoon are also reported.

In addition to water quality (salinity) issues, some of the locations in this region are reportedly water-depleted or water-logged. The districts of Bathinda and Mansa can be termed as water depleted because of the depleting water levels up to 0.2m/a.[39, 40] On the other hand, waterlogging occurs when the moisture saturates the soil pores present in the root zone. This restricts air circulation and develops a low oxygen level, allowing the build-up of carbon dioxide in the root zone. This condition impacts plant growth and degrades water quality. Water-logged soils also limit the removal of salts that are transported through irrigation water and lead to the accumulation of high salt contents through evapotranspiration. The waterlogging scenario is found to be on the rise continuously in the NW part of India, especially in the Muktsar and Faridkot districts of Punjab, where the rise in water levels has been as high as 0.5m/a during the last 10 years.[39] Another report by Singh[40] suggests that 70% of these districts is waterlogged and the water level rise from 0.008m/year to 0.322m/year.

Compiled data from published literature, as explained above, have been used to develop models for the estimation of chloride from TDS measurements. First, the models were developed with the entire data set together, as reported in the literature[23–25] wherein EC was used as a proxy. This dataset is referred to as the Five Districts set henceforth. To examine the suitability of the models and examine any seasonal influence, three models were developed for each set

TABLE 1 Data sources and seasonal water table levels.

District[reference]	Depth of well (m bgl)	Water table (premonsoon) (m bgl)	Water table (postmonsoon) (m bgl)
Faridkot[34]	4.8–24.6	3.50–15.35	1.94–16.1
Muktsar[34]	13–213	3.5–9.0	1.5–6.0
Bathinda[35–37]	9–137	3.4–20.3	2.2–20.7
Amabala[38]	32–152	1.1–10.5	0.7–9.5
Mansa[35]	3.1–30.4	1.82–54.8	1.72–55.86

of data: premonsoon, postmonsoon, and all-season. As explained above, waterlogged and water-depleted conditions prevail in the study area. Therefore, it is worthwhile to explore whether the accuracy of the models could be improved by considering this pertinent factor: waterlogged or water depleted. For that purpose, the districts Mansa and Bhatinda have been grouped under "Water depleted" and the remaining three districts (Faridkot, Ambala, and Muktsar) have been considered under "Water logged." For each of these two sets, three seasonal models, as explained for the "Five district" dataset, were developed. Consequently, results from a total of nine best models were reported as the outcome of the study. The statistical characteristics of the three datasets are listed in Table 2.

Considering the entire dataset together (five districts), the statistical parameters displayed marked variation across seasons for both TDS and chloride. For the water-logged set, the statistical characteristics of the premonsoon, postmonsoon, and all-season TDS match quite closely, whereas, for chloride, the range and standard deviation vary noticeably between the seasonal data sets. For the water-depleted set, almost all parameters for both TDS and chloride varied across seasons. Such variations indicate that the seasonal (premonsoon and postmonsoon) models would yield more accurate estimates than the all-season model. The statistical parameters for the two variables TDS and chloride, when examined for a particular season (premonsoon/postmonsoon) or all seasons across the datasets from water-logged and water-depleted regions show wide variation. Therefore, it is highlighted that in place of the estimation models developed from all available data, as reported for chloride estimation in literature,[23–25] considering seasons and degree of saturation of the soil in the study area could result in better estimation accuracy and deserves exploration.

2.2 Decision tree

A decision tree (DT), which is also known as a model tree or regression tree in the literature, is a data-driven tool suitable for regression models. DT works on the principle of individual regression in subdomains of the model space, and therefore, it ends up in multiple models for different regions of the model space. This method is often found to be

TABLE 2   Statistical parameters of the datasets used for the study.

Dataset	Statistics	All season TDS (mg/L)	All season Chloride (mg/L)	Premonsoon TDS (mg/L)	Premonsoon Chloride (mg/L)	Postmonsoon TDS (mg/L)	Postmonsoon Chloride (mg/L)
Five districts	Number	402	402	182	182	220	220
	Mean	1341.44	176.60	1212.35	162.18	1448.23	188.54
	Standard Deviation	818.41	174.01	644.37	156.82	926.26	186.56
	Coefficient of variation	0.61	0.99	0.53	0.97	0.64	0.99
	Range	5323.15–129.00	1242.50–4.97	4103.75–150.00	1242.50–4.97	5323.15–129.00	1174.30–7.95
Water logged	Number	242	242	121	121	121	121
	Mean	939.99	115.74	944.40	111.78	935.59	119.71
	Standard Deviation	476.69	123.38	467.00	112.35	488.08	133.86
	Coefficient of variation	0.51	1.07	0.49	1.01	0.52	1.12
	Range	2765.00–129.00	1039.44–4.97	2500–150.00	858.39–4.97	2765.00–129.00	1039.44–7.95
Water depleted	Number	160	160	61	61	99	99
	Mean	1948.63	268.66	1743.88	262.15	2074.79	272.66
	Standard Deviation	853.53	197.82	619.95	183.46	951.04	206.98
	Coefficient of variation	0.44	0.74	0.36	0.70	0.46	0.76
	Range	5323.15–1008.35	1242.50–15.40	4103.75–1008.35	1242.50–49.70	5323.15–1018.40	1174.30–15.40

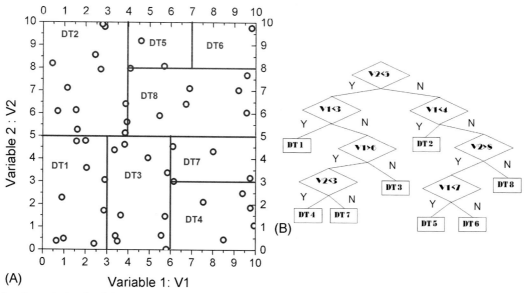

FIG. 2   Sample schematic details of decision tree (A) model domain split and (B) tree structure resulting from subdivision of model space into subdomains shown in (A).

efficient for developing relationships for data with large scatter, for which a single regression model would not result in acceptable accuracy. Considering the wider scatter of points in the data points in certain sets selected for this study (discussed in more detail in Section 2.3), it was deemed worthwhile to examine this tool for application. The name "decision tree" is derived from the tree structure, which is the final outcome of the computations carried out to develop a DT model. The origin of the DT is a decision box called the root node, which branches off into other decision boxes (nodes) or leaves (models). Whether the branching terminates at a leaf node or continues with other decision boxes or nodes depends on the decision output based on a certain criterion; the output is a "yes" or "no." Following this process, the model space is subdivided into subdomains, for each of which a separate model (relationship) is established. To execute the subdivision or domain splitting, algorithms such as minimum entropy in the subdomain, collecting as many samples as possible in the class, or any other algorithm would be employed. A popular M5 algorithm of DT employs the method of minimizing the standard deviation of the class value reaching a node for domain splitting, and the same has been adopted for this study.[41–43]

A sample division of model space is presented in Fig. 2A and the corresponding tree structure of the DT model is represented in Fig. 2B. The selection of the root node is based on the resulting reduction of the standard deviation of the model output, and the root node for which it is minimized is selected. The model development process involves examining many such input divisions and the particular choice which results in maximum reduction of standard deviation is selected for model space division. Subsequently, linear regression models are developed for each subdomain. There are algorithms available in the literature for avoiding DT models with too many domain splits or large discontinuities between the adjacent model spaces.[42–44] Interested readers are directed to textbooks for a more detailed treatment on the development of DT models.[42, 44]

2.3 Methodology

The data scatter (TDS vs Cl⁻) was studied and is presented in Fig. 3 for all nine sets of data, along with the correlation between the variables for each set. The fact that the data points exhibited high scatter (departures from the 1:1 diagonal) along with poor correlation in many cases (particularly Fig. 3A, D, E, and F), suggested that instead of a single model for the entire model space, multiple models for different subdomains could be more effective in improving the accuracy of estimates. The task of subdividing the model space could be performed along with the development of models with a data-driven tool: a decision tree. As explained above, this was the motivation for exploring this tool for the desired objective of estimating the Cl⁻ concentration from the TDS measurements in this study.

The DT models were developed from randomly selected 75% of the data (termed modeling data) and were subsequently evaluated with the remaining 25% of data (termed evaluation data). For each of the nine datasets (as explained in Section 2.1), six different models were developed from the random data divisions and the developed models were

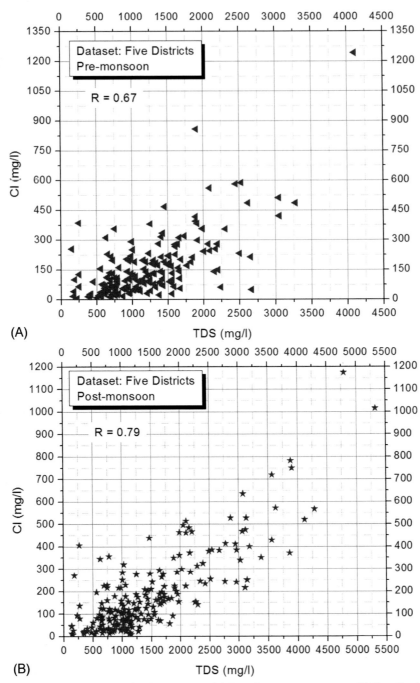

FIG. 3   Scatter plot for the TDS—chloride ion data for the nine cases (A) Five districts: premonsoon; (B) Five districts: postmonsoon;

(Continued)

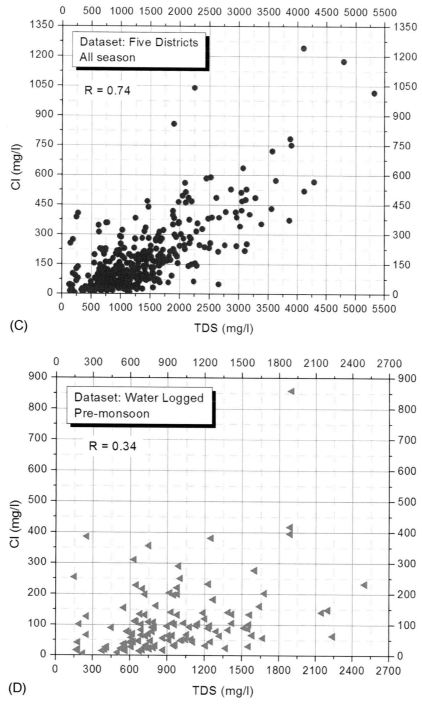

FIG. 3, CONT'D (C) Five districts: all season; (D) Waterlogged: premonsoon;

(Continued)

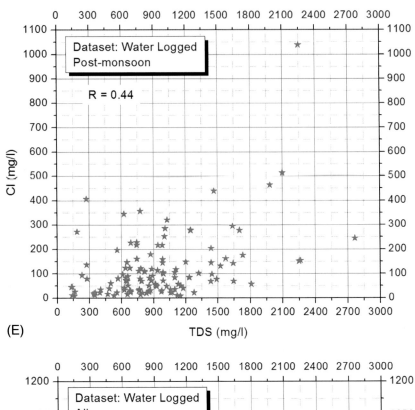

(E)

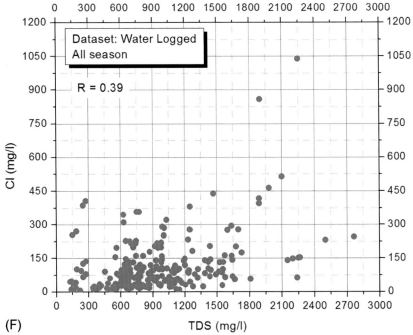

(F)

FIG. 3, CONT'D   (E) Waterlogged: postmonsoon; (F) Waterlogged: all season;

(Continued)

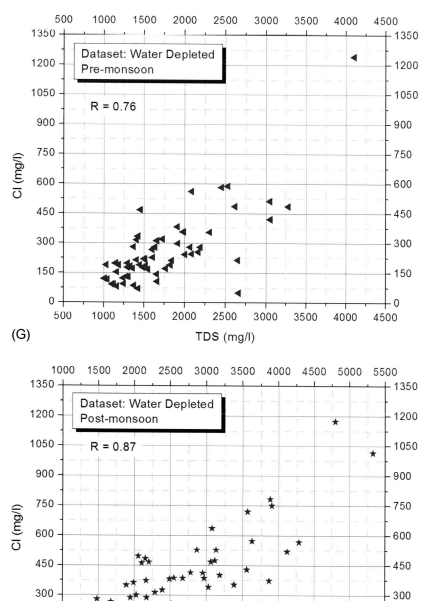

(G)

(H)

FIG. 3, CONT'D (G) Water depleted: Premonsoon; (H) Water depleted: postmonsoon;

(Continued)

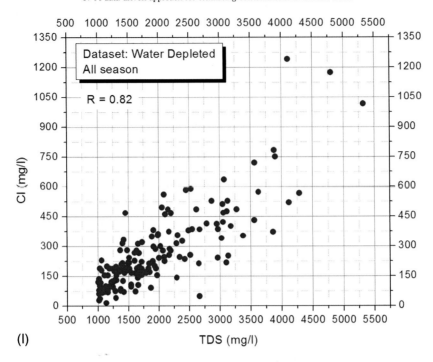

FIG. 3, CONT'D   (I) Water depleted: all season.

evaluated based on their performance to arrive at the best model for a particular dataset. The method, by which the best model was chosen from the six-candidate models, would be explained subsequently. It is needless to mention that the performance evaluation is very important for any data-driven model. Different performance metrics examine different aspects of the accuracy of the developed models and therefore evaluation of such models should be performed in a comprehensive manner for which multiple metrics should be employed,[45–47] depending on the application of the model, as discussed in the literature.[48] For this study, the chosen performance metrics are: correlation coefficient (R); mean absolute relative error (MARE); and root mean squared relative error (RMSRE). Whereas R indicates the degree of linear association of the observations and the estimates, the MARE evaluates the relative errors in the absolute scale, and the RMSRE penalizes the models with estimates which have large relative deviations from the measurements. The combination of these three metrics would help to thoroughly evaluate the developed DT models for suitability of application. It is highlighted here that the performance metrics reported in tables or figures, and discussed in the text of this chapter refer strictly to the estimates of Cl⁻ from the evaluation set only. These performance measures are self-explanatory and basic statistics textbooks can be referred to for further details.

Different models might be best considering different performance measures; therefore, decisions regarding the best model considering all three metrics are necessary. To combine the results of these three performance measures and the selection of the best model from the six-candidate DT models developed for a particular dataset, an intuitive ranking system was adopted in this study. Each model was ranked according to each performance measure, which was equivalent to a score of that model against that performance measure. It is again reiterated that the performance metric under discussion is evaluated from the evaluation set only. Better models are those with a higher correlation coefficient and lower MARE or RMSRE and, therefore, given a higher rank. Thus, a model may have different ranks based on different performance considerations. Subsequently, the total ranks received by the model from the three metrics were calculated. The lowest score, that is, the lowest such total, would indicate the best model or overall rank one while considering the three error measures in a combined fashion. Thus, the best DT model was chosen among the candidate models for each dataset, and the results were reported. The quantitative evaluation is further corroborated by the scatter plots of the measured and estimated chloride content, and helps to appreciate the accuracy that can be achieved with the DT models. For a perfect fit, the points should fall on the 1:1 diagonal line on the scatter plot, but this is seldom the case with the developed models. For this study, the models for which the points fall within an error band of one standard deviation on either side (marked in the plots as +σ and −σ lines on either side of the 1:1 line) in the scatter plot have been considered as acceptable models, as adopted for other studies elsewhere.[49] Other types of researches, however, may have different limits on the errors for acceptance, depending upon the application.

In this study, a new proxy, TDS, was employed to estimate Cl^- concentration. Improvement in methodology in this study for estimation of Cl^- using TDS as a proxy over chloride estimation models using EC as a proxy reported in the literature[23-25] are: employment of separate sets for modeling and evaluation of models, consideration of seasonal influence on the models, consideration of the degree of saturation of soil in the region for model development, comprehensive performance evaluation using multiple metrics, and adoption of an intuitive ranking system for model selection.

3 Results and discussion

3.1 Development of DT models with recorded data from five districts

This dataset consisted of 182 records of premonsoon data, 220 records of postmonsoon data, and combined data from all seasons had 402 entries. The statistical parameters of the data are presented in Table 2. Identification of the best DT model for the premonsoon data, for example, was performed as follows: the performance metrics for the six models are presented in Table 3 (columns 2, 4, and 6) for the evaluation data. Now, according to the correlation (R), the higher is better. Table 3 (column 3) shows the ranking as follows: Models 2 and 6 are jointly ranked 1, Model 4 is rank 3 (as we have two entries in rank 1, the next one is not rank 2, but rank 3), Models 1 and 5 are rank 4, and Model 3 is ranked 6 (as we have two entries in rank 4, this is rank 6 and not 5). Similarly, for MARE and RMSRE, the lower of which would be better, the ranks are ascertained and presented in Table 3 (columns 5 and 7).

The ranks of each model considering the three metrics are added to obtain the total score, and based on the total score (Table 3, column 8), the overall rank (Table 3: column 9), is decided, with the lower total score considered as the better model. DT model 2 with a minimum overall score of 5 (in boldface: Table 3) was selected as rank 1: the best model for this dataset. All subsequent discussions would be based on the results from the best model selected similarly for the respective datasets. In a similar fashion to premonsoon data, the best DT models for the dataset: Five districts are identified for postmonsoon as well as for all-season datasets. The performance evaluation results are presented in Table 4. In Fig. 4, the scatter plots for the finally selected models are presented, wherein it is found that 2.2% of the points fall outside the acceptable limit of $\pm\sigma$ for premonsoon data (Fig. 4A) on the underestimation side. For postmonsoon data (Fig. 4B), 7.3% of the points fall outside the limits, and for all-season data, 6.0% of points fall outside the limits (Fig. 4C), on either side for both cases. For the dataset: Five districts, with a correlation range of 0.67–0.79 (Fig. 3A, B, and C) between the variables (TDS-Cl^-), the DT models offer a range of correlation of 0.79–0.82 between the measurement and corresponding estimations. It can be concluded that for data with higher scatter (lower variable correlations), the DT would be especially effective for enhancing the performance of estimation models owing to the domain-splitting approach. To that extent, the choice of DT for this application is justified by the results. Considering the

TABLE 3 Performance of the DT models for premonsoon data for five districts.

Model No.	R	Rank (R)	MARE	Rank (MARE)	RMSRE	Rank (RMSRE)	Total score	Overall rank
1	0.75	4	0.77	1	1.24	1	6	2
2	**0.79**	**1**	**0.85**	**2**	**1.28**	**2**	**5**	**1**
3	0.71	6	1.51	5	4.06	5	16	6
4	0.78	3	0.95	3	1.56	3	9	3
5	0.75	4	1.18	4	2.12	4	12	4
6	0.79	1	1.66	6	4.57	6	13	5

TABLE 4 Performance of the best DT models for different seasons: five districts.

Season	R	MARE	RMSRE
Premonsoon	0.79	0.85	1.28
Postmonsoon	0.80	0.65	1.06
All-season	0.82	0.74	1.60

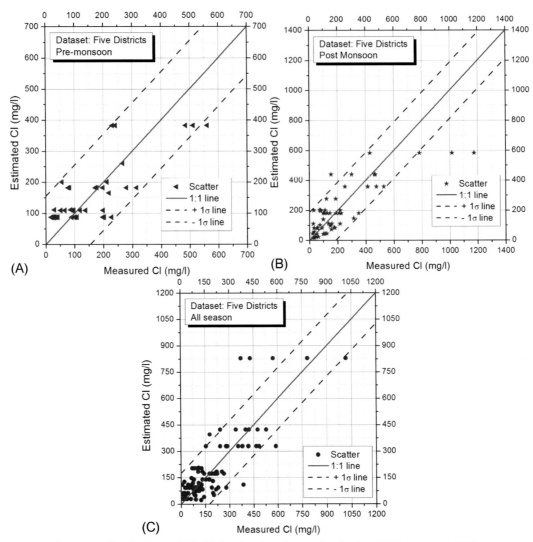

FIG. 4   Scatter plot for the measured and estimated chloride ion data for a dataset: five districts (A) Premonsoon; (B) Postmonsoon; (C) All-season.

performance of seasonal models and the all-season model, the seasonal models (premonsoon and postmonsoon) are marginally better only when RMSRE is considered; otherwise, the performances are comparable for this dataset.

3.2 Development of DT models with recorded data from water logged districts

This dataset consisted of 121 records of premonsoon data, 121 records of postmonsoon data, and for all season's data, there were 242 entries. Like the one explained in Section 3.1, the best models for the premonsoon and postmonsoon data as well as for the all-season data are identified, and their evaluation performance is summarized in Table 5 for the dataset: water-logged districts. Scatter plots of the selected models are presented in Fig. 5. It is noted that 6.7% of the points fall outside the acceptable limit of $\pm\sigma$ for premonsoon data (Fig. 5A), and 10.0% of points fall outside the limits for postmonsoon data (Fig. 5B) on the underestimation side in both cases. In the case of all-season data (Fig. 5C),

TABLE 5   Performance of the best DT models for different seasons: water logged districts.

Season	R	MARE	RMSRE
Premonsoon	0.80	0.90	1.65
Postmonsoon	0.74	0.57	0.84
All-season	0.65	0.70	1.19

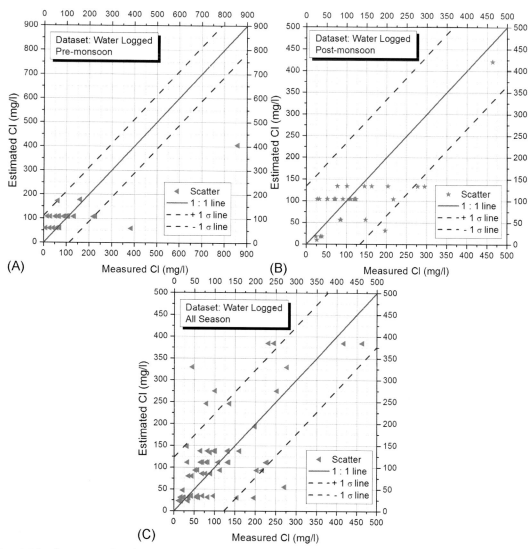

FIG. 5 Scatter plot for the measured and estimated chloride ion data for a dataset: water logged districts (A) Premonsoon; (B) Postmonsoon; (C) All-season.

11.7% of the points fall outside the limits on both sides. Another observation is that from the correlation range of 0.34–0.44 between TDS-Cl$^-$ basic data (Fig. 3D, E, and F), the models for various subdomains help to bring the correlation of the estimations and measurements up to a range of 0.65–0.80; thus, the application of DT has been successful in improving the estimation performance substantially.

Generally, in waterlogged regions, the groundwater quality is highly vulnerable considering the shallow levels of groundwater. Many factors like evaporation, rainfall, irrigation return flow, canal water influence can easily impact the overall quality of groundwater, including its Cl$^-$ content. Therefore, the TDS-Cl$^-$ and are not well correlated (Fig. 3). Barring rainwater influence and irrigation return flow, the rest of the factors become less pronounced during postmonsoon, therefore the dominant controls are restricted to only two (rainwater and irrigation return flow). In this case, a better correlation can be expected between TDS-Cl$^-$ data. In the present case, the TDS-Cl$^-$ correlation coefficient was found to be 0.34 during premonsoon which increases to 0.44 during postmonsoon (Fig. 3D and E), which is in agreement with the above reasoning. Combining both pre- and postmonsoon data should yield an intermediate value of correlation coefficient, which is found to be true (0.39) in the present case. The modeled output shows that samples falling outside one standard deviation are more in the case of postmonsoon compared to premonsoon, but becomes highest at 11.7% when all-season data is taken. The higher correlation of the seasonal model over the one developed for all-season demonstrated that the seasonal influence on the TDS-Cl$^-$ relationship has been captured, to a certain extent, by developing the premonsoon and postmonsoon models separately. Such findings justify the approach adopted for the development of the seasonal DT models.

3.3 Development of DT models with recorded data from water depleted districts

This dataset consisted of 61 records of premonsoon data, 99 records of postmonsoon data, and for all season's data, there were 160 entries. The data characteristics are presented in Table 2 and are discussed in Section 2.1. As explained for the earlier datasets, the best DT models for the dataset water-depleted districts were identified for the different seasons. The evaluation of the performance is summarized in Table 6, and the scatter plots for the selected models are depicted in Fig. 6. From the figure, it can be observed that 13.3% of the points fall outside the acceptable limit of $\pm\sigma$ for premonsoon data (Fig. 6A), 4.0% points fall outside the limits for postmonsoon data (Fig. 6B), and 10.0% points fall outside the limits for all-season data (Fig. 6C), on the underestimation side for all three cases. In this case,

TABLE 6 Performance of the best DT models for different seasons: water depleted districts.

Season	R	MARE	RMSRE
Premonsoon	0.81	0.25	0.31
Postmonsoon	0.89	0.37	0.54
All-season	0.84	0.43	0.67

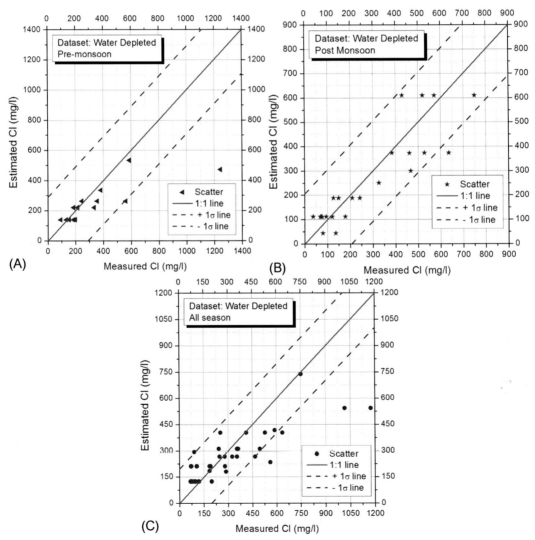

FIG. 6 Scatter plot for the measured and estimated chloride ion data for a dataset: water depleted districts (A) Premonsoon; (B) Postmonsoon; (C) All-season.

as the correlation between the variables (TDS-Cl$^-$) was in the range of 0.76–0.87, the correlations of the estimations with the measurements fell in the range of 0.81–0.89, and the improvement in model domain splitting by DT is not as apparent as that observed for the dataset: Water Logged in Section 3.2.

In the case of groundwater salinization with depleting water levels, it can be envisaged that any surface manifestations take a longer time to reflect in the groundwater levels. Therefore, the rainfall, irrigation return, and other anthropogenic activities would not immediately change the Cl$^-$ content of groundwater. During the vertical travel (infiltration) the Cl$^-$ fluctuations due to variable sources die down leading to a better (less varying) correlation between TDS-Cl$^-$, which is more pronounced in the postmonsoon period. However, the overall correlation between TDS-Cl$^-$ is higher in water-depleted areas compared to water-logged areas (Fig. 3). This observation clearly suggests that the correlations are highly influenced by the hydro-geological and geo-morphological setup of the area. A similar finding was reported by Sharma et al.[50] where uranium contamination was found to be controlled by local geomorphology. The number of samples falling out of $\pm\sigma$ limits is higher in the case of premonsoon compared to postmonsoon. This again corroborates the idea that for water level depleting condition in postmonsoon there is a better correlation between TDS-Cl$^-$ due to reduction in the contribution of other factors barring rainfall and irrigation return flow, which in turn facilitates better estimations of Cl$^-$ from TDS.

3.4 Comparison with previous works

The models reported in literature[23–25] for estimation of Cl$^-$ from EC were developed from well-correlated data, as depicted in the respective plots of the original articles.[23–25] As mentioned earlier, the data scatter (chloride concentration and TDS) in the present study area was quite high for all datasets considered. As an example, the all-season dataset for the water-depleted region was used to estimate chloride concentration using the formulations from literature,[23–25] and the performance of the empirical formulations was compared to the results obtained in this study (Table 7). The results clearly demonstrate that whereas the correlation coefficients of the estimates using expressions from literature[23–25] are similar, the relative errors are greatly reduced for the model presented in this study. The scatter plot of the estimates of Cl$^-$ using empirical expressions from literature[23–25] is presented in Fig. 7, which shows that these expressions grossly overestimate the Cl$^-$ in groundwater for the present study area, leading to the higher relative errors mentioned earlier. This demonstrated that the application of DT for estimation of chloride concentration using proxy: TDS was more accurate than using the expressions available in the literature.[23–25] The improved accuracy could be attributed to either the site-specific model development, the domain splitting approach adopted for the DT model in this study or the use of a different proxy (TDS) or a combination thereof.

3.5 Merits and demerits of the models in this study

The models developed in this study were more accurate than their counterparts in the literature,[23–25] possibly due to the reasons discussed earlier. Furthermore, this study demonstrated the seasonal influence on the TDS-Cl$^-$ relationship, indicating that seasonal models should be considered rather than all-season models for higher accuracy in Cl$^-$ estimation. Consideration of the saturation status of the soil (water-depleted or water-logged) was another factor introduced in this study, which improved the accuracy of the Cl$^-$ estimates. One limitation of the models reported here is that they are site-specific, season-specific, and soil saturation specific for the given study area and therefore cannot be used for Cl$^-$ estimations at other locations or seasons. However, site-specific models can be developed based on the proposed approach for any other site when sufficient data are available. The relative errors of the Cl$^-$ estimates using

TABLE 7   Comparison of performance of the chloride estimation models using a proxy from literature[23–25] and present study for water depleted districts (all seasons).

Season	R	MARE	RMSRE
White et al.[24]	0.83	3.67	5.07
Guan et al.[23]	0.83	2.51	3.67
Bresciani et al.[25]	0.83	2.31	3.42
This study	0.84	0.43	0.67

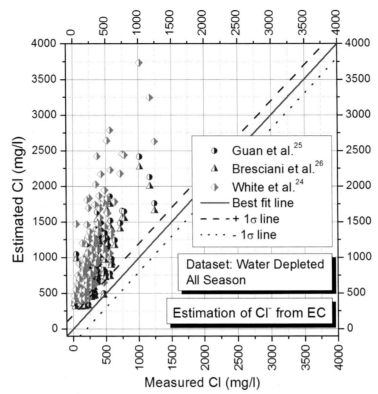

FIG. 7   Scatter plot for the measured and estimated chloride ion data for a dataset: water depleted districts, all-season, using empirical expressions from literature.[23-25]

the DT models developed in this study are much lower than those obtained using models from the literature.[23-25]; however, there is still scope for further reduction of the relative errors by employing a larger data set for model development or using other data-driven tools.

4 Conclusions

In this study, a data-driven tool, DT, was applied for the first time for the development of chloride estimation models using a new proxy, TDS, of groundwater. Acceptable models of reasonable accuracy were developed using the DT approach, and the promise of such applications for the estimation of other groundwater contaminants was highlighted. Starting with a correlation range of input data (0.34–0.87), the estimation performance resulted in increased correlation (0.69–0.89) and reduced the relative errors within the range 0.25 and 2.19 in the various cases. This demonstrates that the performance of the models developed in this study was much better than the models reported in the literature.[23-25]

The DT application improved the prediction performance relatively more when the input data had a wider scatter and lower correlation. Among all the cases examined, only in two cases was the error in the estimation of Cl⁻ was outside the bounds of one standard deviation on either side of the perfect model fit. The consideration of seasonal influences, such as premonsoon and postmonsoon, has helped to improve the performance of the estimation models, highlighting the dominant influence of season on the models. In addition, considering the degree of saturation of the soil in the region (waterlogged or water depleted) also yielded models with more accurate estimates. Therefore, it is emphasized that such pertinent factors should be accounted for while developing contaminant estimation models using proxies. The DT models developed for a region would be especially useful for the rapid estimation of contaminants such as chloride in groundwater from on-site routine TDS measurements. Future studies should focus on the estimation of other contaminants in groundwater using routinely measured proxies.

Acknowledgments

The authors sincerely acknowledge the various sources and authors of the data used in this study. The authors express their gratitude to the critical comments received from the editors/reviewers and this was extremely helpful in improving the quality of the article.

Conflict of interest

The authors declare that there is no known conflict of interest.

Data availability

The data used in this study are available in the literature[34–38] and are listed in Table 1.

References

1. Post VEA. Fresh and saline groundwater interaction in coastal aquifers: is our technology ready for the problems ahead? *Hydrgeol J.* 2005;13(1):120–123.
2. Simmons CT. Variable density groundwater flow: from current challenges to future possibilities. *Hydrgeol J.* 2005;13:116–119.
3. Barlow PM, Reichard EG. Saltwater intrusion in coastal regions of North America. *Hydrgeol J.* 2010;18(1):247–260. https://doi.org/10.1007/s10040-009-0514-3.
4. Custodio E. Coastal aquifers of Europe: an overview. *Hydrgeol J.* 2010;18(1):269–280. https://doi.org/10.1007/s10040-009-0496-1.
5. Werner A. A review of seawater intrusion and its management in Australia. *Hydrgeol J.* 2010;18(1):281–285. https://doi.org/10.1007/s10040-009-0465-8.
6. Bobba AG. Numerical modelling of saltwater intrusion due to human activities and sea level change in the Godavari delta India. *Hydrol Sci J.* 2002;47(S):S67–S80.
7. Maas K. Influence of climate change on a Ghijben Herzberg lens. *J Hydrol.* 2007;347(1–2):223–228.
8. Vandenbohede A, Luyten K, Lebbe L. Impacts of global change on heterogeneous coastal aquifers: case study in Belgium. *J Coast Res.* 2008;24(2B):160–170.
9. OudeEssink GHP. VanBaaren ES, deLouw PGB Effects of climate change on coastal groundwater systems: a modeling study in the Netherlands. *Water Resour Res.* 2010;46:10. https://doi.org/10.1029/2009WR008719.
10. Lindstrom R. A system for modelling groundwater contamination in water supply areas: chloride contamination from road de-icing as an example. *Hydrol Res.* 2006;37(1):41–51.
11. CGWB. *Ground water quality scenario in India, Central Ground Water Board (CGWB), Ministry of Jal Shakti, Department of Water Resources, River Development and Ganga Rejuvenation Government of India;* 2010. http://cgwb.gov.in/wqoverview.html. Accessed 31.01.21.
12. CGWB. *Ground water in shallow aquifers of India, Central Ground Water Board, Ministry of Water Resources, Faridabad, 2010;119;* 2010. http://cgwb.gov.in/wqoverview.html. Accessed 31.01.21.
13. Greene R, Timms W, Rengasamy P, Arshad M, Cresswell R. Soil and aquifer salinization: toward an integrated approach for salinity management of groundwater. In: *Integrated Groundwater Management;* 2016:377–412.
14. Chowdary M, Chandran V, Neeti, et al. Assessment of surface and sub-surface waterlogged areas in irrigation command areas of Bihar state using remote sensing and GIS. *Agric Water Manag.* 2008;95:754–766. https://doi.org/10.1016/j.agwat.2008.02.009.
15. U.S. Environmental Protection Agency. *Secondary Drinking Water Regulations—Guidance for Nuisance Chemicals: EPA 810/K–92–001;* 1992. http://www.epa.gov/safewater/consumer/2ndstandards.html. Accessed 15.12.20.
16. U.S. Environmental Protection Agency. *Ambient water quality criteria for chloride—1988: U.S. Environmental Protection Agency Office of Water Regulations and Standards 440/5–88–001;* 1988:47. http://www.epa.gov/ost/pc/ambientwqc/chloride1988.pdf. Accessed 15.12.20.
17. Fadeeva VK. Effect of drinking water with different chloride contents on experimental animals. *Gigienaisanitarija.* 1971;36(6):11–15. (Dialog Abstract No. 051634) [in Russian].
18. Wesson LG. *Physiology of the Human Kidney.* New York: NY:Grune and Stratton; 1969.
19. Gregory R. Galvanic corrosion of lead solder in copper pipework. *Water Environ J.* 1990;4(2):112–118.
20. Dauji S. Reinforcement corrosion in coastal and marine concrete: a review. *Chall J Concr ResLett.* 2018;9(2):62–70. https://doi.org/10.20528/cjcrl.2018.02.003.
21. Bureau of Indian Standards (BIS). *IS: 456–2000 (Reaffirmed 2005): Plain and Reinforced Concrete—Code of Practice.* New Delhi, India: BIS; 2000.
22. APHA, ed. *Standard Methods for the Examination Of Water And Wastewater.19th Ed.* Washington: APHA; 1998.
23. Guan H, Love AJ, Simmons CT, Hutson J, Ding Z. Catchment conceptualisation for examining applicability of chloride mass balance method in an area with historical forest clearance. *Hydrol Earth Syst Sci.* 2010;14:1233–1245. https://doi.org/10.5194/hess-14-1233-2010.
24. White I, Macdonald BC, Somerville PD, Wasson R. Evaluation of salt sources and loads in the upland areas of the Murray-Darling Basin Australia. *Hydrol Process.* 2009;23(17):2485–2495.
25. Bresciani E, Cranswick RH, Banks EW, Batlle-Aguilar J, Cook PG, Batelaan O. Using hydraulic head, chloride and electrical conductivity data to distinguish between mountain-front and mountain-block recharge to basin aquifers. *Hydrol Earth Syst Sci.* 2018;22:1629–1648. https://doi.org/10.5194/hess-22-1629-2018.
26. Saghebian SM, Sattari MT, Mirabbasi R, Pal M. Ground water quality classification by decision tree method in Ardebil region Iran. *Arab J Geosci.* 2014;7:4767–4777. https://doi.org/10.1007/s12517-013-1042-y.
27. Haselbeck V, Kordilla J, Krause F, Sauter M. Self-organizing maps for the identification of groundwater salinity sources based on hydrochemical data. *J Hydrol.* 2019;576:610–619. https://doi.org/10.1016/j.jhydrol.2019.06.053.

28. Engle MA, Brunner B. Considerations in the application of machine learning to aqueous geochemistry: Origin of produced waters in the northern U.S. Gulf Coast Basin. *Appl Comput Geosci.* 2019;3–4:100012. https://doi.org/10.1016/j.acags.2019.100012.

29. Mosavi A, Hosseini FS, Choubin B, et al. Susceptibility mapping of groundwater salinity using machine learning models. *Environ Sci Pollut Res.* 2020;1–14. https://doi.org/10.1007/s11356-020-11319-5.

30. Norouzi H, Moghaddam AA. Groundwater quality assessment using random forest method based on groundwater quality indices (case study: Miandoab plain aquifer, NW of Iran). *Arab J Geosci.* 2020;13(912):1–13. https://doi.org/10.1007/s12517-020-05904-8.

31. Hem JD. *Study and Interpretation of the Chemical Characteristics of Natural Water, 3rd ed.* USGS, Water Supply Paper 2254, USA: USGS; 1989.

32. Clark ID, Fritz P. *Environmental Isotopes in Hydrogeology.* New York: Lewis Publishers; 1997.

33. Chopra RPS, Krishan G. Assessment of groundwater quality in Punjab India. *Earth Sci Clim Change.* 2014;5:10. https://doi.org/10.4172/2157-7617.1000243.

34. Sharma P. *A comparative study of groundwater quality in parts of Muktsar and Faridkot districts of southwest Punjab India in relation to human health and agriculture.* PhD Thesis, Chandigarh: Faculty of Science, Punjab University; 2015. http://hdl.handle.net/10603/199271;. Accessed 15.12.20.

35. Sharma AD. *Investigations of source of uranium and its geochemical pathways in aquifer systems in parts of Southwest Punjab using environmental isotope techniques.* Unpublished PhD Thesis, Chandigarh: Faculty of Science, Punjab University; 2017.

36. Dhanda DS. *Land Use Study and Assessment of Surface and Ground Water Quality in Talwandi Sabo block of Bathinda, Punjab, India.* PhD Thesis, Chandigarh: Faculty of Science, Punjab University; 2011. http://hdl.handle.net/10603/82110;2011. Accessed 15.12.20.

37. Kumar A. *Hydrogeological and Geochemical Studies of Part of Sirhind Nala Sub Basin of Ghaggar River Basin in Relation to Buried Aravali Delhi Ridge South West Punjab India.* PhD Thesis, Chandigarh: Faculty of Science, Punjab University; 2006. http://hdl.handle.net/10603/82301;2006. Accessed 15.12.20.

38. Ritu S. *Urban hydrogeology of Ambala city and Ambala cantonment area, Haryana, India.* Unpublished PhD Thesis, Chandigarh: Faculty of Science, Punjab University; 2015.

39. Shah M. *Report by the High Level Expert Group on Waterlogging in Punjab, Water Resources and Rural Development.* Government of India Planning Commission, New Delhi: Government of India; 2013.

40. Singh S. Waterlogging and its effect on cropping pattern and crop productivity in south-West Punjab—a case study of Muktsar district. *J Econo & Soc Develop.* 2013;9(1):71–80.

41. Quinlan JR. *C4.5: Programs for Machine Learning.* San Francisco: Morgan Kaufmann; 1992.

42. Rokach L, Maimon O. *Data Mining with Decision Trees: Theory and Applications.* Singapore: World Scientific; 2015.

43. Jekabsons G. *M5PrimeLab: M5' Regression Tree, Model Tree, and Tree Ensemble Toolbox for Matlab/Octave;* 2016. http://www.cs.rtu.lv/jekabsons/;. Accessed 08.05.2016.

44. Witten IH, Frank E. *Data Mining: Practical Machine Learning Tools and Techniques.* SanFrancisco: Morgan Kaufmann; 2000.

45. Dauji S. New approach for identification of suitable vibration attenuation relationship for underground blast. *Engin J.* 2018;22(4):147–159. https://doi.org/10.4186/ej.2018.22.4.147.

46. Ray S, Dauji S. Ground vibration attenuation relationship for underground blast: a case study. *JInstit Eng (India) Series A.* 2019;100:763–775. https://doi.org/10.1007/s40030-019-00382-y.

47. Dauji S. Prediction accuracy of underground blast variables: Decision tree and artificial neural network. *Int J Earthq Impact Eng.* 2020;3(1-2-3):40–59. https://doi.org/10.1504/IJEIE.2020.105382.

48. Dauji S. Re-look into modified scaled distance regression analysis approach for prediction of blast-induced ground vibration. *Int J Geotech Earthq Eng.* 2021;12(1):22–39. https://doi.org/10.4018/IJGEE.2021010103.

49. Rafi A, Dauji S, Bhargava K. Estimation of SPT from Coarse Grid Data by Spatial Interpolation Technique. In: Gali ML, RRP, eds. *Geotechnical Characterization and Modelling, Lecture Notes in Civil Engineering 85.* Singapore: Springer Nature; 2020:1079–1091. https://doi.org/10.1007/978-981-15-6086-6_87.

50. Sharma DA, Keesari T, Rishi MS, et al. Distribution and correlation of radon and uranium and associated hydrogeochemical processes in alluvial aquifers of Northwest India. *Environ Sci Pollut Res.* 2020;27(31):38901–38915.

4

Application of analytical hierarchy process (AHP) in landslide susceptibility mapping for Qazvin province, N Iran

Ebrahim Sharifi Teshnizi[a], Mohsen Golian[b], Shahriar Sadeghi[c], and Ahmad Rastegarnia[d]

[a]Department of Geology, Faculty of Science, Ferdowsi University of Mashhad, Mashhad, Iran [b]Tehran Science and Research Branch, Islamic Azad University, Tehran, Iran [c]Department of Geology, Faculty of Science, Imam Khomeini International University, Qazvin, Iran [d]Department of Geology, Faculty of Science, Ferdowsi University of Mashhad, Mashhad, Iran

1 Introduction

Mass movements are morphodynamic phenomena that are affected by various factors in the mountainous slopes. These movements not only cause changes in the morphology of the slopes but also, in some cases, endanger humans and cause financial risks.

In the last decade, a lot of work conducted around the world on landslide susceptibility various classifications presented (Table 1). Support vector machine (SVM), random forest (RF), multivariate adaptive regression spline (MARS), artificial neural network (ANN), quadratic discriminant analysis (QDA), linear discriminant analysis (LDA), naive Bayes (NB), information content method (ICM), analytical hierarchy process (AHP), random forest (RF), the bee algorithm (Bee), the adaptive neuro-fuzzy inference system (ANFIS), the gray wolf optimizer (GWO), fuzzy gamma (FG), binary logistic regression (BLR), backpropagation artificial neural network (BPANN), and C5 decision tree (C5DT) are the methods applied for this purpose. The use of the AHP method was preferred because of the mathematical simplicity and data extraction in a limited time. An exhaustive statistical evaluation for susceptibility mapping in hilly terrain is essential for testing the analogy of slope failure occurrence and its timely mitigation, which can also be used for other study areas. The objective is to define a relationship between the independent causative factors and the distribution of present landslides in any area. The most common information used for the landslide susceptibility is lithological factors, weathering, slope angle, slope direction (aspect), land cover, distance from the streams, drainage density and distance from the roads.

In this study, after analysis of all information layers of Qazvin province, using the AHP method in GIS software, series of specific steps were performed to diagnose the landslide susceptibility.

2 Study area

Qazvin province is located in the north of Iran, between longitudes of $48° 45'$ to $50° 50'$ E and latitudes of $35° 37'$ and $36° 45'$ N with an area of $15,821 \, km^2$ which is less than 1% of Iran area (Fig. 1). The province's average annual rainfall is about 330 mm. The amount of precipitation varies with topographic conditions, so that mountainous regions are more prevalent than the foothills and lands located in the plain.

TABLE 1   Methods and information used in the previous landslide susceptibility studies.

Author(s)	Method(s)	Information used
Youssef and Pourghasemi[1]	SVM, RF, MARS, ANN, QDA, LDA, and NB	Slope angle, slope aspect, slope length, distance from roads, lithology, distance from wadis, distance from faults, altitude, normalized difference vegetation index (NDVI), plan curvature, profile curvature, and land use/landcover
Yu and Chen[2]	ICM, AHP, and RF	Lithology, slope angle, slope aspect, rainfall, land use, seismic intensity, distance to the river, and distance to the fault
Panahi et al.[3]	ANFIS, SVR, Bee and GWO	Slope, aspect, plan curvature, effective airflow, terrain surface convexity, terrain surface texture, wind exposition index, topographic wetness index, valley depth, the density of the forest, and type of forest, geology, land use
Pourghasemi et al.[4]	Statistical functions	Land use, lithology units, drainage density, plan curvature, slope aspect, altitude, slope angle, profile curvature, slope length, distance from faults, distance from roads, and distance from rivers
Pourghasemi et al.[5]	BRT	Topographical (altitude, slope degree, slope aspect, plan curvature, and profile curvature), hydrological (distance from stream and topographic wetness index (TWI)), geological (lithological formation, distance from fault, and soil texture), land use/land cover (land use, normalized difference vegetation index (NDVI)), and anthropological (distance from roads)
Emami et al.[6]	RF, MARS, and BRT	Elevation, aspect, TWI, slope, plan curvature, convergence index, distance from the river, drainage density, distance from the road, land use, lithology, and distance from the fault
Basu and Pal[7]	AHP	Slope, drainage density, relative relief, rainfall, road, agriculture land, gravity anomaly, geology, NDVI, soil, lineament, peak ground acceleration (PGA), settlement, SPI, STI, TWI, historical landslide locations
Vojteková and Vojtek[8]	AHP	Elevation, slope angle, slope aspect, distance from rivers, distance from faults, geology, and land use
Arsyad and Hamid[9]	AHP	Slope, distance from the road, distance from the drainage, lithology, distance from the fault, precipitation
Pratama et al.[10]	AHP	Precipitation, geology, soil texture, slope, and land use
Vakhshoori et al.[11]	FG, BLR, BPANN, SVM and C5DT	Slope degree, slope aspect, modified sediment transport index (STI—V), stream power index (SPI), distance to rivers network, distance to faults, distance to roads, climate, annual average rainfall, and annual average temperature
Arjmandzadeh et al.[12]	AHP	Geology (land slope-bedding slope, land slope direction-bedding direction, earthquake acceleration, Lithology), geomorphology (land SLOPE, land slope direction), engineering geology (GSI, Is50, saturated density, cohesion, layer slope-international friction angle, weathering conditions, jointing conditions)
Abay et al.[13]	AHP	Slope angle, aspect, proximity to drainage, elevation, proximity to a fault, land use/land cover.
He et al.[14]	AHP	Lithology, distance from faults, NDVI, elevation, slope, aspect, profile curvature, distance from rivers, annual rainfall, distance from roads
El Jazouli et al.[15]	AHP	Slope aspect, slope degree, elevation, land use, distance to drainage network, distance to roads, distance to faults, lithology
Vijith and Dodge-Wan[16]	AHP	Slope, aspect, relative relief, slope length and steepness, curvature, landforms, topographic wetness index (TWI), stream power index (SPI), stream head density, land use/land cover
Nguyen and Liu[17]	AHP	DEM map, geological map, land use map, aspect map, drainage density map, slope map
Abedini and Tulabi[18]	LNRF, FR, and AHP	Lithology, slope, aspect, altitude, distance from the fault, distance from the river, fault land use, rainfall,
Mokarram and Zarei[19]	Fuzzy-AHP	Aspect, the distance of the road, the distance of stream, the distance of fault, slope, DEM, land use, precipitation, lithology
Sharma and Mahajan[20]	FRM, AHP	Lithology, Land cover, Aspect, Slope, Fault buffer, Road buffer, Drainage buffer, Soil type

TABLE 1 Methods and information used in the previous landslide susceptibility studies—cont'd

Author(s)	Method(s)	Information used
Pourghasemi and Rossi[21]	GLM, GAM, MARS, and M-AHP	Distance from rivers, Distance from roads, Distance from faults, Slope angle, Slope aspect, Altitude, Topographic wetness index (TWI), Plan curvature, Profile curvature
Abedini et al.[22]	LR, AHP	Elevation, Slope inclination, Precipitation, Distance to the road, Lithology, Land Use, slope aspect, Distance to the river, Distance to fault
Rahaman and Aruchamy[23]	AHP	Elevation, aspect, slope angle, land use, lineament density, soil depth, precipitation, distance to the road, NDVI
Wu et al.[24]	Logistic regression	Slope angle, slope aspect, curvature, plan curvature, profile curvature, altitude, NDVI, rainfall, STI, distance to roads, distance to rivers, lithology, distance to faults, TWI, and SPI.
Zhao et al.[25]	AHP-fuzzy, entropy-FAHP	Lithology, relief amplitude, slope, aspect, slope morphology, altitude, annual mean rainfall, distance to River.
Wang and Li[26]	AHP	Slope angle, altitude, distance to rivers, distance to roads, distance to faults, rainfall, lithology; and NDVi
Kumar and Anbalagan[27]	AHP	Geology, soil, LULC, aspect, slope, relative relief.
Myronidis et al.[28]	AHP	Land cover, Slope angle, Slope aspect, Altitude, Geologic structure, Soil types, Mean annual rainfall, Proximity to streams, Proximity to roads

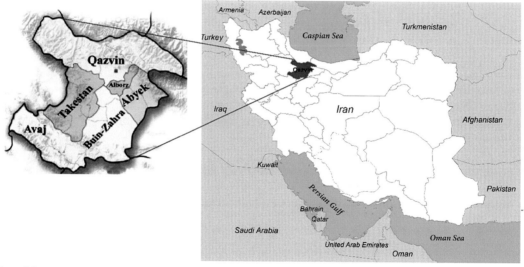

FIG. 1   Location of Qazvin province on Iran map.

More than half of the total area of Qazvin province is mountainous regions of the southern part of the Alborz Mountain range. Various factors such as slope, geology, fault, and human activities are systematically affected the occurrence of landslides in this province. The high potential of a landslide of this area cause hazard for communication lines and villages located in this province.

In this study, considering the factors that influence the occurrence of landslides in Qazvin province using AHP in GIS software, landslide potential zonation is provided and critical areas are determined.

3 Material and methods

3.1 Information used

The information layers used in this study are slope, aspect, distance from the stream, land use and vegetation cover, geology (lithology, distance from fault, difference of slope and layer slope, slope aspect difference with slope direction of layer), distance from the road, point load index, geological strength index (GSI), specific gravity, cohesion, internal friction angle, weathering, and condition of discontinuities

3.2 Steps of AHP analysis

The first step in building an AHP analysis is to prepare a hierarchical tree with the general problem locating on the top and the criterion and subcriterion and other options being in the lower levels. In other words, converting the studies problem to a hierarchical tree converts a complex problem to an AHP format to disentangle the problem. The zero levels of each tree show the goal of the decision-maker. The first level, which is the most important level, presents some criteria upon which the counterpart options are compared. Finally, the least level illustrates and compares the proposed options of the counterpart criterion. The other steps of the AHP process (briefly discussed in the following) are:

- Estimating the weight of criterion
- Estimating the weight of options
- Estimating the overall score of the options
- Evaluating the logical compatibility of the decisions

3.3 Principles of scoring the criterion and subcriterion

There are several methods for determining the importance of coefficients of the criterion and subcriterion. The most important method, however, is the binary comparison through which two criteria are compared together and their values are determined based on their importance and the goal of the analysis. For this purpose, the standard method proposed by[29] can be implemented. Through this procedure, first, a number from the range 1 to 9 is assigned to each binary comparison.[30] Table 2 shows the scores assigned to each criterion and subcriterion according to Saaty's method.

After determining the coefficients of criterion and subcriterion, the importance coefficients of the criterion are determined. In this step, the priority of each option is judged over other subcriterion and directly with the corresponding criterion (if it has no subcriterion). The process involved in obtaining the weight of each option compared to each subcriterion is similar to the process followed for estimating the importance coefficient of criteria with respect to the goal. In both states, the judgments are based on a binary comparison of the criteria using the 9-grade Saaty's scale.

3.4 Estimating the final weight of the criteria

In this step, the weight of criteria and subcriteria are used to determine the ultimate score of each option. To this end, Eq. (1) is used:

$$\text{final score of } j \text{ option} = \sum_{k-1}^{n} \sum_{i-1}^{m} w_k w_i (gij) \tag{1}$$

where, w_k is the importance coefficient of K criterion, w_i is the importance coefficient of i subcriterion, and gij is option score of j in connection with i subcriterion.

TABLE 2   Binary comparison of Saaty's 9-grade scale for criteria and subcriteria.[30]

Description	Score (preference severity)
Same preferred	1
Slightly preferred	3
More preferred	5
Very high preferred	7
Quite preferred	9
One in between preferred	2, 4, 6, and 8

3.5 Evaluating the logical compatibility of the decisions

One of the advantages of AHP is judgment evaluation for determining the importance coefficients of criteria and subcriteria using the compatibility ratio (CR). For an acceptable judgment, CR must be less than 0.1. Using this ratio (Eq. 1) facilitates decision analysis before the ultimate choice.[31]

$$CR = \frac{CI}{RI} \qquad (2)$$

where RI is the average compatibility index estimated using the matrix proposed by[32] and CI is compatibility index calculated using Eq. (3):

$$CI = \frac{(\lambda_{\max} - n)}{(n-1)} \qquad (3)$$

where $\lambda_{\max}$ is the largest eigenvalue of the matrix and n is the matrix order.

The hierarchical tree of decision-making in this research, along with the branches of the main criteria and subcriteria of each branch, according to the results of the statistical surveys carried out on the questionnaires (Fig. 2).

3.6 Prioritization of the effective parameters and preparation of decision-making tree

The most common methods for evaluating the importance of effective factors in landslide zonation are morphometric studies on existing landslides in the field and questionnaires filled by experts. Parameters such as rainfall, lithology of the outcropped rock units, land use, slope angle, road construction factors, streams and rivers, elevation, and slope direction are considered as the factors affecting landslide occurrence and recorded through the field studies. Moreover, considering the importance level of the effective factors on landslide occurrence, the precise detection and prioritization of the factors is also an essential part of this study. This requirement is fulfilled partially using the related questionnaires

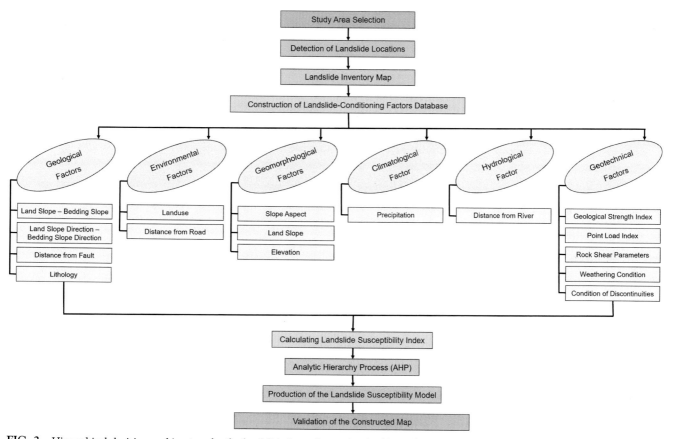

FIG. 2   Hierarchical decision-making tree for the landslide hazard zonation in this study.

and comparing the parameters binary. For this purpose, first, the priority of the effective factors is determined from a questionnaire filled by some experts and prioritizes the factors giving them scores ranging from 1 to 9.

In the present research, using the information from different sources including the experts' remarks, field surveys, and historical landslide data in Qazvin province, 451 landslides were collected, and then their geographical coordinates and a short description of their occurrence mechanisms were imported in GIS (Figs.3 and 4). Next, to prioritize the factors effective in landslide occurrence based on the conditions and triggering factors in the study area, a questionnaire was filled and evaluated based on the remarks of 50 landslide experts including geologists, geotechnical engineers, meteorologists, hydrogeologists, geomorphologists, and rock mechanic engineers. Next, 20 questionnaires were distributed among other landslide experts to determine the factors effective in landslide occurrence and their importance through comparing the obtained results with those from the previous group of experts in SPSS software and statistical analyses on their responses. The decision tree prepared based on the determined main criteria and subcriteria determined through the statistical analyses conducted on the questionnaires are presented in Fig. 2. In the following, each of these criteria and subcriteria are described in details.

FIG. 3   Some images of landslides that occurred in the study area.

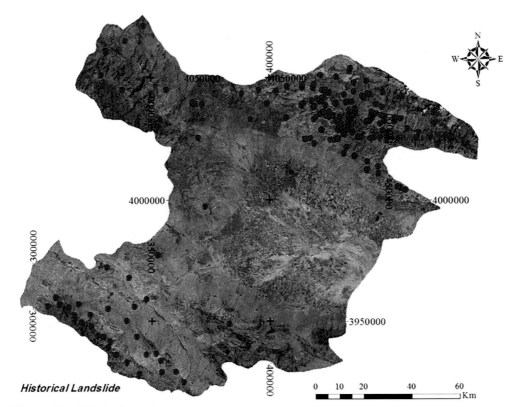

FIG. 4   Location of historical landslide in the study area.

4 Results and discussion

4.1 Geological criteria

Lithology, distance from the fault, the difference between the layer's dip and topography, the difference in a strike of strata, and direction of the topographic relief, are four main subcriterion of geological factor.

Due to the geographical location of the country, Iran has different geological zones and, as a result, a large variety of geological formations. Due to its presence on the active belt of the Alpine Himalayas, these materials have always been under stress and deformation, and various types of geological structures such as folds, faults, joints, and fractures are expanded. Such deformation reduces rock mass strength and consequently increases mass movements.

In general, the geological factor plays an important role in the distribution of landslides, because its geology and its various structures cause differences in the stability and strength of rocks as well as variation in soil texture.[33]

- **Lithology**

In recent studies conducted by researchers, lithology has been considered as the most important factor in landslide hazard zonation.[12, 34, 35] By investigating different formations in different regions of Iran, the sensitivity of various geological materials to landslides has been determined (Table 3). According to this table, the loose Quaternary sediments have more potential for landslide than other geological materials.

Fig. 5 shows the geological map of Qazvin province provided by compiling 1:100000 scaled geological maps. In order to determine the weight of each geological unit of the province in the landslide event, after forming the decision-making matrix of geological units, each unit compared to the other. Considering the sensitivity of different geological materials to landslides (Table 2), the priority between 1 up to 9 is given then the weight of each unit is determined using the AHP program in GIS software, finally a weighted lithology map is prepared (Table 3 and Fig. 6).

- **Slope angle difference with layer's dip**

After plotting the slope of the rock outcrops of Qazvin province (Fig. 7), according to the number of strata dip indicated in the geological map, with the difference of the mentioned map from the topographic slope map and using the Raster Calculator in GIS software, the information layer of the difference of slope angle with layering slope were determined.

In this map, alluvial deposits, sedimentary massive rocks, volcanic rocks, and igneous intrusive masses that lack layering is separately identified and separated. After forming the decision-making matrix, the highest score for the highest difference and the lowest score for the nonlayered outcrops are considered, and finally, the weight of each class is determined using the AHP program. Table 3 shows the decision-making matrix and Fig. 8 shows the weighted map of this subcriterion.

TABLE 3 Classification of various geological materials in terms of slip resistance.

Preference	Lithological description
1	Massive and thick limestone-quartzite-conglomerate with silica cement
2	Dolomite—igneous rocks—sandy limestone with medium layering
3	Limestone with medium layering—sandstone—quartzite with medium layering
4	Schist and metamorphic rocks with joint and layering—weathered basalt and andesite with joints
5	Andesitic tuff—limy shale—weathered limy tuff
6	Coal marl—gypsum marl—alternation of shale and sandstone
7	Marl—shale—weathered tuff—coal
8	Weathered shale and marl
9	Old alluviums-gypsum
10	Young alluviums—flood and glacial deposits—sliding materials of old landslides

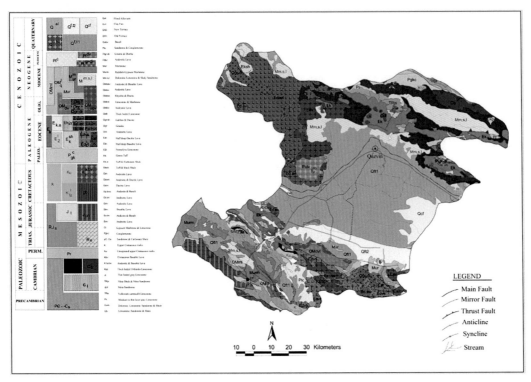

FIG. 5   Geological map of Qazvin province (based on 1:100000 scaled geological maps).

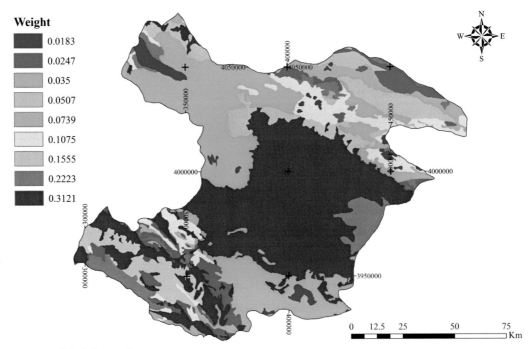

FIG. 6   Weighted map of the lithology factor.

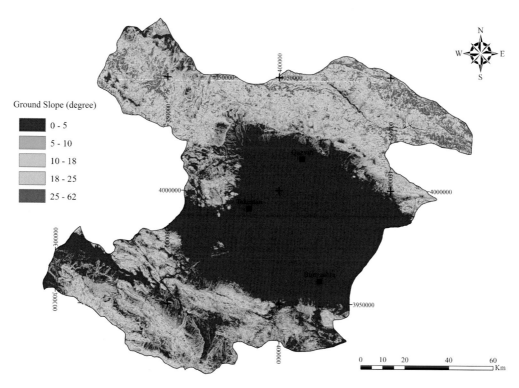

FIG. 7   Layering slope map based on a geological map of 1:100000.

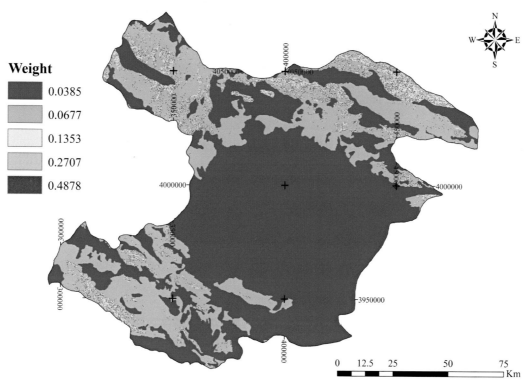

FIG. 8   Weighted map of the subcriterion of slope angle difference with layering slope.

- **Aspect difference of strata and topography**

After preparing the strata aspect map of the rocky outcrops, based on the strata aspect in the geological map of 1:100000, by differentiating the mentioned map from the map to the aspect map using the Raster Calculator, the information layer of aspect difference of strata and topography has been determined (Fig. 9). In this map alluvial deposits, sedimentary mass rocks, volcanic rocks, as well as igneous intrusive masses that are nonlayered are distinct.

After forming the decision-making matrix, the highest score for the difference between zero and 30 degrees and the lowest score for nonlayered outcrops are considered. Finally, the weight of each class is determined using the AHP program (Table 3 and Fig. 10).

- **Distance from fault**

Faults play an important role in creating or re-activating regions with a potential of sliding. Crushing in faulting zones, the penetration of water from these areas into the slopes, the appearance of discontinuity around the fault, and the difference in erosion in the slopes are among the effects that can be noted. The motion of the fault can also be a kind of landslide start on the slopes.[35]

Fig. 11 shows the faults location map of Qazvin province. This map is extracted from the 1: 100000 geological maps. After designing the faults map, the zones around the faults are made using the Buffering tools in GIS. Then by forming the decision-making matrix for each zone was scored based on the proximity to the fault plane (Table 6). Accordingly, the highest score is attributed to the zone up to 0.5 km adjacent to the fault plane (Fig. 12).

- **Summarizing geological criteria**

After weighting to each of the geological subcriteria, by determining the decision-making matrix, the priority of each factor is compared to other factors, and according to the results of the questionnaires, the score is given from 1 to 9. Then, after determining the superiority of geological subcriteria relative to each other, the weight of each subcriterion was determined using the AHP program. Finally, by integrating the geological subcriteria according to their weight, a map of the weighted geological criterion was prepared (Table 4 and Fig. 13).

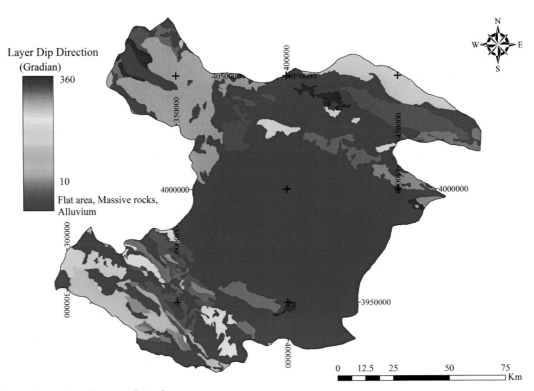

FIG. 9 Strata aspect map based on a geological map.

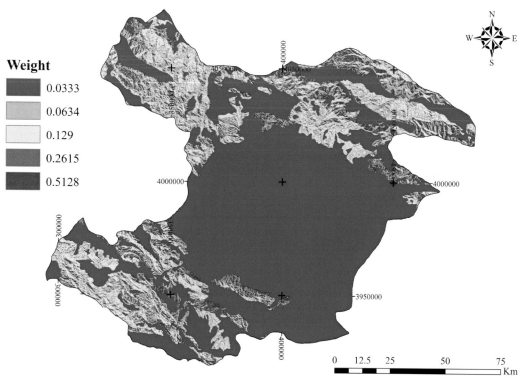

FIG. 10 Weighted map of aspect difference of strata and topography subcriterion.

FIG. 11 Faults map of Qazvin province.

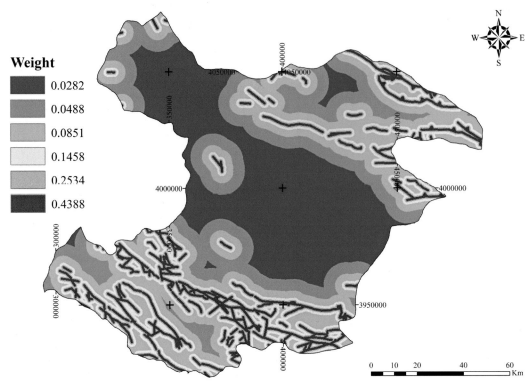

FIG. 12   Weighted map of the distance from the fault subcriterion.

4.2 Environmental criterion

Artificial factors also have a significant effect on landslide occurrence. The effect of these factors, especially in recent decades, has been a significant contribution to land use by humans, is more important than ever before. The discontinuation of forests and their conversion to crops and residential areas is one of the most important triggering factors of landslides.

- **Distance from road**

The construction of roads is one of the synthetic factors affecting instability created by humans.[36, 37] The construction of the road changes the topography and decreases the shear strength in the slope and increases tensile stress also causes water to penetrate the slope and exerts additional pressures due to increased weight. In Fig. 14, the road map of Qazvin province is presented. After locating the main roads, the areas surrounding these roads are zoned using Buffering tools in GIS and then weighted by forming a decision-making matrix based on the proximity of the areas to the road lines (Table 5). Accordingly, the highest score is assigned to areas with a maximum distance of 100 m from the road (Fig. 15). By getting away from the roads, they were assigned less weight.

- **Land use**

In Fig. 16, the land use map of Qazvin province is presented. After the formation of the decision-making matrix (Table 5), because the arid lands and weak vegetation areas are more prone to landslide than other areas the most scores are given to them. Vegetation acts as an effective factor in increasing landslide resistance. The roots of the plants are considered as a reinforcing factor in slope stability, especially in marly rocks. Also, the roots of the plants cause water absorption of the slopes and reduce the potential for slipping. In studies conducted by Leventhal and Kotze,[38] this parameter is used as an important parameter in a landslide zoning map. Subsequently, residential areas will be prone to landslides due to the entry of sewage and the leakage of wells into the slopes, as well as due to the loading of the building on the slopes. Unlike those areas of agriculture, gardens and forests that have enormous vegetation cover are much less sensitive to landslides. Also, rangeland areas with bushes and grass cover have the potential for mass

TABLE 4 Decision-making matrix for geological subcriteria.

Causative factors and classes with in each factor	Pair-wise comparison matrix									Weight	Consistency ratio (CR)
	[1]	[2]	[3]	[4]	[5]	[6]	[7]	[8]	[9]		
Geological criteria											
Lithology											
[1] alluvium	1	2	3	4	5	6	7	8	9	0.3121	0.0345
[2] Shale		1	2	3	4	5	6	7	8	0.2223	
[3] Marl			1	2	3	4	5	6	7	0.1555	
[4] Tuff				1	2	3	4	5	6	0.1075	
[5] Sandstone					1	2	3	4	5	0.0739	
[6] Conglomerate						1	2	3	4	0.0507	
[7] Volcanic lava							1	2	3	0.035	
[8] Limestone								1	2	0.0247	
[9] Igneous rock									1	0.0183	
Ground slope difference with strata slope											
[1] Massive rocks and alluvium	1									0.0385	0.0089
[2] < 0	2	1								0.0677	
[3] 0–10	4	2	1							0.1353	
[4] 10–20	8	4	2	1						0.2707	
[5] 20–35	9	8	4	2	1					0.4878	
Aspect difference between ground slope and strata											
[1] Massive rocks and alluvium	1									0.0333	0.0534
[2] 60–120	3	1								0.0634	
[3] 120–180	5	3	1							0.1290	
[4] 30–60	7	5	3	1						0.2615	
[5] 0–30	9	7	5	3	1					0.5128	
Distance from fault											
[1] 0–0.5	1	3	4	5	7	9				0.4388	0.0363
[2] 0.5–1		1	3	4	5	7				0.2534	
[3] 1–2			1	3	4	5				0.1458	
[4] 2–5				1	3	4				0.0851	
[5] 5–10					1	3				0.0488	
[6] 10 <						1				0.0282	
Geological subcriteria											
[1] Lithology	1	2	3	5						0.4729	0.0194
[2] Land slope—a slope of layer		1	2	4						0.2844	
[3] Land slope direction—slope direction of the layer			1	3						0.1699	
[4] Distance from Fault				1						0.0729	

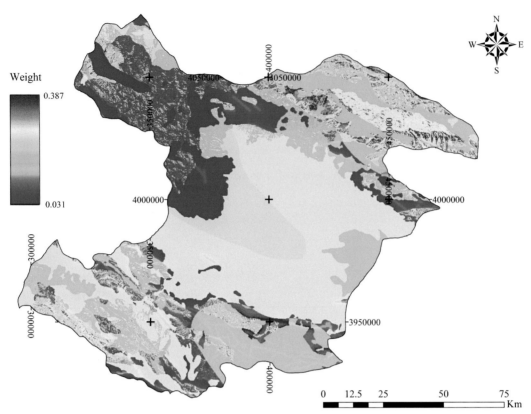

FIG. 13   Map of weights of the geological criterion.

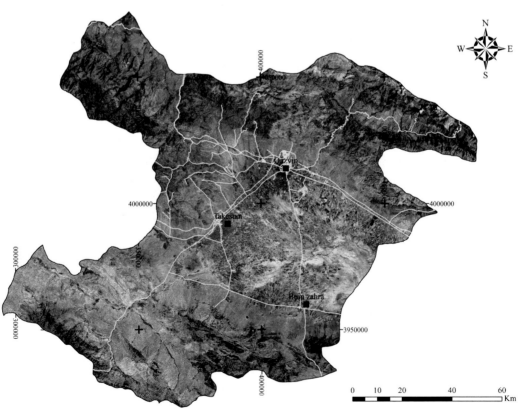

FIG. 14   Road Map of Qazvin province.

TABLE 5 Decision-making matrix for environmental subcriteria.

Causative factors and classes with in each factor		Pair-wise comparison matrix									Weight	Consistency ratio (CR)
		[1]	[2]	[3]	[4]	[5]	[6]	[7]	[8]	[9]		
Environmental criteria												
Distance from road												
[1]	0–100	1	3	5	7	8	9				0.4625	0.0450
[2]	100–200		1	3	5	7	8				0.2589	
[3]	200–500			1	3	5	7				0.1405	
[4]	500–1000				1	3	5				0.0740	
[5]	1000–1500					1	3				0.0401	
[6]	>1500						1				0.0241	
Land use												
[1]	Arid lands	1	2	3	5	6	7	9			0.3725	0.0297
[2]	residential area		1	2	3	5	6	8			0.2467	
[3]	grassland			1	2	3	5	6			0.1566	
[4]	Agricultural lands				1	2	3	5			0.0966	
[5]	Garden					1	2	4			0.0624	
[6]	forest						1	3			0.0416	
[7]	Watery areas							1			0.0236	
Environmental subcriteria												
[1]	Land use	1	3								0.7500	–
[2]	Distance from road		1								0.2500	

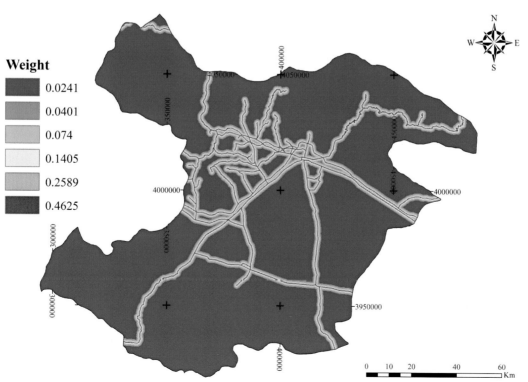

FIG. 15 Weighted map of distance from road subcriterion.

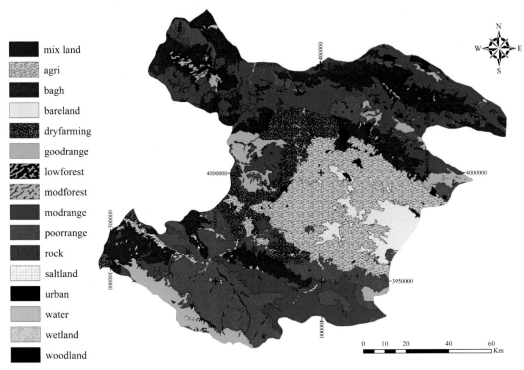

FIG. 16 Land use map of Qazvin province.

movements more than gardens, crops, and forests. Water bodies, which include ponds, wetlands, and artificial and natural lakes, are irrelevant in the landslide phenomenon. Finally, after determining the weight of each area, a weighted map of land use subcriterion was provided (Fig. 17).

- **Summarizing environmental criteria**

After weighing each of the environmental subcriteria, the decision-making matrix was made up (Table 5), the priority of each factor was compared to the other and based on the results of the questionnaires, and an appropriate score was given. Then, after determining the superiority of the subcriteria to each other, the weight of each subcriterion is determined and finally, the environmental weighted map is obtained according to the weight of each subcriterion (Fig. 18).

4.3 Geomorphological criteria

Slope aspect, slope angle, and topography elevation are the most important factors in geomorphology, which influence slope stability.

- **Slope angle**

Slope angle is one of the most important factors in creating instability of slopes. Based on morphology, each study area may have different slopes. The higher the angle of a slope, the greater the unstable force. Therefore, theoretically, and assuming that other factors are identical, the risk of landslides is higher in more steep slopes.[39] Considering the importance of this parameter in landslide zonation in different studies, this parameter has been used for zoning the risk of landslides. Flat surfaces (slope with an angle less than 10%) also have no-slip.[40, 41]

In Fig. 19, the map of the slope angle of Qazvin province is presented. In this regard, after the formation of the decision-making matrix, the highest weight for steep slopes and the lowest weight for flat surfaces are considered (Table 6). Fig. 20 shows the weighted map of the slope angle subcriterion.

- **Slope aspect**

The direction of the slope shows the different effects of getting sunlight, hot and dry winds, and rainfall in different directions. Many studies on the effect of slope aspect on landslides indicated that slopes that are more exposed to sunlight are more stable than those that are less time-consuming than sunlight. The reason for this phenomenon can be

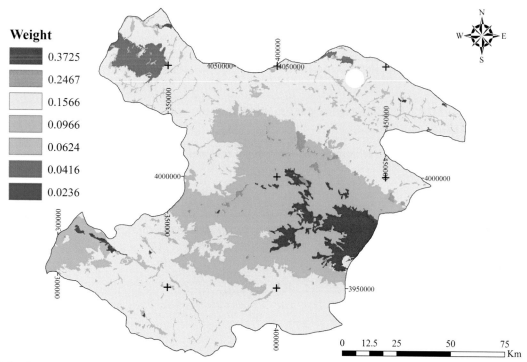

FIG. 17 Weighted map of land use subcriterion.

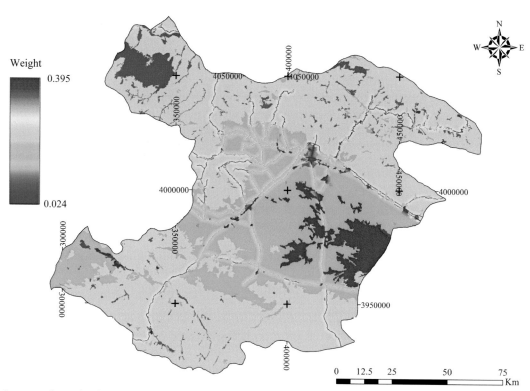

FIG. 18 Environmental weighted map.

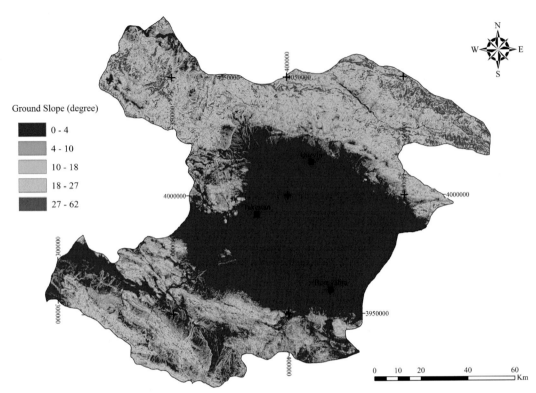

FIG. 19   Map of topographic slope angle of Qazvin province.

attributed to the amount of evaporation and less moisture content of these slopes. In the northern hemisphere, the southern slopes are more exposed to sunlight and can be expected these slopes to be more stable than the northern ones.[42]

In Fig. 21, the aspect map of Qazvin province's topography is presented and in Table 5, the decision-making matrix for the aspect is presented. Aspect classification is made based on the presence of different factors in different directions of slope angle and the difference in the expansion of the slopes. Due to the presence of Iran in the northern hemisphere, the southern slopes compared to the northern slopes receive more sunlight throughout the day and have a colder climate; thus, the northern slopes provide good conditions for chemical weathering, and since weathering intensity has an inverse relationship with adhesive strength, they have the greatest potential for a landslide. On the other hand, due to the general direction of the winds from the west in the province, atmospheric rainfall in the western slopes is higher than other slopes. Fig. 22 shows the aspect subcriterion map.

- **Topography elevation**

Fig. 23 shows the topographic map of Qazvin province using the SRTM (the Shuttle Radar Topography Mission) 90 m DEM's (Digital Elevation Model). It is shown that the height factor due to the effect on slope, dispersion, and density of the network of streams is considered as an effective factor in the occurrence of mass movements of the slopes. The highest risk of landslide is at higher altitudes. After the decision-making matrix and zoning of Qazvin province based on altitude, the highest weight was assigned to the zones with higher topographic levels. In addition, with a decrease in the amount of topographic elevation, less weight is assigned to the zones (Table 6). Fig. 24 shows the weighted map of the subcriterion of topographic elevation.

- **Summarizing the geomorphological criteria**

After the weighting of each of the subcriteria of geomorphology (Table 6), by determining the decision-making matrix, the preferences of each subcriterion are compared to the other, and according to the results of the questionnaires, an appropriate score has been given. After determining the superiority of the subcriteria to each other, the weight of each subcriterion was determined using the AHP program, and finally, the weighted map of the environmental subcriterion was prepared (Fig. 25).

TABLE 6 Decision-making matrix for geomorphological criteria.

Causative factors and classes with in each factor		Pair-wise comparison matrix									Weight	Consistency ratio (CR)
		[1]	[2]	[3]	[4]	[5]	[6]	[7]	[8]	[9]		
Geomorphological criteria												
Slope angle (in degree)												
[1]	0–10	1									0.0261	0.0546
[2]	10–20	3	1								0.0486	
[3]	20–30	5	3	1							0.0894	
[4]	30–40	6	4	3	1						0.1522	
[5]	40–50	8	5	4	3	1					0.2753	
[6]	>50	9	7	5	4	2	1				0.4085	
Slope aspect												
[1]	North	1									0.0451	0.0245
[2]	East	3	1								0.0993	
[3]	West	7	4	1							0.3200	
[4]	South	9	6	2	1						0.5356	
Topography level (in meter)												
[1]	1000>	1									0.0333	0.0534
[2]	1000–2000	3	1								0.0634	
[3]	2000–3000	5	3	1							0.1290	
[4]	3000–4000	7	5	3	1						0.2615	
[5]	4000<	9	7	5	3	1					0.5128	
Geomorphological subcriteria												
[1]	Slope angle	1	3	7								0.0061
[2]	Slope aspect		1	3								
[3]	Topography level			1								

4.4 Hydrological criteria

For this criterion, the only subcriterion that can be counted is the distance from the river (river boundary). Due to the presence of drainage of water and sloping walls, slope earoding and increasing saturation of sloping materials, rivers tend to have a higher landslide.[40]

Some studies have shown that the dispersion of landslides is closely related to the system of streams. This phenomenon can be attributed to two factors:

1. Rivers indicate the presence of a large number of slopes and, consequently, create more landslides.
2. The river causes the slope to be removed and disturbs the slope balance and increases the chance of slipping.

In Fig. 26, the map of the rivers of Qazvin province is presented. Due to the position of the streams, the surrounding areas were zoned using the Buffering tools in GIS, and then the decision-making matrix was weighted based on the proximity to the river (Table 7). Accordingly, the highest score is given to the zones with a maximum distance of 100 m from the river. By distance from the rivers, the zones are less weighed. In Fig. 27, the map of the hydrological criterion is presented.

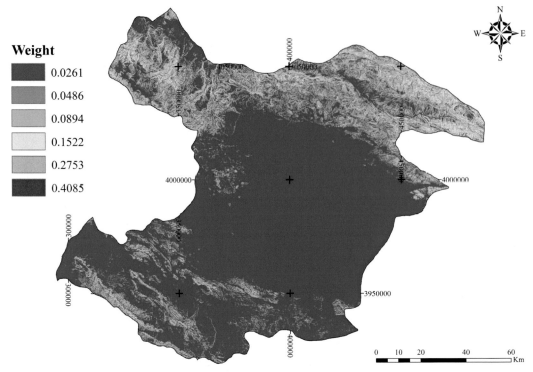

FIG. 20   Weighted map of slope angle subcriterion.

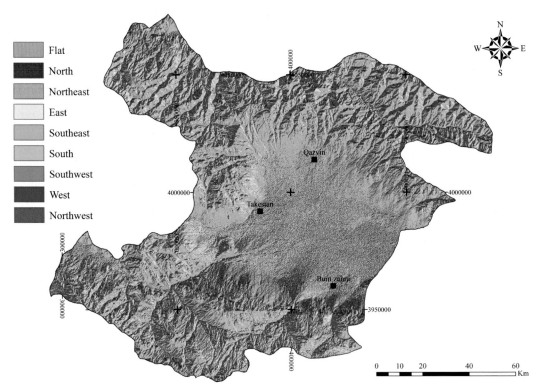

FIG. 21   Aspect map of the Qazvin province.

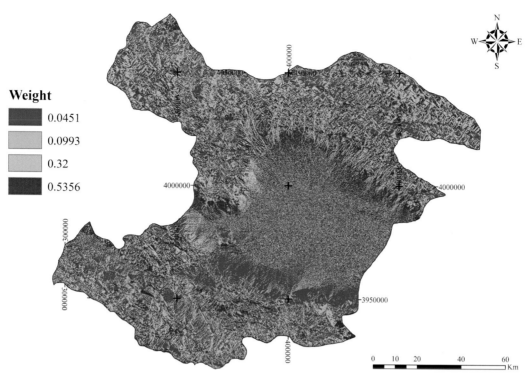

FIG. 22 Weighted map of the aspect subcriterion.

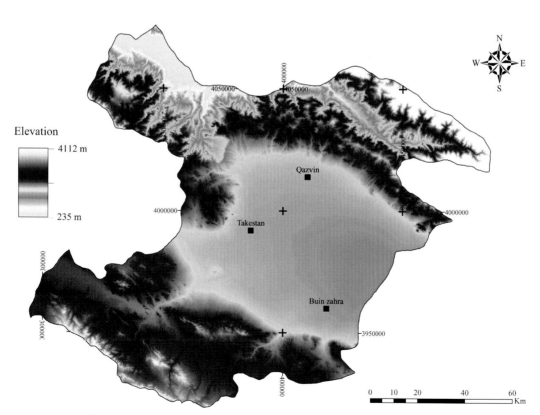

FIG. 23 Topographic Map of Qazvin province.

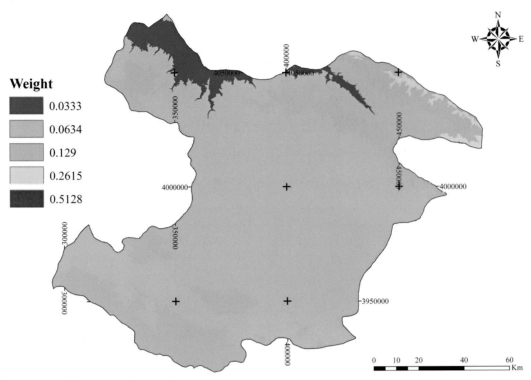

FIG. 24 Weighted map of the topographic subcriterion.

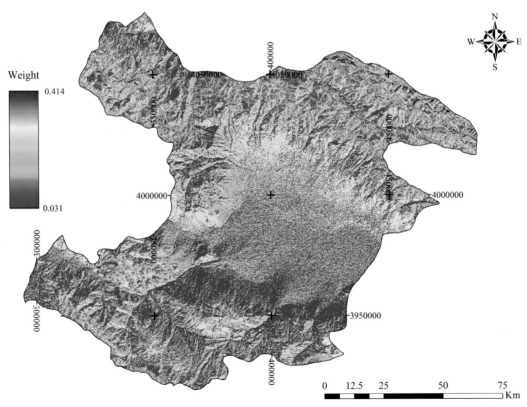

FIG. 25 Weighted map of geomorphology.

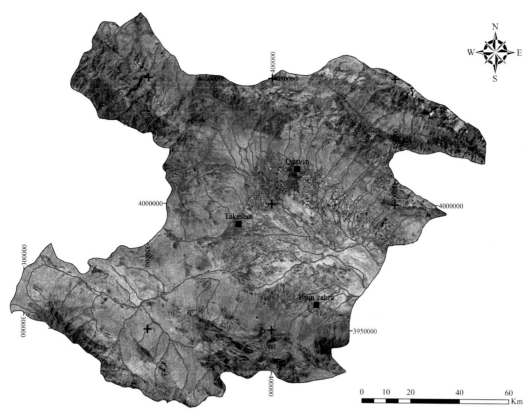

FIG. 26 Map of the rivers of Qazvin province.

TABLE 7 Decision-making matrix of rainfall subcriterion (in millimeters per year).

Causative factors and classes with in each factor		Pair-wise comparison matrix									Weight	Consistency ratio (CR)
		[1]	[2]	[3]	[4]	[5]	[6]	[7]	[8]	[9]		
Hydrological criteria												
Distance from river (in meter)												
[1]	0–100	1	3	6	9						0.5912	0.0307
[2]	100–200		1	3	6						0.2571	
[3]	200–300			1	3						0.1052	
[4]	>300				1						0.0466	
Climatological criteria												
Rainfall (in millimeters per year)												
[1]	200–260	1									0.0333	0.0296
[2]	260–320	3	1								0.0634	
[3]	320–380	5	3	1							0.1290	
[4]	380–450	7	5	3	1						0.2615	
[5]	> 450	9	7	5	3	1					0.5128	

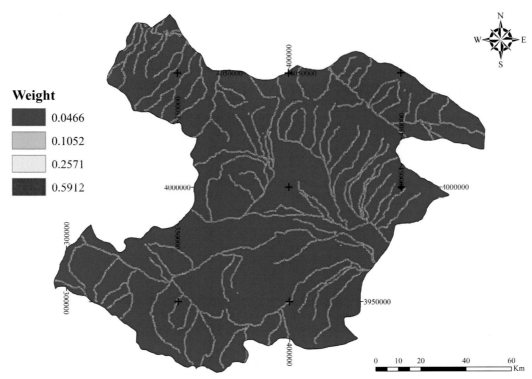

Weight

■ 0.0466
□ 0.1052
□ 0.2571
■ 0.5912

FIG. 27 Weighted map of hydrological criteria.

4.5 Climatological criteria

For this criterion, the only subcriterion to be considered is the amount of precipitation. Infiltrated water in the slopes increases the pore pressure of the water, the rise of the groundwater level, and consequently the saturation of the slopes, decreasing the shear strength of the masses of the soil and rock, increasing the slope weight, fluctuating the river water level and slope scour. This condition causes the slippery materials to reach impenetrable levels and increases the talent of the landslide event. Experience has shown that the highest number of rupture slopes occurs after heavy rainfall or melting of snow in the spring due to water penetration in the cracks.[43] In Fig. 28, the rainfall map of the Qazvin province is presented. After zoning, based on the amount of annual precipitation (in mm), the decision-making matrix of each zone's priority was determined relative to the other zones, and based on the results of the questionnaires, the score was given between 1 and 9. Table 7 presents the decision-making matrix of the rainfall subcriterion. As seen, the highest score is attributed to areas with a minimum annual rainfall of 450 mm. Finally, after determining the preference of the factors relative to each other, the weight of each zone was determined by the AHP program and based on the map of the weighted precipitation subcriterion was obtained (Fig. 29).

4.6 Geotechnical criteria

Due to the effect of geotechnical characteristics on the landslide hazard, several types of research have been conducted.[44]

Geotechnical characteristics include point load index, GSI (Geological Strength Index), saturated specific gravity, cohesion, internal friction angle, weathering level, and condition of discontinuities. The corresponding values of these parameters are presented in Table 8 for rock and deposits of Qazvin province. Table 9 presents the decision-making matrix of the geotechnical subcriterion.

- **Geological strength index (GSI)**

Geological strength index (GSI) is a pure geological index that is controlled by the lithology, structures, and surface conditions of rock mass discontinuities. This index, which was firstly proposed by Marinos et al.[45] for rock mass classification, is estimated through observing rock mass outcrops through the surface and subsurface excavations and core boxes. This classification has a stronger emphasis on geological observations on rock mass properties and reflects the

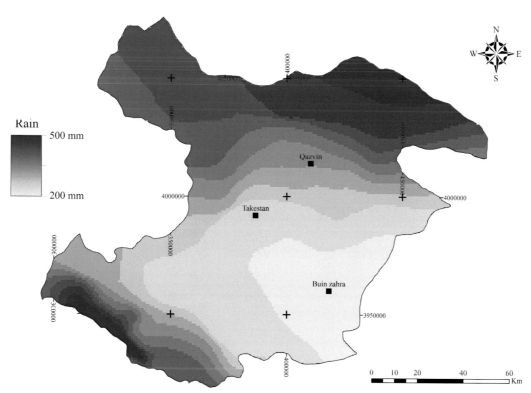

FIG. 28 Precipitation map of Qazvin province.

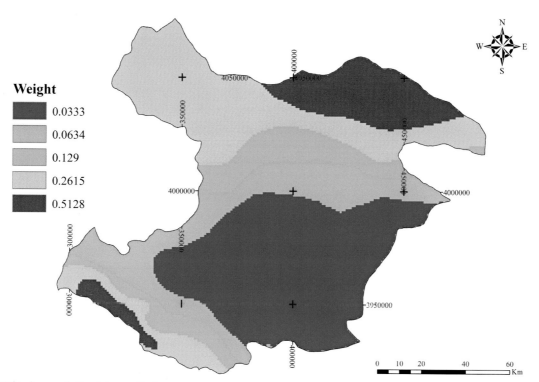

FIG. 29 Weighted map of climatology criterion.

TABLE 8 Engineering geology characteristics deposits of Qazvin province.

Lithology	GSI	Point load index (MPa)	Saturation density (gr/cm³)	Cohesion (MPa)	Internal friction angle (dge)	Weathering	Jointing conditions
New Alluviums	–	>1	1.9–1.92	0.03–0.05	22–25	W3–W4	–
Old alluviums	–		2.1–2.2	0.05–0.07	25–30	W1–W2	–
Tuff	20–30	1–1.5	2.4–2.45	0.6–1.1	23–26	W3–W4	D3–D4
Volcanic lava	25–35	1.5–2	2.45–2.5	1.1–1.9	28–31	W3–W4	D3–D4
Granite	60–70	6–8	2.7–2.75	9–14	43–47	W0–W1	D0–D1
Gabbro	55–65	6–8	2.9–2.92	8–13	40–45	W1–W2	D0–D1
Sandstone	40–45	1.5–3	2.68–2.73	1.5–3.2	32–35	W2–W3	D2–D3
Marl	25–30	1–1.5	2.45–2.55	0.5–0.9	20–22	W3–W4	D3–D4
Limestone	50–60	3–5	2.7–2.75.	2.8–5.7	30–35	W1–W2	D1–D2
Shale	25–35	1–1.5	2.4–2.55	0.5–0.9	20–25	W3–W4	D3–D4
Conglomerate	45–55	2–4	2.6–2.7	2.3–5.3	35–40	W2–W3	D2–D3

TABLE 9 Decision-making matrix for Geotechnical criteria.

Causative factors and classes within each factor	Pair-wise comparison matrix										Weight	Consistency Ratio (CR)
	[1]	[2]	[3]	[4]	[5]	[6]	[7]	[8]	[9]			
Geotechnical criteria												
Geological strength index (GSI)												
[1]	<20	1	3	5	7						0.5650	0.0439
[2]	20–40		1	3	5						0.2622	
[3]	40–60			1	3						0.1175	
[4]	60–80				1						0.0553	
The point load index (IS50)												
[1]	<1	1	3	5	7						0.5650	0.0439
[2]	1–3		1	3	5						0.2622	
[3]	3–5			1	3						0.1175	
[4]	5–7				1						0.0553	
Saturated specific gravity (Gsat)												
[1]	<2	1									0.0336	0.0090
[2]	2–2.2	2	1								0.0540	
[3]	2.2–2.4	3	2	1							0.0896	
[4]	2.4–2.6	5	3	2	1						0.1532	
[5]	2.6–2.8	7	5	3	2	1					0.2547	
[6]	>2.8	9	7	5	3	2	1				0.4148	

TABLE 9 Decision-making matrix for Geotechnical criteria—cont'd

Causative factors and classes within each factor	Pair-wise comparison matrix									Weight	Consistency Ratio (CR)
	[1]	[2]	[3]	[4]	[5]	[6]	[7]	[8]	[9]		
Cohesion strength (C)											
[1]	<0.1	1	2	3	4	5	6	9			0.3609	0.0204
[2]	0.1–1		1	2	3	4	5	6			0.2383	
[3]	1–2			1	2	3	4	5			0.1574	
[4]	2–4				1	2	3	4			0.1027	
[5]	4–7					1	2	3			0.0669	
[6]	7–10						1	2			0.0444	
[7]	>10							1			0.0294	
Internal friction angle (φ)											
[1]	Massive rocks and alluvium	1									0.0277	0.0163
[2]	0	2	1								0.0415	
[3]	0–10	3	2	1							0.0655	
[4]	10–20	4	3	2	1						0.1005	
[5]	20–30	5	4	3	2	1					0.1543	
[6]	30–40	8	5	4	3	2	1				0.2422	
[7]	>40	9	8	5	4	3	2	1			0.3683	
Weathering condition (W)											
[1]	W0-W1	1									0.0553	0.0439
[2]	W1-W2	3	1								0.1175	
[3]	W2-W3	5	3	1							0.2622	
[4]	W3-W4	7	5	3	1						0.5650	
Geotechnical criteria											
[1]	GSI	1	2	3	4	5	6	7			0.3543	0.0297
[2]	C		1	2	3	4	5	6			0.2399	
[3]	φ			1	2	3	4	5			0.1587	
[4]	D				1	2	3	4			0.1036	
[5]	Gsat					1	2	3			0.0676	
[6]	IS50						1	2			0.0448	
[7]	W							1			0.0312	

type, structure, and geological history of rock masses. The variations in the GSI score of various rock units and their decision-making matrix are presented in Fig. 30 and Table 9, respectively. Also, the weight map of this subcriterion is illustrated in Fig. 31.

- **The point load index (IS50)**

Strength decline along the slide surface is among the key factors in landslide occurrence. Landslide phenomenon may occur in slopes with an alternative sequence of marls, conglomerates, and sandstones along their largest slope angle, but not necessarily in the contact of marl and sandstone or conglomerate; rather, it occurs in the marl layers with reduced shear strength.[46]

Through the field surveys performed in the study area, rock blocks were collected from different rock formations. After the transfer of the samples to a laboratory and suturing them, point load tests conducted on the rock blocks and IS50 were determined to determine rock strength, as a variable effective in slide occurrence. The IS50 map for geological materials of Qazvin province and decision-making matrix for this subcriterion is presented in Fig. 32 and Table 9, respectively. Moreover, the weight map of the IS50 subcriterion for the study area is presented in Fig. 33.

- **Saturated specific gravity (Gsat)**

Water is the most important factor in landslide occurrence. The majority of landslides occur after heavy precipitations. Water accumulation increases pore pressure and bulk density of the ground materials, which are in favor of landslide occurrence. Water absorption by rocks such as shale and marl increases the volume of these rocks, creates a lubricant, and thus leads to the slide of overlain rocks.[47]

To perform saturated specific gravity tests in the present work, the collected samples were saturated and then the tests were carried out following the ISRM standard.[48] Fig. 34 illustrates Gsat map of the geological materials in Qazvin province and Table 2 presents the decision-making matrix for Gsat subcriterion. As shown by these results, an increase in Gsat reduces the slope stability. Fig. 35 shows the weight map of Gsat subcriterion for geological materials of Qazvin province.

- **Rock mass shear strength parameters**

Most of the analytical techniques for slope design apply cohesion strength (C) and internal friction angle (ϕ) as the shear strength parameters. C and ϕ of rock mass are among the most important factors in rock mass slide.[49] The Hoek-

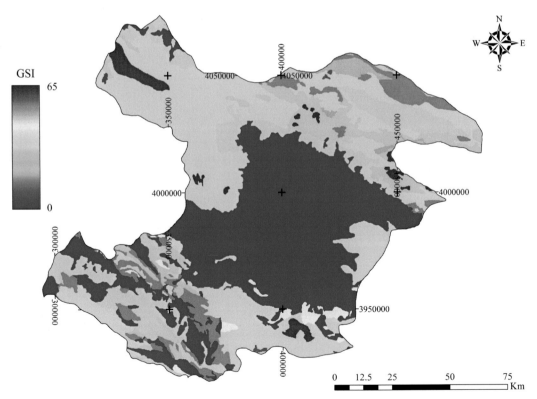

FIG. 30   GSI map of rock outcrops in Qazvin province.

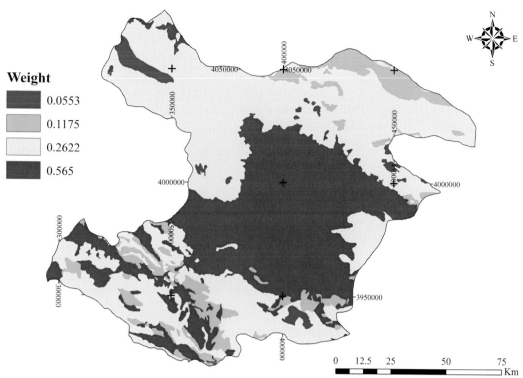

FIG. 31 Weight map of GSI subcriterion.

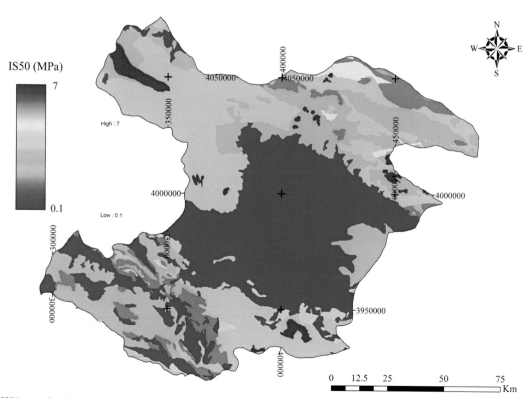

FIG. 32 An IS50 map for the geological materials of Qazvin province.

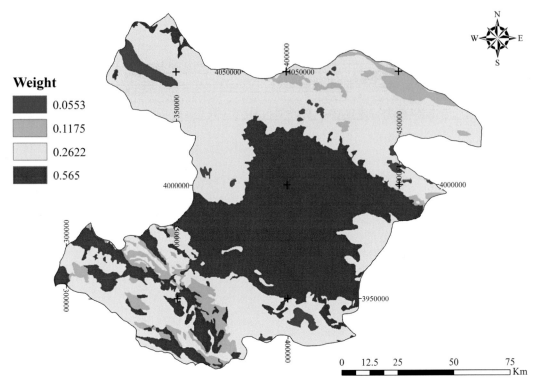

FIG. 33 Weight map of IS50 subcriterion.

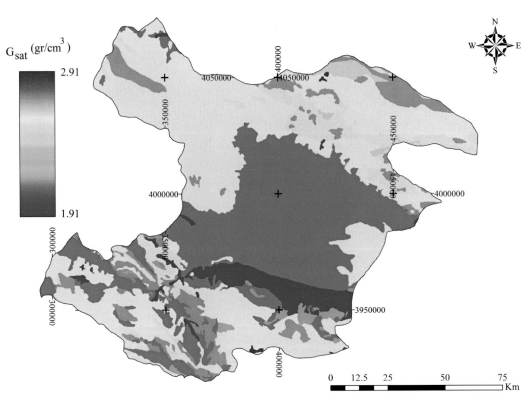

FIG. 34 Gsat map of geological materials in Qazvin province.

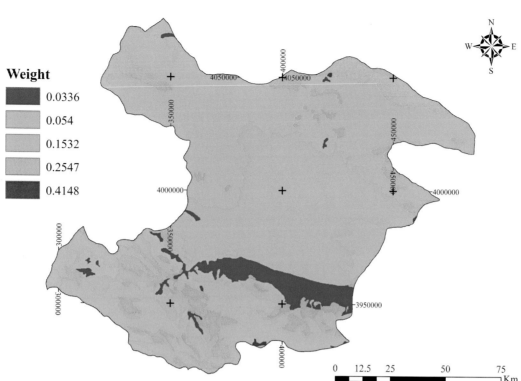

FIG. 35   Weight map of Gsat subcriterion.

Brown failure criterion is one of the most practical criteria for the determination of rock mass C and φ values under in situ conditions.

To obtain parameters of Mohr-Coulomb criterion (i.e., C and φ), first, using the laboratory parameters, field surveys, and Hoek-Brown criterion, the rock failure envelope was plotted and then C and φ of rock mass were determined by solving Mohr-Coulomb relation in the σ1–σ3 space and fit model of Mohr-Coulomb and Hoek-Brown envelopes.

Based on the field surveys, C and φ for the rock masses of the study area were determined using the equations proposed by Hoek and Diederichs[50] and Hoek et al..[51] The C map and decision-making of the C subcriterion are shown in Fig. 36 and Table 9, respectively. Also, Fig. 37 presents the weight map of the C subcriterion.

In addition, the φ map and decision-making matrix of difference between strata slope and φ for the geological materials in Qazvin province are presented in Fig. 38 and Table 9, respectively. Fig. 39 presents the weight map of the difference between strata slope and φ for the study area.

- **Weathering level**

The outcrops weathering was determined according to Look[52] and shown in Table 10. The results show that an increase in weathering leads to rock mass strength reduction. The type and rate of rock weathering are a function of atmospheric conditions such as temperature and its variations and atmospheric precipitations.

Based on the field surveys, rock mass weathering was one of the most important factors effective in slope stability conditions. Table 9 and Fig. 40 present weathering conditions of geological materials and the decision-making matrix for this parameter for the geological materials of Qazvin province, respectively. Moreover, Fig. 41 presents a weight map of weathering condition subcriterion.

- **Condition of discontinuities**

The condition of discontinuities and geological structures are among the important factors effective on the stability of rock slopes. An increase in the number and severity of jointing leads to rock mass strength reduction and provides better conditions for rockslides. As a consequence, based on the discontinuities' orientation with respect to the slope, different types of slope stability may occur. When the slopes are inclined inward the slope body, the slope is more stable since under such conditions the potential failure is controlled by the shear strength parameters of intact rock

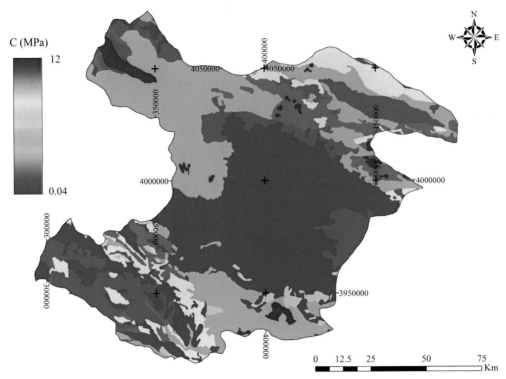

FIG. 36   C map of the geological materials in Qazvin province.

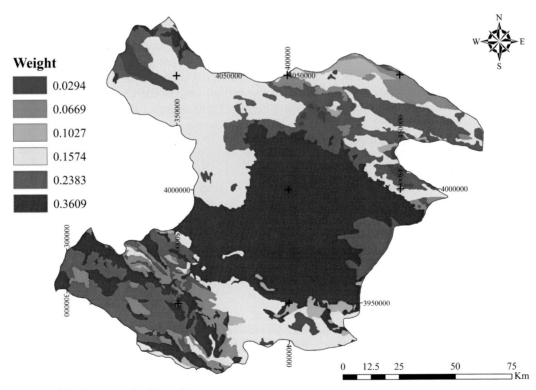

FIG. 37   Weight map of C subcriterion for the study area.

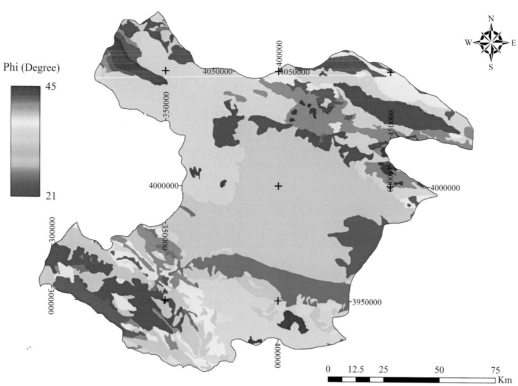

FIG. 38   The ϕ map for the geological materials of Qazvin province.

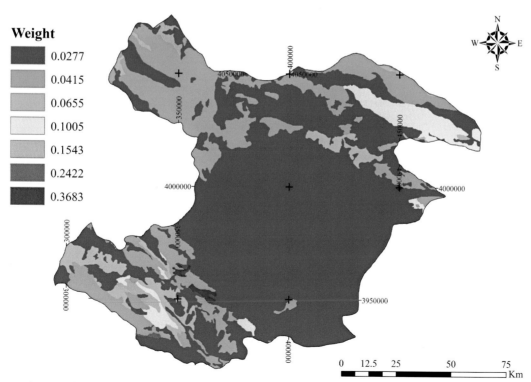

FIG. 39   The weight map of the difference between strata slope and ϕ.

TABLE 10   Description of rock weathering.[52]

Grade	Term	Description
w0–w1	Fresh	No visible sign of rock material weathering, perhaps slight discoloration on major discontinuity surfaces
w1–w2	Slightly weathered	Discoloration indicates weathering of rock material and discontinuities surface. All the rock material may be discolored by weathering and the external surface may be somewhat weaker than in its fresh condition
w3	Moderately weathered	Less than half of the rock material are decomposed and/or disintegrated into the soil. Fresh or discolored rock is present either as continuous framework or as corestones
w4	Highly weathered	More than half of the rock materials are discomposed and/or disintegrated into the soil. Fresh or discolored rock is present either as a discontinuous framework or as corestones
w5	Completely weathered	All rock material is decomposed and/or disintegrated into the soil. The original mass structure is still largely intact
w6	Residual soils	All rock material is converted to the soil. The mass structure and material fabric are destroyed. There is a large change in volume, but the soil has not been significantly transported

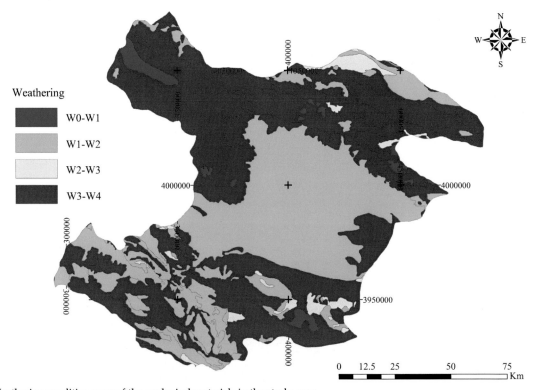

FIG. 40   Weathering condition map of the geological materials in the study area.

material. On the other hand, when the slopes are inclined outward the slope body, they are less stable since discontinuities control the rock stability instead of intact rock parameters.[49]

The condition of discontinuities in different rock units is among the most important factors recorded through the field surveys. The discontinuities in each rock unit were collected from each station and analyzed according to the ISRM standard. Table 11 shows the condition of discontinuities according to the method proposed by Bieniawski.[53]

Based on the surveys and analyses performed in this study, Fig. 42 and Table 9 present the discontinuity conditions or rock outcrops and decision-making matrix for discontinuity condition subcriterion, respectively. Moreover, Fig. 43 presents a weight map of discontinuity conditions subcriterion.

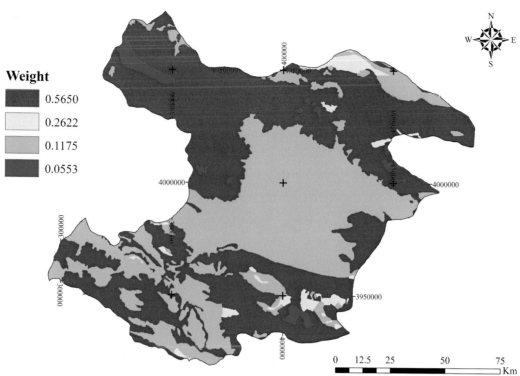

FIG. 41 Weight map of weathering condition subcriterion.

TABLE 11 Condition of discontinuities.[53]

Description	Rating
Very rough and weathered, wall rock tight and discontinuous, no separation	D0–D1
Rough and slightly weathered, wall rock surface separation <1 mm	D1–D2
Slightly rough and moderately to highly weathered, wall rock surface separation <1 mm	D2–D3
Slickensides wall rock surface or 1–5 ram thick gouge or 1–5 ram wide continuous discontinuity	D3–D4
5 mm thick soft gouge, 5 mm wide continuous discontinuity	D4

- **Summarizing the geotechnical criteria**

The decision-making matrix for geotechnical subcriteria and weight map of geotechnical criteria are presented in Table 9 and Fig. 44, respectively. As can be seen, the maximum and minimum weights are for GSI and weathering conditions, respectively.[49]

4.7 Landslide hazard zoning in Qazvin province

After weighing each criterion, with the formation of the decision-making matrix, the priority of each factor relative to other factors score from 1 to 9 is given according to the results obtained from the opinions of the specialists (Table 12). Then, after determining the priority of the criteria against each other, the weight of each criterion was determined. Finally, by integrating the criteria according to their weight, a weighted map of landslide hazard of Qazvin province was determined (Fig. 45). Since the landslide does not occur in the flat area, applying the zero coefficient in zones with slopes less than 10 degrees using the Raster Calculator in GIS has virtually eliminated the value of these areas in the occurrence of landslides. It should be noted that this is the only possible way to reduce the landslide of the flat terrain and cannot be considered any solution in AHP evaluation steps for it.

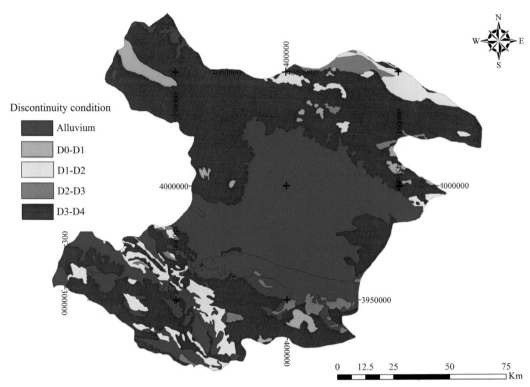

FIG. 42　Discontinuity condition map of rock outcrops in Qazvin province.

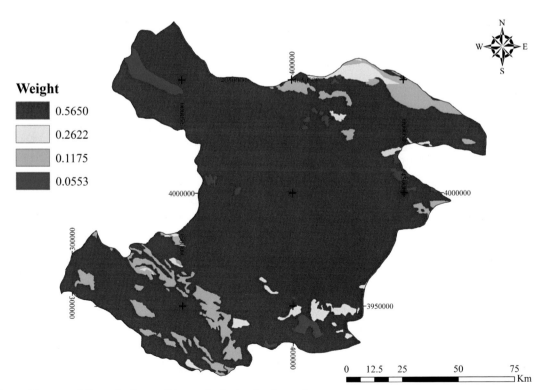

FIG. 43　The weight map of discontinuity conditions subcriterion for the study area.

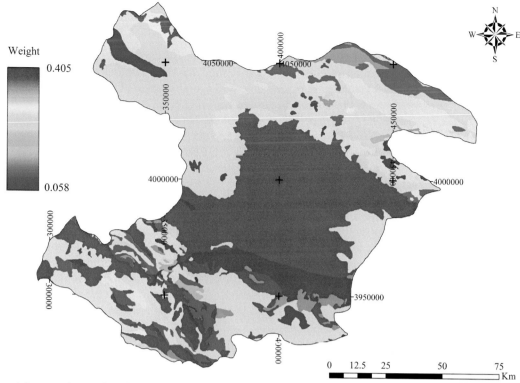

FIG. 44 The weight map of geotechnical criterion.

TABLE 12 Classification of landslide risk potential based on the final weights of AHP.

Risk level	Weight range	Risk level	Weight range
Medium	0.15–0.2	Negligible	0–0.05
High	0.2–0.25	Very low	0.05–0.1
Very high	0.25–0.32	Low	0.1–0.15

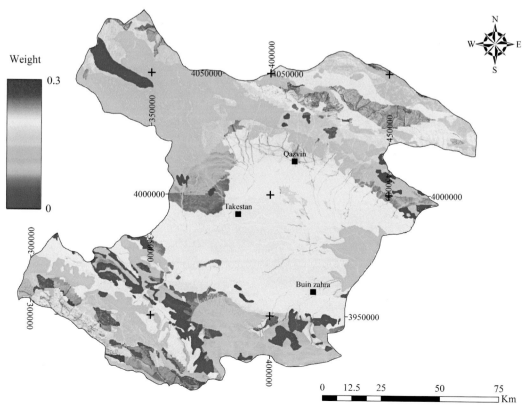

FIG. 45 Primary weighted map of risk in Qazvin province.

Fig. 46 shows an edited map of the landslide hazard in Qazvin province. As can be seen, the final weight varies from zero to a maximum of 0.3. Therefore, considering the final weight range, the risk of landslide occurrence in the province is divided into 6 categories, the riskiest category is from 0.25 to 0.3 weight range (Table 12). Fig. 47 shows the final classification results of the potential hazard of landslide occurrence in Qazvin province.

In order to verify the accuracy of the hazard map, the location of the historic landslides in the province is plotted on the map. The fairly complete compatibility of the hazardous areas with the historical landslides of the province indicates the high accuracy of the map and the high accuracy of the AHP method. The efficiency of the AHP method is confirmed by other researchers.[54–56]

Figs. 30 and 33 show the highest concentration of landslides. In this area (on the right and top of the map), the strata slope is 30 to 40 degrees. The lithology of the land in this area is more of a type of dolomitic limestone and sandstone with shale in the Miocene age. On average, basaltic lava is also seen. In this area, the gradient slope of Qazvin province is 5–20 degrees and precipitation are more than 400 mm per year. There is a fairly good correlation between the distance from the fault and the landslides in this area. Less than 0.5 km the greatest number of landslides have occurred. Among land use criteria, mordrange lands and mixed land areas were the most susceptible areas to the landslide.

In Fig. 48, the area of each specified zone is shown based on the classification for landslide hazard in Qazvin province, and Fig. 49 the percentage of each specified zone is presented. According to these figures, the largest area of the negligible zone is in landslide zoning and the least is related to the very high zone in landslide zoning.

5 Conclusions

Considering the factors affecting landslide and landslide distribution map in Qazvin province, some important results are obtained. Most of the mass movements occurred, where slopes are between 5 and 20 degrees. In slopes, less than 5 degrees (about 10%), the probability of landslide is reduced because of a decrease in the effect of gravity, however on slopes of higher than 20 degrees it is reduced because of the soil process and consequently the absence of susceptible materials for a landslide. The topography aspect did not show a logical correlation with a landslide for the study area.

Among the geological criteria, after the slope angle, the lithology has the greatest effect on landslide. Most of the landslides are compatible with dolomitic limestones and shaly sandstones that are Miocene in age. Land use has the

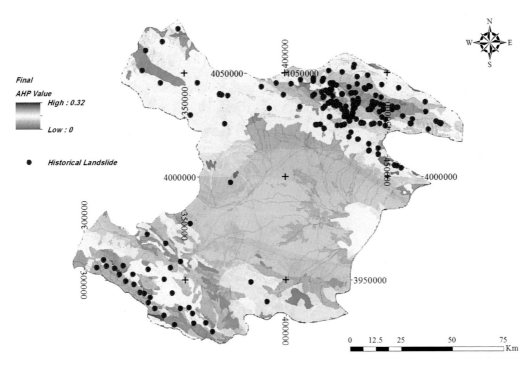

FIG. 46   Modified weighted map of the landslide risk of Qazvin province by eliminating the weight of flat area.

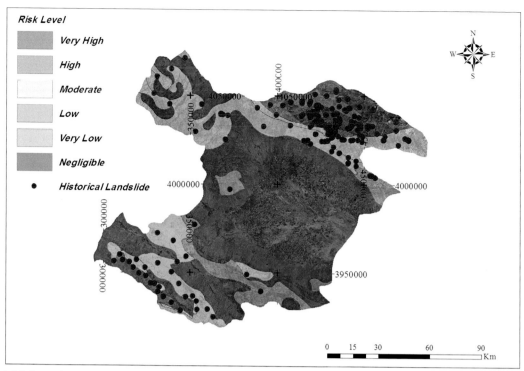

FIG. 47 Landslide hazard zonation map of Qazvin province along with historical landslides of the province.

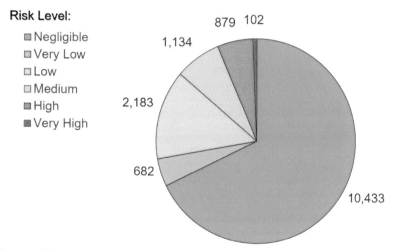

FIG. 48 The area of hazard potential classification of landslide occurrence in the Qazvin province.

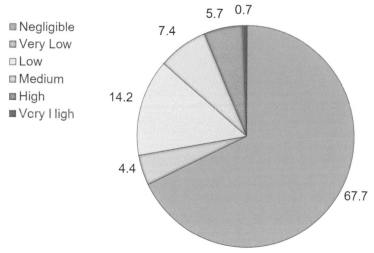

FIG. 49 The percentage of each class of landslide hazard potential in Qazvin province.

highest landslide in mordrange lands and mixed land, while forest soils and rangelands have less landslide rates. The minimum distance from the road and faults in the area for nonslipping are 200 and 500 m, respectively. At intervals greater than these values, the probability of the landslide decreases sharply. At less than 100 m from the streams, the probability of occurrence of the landslides increases extremely. The most landslide has occurred in slopes with the direction of the south-east, south-west. For the occurrence of landslides, the amount of rainfall should be sufficient to penetrate the soil and increase the weight of the mass of the soil on the slopes, which this precipitation rate (more than 400 mm per year) only occurred in areas with slipping areas overlapping. Finally, given the overlapping of historical landslides and the results of the AHP method, it can be concluded that AHP has a good performance for landslide hazard zoning. The results of this study in the future can be applied for environmental planning in terms of building infrastructure, communication lines, building residential areas, etc.

References

1. Youssef AM, Pourghasemi HR. Landslide susceptibility mapping using machine learning algorithms and comparison of their performance at Abha Basin, Asir region, Saudi Arabia. *Geosci Front*. 2021;12:639–655. https://doi.org/10.1016/j.gsf.2020.05.010.
2. Yu C, Chen J. Application of a gis-based slope unit method for landslide susceptibility mapping in Helong city: comparative assessment of icm, ahp, and rf model. *Symmetry (Basel)*. 2020;12:1–21. https://doi.org/10.3390/sym12111848.
3. Panahi M, Gayen A, Pourghasemi HR, et al. Spatial prediction of landslide susceptibility using hybrid support vector regression (SVR) and the adaptive neuro-fuzzy inference system (ANFIS) with various metaheuristic algorithms. *Sci Total Environ*. 2020;741:139937. https://doi.org/10.1016/j.scitotenv.2020.139937.
4. Pourghasemi HR, Kariminejad N, Gayen A, Komac M. Statistical functions used for spatial modelling due to assessment of landslide distribution and landscape-interaction factors in Iran. *Geosci Front*. 2020;11:1257–1269. https://doi.org/10.1016/j.gsf.2019.11.005.
5. Pourghasemi HR, Kornejady A, Kerle N, Shabani F. Investigating the effects of different landslide positioning techniques, landslide partitioning approaches, and presence-absence balances on landslide susceptibility mapping. *Catena*. 2020;187:104364. https://doi.org/10.1016/j.catena.2019.104364.
6. Emami SN, Yousefi S, Pourghasemi HR, et al. A comparative study on machine learning modeling for mass movement susceptibility mapping (a case study of Iran). *Bull Eng Geol Environ*. 2020;79:5291–5308. https://doi.org/10.1007/s10064-020-01915-7.
7. Basu T, Pal S. *A GIS-Based Factor Clustering and Landslide Susceptibility Analysis Using AHP for Gish River Basin*. Springer Netherlands: India; 2020.
8. Vojteková J, Vojtek M. Assessment of landslide susceptibility at a local spatial scale applying the multi-criteria analysis and GIS: a case study from Slovakia. *Geomatics, Nat Hazards Risk*. 2020;11:131–148. https://doi.org/10.1080/19475705.2020.1713233.
9. Arsyad A, Hamid W. Landslide susceptibility mapping along road corridors in West Sulawesi using GIS-AHP models. *IOP Conf Ser Earth Environ Sci*. 2020;419. https://doi.org/10.1088/1755-1315/419/1/012080.
10. Pratama RP. Application of GIS for the mapping of landslide-vulnerable areas by through android-based analytical hierarchy process (AHP) method in Bantul Regency. *IOP Conf Ser Earth Environ Sci*. 2019;245(1):1–11. https://doi.org/10.1088/1755-1315/245/1/012008.
11. Vakhshoori V, Pourghasemi HR, Zare M, Blaschke T. *Landslide Susceptibility Mapping Using GIS-Based Data Mining Algorithms*. vol. 11. 11th. MDPI (Water); 2019:2–30. https://doi.org/10.3390/w11112292.
12. Arjmandzadeh R, Sharifi Teshnizi E, Rastegarnia A, et al. GIS - based landslide susceptibility mapping in Qazvin province of Iran. *Iran J Sci Technol Trans Civ Eng*. 2019. https://doi.org/10.1007/s40996-019-00326-3.
13. Abay A, Barbieri G, Woldearegay K. GIS-based landslide susceptibility evaluation using analytical hierarchy process (AHP) approach: the case of Tarmaber District, Ethiopia. *Momona Ethiop J Sci*. 2019;11:14. https://doi.org/10.4314/mejs.v11i1.2.
14. He H, Hu D, Sun Q, et al. A landslide susceptibility assessment method based on GIS technology and an AHP-weighted information content method: a case study of southern Anhui, China. ISPRS. *Int J Geo-Information*. 2019;8. https://doi.org/10.3390/ijgi8060266.
15. El Jazouli A, Barakat A, Khellouk R. GIS-multicriteria evaluation using AHP for landslide susceptibility mapping in Oum Er Rbia high basin (Morocco). *Geogr Discuss*. 2019;6. https://doi.org/10.1186/s40677-019-0119-7.
16. Vijith H, Dodge-Wan D. Modelling terrain erosion susceptibility of logged and regenerated forested region in northern Borneo through the analytical hierarchy process (AHP) and GIS techniques. *Geogr Discuss*. 2019;6. https://doi.org/10.1186/s40677-019-0124-x.
17. Nguyen TTN, Liu CC. A new approach using AHP to generate landslide susceptibility maps in the chen-yu-lan watershed. *Taiwan Sensors (Switzerland)*. 2019;19. https://doi.org/10.3390/s19030505.
18. Abedini M, Tulabi S. Assessing LNRF, FR, and AHP models in landslide susceptibility mapping index: a comparative study of Nojian watershed in Lorestan province, Iran. *Environ Earth Sci*. 2018;77:0. https://doi.org/10.1007/s12665-018-7524-1.
19. Mokarram M, Zarei AR. Landslide susceptibility mapping using fuzzy-AHP. *Geotech Geol Eng*. 2018;36:3931–3943. https://doi.org/10.1007/s10706-018-0583-y.
20. Sharma S, Mahajan AK. Comparative evaluation of GIS-based landslide susceptibility mapping using statistical and heuristic approach for Dharamshala region of Kangra Valley, India. *Geogr Discuss*. 2018;5. https://doi.org/10.1186/s40677-018-0097-1.
21. Pourghasemi HR, Rossi M. Landslide susceptibility modeling in a landslide prone area in Mazandarn Province, north of Iran: a comparison between GLM, GAM, MARS, and M-AHP methods. *Theor Appl Climatol*. 2017;130:609–633. https://doi.org/10.1007/s00704-016-1919-2.
22. Abedini M, Ghasemyan B, Rezaei Mogaddam MH. Landslide susceptibility mapping in Bijar city, Kurdistan Province, Iran: a comparative study by logistic regression and AHP models. *Environ Earth Sci*. 2017;76. https://doi.org/10.1007/s12665-017-6502-3.
23. Rahaman SA, Aruchamy S. Geoinformatics based landslide vulnerable zonation mapping using analytical hierarchy process (AHP), a study of Kallar river sub watershed, Kallar watershed, Bhavani basin, Tamil Nadu. *Model Earth Syst Environ*. 2017;3:0. https://doi.org/10.1007/s40808-017-0298-8.
24. Wu Z, Wu Y, Yang Y, et al. A comparative study on the landslide susceptibility mapping using logistic regression and statistical index models. *Arab J Geosci*. 2017;10. https://doi.org/10.1007/s12517-017-2961-9.

25. Zhao H, Yao L, Mei G, et al. A fuzzy comprehensive evaluation method based on AHP and entropy for a landslide susceptibility map. *Entropy*. 2017;19:1–16. https://doi.org/10.3390/e19080396.

26. Wang Q, Li W. A GIS-based comparative evaluation of analytical hierarchy process and frequency ratio models for landslide susceptibility mapping. *Phys Geogr*. 2017;38:318–337. https://doi.org/10.1080/02723646.2017.1294522.

27. Kumar R, Anbalagan R. Landslide susceptibility mapping using analytical hierarchy process (AHP) in Tehri reservoir rim region, Uttarakhand. *J Geol Soc India*. 2016;87:271–286. https://doi.org/10.1007/s12594-016-0395-8.

28. Myronidis D, Papageorgiou C, Theophanous S. Landslide susceptibility mapping based on landslide history and analytic hierarchy process (AHP). *Nat Hazards*. 2016;81:245–263. https://doi.org/10.1007/s11069-015-2075-1.

29. Saaty TL, Vargas LG. *Models, Methods, Concepts & Applications of the Analytic Hierarchy Process*; 2012 [Springer Science & Business Media].

30. Çimren E, Çatay B, Budak E. Development of a machine tool selection system using AHP. *Int J Adv Manuf Technol*. 2007;35:363–376. https://doi.org/10.1007/s00170-006-0714-0.

31. Saaty TL. Rank from comparisons and from ratings in the analytic hierarchy/network processes. *Eur J Oper Res*. 2006;168:557–570.

32. Saaty TL. A scaling method for priorities in hierarchical structures. *J Math Psychol*. 1977;15:234–281.

33. Huang H, Lin B. The application of Gis-based logistic regression for landslide susceptibility mapping in the Shihmen. *Geomorphology*. 2005;65:428–434.

34. Poorbehzadi K, Yazdi A, Sharifi Teshnizi E, Dabiri R. Investigating of geotechnical parameters of alluvial Foundation in Zaram-Rud dam Site, North Iran. *Int J Min Eng Technol*. 2019;1:33–44.

35. Pourghasemi HR, Pradhan B, Gokceoglu C. Application of fuzzy logic and analytical hierarchy process (AHP) to landslide susceptibility mapping at Haraz watershed, Iran. *Nat Hazards*. 2012;63:965–996. https://doi.org/10.1007/s11069-012-0217-2.

36. Corominas J, Leroi E, Savage WZ, et al. Landslide susceptibility and hazard mapping in australia for land use planning—with reference to challenges in metropolitan suburbia. *Eng Geol*. 2008;102:238–250. https://doi.org/10.1016/j.enggeo.2008.03.021.

37. Yalcin A. GIS-based landslide susceptibility mapping using analytical hierarchy process and bivariate statistics in Ardesen (Turkey): comparisons of results and confirmations. *Catena*. 2008;72:1–12. https://doi.org/10.1016/j.catena.2007.01.003.

38. Longpré MA, Del Potro R, Troll VR, Nicoll GR. Engineering geology and future stability of the El Risco landslide, NW-Gran Canaria, Spain. *Bull Eng Geol Environ*. 2008;67:165–172. https://doi.org/10.1007/s10064-007-0119-9.

39. Wang WD, Xie CM, Du XG. Landslides susceptibility mapping based on geographical information system, GuiZhou, south-West China. *Environ Geol*. 2009;58:33–43. https://doi.org/10.1007/s00254-008-1488-5.

40. Geology E. GIS-based landslide susceptibility mapping for a problematic segment of the natural gas GIS-based landslide susceptibility mapping for a problematic segment of the natural gas pipeline, Hendek (Turkey). *Environ Geol*. 2014;44:949–962. https://doi.org/10.1007/s00254-003-0838-6.

41. Hasekioğulları GD, Ercanoglu M. A new approach to use AHP in landslide susceptibility mapping: a case study at Yenice (Karabuk, NW Turkey). *Nat Hazards*. 2012;63:1157–1179. https://doi.org/10.1007/s11069-012-0218-1.

42. Chacón J, Irigaray C, Fernández T. *Large to Middle Scale Landslides Inventory, Analysis and Mapping with Modelling and Assessment of Derived Susceptibility, Hazards and Risks in a GIS*. In: International congress International Association of Engineering Geology; 1994:4669–4678.

43. Cornforth D. *Landslides in Practice: Investigation, Analysis, and Remedial/Preventive Options in Soils*; 2005:620.

44. Gao FZÆ, Chen LÆW, Bai WHÆS. Engineering geology and stability of the Jishixia landslide, Yellow River, China. *Bull Eng Geol Environ*. 2010;69:99–103. https://doi.org/10.1007/s10064-009-0224-z.

45. Marinos P, Hoek E, others. *GSI: A Geologically Friendly Tool for Rock Mass Strength Estimation*. In: ISRM International Symposium; 2000.

46. Eberhardt E, Thuro K, Luginbuehl M. Slope instability mechanisms in dipping interbedded conglomerates and weathered marls—the 1999 Rufi landslide, Switzerland. *Eng Geol*. 2005;77:35–56. https://doi.org/10.1016/j.enggeo.2004.08.004.

47. Ghobadi MH. *Engineering Geologic Factors Influencing the Stability of Slopes in the Northern Illawarra Region*; 1994.

48. Aydin A. *The ISRM Suggested Methods for Rock Characterization, Testing and Monitoring: 2007-2014*; 2015:2007–2014. https://doi.org/10.1007/978-3-319-07713-0.

49. Hoek E, Bray JD. *Rock Slope Engineering*. CRC Press; 1981.

50. Hoek E, Diederichs MS. Empirical estimation of rock mass modulus. *Int J Rock Mech Min Sci*. 2006;43:203–215. https://doi.org/10.1016/j.ijrmms.2005.06.005.

51. Hoek E, Carranza C, Corkum B. *Hoek-Brown Failure Criterion–2002 Edition*. Narms-Tac; 2002:267–273. https://doi.org/10.1016/0148-9062(74)91782-3.

52. Look BG. *Handbook of Geotechnical Investigation and Design Tables*. CRC Press; 2014.

53. Bieniawski ZT. *Engineering Rock Mass Classifications: A Complete Manual for Engineers and Geologists in Mining, Civil, and Petroleum Engineering*. John Wiley & Sons; 1989.

54. Feizizadeh B, Jankowski P, Blaschke T. A GIS based spatially-explicit sensitivity and uncertainty analysis approach for multi-criteria decision analysis. *Comput Geosci*. 2014;64:81–95. https://doi.org/10.1016/j.cageo.2013.11.009.

55. Hadmoko DS, Lavigne F, Samodra G. Application of a semiquantitative and GIS-based statistical model to landslide susceptibility zonation in Kayangan catchment, Java, Indonesia. *Nat Hazards*. 2017;87:437–468. https://doi.org/10.1007/s11069-017-2772-z.

56. Nefeslioglu HA, Sezer EA, Gokceoglu C, Ayas Z. A modified analytical hierarchy process (M-AHP) approach for decision support systems in natural hazard assessments. *Comput Geosci*. 2013;59:1–8. https://doi.org/10.1016/j.cageo.2013.05.010.

5

Assessment of machine learning algorithms in land use classification

Hassan Khavarian Nehzak[a], Maryam Aghaei[a], Raoof Mostafazadeh[b], and Hamidreza Rabiei-Dastjerdi[c],*

[a]School of Physical Geography, University of Mohaghegh Ardabili, Ardabil, Iran [b]School of Natural Resources Management, University of Mohaghegh Ardabili, Ardabil, Iran [c]School of Computer Science and CeADAR, University College Dublin (UCD), Dublin, Belfield, Dublin 4, Ireland

1 Introduction

Satellite images allow the observation of human impacts on natural resources and obtain data to analyze the spatiotemporal changes and planning objectives.[1,2] Classification of an image is a process to identify different spectral classes and connect them with ground feature classes. Extracted land use maps from satellite data could be highly valuable in many applications such as monitoring environmental change and spatial planning.[3,4] Both artificial neural networks (ANNs) and support vector machines (SVMs) are nonparametric statistical learning methods, i.e., there is no need to assume the normal distribution of the data, which are used for land use classification.[5,6] ANN is a black-box model, which is trained by a backpropagation algorithm and is functioned like a human nervous system.[6] ANN comprises an input, hidden, and output layers and employs an activation function. SVM is a supervised learning technique and creates a binary classification by creating an optimized hyperplane between the desired class and other classes.[7] SVM is developed for nonlinear classification by the nonlinear kernel for internal displacement and optimization.[8] Yuan et al.[9] assessed the performance of SVM and ANN methods in land use classification. They showed ANN has better accuracy in land use classification. Deilmai et al.[10] compared maximum likelihood classification and SVM methods in land use extraction. The results of their study showed the support vector machine method is suitable for land use classification. Also, Aziz et al.[11] analyzed the efficiency of the ANN method on the classification of land cover.

This paper aims to compare the ANN and SVM (radial basis, linear, polynomial, and sigmoid kernels) methods to classify Landsat 8 satellite imagery using thermal bands and slope data in Kozehtopraghi watershed to obtain high precision land use maps.

In this paper, we will answer the following question:

1. Could elevation, thermal, slope, and aspect information increase the classification accuracy?

2 Material and method

2.1 Study area

Kozehtopraghi watershed is located in the southern part of Ardabil province in Iran. It has an area of 812 km^2. The minimum and maximum elevation of the study area are 1384 and 2385 m, respectively. According to the 40-years

*Hamidreza Rabiei-Dastjerdi is a Marie Skłodowska-Curie Career-FIT Fellow at the UCD School of Computer Science and CeADAR (Ireland's National Centre for Applied Data Analytics & AI). Career-FIT has received funding from the European Union's Horizon 2020 research and innovation programme under the Marie Skłodowska-Curie grant agreement No. 713654.

97

FIG. 1 Location of the study area (shown with *red color*) in Iran.

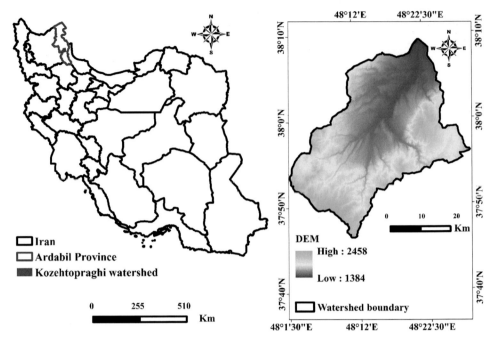

TABLE 1 Landsat 8 bands specifications.

Band number	Wavelength (μm)	Band number	Wavelength (μm)	Band number	Wavelength (μm)
1	0.43–0.45	5	0.85–0.88	9	1.36–1.38
2	0.45–0.51	6	1.57–1.65	10	10.60–11.19
3	0.53–0.59	7	2.11–2.29	11	11.50–12.51
4	0.64–0.67	8	0.50–0.68	–	–

recorded meteorological data, the average annual rainfall and annual temperature are 300 mm and 6.95°C, respectively. Fig. 1 shows the location of the study area in Iran. Also, the characteristics of the satellite imagery used in this paper are presented in Table 1.

In this paper, Landsat 8 satellite imagery of 2018 has been used to obtain a land use map of the study area using SVM (applying radial basis, linear, polynomial, and sigmoid kernels) and ANN methods.[12,13] Table 1 shows spectral charactristics of Landsat 8. We, also, used a DEM of the study area, which originally was extracted from PALOSAR data, with 12.5 m resolution.

2.2 Preprocessing

The first step in land use classification is image preprocessing.[14] This operation could be divided into atmospheric and geometric corrections.[15]

2.2.1 Atmospheric correction

The aim of the atmospheric correction is the conversion of radiance at the top of the atmosphere (received by the sensor) to output radiance of the earth's surface.[16] In this study, the atmospheric correction has been carried out using FLAASH (Fast Line of sight Atmospheric Analysis of Spectral Hypercubes) Module in ENVI image processing Software. The obtained information from the metadata including flight and sensor information are imported in FLAASH.

The FLAASH technique uses the MODTERAN 4 model to correct atmospheric effects in the range of shortwave of solar radiation according to Eq. (1):

$$I = \left(\frac{A_\rho}{1 - \rho_e S}\right) + \left(\frac{B_{\rho_e}}{1 - \rho_e S}\right) + L_a \tag{1}$$

Where,

L is the spectral radiance at the sensor,
p is the surface reflectance at the pixel level,
pe is the mean surface reflectance for the surrounding area,
S is the atmospheric spherical albedo,
A and B are the surface-independent coefficients that vary with atmospheric and geometric conditions.[17]

To increase the spatial resolution of the data, the Gram Schmith pansharpening method has been employed.[18] The spatial resolution of multispectral bands was increased from 30 to 15 m. Gram-Schmith pansharpening is the fusion of a low-resolution multispectral (MS) image and a high-resolution panchromatic (PAN) image to achieve a high-resolution MS image,[19] according to Eq. (2) as follows:

$$g_x = \frac{\mathrm{cov}\left(\widetilde{MS}_{k,} \; I_L\right)}{\mathrm{var}(I_L)}, k = 1, \ldots, N \tag{2}$$

Where,

cov and var are covariance and variance, N is the number of bands, $\widetilde{MS}_{k,}$ denotes the pan-sharpened image, I_L is the image synthesized from the MS image bands with weight factors, and g_x is a gain factor determined by the utilized transformation in the method.[20]

2.3 Artificial neural network

ANN is a well-developed method and simulates the patterns based on human brain function. ANN uses a standard backpropagation algorithm in ENVI software. ANN is a method that simulates the function of the human brain based on a neural network composed of an input layer, hidden layers, and output layer. The training was done by adjusting each node's weight based on minimum differences between the input and output nodes. The ANN can adapt to the rich texture and high spectrum confusion of remote sensing, especially by setting the nodes in the hidden layers, the problem of "homogeneous spectrum" and "foreign matter" can be solved in the process of land use classification.[21,22] The nodes in each layer are connected with a particular weight to the next layers' nodes, and training comes after each data processing time by the change in weights.[23] In this study, logistic ANN was used for classification. Logistic is a nonlinear classifier, but it can be made as a linear classifier with a simple transformation. This type of ANN based on spatial predictor's output provides the probability of belonging class for each cell.[24] Logistic parameters consist of the learning rate, learning momentum, and the number of training iteration. Learning rate acts similar to a low-pass filter and allows the network to ignore small features on surface error. Its range varies between 0 and 1. The lower value requires more training iteration. Learning momentum is the additional correction of learning rate to setting weights, and its range is 0.1–0.9. the number of training iteration is defined based on the training errors of the neural network system.[25] Kavzoglu and Mather[26] considered 0.2 and 0.1 values for training rate, and 0.5 and 0.6 for training momentum of the ANN algorithm and compared the results. They concluded that these values could lead to higher accuracy of classification. On this basis, in this paper, the proposed values by these researchers and the default value of ENVI software were used.

2.4 Support vector machine

SVM is a binary classifier that uses a straight line to separate the data for classification from a statistical learning theory[27]; but in some datasets, it is impossible to separate from a straight line. To solve this problem, some of the kernels have been developed. Kernel defines data in a multidimensional space, and so the data are easily separated by a straight line.[28] In addition, in most cases, the hyperplane may not be located precisely between the two classes. So, the error value is manipulated by trial and error between the maximum and minimum boundaries of the training samples

TABLE 2 Equations of four kernels of SVM.[29]

Radial basis	$K(x_i, x_j) = \exp(-g \parallel x_i - x_j \parallel^2), g > 0$
Linear	$K(x_i, x_j) = x_i{}^T x_j$
Polynomial	$K(x_i, x_j) = (gx_i{}^T x_j + r)^{d,} \ g > 0$
Sigmoid	$K(x_i, x_j) = \tanh(gx_i{}^T x_j + r)$

located in the wrong areas.[8] In this paper SVM (radial basis, linear, polynomial, and sigmoid kernels) method was used for land use classification. The SVM kernels equations are given in Table 2.

Where,

g is gamma parameter defined by the user as the width of a kernel,

x_i and x_j a collection of training data,

d is the polynomial degree for the polynomial kernel,

and

r is a bias for polynomial and sigmoid kernels.

2.5 Accuracy assessment

Accuracy assessment evaluates the classification algorithm and is the error level estimation that may exist in the classified image. The accuracy of each classifier is expressed in the error matrix.[30] In this paper, overall accuracy and kappa coefficient were used to examine the classification accuracy. Error matrix consists of n * n dimension, where n is the number of classes in the reference map. The rows of the matrix denote the actual classes or the information in the reference map, while the columns of the matrix shown the classified classes in the classified map. Overall accuracy is determined by the following equation:

$$OA = \frac{1}{N} \Sigma P_{ii} \tag{3}$$

Where:

OA is overall accuracy,

N is the number of training pixels,

and ΣP_{ii} is the sum of the elements in the main diagonal of the error matrix.

Kappa coefficient determines the difference between the observed values of two maps, as in the diagonal numbers in the error matrix and the agreement that could be achieved only by chance on two maps.[31] The kappa coefficient is calculated based on Eq. (4):

$$Kappa = \frac{P_o - P_C}{1 - P_C} \times 100 \tag{4}$$

Where,

P_O is observed accuracy and P_C is excepted agreement.[32,33]

3 Results

In this study, the reflective and thermal bands and topography information were used to carry out the land use classification. Fig. 2 shows the land use maps resulting from the SVM with linear kernel (on the left) and ANN (on the right) methods and considering reflective, thermal and slop information.

Based on the results, the most part of the study area has been covered by pasture and dry farming land uses. Thus, the overall accuracy of the produced land use maps are affected by the accuracy of these two classes. To consider the effects of the thermal bands and the slope information, we calculated the mean values of the brightness temperature and slope in each class (Table 3).

As Table 3 shows, the waterbody and dry farming land use classes had the lowest and the highest brightness temperature, respectively. Also, considering the mean slope value for each land use, the water body and the pasture land

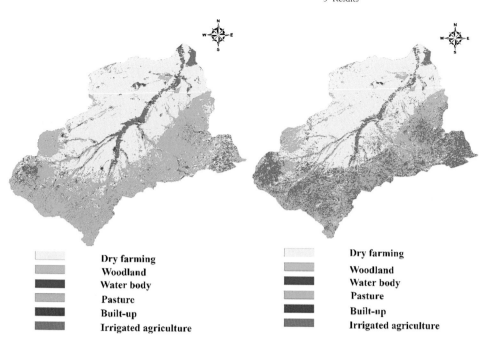

FIG. 2 The produced land use maps using SVM (on the *left*) and ANN (on the *right*) methods, considering all bands and slope data.

Dry farming
Woodland
Water body
Pasture
Built-up
Irrigated agriculture

Dry farming
Woodland
Water body
Pasture
Built-up
Irrigated agriculture

use classes showed the lowest and highest slope value, respectively. Thus, it is expected to find a better result in the results of the classification if we take into the account these two variables. Some studies, also, showed that considering thermal data could increase the accuracy of the resulted land use map.[34] In this paper, the effect of thermal bands and slope information is analyzed. The accuracy of each land use classification using reflective, thermal, and slope are shown in Table 4.

First, the results from the ANN method is assessed. Based on the results presented in Table 4, the land use classification using the ANN method and considering the reflective bands only, dry farming and built-up land use classes had the highest and lowest accuracy, respectively. Some other studies showed the similar results.[35] The ANN classifier could not separate the built-up area from the fallow agricultural lands well in the study area. The land use classification using the ANN method and considering the thermal and reflective information increased the accuracy in the woodland and irrigated agriculture land uses. Some studies used thermal data to improve the accuracy of the classification in woodland, irrigated agricultural lands, water bodies, and built-up areas.[36–38] Adding the slope information to the reflective bands and using the ANN classification method increased the accuracy of water body but decreased the accuracy of built-up area class. The results of the land use classification by the SVM method is assessed based on the different kernels and different input data. In the classification by the SVM classification method and the reflective bands, the built-up and dry farming classes had the highest accuracy. Among all four kernels, the SVM with linear kernel showed the highest accuracy in the classification of the land use map and the SVM with sigmoid kernel showed the lowest accuracy. By adding the thermal information to the reflective bands improved the accuracy of the classification of all land use classes in sigmoid kernel method. The other kernels showed a slight improvement in the

TABLE 3 The mean values of brightness temperature and slope for each land use.

Land use	Brightness temperature (K)	Slope mean (degree)
Dry farming	317.84	7.2
Woodland	311.68	11.41
Waterbody	303.1	3.7
Pasture	316.84	12.2
Built-up	314.81	5.4
Irrigated agriculture	314.20	4.86

TABLE 4   The obtained land use classification accuracies.

Row	Method	Irrigated agriculture	Built-up	Waterbody	Pasture	Woodland	Dry farming
1	ANN and reflective bands	92.14	83.73	97.59	98.91	92.71	99.66
2	SVM (RBF) and reflective bands	94.50	90.40	97.07	98.58	97.74	99.55
3	SVM (polynomial) and reflective bands	94.38	90.67	97.07	98.50	97.70	99.50
4	SVM (sigmoid) and reflective bands	93.47	70.13	97.14	96.19	97.55	97.49
5	SVM (linear) and reflective bands	94.81	91.73	99.41	98.43	97.70	99.41
6	ANN and reflective and thermal bands	95.26	73.44	97.32	98.33	97.13	98.68
7	SVM (RBF) and reflective and thermal bands	95.31	82.81	95.08	97.59	96.21	99.18
8	SVM (polynomial) and reflective and thermal bands	95.34	83.33	95.08	97.71	96.21	99.27
9	SVM (Sigmoid) and reflective and thermal bands	96.40	75.88	98.35	96.88	93.59	98.92
10	SVM (Linear) and reflective and thermal bands	95.92	86.98	95.97	97.97	96.34	99.61
11	ANN and reflective bands and slope	91.95	78.88	99.33	83.09	90.51	92.86
12	SVM (RBF) and reflective bands and slope	94.55	89	97.75	98.73	98.26	99.52
13	SVM (Polynomial) and reflective bands and slope	97.51	89	97.51	98.73	98.26	99.47
14	SVM (Sigmoid) and reflective bands and slope	97.51	89	97.51	98.73	98.26	99.47
15	SVM (Linear) and reflective bands and slope	98.09	95.44	98.09	98.77	98.22	99.44
16	ANN and reflective and thermal bands and slope	98.67	78.65	93.74	95	95.95	95.06
17	SVM (RBF) and reflective and thermal bands and slope	95.31	82.81	95.08	97.59	96.21	99.18
18	SVM (polynomial) and reflective and thermal bands and slope	94.43	82.81	95.97	99.08	97.74	99.65
19	SVM (sigmoid) and reflective and thermal bands and slope	94.34	79.71	94.85	96.64	96.87	97.95
20	SVM (linear) and reflective and thermal bands and slope	94.92	86.72	99.72	99.05	96.24	99.41
21	SVM and reflective and thermal bands and slope and elevation and aspect	96.03	82.72	98.95	96.88	95.81	99.80

irrigated agricultural lands but a lower accuracy in the other land use classes. By adding the slope information to the reflective bands, the accuracy of the produced land use map was assessed in each SVM kernels. There was no changes in the results from the RBF kernel, an improvement in the classification of the irrigated agriculture by the polynomial kernel, a higher accuracy in the classification of built-up areas and irrigated agricultural lands by linear kernel. The result of the classification by the sigmoid kernel showed more improvement than the others. There was a significant improvement in the classification of the built-up area and also an increase in the accuracy of the dry farming, pasture and irrigated agricultural lands. By using the reflective and thermal bands and slope data and the SVM method, the results of the classified land use were reviewed. There was a significant deterioration in the classification of the built-up areas and also a decrease in the accuracy of the other classes when the RBF, polynomial and linear kernels were applied. On the other hand, there was a significant improvement in the sigmoid kernel's results. Some studies showed that there was different results in the accuracy of land use classification when different SVM kernels were applied.[27,39,40]

The results of the overall accuracy and kappa coefficient of each method are shown in Table 5.

The maximum overall accuracy of the land use map was produced by the SVM method with linear kernel when the thermal and reflective bands and the slope data were used. The highest overall accuracy by the ANN method was obtained when the reflective and thermal bands were used. Similar studies, also, approved the higher accuracy of the results obtained by the ANN in comparisons with the ANN.[41] As the parameters, such as training information and hidden layer parameters, defined by the user could affect the accuracy of the SVM and ANN method,[42] different values for each parameter were used. The results showed that the default values used by ENVI software produced the best results. Prevez et al.[43]; Kalantar et al.[44]; and Prasad et al.[45] used the same methods for land use classification.

TABLE 5 Overall accuracy values and kappa coefficients of 20 methods.

Method	Kappa coefficient	Overall accuracy	Method	Kappa coefficient	Overall accuracy	Method	Kappa coefficient	Overall accuracy
1	0.95	96.7	8	0.96	97.6	15	0.97	97.7
2	0.96	97.5	9	0.97	97.7	16	0.92	94.3
3	0.96	97.4	10	0.97	97.7	17	0.96	97.1
4	0.94	95.7	11	0.85	89	18	0.97	98.3
5	0.96	97.5	12	0.96	97.5	19	0.95	96.3
6	0.96	97.1	13	0.96	97.5	20	0.98	98.7
7	0.96	97.1	14	0.96	97.5	21	0.97	98

4 Conclusion

The results of this paper showed that both the SVM and ANN machine learning algorithm produced a land use map with a high overall accuracy. However, these two methods produced different results in different land use classes. The SVM classifier, other than the sigmoid kernel, resulted in a higher accuracy map in the built-up areas. Comparing different SVM kernels showed that they computed different accuracies in different land use classes. However, in general, the linear kernel produced better results. Comparing the results of using different input data showed that the accuracy of the resulting map varied in each different land use class. Considering both the thermal and reflective bands resulted in a slightly better results in classification of agricultural areas but decreased the accuracy in the other land uses. Adding the slope information significantly increased the accuracy of the resulting land use maps when the SVM method with sigmoid kernel was used. The slope information also slightly increased the accuracy of the results from the linear SVM kernel but it had a negative effect in the other methods of the classification used. In this study we only considered the role of a few factors and the results could change when considering the other effective factors such as satellite data with different spatial and spectral characteristics, multiple temporal remote sensing data, geographical area with different land uses, and other climatology and hydrology ancillary data.

References

1. Kantakumar LN, Neelamsetti P. Multi-temporal land use classification using hybrid approach. *Egypt J Remote Sens Space Sci*. 2015;18(2):289–295. https://doi.org/10.1016/j.ejrs.2015.09.003.
2. Alimohammadi A, Rabiei HR, Firouzabadi PZ. A new approach for modeling uncertainty in remote sensing change detection process. In: *Proceedings of the 12th International Conference onGeoinformatics-Geospatial Information Research: Bridging the Pacific and Atlantic*; 2004:07–09.
3. van Leeuwen B, Tobak Z, Kovács F. Machine learning techniques for land use/land cover classification of medium resolution optical satellite imagery focusing on temporary inundated areas. *J Environ Geol*. 2020;13(1–2):43–52. https://doi.org/10.2478/jengeo-2020-0005.
4. Zeaiean P, Rabiei HR, Alimohamadi A. Detection of land use/cover changes of Isfahan by Agricultural Lands around urban area using remote sensing and GIS technologies. *J Spat Plan*. 2005;9(4):41–54.
5. Hermes L, Frieauff D, Puzicha J, Buhmann JM. Support vector machines for land usage classification in Landsat TM imagery. In: *IEEE 1999 International Geoscience and Remote Sensing Symposium. IGARSS'99 (Cat. No.99CH36293). IEEE 1999 International Geoscience and Remote Sensing Symposium. IGARSS'99. Hamburg, Germany, 28 June-2 July 1999*: IEEE; 1999:348–350.
6. Talukdar S, Singha P, Mahato S, et al. Land-use land-cover classification by machine learning classifiers for satellite observations—a review. *Remote Sens*. 2020;12(7):1135. https://doi.org/10.3390/rs12071135.
7. Rudrapal D, Subhedar M. Land cover classification using support vector machine. *IJERT*. 2015;V4(09). https://doi.org/10.17577/IJERTV4IS090611.
8. Shi D, Yang X. Support vector machines for land cover mapping from remote sensor imagery. In: Li J, Yang X, eds. *Monitoring and Modeling of Global Changes: A Geomatics Perspective*. Dordrecht: Springer Netherlands; 2015:265–279. [Springer Remote Sensing/Photogrammetry].
9. Yuan H, van der Wiele C, Khorram S. An automated artificial neural network system for land use/land cover classification from Landsat TM imagery. *Remote Sens (Basel)*. 2009;1(3):243–265. https://doi.org/10.3390/rs1030243.
10. Deilmai BR, Ahmad BB, Zabihi H. Comparison of two classification methods (MLC and SVM) to extract land use and land cover in Johor Malaysia. *IOP Conf Ser: Earth Environ Sci*. 2014;20:12052. https://doi.org/10.1088/1755-1315/20/1/012052.
11. Aziz N, Minallah N, Junaid A, Gul K, eds. *Performance Analysis of Artificial Neural Network Based Land Cover Classification*. International Scholary and Scientific Research: IEEE; 2017.
12. Khodabandehlou B, Khavarian Nehzak H, Ghorbani A. Change detection of land use/land cover using object oriented classification of satellite images (Case study: Ghare Sou basin, Ardabil province). *J RS GIS Nat Resour*. 2019;10(3):76–92.
13. Aghaei M, Khavarian H, Mostafazadeh R. Prediction of land use changes using the CA-Markov and LCM models in the Kozehtopraghi Watershed in the Province of Ardabil. *Watershed Manag Res J*. 2020;33(3):91–107.

14. Chrystal JB, Joseph S. Land cover classification of satellite images using artificial neural network and support vector machine. *Int J Adv Elect Comput Sci*. 2015;2(6):83–85.

15. Hu Y, Zhang Q, Zhang Y, Yan H. A deep convolution neural network method for land cover mapping: a case study of Qinhuangdao, China. *Remote Sens*. 2018;10(12):2053. https://doi.org/10.3390/rs10122053.

16. Hadjimitsis DG, Papadavid G, Agapiou A, et al. Atmospheric correction for satellite remotely sensed data intended for agricultural applications: impact on vegetation indices. *Nat Hazards Earth Syst Sci*. 2010;10:89–95.

17. Lin C, Wu C-C, Tsogt K, Ouyang Y-C, Chang C-I. Effects of atmospheric correction and pansharpening on LULC classification accuracy using WorldView-2 imagery. *Inf Process Agri*. 2015;2(1):25–36. https://doi.org/10.1016/j.inpa.2015.01.003.

18. Palubinskas G. Fast, simple, and good pan-sharpening method. *J Appl Remote Sens*. 2013;7(1):73526. https://doi.org/10.1117/1.JRS.7.073526.

19. Ma L, Liu Y, Zhang X, Ye Y, Yin G, Johnson BA. Deep learning in remote sensing applications: a meta-analysis and review. *ISPRS J Photogramm Remote Sens*. 2019;152:166–177. https://doi.org/10.1016/j.isprsjprs.2019.04.015.

20. Kahraman S, Erturk A. Review and performance comparison of Pansharpening algorithms for RASAT images. *IU-JEEE*. 2018;18(1):109–120. https://doi.org/10.5152/iujeee.2018.1817.

21. Chen Y, Dou P, Yang X. Improving land use/cover classification with a multiple classifier system using AdaBoost integration technique. *Remote Sens*. 2017;9(10):1055. https://doi.org/10.3390/rs9101055.

22. Pakale GK, Gutpa PK. Comparison of advanced pixel based (ANN and SVM) and objected-oriented classification approaches using Landsat 7 ETM+ data. *Eng Technol*. 2010;2(4):245–251.

23. Omer G, Mutanga O, Abdel-Rahman EM, Adam E. Performance of support vector machines and artificial neural network for mapping endangered tree species using WorldView-2 data in Dukuduku Forest, South Africa. *IEEE J Sel Top Appl Earth Obs Remote Sens*. 2015;8(10):4825–4840. https://doi.org/10.1109/JSTARS.2015.2461136.

24. Tayyebi A. *Simulating Land Use Land Cover Change Using Data Mining and Machine Learning Algorithms*; 2013. Purdue university thesis, philosophy.

25. Ozyavuz M. *Landscape Planning*. InTech; 2012.

26. Kavzoglu T, Mather PM. The use of backpropagating artificial neural networks in land cover classification. *Int J Remote Sens*. 2003;24 (23):4907–4938. https://doi.org/10.1080/0143116031000114851.

27. Kesikoglu MH, Atasever UH, Dadaser-Celik F, Ozkan C. Performance of ANN, SVM and MLH techniques for land use/cover change detection at sultan marshes wetland, Turkey. *Water Sci Technol J Int Assoc Water Pollut Res*. 2019;80(3):466–477. https://doi.org/10.2166/wst.2019.290.

28. Pahwa S, Sinwar D. Comparison of various kernels of support vector machine. *IJRASET*. 2015;3:532–536.

29. Pradhan B, Sameen MI. Manifestation of SVM-based rectified linear unit (ReLU) kernel function in landslide modelling. In: Suparta W, Abdullah M, Ismail M, eds. *Space Science and Communication for Sustainability*. Singapore: Springer Singapore; 2018:185–195.

30. Tatti A, Sarmadian F, Mousavi A, Taghati Hossein Pour C, Esmaile Shahir AH. Land use classification using support vector machine and maximum likelihood algorithms by Landsat 5 TM images. *Eng Phys Sci*. 2014;12(8):681–687.

31. Kim C. Land use classification and land use change analysis using satellite images in Lombok Island, Indonesia. *For Sci Technol*. 2016;12 (4):183–191. https://doi.org/10.1080/21580103.2016.1147498.

32. Islam K, Jashimuddin M, Nath B, Nath TK. Land use classification and change detection by using multi-temporal remotely sensed imagery: the case of Chunati wildlife sanctuary, Bangladesh. *Egypt J Remote Sens Space Sci*. 2018;21:37–47.

33. Sarkar A. Accuracy assessment and analysis of land use land cover change using Geoinformatics technique in Raniganj coalfield area, India-Technique in Raniganj coalfield area, India. *Int J Environ Sci Nat Resour*. 2018;11(1):25–34.

34. Sun L, Schulz K. The improvement of land cover classification by thermal remote sensing. *Remote Sens*. 2015;7(7):8368–8390. https://doi.org/10.3390/rs70708368.

35. Tan KC, Lim HS, MatJafri MZ, Abdullah K. Landsat data to evaluate urban expansion and determine land use/land cover changes in Penang Island, Malaysia. *Environ Earth Sci*. 2010;60(7):1509–1521. https://doi.org/10.1007/s12665-009-0286-z.

36. Reis S. Analyzing land use/land cover changes using remote sensing and GIS in Rize, north-East Turkey. *Sensors*. 2008;8(10):6188–6202. https://doi.org/10.3390/s8106188.

37. Liberti M, Simoniello T, Carone MT, Coppola R, D'Emilio M, Macchiato M. Mapping badland areas using LANDSAT TM/ETM satellite imagery and morphological data. *Geomorphology*. 2009;106(3–4):333–343. https://doi.org/10.1016/j.geomorph.2008.11.012.

38. Mushore TD, Mutanga O, Odindi J, Dube T. Assessing the potential of integrated Landsat 8 thermal bands, with the traditional reflective bands and derived vegetation indices in classifying urban landscapes. *Geocarto Int*. 2017;32(8):886–899. https://doi.org/10.1080/10106049.2016.1188168.

39. Debojit B, Hitesh J, Manoj KA, Balasubramanian R. Microsoft word - #020410455. *Int J Earth Sci Eng*. 2011;4(6):985–988.

40. Li C, Wang J, Wang L, Hu L, Gong P. Comparison of classification algorithms and training sample sizes in urban land classification with Landsat thematic mapper imagery. *Remote Sens (Basel)*. 2014;6(2):964–983. https://doi.org/10.3390/rs6020964.

41. Kolios S, Stylios CD. Identification of land cover/land use changes in the greater area of the Preveza peninsula in Greece using Landsat satellite data. *Appl Geogr*. 2013;40:150–160. https://doi.org/10.1016/j.apgeog.2013.02.005.

42. Li J, Yang X, eds. *Monitoring and Modeling of Global Changes: A Geomatics Perspective*. Dordrecht: Springer Netherlands (Springer Remote Sensing/Photogrammetry); 2015.

43. Pervez W, Uddin V, Khan SA, Khan JA. Satellite-based land use mapping: comparative analysis of Landsat-8, Advanced Land Imager, and big data Hyperion imagery. *J Appl Remote Sens*. 2016;10(2):26004. https://doi.org/10.1117/1.JRS.10.026004.

44. Kalantar B, Pradhan B, Naghibi SA, Motevalli A, Mansor S. Assessment of the effects of training data selection on the landslide susceptibility mapping: a comparison between support vector machine (SVM), logistic regression (LR) and artificial neural networks (ANN). *Geomat Nat Haz Risk*. 2018;9(1):49–69. https://doi.org/10.1080/19475705.2017.1407368.

45. Prasad SVS, Savithri TS, Iyyanki VMK. Comparison of accuracy measures for RS image classification using SVM and ANN classifiers. *IJECE*. 2017;7(3):1180. https://doi.org/10.11591/ijece.v7i3.pp1180-1187.

6

Evaluation of land use change predictions using CA-Markov model and management scenarios

Hassan Khavarian Nehzak[a], Maryam Aghaei[a], Raoof Mostafazadeh[b], and Hamidreza Rabiei-Dastjerdi[c]

[a]School of Physical Geography, University of Mohaghegh Ardabili, Ardabil, Iran [b]School of Rangeland and Watershed Management, University of Mohaghegh Ardabili, Ardabil, Iran [c]School of Computer Science and CeADAR, University College Dublin (UCD), Dublin, Belfield, Dublin 4, Ireland

1 Introduction

Land-use change modeling and prediction is one of the most important decision tools for global and regional planning,[1–3] and the development of land-use scenarios can be suitable to identify underlying land change processes in the future.[4] Remotely sensed data provide valuable information about the process, relation, location, and natural process of land-use change pattern.[5–7] Some dynamic simulation models can help regarding the simulation of future scenarios. Examples of such models can be referred to as Cellular Automata[8] and Markov models.[9] Markov model produces only temporal dynamics and does not provide any information about the location of the changes. Compared to the Markov model, the cellular automata (CA) consider the spatial component of changes, which produces changes in the future by specific laws of neighboring cells. It should be noted that the Markov model output is non-spatial in nature, in which there is no knowledge of the geographical location of land use. The main concept of CA is that land use change in each cell can be described by its current phase and change in its neighboring cells.[10, 11] In the CA-Markov model, the status of each cell depends on both spatial and temporal conditions of its neighbors.[12, 13] CA-Markov model, considering the trend of land use changes is useful in generating land-use spatiotemporal changes[14]. For instance, Bello et al.[15] Gollnow et al.[16] Nguyen and Ngo[17] and Sadoddin et al.[18] predicted land-use change using the CA-Markov model and scenario-based approach. In addition, to define land-use change scenarios, the scenarios may involve a certain amount of agricultural land use escalation[16] or development of forests and pastures and converting agricultural land into pastures that have a positive impact on total runoff reduction.[19] In this regard,[20] and[21] studied the effects of land-use change on runoff amount. Meanwhile, agricultural land-use covers larger areas and their management can affect the final runoff intensity.[22] Furthermore, woodlands can increase evaporation and reduce surface runoff.[23] However, the loss of natural habitats, due to the development of agriculture, has been the main cause of the decline of biodiversity to date.[24]

This study aims to evaluate the spatiotemporal pattern of land-use changes using management scenarios and the CA-Markov model. Therefore, the spatiotemporal patterns of land-use changes in the Kozehtopraghi watershed in Ardabil province have been identified using remote sensing data and GIS. In this regard, the reclamation and destruction land-use change scenarios based on possible impacts on runoff are formulated and their results are compared with the results of the CA-Markov model.

In this chapter we will answer the following questions:

1. Is this spatial trend of land-use changes in the study area based on the reclamation or destruction scenarios?
2. Is the prediction of land-use change scenarios consistent with the results of the CA-Markov model?

2 Definitions and background

2.1 CA-Markov model

CA-Markov consists of Markov Chain and Cellular Automata (CA) models [25]. Markov chain is a sequence of random variables X_1, X_2, X_3, and X_n, that has a stochastic process. In Eq. (1), $X[k]$ is a Markov chain and $P_{i,j}$ is the probability of transmission from state i to j at a given time.

$$P_{i,j} = \Pr(X[k+1] = j \mid X[k] = i) \tag{1}$$

When a Markov chain has a finite number of states, it is possible to obtain the transition probability matrix using Eq. (2).[26]

$$\begin{bmatrix} p_{1,1}p_{1,2}\ldots\ldots\ p_{1,n} \\ p_{2,1}p_{2,2}\ldots\ldots\ p_{2,n} \\ \ldots\ \ldots\ \ldots\ \ldots \\ p_{n,1}p_{n,2}\ldots\ldots\ p_{n,n} \end{bmatrix} \tag{2}$$

In transition probability matrix, the change from one class to another is described. Based on the values of the transition probability matrix, the data are obtained about the converting of pixels or classes to the another cell or class.[27] This is achieved by constructing a probability matrix from time to time, which is the basis of the planning for the future.[28][29] CA consists of a grid space, a set of states, a set of transition rules determine the state transition for each cell based on the position of neighboring cells and a sequence of discrete time steps.[11] The cellular automata model can be expressed as follows (Eq. 3):

$$S(t, t+1) = f(S(t), N) \tag{3}$$

where

S is the set of discrete cellular states,
n is the cell field, and $t+1$ at different times,
and f is the law of cell state change in the local space.

CA-Markov model is based on the theoretical principles of both CA and Markov models. In other words, while the Markov process controls the temporal dynamics between land-use classes, the cellular automata have a local mechanism and proceed local laws associated with neighborhood configuration aligns with the possibility of transferring the spatial dynamics of land-use types.[12][30]

2.2 Land use change scenario

There are several definitions for a scenario including, the Intergovernmental Panel on Climate Change (IPCC) defined the scenario as a coherent, consistent, and rational description of the possible future situation.[31] The main idea of the land-use change scenario is to use multiple views to explore a particular problem. Land-use change scenarios can be identified in two main explorative and predictive categories. Explorative scenarios consider uncertainties and drivers of change and predictive scenarios consider the results and effects of future events.[32] The improper use of land cover intensifies the runoff volume and the amount of sediment yield. However, it is not easy to determine the relationship between runoff and land use. Detecting the effects of land-use change on runoff can be complicated as it requires to collect information about land use in long term.[33] By using the management scenarios (reclamation and destruction) of land-use, it is possible to investigate the effects of land use changes on runoff components. Reclamation scenarios have positive impacts on reducing the speed and volume of runoff. In addition, destruction land-use change scenarios increase the runoff and the hydrological risks in a watershed (e.g., flood, erosion, and so on). In order to deal with the risk of flooding, it is necessary to improve protection measures, promote proper land use, and control land-use practices in short term.[34] Scenarios provide a framework for predicting future changes when there are many possibilities. Scenarios do not predict the future but predict a set of possible future alternative stories which can provide insight into a wide range of outcomes.[35] Scenarios are also capable of responding to questions about what can happen while dealing with uncertainty and complexity.[36] An essential element of the scenarios is to provide a structural framework for the discovery of future alternatives that reveals the conditions of the future on "if-then" basis. In addition, scenarios have the ability to capture both qualitative and quantitative elements[37] and combine field information, GIS-based land-use change analysis, especially focusing on group discussions and specialized interviews with local authorities.[38] So, using land-use change scenarios analysis can improve our knowledge in environmental management plans. In addition, Land-use changes can have positive (reclamation) or negative (destruction) impacts on environmental components and responses.[19, 39, 40]

3 Material and method

3.1 Study area

The center of Kozehtopraghi watershed has the geographical coordinates of 48°28 01″ longitude and 38°07 28″ latitude. This watershed is located in the southern part of Ardabil city and the northern part of Kosar city in Iran, with an area of 812 km². The minimum and maximum elevations of the watershed are 1384 and 2485 m, respectively. According to annual climate data, the mean annual rainfall and temperature in the watershed are 300 mm and 6.59°C, respectively. Fig. 1 shows the location of the study area in Iran.

3.2 Image preprocessing and classification

In this research, three Landsat satellite imageries in 2000, 2010, and 2018, downloaded from USGS website,[a] were used to model land-use changes and to define land-use scenarios. The June and July imageries were cloud-free, and the vegetation cover of the study area was fully grown. The Digital Elevation Model of the study area, obtained from ALOS PALSAR[b] sensor, was used to produce the slope. The aim of the atmospheric correction is the conversion of radiance at the top of the atmosphere (received by the sensor) to output radiance of the earth's surface. In this study,

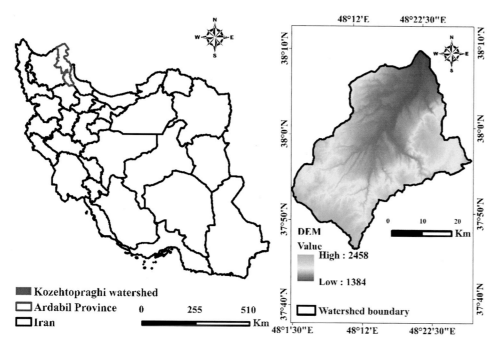

FIG. 1 Location of Kozehtopraghi watershed in Iran.

[a] earthexplorer.usgs.gov

[b] asf.alaska.edu

the atmospheric correction was carried out by FLAASH[c] algorithm and SVM supervised classification method, as suggested by some studies,[41] was used for land use classification in the study area. The land-use classes in the study area consist of dry farming, irrigated agriculture, pasture, garden, waterbody, and built-up.

3.3 Land-use change model of 2018 with CA-Markov

To predict the 2018 land use map, first land use maps of 2000 and 2010 were used to produce the transition area matrix of the Markov model. Markov model inputs consist of 2000 and 2010 classified images, and its output is land-use change prediction of 2018. The accuracy of the past and present analysis plays an important role in the quality of predicted changes.[42] The 2018 classified image and predicted image were used to validate the CA-Markov model.

3.4 Suitability map

Suitability maps were obtained using multicriteria evaluation.[43] These maps demonstrate the suitability of converting each image cell from one land use to another land uses. The values of these images are in the range of 0 to 255; the zero value indicates the lack of competence, and the value of 255 maximum means competency to convert the land use from one class to another. In this study, six suitability land use maps have been prepared. In multicriteria evaluation, the constraints were chosen based on the suitability analysis of each land use conversion. The constraints that were defined for producing suitability maps in this study are listed in Table 1. In this study multicriteria Boolean method, as suggested by some studies,[48] was used to produce suitability maps.

3.5 Land-use change model of 2036

After determining the accuracy for the 2018 predicted map, the prediction was made for 2036. The classified maps of 2000 and 2018 were used to obtain the transition area matrix. The value of the proportional error was chosen 0.6. Also, the time interval for prediction was determined 18 years. Total numbers of iterations are based on the time determined to predict land-use change.[28] Land use change scenarios prediction

To predict future land-use change scenarios in Kozehtopraghi watershed, 10 managerial scenarios (five reclamation scenarios that have positive implications and five destruction scenarios that have negative implications on land-use changes) based on different land use activities were defined. The management scenarios and their formulation rules are listed in Table 2.

In this paper, the reclamation and destruction scenarios of land-use change were formulated and defined by the emphasis on dry farming, pasture, woodland, and irrigated agriculture land uses. To consider the impacts of reclamation and destruction land-use change scenarios on runoff, the land use reclamation scenarios were formulated

TABLE 1 Land-use suitability map constraints.

Land use	Constrains	References
Dry farming	Distances greater than 10 m from rivers, distances greater than 10 m from roads, slope more than 25%, and built-up area	44, 45
Irrigated agriculture	Distances greater than 10 m from rivers, distances greater than 10 m from roads, and slope more than 8%	14
Garden	Distances greater than 5 m from rivers, distances greater than 25 m from roads, slope more than 30%, and built-up area	27
Built-up	Distances greater than 200 m from rivers, distances greater than 70 m from roads, and slope more than 12%	44, 46
Waterbody	Distances greater than 20 m from waterbody and slope more than 10%	17
Pasture	Built-up area, distances greater than 10 m from waterbody (10 m), distances greater than 10 m from built-up area, distances greater than 5 m from roads, irrigated agriculture, and woodland	47

[c] Fast line-of-sight atmospheric analysis of hypercubes.

TABLE 2 Land use change management scenarios.

Managerial scenarios	Activity title	Rules of formulation and explanation
Existing condition	Existing land use	Base scenario (2018)
Scenario1 (reclamation)	Reclamation low slope pastures in dry farming lands	Dry farming in 12%–20% slopes
Scenario2 (reclamation)	Reclamation of plowed pastures	Dry farming land use in slopes greater than 25%, a buffer of pastures (500 m)
Scenario3 (reclamation)	Construction gardens in irrigated agriculture	Irrigated agriculture in slopes less than 25%, a buffer of woodland (500 m)
Scenario4 (reclamation)	Combination of scenario 2 and 3	Rules of scenario 2 and 3
Scenario5 (reclamation)	Construction of walnut, almond and rose in plowing pastures	Dry farming in 12%–20% slopes
Scenario6 (destruction)	Convert from low-slope pastures to dry farming	Pasture in 12%–20% slopes, a buffer of dry farming (100 m)
Scenario7 (destruction)	Convert from high-slope pasture to dry farming	Pasture in 20% 35% slopes, a buffer of dry farming (100 m)
Scenario8 (destruction)	Convert from low-slope pasture to dry farming	Pasture in 12%–20% slopes, a buffer of dry farming (150 m)
Scenario9 (destruction)	Convert from high-slope pasture to dry farming	Pasture in 20%–35% slopes, a buffer of dry farming (150 m)
Scenario10 (destruction)	Convert from low-slope pasture to uncover lands	Pastures in 12%–20% slopes, a buffer of dry farming (100 m)

based on an increase in the extent of pasture and woodland and a reduction in the extent of agriculture land use. This could reduce the runoff through an increase in the soil permeability. In addition, land use degradation scenarios were formulated with the possibility of increasing dry farming land use and decreasing woodland and pasture land uses. Destructive land use change scenarios increase the intensity and volume of the runoff because of the reduction of soil permeability. In Kozehtopraghi watershed, dry farming covers 44.9%, and irrigated agriculture covers 8% of the total area. Considering the extent of dry farming, pasture, irrigated agriculture, and woodland as a large proportion in the Kozehtopraghi watershed, all possible contingencies can be considered in the future in land-use changes. In most studies, the slope factor or digital elevation model was used. Finally, land use maps derived from defined scenarios were compared with the predicted land use map of 2036 from the CA-Markov model to identify future land-use scenarios in the Kozehtopraghi watershed. The sequential steps of this study are shown in Fig. 2.

4 Results and discussion

In this paper, the accuracy of each land use map was analyzed. Based on the results of land use classification in three time periods using SVM method, the obtained overall accuracies were 96, 95.8, and 97.5, and the obtained kappa coefficients were 0.95, 0.94, and 0.96 for 2000, 2010, and 2018, respectively. The classification maps of 2000, 2010, and 2018 are shown in Fig. 3. So, we used these land use maps to predict the land use changes and produce the land-use change scenarios maps.

To validate the CA-Markov model, the land use classified map of 2018 was compared with the predicted land use map of 2018. As the Kappa coefficients (Kno (0.88), Klocation (0.94), Klocationstrata (0.94), and Kstandard (0.84)) indicated, the accuracy of the CA-Markov model was high. To model the land use map of 2036, the transition area matrix of 2036 was obtained from land use maps of 2000 and 2018. The transition area matrix and transition probability matrix of 2036 are shown in Tables 3 and 4. The probability transition matrix is describing the probability of converting any land use class to another land use. Also, the transition area matrix represents the number of pixels that are chosen by the Markov model for conversion.[49]

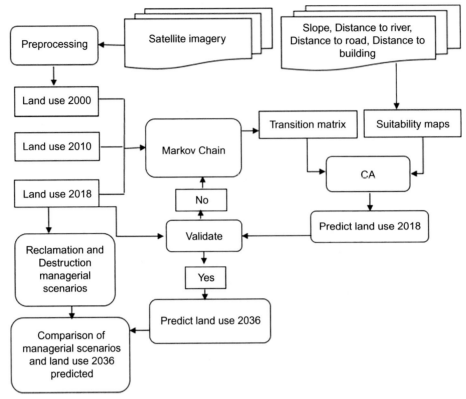

FIG. 2   Flowchart of the research.

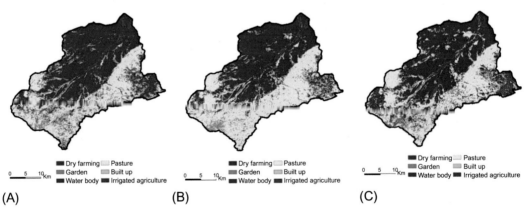

FIG. 3   Land use maps of (A) 2000, (B) 2010, and (C) 2018.

According to Table 3, the highest transition area from 2000 to 2018 were in dry farming, pasture, irrigated agriculture, waterbody, built-up, and garden, respectively.

The transition probability matrix represents the probability of change each land-use type in the future given time for any pixel in the image. In this matrix, the rows showed the start year (2000) and columns showed the end year (2018). The results of Table 4 showed that the highest transition probability land-use change is related to the garden, waterbody, built-up, irrigated agriculture, pasture, and dry farming. The obtained transition area matrix by the Markov model was used as the input of a CA-Markov model to prepare a change map for the year 2036.

The predicted maps of 2018 and 2036 for CA-Markov are shown in Fig. 4.

The land-use change scenarios maps (reclamation and destruction) are represented in and Fig. 5.

TABLE 3 Transition area matrix of 2036 land use.

Land use	Dry farming	Garden	Waterbody	Pasture	Built-up	Irrigated agriculture
Dry farming	1573	928	0	2636	85	9617
Garden	482	2	0	111	3	88
Water body	7	0	14	5	4	3
Pasture	553	1012	0	530	34	2145
Built-up	41	0	0	21	2	70
Irrigated agriculture	609	1625	0	1828	0	353

TABLE 4 Transition probability matrix for 2036.

Land use	Dry farming	Garden	Waterbody	Pasture	Built-up	Irrigated agriculture
Pasture	0.106	0.063	0.000	0.178	0.006	0.006
Irrigated agriculture	0.701	0.003	0.000	0.162	0.005	0.128
Dry farming	0.389	0.000	0.000	0.267	0.188	0.156
Built-up	0.129	0.237	0.000	0.124	0.008	0.502
Garden	0.278	0.000	0.096	0.142	0.139	0.470
Waterbody	0.138	0.368	0.000	0.414	0.000	0.080

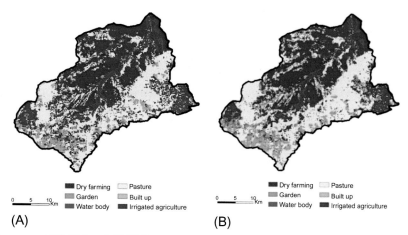

(A)   (B)

FIG. 4 Land use prediction map of (A) 2018 and (B) 2036.

According to Fig. 5 , in scenarios 1–5 (reclamations), dry farming, and irrigated agriculture land use will be reduced, and pasture and gardens will be increased. This increase in the woodland and pasture land uses in the study area will reduce the damages caused by flooding. The results of Fig. 6A–D indicate that the pasture area will be reduced and will be added to the area of dry farming and irrigated agricultural land. Also, because of the intensive use of pastures, they will be converted to bare lands. This reduction in the extent of pasture would increase damages and effects on flooding and erosion in the study area. The results refer to identify the dynamics of land use, simulation, and prediction of temporal–spatial patterns and land use distribution in different scenarios in the future.[28] Sadoddin et al.,[18] and Tornquist and Silva,[50] have used a similar approach to formulate of management land-use scenarios. Mukhopdhaya[29] and Koo et al.[51] in their studies, predicted land-use scenarios with the CA-Markov model and pointed out the scenario of converting agricultural land to pasture and garden. In this study, the CA-Markov model predicted the conversion from

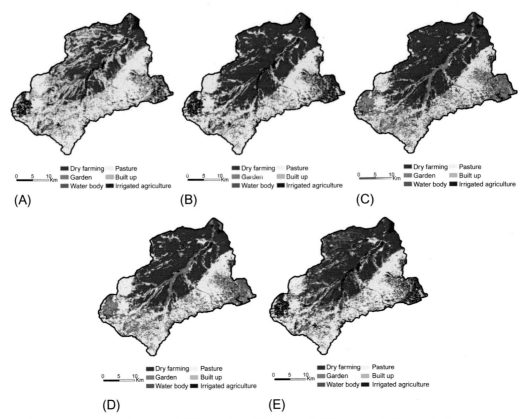

FIG. 5   Land use change reclamation (A) scenario 1, (B) scenario 2, (C) scenario 3, (D) scenario 4 (E) scenario 5.

dry farming and irrigated agriculture to garden that fits within scenario 2 (reclamation of plowing pastures). Also, the model predicted conversation of dry farming and irrigated agriculture to garden which has an agreement with scenario 3 (construction of gardens in irrigated agriculture) and scenario 5 (construction of walnut, almond and rose in plowed pastures). In destruction scenarios, we predicted convert from pasture to dry farming that it partly is in agreement with scenario 6 (converting from low-slope pastures to dry farming), but in addition to it, model predicted convert of pasture to irrigated agriculture.

5 Conclusion

The results of comparing the predicted land use map and 10 scenarios map from this study can be a good guideline for managers and planners in the natural resource sector. In addition, the predicted land use map and maps of different land-use scenarios can be used as a warning system for future impacts of land-use changes over the study area. Emphasizing that, usually, the purpose of prediction land-use change is to evaluate the consequences of different scenarios, especially the continuation of the exiting trend; the results of this prediction, however, maybe an environmental warning for land use in the future. Also, one of the most important factors influencing the occurrence of floods and crosion is land-use change. Therefore, it is important to prepare plans to simulate land-use changes to determine where, how, and when the changes have occurred to help planning and decision process for managers. The presented land use maps can be used as a useful tool in decision-making under uncertain conditions. The restoration and degradation

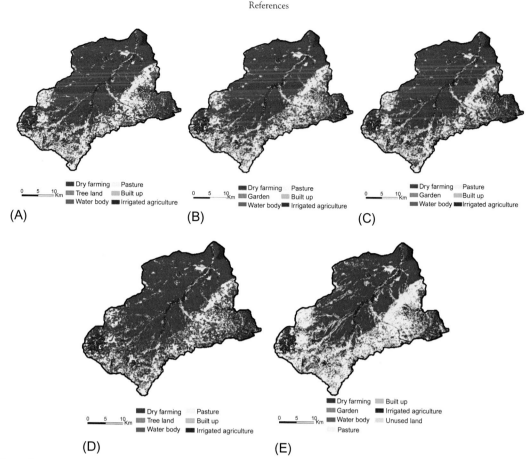

FIG. 6 Land use change destruction (A) scenario 6, (B) scenario 7, (C) scenario 8, (D) scenario 9, (E) scenario 10.

scenarios provide a range of possible conditions regarding the restoration of the watershed and implementation of management activities to improve the integrity of natural and low-impact land use at the watershed scale.

References

1. Sundara Kumar K, Udaya Bhaskar P, Padmakumari K. Application of land change modeling for prediction of future of land use land cover: a case study of Vijayawada city. *Adv Technol Eng Sci*. 2015;3(1):773–783.
2. Zeaiean P, Rabiei HR, Alimohamadi A. Detection of land use/cover changes of Isfahan by agricultural lands around urban area using remote sensing and GIS technologies. *J Spat Plan*. 2005;9(4):41–54.
3. Alimohammadi A, Rabiei HR, Firouzabadi PZ. A new approach for modeling uncertainty in remote sensing change detection process. In: *Proceedings of the 12th International Conference on Geoinformatics-Geospatial Information Research: Bridging the Pacific and Atlantic*; 2004:7–9.
4. Armenteras D, Murcia U, González TM, Barón OJ, Arias JE. Scenarios of land use and land cover change for NW Amazonia: impact on forest intactness. *Glob Ecol Conserv*. 2019;17. https://doi.org/10.1016/j.gecco.2019.e00567, e00567.
5. Munthali MG, Botai JO, Davis N, Ade la AM. Multi-temporal analysis of land use and land cover change detection for Dedza district of Malawi using geospatial techniques. *Appl Eng*. 2019;14(5):1151–1162.
6. Abdolalizadeh Z, Ebrahimi A, Mostafazadeh R. Landscape pattern change in Marakan protected area, Iran. *Reg Environ Chang*. 2019;19 (6):1683–1699. https://doi.org/10.1007/s10113-019-01504-9.
7. Khodabandehlou B, Khavarian Nehzak H, Ghorbani A. Change detection of land use/land cover using object oriented classification of satellite images (Case study: Ghare Sou basin, Ardabil Province). *J RS GIS Nat Resour*. 2019;10(3):76–92.
8. Abuelaish B, Camacho O MT. Scenario of land use and land cover change in the Gaza strip using remote sensing and GIS models Golestan, Iran,: the implications and way-out. *Arab J Geosci*. 2016;9(274):1–14.

9. Samie A, Deng X, Jia S, Chen D. Scenario-based simulation on dynamics of land-use-land-cover change in Punjab Province, Pakistan. *Sustainability*. 2017;9(8):1285. https://doi.org/10.3390/su9081285.

10. Qiu Y, Lu J. Dynamic simulation of Spartina alterniflora based on CA-Markov model—a case study of Xiangshan bay of Ningbo City, China. *Aquat Invasions*. 2018;13(2):299–309. https://doi.org/10.3391/ai.2018.13.2.10.

11. Hua AK. Application of ca-MARKOV model and land use/land cover changes in MALACCA river watershed, MALAYSIA. *Appl Ecol Environ Res*. 2017;15(4):605–622. https://doi.org/10.15666/aeer/1504_605622.

12. Reddy CS, SINGH S, DADHWAL VK, JHA CS, RAO NR, DIWAKAR PG. Predictive modelling of the spatial pattern of past and future forest cover changes in India. *J Earth Syst Sci*. 2017;126(1). https://doi.org/10.1007/s12040-016-0786-7.

13. Aghaei M, Khavarian H, Mostafazadeh R. Prediction of land use changes using the CA-Markov and LCM models in the Kozehtopraghi Watershed in the Province of Ardabil. *Watershed Manag Res J*. 2020;33(3):91–107.

14. Mujiono, Indra TL, Harmantyo D, Rukmana IP, Nadia Z. Simulation of land use change and effect on potential deforestation using Markov Chain—cellular automata. In: *AIP Conference Proceedings*; 2017:30177. https://doi.org/10.1063/1.4991281.

15. Bello HO, Ojo OI, Gbadegesin AS. Land use/land cover change analysis using Markov-based model for Eleyele reservoir. *J Appl Sci Environ Manag*. 2019;22(12):1917. https://doi.org/10.4314/jasem.v22i12.8.

16. Gollnow F, Göpel J, de Barros Viana Hissa L, Schaldach R, Lakes T. Scenarios of land-use change in a deforestation corridor in the Brazilian Amazon: combining two scales of analysis. *Reg Environ Chang*. 2018;18(1):143–159. https://doi.org/10.1007/s10113-017-1129-1.

17. Nguyen TTH, Ngo TTP. Land use/land cover change prediction in Dak Nong Province based on remote sensing and Markov chain model and cellular automata. *Adv Online Publ*. 2018. https://doi.org/10.13141/JVE.VOL9.NO3.PP132-140.

18. Sadoddin A, Sheikh V, Mostafazadeh R, Halili MG. Analysis of vegetation-based management scenarios using MCDM in the Ramian watershed, Golestan, Iran. *Int J Plant Prod*. 2010;4(151–62).

19. Divín J, Mikita T. Effects of land use changes on the runoff in the landscape based on hydrological simulation in HEC-HMS and HEC-RAS using different elevation data. *Acta Universitatis Agriculturae Et Silviculturae Mendelianae Brunensis*. 2016;64(3):759–768. https://doi.org/10.11118/actaun201664030759.

20. Appollonio C, Balacco G, Novelli A, Tarantion E, Ferruccio Piccinni A. Land use change impact on flooding areas. The case study of Carraro basin. *Sustainability*. 2018;8(996):1–18.

21. Schilling KE, Gassman PW, Kling CL, et al. The potential for agricultural land use change to reduce flood risk in a large watershed. *Hydrol Process*. 2014;28(8):3314–3325. https://doi.org/10.1002/hyp.9865.

22. Halounova L, Holubec V. Assessment of flood with regards to land cover changes. *Procedia Econom Bus Adm*. 2014;18:940–947. https://doi.org/10.1016/S2212-5671(14)01021-1.

23. Schaldach R, Alcamo J, Heisterman M. The multiple-scale land use change model Landshift: a scenario analysis of land use change consequence in Africa. *Environ Model Software*. 2006;196:1–7.

24. Powell TWR, Lenton TM. Scenarios for future biodiversity loss due to multiple drivers reveal conflict between mitigating climate change and preserving biodiversity. *Environ Res Lett*. 2013;8(2):25024. https://doi.org/10.1088/1748-9326/8/2/025024.

25. Nouri J, Gharagozlou A, Arjmandi R, Faryadi S, Adl M. Predicting urban land use changes using a CA–Markov model. *Arab J Sci Eng*. 2014;39 (7):5565–5573. https://doi.org/10.1007/s13369-014-1119-2.

26. Meneses BM, Reis E, Vale MJ, Reis R. Modelling the land use and land cover changes in PORTUGAL: a multi-scale and multi-temporal approach. *Finisterra*. 2018;53(107). https://doi.org/10.18055/Finis12258.

27. Patil SP, Jamgade MB. VOLUME-8 ISSUE-10, august 2019, regular issue. *Int J Innov Technol Explor Eng*. 2019;8(10):484–490. https://doi.org/10.35940/ijitee.I8532.0881019.

28. Surabuddin Mondal M, Sharma N, Kappas M, Garg PK. CA Markov modeling of land use land cover dynamics and sensitivity analysis to identify sensitive parameter(s). *Int Arch Photogramm Remote Sens Spat Inf Sci*. 2019;XLII-2(W13):723–729. https://doi.org/10.5194/isprs-archives-XLII-2-W13-723-2019.

29. Mukhopdhaya S. *Land Use and Land Cover Change Modelling Using CA-Markov Case Study: Deforestation Analysis of Doon Valley*; 2016.

30. Onwuka S, Eneche P, Ismail N. Geospatial Modeling and Prediction of lLand Use/Cover Dynamics in Onitsha Metropolis, Nigeria: A Sub-pixel Approach. *Current J Appl Sci TechnolJournal of Applied Science and Technology*. 2017;22(6):1–18. https://doi.org/10.9734/CJAST/2017/35294.

31. Rounsevell MDA, Ewert F, Reginster I, Leemans R, Carter TR. Future scenarios of European agricultural land use. *Agr Ecosyst Environ*. 2005;107 (2–3):117–135. https://doi.org/10.1016/j.agee.2004.12.002.

32. Godet M. The art of scenarios and strategic planning: tools and pitfalls. *Technol Forecast Soc Chang*. 2000;65:3–22.

33. Yang X, Chen H, Wang Y, Xu C-Y. Evaluation of the effect of land use/cover change on flood characteristics using an integrated approach coupling land and flood analysis. *Hydrol Res*. 2016;47(6):1161–1171. https://doi.org/10.2166/nh.2016.108.

34. Barredo JI, Engelen G. Land use scenario modeling for flood risk mitigation. *Sustainability*. 2010;2(5):1327–1344. https://doi.org/10.3390/su2051327.

35. Thorn AM, Wake CP, Grimm CD, Mitchell CR, Mineau MM, Ollinger SV. Development of scenarios for land cover, population density, impervious cover, and conservation in New Hampshire, 2010–2100. *Ecol Soc*. 2017;22(4). https://doi.org/10.5751/ES-09733-220419.

36. Rega C, Helming J, Paracchini ML. Environmentalism and localism in agricultural and land-use policies can maintain food production while supporting biodiversity Findings from simulations of contrasting scenarios in the EU. *Land Use Policy*. 2019;87:103986. https://doi.org/10.1016/j.landusepol.2019.05.005.

37. Sleeter BM, Sohl TL, Bouchard MA, et al. Scenarios of land use and land cover change in the conterminous United States: utilizing the special report on emission scenarios at ecoregional scales. *Glob Environ Chang*. 2012;22(4):896–914. https://doi.org/10.1016/j.gloenvcha.2012.03.008.

38. Lippe M, Hilger T, Sudchalee S, Wechpibal N, Jintrawet A, Cadisch G. Simulating stakeholder-based land-use change scenarios and their implication on above-ground carbon and environmental Management in Northern Thailand. *Landarzt*. 2017;6(4):85. https://doi.org/10.3390/land6040085.

39. Zhang L, Zhang S, Huang Y, et al. Prioritizing abandoned mine lands rehabilitation: combining landscape connectivity and pattern indices with scenario analysis using land-use modeling. *ISPRS Int J Geo Inf*. 2018;7(8):305. https://doi.org/10.3390/ijgi7080305.

40. Riyando Moe I, Kure S, Fajar Januriyadi N, et al. Future projection of flood inundation considering land-use changes and land subsidence in Jakarta, Indonesia. *Hydrol Res Lett.* 2017;11(2):99–105. https://doi.org/10.3178/hrl.11.99.

41. Siregar VP, Prabowo NW, Agus SB, Subarno T. The effect of atmospheric correction on object based image classification using SPOT-7 imagery: a case study in the Harapan and Kelapa Islands. *IOP Conf Ser Earth Environ Sci.* 2018;176:12028. https://doi.org/10.1088/1755-1315/176/1/012028.

42. Hamad R, Balzter H, Kolo K. Predicting land use/land cover changes using a CA-Markov model under two different scenarios. *Sustainability.* 2018;10(10):3421. https://doi.org/10.3390/su10103421.

43. Abuelaish B, Olmedo MTC. Scenario of land use and land cover change in the Gaza strip using remote sensing and GIS models. *Arab J Geosci.* 2016;9(4). https://doi.org/10.1007/s12517-015-2292-7.

44. Li X, Wang M, Liu X, Chen Z, Wei X, Che W. MCR-modified CA–Markov model for the simulation of urban expansion. *Sustainability.* 2018;10(9):3116. https://doi.org/10.3390/su10093116.

45. Omar NQ, Sanusi SAM, Hussin WMW, Samat N, Mohammed KS. Markov-CA model using analytical hirareny process and multi regression technique. *Earth Environ Sci.* 2014;20:1–18.

46. Ignacio B, Bosque-Sendra J. Comparison of multi criteria evaluation methods in geographical methods integrated in Geographical Information System to allocate urban areas systems to allocate urban areas. *Geogr Syst.* 1998;1–12.

47. Behera M, Borate D, Panda SN, Behera SN, Roy PR. Modeling and analyzing the watershed dynamics using cellular automata CA-Markov model a Geo-Information based model. 2012;121(4):1011–1024.

48. Hadi SJ, Shafri HZM, Mahir MD. Modelling LULC for the period 2010-2030 using GIS and remote sensing: a case study of Tikrit, Iraq. *IOP Conf Ser Earth Environ Sci.* 2014;20:12053. https://doi.org/10.1088/1755-1315/20/1/012053.

49. Viana CM, Rocha J. Evaluating dominant land use/land cover changes and predicting future scenario in a rural region using a memoryless stochastic method. *Sustainability.* 2020;12(10):4332.

50. Tornquist CG, Silva, D. S. d. Current and future land use and land cover scenarios in the Arroio Marrecas watershed. *Revista Brasileira De Engenharia Agrícola E Ambiental.* 2019;23(3):215–222. https://doi.org/10.1590/1807-1929/agriambi.v23n3p215-222.

51. Koo H, Kleemann J, Fürst C. Land use scenario modeling based on local knowledge for the provision of ecosystem Services in Northern Ghana. *Landarzt.* 2018;7(2):59. https://doi.org/10.3390/land7020059.

7

Topographical features and soil erosion processes

Mahboobeh Kiani-Harchegani[a], Ali Talebi[a], Ebrahim Asgari[a],
and Jesús Rodrigo-Comino[b]

[a]Department of Watershed Management Engineering, Faculty of Natural Resources, Yazd University, Yazd Province, Iran
[b]Departamento de Análisis Geográfico Regional y Geografía Física, Facultad de Filosofía y Letras, Campus Universitario de Cartuja, University of Granada, Granada, Spain

1 Introduction

Soil erosion is a serious environmental issue with supposing a key challenge for humankind because of the potential hazards and negative impacts.[1, 2] Statistics on the degree of erosion and sedimentation and associated damages (e.g., destruction and reduction of soil fertility, a decrease in the capacity of waterways and reservoirs, and environmental pollution) worldwide have proven the phenomenon of soil erosion as a critical concern. Hence, identifying, determining, predicting, and simulating this natural process can act as a guide for soil conservation management and planning.[3]

Soil erosion is the process of separating soil particles from their beds and transporting them to another place by the force of a transporting agent. The separated soil particles are then transported by water upon entering the stream.[4–6] Some governing factors such as soil type,[7] land cover,[8] roughness,[9] rainfall intensity, and slope[10] have been considered in human and natural ecosystems to understand erosion processes. Rainfall intensity and topography variables have also received considerable attention in many risk assessment models and techniques.[11, 12]

Topographic features are crucial factors to consider in hydrological processes at different spatial and temporal scales. It is helpful to determine the impact of these factors in the proper planning and management of water and soil resources and control the evolutionary process of soil erosion.[13] The slope and its length, the difference in altitude, and the shape of the slope are influential topographic factors in hydrological processes. The shape of the slope and the process by which its dimensions expand in various directions remarkably impact the runoff start threshold, time to reach the peak, time of concentration, volume and intensity of flow, and variation in their erosion power.[14] Therefore, mutual understanding and feedback among various forms of hillslopes and hydrological processes can be useful in watershed management.[15] Furthermore, topography and other associated data are influential information in the correctness and performance of various models developed in the field of soil erosion.[16, 17]

Numerous hydrological models have been developed to better understand the factors influencing the hydrological cycle in watersheds. These models are popular to solve diverse water resource management issues, including rainfall-runoff modeling, sediment management, and evaluation of the impact of climate change.[18]

The prediction of the hydrological behavior of watersheds is essential for better evaluation of water projects, development of soil conservation measures, and management of water resources. This is due to the low probability of uniformly shaped slopes in nature and the fact that watersheds are formed of a set of complex hillslopes and connectivity processes.[19, 20] The hydrological reactions of hillslopes are mostly affected by the longitudinal profile and shape of the plan. Therefore, most of the models proposed by experts are less adaptable and fail to provide reliable modeling of hydrological processes in hillslopes.[14]

Various techniques have been proposed to study hydrological processes in watersheds. However, the effect of topography in these processes is deemed simply by applying the land slope and their true geometric shape (e.g., concavity or convexity in the longitudinal profile) is not covered. The dynamic reaction of a hillslope depends greatly on the plan shape and the profile curvature, and the angle of the slope; therefore, dynamic and 3D models are required to study the hydrological processes of complex hillslopes.[21, 22] In most hydrological studies, the shape of the hillslopes geometry has been considered rectangular or planar.[23–25] The majority of studies on different soil erosion processes have only focused on the degree and length of the slope.[12, 26–30] Among similar studies conducted in recent years, Sadeghi et al.[31] and Kiani-Harchegani et al.[13] examined the components of runoff and sediment under various slope degrees with smooth geometric shapes and parallel slopes under different rainfall intensities at the plot scale. The results showed dynamic variations in the concentration and size distribution of sediment particles with different slopes and intensities of rainfall. However, natural hillslopes have a complex topography and their shapes are marked by contour lines of irregular distance.

Therefore, efforts have been made over the past few decades to consider the influence of various slope shapes including plan shape (convergent, divergent, and parallel) and longitudinal profile of the slope in three straight, convex, and concave modes on the behavior and creation of surface and subsurface runoff and soil loss during the soil erosion processes in hydrological models.[14, 21, 32, 33] For example, Evans[34] studied the effect of four types of slopes (i.e., uniform, convex, concave, and combined) in the study area and found that concave and convex slopes have the lowest and highest rates of soil loss and erosion, respectively, and uniform slopes fall between these two types of slopes.

To determine the leading and determining factors in subsurface flows, Troch et al.[33] worked with field and numerical methods and found that the longitudinal profile of the bedrock and the shape of the hillslope are critical factors influencing subsurface flows. For example, Aryal et al.[35] studied the plan shape of complex hillslopes (convergent to divergent) and the curvature of the profile (concave and convex) and concluded that the convergent hillslopes are more saturated. In addition, Troch et al.[22] developed a dynamic 3D model to study the motion of water in an unsaturated environment in which all forms of slopes and variations in slope width and curvature of the hillslope were covered. According to Agnese,[36] at the same convergent hillslopes, the time required to drain surface water is more than that required for divergent hillslopes and convex hillslopes have the shortest surface flow time. Talebi et al.[37] investigated the effect of the shape and geometry of complex hillslopes on slope stability and found that convergent and concave hillslopes reach instability much earlier than divergent and convex hillslopes and their stability coefficients is lower. Similar research has been carried out on the development of hydrological models in complex hillslopes in recent years, but no clear consensus has been reached.[15, 38–40]

2 Terminology and definition of hillslope geometry

Hillslopes can be considered as the basic landscape element to assess watersheds with sufficient accuracy.[21] The natural or human-affected hillslopes of a specific watershed show a wide variety of shapes and are not always straight curvature and/or rectangular (Fig. 1). Whether flow lines are convergent or divergent along the flow path is an indicator for classifying domains as convergent, parallel, or divergent. In contrast to divergent hillslopes, in convergent hillslopes, the width decreases from upstream to downstream. The curvature of the hillslopes has three potential shapes: concave, convex, or straight. For example, in Fig. 1, sub-watershed No. 2 is of the convex-divergent type, sub-watershed No. 7 is of the convex-convergent type, and sub-watershed No. 15 is of the convex-parallel type. Therefore, it is necessary to first define the geometry of the hillslopes in three dimensions, including the curvature of the profile and the shape of the plan. The Slope curvature indicates the curvature in the direction of the flow gradient, which is concave, straight, or convex, and the plane shape is the variation in the transverse slope in the direction perpendicular to the flow lines. Studies on the geometry of the slopes consider the shape of the plan, similar to the convergence and divergence of the plan.[38]

The method proposed by Troch et al.[33] is applied to introduce a suitable function that can show the complex hillslopes geometry. These equations are based on the Evans model. Evans[34] proposed the application of a quadratic two-variable function as the following equation:

$$z = ax^2 + by^2 + cxy + dx + ey + f \tag{1}$$

where, z is the height, x is the horizontal distance in the direction of the length of hillslope towards the end of the basin, y is the horizontal distance from the center of the gradient in the vertical direction towards the length of the hillslope, and the other parameters are fixed. Di Stefano et al.[41] proposed a basic curvature function of the profile in Eq. (2), and Troch et al.[33] proposed Eq. (3).

$$z(x) = E + H \left(1 - \frac{x}{L}\right)^n \tag{2}$$

$$z(x, y) = E + H \left(1 - \frac{x}{L}\right)^a + \omega y^2 \tag{3}$$

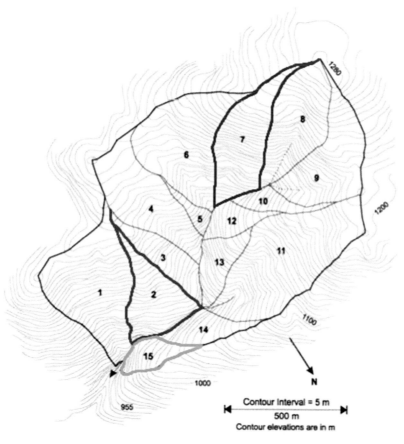

FIG. 1 Contour and sub-watersheds of Bushrangers watershed in Australia. *Modified from Aryal SK, O'Loughlin EM, Mein, RG. A similarity approach to determine response times to steady-state saturation in landscapes. Adv Water Resour. 2005; 28(2), 99–115.*

where E and H are the minimum and maximum heights relative to the baseline, respectively. L is the length of the hillslope in meters, and a is the profile curvature parameter which has no dimension and ω is the plan shape parameter. Allowing profile curvature (a) to assume values $a > 1$, $a = 1$ and $a < 1$ less than, and plan curvature (ω) to assume either a positive, zero or negative value, one can define nine geometric hillslope forms.[14, 32]

Table 1 shows nine different hillslope geometry types that are formed by combining three profile curvatures and three plan shapes.[33] Fig. 2 also illustrates the 3-dimensional view of a convex-convergent hillslope.[39]

3 Three-dimensional hillslopes characteristics in hydrological models

The topographic structure of watersheds depends on the interaction between the hillslope geometry and the processes along the channel. However, most models have provided a very simple representation of topographic properties to date in landslides, runoff, sediment production, and vegetation cover.[42] Therefore, in recent decades, studies have attempted to cover 3-dimensional hillslope characteristics such as curvature profiles and shape plans in quantitative hydrological models of surface and subsurface flow, saturation permeability, time of concentration, lag time, and soil erosion models such as the USLE model (e.g., Refs. 14, 15, 21, 32, 33, 36, 38, 39, 43–45).

3.1 The effect of hillslopes geometry on surface and subsurface runoff

Runoff consists of three sections: surface runoff, subsurface runoff, and baseflow. Surface runoff is usually defined as the amount of rainfall that is not absorbed by infiltration. When the infiltration rate is higher than the rainfall intensity, the subsurface flow replaces the surface flow.[46] In recent years, some authors have considered the shape of hillslopes as an influential factor in the behavior of subsurface and surface flow. A mutual understanding between

TABLE 1 Geometric characteristics of complex hillslopes.

No	Longitudinal profile	Plan shape	Complex hillslopes
1	Concave	Convergent	
2		Parallel	
3		Divergent	
4	Straight	Convergent	
5		Parallel	
6		Divergent	
7	Convex	Convergent	
8		Parallel	
9		Divergent	

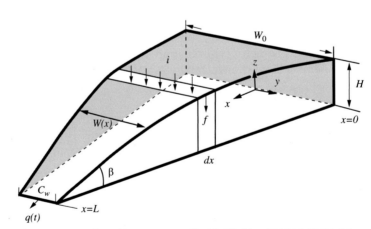

FIG. 2 Three-dimensional view of a convergent plan with a convex profile. *Modified from Talebi A, Hajiabolghasemi R, Hadian MR, Amanian N. Physically based modelling of sheet erosion (detachment and deposition processes) in complex hillslopes. Hydrol. Process. 2016; 30(12): 1968–1977.*

the hillslope's geometry and the rainfall-runoff processes can be useful in the management of urban and natural catchments.[43, 47]

Hillslope shape is a powerful tool for investigating the complex effects of topography on surface runoff, subsurface runoff, and travel time. In this regard, Agnese[36] showed that convex hillslopes have more surface runoff than other profiles (in a fixed plane), while convergent hillslopes have more surface runoff than parallel and divergent hillslopes

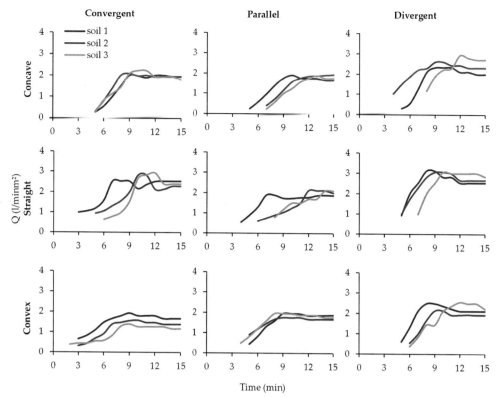

FIG. 3 Hydrographs of rainfall-runoff processes at different roughness coefficients in complex hillslopes. *Modified from Meshkat M, Amanian N, Talebi A, Kiani-Harchegani M, Rodrigo-Comino J. Effects of roughness coefficients and complex hillslope morphology on runoff variables under laboratory conditions. Water. 2019; 11(12): 2550.*

(in a fixed profile). Furthermore, according to Geranian et al.[38] in convergent hillslopes, the time required to drain surface water is greater than that required for divergent hillslopes. Meshkat et al.[47] investigated the effect of the surface roughness coefficient in three types of soils with different roughness coefficients on surface flow components in complex hillslopes. The results showed that the minimum start time of runoff in the convex-convergent slope was achieved after 71 s and the least time of concentration in the convex-divergent hillslope was obtained after 162 s for soil No. 1 with a roughness coefficient of 0.015. In the concave-divergent hillslope and soil No. 3, with a roughness coefficient of 0.018, the maximum peak discharge was obtained.

The highest peak discharge in the concave-divergent hillslope was also obtained in soil No. 3, with a roughness coefficient of 0.018. Furthermore, the results of the soil hydrograph with different roughness coefficients were drawn for complex hillslopes.

Fig. 3 illustrates that parallel hillslope has a constant rate of head change due to the constant upward and downward plot, while divergent hillslopes have a decreasing rate of change due to direct and parallel motion of flow towards the outlet. However, the increase rate in the flow peak also increased in convergent hillslopes because of the greater area of the plot upward. The study of hillslopes geometry also shows that the time to concentration and start time of runoff could increase with increasing surface roughness coefficient. So that, in convex profiles, owing to the steep slope of the plot downward, with increasing surface roughness coefficient and consequently increasing its permeability, the flow is discharged earlier and the time of concentration is reduced. Therefore, the hydrographs of different soil types 1, 2, and 3 almost overlap.

Subsurface flow can be defined as the water that penetrates the soil.[48] Subsurface flow may face impermeable layers or obstacles in its path which hinder it from deeper penetration. In this case, water collects on these layers, takes on lateral movements, and moves along the slope.[49–54] For this, a related study was carried out by Troch et al.[22] and Hilberts et al.[55] in laboratory conditions. Sabzevari et al.[40] modeled a saturation zone extension in unsteady-state conditions based on rainfall temporal distributions by developing proposed models for complex hillslopes. The results of these studies showed that the average subsurface runoff in divergent hillslopes use to be higher than the convergent ones and the corresponding values in convex hillslopes are higher than concave hillslopes. Thus, the lowest and highest values were observed respectively in the concave-convergent (0.04) and parallel convex (34) hillslopes (Fig. 4).

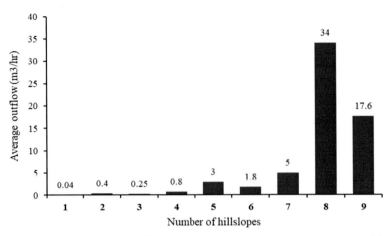

FIG. 4   Average subsurface flow in complex hillslopes ($N = 30\,mm/day$). *Modified from Sabzevari T, Talebi A., Ardakanian R, Shamsai A. A steady-state saturation model to determine the subsurface travel time (STT) in complex hillslopes. Hydrol. Earth Syst. Sci. 2010; 14(6): 891–900.*

Amanian et al.[15] also reported that the influence of the hillslope plan on the threshold of surface and subsurface flow is much more than the hillslope profile and stated that the hillslope plan can start a different flow threshold in different plans by altering the storage volume at the outlet of the hillslope. Fariborzi et al.[56] estimated subsurface runoff on complex hillslopes at three slopes and three rainfall intensities using the time-area method. The results showed a great estimate of surface runoff using the time-area method with a coefficient of determination of 0.98.

3.2 The effect of hillslopes geometry on time of concentration and time to equilibrium

Most rainfall-runoff models require parameters to accurately describe the response time in catchments. The most popular response time parameters employed in hydrological models are the lag time, time of concentration, and time to equilibrium. Slope steepness is one of the parameters in most equations related to the time of concentration. However, most researchers have formulated straight hills, which should include the impact of the profile curvature. Several authors, such as Troch et al.,[22] and Hilberts et al.[55] showed that the underlying hydrological processes are influenced by hillslope geometry. Sabzevari et al.[39] developed the equilibrium time equation proposed by Agnese[36] concerning the 3-dimensional geometry of hillslopes and proposed an equation for calculating the time of concentration of complex hillslopes covering the plane shape and curvature of hillslopes. The results showed that convergent hillslopes with different profile curvatures had a long time concentration-time than parallel and divergent hillslopes. The mean time of concentration of the convergent hillslopes was 2.4 times that of the divergent hillslopes and 1.7 times that of the parallel hillslopes. Therefore, considering that the time of concentration is directly related to the plan shape, variations in the time of concentration for rainfall intensities greater than $10\,mm/h$ are shown in Fig. 5. According to Fig. 5, the shortest time of concentration is for divergent convex hillslopes ($8.6\,min$) and the longest is for convergent convex hillslopes ($23.9\,min$). Furthermore, the equilibrium time in convergent hillslopes is lower than that of divergent hillslopes and convex hillslopes have more equilibrium time than concave hillslopes. In convergent hillslopes, the area engaged in runoff production was low at the onset of rainfall and increased as the equilibrium time was reached, but the process was reversed in divergent hillslopes.

3.3 The effect of hillslopes geometry on different soil erosion processes

Soil erosion is a complex naturally occurring phenomenon that causes many problems in the world, including considerable financial costs and major environmental problems.[5, 57] If planning to avoid these damages, it is crucial to consider hillslope erosion. Furthermore, by understanding and analyzing the processes that are affected by the hillslopes geometry, the rate of surface erosion and sediment production could be predicted and then estimated.[21] Sheet or interrill erosion is a fundamental type of erosion in forests, agriculture, and rangeland ecosystems.[10, 58, 59] Therefore, it is often difficult to study them in combination with other interrelated factors, such as rainfall, runoff, and erosion, when humans play a key role.

The study of mathematical physics-based models developed by some researchers to investigate sheet erosion due to rain often considers 1-dimensional topography.[60, 61] These models were developed by Talebi et al.[21] with

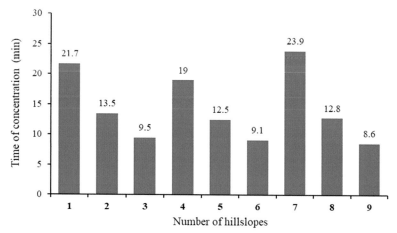

FIG. 5 Time of concentration in complex hillslopes ($r = 10\,\text{mm/h}$). *Modified from Sabzevari T, Saghafian B, Talebi A, Ardakanian R. Time of concentration of surface flow in complex hillslopes. J. Hydrol. Hydromech. 2013; 61(4): 269–277.*

3-dimensional consideration of hillslope topography, who used a numerical model based on the equation of sediment continuity and the detachment and deposition rate of sediments due to sheet erosion in different profiles and plans of complex hillslopes were estimated for two slope lengths (2 and 4.53 m), two slope steepness (5.7% and 15%) and tow rainfall intensities (57 and 93 mm/h) (Figs. 6–9). The results showed that variations in slope length, degree, and rainfall intensity did not change the shape of the sediment detachment/deposition diagrams due to sheet erosion. However, the hillslope geometry has a greater effect on variations in sediment detachment/deposition, so that the outcome of the profile (concave and convex) in sheet erosion was greater than the plan (divergence and convergence) of complex hillslopes.

In another study, Sabzevari and Talebi[18] investigated the rate of erosion on complex hillslopes using the universal soil loss equation (USLE) and the topographic factor (LS) was developed as a function of the profile curvature and plan shape. The mean erosion in convex hillslopes was 1.43 times that of concave hillslopes and 1.19 times that of straight hillslopes (Fig. 10) and the impact of the profile curvature on erosion was much greater than that of the plan shape. The largest and least erosion was related to the convex-divergent and concave-divergent hillslopes, respectively.

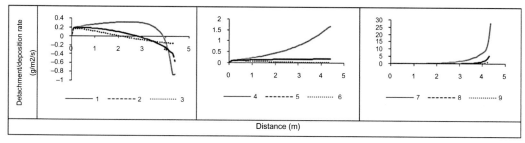

FIG. 6 Detachment/deposition in 5.7% slope and 4.58 m length under 57 mm/h rainfall intensity in complex hillslopes.

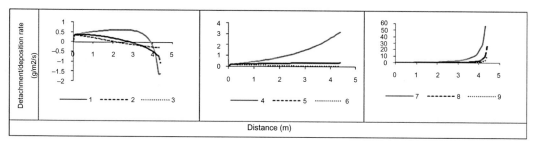

FIG. 7 Detachment/deposition in 5.7% slope and 4.58 m length under 93 mm/h rainfall intensity in complex hillslopes.

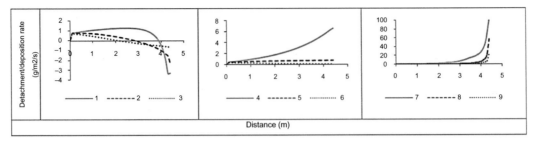

FIG. 8 Detachment/deposition in 15% slope and 4.58 m length under 57 mm/h rainfall intensity in complex hillslopes.

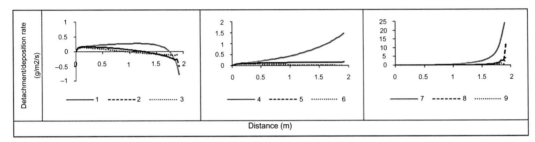

FIG. 9 Detachment/deposition in 5.7% slope and 2 m length under 57 mm/h rainfall intensity in complex hillslopes.

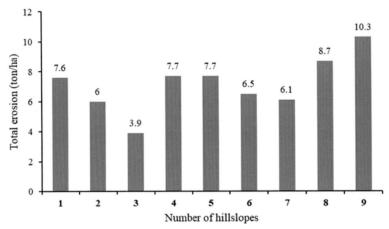

FIG. 10 The total erosion in complex hillslopes. *Modified from Sabzevari T, Talebi A. Effect of hillslope topography on soil erosion and sediment yield using USLE model. Acta Geophys. 2019; 67(6): 1587–1597.*

Considering that soil erosion needs to be measured, monitored, and evaluated and since mathematical and computer models are an easy-to-use and flexible tools to organize data and maps and they can provide extensive information for different regions and temporal scales. Therefore, combining the use of different models related to soil erosion, especially the USLE-type models, with computer software and programs can help managers, planners, and experts in conserving water and soil resources and controlling their loss.

Since modeling soil erosion, only represent reality, not reality itself. Therefore, understanding the complexities of nature and considering them in models can lead to a more accurate understanding of system performance and how the interaction affects the elements of a system.[18, 62–64]

4 Conclusion

In this chapter, we demonstrated that, among the factors related to topography, there are other keys (e.g., longitudinal profile and plan shape) influencing factors that alter hydrological processes at the hillslope scale, besides the degree and slope length. Hydrological processes under the influence of each of the hillslope forms, that is, three plan shapes and three profile curvatures, as well as their interaction, show special connections, as confirmed by the behavioral complexity of runoff and sediment components. The effect of the plan on the peak flow rate is much greater than that of the profile, and the effect of the curvature of the profile is much greater than that of the plan on the erosion process. Therefore, in hydrological models, these factors must be considered to obtain correct results from these models and reduce their uncertainties. A mutual understanding between different forms of hillslopes and hydrological processes can be very useful and efficient in watershed management, as well as in many cases of water resource management issues, including rainfall-runoff modeling and sediment management.

Acknowledgments

The present study was conducted during the postdoctoral period of the corresponding author at Yazd University. The authors wish to thank Yazd University for its support of this research (Project No. 97.50.2374).

References

1. Rodrigo-Comino J, López-Vicente M, Kumar V, et al. Soil science challenges in a new era: a transdisciplinary overview of relevant topics. *Air Soil Water Res.* 2020;13. https://doi.org/10.1177/1178622120977491.
2. Rodrigo-Comino J, Terol E, Mora G, Giménez-Morera A, Cerdà A. Vicia sativa Roth. Can reduce soil and water losses in recently planted vineyards (Vitis vinifera L.). *Earth Syst Environ.* 2020;4:827–842.
3. Panagos P, Ballabio C, Poesen J, et al. A soil Erosion Indicator for supporting agricultural, environmental and climate policies in the European Union. *Remote Sens.* 2020;12:1365. https://doi.org/10.3390/rs12091365.
4. Keesstra S, Nunes JP, Saco P, et al. The way forward: can connectivity be useful to design better measuring and modelling schemes for water and sediment dynamics? *Sci Total Environ.* 2018;644:1557–1572.
5. Kiani-Harchegani M, Sadeghi SH, Asadi H. Inter-storm variability of coefficient of variation of runoff volume and soil loss during rainfall and erosion simulation replicates. *J Ecohydrology.* 2017;4(1):191–199 [In Persian].
6. Kiani-Harchegani M, Sadeghi SH. Practicing land degradation neutrality (LDN) approach in the Shazand watershed, Iran. *Sci Total Environ.* 2020;698:134319.
7. Ben-Hur M, Wakindiki IIC. Soil mineralogy and slope effects on infiltration, interrill erosion, and slope factor. *Water Resour Res.* 2004;40(3).
8. López-Vicente M, Calvo-Seas E, Álvarez S, Cerdà A. Effectiveness of cover crops to reduce loss of soil organic matter in a rainfed vineyard. *Land.* 2020;9:230. https://doi.org/10.3390/land9070230.
9. Eitel JU, Williams CJ, Vierling LA, Al-Hamda OZ, Pierson FB. Suitability of terrestrial laser scanning for studying surface roughness effects on concentrated flow erosion processes in rangelands. *Catena.* 2011;87(3):398–407.
10. Sun L, Fang H, Qi D, Li J, Cai Q. A review on rill erosion process and its influencing factors. *Chin Geogr Sci.* 2013;23(4):389–402.
11. Armstrong A, Quinton JN, Heng BCP, Chandler JH. Variability of interrill erosion at low slopes. *Earth Surf Process Landf.* 2011;36(1):97–106.
12. Chaplot V, Le Bissonnais Y. Field measurements of interrill erosion under different slopes and plot sizes. *Earth Surf Process Landf.* 2000;25(2):145–153.
13. Kiani-Harchegani M, Sadeghi SH, Singh VP, Asadi H, Abedi M. Effect of rainfall intensity and slope on sediment particle size distribution during erosion using partial eta squared. *Catena.* 2019;176:65–72.
14. Talebi A, Troch PA, Uijlenhoet R. A steady-state analytical slope stability model for complex hillslopes. *Hydrol Process.* 2008;22(4):546–553.
15. Amanian N, Geranian M, Talebi A, Hadian MR. The effect of plan and slope profile on runoff initiation threshold. *Watershed Manag Sci Eng.* 2018;11(39):105–108 [In Persian].
16. Arabameri A, Cerda A, Pradhan B, Tiefenbacher JP, Lombardo L, Bui DT. A methodological comparison of head-cut based gully erosion susceptibility models: combined use of statistical and artificial intelligence. *Geomorphology.* 2020;359:107136. https://doi.org/10.1016/j.geomorph.2020.107136.
17. Pourghasemi HR, Kariminejad N, Amiri M, et al. Assessing and mapping multi-hazard risk susceptibility using a machine learning technique. *Sci Rep.* 2020;10(1):1–11.
18. Sabzevari T, Talebi A. Effect of hillslope topography on soil erosion and sediment yield using USLE model. *Acta Geophys.* 2019;67(6):1587–1597.
19. Bracken LJ, Turnbull L, Wainwright J, Bogaart P. Sediment connectivity: a framework for understanding sediment transfer at multiple scales. *Earth Surf Process Landf.* 2015;40(2):177–188.
20. Jencso KG, McGlynn BL. Hillslope hydrologic connectivity controls riparian groundwater turnover: implications of catchment structure for riparian buffering and stream water sources. *Water Resour Res.* 2010;46. https://doi.org/10.1029/2009WR008818, W10524.
21. Talebi A, Hajiabolghasemi R, Hadian MR, Amanian N. Physically based modelling of sheet erosion (detachment and deposition processes) in complex hillslopes. *Hydrol Process.* 2016;30(12):1968–1977.
22. Troch PA, Paniconi C, Emiel van Loon AE. Hillslope-storage Boussinesq model for subsurface flow and variable source areas along complex hillslopes: 1. Formulation and characteristic response. *Water Resour Res.* 2003;39(11).
23. Baiamonte G, D'Asaro F, Grillone G. Simplified probabilistic-topologic model for reproducing hillslope rill network surface runoff. *J Irrig Drain Eng.* 2015;141(7), 04014080.
24. Baiamonte G, Singh VP. Overland flow times of concentration for hillslopes of complex topography. *J Irrig Drain Eng.* 2016;142(3), 04015059.
25. Reggiani P, Todini E, Meißner D. Analytical solution of a kinematic wave approximation for channel routing. *Hydrol Res.* 2014;45(1):43–57.
26. Chen H, Zhang X, Abla M, et al. Effects of vegetation and rainfall types on surface runoff and soil erosion on steep slopes on the Loess Plateau, China. *Catena.* 2018;170:141–149.
27. Kinnell PIA. The effect of slope length on sediment concentrations associated with side-slope erosion. *Soil Sci Soc Am J.* 2000;64(3):1004–1008.

28. Panagos P, Borrelli P, Meusburger K. A new European slope length and steepness factor (LS-factor) for modeling soil erosion by water. *Geosciences*. 2015;5(2):117–126.

29. Sadeghi SH, Kiani-Harchegani M, Hazbavi Z, et al. Field measurement of effects of individual and combined application of biochar and polyacrylamide on erosion variables in loess and marl soils. *Sci Total Environ*. 2020;138866.

30. Zhao B, Zhang L, Xia Z, et al. Effects of rainfall intensity and vegetation cover on erosion characteristics of a soil containing rock fragments slope. *Adv Civ Eng*. 2019;1–14. https://doi.org/10.1155/2019/7043428. 7043428.

31. Sadeghi SH, Singh VP, Kiani-Harchegani M, Asadi H. Analysis of sediment rating loops and particle size distributions to characterize sediment source at mid-sized plot scale. *Catena*. 2018;167:221–227.

32. Talebi A, Uijlenhoet R, Troch PA. A low-dimensional physically based model of hydrologic control of shallow landsliding on complex hillslopes. *Earth Surf Process Landf*. 2008;33(13):1964–1976.

33. Troch P, Van Loon E, Hilberts A. Analytical solutions to a hillslope-storage kinematic wave equation for subsurface flow. *Adv Water Resour*. 2002;25(6):637–649.

34. Evans IS. An integrated system of terrain analysis and slope mapping. *Z Geomorphol*. 1980;36:274–295.

35. Aryal SK, O'Loughlin EM, Mein RG. A similarity approach to determine response times to steady-state saturation in landscapes. *Adv Water Resour*. 2005;28(2):99–115.

36. Agnese C, Baiamonte G, Corrao C. Overland flow generation on hillslopes of complex topography: analytical solutions. *Hydrol Process*. 2007;21:1308–1317.

37. Talebi A, Uijlenhoet R, Troch PA. Soil moisture storage and hillslope stability. *Nat Hazards Earth Syst Sci*. 2007;7(5):523–534.

38. Geranian M, Amanian N, Talebi A, Hadian MR, Zeini M. Laboratorial investigation of effect of plan shape and profile curvature on variations of surface flow in complex hillslopes. *Water Resour Res*. 2013;9(2):64–72 [In Persian].

39. Sabzevari T, Saghafian B, Talebi A, Ardakanian R. Time of concentration of surface flow in complex hillslopes. *J Hydrol Hydromech*. 2013;61(4):269–277.

40. Sabzevari T, Talebi A, Ardakanian R, Shamsai A. A steady-state saturation model to determine the subsurface travel time (STT) in complex hillslopes. *Hydrol Earth Syst Sci*. 2010;14(6):891–900.

41. Di Stefano C, Ferro V, Porto P, Tusa G. Slope curvature influence on soil erosion and deposition processes. *Water Resour Res*. 2000;36(2):607–617.

42. Tucker GE, Bras RL. Hillslope processes, drainage density, and landscape morphology. *Water Resour Res*. 1998;34(10):2751–2764.

43. Afshar Ardekani A, Sabzevari T. Effects of hillslope geometry on soil moisture deficit and base flow using an excess saturation model. *Acta Geophys*. 2020;1–10.

44. Boll J, Brooks ES, Crabtree B, Dun S, Steenhuis TS. Variable source area hydrology modeling with the water erosion prediction project model. *J Am Water Resour Assoc*. 2015;51(2):330–342.

45. de Lima JL, Isidoro JM, de Lima MIP, Singh VP. Longitudinal hillslope shape effects on runoff and sediment loss: laboratory flume experiments. *J Environ Eng*. 2018;144(2), 04017097.

46. Sabzevari T, Noroozpour S. Effects of hillslope geometry on surface and subsurface flows. *Hydrogeol J*. 2014;22(7):1593–1604.

47. Meshkat M, Amanian N, Talebi A, Kiani-Harchegani M, Rodrigo-Comino J. Effects of roughness coefficients and complex hillslope morphology on runoff variables under laboratory conditions. *Water*. 2019;11(12):2550.

48. Hewlett JD, Hibbert AR. Factors affecting the response of small watersheds to precipitation in humid areas. In: *Forest Hydrology*; 1967:275–290.

49. Beven K. On subsurface stormflow: an analysis of response times. *Hydrol Sci J*. 1982;27(4):505–521.

50. Fiori A, Romanelli M, Cavalli DJ, Russo D. Numerical experiments of streamflow generation in steep catchments. *J Hydrol*. 2007;339(3–4):183–192.

51. Freeze RA. Role of subsurface flow in generating surface runoff: 2. Upstream source areas. *Water Resour Res*. 1972;8(5):1272–1283.

52. Freeze RA. Role of subsurface flow in generating surface runoff: 1. Base flow contributions to channel flow. *Water Resour Res*. 1972;8(3):609–623.

53. Montgomery DR, Dietrich WE. Runoff generation in a steep, soil-mantled landscape. *Water Resour Res*. 2002;38(9):7-1.

54. Morbidelli R, Saltalippi C, Flammini A, Cifrodelli M, Corradini C, Govindaraju RS. Infiltration on sloping surfaces: laboratory experimental evidence and implications for infiltration modeling. *J Hydrol*. 2015;523:79–85.

55. Hilberts AG, Troch PA, Paniconi C, Boll J. Low-dimensional modeling of hillslope subsurface flow: relationship between rainfall, recharge, and unsaturated storage dynamics. *Water Resour Res*. 2007;43(3).

56. Fariborzi H, Sabzevari T, Noroozpour S, Mohammadpour R. Prediction of the subsurface flow of hillslopes using a subsurface time-area model. *Hydrogeol J*. 2019;27(4):1401–1417.

57. Sidle R, Ochiai H. *Processes, Prediction, and Land Use. Water Resources Monograph*. American Geophysical Union, Washington; 2006. 308 p.

58. Kinnell PIA. Modeling of the effect of flow depth on sediment discharged by rain-impacted flows from sheet and interrill erosion areas: a review. *Hydrol Process*. 2013;27:2567–2578.

59. Tayfur G. Modeling two-dimensional erosion process over infiltrating surfaces. *J Hydrol Eng*. 2001;6(3):259–262.

60. GopalNaik M, Rao EP, Eldho TI. A kinematic wave based watershed model for soil erosion and sediment yield. *Catena*. 2009;77(3):256–265.

61. Kinnell PIA. Applying the RUSLE and the USLE-M on hillslopes where runoff production during an erosion event is spatially variable. *J Hydrol*. 2014;519:3328–3337.

62. Alewell C, Borrelli P, Meusburger K, Panagos P. Using the USLE: chances, challenges and limitations of soil erosion modelling. *Int Soil Water Conserve Res*. 2019;7(3):203–225.

63. Igwe PU, Onuigbo AA, Chinedu OC, Ezeaku II, Muoneke MM. Soil erosion: a review of models and applications. *Int J Adv Engin Res Sci*. 2017;4(12):237341.

64. Pampalone V, Ferro V. Estimating soil loss of given return period by USLE-M-type models. *Hydrol Process*. 2020;34(11):2324–2336.

8

Mapping the NDVI and monitoring of its changes using Google Earth Engine and Sentinel-2 images

Mahdis Amiri[a] and Hamid Reza Pourghasemi[b]

[a]Department of Watershed and Arid Zone Management, Gorgan University of Agricultural Sciences and Natural Resources, Gorgan, Iran [b]Department of Natural Resources and Environmental Engineering, College of Agriculture, Shiraz University, Shiraz, Iran

1 Introduction

Vegetation changes over time are important because they are widely used in many studies, including the management of natural resources, water resources, agriculture activities, and drought. In recent years, remote sensing has been one of the most important sources of spatial information gathering because of its benefits such as wide coverage of the study area, ability to capture data in a regular sequence, a digital format suitable for computer processing, and cheaper than field survey techniques. It has always been one of the most important sources of spatial information gathering and has attracted the attention of many experts in the field of natural resource monitoring.[1] In general, plant cover changes over time due to various causes, including natural or human factors that affect the condition and performance of the ecosystem. Therefore, it is essential to detect, predict, and manage such changes in an ecosystem. One of the most important problems in the study of vegetation changes is the lack of accurate information from the past. Satellite imagery and remote sensing (RS) technology provide an opportunity to rely on the gathered information to improve environmental management. Remote sensing is a useful technology that can be used to obtain information layers from soil and vegetation cover.[2] Features such as providing a vast and integrated view of an area, repeatability of the land cover, easy access to information, high accuracy of the information obtained, and time-saving are features that make a preference for such information to evaluate vegetation cover and control of its changes over other methods.[3] Accordingly, many researchers have used remote sensing data to study vegetation cover and have stated that this technique is suitable for such studies. On the other hand, The NDVI has been used as an appropriate vegetation indicator for studying the effects of climate change since the early 80s.[4] The NDVI series can be used to detect changes in health and vegetation growth, especially in semi-arid regions.[5] On the other hand, the vegetation index is based on the fact that chlorophyll in the plant structure can absorb red light and the leaf mesophilic layer, reflecting near-infrared light.[6] The NDVI time series can be examined in two ways: annual or seasonal parameters, their long-term trend, and sudden changes between successive years.[7]

Owing to the importance of vegetation cover and its various applications, several methods have been developed. In some studies, a close relationship was observed between monthly vegetation changes and climate change. However, this relationship is weaker for vegetation data with shorter durations, such as weekly.[8,9] In recent decades, some studies have suggested that NDVI has often been used to measure plant biomass and production by measuring nitrogen, leaf area, and sunlight intake.[10–12] In addition, temporal evaluation of vegetation processes due to their simple performance and high importance due to the comparative issue during different times in land management and agricultural management programs is widely used.[11,13]

Focus on dynamic vegetation parameters such as seasonal assessment, gradual trends, and sudden changes in various studies can be mentioned.[7,14–16] In a series of recent studies, the importance and efficiency of NDVI have been proven and in management measures, the use and spatial and temporal study of this index has been strongly recommended.[6,17–19]

So, the current research tried to use the GEE-Google Earth Engine Platform. The GEE is a cloud computing platform designed to reserve and process large amounts of data (based on byte scale) for future analyses and decision-making.[20] The GEE provides a suitable environment for interactive data and algorithm development.

Therefore, the purpose of this study was to evaluate the trend of vegetation changes using the NDVI index over 14 months from September 2017 to October 2018 and its mapping using Sentinel-2 satellite images and the GEE platform. By examining the temporal and spatial trends of vegetation, the extent of land degradation and vegetation can be assessed. It is also necessary to understand the causes of destruction and loss of vegetation from climatic, soil, and other conditions. Therefore, due to the time and cost of field visits, it is recommended to use remote sensing methods and satellite images with high accuracy.

2 Doroudzan Dam Watershed

The study area is located in the southwest of Iran and also in the northwest of Fars Province between the geographical longitudes of 51° 40′ to 52° 55′ east and latitudes of 30° 7′ to 30° 55′ north. The Doroudzan Dam Watershed is part of the Kor River Basin, with a total area of 4522 km^2, which is part of the Maharlou-Bakhtegan large watershed. The highest elevation of the Doroudzan Dam Watershed is 3749 m from the mean sea level and is located northwest of the watershed, whereas the average elevation is about 2167 m. The yearly average rainfall in this watershed/catchment area was 587 mm. The total volume and storage of dam reservoirs were 993 and 133, respectively.[21] The position of the watershed in the Fars Province and Iran is shown in Fig. 1.

3 Methodology

The research methodology of this research was conducted in four stages (Fig. 2):

(1) Selecting the area of interest (AOI) in the GEE,
(2) Encoding NDVI using Sentinel-2 satellite imagery,
(3) Transfer the final NDVI maps to ArcGIS 10.6.1 and convert them to the ".shp" file,
(4) Preparing NDVI maps for 14 months between 2017 and 2018.

FIG. 1 Location of Doroudzan Dam Watershed in Fars Province, Iran.

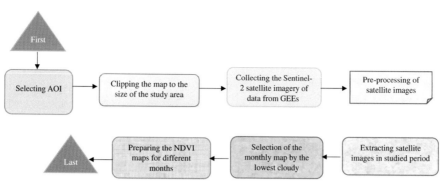

FIG. 2 Flowchart of the used methodology in the Doroudzan Dam Watershed.

3.1 NDVI

Vegetation indices are widely used as criteria for analyzing land cover changes and monitoring trends.[19] This factor is widely used to identify areas with unhealthy vegetation and vegetation-free areas.[22] The numerical value of the NDVI varies from −1 to 1.[14] Positive numerical values are related to dense vegetation and zero and near those values are for non-vegetated areas and wet and water locations have close to −1 digits. This index was calculated using Eq. (1)[23]:

$$NDVI = B_1 - R/B_1 + R \tag{1}$$

where in $B_1 = $ NIR, $R = $ RED, NIR = near-infrared band, and RED = red band.

4 Results

In the present study, 14 NDVI maps were prepared using encoding in the GEE in the Doroudzan Dam watershed (Fig. 3). The NDVI is divided into 5 classes as follows (Fig. 3, Table 1): smaller or equal to zero ($\leq$ 0) "water, snow and ice," 0–0.05 "bare soil," 0.05–0.1 "sparse vegetation," 0.1–0.5 "compact vegetation" and >0.5 "forest cover".[24,25] The Vegetation index is one of the most applicable vegetation indices, and its efficiency has been reported by many researchers in many studies, ranging from −1 to +1.[26–28]

The first picture (September 2017) shows that the southwest and central parts of the DDW have NDVI values smaller or equal to zero, that is, there is water in the aforementioned areas, and most regions, there is dense vegetation (NDVI = 0.1–0.5). Even in some parts of the southwest and southeast, there is forest cover (NDVI >0.5). The second month (October 2017) demonstrates that most parts of the southeast, northeast, and center have low-density vegetation (NDVI = 0.05–0.1). in most other areas, compact vegetation is visible (NDVI = 0.1–0.5). The third month (November) of 2017 showed that bare soil (<0) was found in the southwest and most parts of the southeast, central, and northeast, there was sparse vegetation (NDVI = 0.05–0.1) and southwest compact vegetation (NDVI = 0.1–0.5). The latest image from the first year (December 2017) indicates that in the western, central, and northwestern parts, bare soil is more abundant than other areas, and there is a water spot in the southwest of the study area (DDW). In the same year and month, sparse vegetation (0.05–0.1) was observed in most of the remaining parts of the Doroudzan Dam Watershed. On the other hand, in the first month of the crop year (January 2018), there is a low-density NDVI (0.05–0.1) and high-density vegetation (0.1–0.5) in all parts of the Doroudzan Dam Watershed, and in January 2018, bare soil (NDVI = 0–0.05) was found in the west, southwest, and northwest. In February 2018, in the eastern parts, the vegetation is low-density (0.05–0.1) and in the other parts is high-density vegetation (0.1–0.5), in the SW, SE, and S The spots are bare soil (0–0.05) and also had the highest and lowest NDVI values of 0.78 and −0.07, respectively. In the following month's image, the year 2018 is similar to the previous images: there is water in the southwest, and the southeast and southwest there are low density (NDVI = 0.05–0.1) and high density (NDVI = 0.1–0.5), respectively; in April, May, and June, the vegetation density was abundant for three months (NDVI = 0.1–0.5), and bare soil in the southwest of the case study on April and May months and on June month, there is seen water; in general, forest cover (>0.5) was densely observed. Sparse vegetation appears as a patch in the center of the study area in three months (April, May, and June). Also, in July 2018 as in previous months, there is overcrowded vegetation is found throughout the DDW, in the southwest and eastern part of the Doroudzen Dam, sparse vegetation and forest cover were observed.

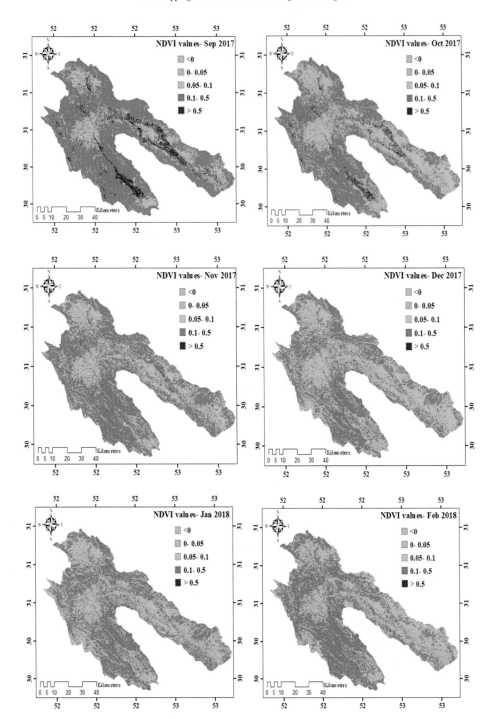

FIG. 3   NDVI mapping during different months in the first and second years.

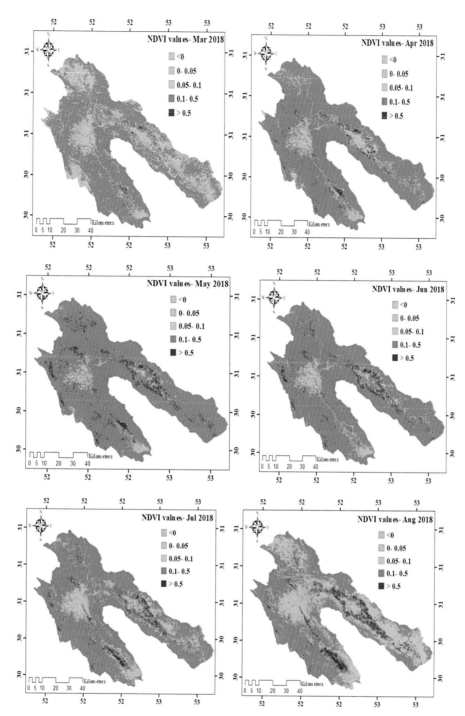

FIG. 3, CONT'D

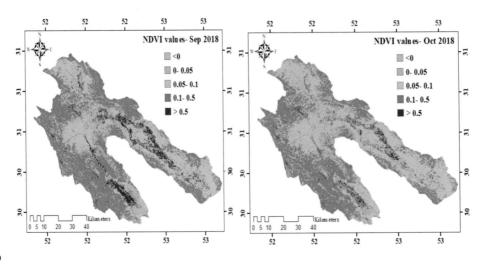

FIG. 3, CONT'D

In August and September, in both months, the vegetation is similar, for example, water cover is in central and southwest, low-density vegetation is in center, northeast, and southeast, forest cover is in southwest and east in August and September. In addition, dense vegetation was observed in the study area. On the other hand, in October, bare soil spots were evident in the central and southwestern regions, and in the last month of the crop year, vegetation deficiency spots were observed to be proportional to the whole area of the DDW. Also, the Percentage of vegetation changes for different months is given in Table 1.

TABLE 1 Percentage of vegetation changes for different months.

Month-Year	Percentage of vegetation in each class
Sep 2017	1; 0.7833 2; 0.5300 3; 15.2964 4; 79.1436 5; 4.2465
Oct 2017	1; 0.8006 2; 0.8933 3; 31.3526 4; 65.6913 5; 1.2619
Nov 2017	1; 0.0488 2; 1.6445 3; 26.3818 4; 71.7163 5; 0.2085
Dec 2017	1; 0.7688 2; 4.2103 3; 36.7856 4; 58.2082 5; 0.0269
Jan 2018	1; 0.3222 2; 4.4109 3; 32.8400 4; 62.4205 5; 0.0061
Feb 2018	1; 0.0845 2; 4.9637 3; 31.3180 4; 63.6095 5; 0.0241

TABLE 1 Percentage of vegetation changes for different months—cont'd

Month-Year	Percentage of vegetation in each class
Mar 2018	1; 0.0125 2; 1.3818 3; 8.5423 4; 88.6713 5; 1.3919
Apr 2018	1; 0.0125 2; 1.3818 3; 8.5423 4; 88.6713 5; 1.3919
May 2018	1; 0.1457 2; 0.7405 3; 3.4651 4; 92.1623 5; 3.4861
Jun 2018	1; 0.7923 2; 0.5431 3; 5.8916 4; 90.6096 5; 2.1631
July 2018	1; 0.7563 2; 0.7535 3; 12.1487 4; 84.4175 5; 1.9238
Aug 2018	1; 0.4266 2; 0.9733 3; 29.3957 4; 65.9464 5; 3.2577
Sep 2018	1; 0.6924 2; 0.8992 3; 34.7394 4; 60.2029 5; 3.4657
Oct 2018	1; 0.0002 2; 1.9053 3; 43.3024 4; 53.7818 5; 1.0101

5 Discussion

In general, knowing the temporal and spatial changes in the vegetation index is necessary and important for the management of ecosystems, especially in ecosystems sensitive to degradation.[29] NDVI is a vegetation index that has many applications in the study of monitoring vegetation changes over time. It should be noted that this indicator requires continuous and complete-time data over a given time interval. The main objective of the present study was to investigate the temporal and spatial vegetation indices during the 2017 and 2018 crop years in the Doroudzan Dam Watershed (DDW). In the first months (September, October, November, and December), the 2017 crop year was moderate at all levels of vegetation and occurred in the southwest, west, east, and center of the vegetation, but very low, which may be due to disturbances in biophysical and meteorological factors,[30] and inappropriate vegetation may also be caused by hazards such as drought, floods, and fires.[31,32] On the other hand, some areas with vegetation indicate good rainfall and high humidity during the study season.[33] In January, February and April, there was minimal vegetation cover at all levels of the area. In other months, including March, June, July, August, and September, the vegetation index was averaged over almost all the Doroudzan watershed, and the minimum and maximum vegetation cover were −0.34, 0.83, −0.14, 0.83, −0.39, 0.83, −0.31, 0.83, −0.48, and 0.83, respectively. In October 2018, a sudden change in Fig. 3 was visible, with a marked reduction in vegetation cover. It is sometimes observed in studies that a sudden change is seen due to a lack of vegetation; for example, there is usually a natural leaf fall from July.[30,34] In the present study, this issue is well understood. In general, the strong dependence of vegetation on the season and crop and climates, such as rainfall, temperature, and humidity, has been proven in previous studies.[35,36] In this study, we examined the seasonal variability of green vegetation density using NDVI time-series data, and several case studies have shown that trees and grasses are found in different locations, such as the plains of South Africa, from the beginning to the end of the growing season, and show heterogeneous phenological products.[14,34,37]

Currently, gaining knowledge and knowledge about vegetation and its health plays an important role in soil management. On the other hand, vegetation cover in agricultural and rangelands is associated with applied corrected crops directly in each region, water and soil management. Accordingly, the use of satellite imagery in this regard provides accurate information on the application of operations and management. Vegetation indices are one of the most commonly used examples of banding calculations used to calculate vegetation percentages, study vegetation types, and vegetation status of an area during different periods. GEE is a cloud computing/calculation operative apparatus deliberated/ designed to stock/store and procedure/process big/large datasets "petabyte-scale" for final analysis and decision making.[20] It is noteworthy that the following advantages, such as the availability and free availability of "Landsat Series in 2008," Google archives all data sets and can be connected to the cloud computing engine for open-source use. Current data archiving includes other satellites as well as geographic information systems (GIS), social, demographic, climate, digital elevation models, and aerial data layers.[38] The results of this study showed that October was the year with the highest vegetation deficiency in the year, the average vegetation scattered had seen in the other months, and a bold blue spot indicating a lack of vegetation in the southwest in all months. However, as a final point, it should be noted that the vegetation of an area reflects the climate, soil type, socio-economic status, and even the history of rangeland use.[31] Management factors have a great impact on the sustainability of vegetation in the region and generally show the relationship between humans, the environment, and plants within the region, as well as an important factor in energy conversion, especially radiant energy in biomass; therefore, human livelihood depends on it; more studies in this area are recommended by comparing more vegetation indices and wider and newer satellite images over many years. Because prevention and prevention of degradation of pastures and vegetation should be necessary by managers and planners to carry out accurate management strategies.

6 Conclusion

Iran is a land with a hot and dry climate in the middle latitude. The most important environmental factor is the lack of water resources, which strongly affects the distribution of vegetation in the country. In other words, vegetation changes are a function of the temporal distribution of temperature and rainfall. Since Iran is poor in terms of vegetation, this lack of cover has caused the land to lead to soil erosion, abundant deserts or desertification, and ultimately the destruction of water resources, especially groundwater. Therefore, efforts to maintain vegetation and prevent harmful consequences are necessary and inevitable. To plan for the conservation, expansion, and even restoration of vegetation, it is essential to identify the behavior of its changes in the country. Therefore, a comprehensive, low-cost, and fast method for studying vegetation is the use of satellite vegetation data. Therefore, in the present study, an attempt was made to use the monthly vegetation data for one crop year, that is, from September 2017 to October 2018, to monitor the temporal and spatial changes of the NDVI index using Sentinel 2 images in the Google Earth Engine platform. The results of the trend of different months are explained in detail in the previous section and the distribution of vegetation in accordance with the seasons in the (DDW) is known clearly. However, by examining all the months, each of the vegetation categories in terms of density, evaluating the trend of vegetation reduction in the study area and identifying low-coverage areas, preventing their further destruction and in the next stage, their restoration is possible. These studies are necessary because the destruction of vegetation is the destruction of ecosystems.

References

1. Lu D, Mausel P, Brondizio E, Moran E. Change detection techniques. *Int J Remote Sens.* 2004;25(12):2365–2407. https://doi.org/10.1080/0143116031000139863.

2. French AN, Hunsaker DJ, Sanchez CA, Saber M, Gonzalez JR, Anderson R. Satellite-based NDVI crop coefficients and evapotranspiration with eddy covariance validation for multiple durum wheat fields in the US Southwest. *Agric Water Manag.* 2020;239:106266. https://doi.org/10.1016/j.agwat.2020.106266.

3. Nse OU, Okolie CJ, Nse VO. Dynamics of land cover, land surface temperature and NDVI in Uyo City, Nigeria. *Sci Afr.* 2020;10. https://doi.org/10.1016/j.sciaf.2020.e00599, e00599.

4. An L, Che H, Xue M, et al. Temporal and spatial variations in sand and dust storm events in East Asia from 2007 to 2016: relationships with surface conditions and climate change. *Sci Total Environ.* 2018;633:452–462.

5. Measho S, Chen B, Trisurat Y, et al. Spatio temporal analysis of vegetation dynamics as a response to climate variability and drought patterns in the semiarid region. *Eritrea Remote Sens.* 2019;11:724.

6. Ghebrezgabher MG, Yang T, Yang X, Eyassu Sereke T. Assessment of NDVI variations in responses to climate change in the Horn of Africa. *Egypt J Remote Sens Space Sci.* 2020;23(3):249–261. https://doi.org/10.1016/j.ejrs.2020.08.003.

7. Wingate VR, Phinn SR, Kuhn N. Mapping precipitation-corrected NDVI trends across Namibia. *Sci Total Environ.* 2019;684:96–112. https://doi.org/10.1016/j.scitotenv.2019.05.158.

8. Hou G, Zhang H, Wang Y. Vegetation dynamics and its relationship with climatic factors in the Changbai Mountain Natural Reserve. *J Mountain Sci.* 2011;8(6):865–875. https://doi.org/10.1007/s11629-011-2206-4.

9. Piao J, Chen W, Zhang Q, Hu P. Comparison of moisture transport between Siberia and Northeast Asia on annual and interannual time scales. *J Climate.* 2018;31(18):7645–7660. https://doi.org/10.1175/jcli-d-17-0763.1.

10. Esquerdo JCDM, Zullo Júnior J, Antunes JFG. Use of NDVI/AVHRR time-series profiles for soybean crop monitoring in Brazil. *Int J Remote Sens.* 2011;32(13):3711–3727. https://doi.org/10.1080/01431161003764112.

11. Meng J, Du X, Wu B. Generation of high spatial and temporal resolution NDVI and its application in crop biomass estimation. *Int J Digit Earth.* 2013;6(3):203–218. https://doi.org/10.1080/17538947.2011.623189.

12. Zhou Z, Jabloun M, Plauborg F, Andersen MN. Using ground-based spectral reflectance sensors and photography to estimate shoot N concentration and dry matter of potato. *Comput Electron Agric.* 2018;144:154–163. https://doi.org/10.1016/j.compag.2017.12.005.

13. Wu B, Meng J, Li Q, Yan N, Du X, Zhang M. Remote sensing-based global crop monitoring: experiences with China's CropWatch system. *Int J Digit Earth.* 2014;7(2):113–137. https://doi.org/10.1080/17538947.2013.821185.

14. Cho MA, Ramoelo A. Optimal dates for assessing long-term changes in tree-cover in the semi-arid biomes of South Africa using MODIS NDVI time series (2001–2018). *Int J Appl Earth Observ Geoinf.* 2019;81:27–36. https://doi.org/10.1016/j.jag.2019.05.014.

15. Gholamnia M, Khandan R, Bonafoni S, Sadeghi A. Spatiotemporal analysis of MODIS NDVI in the semi-arid region of Kurdistan (Iran). *Remote Sens (Basel).* 2019;11(14):1723. Retrieved from https://www.mdpi.com/2072-4292/11/14/1723.

16. Muradyan V, Tepanosyan G, Asmaryan S, Saghatelyan A, Dell'Acqua F. Relationships between NDVI and climatic factors in mountain ecosystems: a case study of Armenia. *Remote Sens Appl Soc Environ.* 2019;14:158–169. https://doi.org/10.1016/j.rsase.2019.03.004.

17. He P, Xu L, Liu Z, Jing Y, Zhu W. Dynamics of NDVI and its influencing factors in the Chinese Loess Plateau during 2002–2018. *Region Sustain.* 2021;2(1):36–46. https://doi.org/10.1016/j.regsus.2021.01.002.

18. Sotille ME, Bremer UF, Vieira G, Velho LF, Petsch C, Simões JC. Evaluation of UAV and satellite-derived NDVI to map maritime Antarctic vegetation. *Appl Geogr.* 2020;125:102322. https://doi.org/10.1016/j.apgeog.2020.102322.

19. Zhe M, Zhang X. Time-lag effects of NDVI responses to climate change in the Yamzhog Yumco Basin, South Tibet. *Ecol Indic.* 2021;124:107431. https://doi.org/10.1016/j.ecolind.2021.107431.

20. Kumar L, Mutanga O. Google earth engine applications since inception: usage, trends, and potential. *Remote Sens (Basel).* 2018;10:1509.

21. Goodarzi E, Shui LT, Ziaei M. Risk and uncertainty analysis for dam overtopping – case study: the Doroudzan Dam, Iran. *J Hydro-Environ Res.* 2014;8(1):50–61. https://doi.org/10.1016/j.jher.2013.02.001.

22. Kong J, Ryu Y, Huang Y, et al. Evaluation of four image fusion NDVI products against in-situ spectral-measurements over a heterogeneous rice paddy landscape. *Agric For Meteorol.* 2021;297:108255. https://doi.org/10.1016/j.agrformet.2020.108255.

23. Rouse JW, Haas RH, Schell JA, Deering DW. Monitoring vegetation systems in the great plains with ERTS. In: *3rd ERTS Symposium, NASA SP-351 I*; 1973:309–317.

24. Carlson TN, Ripley DA. On the relation between NDVI, fractional vegetation cover, and leaf area index. *Remote Sens Environ.* 1997;62(3):241–252. https://doi.org/10.1016/S0034-4257(97)00104-1.

25. Liu L, Wang Y, Wang Z, et al. Elevation-dependent decline in vegetation greening rate driven by increasing dryness based on three satellite NDVI datasets on the Tibetan Plateau. *Ecol Indic.* 2019;107:105569. https://doi.org/10.1016/j.ecolind.2019.105569.

26. Han J-C, Huang Y, Zhang H, Wu X. Characterization of elevation and land cover dependent trends of NDVI variations in the Hexi region, Northwest China. *J Environ Manage.* 2019;232:1037–1048. https://doi.org/10.1016/j.jenvman.2018.11.069.

27. Peng W, Kuang T, Tao S. Quantifying influences of natural factors on vegetation NDVI changes based on geographical detector in Sichuan, western China. *J Clean Prod.* 2019;233:353–367. https://doi.org/10.1016/j.jclepro.2019.05.355.

28. Zhang Y, Ling F, Foody GM, et al. Mapping annual forest cover by fusing PALSAR/PALSAR-2 and MODIS NDVI during 2007–2016. *Remote Sens Environ.* 2019;224:74–91. https://doi.org/10.1016/j.rse.2019.01.038.

29. Gaughan AE, Holdo RM, Anderson TM. Using short-term MODIS time-series to quantify tree cover in a highly heterogeneous African savanna. *Int J Remote Sens.* 2013;34(19):6865–6882. https://doi.org/10.1080/01431161.2013.810352.

30. Fang X, Zhu Q, Ren L, Chen H, Wang K, Peng C. Large-scale detection of vegetation dynamics and their potential drivers using MODIS images and BFAST: a case study in Quebec, Canada. *Remote Sens Environ.* 2018;206:391–402. https://doi.org/10.1016/j.rse.2017.11.017.

31. Verbesselt J, Zeileis A, Herold M. Near real-time disturbance detection using satellite image time series. *Remote Sens Environ.* 2012;123:98–108.

32. Watts LM, Laffan SW. Effectiveness of the BFAST algorithm for detecting vegetation response patterns in a semi-arid region. *Remote Sens Environ.* 2014;154:234–245.

33. Chen L, Michishita R, Xu B. Abrupt spatiotemporal land and water changes and their potential drivers in Poyang Lake, 2000–2012. *ISPRS J Photogramm Remote Sens.* 2014;98:85–93. https://doi.org/10.1016/j.isprsjprs.2014.09.014.

34. Geng L, Che T, Wang X, Wang H. Detecting spatiotemporal changes in vegetation with the BFAST model in the Qilian Mountain region during 2000–2017. *Remote Sens (Basel).* 2019;11(2):103. Retrieved from https://www.mdpi.com/2072-4292/11/2/103.

35. Janecke BB, Smit GN. Phenology of woody plants in riverine thicket and its impact on browse availability to game species. *Afr J Range Forage Sci.* 2011;28(3):139–148. https://doi.org/10.2989/10220119.2011.642075.

36. Nabavi SO, Haimberger L, Samimi C. Climatology of dust distribution over West Asia from homogenized remote sensing data. *Aeolian Res.* 2016;21:93–107. https://doi.org/10.1016/j.aeolia.2016.04.002.

37. Naidoo L, Mathieu R, Main R, Wessels K, Asner GP. L-band synthetic aperture radar imagery performs better than optical datasets at retrieving woody fractional cover in deciduous, dry savannahs. *Int J Appl Earth Observ Geoinf.* 2016;52:54–64. https://doi.org/10.1016/j.jag.2016.05.006.

38. Mutanga O, Kumar L. Google Earth engine applications. *Remote Sens.* 2019;11(5):591. https://doi.org/10.3390/rs11050591.

9

Spatiotemporal urban sprawl and land resource assessment using Google Earth Engine platform in Lahore district, Pakistan

Adeel Ahmad[a], Hammad Gilani[b], Safdar Ali Shirazi[a], Hamid Reza Pourghasemi[c], and Ifrah Shaukat[d]

[a]Department of Geography, University of the Punjab, Lahore, Pakistan [b]Department of Space Science, Institute of Space Technology, Islamabad, Pakistan [c]Department of Natural Resources and Environmental Engineering, College of Agriculture, Shiraz University, Shiraz, Iran [d]Geography of Environmental Resources and Human Security, United Nations University, Bonn, Germany

1 Introduction

Remote sensing is the science of acquiring information on features, processes, and systems of the earth's surface using satellite and airborne sensors that are not in actual physical contact with the areas under investigation. It is a technology and art for recording, analyzing, and interpreting the connected events of acquired information for under-standing and decision-making related to our earth and environment.[1] Furthermore, remote sensing has a wide range of applications in various domains, including terrestrial and aquatic ecosystems and especially in urban studies. The importance of remote sensing can be assessed through its wide applications in "urban geography," a science that deals with the impact assessment of human factors (economy and population) and physical factors (infrastructure and environment) over urban areas.[2] The urban growth trend is a very important indicator of urban geography and is regarded as the concentration of the human population in a specific area, in terms of demography and spatial extent over a specific period.[3] Global urbanization is one of the most anticipated challenges our world faces and will face more drastically shortly. This challenge could be coped with knowledge and timely strategy; however, knowledge on the spatial distribution and quantification of urban areas is still lacking, especially in the global north. From 1950 to 2008, the urban population has increased from half to more in number in comparison to the rural population of the world.[4] This rapid increase in the urban population has led to drastic temporal changes in land cover, which are a result of the uncontrolled growth of urban areas.

Remote sensing provides a suitable platform for monitoring these changes in land covers over time.[5] To quantify various parameters for urban sprawl and land resource assessment, some specific properties of satellite images are considered, including spatial, radiometric, spectral, and temporal resolutions.[1] Analyzing urban growth through remote sensing science depends precisely on image processing algorithms and improved satellite imagery.[6] Improved satellite imagery refers to better spatial resolution, more spectral channels a satellite sensor can capture, improved depth of shades in terms of radiometric resolution, and better temporal resolution for timely access to information.[7–9]

Google Earth Engine (GEE) is an advanced and sophisticated platform that offers processing and classification of multi-temporal and multi-resolution satellite image datasets to be classified according to the user's choice over a selected area (large or small scale).[10] GEE is an online web-based geospatial analysis platform that can be used to perform spatial analysis on an archive of datasets available free of charge for research purposes.[11] Landsat satellite images, which are medium-resolution images, provide an opportunity to analyze Earth surface trends from 1972 to to-date and was made publicly available via the internet in 2008.[12] Several studies have utilized Landsat data to quantify and map urban areas using various classification approaches, including pixel-based, sub-pixel-based, and object-based analysis.[10,13–17]

The process of rapid urbanization leads to an increase in temperature in both terrestrial and marine ecosystems and has great implications for the urban climate.[18] Remote sensing sensors have been designed to monitor electromagnetic radiation at various wavelengths, including thermal infrared. Satellite sensors ranging from the Advanced Very-High-Resolution Radiometer (AVHRR) for more precise thermal measurements to MODerate resolution Imaging Spectro-radiometer (MODIS) for regional-level studies related to spatial and temporal thermal variations.[6] Landsat satellite series (notably Landsat 4–5 TM, Landsat 7 ETM+ and Landsat 8 OLI) have specific bands to monitor the thermal radiation of the earth's surface. It is very important to study the spatiotemporal variations in temperature and vegetation in an urban area for administration/ local governments to devise policies for urban areas.[18]

1.1 Literature review

Several studies have discussed urban, transportation, and environmental issues in Lahore. Spatiotemporal assessment of land resources in mega metropolitans, such as Lahore, provides important information for decision-makers and policymakers. The first study, which used remotely sensed data to assess urban sprawl in Lahore, was conducted by Almas et al.[19] for 1973, 1992, and 2001. They used supervised image classification to assess land-cover changes using Landsat data. This was followed by Shirazi,[20] who assessed land use and land cover (LULC) changes in Lahore for 1992, 2001, and 2009. In this study, the data were accessed from a freely available online global land cover facility (GLCF), and ground control points (GCPs) were used to classify the Landsat images through a visual interpretation technique. Five LULC classes were mapped in this study, including built-up areas, vegetative cover, open areas, water bodies, and mixed land use. Another similar study by Shirazi and Kazmi[21] reported LULC change assessment of Lahore district, on union council (UC) level administrative units, using Landsat data. The maximum likelihood image classification technique was used for classifying the Landsat data, while Quickbird and GeoEye, high-resolution images, taken from Google Earth, were used to extract GCPs of required LULC classes including vegetation/ cultivated land, built-up/ urban area, open area/land, water and mixed from 1973 to 2009. This study reported a substantial decrease in the vegetative cover of the Lahore district. Shirazi and Kazmi[22] conducted another study focusing on urban development in the Lahore district (1973–2009). They identified potential UCs in the Lahore district for future urban development. For almost the same time, i.e., 1972–2009, another study was conducted by Riaz et al.[23] to model the land-use patterns of Lahore. They used Landsat data for 1972, 1981, 1992, 2000, and 2009 after stacking the layers of Landsat's multispectral scanner (MSS) and thematic mapper (TM) sensors. The Parallelepiped algorithm with the maximum likelihood classification method was used to classify five land cover types, including urban land use, agricultural land, forest land, water bodies, and others. Saleem[24] used remote sensing and geographic information system (GIS) techniques to detect urbanization in Lahore from 1992 to 2010. An unsupervised image classification technique, along with a band differencing method on Landsat data (red and green band differencing), was used to detect urban changes. Saleem and Saleem[25] used calibrated Landsat data to evaluate the spatiotemporal vegetation cover of Lahore from 1994 to 2014. This study utilized Landsat images from 1994, 2003, and 2014 to extract vegetation cover using the normalized difference vegetation index (NDVI). Similarly, a socio-environmental study of Lahore was conducted by Shirazi and Kazmi[26] who focused on analyzing loss in tree and vegetation cover using Landsat data. They identified population and urbanization as the major driving factors of tree cover loss. Their study was supported by public perception and responses through a questionnaire survey conducted in 2009. Likewise, the work done by Asad et al.[27] introduced a comparison of Normalized Difference Built-up Index (NDBI), supervised image classification, and Object-Based Image Analysis (OBIA) to map the impervious surface (land cover) of Lahore. In their research, only visual accuracy assessment was performed without using a concrete rationale approach to discuss accuracy statistically. Farhat et al.[28] reported the land-use dynamics of Lahore from 2000 to 2015 using supervised image classification. The main objective of this work was to identify urban sprawl in the study area and correlate it with demographic trends. Shakrullah et al.[29] used geospatial techniques to spatially compare land cover changes in Lahore and New Delhi (India). They used Landsat images from 1972, 1980, 1993, 2000, 2010, and 2015 to identify these changes. For this study, Landsat MSS, TM, ETM+ and OLI/ TIRS sensor data were processed to calculate the NDVI, normalized difference water index (NDWI) and soil adjusted vegetation index (SAVI). Another study conducted by Akbar et al.[30] used Landsat-5 TM (for years 1988 and 2001) and Landsat-8 OLI (for 2016) to predict existing and future LULC in the Lahore district. They used a Markov model to predict urbanization until 2040 in the study area. NDVI, with precise class value ranges, was used to extract six land cover classes, including water, built-up land, barren land, shrub and grassland, sparse vegetation, and dense vegetation. Imran and Mehmood[31] analyzed the spatial variability of Land Surface Temperature (LST) caused by LULC changes in Lahore from 1996 to 2016. Landsat data from the TM, ETM+, and OLI sensors were used to calculate the NDVI and NDBI. The MultiLayer Perceptron-Markov Chain Analysis (MLP-MCA) model was used to predict LULC changes until 2035. They correlated LULC changes (current and future) with the future local climate of the study area. The latest study in this regard by Mumtaz et al.[32] focused on monitoring spatiotemporal LULC changes in the cities of Lahore and Peshawar. This study used the CA-Markov model to predict LULC changes for the years 2023 and 2028 based on the results of Landsat image classification for both study areas from 1998 to 2018. The study focused on the impact of these LULC (current and future) changes on LST. Supervised image classification was used to extract four LULC classes, including vegetation, water bodies, built-up land, and barren land.

In addition to these studies, which utilized remotely sensed data to study the spatial behavior of Lahore, several other studies, focused on different themes, were also conducted. First, among these studies, Bajwa et al.[33] identified demand and supply gaps regarding housing requirements for people living in Lahore. A shortage of housing compels people to live in substandard housing with poor management. Despite the introduction of new housing schemes by the public sector, the requirements of the people did not meet because of the high prices that people cannot afford. In the peri-urban area of Lahore, agriculture has been converted to urban areas over time because of this mobility. Spatiotemporal climatic variations in Lahore were analyzed by Qureshi et al.,[34] who examined temperature variations from 1950 to 2010. Using Landsat and ASTER satellite images, they observed an increase in urban areas from 1973 to 2006. Ahmad et al.[35] identified and discussed the role of urban development authorities in Lahore and their failure to manage the city's urban sprawl and maintain a green environment. The main findings of their research depict issues such as delaying the preparation of city plans by the Lahore Development Authority (LDA). The following research conducted by Shah and Abbas[36] discussed the socio-economic situation of people living in Lahore City. They used a questionnaire survey to identify why people migrated to Lahore. Likewise, various studies for the assessment of different kinds of pollution were conducted by.[37–41] Butt[39] discussed light pollution using remote-sensing techniques. His study concluded that the major source of light pollution in Lahore is artificial nightlights. An increase in the amount of sulfur dioxide (SO_2) due to transportation of volcanic gases from the west and high concentration of nitrogen dioxide (NO_2) in the built environment of Lahore were reported by Ul-Haq et al.[41] and Andleeb et al.[38] respectively.

1.2 Rationale of the study

Based on the literature review we analyzed, although several studies have been conducted on land cover dynamics changes in Lahore, none of the studies conducted in Lahore used landscape metrics or other methods to assess the type of urban agglomeration. For the landscape analysis, a land cover map is essential, and already developed and published land cover maps were not available for this study. This chapter aims to analyze, quantify, and map the land cover of Lahore using temporal Landsat imagery through a machine-learning classifier in GEE. Spatiotemporal temperature variations were assessed using the thermal bands derived from Landsat imagery. The shape, structure, and complexity of urban patches in the study area were also analyzed using landscape metrics that were not reported in the literature review.

2 Study area

Lahore, located between 31.20°N to 31.71°N latitude and 74.00°E to 74.65°E longitude, is a district in the Punjab Province of Pakistan. It is located on the left bank of the Ravi River and is the second-largest metropolitan city in the country (Fig. 1). This city is one of the oldest cities in the world with written history, dating back to approximately 1000 years. For a longer period, Lahore was regarded as the city of the gardens due to its affiliation with the royal dynasty and Mughal emperors, but these gardens have long gone.[42] The Hindus, Sikhs, Mughals, and British empires ruled this ancient city. The city of Lahore was founded about 2000 years ago by "*Loh*" who was the son of Rama Chandra.[42] Lahore city has experienced intense spatial urban changes and rapid urban sprawl over the last few decades.[43] The city has faced a paradigm shift in terms of urban growth and its settings.[42]

The Physical geography and climate conditions of the Lahore district make its soil suitable for various crops, including cotton, sugarcane, and other vegetables, with some connection with urban land at its periphery.[23] The population of Lahore district increased from 6,318,745 persons in 1998[44] to 11,126,285 persons in 2017.[45]

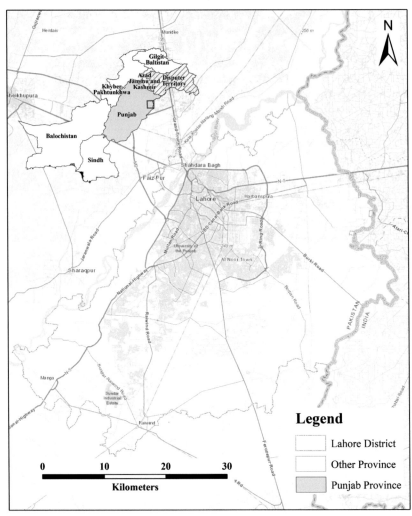

FIG. 1 Lahore district map.

3 Methodology

Fig. 2 presents the methodological framework for this study. The methodology was divided into three categories to assess the spatiotemporal patterns of land cover, LST, and landscape metrics.

The recorded information from satellite images is assigned as a digital number (DN) at the backend of each pixel of the grid. This digital number or information needs to be unlocked in various ways. Generally, two methodologies have been used to classify the data contained in satellite imagery: supervised and unsupervised image classification.[46] A wide range of satellite image analysis softwares are available for performing different types of image classification. GEE has been utilized to perform image classification, using the machine-learning technique of Lahore district for the years 1990, 2000, 2010, and 2017 using 30 m spatial resolution Landsat 5 TM and Landsat 8 OLI (Table 1). Machine-learning algorithm-based image classification may require, at a time, a higher number of training data (land cover samples). However, the accuracy achieved using machine-learning classifiers is usually higher than that of conventional image classification techniques.[47] Several machine learning classifiers are available, including classification and regression trees (CART), random forests (RF), artificial neural networks (ANNs), and support vector machines (SVMs).[47]

Machine-learning image classification by training the CART algorithm through training samples was used in the present study. CART is a binary decision tree that is based on "if-then" questions at every node of the binary tree.[10] Land cover mapping using the CART classifier yields very good results with accuracies around 75% or higher.[47] A total of five land cover classes were extracted from the satellite data, including greenery/agricultural land, water bodies,

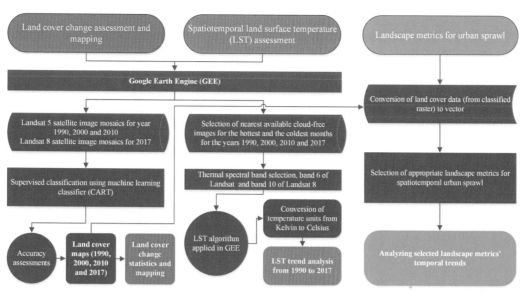

FIG. 2 Methodological framework of the current study.

urban areas, tree cover, and barren areas. Out of all training samples, 70% were used for the classification of the satellite image, while the remaining 30% were used for validation of the land cover product.

LST is usually referred to as the skin surface temperature of the earth's surface.[48] Using band 6 of Landsat 5 and band 10 of Landsat 8, LST was calculated using the GEE. Single and multiple Landsat images from a single month (coldest and hottest) were used to calculate the LST in GEE (Table 2). For multiple images in a single month, the average was used to calculate surface temperature. The Nearest available cloud-free images for the years 1990, 2000, 2010, and 2017 were taken for the coldest (December–January) and hottest (May–June) months. The temperature from the thermal bands was converted from kelvin (K) to Celsius (°C) by subtracting −273.15.

Landscape metrics are useful for the quantification of configurations and trends assessment of urban sprawl over time. Remote sensing data deals with grid data in the form of pixels or cells. Hence, it is quite challenging for raster data (pixel-based data) to effectively describe the changes occurring in the structures and shapes of urban areas in depth. On the other hand, vector data (point, line, and polygon) are very effective for analyzing and describing these changes in detail. For this purpose, landscape metrics, available as patch analyst extension in ArcGIS software as well as statistical software of Fragstats, have been used to quantify these changes in urban patches from 1990 to 2017.[49] There are several indicators available to extract for analyzing landscape trends over time, but the indicators presented in Table 3 were selected for the current study.

4 Results and discussion

4.1 Land cover change assessment and mapping

The overall classification accuracy for all years ranged from 84% to 96%. In parallel to the overall classification accuracies, producer and user accuracies were also assessed and reported (Table 4).

TABLE 1 Specifications of the datasets used for classification in GEE.

Year	Landsat datasets properties		
	Satellite/sensor	Date of acquisition	Path/row
1990	Landsat 5 (TM)	09 March 1990	148/38
2000	Landsat 5 (TM)	24 February 2000	149/38
2010	Landsat 5 (TM)	02 February 2010	149/38
2017	Landsat 8 (OLI)	26 March 2017	149/38

TABLE 2 Specifications of the datasets used for LST calculation in GEE.

Year	Landsat datasets properties		
	Satellite/sensor	Date of acquisition	Path/row
1991 (coldest month)	Landsat 5 (TM)	23 January 1991	148/38
		30 January 1991	
1989 (hottest month)	Landsat 5 (TM)	01 June 1989	149/38
1998 (coldest month)	Landsat 5 (TM)	17 January 1998	149/38
2000 (hottest month)	Landsat 5 (TM)	14 May 2000	149/38
		30 May 2000	
2009 (coldest month)	Landsat 5 (TM)	01 December 2009	149/38
		17 December 2009	
2010 (hottest month)	Landsat 5 (TM)	04 June 2010	148/38
		20 June 2010	
2017 (coldest month)	Landsat 8 (OLI)	14 January 2017	148/38
		30 January 2017	
2017 (hottest month)	Landsat 8 (OLI)	07 June 2017	148/38
		23 June 2017	

The producer's accuracy refers to the percentage of pixels in training samples of ground truth that are accurately identified in the classified map for the same relative classes and presents the prospect of the producer.[51] On the other hand, user accuracy refers to the percentage of classified pixels that accurately overlay the ground truth training samples for each land cover class and presents the user's prospect.[51] Both these accuracies were calculated against each land cover class, as shown in Table 4. The producer's accuracy for the urban area class ranged between 83% and 96%, showing high to very high accuracy. Similarly, the user's accuracy for this class ranged between 75% and 96%, which is also good to very high for all years. If the user's accuracy is higher than the producer's accuracy, then the land cover class would be under-classified and vice versa.[11] The Kappa value is a statistical index value that refers to the agreement level of validation of the results obtained.[52,53] Ranging from +1 (perfect agreement) to −1 (complete disagreement) and 0 (for no agreement above that was expected).[53] Kappa value was obtained for all classified images ranging from 0.77 (for 1990) to 0.94 (for 2000) showing a promising value for all classified images.

TABLE 3 Details of indicators used in landscape matrics.

Indicator	Acronym	Description
Area-weighted mean shape index	AWMSI	For assessing the merging of non-built-up areas into urban polygons[49]
Mean shape index	MSI	For measuring the geometric complexity of urban patches[49]
Mean perimeter-area ratio	MPAR	For measuring shape complexity in meters/hectares[49]
Area weighted mean patch fractal dimension	AWMPFD	For assessing the complexity of shapes of urban patches[50]
Total edge	TE	For measuring the total perimeter of all urban patches[49]
Edge density	ED	For calculating the complexity of urban patches[50]
Mean patch edge	MPE	For measuring the average amount of edge of individual urban patches[49]
Mean patch size	MPS	For measuring the ratio of patch number and the total area covered by urban patches[49]
Number of patches	NumP	For measuring disjoint individual urban patches[50]
Patch size coefficient of variation	PSCoV	For measuring the standardized spread of sizes of urban patches[49]
Patch size standard deviation	PSSD	For measuring the spread of patch sizes[49]

TABLE 4 Classification accuracies achieved after classification using CART classifier.

Land cover class	1990		2000		2010		2017	
	Producer's accuracy	User's accuracy	Producer's accuracy	User's accuracy	Producer's accuracy	User's accuracy	Producer's accuracy	User's accuracy
Greenery/ agricultural land	60%	100%	100%	100%	83%	100%	100%	75%
Waterbody	70%	100%	86%	100%	67%	100%	100%	83%
Urban area	83%	75%	89%	89%	96%	96%	96%	100%
Tree cover	94%	81%	100%	83%	87%	78%	67%	100%
Barren area	91%	91%	100%	100%	100%	75%	86%	100%
Overall accuracy	84%		95%		91%		93%	
Kappa value	0.77		0.94		0.86		0.90	

It is evident from the classified images of the Lahore district over a span of 37 years, urbanization has increased at a rapid pace. Major land cover changes in the Lahore district are due to urbanization, which affected other land cover classes, including agricultural land, tree cover, and barren land (Fig. 3).

Only during the interval of 2000–2010, greenery/agricultural area increased at a rate of approximately 2 km² per year. The urban area did not decrease during any interval from 1990 to 2017, while tree cover did not increase in Lahore from 1990 to 2017, as shown in Fig. 4.

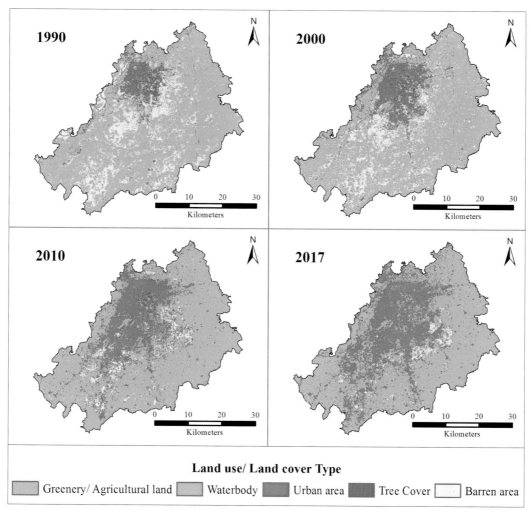

FIG. 3 Land cover map of Lahore district for the year 1990, 2000, 2010 and 2017.

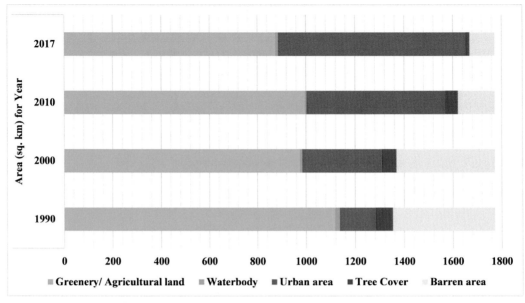

FIG. 4   Total area gains per land cover class for each year.

Table 5 shows the annual rate of change of each land cover class from 1990 to 2017. The table suggests that rapid urban growth has occurred during the last 27 years in Lahore. The urban area of Lahore has increased at an enormous rate of 23 km^2 per year from 1990 to 2017 with the most rapid increase between the intervals from 2000 to 2010. On the other hand, barren land decreased at the most rapid rate of around 24 km^2 per year during the same year interval i.e., 2000–2010.

It can be observed that through urban temporal change (1990–2017), Lahore's urban area has expanded along the major roads, as shown in Fig. 5. The transport network of an area influences the urban expansion pattern, as reported by.[54] In the case of the Lahore district, this expansion can be observed along the Grand Trunk (GT) road, Ferozepur road, Barki road, and National Highway (N-5). The research by Duranton and Turner[55] showed that a region's expansion along highways also contributes to more employment opportunities in the area.

4.2 Spatiotemporal land surface temperature

January and June are the coldest and hot months in the Lahore district[48] respectively. The average LST for both months was calculated separately for the entire Lahore district. Fig. 6 shows the spatial distribution of temperature distribution for the month of January over the Lahore district. Figs. 6 and 7A show an increase in the lowest and highest range of temperatures for the coldest month, which is an indicator of the strong correlation between LST and urban growth in the Lahore district.

The minimum temperature in the month of January increased from a minimum of 8.9°C (1990) to 15.2°C (2017) to a maximum of 24.5°C (1990) to 26.7°C (2017).

TABLE 5   Land cover change in sq. km per year during various time intervals.

Land cover class	Change in area (sq. km/ year)			
	1990–2000	2000–2010	2010–2017	1990–2017
Greenery/agricultural land	−14.23	2.09	−11.95	−8.92
Waterbody	−0.94	−0.37	0.47	−0.31
Urban area	17.77	24.38	19.96	23.00
Tree cover	−1.11	−0.73	−3.71	−2.06
Barren area	−1.49	−25.37	−4.76	−11.71

FIG. 5 Urban change over Lahore district (1990–2017).

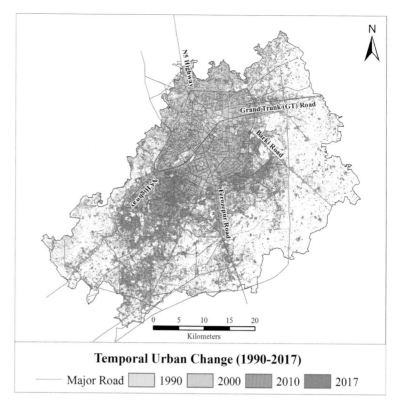

Temporal Urban Change (1990-2017)

—— Major Road | 1990 | 2000 | 2010 | 2017

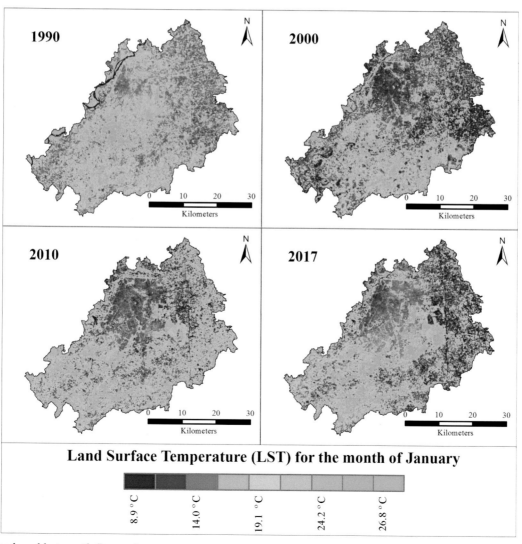

Land Surface Temperature (LST) for the month of January

8.9 °C   14.0 °C   19.1 °C   24.2 °C   26.8 °C

FIG. 6 LST for the coldest month (January).

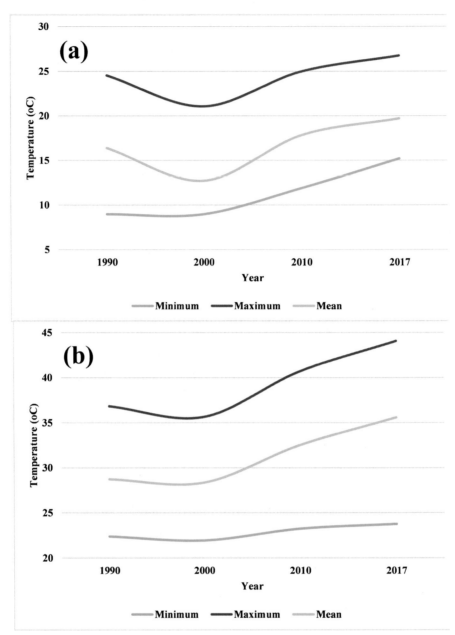

FIG. 7 (A) LST trend (minimum, maximum and mean) for the coldest month (January). (B) LST trend (minimum, maximum and mean) for the hottest month (June).

A similar trend can be seen in Fig. 7B and 8, which show a similar trend in the hottest month as well. The LST increased from a minimum of 22.4°C (1990) to 23.8°C (2017) to a maximum of 36.8°C (1990) to 44.1°C (2017). Table 6 shows that not only minimum and maximum LST has increased over the study area, but also the average (mean) temperature has increased from 16.4°C to 19.7°C and from 28.7°C to 35.6°C for winters and summers, respectively.

4.3 Landscape metrics for urban sprawl

The landscape metrics indices were selected to study the absolute and relative size and complexity of urban patches in the Lahore district. To measure and analyze the absolute size of the urban patches, CA, TE, MPE, and NumP were assessed and quantified, as shown in Fig. 9. Where CA refers to the size of an urban patch that tends to increase continuously under the process of urban sprawl over time, TE is the sum of perimeters of all patches in an urban class, MPE

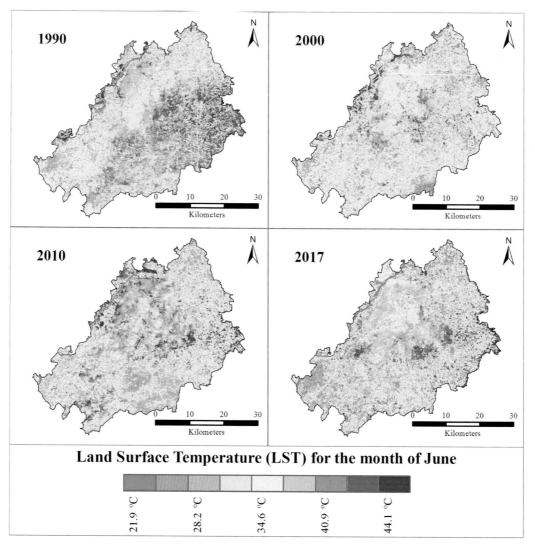

FIG. 8 LST for the hottest month (June).

is the mean of the number of edges per patch and NumP is the sum of the total number of patches in a single year for urban class.[49] For relative size quantification, MPS, ED, PSSD, and PSCoV were calculated. MPS refers to a function that calculates the number of patches and the size of the total size of the area covered by the urban class, ED is the function for calculating the total edge relative to the total area of the urban class, PSSD calculates the standard deviation of patch sizes and calculates the spread and variation of patch size of the urban class and PSCoV is the coefficient of variation of urban patch sizes.[49]

For the second characteristic of relative complexity, the AWMPFD, MSI, and MPAR were quantified. AWMPFD refers to the degree of complexity or irregularity of patches of urban class, MSI measures the complexity of shapes and MPAR is also used to measure the shape complexity of urban patches in meters/hectare.[49]

TABLE 6 LST for the coldest and hottest months over Lahore district (1990–2017).

Land surface temperature (LST)	Coldest month (January)				Hottest month (June)			
	1990	2000	2010	2017	1990	2000	2010	2017
Minimum (°C)	9.0	9.0	11.9	15.2	22.4	21.9	23.2	23.8
Maximum (°C)	24.5	21.1	25.0	26.8	36.8	35.7	40.7	44.1
Mean (°C)	16.4	12.7	17.8	19.7	28.7	28.4	32.5	35.6

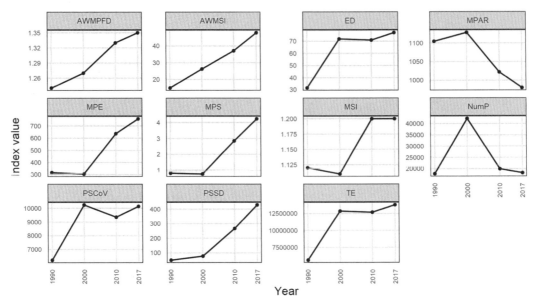

FIG. 9   Landscape metrics trend for urban areas from 1990 to 2017.

The value of AWMSI increased from 14.84 to 47.88, indicating the complexity of urban patches in the Lahore district from 1990 to 2017 and it also shows consistency with the increase in an urban sprawl over the same period. There is also an increase in the value of MSI from 1.12 to 1.20, which shows that urban patches are becoming more complex over time. On the other hand, the index MPAR shows a declining trend from 1104.54 to 980.24 from 1990 to 2017. This refers that the individual urban patches merge over time due to urban sprawl and trending towards the larger patches, thereby reducing the complexity of individual urban patches. It should be noted here that, from a broader perspective, complexity increases relative to other land cover classes, but at a detailed level, the relative complexity of individual patches has declined over time. Similarly, AWMPFD had a value of 1.24 in 1990 which increased to 1.35 in 2017, indicating that urban sprawl is somewhat irregular and less symmetrical in nature. The values of TE, ED, PSCoV, and PSSD also increased significantly from 1990 to 2017. All these indices were used as determinants of the urban patch size. MPS increased from 0.81 to 4.23 referring that because of urbanization the patch sizes will increase and will agglomerate with other urban patches in the neighborhood.

5   Conclusion

This study sheds new light on the systematic land cover dynamics of the Lahore district over a period of around three decades, underlining a major decrease in tree cover along with a significant increase in urban area/settlements. The findings from the analysis of land cover changes are aligned well and agreed overall with previously published work along with the landscape metrics analysis. Cloud computing, the computation of geospatial products, is vital to understanding the rate of land change, historical land consumption traditions with increasing LST and to guide decision and policymakers on planned urban expansion while ensuring the protection of environmental, social, and economic assets.

Both natural and anthropogenic activities are responsible for land cover change and increasing LSTs in Lahore. In terms of landscape, the urban landscape ecology of Lahore is continuously threatened by the encroachment of housing societies and influential business communities. The findings from urban landscape matrix analysis in this study could help the Lahore Development Authority (LDA), Environment Protection Department (EPD), and other relevant departments in making strategic decisions to prevent tree loss and encroachment in the urban landscape. The concerned departments and key stakeholders should plan future urban growth keeping with the use of remote sensing imagery and geospatial analysis. The methodology outlined in this chapter is cost-effective and can be easily replicated by integrating freely available satellite images with a 5–10 years interval.

References

1. Yang X. *Urban remote sensing.* John Wiley & Sons, Ltd.; 2011. https://doi.org/10.1002/9780470979563.
2. Malczewski J, Rinner C. *Development of GIS-MCDA;* 2015. https://doi.org/10.1007/978-3-540-74757-4_3.
3. Wilson A. *The science of cities and regions.* Springer; 2012. https://doi.org/10.1007/978-94-007-2266-6.
4. Deng C, Lin W. In: Weng Q, Quattrochi D, Gamba P, eds. *City in desert: mapping subpixel urban impervious surface area in a desert environment using spectral unmixing and machine learning methods.* 2nd ed. CRC Press (Taylor & Francis); 2018. https://doi.org/10.1201/9781138586642.
5. Belal AA, Moghanm FS. Detecting urban growth using remote sensing and GIS techniques in Al Gharbiya governorate, Egypt. *Egypt J Remote Sens Sp Sci.* 2011;14(2):73–79. https://doi.org/10.1016/j.ejrs.2011.09.001.
6. Xian G. *Remote sensing applications for the urban environment.* Taylor & Francis; 2015. https://doi.org/10.1007/s00247-007-0582-2.
7. Sakhre S, Dey J, Vijay R, Kumar R. Geospatial assessment of land surface temperature in Nagpur, India: an impact of urbanization. *Environ Earth Sci.* 2020;79(10):1–13. https://doi.org/10.1007/s12665-020-08952-1.
8. Rocchini D. Effects of spatial and spectral resolution in estimating ecosystem α-diversity by satellite imagery. *Remote Sens Environ.* 2007;111(4):423–434. https://doi.org/10.1016/j.rse.2007.03.018.
9. Chen S, Li WB, Du YD, Mao CY, Zhang L. Urbanization effect on precipitation over the Pearl River Delta based on CMORPH data. *Adv Clim Chang Res.* 2015;6(1):16–22. https://doi.org/10.1016/j.accre.2015.08.002.
10. Goldblatt R, You W, Hanson G, Khandelwal AK. Detecting the boundaries of urban areas in India: a dataset for pixel-based image classification in Google Earth Engine. *Remote Sens.* 2016;8(8). https://doi.org/10.3390/rs8080634.
11. Sidhu N, Pebesma E, Câmara G. Using Google earth engine to detect land cover change: Singapore as a use case. *Eur J Remote Sens.* 2018;51(1):486–500. https://doi.org/10.1080/22797254.2018.1451782.
12. Dwyer JL, Roy DP, Sauer B, Jenkerson CB, Zhang HK, Lymburner L. Analysis ready data: enabling analysis of the Landsat archive. *Remote Sens (Basel).* 2018;10(9):1–19. https://doi.org/10.3390/rs10091363.
13. Chen THK, Prishchepov AV, Fensholt R, Sabel CE. Detecting and monitoring long-term landslides in urbanized areas with nighttime light data and multi-seasonal Landsat imagery across Taiwan from 1998 to 2017. *Remote Sens Environ.* 2019;225:317–327. https://doi.org/10.1016/j.rse.2019.03.013.
14. Deng Z, Zhu X, He Q, Tang L. Land use/land cover classification using time series Landsat 8 images in a heavily urbanized area. *Adv Sp Res.* 2019;63(7):2144–2154. https://doi.org/10.1016/j.asr.2018.12.005.
15. Goldblatt R, Stuhlmacher MF, Tellman B, et al. Using Landsat and nighttime lights for supervised pixel-based image classification of urban land cover. *Remote Sens Environ.* 2018;205(November 2017):253–275. https://doi.org/10.1016/j.rse.2017.11.026.
16. Phiri D, Morgenroth J. Developments in Landsat land cover classification methods: a review. *Remote Sens (Basel).* 2017;9(9):2–25. https://doi.org/10.3390/rs9090967.
17. Sun Y, Zhang X, Zhao Y, Xin Q. Monitoring annual urbanization activities in Guangzhou using Landsat images (1987–2015). *Int J Remote Sens.* 2017;38(5):1258–1276. https://doi.org/10.1080/01431161.2016.1268283.
18. Bonafoni S, Keeratikasikorn C. Land surface temperature and urban density: multiyear modeling and relationship analysis using MODIS and Landsat data. *Remote Sens (Basel).* 2018;10(9):1471. https://doi.org/10.3390/rs10091471.
19. Almas AS, Rahim CA, Butt MJ, Shah TI. Metropolitan growth monitoring and landuse classification using geospatial techniques. In: *Proceedings of international workshop on service and application of spatial data infrastructure, Hangzhou, China;* 2005:277–282.
20. Shirazi SA. Temporal analysis of land use and land cover changes in Lahore-Pakistan. *Pakistan Vis.* 2012;13(1):187–206. http://pu.edu.pk/images/journal/studies/PDF-FILES/5%20-%20Shirazi%20Geography%20Artical-7_v13No1.pdf.
21. Shirazi S, Kazmi S. Appraisal of the change in spatio-temporal patterns of Lulc and its impacts on the vegetation of Lahore, Pakistan. *Pakistan Vis.* 2013;14(1):25–44. http://pu.edu.pk/images/journal/studies/PDF-FILES/Artical-2_Vol_14_No1.pdf.
22. Shirazi SA, Kazmi SJH. Analysis of population growth and urban development in Lahore-Pakistan using geospatial techniques: suggesting some future options. *South Asian Stud.* 2014;29(July 2014):269–280. http://pu.edu.pk/images/journal/csas/PDF/20%20Safdar%20Shirazi_29_1.pdf.
23. Riaz O, Ghaffar A, Butt I. Modelling land use patterns of Lahore (Pakistan) using remote sensing and GIS. *Glob J Sci Front Res Environ Earth Sci.* 2014;14(1):25–30.
24. Saleem MU. Urban change detection of Lahore (Pakistan) using the Thematic Mapper Images of Landsat since 1992–2010. In: *Fourth international conference on aerospace science & engineering (ICASE 2015).* Institute of Space Technology; 2015:38–43.
25. Saleem SS, Saleem SS. Spatial and temporal evolution of vegetation cover in Lahore, Pakistan. *Bull Environ Sci.* 2016;1(3):81–87.
26. Shirazi SA, Kazmi SJH. Analysis of socio-environmental impacts of the loss of urban trees and vegetation in Lahore, Pakistan: a review of public perception. *Ecol Process.* 2016;5(1). https://doi.org/10.1186/s13717-016-0050-8.
27. Asad M, Ahmad SR, Ali F, Mehmood R, Butt MA, Rathore S. Use of remote sensing for urban impervious surfaces : a case study of Lahore. *Int J Eng Appl Sci.* 2017;4(8):86–92.
28. Farhat K, Waseem LA, Ahmad KA, Baig S. Spatiotemporal demographic trends and land use dynamics of metropolitan Lahore. *J Hist Cult Art Res.* 2018;7(5):92–102. https://doi.org/10.7596/taksad.v7i5.1774.
29. Shakrullah K, Shirazi SA, Sajjad SH. An assessment of land use and land cover changes in Lahore (Pakistan) and New Delhi (India) using geospatial techniques. *Pak J Sci.* 2019;71(4):249–253.
30. Akbar TA, Hassan QK, Ishaq S, Batool M, Butt HJ, Jabbar H. Investigative spatial distribution and modelling of existing and future urban land changes and its impact on urbanization and economy. *Remote Sens.* 2019;11(2). https://doi.org/10.3390/rs11020105.
31. Imran M, Mehmood A. Analysis and mapping of present and future drivers of local urban climate using remote sensing: a case of Lahore, Pakistan. *Arab J Geosci.* 2020;13(6):1–14. https://doi.org/10.1007/s12517-020-5214-2.
32. Mumtaz F, Tao Y, de Leeuw G, et al. Modeling spatio-temporal land transformation and its associated impacts on land surface temperature (LST). *Remote Sens (Basel).* 2020;12(18):1–23. https://doi.org/10.3390/RS12182987.
33. Bajwa IU, Ahmad I, Khan Z. Urban housing development in Pakistan: a case study of Lahore Metropolitan area. *J Pakistan Eng Congr.* 2000. https://pecongress.org.pk/images/upload/books/Paper248.pdf.
34. Qureshi J, Mahmood SA, Almas AS, Irshad R, Rafique HM. Monitoring spatiotemporal and micro-level climatic variations in Lahore and suburbs using satellite imagery and multi-source data. *J Fac Eng Technol.* 2012;19:51–65. https://www.pu.edu.pk/journals/index.php/jfet/index.

35. Ahmad I, Mayo SM, Bajwa IU, Rahman A, Mirza AI. Role of development authorities in managing spatial urban growth; a case study of Lahore development authority, Pakistan. *Pakistan J Sci.* 2013;65(4):546–549. https://www.researchgate.net/profile/Ali_Mirza4/publication/283497144_Role_of_development_authorities_in_managing_Spatial_Urban_Growth_a_case_study_of_Lahore_Development_Authority_Pakistan/links/563b21ae08aeed0531dcd1f3.pdf.
36. Shah QA, Abbas H. *Livelihoods and access to services: an analysis of peri-urban areas of Lahore, Pakistan;* 2015.
37. Tariq S, Ul-Haq Z. Investigating the aerosol optical depth and angstrom exponent and their relationships with meteorological parameters over Lahore in Pakistan. *Proc Natl Acad Sci India Sect A Phys Sci.* 2019;90. https://doi.org/10.1007/s40010-018-0575-6.
38. Andleeb S, Ali Z, Afzal F, et al. Exposure to NO$_2$ in occupationalbuilt environmnets in urban Centre in Lahore. *J Anim Plant Sci.* 2015;25(3):656–659.
39. Butt MJ. Estimation of light pollution using satellite remote sensing and geographic information system techniques. *GISci Remote Sens.* 2012;49(4):609–621. https://doi.org/10.2747/1548-1603.49.4.609.
40. Sherwani RK, Shahid Z, Asim M. Impact of traffic pollution on historic buildings in Lahore. *Int J Res Chem Metall Civ Eng.* 2016;3(2):270–273. https://doi.org/10.15242/ijrcmce.ae0916304.
41. Ul-Haq Z, Tariq S, Ali M, Mahmood K, Rana AD. Sulphur dioxide loadings over megacity Lahore (Pakistan) and adjoining region of Indo-Gangetic Basin. *Int J Remote Sens.* 2016;37(13):3021–3041. https://doi.org/10.1080/01431161.2016.1192701.
42. Shirazi SA, Kazmi SJH. Analysis of population growth and urban development in Lahore-Pakistan using geospatial techniques: suggesting some future options. *South Asian Stud.* 2014;29(1):269–280. http://pu.edu.pk/images/journal/csas/PDF/20%20Safdar%20Shirazi_29_1.pdf.
43. Riaz O. Urban change detection of Lahore (Pakistan) using a time series of satellite images since 1972. *Asian J Nat Appl Sci.* 2013;2(4):101–105.
44. Pakistan Bureau of Statistics. *Census 1998;* 2005. http://www.pbs.gov.pk/.
45. Pakistan Bureau of Statistics. *Census 2017;* 2018. http://www.pbs.gov.pk/.
46. Cheng L. *Monitoring and analysis of urban growth process using remote sensing, GIS and cellular automata modeling: a case study of Xuzhou City, China;* 2014 [September].
47. Tsai YH, Stow D, Chen HL, Lewison R, An L, Shi L. Mapping vegetation and land use types in Fanjingshan National Nature Reserve using Google Earth Engine. *Remote Sens (Basel).* 2018;10(6):1–14. https://doi.org/10.3390/rs10060927.
48. Minallah MN. *Assessing the impact of urban expansion on land surface temperature in lahore using remote sensing Techniques;* 2017.
49. Magidi J, Ahmed F. Assessing urban sprawl using remote sensing and landscape metrics: a case study of city of Tshwane, South Africa (1984–2015). *Egypt J Remote Sens Sp Sci.* 2018;22. https://doi.org/10.1016/j.ejrs.2018.07.003.
50. Abebe GA. *Quantifying urban growth pattern in developing countries using remote sensing and spatial metrics: a case study in Kampala, Uganda;* 2013.
51. Brovelli MA, Molinari ME, Hussein E, Chen J, Li R. The first comprehensive accuracy assessment of GlobeLand30 at a National Level: methodology and results. *Remote Sens (Basel).* 2015;7(4):4191–4212. https://doi.org/10.3390/rs70404191.
52. Megahed Y, Cabral P, Silva J, Caetano M. Land cover mapping analysis and urban growth modelling using remote sensing Techniques in greater Cairo region—Egypt. *ISPRS Int J Geo-Inf.* 2015;4(3):1750–1769. https://doi.org/10.3390/ijgi4031750.
53. Paudel S, Yuan F. Assessing landscape changes and dynamics using patch analysis and GIS modeling. *Int J Appl Earth Obs Geoinf.* 2012;16(1):66–76. https://doi.org/10.1016/j.jag.2011.12.003.
54. Mundia CN, Aniya M. Analysis of land use/cover changes and urban expansion of Nairobi city using remote sensing and GIS. *Int J Remote Sens.* 2005;26(13):2831–2849. https://doi.org/10.1080/01431160500117865.
55. Duranton G, Turner MA. Urban growth and transportation. *Rev Econ Stud.* 2012;79(4):1407–1440.

10

Using OWA-AHP method to predict landslide-prone areas

Marzieh Mokarram[a] *and Hamid Reza Pourghasemi*[b]

[a]Department of Range and Watershed Management, College of Agriculture and Natural Resources of Darab, Shiraz University, Shiraz, Iran [b]Department of Natural Resources and Environmental Engineering, College of Agriculture, Shiraz University, Shiraz, Iran

1 Introduction

Landslides are the mass movements of materials that occur naturally under various factors, especially gravity on sloping surfaces. The occurrence of landslides in many parts of the world leads to the destruction of vegetation cover and even human fatalities.

Landslides are known as one of the natural hazards that effect under the influence of various internal and external factors by a lot of financial losses and possible loss of life. In recent years, with high landslide damage, human beings have attempted to reduce their damage. Predicting the exact time of landslide occurrence is beyond human scientific ability, identifying landslide-prone areas-LPA and ranking them can decrease its risk.[1] Therefore, recognizing this phenomenon and the factors affecting it can be helpful in basic planning for development projects and reducing dangerous areas, and lead to the presentation of appropriate solutions to stabilize them.

Because of the importance of this subject, various methods have been applied to predict areas prone to landslides, such as AHP,[2, 3] logistic regression,[4] and artificial neural networks (ANN).[5] Among the various methods, GIS-based multicriteria decision-making (MCDM) is very suitable for preparing landslide maps based on experts' opinions in the results.[6]

GIS-based MCDM is the main method of spatial analysis that applies spatial and nonspatial data to identify LPA in a region.[7] GIS with MCDM methods provides a spatial framework for organizing different thematic layers in a hierarchical structure.[8] Today, with the development of geographical sciences, new methods such as automatic and semi-automatic methods have been applied for landslide susceptibility and hazard mapping.[9] Many efforts have been made to apply a landslide susceptibility map based on qualitative methods.[10] Mokarram and Zarei[11] used the MCDM (AHP and fuzzy) method to prepare an LPA map.

Given the importance of landslides and studies that have been conducted so far with different methods, it is important to use a method that predicts landslide-prone areas with different levels of susceptibility. In fact, by preparing maps with different levels of susceptibility based on the importance of the region and the amount of budget, the necessary measures to prevent possible damage in the region can be considered. One of the methods to create a landslide map with different levels of susceptibility is ordered weighted averaging (OWA).

The OWA method is a set of multicriteria evaluations, which is a suitable method for investigating and deciding on natural processes.[12–16]

This study uses the OWA-AHP approach to produce a landslide susceptibility map with different levels of susceptibility. This helps users (e.g., managers) select one of the landslide maps with one level of susceptibility based on the condition of the region.

It is worth mentioning that one of the major innovations of this research is the combination of two AHP-OWA methods to prepare landslide susceptibility maps with different levels of risk. So that, according to the conditions of the region, one of these maps is used to manage landslides. Therefore, the combination method can make a powerful spatial decision for LPA.

2 Study area

This study was conducted in northern Lorestan Province, Iran. The area in the region is 11,712 km^2 and is located at longitudes of 33° 11–34° 18 N and latitudes of 46° 51–48° 40 E (Fig. 1). The study area is a part of the folded Zagros Mountains that are in the form of anticlines and synchronous structures with axial plates northwest-southeast and spiral-shaped, which have created elevations and subsidence, respectively. The material of the anticlines of the study area and its surroundings is usually limestone from the Sarvak Formation.

Most of the study area is located in the Zagros highlands. The highest altitude of the study area is 3487 m and its lowest altitude is located in the southernmost region (682 m). It is the third most water-rich province in Iran and accounts for 12% of the country's water. The maximum recorded temperature was 47.4°C, the minimum absolute temperature was −35°C and the average annual rainfall was 550–600 mm. Because of geological characteristics such as lithology, tectonics, seismicity, and climatic conditions, this region is very sensitive to landslides in Iran.[17]

3 Material and methods

In this study, to determine LPA and landslides mapping with different levels of susceptibility, after preparing the most important effective factors such as DEM, lithology, slope, land use, distance to a road (DTR), distance to stream (DTS), distance to fault (DTF) and rainfall, the incremental and decremental membership functions applied to prepare fuzzy maps for each factor.

The OWA-AHP method was used for weighting thematic layers, whereas the ANFIS method was used to predict landslide susceptibility classes. For determining the most important factors, the best subset regression method is applied. The steps of the research method used in this study are shown in Fig. 2.

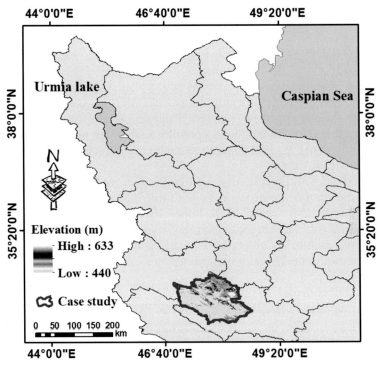

FIG. 1 The study area in Iran.

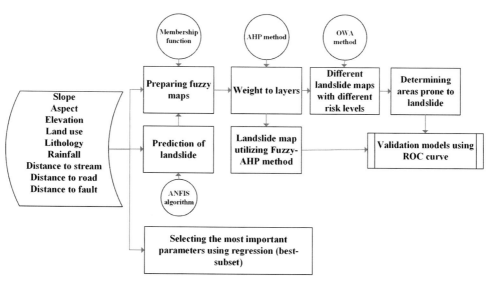

FIG. 2 Flowchart of steps to determine LPA in the region.

3.1 Data used

In this study, 82 landslides were used to map landslides in the region. After determining the effective factors and determining the required spatial and descriptive information, the data were collected based on the specifications of the region. The Data sources, scale, and years of data preparation are listed in Table 1. According to Table 1, it is clear that the coordinate system, scale, and resolution of data are different. Therefore, at this stage, the data were edited and prepared in the same format and resolution in ArcGIS V.10.8.

Using the digital elevation model (DEM 30×30 m), altitude, slope, and aspect maps were prepared (Fig. 3). According to Fig. 3, it is clear that many northeastern regions have altitudes of more than 3000 m that are prone to landslides. The aspect value is between -1 and 360 degrees, where the south and west directions are sensitive to landslides.[18] In addition, the maximum and minimum slope values were 0 and 66.11 degrees, respectively, and the maximum slope value was observed in the northeastern and central regions. Distance to roads, faults, and rivers was prepared based on basic layers such as topographic and geological maps (Table 2 and Fig. 3).

A rainfall map of this region is shown in Fig. 3. Fore preparing rainfall map as a climatic parameter, 15 meteorological stations were used, and an "inverse distance weighting (IDW)" method was applied.

TABLE 1 The data source of the region to prepare landslide map.

Parameters	Format	Scale	Source	Year
Land use	Polygon	1:250,000	Agriculture	2018
Rainfall (mm)	Polygon	15 weather stations	Meteorological organization	2019
Distance of road (m)	Polyline			
Distance of fault (m)	Polyline	1:100,000	Geology organization	2015
Distance of stream (m)	Polyline			
Lithology (formation susceptibility to erosion)	Polygon	1:100,000	Geology organization	2015
Aspect	Raster	Resolution 30 m	USGS	2019
DEM (m)				
Slope (degree)				

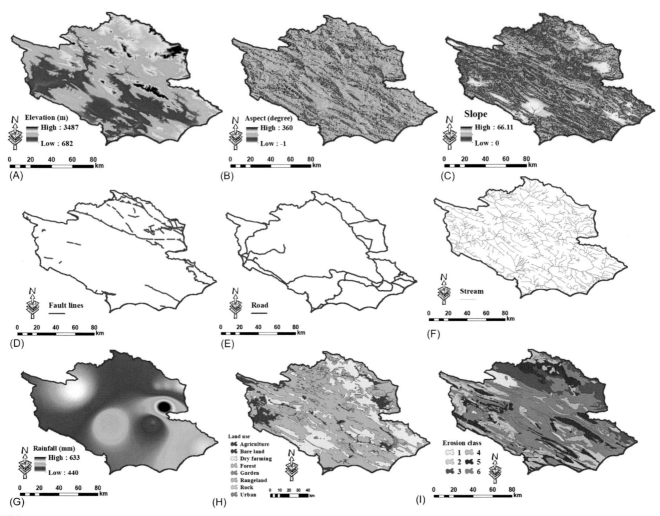

FIG. 3   Maps of the effective factors on landslides event in the study area. (A) Elevation, (B) aspect, (C) slope, (D) fault lines, (E) road, (F) stream, (G) rainfall, (H) land use, and (I) erosion class.

TABLE 2   Distance to fault, stream, and roads.

Feature				
Distance of fault (m)				
0–1000	1000–2000	2000–3000	3000–4000	>4000
Distance of stream (m)				
0–50	50–100	100–150	150–200	>200
Distance of road (m)				
0–25	25–50	50–75	75–100	>100

According to Fig. 3, the rainfall in the study area was between 440 and 633 mm. The southern and eastern parts had more rainfall than the other parts of the study area. Land use in the region includes forest, agricultural, pasture, bare land, rock, and urban areas (Fig. 3). To determine the effectiveness of lithology on landslides, a map of formation sensitive to water erosion was used. As shown in Fig. 3, the areas located in the center and east of the study area were more sensitive to erosion. Table 3 shows the lithological units and the features of the study area.

TABLE 3 Explanation of lithological features and their sensitive.

Classes	Geology units	Description
1	Kbgp	Limestone
2	EMas, PeEtz	Massive of fossiliferous and limestone
3	Trjvm, TRkurl, K1b1	Shale and chert bedded to massive orbitolina limestone
4	Omq, EKN, ekN, Plbk	Limestone and calcareous shale, sandstone
5	Kpeam	Piedmont conglomerate and sandstone, shale, and limestone
6	Qft2	Limestone, low level of pediment fan, and valley terrace deposits

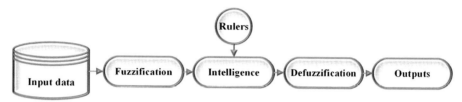

FIG. 4 Relationship of different parts in fuzzy logic.

3.2 Fuzzy logic

The term fuzzy logic was first used by Professor Zadeh[19] at Berkeley University. Fuzzy logic has four main parts; these parts and their relations are shown in Fig. 4.

In general, the steps of the fuzzy method include:

Basic rules: In the first stage, fuzzy rules are determined by utilizing the opinions of experts and studies conducted in this field.

Fuzzification: In this step, the data are converted to values between 0 and 1 by the membership functions, depending on their effect on the target.

Inference or intelligence engine: In this step, the output of this layer is the multiplication of input signals that is equivalent to "if" in the fuzzy method actually.

Defuzzification: In the last step, the results of fuzzy results, which are in the form of fuzzy sets, are converted into quantitative data and information.

3.3 Analytical hierarchy process (AHP)

The AHP is one of the most efficient multicriteria decision-making techniques that was first presented by Saaty and Vargas.[20] This method is based on pairwise comparisons of factors and allows decision-makers to consider different scenarios. This technique allows the hierarchical formulation of complex natural problems. It is also possible to consider different quantitative and qualitative criteria for the problem.

The first step in the hierarchical analysis was to formulate the structure. In general, the hierarchical structure can be considered as one of the following issues[21]:

(1) Objective, criterion, subcriterion, option
(2) Purpose, agent, subagent, option

In this study, to create a hierarchical structure for LPA, the first structure (objective, criterion, subcriterion, option) is used and the relationships between the criteria and subcriteria are shown in Fig. 5.

In the AHP, judgments are converted by Saaty[22] to small values between 1 and 9, which are listed in Table 4.

After preparing the pairwise comparison matrix and determining the degree of preference for each factor, the relevant information is extracted and the degree of compatibility of the comparisons is determined. After ensuring that the obtained priorities are acceptable, the weight of each parameter is determined using the geometric mean.[22]

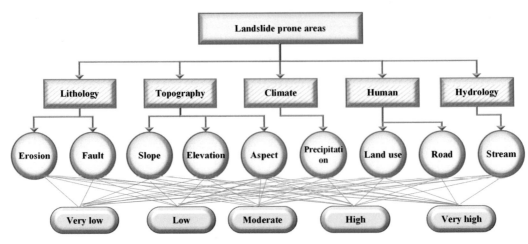

FIG. 5   Relation of criteria and subcriteria for weighting the parameters.

TABLE 4   The number of preferences for pairwise comparison of factors.[22]

Number value	Description	Number value	Description
9	Extremely preferred	3	Moderately preferred
7	Very strongly preferred	1	Equally preferred
5	Strongly preferred	2, 4, 6, and 8	Intermediate

3.4 OWA-AHP method

The OWA and AHP methods were not implemented at the same level. In the AHP method, a hierarchical model is created based on the degree of importance of each meter for the desired goal, in which a simple linear combination is created to weigh the desired parameters, which also provides a general framework for performing processes such as AHP. The nature and structure of these two algorithms are such that their combination can be used to make a more powerful spatial decision tool.[23] To achieve this framework, the formation of the hierarchical structure and the calculation of the relative weights of the criteria by performing a pairwise comparison of the criteria were calculated. In the next step, the problem is processed using OWA-guided quantifiers.[24] The procedure at this stage consisted of three main steps.

(A) determining the linguistic quantifier (Q)
(B) obtaining rank weights by a linguistic quantifier (Q)
(C) to calculate the total evaluations for each location of each level of the hierarchical structure using the OWA composition function. The overall score of *option i* can be calculated in two steps. In the first step, the score of the first option for each criterion was calculated based on Eq. (1), and in the second step, the weighting is calculated using Eq. (2).

$$s_{iq} = \sum_{k=1}^{l} v_{k(q)} \cdot z_{jk(q)} \quad i=1,2,...,m, \ q=1,2,...p \tag{1}$$

$$v_{k(q)} = \left(\sum_{k=1}^{l} u_{k(q)} \right)^{\alpha(q)} - \left(\sum_{k=1}^{l-1} u_{k(q)} \right)^{\alpha(q)} \tag{2}$$

where, $Z_{ik(q)}$ by reclassifying the values of the attributes related to q, which is the second target $u_{k(q)}$ and $x_{ik(q)}$ is reclassifying the qth weight of the attribute corresponding to k of this criterion. $\alpha(q)$ is a factor related to linguistic quantifiers that relate to the q target, which is a class of fuzzy language quantifiers known as regular increasing monotone—RIM (Table 5).[25]

TABLE 5 Linguistic quantifiers and corresponding α values.[24]

Linguistic quantifier (Q)	At least one	Little	Some	Half	Many	Most	All
α	0.0001	0.01	0.5	1	?	10	100

The total score of each option in relation to the final goal was obtained based on Eq. (3):

$$\text{OWA} - \text{AHP}_{(i)} = \sum_{k=1}^{l} v_k z_{iq} \quad i = 1, 2, \dots, m. \tag{3}$$

V_a is calculated according to Eq. (4):

$$v_q = \left(\sum_{q=1}^{p} u_q \right)^{\alpha_g} - \left(\sum_{k=1}^{p-1} u_q \right)^{\alpha_g} \tag{4}$$

Z_{iq} by re-ranking the values of the options at the qth target level of S_{iq} and u_q is the weight q ranking of the target. α_g is the parameter linking linguistic quantifiers to the ultimate goal of spatial decision-making based on a hierarchical structure.

The OWA method is introduced as a decision-making method that can consider the priorities and mental evaluations of the decision-maker. This method can consider susceptibility-taking and susceptibility-averse decision-making in the decision-making process and can make the final decision based on susceptibility-taking or susceptibility-averse decision-making.[26] The OWA operator consists of two main characteristics that express the behavior of this operator: the degree of susceptibility between the indicators (degree of ORness) that indicates the position of the OWA operator between the AND and OR relationships.[27] This degree indicates the degree of decision-makers of emphasis on better or worse values of a set of indicators or the same susceptibility-taking and susceptibility-averse decision-making and is defined by Eq. (5):

$$\text{ORness} = \frac{1}{n-1} \sum_{i=1}^{n} (n-i) w_i \quad 0 \leq \text{ORness} \leq 1 \tag{5}$$

The higher the ORness, the greater the optimism or susceptibility taking of the decision-maker, and vice versa. The second characteristic is the amount of compromise or trade-off, which shows the degree of influence or exchange of one index from other indices and is defined as follows (Eq. 6):

$$\text{Trade} - \text{off} = 1 - \sqrt{\frac{n}{n-1} \sum_{i=1}^{n} \left(w_i - \frac{1}{n} \right)^2} \quad 0 \leq \text{ORness} \leq 1 \tag{6}$$

Different levels of susceptibility range from 1 to 6 based on trade-off values, as shown in Fig. 6.[28]

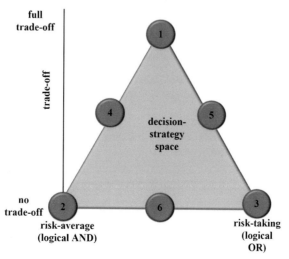

FIG. 6 Susceptibility status based on trade-off values.

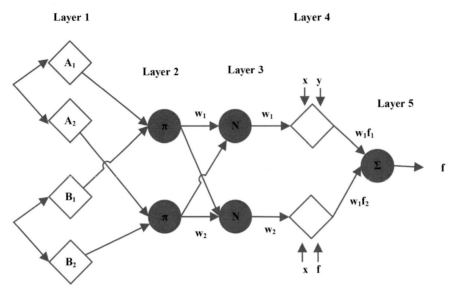

FIG. 7 The general structure of the ANFIS model.

3.5 Adaptive neuro-fuzzy inference system (ANFIS)

ANFIS is an artificial neural network (ANN) that is based on the Takagi-Sugino fuzzy system. This method was first proposed by Jang.[29] Because this system integrates ANNs and concepts of fuzzy concepts, it can take advantage of the possibilities of two methods in one method. Its inference system conforms to a set of fuzzy rules that can be learned to approximate nonlinear functions.[29] Fig. 7 shows the general structure of the ANFIS model with two inputs and one output (Eqs. 7 and 8):

$$\text{Rule 1}: \text{if } x \text{ is } A_1 \text{ and } y \text{ is } B_1 \text{ then } f_1 = p_1 x + q_1 y + r_1 \tag{7}$$

$$\text{Rule 1}: \text{if } x \text{ is } A_2 \text{ and } y \text{ is } B_2 \text{ then } f_2 = p_2 x + q_2 y + r_2 \tag{8}$$

To assess the precision of the model (subclustering, grid partitioning, fuzzy c-means (FCM)), root mean square error (RMSE), mean absolute error (MAE), and correlation coefficient (R) were used. The RMSE evaluates the variance of the errors independent of the sample size, which is calculated using Eq. (9):

$$RMSE = \sqrt{\frac{1}{p}\sum_{p=1}^{p}\left(f^{(p)} - t^{(p)}\right)^2} \tag{9}$$

MAE shows the average of all individual errors calculated using Eq. (10):

$$MSE = \frac{1}{p}\sum_{p=1}^{p}\left|f^{(p)} - t^{(p)}\right| \tag{10}$$

The correlation coefficient (R) is a statistical concept that shows the extent to which the predicted values correspond to the real data and is obtained through Eq. (11):

$$R = \frac{\sum_{p=1}^{p}\left(f^{(p)} - \bar{f}^{(p)}\right)^2\left(t^{(p)} - \bar{t}\right)}{\sqrt{\sum_{p=1}^{p}\left(f^{(p)} - \bar{f}\right)^2\sum_{p=1}^{p}\left(t^{(p)} - \bar{t}\right)^2}} \tag{11}$$

TABLE 6 Maximum and minimum of m, n to define the MF for each of the parameters.[18]

Parameters	Minimum	Maximum
Land use	Forest, agriculture	Rocky and bare lands
Precipitation (mm)	<250	>500
Distance of road (m)	>100	<25
Distance of fault (m)	>4000	0–1000
Distance of stream (m)	>200	0–50
Sensitive	5	6
Aspect	South	North
DEM (m)	>3000	<1200
Slope (degree)	0–10	>40

4 Results and discussion

4.1 Fuzzy logic

According to Table 6, it is determined that the rocky lands and bare lands, rainfall of more than 500 mm, distance to roads less than 25 m, distance to faults lower than 1000 m, distance to rivers less than 50 m, lithological formations with high susceptibility to water erosion (Class 6), northern slope facing, altitudes less than 1200 m, and slopes greater than 40 degrees lead to intensification of landslides in the region where the membership function receives values close to 1.[18]

According to Table 6, a fuzzy map was created for each of the factors (Fig. 8). Values close to 1 represent the appropriate factor conditions for making landslides and vice versa. North slopes, more sensitive classes (most of the region), bare lands (areas located in the northeastern, eastern, and western parts shown in green), areas with more rainfall (located in the northeast and northwest), areas close to the road, waterways, and faults (northeastern parts) and higher elevations (northeastern areas) are prone to landslides and have values of 1 and close to 1 in fuzzy maps.

4.2 Computation of criterion weight

After prioritizing each factor based on the opinions of experts and different studies (such as[18]), the weight of each factor was calculated using the AHP method (Table 7).

According to Table 7, it is determined that the lithology (formation susceptibility to water erosion) and rainfall are the most important and altitude has the least importance in creating landslides in the region.

Mansour et al.[30] concluded that rainfall plays a very important role in landslide occurrence. Nguyen et al.[31] in the Atlantic Island of Madeira concluded that rainfall led to intensified landslides in the area so that in 2 days of rainfall events, about 120 landslides occurred in the area.

Then, according to Table 7, the final map of the LPA is prepared, as shown in Fig. 9. According to Fig. 9, approximately 47% of the area is highly susceptible to landslides. By the way, it is clear that the areas located in the north and east of the region have a higher susceptibility to landslides than other areas.

Areas located in the north and east of the region are more sensitive to landslides, as approximately 47% of the area has the most landslides.

In the present study, six weights arranged according to nine input parameters (different α values) were used for the weighted factors using the AHP method (Table 8). Considering the use of six types of α values from at least one (($\alpha = 0$) to all ($\alpha = \infty$)), six types of landslide maps with different levels of susceptibility are made such that the different weights of each are shown in Table 8. A Description of each α value is shown in Table 9.

The final landslide maps with different levels of susceptibility levels are shown in Fig. 10. Based on Fig. 10, it is clear that by reducing the susceptibility, only the areas that are in the priority of landslides are identified as high-susceptibility areas. As shown in Fig. 10, high-susceptibility areas were located in the northeast of the area. These areas has a high formation of sensitivity to water erosion, high rainfall, and high slope, making the prone region prone to landslides. Conversely, with increasing susceptibility, landslides that are less priority in the region are identified as

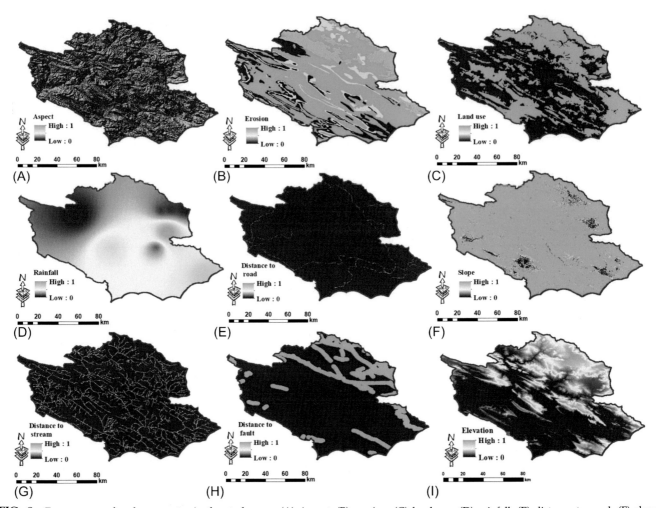

FIG. 8 Fuzzy maps of each parameter in the study area. (A) Aspect, (B) erosion, (C) land use, (D) rainfall, (E) distance to road, (F) slope, (G) distance to stream, (H) distance to fault, and (I) elevation.

TABLE 7 Pairwise comparison of each factor using the AHP method.

Parameter	Lithology	Precipitation	Land use	Slope	Distance to fault	Distance to stream	Distance to road	Aspect	DEM	Weight
Lithology	1	2	3	4	5	6	7	8	9	0.31
Precipitation	1/2	1	2	3	4	5	6	7	8	0.22
Land use	1/3	1/2	1	2	3	4	5	6	7	0.15
Slope	1/4	1/3	1/2	1	2	3	4	5	6	0.11
Distance to fault	1/5	1/4	1/3	1/2	1	2	3	4	5	0.08
Distance to stream	1/6	1/5	1/4	1/3	1/2	1	2	3	4	0.05
Distance to road	1/7	1/6	1/5	1/4	1/3	1/2	1	2	3	0.04
Aspect	1/8	1/7	1/6	1/5	1/4	1/3	1/2	1	2	0.03
DEM	1/9	1/8	1/7	1/6	1/5	1/4	1/3	1/2	1	0.02

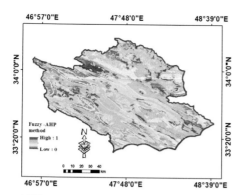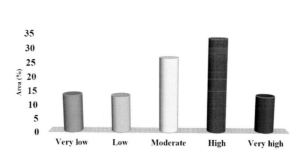

FIG. 9 Fuzzy-AHP maps and area of each class.

TABLE 8 Weights calculated for each of the factors at different levels of susceptibility.

j	Quantifier	Criterion weights u_k	$\left(\sum_{k=1}^{j} u_k\right)^{\propto}$	$\left(\sum_{k=1}^{j} u_k\right)^{\propto} - \left(\sum_{k=1}^{j-1} u_k\right)^{\propto}$
Lithology	**(a)** At least one ($\alpha \to = 0$)	0.31	1	1
Precipitation		0.22	1	0
Land use		0.15	1	0
Slope		0.11	1	0
Distance to fault		0.08	1	0
Distance to stream		0.05	1	0
Distance to road		0.04	1	0
Aspect		0.03	1	0
DEM		0.18	1	0
Lithology	**(b)** At least a few ($\alpha \to = 0.1$)	0.31	0.88948	0.8894
Precipitation		0.22	0.938486	0.04900
Land use		0.15	0.962168	0.0145
Slope		0.11	0.976703	0.0094
Distance to fault		0.08	0.98617	0.0055
Distance to stream		0.05	0.991697	0.0042
Distance to road		0.04	0.995926	0.0030
Aspect		0.03	0.998995	0.0019
DEM		0.18	1	
Lithology	**(c)** A few ($\alpha \to = 0.5$)	0.31	0.56	0.56
Precipitation		0.22	0.73	0.17
Land use		0.15	0.82	0.10
Slope		0.11	0.89	0.06
Distance to fault		0.08	0.93	0.04
Distance to stream		0.05	0.96	0.03
Distance to road		0.04	0.98	0.02
Aspect		0.03	1	0.02
DEM		0.18	1	0
Lithology	**(d)** Half (identity) ($\alpha \to = 1$)	0.31	0.31	0.31
Precipitation		0.22	0.53	0.22
Land use		0.15	0.68	0.15
Slope		0.11	0.79	0.11
Distance to fault		0.08	0.87	0.08
Distance to stream		0.05	0.92	0.05
Distance to road		0.04	0.96	0.04
Aspect		0.03	0.99	0.03
DEM		0.18	1	0.018
Lithology	**(e)** Most ($\alpha \to = 2$)	0.31	0.0961	0.0961
Precipitation		0.22	0.2809	0.1848
Land use		0.15	0.4624	0.1815
Slope		0.11	0.6241	0.1617
Distance to fault		0.08	0.7569	0.1328

Continued

TABLE 8 Weights calculated for each of the factors at different levels of susceptibility—cont'd

j	Quantifier	Criterion weights u_k	$\left(\sum_{k=1}^{j} u_k\right)^{\infty}$	$\left(\sum_{k=1}^{j} u_k\right)^{\infty} - \left(\sum_{k=1}^{j-1} u_k\right)^{\infty}$
Distance to stream		0.05	0.8464	0.0895
Distance to road		0.04	0.9216	0.0752
Aspect		0.03	0.9801	0.0585
DEM		0.18	1	0.037981
Lithology	**(f)** All $(\alpha \to \infty)$	0.31	0	0
Precipitation		0.22	0	0
Land use		0.15	0	0
Slope		0.11	0	0
Distance to fault		0.08	0	0
Distance to stream		0.05	0	0
Distance to road		0.04	0	0
Aspect		0.03	0	0
DEM		0.18	1	1

TABLE 9 The α values of each susceptibility level and trade-off.

Code	α value	Description
(a)	0	Low level of susceptibility and no trade-off (LLR-NTO)
(b)	0.1	Low level of susceptibility and average trade-off (LLR-ATO)
(c)	0.5	Average level of susceptibility and full trade-off (ALR-FTO)
(d)	1	Average level of susceptibility and no trade-off (ALR-NTO)
(e)	2	High level of susceptibility and average trade-off (HLR-ATO)
(f)	∞	High level of susceptibility and no trade-off (HLR-NTO)

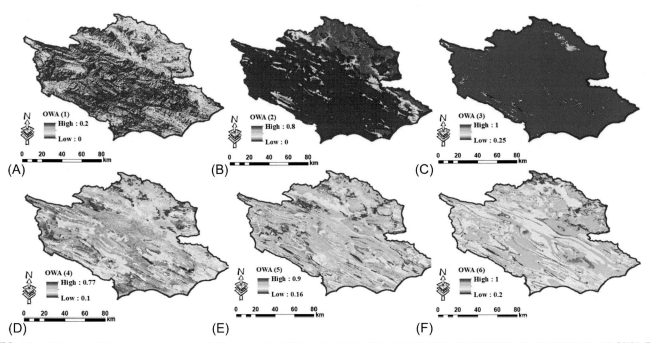

FIG. 10 LPA with different levels of susceptibility using the OWA method. (A) OWA (1), (B) OWA (2), (C) OWA (3), (D) OWA (4), (E) OWA (5), and (F) OWA (6).

TABLE 10 Description of each class in OWA-AHP method.

Range	Description
0–0.25	Very low
0.25–0.5	Low
0.5–0.75	Medium
0.75–1	High

high-susceptibility areas; thus, according to Fig. 10, most areas are in the high susceptibility class. Therefore, according to the conditions of the region and the amount of budget, one of these maps can be used as a landslide susceptibility map of the region. For example, by reducing the budget, maps with lower susceptibility levels can be used for regional management, and vice versa.

Based on Table 10, the LPA classification maps are shown in Fig. 11. The area of each class is shown in Fig. 12. According to Figs. 11 and 12, it is found that HLR (class 4 (0.75–1)) is higher than LLR (class 1 (0–0.25)). As shown in Fig. 12 and Table 10, LLR (2) shows that the largest area is in the very low class (0–0.25). the HLR in the very high class (0.75–1) is high.

To validate the methods (Fuzzy-AHP and OWA-AHP with average trade-off methods), 15 points where landslides occurred were selected as sample points. An example of landslides that occurred in some parts of the sample is shown in Fig. 13. Given that the ROC curve is the most efficient method for calculating the accuracy of the method,[32] the curve was used to evaluate the accuracy of the methods. The results are presented in Fig. 14. Owing to the high values of the Area under the curve (AUC) in the two methods ($AUC_{Fuzzy-AHP} = 90\%$, $AUC_{ALOR-NTO} = 88\%$), the two methods have high accuracy in determining landslides in the region.

Methods such as fuzzy and AHP have been used to determine landslide-prone locations.[33-35] As a result, only a map of the landslide situation has been prepared without considering the level of risk. As part of the study, the advantages of the AHP and OWA methods were applied and after prioritizing the data using the AHP method, linguistic quantifier, and ranking weights required by the OWA method, maps indicating different levels of risk-taking were prepared. With these maps and the different levels of risk, managers can make the right decisions based on different

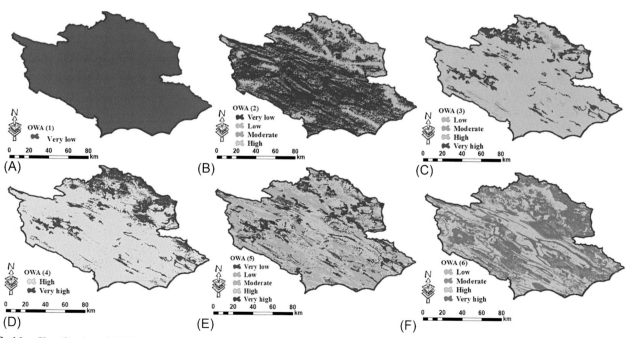

FIG. 11 Classification of OWA maps for landslide susceptibility. (A) OWA (1), (B) OWA (2), (C) OWA (3), (D) OWA (4), (E) OWA (5), and (F) OWA (6).

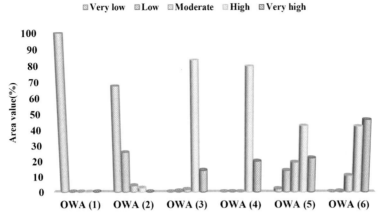

FIG. 12   Area of five classes determined using the OWA method.

FIG. 13   Examples of landslides in the study area.

financial situations. This multicriteria assessment method is often used to investigate environmental issues.[24] A lot of natural scientists are new to OWA and use it for the first time for landslide sensitivity.

4.3 ANFIS method

Then, the ANFIS method was applied to predict landslide susceptibility. For this purpose, the altitude, lithology, slope, land use, distance to road, distance to stream, distance to fault, and rainfall factors were used as input data. In the ANFIS method, subclustering, grid partitioning, and fuzzy c-means (FCM) with two models of backpropagation and hybrid are applied. The results for each method are listed in Table 11.

According to Table 11, it is clear that the hybrid model has high accuracy for the three methods. As shown in Table 11, the three hybrid methods had the highest values of R ($R_{FCM} = 0.99$, $R_{Subclustering} - 1$, $R_{Grid\ partitioning} - 1$) and the lowest RMSE values ($RMSE_{FCM} = 0.06$, $RMSE_{Subclusetring} = 0.00$, $RMSE_{Grid\ partitioning} = 0.00$) and the lowest MSE values ($MSE_{FCM} = 0.01$, $MSE_{Subclustering} = 0.00$, $MSE_{Grid\ partitioning} = 0.02$).

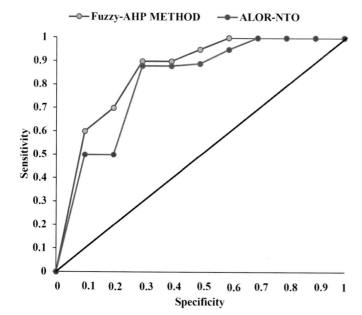

FIG. 14   The ROC curve results.

TABLE 11   Accuracy of each model of the ANFIS method for landslide prediction.

Method	Model	Test			Train		
		R	RMSE	MSE	R	RMSE	MSE
FCM	Back propagation	0.20	7.30	53.50	0.08	8.12	66.01
	Hybrid	0.99	0.00	0.01	0.99	0.01	0.00
Subclustering	Back propagation	0.11	84.60	9.20	0.35	125.70	11.21
	Hybrid	1.00	0.02	0.01	0.99	0.00	0.00
Grid	Back propagation	0.25	0.77	0.61	0.10	0.81	0.65
	Hybrid	1.00	0.02	0.02	0.99	0.01	0.00

The membership functions used for each of the input parameters and the results of each ANFIS method are shown in Fig. 15.

4.4 Determining the optimal regression model

Then, an all-subset regression is used to evaluate the data status and landslide prediction and select the most important effective parameters in a landslide. According to nine input factors, nine regression models are developed. The results of the best subset regression model are shown in Fig. 16. Studies have shown that smaller C_p values indicate the accuracy of the model for predicting landslides.[36] Therefore, the smallest C_p was selected as the optimal model in this study. The results show that distance to road, distance to stream, distance to fault, precipitation parameters with $R^2 = 96.9$, $AdjR^2 = 96.2$ and $C_p = 0.6$ are selected as the optimal model.

In landslide-prone areas in the study area, various methods can be used to improve the condition of unstable slopes, such as changing the shape of the slope. Improving surface drainage, blocking crevices, creating ditches and water transfer channels above the slopes, using satellite data and models for landslide susceptibility prediction, and preparing landslide maps.

The AHP method leads to data prioritization and final landslide mapping by paired data comparison and prevents ambiguous data evaluation.[37] On the other hand, preparing maps of landslide-prone areas with different susceptibility

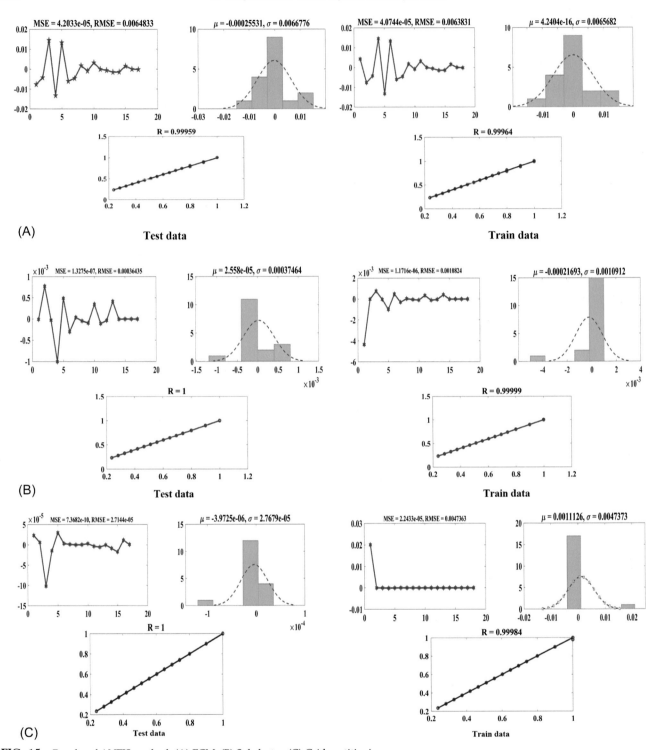

FIG. 15   Results of ANFIS method. (A) FCM, (B) Subclaster, (C) Grid partitioning.

levels, compared to the AHP method, OWA method leads to preparing multiple landslide maps that based on the situation in the area, managers can use one of these maps to control landslides in the area.[38] In this region, maps with different susceptibility levels were created using the OWA method. On the other hand, using the fuzzy method to prioritize the values of each factor and prepare a fuzzy map for each factor[39] and multicriteria decision making was utilized to weigh the layers, leading to an increase in the accuracy of the final maps for the decision.[37]

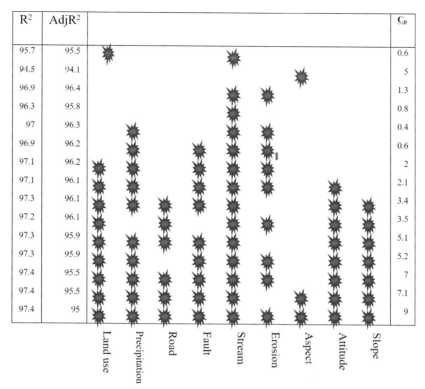

FIG. 16 Best-subset regression results for the most important parameters affecting moat erosion.

5 Conclusions

The occurrence of landslides in susceptible areas is a major problem in northwestern Iran. Therefore, creating a regional strategy to protect human and natural resources and reduce the damage caused by their occurrence is very important and necessary. Providing a suitable model and preparing a landslide susceptibility map can be of great help in planning and managing the environment of the study area. GIS technique, due to its ability to manage large volumes of spatial information, is a powerful tool for this type of initial study.

In addition, the AHP model is used by planners to solve complex management problems. Therefore, the combination of the GIS technique and the AHP model can be used as a powerful method to provide a prediction model and prepare a landslide map for the study area. In this study, maps with different levels of susceptibility were prepared using the combined method of OWA-AHP to determine the LPA. According to the results, it is clear that based on the conditions of the region, one of the maps prepared with certain levels of susceptibility should be used to control landslides and reduce damage.

Acknowledgments

The authors are very grateful to Shiraz University for providing financial support for this research. We also thank the Agricultural Jihad of Lorestan Province for their cooperation.

References

1. Vranken L, Van Turnhout P, Van Den Eeckhaut M, Vandekerckhove L, Poesen J. Economic valuation of landslide damage in hilly regions: a case study from Flanders, Belgium. *Sci Total Environ*. 2013;447:323–336.
2. El Jazouli A, Barakat A, Khellouk R. GIS-multicriteria evaluation using AHP for landslide susceptibility mapping in Oum Er Rbia high basin (Morocco). *Geoenviron Disast*. 2019;6(1):3.
3. He H, Hu D, Sun Q, Zhu L, Liu Y. A landslide susceptibility assessment method based on GIS technology and an AHP-weighted information content method: a case study of southern Anhui, China. *ISPRS Int J Geo Inf*. 2019;8(6):266.
4. Jibson RW. Regression models for estimating coseismic landslide displacement. *Eng Geol*. 2007;91(2–4):209–218.
5. Huang F, Zhang J, Zhou C, Wang Y, Huang J, Zhu L. A deep learning algorithm using a fully connected sparse autoencoder neural network for landslide susceptibility prediction. *Landslides*. 2020;17(1):217–229.

6. Othman AN, Naim WM, Noraini S. GIS based multi-criteria decision making for landslide hazard zonation. *Procedia Soc Behav Sci.* 2012;35:595–602.

7. Niu Q, Dang X, Li Y, Zhang Y, Lu X, Gao W. Suitability analysis for topographic factors in loess landslide research: a case study of Gangu county, China. *Environ Earth Sci.* 2018;77:294.

8. Ghorbanzadeh O, Feizizadeh B, Blaschke T. Multi-criteria risk evaluation by integrating an analytical network process approach into GIS-based sensitivity and uncertainty analyses. *Geomat Nat Haz Risk.* 2018;9(1):127–151.

9. Czikhardt R, Papco J, Bakon M, Liscak P, Ondrejka P, Zlocha M. Ground stability monitoring of undermined and LPA by means of sentinel-1 multi-temporal InSAR, case study from Slovakia. *Geosciences.* 2017;7:87.

10. Trigila A, Frattini P, Casagli N, et al. Landslide susceptibility mapping at national scale: the Italian case study. In: *Landslide Science and Practice.* Berlin/Heidelberg, Germany: Springer; 2013:287–295.

11. Mokarram M, Zarei AR. Landslide susceptibility mapping using fuzzy-AHP. *Geotech Geol Eng.* 2018;36(6):3931–3943.

12. Makropoulos CK, Butler D, Maksimovic C. Fuzzy logic spatial decision support system for urban water management. *J Water Resour Plan Manag.* 2003;129(1):69 77.

13. Mokarram M, Aminzadeh F. GIS-based multicriteria land suitability evaluation using ordered weight averaging with fuzzy quantifier: a case study in Shavur Plain, Iran. *Int Arch Photogramm Remote Sens Spat Inf Sci.* 2010;38(2):508–512.

14. Mokarram M, Hojati M. Using ordered weight averaging (OWA) for multicriteria soil fertility evaluation by GIS (case study: Southeast Iran). *J Solid Earth.* 2016;10.

15. Zeng S, Baležentis T, Zhang C. A method based on OWA operator and distance measures for multiple attribute decision making with 2-tuple linguistic information. *Informatica.* 2012;23(4):665–681.

16. Jiang H, Eastman JR. Application of fuzzy measures in multi-criteria evaluation in GIS. *Int J Geo Inform Sci.* 2000;4(2):173–184.

17. General Department of Meteorology of Lorestan Province (GDMLP); 2019. https://www.irimo.ir/far/wd/701.

18. Feizizadeh B, Blaschke T. GIS-multicriteria decision analysis for landslide susceptibility mapping: comparing three methods for the Urmia lake basin, Iran. *Nat Hazards.* 2013;65(3):2105–2128.

19. Zadeh LA. Information and control. *Fuzzy. Sets.* 1965;8(3):338–353.

20. Saaty TL, Vargas LG. Diagnosis with dependent symptoms: Bayes theorem and the analytic hierarchy process. *Oper Res.* 1998;46(4):491–502.

21. Bowen WM. Subjective judgments and data environment analysis in site selection. *Comput Environ Urban Syst.* 1990;14:133–144.

22. Saaty TL. Axiomatic foundation of analytical hierarchy process. *J Manag Sci.* 1986;31(7):841–855.

23. Yager RR. On ordered weighted averaging aggregation operators in multi-criteria decision making. *IEEE Trans Syst Man Cybern.* 1988;18 (1):183–190.

24. Malczewski J. Ordered weighted averaging with fuzzy quantifiers: GIS-based multicriteria evaluation for land-use suitability analysis. *Int J Appl Earth Obs Geoinf.* 2006;8:270–277.

25. Liu X, Han S. Orness and parameterized RIM quantifier aggregation with OWA operators: a summary. *Int J Approx Reason.* 2008;48(1):77–97.

26. Zhou LG, Chen HY. Continuous generalized OWA operator and its application to decision making. *Fuzzy Sets Syst.* 2011;168(1):18–34.

27. Carbonell M, Mas M, Mayor G. On a class of monotonic extended OWA operators. In: *Proceedings of 6th International Fuzzy Systems Conference.* vol. 3. IEEE; 1997:1695–1700.

28. Drobne S, Lisec A. Multi-attribute decision analysis in GIS: weighted linear combination and ordered weighted averaging. *Informatica.* 2009;33:459–474.

29. Jang JS. ANFIS: adaptive-network-based fuzzy inference system. *IEEE Trans Syst Man Cybern.* 1993;23(3):665–685.

30. Mansour MF, Morgenstern NR, Martin CD. Expected damage from displacement of slow-moving slides. *Landslides.* 2011;8(1):117–131.

31. Nguyen HT, Wiatr T, Fernández-Steeger TM, Reicherter K, Rodrigues DM, Azzam R. Landslide hazard and cascading effects following the extreme rainfall event on Madeira Island (February 2010). *Nat Hazards.* 2013;65(1):635–652.

32. Swets JA. Measuring the accuracy of diagnostic systems. *Science.* 1988;240:1285–1293.

33. Bahrami Y, Hassani H, Maghsoudi A. Landslide susceptibility mapping using AHP and fuzzy methods in the Gilan province, Iran. *GeoJournal.* 2020;1–20.

34. Sur U, Singh P, Meena SR. Landslide susceptibility assessment in a lesser Himalayan road corridor (India) applying fuzzy AHP technique and earth-observation data. *Geomat Nat Haz Risk.* 2020;11(1):2176–2209.

35. Zhou S, Zhou S, Tan X. Nationwide susceptibility mapping of landslides in Kenya using the fuzzy analytic hierarchy process model. *Landarzt.* 2020;9(12):535.

36. Mallows CL. More comments on Cp. *Technometrics.* 1995;37(4):362–372.

37. Duru O, Bulut E, Yoshida S. Regime switching fuzzy AHP model for choicevarying priorities problem and expert consistency prioritization: a cubic fuzzypriority matrix design. *Expert Syst Appl.* 2012;39:4954–4964.

38. Khakzad H. OWA operators with different Orness levels for sediment management alternative selection problem. *Water Supply.* 2020;20 (1):173–185.

39. Chen VYC, Pang Lien H, Liu CH, Liou JJH, Hshiung Tzeng G, Yang LS. Fuzzy MCDM approach for selecting the best environment-watershed plan. *Appl Soft Comput.* 2011;11:265–275.

11

Multiscale drought hazard assessment in the Philippines

Arnold R. Salvacion

Department of Community and Environmental Resource Planning, College of Human Ecology, University of the Philippines Los Baños, Los Baños, Laguna, Philippines

1 Introduction

Drought is a natural hazard that has a creeping, long-lasting, and severe impact on human society and the environment.[1–3] It is an extreme condition of the hydrologic cycle, which is characterized by drying and loss of stored water resources caused by a prolonged shortage of available water due to below-normal or inadequate precipitation, exceptionally high temperatures, and low humidity.[4–7] Drought can vary in magnitude, intensity, and spatial extent, which can last for weeks to years.[4, 5] It can occur almost everywhere and is considered a normal and recurrent feature of climate.[8, 9] Drought events can result in agricultural losses, water shortages, reduced hydropower supply, fires, migration, reduced labor or productivity, and famines.[1–3, 10–12]

Meteorological drought is the lack of precipitation over a period resulting from persistent or stagnant atmospheric high-pressure systems over a region.[9, 13] Meteorological drought is measured as a rainfall deficit from the long-term average or cumulative precipitation shortages.[9, 13] Although considered to have the least disaster potential, assessment of meteorological drought is important because it is the root cause of other droughts such as (1) hydrological, (2) agricultural, and (3) socio-economic.[9, 13–17] For example, the duration of meteorological drought lingers can transition to hydrological drought.[9] Hydrological drought can then transition to agricultural drought.[9] Finally, agricultural droughts can transition to socio-economic drought.[9]

In drought hazard assessment, the availability of observed data from meteorological stations is imperative.[18] The availability and accessibility of long-term records of climatic data are limited and vary from one country to another.[18–20] This is true for the Philippines, where long-term climatic records are still inaccessible and unavailable for most researchers and the general public.[21] Meanwhile, the recent development of high-resolution gridded climate data products can provide a solution to address the issues of data availability and accessibility. The TerraClimate dataset has a high spatial resolution (~4 km) monthly climate and climatic water balance (from 1958 to 2019) for the global terrestrial surface derived by interpolating and combining high spatial resolution climatological normal with time-varying (monthly) coarser resolution data.[22] According to Abatzoglou et al.,[22] TerraClimate consists of monthly data on rainfall, temperature (maximum and minimum), solar radiation, wind speed, vapor pressure, and surface water balance.

Several drought studies have been conducted in the Philippines. However, most of these studies are limited to identifying drought occurrences and hazards over a limited spatial extent or a shorter period.[23–26] This chapter aims to assess multiscale drought hazards in the Philippines using the SPI and TerraClimate datasets. It also demonstrates the use of PCA to develop a composite drought hazard map for the Philippines.

Computers in Earth and Environmental Sciences
https://doi.org/10.1016/B978-0-323-89861-4.00024-5

2 Methodology

2.1 Standardized precipitation index

The Standardized Precipitation Index (SPI) is a widely adopted tool to identify the magnitude and extent of drought in an area using long-term records (30 years or greater) of monthly rainfall data.[13, 18, 27–31] The SPI values were derived by fitting a gamma probability distribution function (Eq. 1) to the long series of monthly rainfall that is later converted to a normal distribution to calculate the number of standard deviations (Eq. 2), the observed monthly rainfall deviates from the long-term mean.[27, 30, 31] The calculated SPI is then classified into different categories (Table 1), where months with SPI values less than −1 are considered dry months.[27, 30, 31] According to He et al.,[8] the main advantage of SPI is its simplicity and the ability to provide a regional drought comparison. Lastly, the SPI can be calculated at time scales such as 1-, 3-, 6-, and 12-month periods to describe drought conditions that are imperative for meteorological, agricultural, and hydrological use.[30, 31]

$$G(x) = \frac{1}{\beta^\alpha \Gamma(\alpha)} x^{\alpha-1} e^{\frac{-x}{\beta}} \text{ for } x > 0 \tag{1}$$

where $\alpha > 0$ is the shape parameter, $\beta > 0$ is the scale parameter, $x > 0$ is the amount of precipitation, and $\Gamma(\alpha)$ is the gamma function.

$$SPI = \frac{x_i - \bar{x}}{\sigma} \tag{2}$$

where x_i is the precipitation during month i, and $\bar{x}$ and σ are the mean and standard deviation of the long-term precipitation record, respectively.

2.2 Principal component analysis

Principal component analysis (PCA) was applied to display patterns in multivariate data using distance-based ordination.[32] It aims to reduce the number of variables in a dataset into smaller dimensions for further analysis, which includes the development of a predictive model or construction of composite indicators that maximize the explainable variation from the original data set.[33–39] PCA transforms the dataset of potentially correlated variables into a combination of uncorrelated principal components (PCs) via an orthogonal method.[34, 40, 41] According to Vyas and Kumaranayake,[39] each PC is a linear weight combination of the initial variables. For example, in a set of variables, PCs are derived using Eq. (3).

$$
\begin{aligned}
PC_1 &= a_{11}X_1 + a_{12}X_2 + \cdots + a_{1n}X_n \\
&\vdots \\
PC_m &= a_{m1}X_1 + a_{m2}X_2 + \cdots + a_{mn}X_n
\end{aligned}
\tag{3}
$$

where a_{mn} is the weight for the mth PC for the nth variable.

2.3 Data and data processing

Monthly gridded rainfall data for the Philippines from 1989 to 2019 (30-year period) were downloaded from the TerraClimate webpage.[42] Calculation of SPI values for each month of the study period was performed using

TABLE 1 Classification of SPI values.

Index value	Classification
2.0 and above	Extremely wet
1.5 to 1.99	Very wet
1.0 to 1.49	Moderately wet
0.99 to −0.99	Near normal
−1.0 to −1.49	Moderately dry
−1.5 to −1.99	Severely dry
−2.0 and below	Extremely dry

the *SPEI* package in R software.[43, 44] Four (4) drought characteristics were calculated: (1) drought severity (Eq. 4), (2) total drought events (Eq. 5), (3) drought intensity (Eq. 6), and (4) drought frequency (Eq. 7).[28, 30] These four (4) drought characteristics were then combined into a single drought hazard index using PCA. The drought hazard index was calculated by summing up the factor loadings of the principal components with eigenvalues >1.[35, 45, 46] Finally, the computed drought hazard index was normalized using the min-max rescaling transformation (Eq. 8) for easier spatial comparison. Fig. 1 illustrates the methodology used in this study.

$$S = -\left(\sum_{j=1}^{x} SPI_{ij}\right),$$ (4)

where S is the drought severity, j is the first month when SPI_i is less than the threshold (-1) and continues for x months until SPI_i is greater than the threshold (-1). i represents the scale (i.e., 1-, 3-, 6-, and 12-month) used for the *SPI* calculation.

$$TD = \left(\sum_{j=1}^{x} DM_{ij}\right),$$ (5)

where TD is the total number of dry months (DM) ($SPI \leq -1$). i represents the (i.e., 1-, 3-, 6-, and 12-month) used for the *SPI* calculation.

$$DI = \frac{S}{TD}$$ (6)

where DI is the drought intensity, S is the drought severity, and TD is the total number of dry months.

$$DF = \frac{TD}{TM}$$ (7)

where DF is the drought frequency, TD is the sum of dry months, and TM is the total months of the study period.

$$NV = \frac{X_i - X_{min}}{X_{max} - X_{min}},$$ (8)

where NV is the normalized value

X_i is the actual value of the index
X_{min} is the minimum value of the index
X_{max} is the minimum value of the index

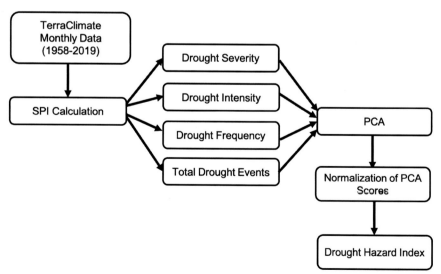

FIG. 1   Methodology flow diagram for multiscale drought hazard assessment.

3 Results

3.1 Multiscale drought characteristics

Fig. 2 shows the map of drought severity in the Philippines at different temporal scales. The difference in drought severity varies across countries and time scales. Higher drought severity was observed in the central and southern regions of the Philippines. Similarly, a higher number of drought events were also observed in these regions across different time scales (Fig. 3). However, in terms of intensity, the higher intensity was observed in the northern portion

FIG. 2 Drought severity map of the Philippines based on (A) 1-month; (B) 3-month; (C) 6-month; and (D) 12-month scale.

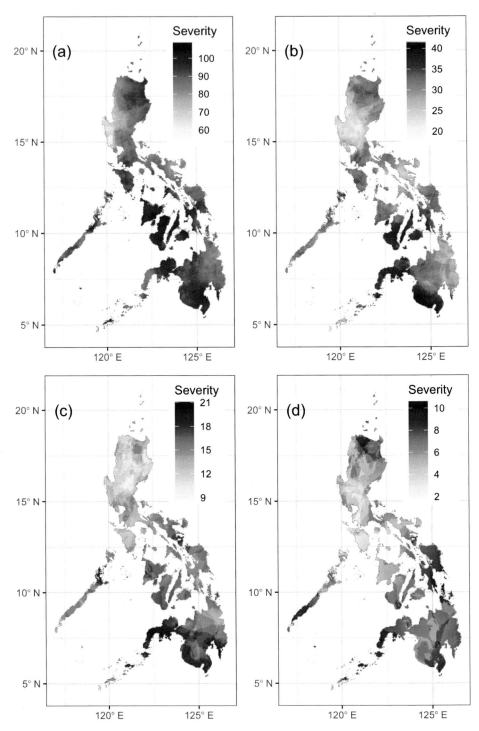

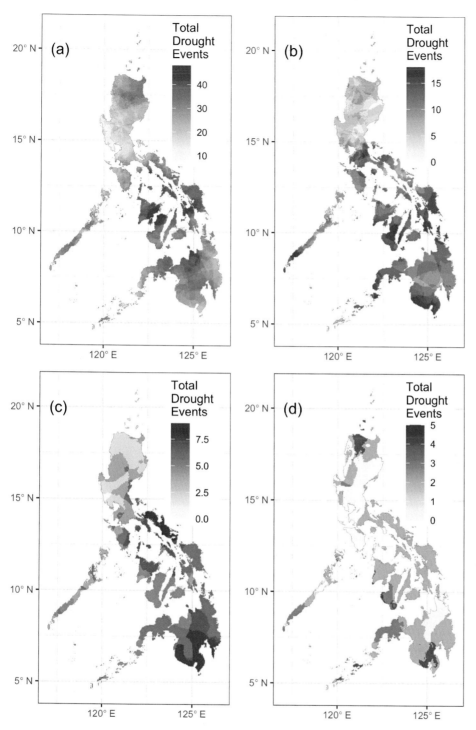

FIG. 3 Overall drought event map of the Philippines based on (A) 1-month; (B) 3-month; (C) 6-month; and (D) 12-month scale.

of the country at 1-, 3-, and 6-month time scales, except for the 12-month time scale (Fig. 4). A similar pattern of higher drought frequency was also observed for the central to the southern portion of the country for the 1-, 3-, and 6-month time scales, except for the 12-month time scale (Fig. 5).

3.2 Drought hazard

The eigenvalues from PCA showed that there are four (4) principal components (Fig. 6) that can be used to develop the drought hazard index for the Philippines. These four (4) principal-components can explain 84% of the total variation for the different drought characteristics at different scales. Fig. 7 shows a drought hazard map for the Philippines. Based on

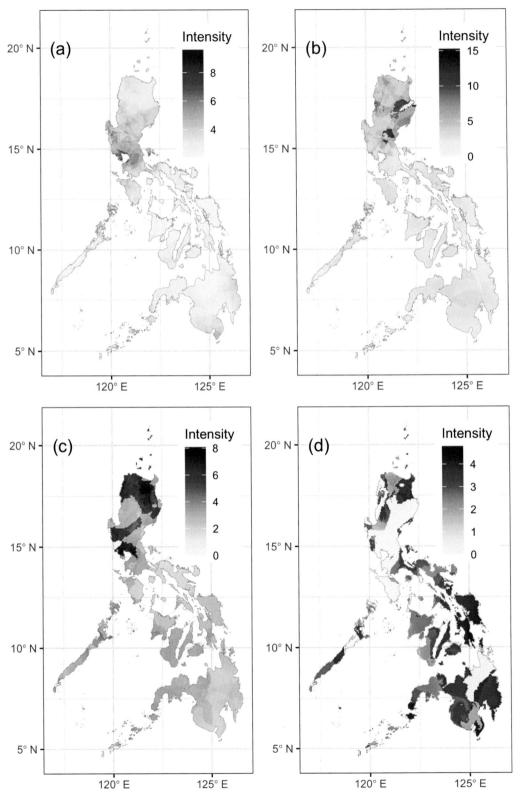

FIG. 4   Drought intensity map of the Philippines based on (A) 1-month; (B) 3-month; (C) 6-month; and (D) 12-month scale.

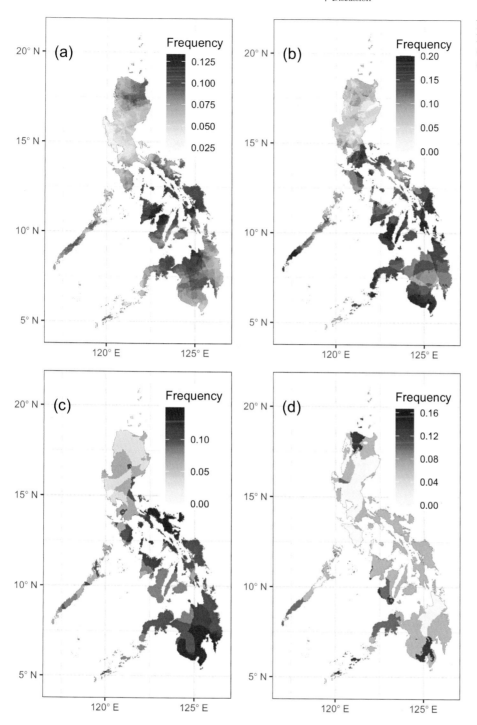

FIG. 5 Drought frequency map of the Philippines based on (A) 1-month; (B) 3-month; (C) 6-month; and (D) 12-month scale.

this map, approximately 72% of the country is under high (40%) and very high (32%) drought hazards. Meanwhile, around 21%, 6%, and less than 1% were under medium, low, and very low drought hazards, respectively.

4 Discussion

According to Lyon et al.,[47] seasonal precipitation in the Philippines is highly influenced by the El Niño Southern Oscillation (ENSO). Low precipitation results from warm ENSO events while cold ENSO events bring too much rainfall.[47] Lyon et al.[47] reported that during warm ENSO, rainfall in most weather stations in the country fell below

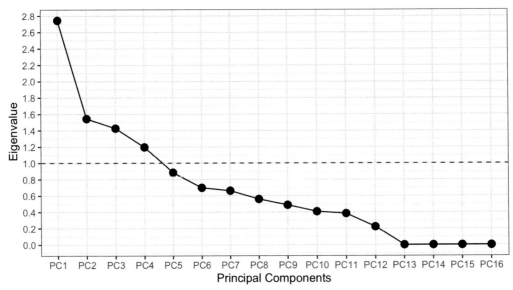

FIG. 6 Scree plot of different principal components for drought hazard index.

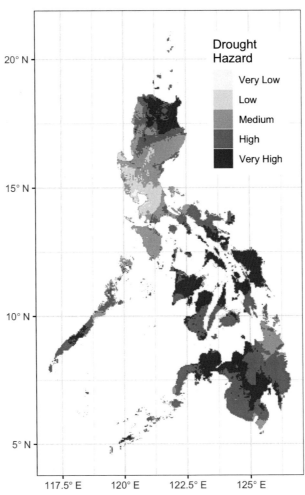

FIG. 7 Drought hazard map of the Philippines.

the observed median rainfall for that particular location with varying levels of significance. However, according to Delos Reyes and David,[48] the effect of ENSO on Philippine rainfall varies with location, duration, magnitude, and spatial extent sea surface temperature (SST) anomalies. Hilario et al.[49] showed that rainfall deviations from the normal during ENSO events vary across time and space. These observations support the results of this study, where the calculated drought characteristics in the Philippines vary across time scales and spaces. Such observations are evident in the maps of different drought characteristics at different temporal scales (e.g., 1-, 3-, 6-, and 12-month). Mapping these characteristics and combining them into a single index revealed areas in the Philippines where drought occurs longer, severely, and more frequently. This information can help formulate strategies to combat drought impacts.[50] In the era of climate change, drought trends and hazard mapping have become indispensable for mitigating its potential impact.[51] Understanding drought hazards is an integral component of drought risk assessment, which is the initial step in managing drought risk and impact minimization at different time scales and thematic foci.[52–54]

According to Hayes et al.,[55] the use of SPI has several advantages and disadvantages compared to other drought indices. SPI is simple and flexible, and requires only precipitation as input data, and can calculate drought at different time scales.[55] Next, SPI is not highly influenced by topography. Lastly, severe and extreme drought classification using SPI is consistent for any location and time scale.[55] Meanwhile, the output of the SPI is highly reliable for the quality of its input data.[55] The use of SPI alone cannot be used to identify drought-prone regions.[55] In addition, according to Vicente-Serrano et al.[56] SPI cannot be used for future drought under climate change conditions because it does not account for temperature increase. Similarly, the use of gridded climate data, such as TerraClimate data, also has advantages and disadvantages. One advantage of TerraClimate is that it can provide climatic data for areas with limited spatial and temporal climatic records.[22] On the other hand, the methodology (i.e., interpolation and extrapolation) and changes in input data (i.e., homogeneities and availability) used to produce the TerraClimate data can introduce inherent uncertainties to the final data product. However, according to Abatzoglou et al.,[22] these measures of uncertainty are provided along with the TerraClimate data for users to determine the robustness of the dataset for specific geographic regions.

5 Conclusion

This chapter demonstrates how open-geospatial data, standard precipitation index, and principal component analysis are combined to develop a drought hazard map for the Philippines. The results show that the drought characteristics of the country vary across time and space. These drought hazard maps can help focus efforts on drought hazard mitigation and impact minimization in the country. Meanwhile, caution should be exercised when using these results because of the inherent uncertainty in the input data and the methodology used.[22, 57–59] These approaches serve as an alternative to address the existing limitations of climate data availability and accessibility. The data and techniques used in this study can also be applied to other countries with climate data limitations similar to those of the Philippines.

References

1. Meza I, Siebert S, Döll P, et al. Global-scale drought risk assessment for agricultural systems. *Nat Hazards Earth Syst Sci.* 2020;20(2):695–712. https://doi.org/10.5194/nhess-20-695-2020.
2. Schwalm CR, Anderegg WRL, Michalak AM, et al. Global patterns of drought recovery. *Nature.* 2017;548(7666):202–205. https://doi.org/10.1038/nature23021.
3. Touma D, Ashfaq M, Nayak MA, Kao S-C, Diffenbaugh NS. A multi-model and multi-index evaluation of drought characteristics in the 21st century. *J Hydrol.* 2015;526:196–207. https://doi.org/10.1016/j.jhydrol.2014.12.011.
4. Bullock JA, Haddow GD, Coppola DP. Hazards. In: Bullock JA, Haddow GD, Coppola DP, eds. *Homeland Security.* 2nd ed. Butterworth-Heinemann; 2018:45–66. https://doi.org/10.1016/B978-0-12-804465-0.00003-0 [chapter 3].
5. Haddow GD, Bullock JA, Coppola DP. Natural and technological hazards and risk assessment. In: Haddow GD, Bullock JA, Coppola DP, eds. *Introduction to Emergency Management.* 6th ed. Butterworth-Heinemann; 2017:33–77. https://doi.org/10.1016/B978-0-12-803064-6.00002-0 [chapter 2].
6. He X, Estes L, Konar M, et al. Integrated approaches to understanding and reducing drought impact on food security across scales. *Curr Opin Environ Sustain.* 2019;40:43–54. https://doi.org/10.1016/j.cosust.2019.09.006.
7. Lybbert TJ, Carter MR. Bundling drought tolerance and index insurance to reduce rural household vulnerability to drought. In: Balisacan AM, Chakravorty U, Ravago M-LV, eds. *Sustainable Economic Development.* Academic Press; 2015:401–414. https://doi.org/10.1016/B978-0-12-800347-3.00022-4 [chapter 22].

8. He B, Lü A, Wu J, Zhao L, Liu M. Drought hazard assessment and spatial characteristics analysis in China. *J Geogr Sci*. 2011;21(2):235–249. https://doi.org/10.1007/s11442-011-0841-x.
9. Şen Z. *Applied Drought Modeling, Prediction, and Mitigation*. 1st ed. Elsevier; 2015. http://www.sciencedirect.com/science/article/pii/B9780128021767099946.
10. Edwards B, Gray M, Hunter B. The impact of drought on mental health in rural and regional Australia. *Soc Indic Res*. 2015;121(1):177–194. https://doi.org/10.1007/s11205-014-0638-2.
11. OBrien LV, Berry HL, Coleman C, Hanigan IC. Drought as a mental health exposure. *Environ Res*. 2014;131:181–187. https://doi.org/10.1016/j.envres.2014.03.014.
12. Wu B, Ma Z, Yan N. Agricultural drought mitigating indices derived from the changes in drought characteristics. *Remote Sens Environ*. 2020;244:111813. https://doi.org/10.1016/j.rse.2020.111813.
13. Mishra AK, Singh VP. A review of drought concepts. *J Hydrol*. 2010;391(1):202–216. https://doi.org/10.1016/j.jhydrol.2010.07.012.
14. Hollins S, Dodson J. Drought. In: Bobrowsky Peter T, ed. *Encyclopedia of Natural Hazards*. Springer Netherlands; 2013:189–197. Encyclopedia of Earth Sciences Series; https://doi.org/10.1007/978-1-4020-4399-4_98.
15. Ma B, Zhang B, Jia L, Huang H. Conditional distribution selection for SPEI-daily and its revealed meteorological drought characteristics in China from 1961 to 2017. *Atmos Res*. 2020;246:105108. https://doi.org/10.1016/j.atmosres.2020.105108.
16. Senay GB, Velpuri NM, Bohms S, et al. Drought monitoring and assessment: remote sensing and modeling approaches for the famine early warning systems network. In: JFSPD B, ed. *Hydro-Meteorological Hazards, Risks and Disasters*. Elsevier; 2015:233–262. http://www.sciencedirect.com/science/article/pii/B9780123948465000096. [chapter 9].
17. Zargar A, Sadiq R, Naser B, Faisal K. A review of drought indices. *Environ Rev*. 2011. https://doi.org/10.1139/a11-013.
18. Duan K, Xiao W, Mei Y, Liu D. Multi-scale analysis of meteorological drought risks based on a Bayesian interpolation approach in Huai River basin, China. *Stoch Env Res Risk A*. 2014;28(8):1985–1998. https://doi.org/10.1007/s00477-014-0877-4.
19. Moreno A, Hasenauer H. Spatial downscaling of European climate data. *Int J Climatol*. 2016;36(3):1444–1458. https://doi.org/10.1002/joc.4436.
20. Wijngaard JB, Tank KGAM, Können GP. Homogeneity of 20th century European daily temperature and precipitation series. *Int J Climatol*. 2003;23(6):679–692. https://doi.org/10.1002/joc.906.
21. Salvacion AR, Magcale-Macandog DB, Cruz PCS, Saludes RB, Pangga IB, CJR C. Evaluation and spatial downscaling of CRU TS precipitation data in the Philippines. *Model Earth Syst Environ*. 2018;4(3):891–898. https://doi.org/10.1007/s40808-018-0477-2.
22. Abatzoglou JT, Dobrowski SZ, Parks SA, Hegewisch KC. TerraClimate, a high-resolution global dataset of monthly climate and climatic water balance from 1958–2015. *Sci Data*. 2018;5(1):1–12. https://doi.org/10.1038/sdata.2017.191.
23. Jaranilla-Sanchez PA, Wang L, Koike T. Modeling the hydrologic responses of the Pampanga River basin, Philippines: a quantitative approach for identifying droughts: drought quantification in the Philippines. *Water Resour Res*. 2011;47(3). https://doi.org/10.1029/2010WR009702.
24. Hasegawa A, Gusyev M, Ushiyama T, Magome J, Iwami Y. Drought assessment in the Pampanga river basin, the Philippines—part 2: a comparative SPI approach for quantifying climate change hazards. In: *Proceedings—21st International Congress on Modelling and Simulation, MODSIM 2015*; 2015:2388–2394.
25. Perez GJ, Macapagal M, Olivares R, Macapagal EM, Comiso JC. Forecasting and monitoring agricultural drought in the Philippines. In: *The International Archives of the Photogrammetry, Remote Sensing and Spatial Information Sciences*. vol. 41; 2016:1263–1269. https://doi.org/10.5194/isprsarchives-XLI-B8-1263-2016.
26. Valete MAP, Perez GJP, Enricuso OB, Comiso JC. Spatiotemporal evaluation of historical drought in the Philippines. In: *40th Asian Conference on Remote Sensing, ACRS 2019: Progress of Remote Sensing Technology for Smart Future*; 2020.
27. McKee TB, Doesken NJ, Kleist J. *Relationship of drought frequency and duration to time scales*; 1993.
28. Awchi TA, Kalyana MM. Meteorological drought analysis in northern Iraq using SPI and GIS. *Sustain Water Resour Manag*. 2017;3(4):451–463. https://doi.org/10.1007/s40899-017-0111-x.
29. Santos JF, Portela MM, Pulido-Calvo I. Regional frequency analysis of droughts in Portugal. *Water Resour Manag*. 2011;25(14):3537. https://doi.org/10.1007/s11269-011-9869-z.
30. Singh GR, Jain MK, Gupta V. Spatiotemporal assessment of drought hazard, vulnerability and risk in the Krishna River basin, India. *Nat Hazards*. 2019;99(2):611–635. https://doi.org/10.1007/s11069-019-03762-6.
31. Tirivarombo S, Osupile D, Eliasson P. Drought monitoring and analysis: standardised precipitation evapotranspiration index (SPEI) and standardised precipitation index (SPI). *Phys Chem Earth Parts A/B/C*. 2018;106:1–10. https://doi.org/10.1016/j.pce.2018.07.001.
32. Syms C. Principal components analysis. In: Fath B, ed. *Encyclopedia of Ecology*. 2nd ed. Elsevier; 2019:566–573. https://doi.org/10.1016/B978-0-12-409548-9.11152-2.
33. Abdelaziz AEM, Leite GB, Hallenbeck PC. Addressing the challenges for sustainable production of algal biofuels: II. Harvesting and conversion to biofuels. *Environ Technol*. 2013;34(13–14):1807–1836. https://doi.org/10.1080/09593330.2013.831487.
34. Li T, Zhang H, Yuan C, Liu Z, Fan C. A PCA-based method for construction of composite sustainability indicators. *Int J Life Cycle Assess*. 2012;17(5):593–603. https://doi.org/10.1007/s11367-012-0394-y.
35. Rabby YW, Hossain MB, Hasan MU. Social vulnerability in the coastal region of Bangladesh: an investigation of social vulnerability index and scalar change effects. *Int J Disast Risk Reduct*. 2019;41:101329. https://doi.org/10.1016/j.ijdrr.2019.101329.
36. Roessner U, Nahid A, Chapman B, Hunter A, Bellgard M. 1.31—Metabolomics—the combination of analytical biochemistry, biology, and informatics. In: Moo-Young M, ed. *Comprehensive Biotechnology*. 3rd ed. Pergamon; 2011:435–447. https://doi.org/10.1016/B978-0-444-64046-8.00027-6.
37. Santos M, Fragoso M, Santos JA. Regionalization and susceptibility assessment to daily precipitation extremes in mainland Portugal. *Appl Geogr*. 2017;86:128–138. https://doi.org/10.1016/j.apgeog.2017.06.020.
38. Tripathi M, Singal SK. Use of principal component analysis for parameter selection for development of a novel water quality index: a case study of river ganga India. *Ecol Indic*. 2019;96:430–436. https://doi.org/10.1016/j.ecolind.2018.09.025.
39. Vyas S, Kumaranayake L. Constructing socio-economic status indices: how to use principal components analysis. *Health Policy Plan*. 2006;21(6):459–468. https://doi.org/10.1093/heapol/czl029.
40. Basu T, Das A. Identification of backward district in India by applying the principal component analysis and fuzzy approach: a census based study. *Socio Econ Plan Sci*. 2020. https://doi.org/10.1016/j.seps.2020.100915, 100915.

41. Pearson K. LIII. On lines and planes of closest fit to systems of points in space. *The London Edinburgh Dublin Philos Mag J Sci.* 1901;2(11):559–572. https://doi.org/10.1080/14786440109462720.

42. TerraClimate. Climatology Lab; 2021. Accessed March 23, 2021 http://www.climatologylab.org/terraclimate.html.

43. Beguería S, Vicente-Serrano SM. *SPEI: Calculation of the Standardised Precipitation-Evapotranspiration Index*; 2017.

44. Ihaka R, Gentleman R. R: a language for data analysis and graphics. *J Comput Graph Stat.* 1996;5(3):299–314. https://doi.org/10.1080/10618600.1996.10474713.

45. Abson DJ, Dougill AJ, Stringer LC. Using Principal Component Analysis for information-rich socio-ecological vulnerability mapping in Southern Africa. *Appl Geogr.* 2012;35(1):515–524. https://doi.org/10.1016/j.apgeog.2012.08.004.

46. Ravago M-LV, Mapa CDS, Aycardo AG, Abrigo MRM. Localized disaster risk management index for the Philippines: is your municipality ready for the next disaster? *Int J Disast Risk Reduct.* 2020;51:101913. https://doi.org/10.1016/j.ijdrr.2020.101913.

47. Lyon B, Cristi H, Verceles ER, Hilario FD, Abastillas R. Seasonal reversal of the ENSO rainfall signal in the Philippines. *Geophys Res Lett.* 2006;33 (24). https://doi.org/10.1029/2006GL028182.

48. de los Reyes RB, David WP. Spatial and temporal effects of El Niño on Philippine rainfall and cyclones. *Philipp Agric Sci.* 2006;89(4):296–308.

49. Hilario FD, De Guzman R, Ortega D, Hayman P, Alexander B. El Niño southern oscillation in the Philippines: impacts, forecasts, and risk management. *Philipp J Dev.* 2009;34(1):9–34.

50. Zagade ND, Umrikar BN. Drought severity modeling of upper Bhima river basin, western India, using GIS–AHP tools for effective mitigation and resource management. *Nat Hazards.* 2021;105(2):1165–1188. https://doi.org/10.1007/s11069-020-04350-9.

51. Pandey V, Srivastava PK, Singh SK, Petropoulos GP, Mall RK. Drought identification and trend analysis using long-term CHIRPS satellite precipitation product in Bundelkhand, India. *Sustainability.* 2021;13(3):1042. https://doi.org/10.3390/su13031042.

52. Blauhut V. The triple complexity of drought risk analysis and its visualisation via mapping: a review across scales and sectors. *Earth Sci Rev.* 2020;210. https://doi.org/10.1016/j.earscirev.2020.103345.

53. Rossi G. Drought mitigation measures: a comprehensive framework. In: Vogt JV, Somma F, eds. *Drought and Drought Mitigation in Europe.* Springer Netherlands; 2000:233–246. Advances in Natural and Technological Hazards Research;.

54. Yang T-H, Liu W-C. A general overview of the risk-reduction strategies for floods and droughts. *Sustainability.* 2020;12(7):2687. https://doi.org/10.3390/su12072687.

55. Hayes MJ, Svoboda MD, Wiihite DA, Vanyarkho OV. Monitoring the 1996 drought using the standardized precipitation index. *Bull Am Meteorol Soc.* 1999;80(3):429–438. https://doi.org/10.1175/1520-0477(1999)080<0429:MTDUTS>2.0.CO;2.

56. Vicente-Serrano SM, Beguería S, López-Moreno JI. A multiscalar drought index sensitive to global warming: the standardized precipitation evapotranspiration index. *J Clim.* 2010;23(7):1696–1718. https://doi.org/10.1175/2009JCLI2909.1.

57. Dunn RJH, Donat MG, Alexander LV. Investigating uncertainties in global gridded datasets of climate extremes. *Clim Past.* 2014;10(6):2171–2199. https://doi.org/10.5194/cp-10-2171-2014.

58. Parkes B, Higginbottom TP, Hufkens K, Ceballos F, Kramer B, Foster T. Weather dataset choice introduces uncertainty to estimates of crop yield responses to climate variability and change. *Environ Res Lett.* 2019;14(12):124089. https://doi.org/10.1088/1748-9326/ab5ebb.

59. Prein AF, Gobiet A. Impacts of uncertainties in European gridded precipitation observations on regional climate analysis. *Int J Climatol.* 2017;37 (1):305–327. https://doi.org/10.1002/joc.4706.

12

Selection of the best pixel-based algorithm for land cover mapping in Zagros forests of Iran using Sentinel-2A data: A case study in Khuzestan province

Saeedeh Eskandari[a], Sajjad Ali Mahmoudi Sarab[b], Mehdi Pourhashemi[a], and Fatemeh Ahmadloo[c]

[a]Forest Research Division, Research Institute of Forests and Rangelands, Agricultural Research, Education and Extension Organization (AREEO), Tehran, Iran [b]Khuzestan Agricultural and Natural Resources Research and Education Center, Agricultural Research, Education and Extension Organization (AREEO), Tehran, Khuzestan, Iran [c]Poplar and Fast-growing Trees Research Division, Research Institute of Forests and Rangelands, Agricultural Research, Education and Extension Organization (AREEO), Tehran, Iran

1 Introduction

The Zagros forests of Iran with an area of about 6.07 million ha in the west of the country,[1] were created more than 5500 years ago.[2] Plant and animal biodiversity in these forests and rangelands are unique in western Iran, which makes it one of the most important biodiversity hotspots in the west of the country.[3] In addition, Zagros forests and rangelands play important roles in water and soil conservation in mountainous landscapes.[4] Unfortunately, these forests and rangelands in Khuzestan province in southwestern Iran have been destroyed by natural events (climate change, drought events, fire occurrence, etc.) or human-made factors (grazing, agricultural development, urbanization, etc.) in recent years.[5–7] Despite the wide destruction of these ecosystems, there is no complete planning for the protective management of these forests. The first step for optimum management is to provide up-to-date information on the situation of these forests. Therefore, spatial and temporal mapping of these forests and rangelands is necessary to manage these ecosystems.[8] In this case, land cover mapping with accurate algorithms can provide valuable information about the distribution and area of different land covers inside these forests. Satellite imagery can provide temporal and spatial information from forest resources, especially in wide and mountainous areas such as the Zagros forests.

To date, some studies have been performed for land cover mapping using different satellite images and various algorithms in western Iran.[8–14] Despite these studies performed in the west of the country, land cover mapping at a spatial resolution of 10 m and novel pixel-based algorithms have not been performed in the forests and rangelands of Khuzestan province; in contrast, Sentinel-2 satellite images have been successfully applied for mapping different characteristics of forests around the world in recent years. For example, the Sentinel-2 satellite image is a very efficient data for vegetation and land cover mapping,[15–17] mapping the different crop type,[18, 19] mapping the tree species,[19] mapping the vegetation type,[20] mapping the forest type,[21] estimation of forest succession,[22] and investigation of defoliation of forests.[23] The results of these studies have confirmed that Sentinel-2 data have high efficiency for land cover and forest mapping in natural areas.

Regarding the wide destruction of Zagros forests of Iran in Khuzestan province, new research for the selection of the best pixel-based algorithm for land cover mapping in the forest area will be very useful for optimum management of these forests. This study aimed to map the current natural and human-made land cover in a part of the Zagros forests in Khuzestan province using Sentinel-2A data with a moderate spatial resolution (10m). However, the main purpose is to compare different pixel-based algorithms to select the most accurate algorithm for land cover mapping in the Zagros forests of Khuzestan province. The results of this study will help forest managers apply the most accurate method for periodic land cover mapping at 10m resolution in the entire Zagros forests in Khuzestan province. Land cover mapping at a spatial resolution of 10m and novel pixel-based algorithms have not been performed in the forests and rangelands of Khuzestan province which is the novelty of this study.

2 Material and methods

2.1 Study area

This study was performed in one 1:25,000 sheet, which is a part of the Zagros forest in Khuzestan province in southwestern Iran (Fig. 1A). The study area has been located between 382,097 to 393,763 east longitude and 3,541,113 to 3,555,079 north latitude. This sheet was selected as the study area, which included different land covers (dense forest, semidense forest, sparse forest, rangeland, garden, agriculture, bare soil-urban area, and water) (Fig. 1B). To select this sheet as the study area, a map of 1:25,000 sheets in Khuzestan province (obtained from Iranian Mapping Organization) was imported to Google Earth Pro.7 and a proper sheet including all land covers was selected as the study area (Fig. 1B). Then, the Sentinel-2A satellite image of this sheet with an area of approximately 16,000 ha was clipped in GIS (Fig. 2). This sheet (study area) was located in Izeh County in northeastern Khuzestan province in Zone 38 (Northern Hemisphere) of the UTM WGS 1984 coordinate system (Fig. 1). The climate of the study area is semihumid to humid in winter and semiarid in summer. The annual rainfall mean is 760mm in the study area.

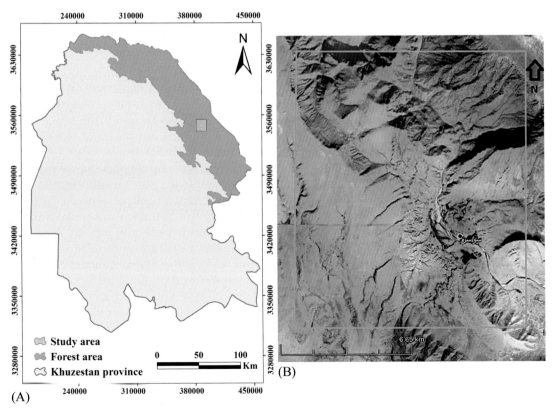

FIG. 1 Location of the study area in Zagros forest area in Khuzestan province (A), and landscape of the study area in Google Earth imagery (2020) (B).

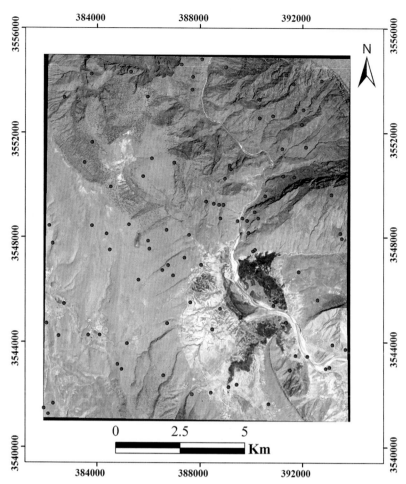

FIG. 2 Sentinel-2A satellite image of the study area and the random points for accuracy assessment.

2.2 Data

Sentinel-2A is a multispectral imaging mission developed by European Space Agency (ESA) as part of the Copernicus Programme, supporting the Copernicus Land Monitoring services, including the monitoring of vegetation, soil, and water cover, as well as the observation of inland waterways and coastal areas.[24] Sentinel-2A has 13 spectral bands with a spatial resolution of 10–60 m.

For this research, a Sentinel-2A Level-1C image for the study area (July 28, 2017) was downloaded from the ESA website (https://scihub.copernicus.eu).[24] The map of Khuzestan province was also provided by the Forest, Rangeland, and Watershed Organization of Iran. The Boundary of the Zagros forests in Khuzestan province (northeast of the province) was digitized using Google Earth Pro. 7 (the green polygon in Fig. 1A). In addition, a map of 1:25,000 sheets covering the Khuzestan province was prepared from the Iranian Mapping Organization. Then, one sheet covering the Zagros forests in Khuzestan province was separated as the study area (pink square in Fig. 1A).

2.3 Methodology

After selecting a sheet as the study area (Fig. 1B), some minor corrections were performed on the Sentinel-2A image of this sheet to reveal different land covers. Then, a layer stack of bands 2, 3, and 4 (blue, green, red) and 8 (NIR) was performed to obtain a color satellite image of the study area (Fig. 2). All bands used for layer stacking have 10 m spatial resolution.

2.3.1 Supervised classification of Sentinel-2A image

In this study, five pixel-based algorithms, including maximum likelihood (ML), minimum distance (MD), Mahalanobis distance (MaD), support vector machine (SVM), and neural network (NN), were used for supervised classification of the Sentinel-2A image in Envi 5.3. These algorithms were selected because of their ability to perform land cover classification in previous studies.[8] In this study, for pixel-based supervised classifications of Sentinel-2A, 20 areas of interest (AOI) as pixel-based training areas were selected for each class of land cover (dense forest, semidense forest, sparse forest, rangeland, garden, agriculture, bare soil-urban area, and water) in Envi 5.3. Then, different pixel-based algorithms were implemented on the Sentinel-2A image in Envi 5.3 software and the final land cover maps were constructed. The algorithms used in this study are described below.

Maximum likelihood (ML)

The most common supervised classification algorithm used in applications of remote sensing applications is the maximum likelihood, which is a parametric statistical method.[25, 26] It computes a probability density function considering the spectral distribution of the data to determine the probability of a pixel belonging to a specific class.[26] This method assigns all unclassified pixels to the class with the highest probability.[25, 27]

Minimum distance (MD)

A Minimum distance algorithm is a supervised classification method that classifies all pixels to the nearest class unless a standard deviation or distance threshold is specified. In fact, this method calculates the Euclidean distance between the values of pixels (xp,yp) and the mean values for the classes. Then it allocates the pixel to that class with the shortest Euclidean distance.[26, 28]

Mahalanobis distance (MaD)

The Mahalanobis distance classification is like the maximum likelihood classifier, but assumes that all class co-variances are equal; therefore, it is a faster method.[29] All pixels are classified to the closest region of interest (ROI) class unless a distance threshold is specified, in which case some pixels may be unclassified if they do not meet the criteria.[30]

Support vector machine (SVM)

Support vector machine (SVM) algorithms are nonlinear binary classifiers that find a threshold to divide the pixels into predefined classes using training area.[31] Optimal separation is applied to minimize misclassifications, which usually occur in the training step.[23, 32] SVMs minimize the upper bound of the expected errors by maximizing the margin between the separating hyperplane and the data.[33, 34] The concept of margin indicates the capability of SVMs.[35, 36]

Neural network (NN)

A neural network classifier is another supervised classification algorithm. The multilayer perceptron is the best neural network classifier for remote sensing applications. Feedforward or multilayer perceptrons (MLP) may have one or more hidden layers of neurons between the input and output layers.[27] This method is independent of the statistical parameters of a particular class and accepts qualitative and quantitative data to be introduced as input data.

2.3.2 Accuracy assessment of pixel-based algorithms for land cover mapping

In this study, the accuracy of the land cover maps was assessed using Google Earth imagery. For this purpose, a set of random points was created in the software and the accuracy of the classified land cover maps was assessed using these points. In this study, 100 stratified random points were created (Fig. 2). The stratified random distribution (stratification method) was used to assign a sufficient number of points to each class of land cover.[37] In this method, the number of points assigned to each class is based on the area of that class and the points are randomly distributed in each class. Based on previous studies, the stratification method has shown more accuracy than other methods for forest inventory in estimating the forest area in Zagros forests.[8, 38] Stratified random sampling has also been recommended in other studies for reference data collection because the reference data should cover the full range of the study area and all the land covers.[39] Therefore, this method was applied in this study for the accuracy assessment

Finally, all the random points with certain geographical coordinates were imported to ArcMap 10.4. These points were found on Google Earth Pro. 7, and their current land cover was determined at the pixel area (10 m × 10 m) on Google Earth (Fig. 3). This means that the current cover of each random point was checked in a square plot with dimensions of 10 m × 10 m, which was by the pixel size (10 m × 10 m) in the Sentinel-2A satellite image (Fig. 3).

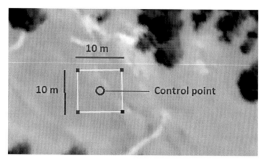

FIG. 3 A control point in a plot area for checking the land cover in Google Earth (an example of rangeland cover).

For each square plot (Fig. 3), the real land cover class (in Google Earth) and the classified land cover class (in the classified map) were compared to evaluate the accuracy of different supervised algorithms. The overall accuracy (*OA*) and Kappa index (*k*) were used for accuracy assessment.[40, 41] The Overall accuracy was calculated using Eq. (1):

$$OA = \frac{\sum_{i=1}^{j} n_{ii}}{n} \times 100 \tag{1}$$

The Kappa Index (*k*) was obtained from Eq. (2)[40, 42, 43]:

$$k = \frac{n\sum_{i=1}^{j} n_{ii} - \sum_{i=1}^{j} n_{i+}n_{+i}}{n^2 - \sum_{i=1}^{j} n_{i+}n_{+i}} \tag{2}$$

n: number of the reference pixels (real pixels) in the error matrix
n_{ii}: sum of the pixels in the main diamond (correct classification) in the error matrix.
n_{i+}: number of pixels in row i of the error matrix
n_{+i}: number of pixels in column i of the error matrix
j: number of classes

The value of overall accuracy (*OA*) ranges from 0 to 100, where values less than 50%, mean undesirable classification, and values close to 100 indicate perfect classification results. The value of the Kappa Index (*k*) ranges from 0 to 1, where values close to 0 indicate the low accuracy of the classification method and values close to 1 indicates the high accuracy of the classification algorithm.[40, 44]

3 Results

3.1 The land cover maps by pixel-based algorithms

The land cover maps obtained from the supervised classification of Sentinel-2A images using pixel-based algorithms are shown in Fig. 4. In addition, the areas of different land covers in the land-cover maps by different algorithms are shown in Table 1.

3.2 Accuracy assessment of pixel-based algorithms for land cover mapping

The overall accuracy and kappa index were applied for the accuracy assessment of land cover maps obtained from different algorithms in this study. The results of the accuracy assessment of the pixel-based algorithms are given in Table 2. The results show that the support vector machine (SVM) algorithm created the most accurate land cover map in the study area (Fig. 4D, Table 2).

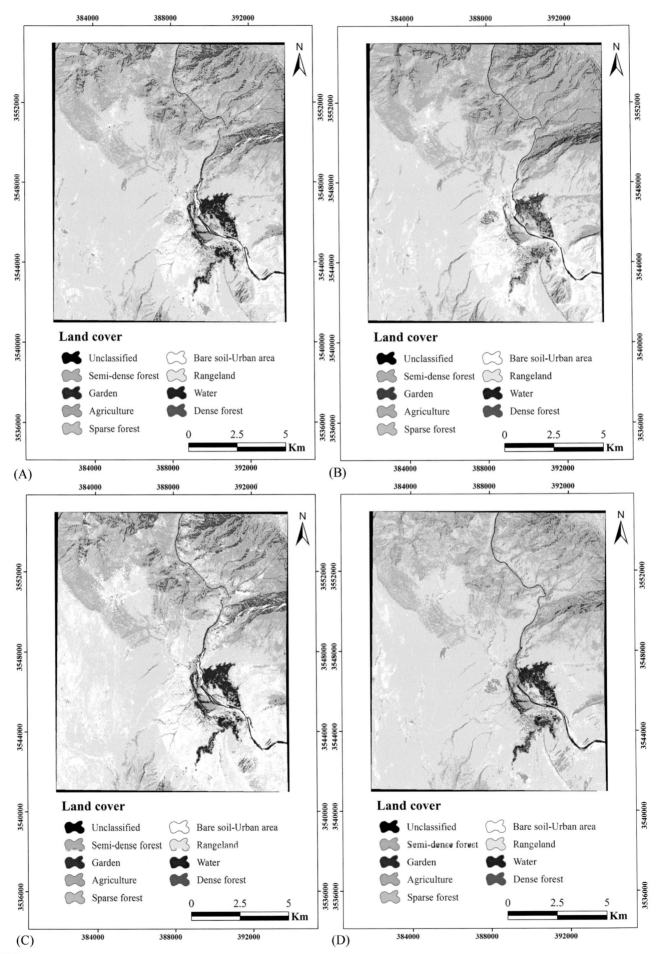

FIG. 4

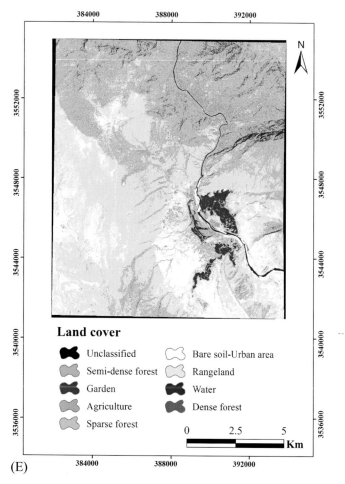

(E)

FIG. 4, CONT'D The supervised classification of Sentinel-2A image by pixel-based algorithms. (A) Support vector machine, (B) maximum likelihood, (C) Mahalanobis distance, (D) minimum distance, and (E) neural network.

TABLE 1 The areas of different land covers in the land cover maps by pixel-based algorithms.

Area of land cover (ha)	Pixel-based algorithm				
	Support vector machine	Maximum likelihood	Mahalanobis distance	Minimum distance	Neural network
Dense forest	812.41	632.88	834.29	418.19	361.96
Semidense forest	533.1	985.44	363.24	569.69	838.13
Sparse forest	3609.47	3601.03	3675.83	2915.77	3288.51
Rangeland	8859.84	8433.98	5865.77	10,260	6431.14
Garden	348.23	137.59	369.6	167.92	360.71
Agriculture	441.28	845.03	41.01	955.06	2665.36
Bare soil-urban area	1663.61	1629.95	5118.43	970.71	2298.67
Water	68.46	70.5	68.23	79.06	91.92
Total	16,336.4	16,336.4	16,336.4	16,336.4	16,336.4

TABLE 2 The accuracy assessment of pixel-based algorithms.

Pixel-based algorithm	Accuracy assessment index	
	Overall accuracy (percent)	Kappa index
Support vector machine	88	0.81
Maximum likelihood	78	0.66
Mahalanobis distance	70	0.58
Minimum distance	71	0.52
Neural network	66	0.51

4 Discussions

Regarding the extensive destruction of Zagros forests in Khuzestan province in southwestern Iran in recent years, this research was performed to map different land covers in a part of these forests using Sentinel-2A data at a spatial resolution of 10 m. In addition, the accuracy of different pixel-based algorithms for providing the land cover map for the entire Khuzestan province was compared in this study. The purpose of this study was to select the best pixel-based algorithm for land cover mapping in the Zagros forests of Khuzestan province by Sentinel-2A data.

The Results demonstrated that the SVM was superior to the other methods because it showed higher accuracy. Based on the results of this study, this algorithm, with an overall accuracy of 88% and a Kappa index of 0.81 (40 of 42), had the highest accuracy in image classification and land cover mapping in the study area in Khuzestan province (Table 2). The results of other studies performed in heterogeneous areas of Iran[8, 45–47] and around the world[15, 17, 20, 23, 48] also showed the highest precision of the SVM method for land cover mapping in natural ecosystems. This classifier is very useful for classifying the satellite image, especially in heterogeneous areas. As the study area has a very heterogeneous structure (Fig. 1B), SVM has shown high accuracy in land-cover mapping in the study area.

After SVM, maximum likelihood with an Overall accuracy of 0.78 and kappa index of 0.66, had the highest accuracy in land cover mapping in the forest area of Khuzestan province. Stefanov et al.,[49] Erbek et al.,[50] Mustapha et al.,[27] Mahdavi and Fallah Shamsi,[9] Heydarian et al.,[51] and Borzafkan et al.[52] also showed that the Maximum likelihood method has high precision in detecting the forest area, which is following the results of the current study.

Investigation of the land cover map constructed by the SVM algorithm indicated that from the total area of the study area (16,336.4 ha), the dense forest covers 812.41 ha, the semidense forest covers 533.1 ha, and the sparse forest covers 3609.47 ha of the study area. In addition, areas of rangelands and gardens have been 8859.84 ha and 348.23 ha, respectively. Agriculture covered 441.28 ha of the study area. Furthermore, 1663.61 ha of the study area has been covered by bare soil-urban areas and 68.46.ha of the study area was covered with water (Table 1).

Based on these results, measures were taken to protect the natural land covers (forests and rangelands) of the study area and to control the human-made land covers (agricultural lands, gardens, and urban areas) in the landscape of Zagros forests in Khuzestan Province are very important. Proper management of the Zagros forests in Khuzestan province is essential for increasing the area from dense forests (812.41 ha) to sparse forests (3609.47 ha) in the study area. For management of the remained forests, a protective plan should be performed. These forests are an important habitat for endemic fauna and flora and should be protected against human activities. In addition, they are important for water and soil conservation. If they are not protected, they may be converted to agricultural lands, bare soils, or urban areas in the near future. Therefore, these forests should be protected and rescued by plantations and forestation. Fortunately, rangelands have covered a wide area of the study area (8859.84 ha), which are important habitats for other fauna and flora in the Zagros Mountains. These rangelands should be protected to prevent their conversion to human-made land cover.

Monitoring human-made land covers (agricultural lands, gardens, and urban areas) should be performed by natural resource managers. Prevention of more development of human-made covers inside Zagros forests in Khuzestan province is very important to conserve forests and rangelands in the study area.

Based on the results of this study, the SVM method is the best classifier for land cover mapping in the entire Zagros forests in Khuzestan province. The most important advantage of the SVM algorithm is that it applies an optimal separation of pixel clusters to minimize misclassifications usually occurring in the training phase.[23, 32] Therefore, this classifier is suitable for heterogeneous areas with several land covers. Whereas the study area has a very heterogeneous structure (Fig. 1B), SVM has shown high accuracy in land cover mapping in the study area. Therefore, it is suggested that the land cover mapping be performed in the entire Zagros forests in this province using Sentinel-2A data and SVM algorithm in a certain interval time series.

5 Conclusions

This research aimed to select the best pixel-based algorithm for land cover mapping in the Zagros forests of Khuzestan province using Sentinel-2A data at a spatial resolution of 10 m. In this study, the SVM algorithm showed the highest accuracy for land cover mapping in the Zagros forests of Khuzestan province. Therefore, mapping different land cover inside Zagros forests in this province using Sentinel-2A satellite image and SVM algorithm will provide useful data for forest managers to protect the natural resources of Khuzestan province and to control the human-made land covers.

References

1. *Forests, Rangelands and Watershed Organization of Iran*. Forests of Iran; 2016. www.frw.org.ir.
2. Potts DT. *The Archaeology of Elam, Formation and Transformation of an Ancient Iranian State*. 2rd ed. Cambridge: Cambridge University Press; 2016. 501p.
3. Farashi A, Shariati M, Hosseini M. Identifying biodiversity hotspots for threatened mammal species in Iran. *Mamm Biol*. 2017;87:71–88.
4. Fattahi M. *Study of Zagros Forests and the Most Important Degradation Factors*. Research Institute of Forests and Rangelands: Tehran; 1994. 63p.
5. Fattahi M, Ansari N, Abbasi HR, Khanhasani M. *Zagros Forest Management*. Research Institute of Forests and Rangelands: Tehran; 2001. 471p.
6. Sadeghi M, Malekian M, Khodakarami L. Forest losses and gains in Kurdistan province, western Iran: where do we stand? *Egypt J Remote Sens Space Sci*. 2017;20:51–59.
7. Khalili F, Malekian M, Hemami MR. Habitat suitability modelling of Persian squirrel (*Sciurus anomalus*) in Zagros forests, western Iran. *J Wildl Biodivers*. 2018;2(2):56–64.
8. Eskandari S, Jaafari MR, Oliva P, Ghorbanzadeh O, Blaschke T. Mapping land cover and tree canopy cover in Zagros forests of Iran: application of Sentinel-2, Google Earth, and field data. *Remote Sens*. 2020;12(12):1912. 1–32.
9. Mahdavi A, Fallah Shamsi SR. Mapping forest cover change, using aerial photography and IRS-LISSIII imagery (Case study: Ilam Township). *J Wood For Sci Technol*. 2012;19:77–91.
10. Eskandari S, Moradi A. Investigation of land use and the analysis of landscape elements in Sivar Village from environmental viewpoint. *J Environ Stud*. 2012;38:35–44.
11. Arkhi S, Niazi Y, Ebrahimi H. Comparison of efficiency of artificial neural network and decision tree algorithms in provision of land use map using ETM+ data, case study: Darreshahr Watershed Basin in Ilam province. *Geogr Space*. 2014;13:47–72.
12. Fathizad H, Fallah Shamsi SR, Mahdavi A, Arkhi S. Comparison of two classification methods of maximum probability and artificial neural network of fuzzy Artmap to produce rangeland cover maps (Case study: Rangeland of Doviraj, Dehloran). *Iran J Range Desert Res*. 2015;22:59–72.
13. Mahdavi A, Rangin S, Mehdizadeh H, Mirzaei ZV. Assessment of forest cover change trends and determination of the main physiographic factors on forest degradation in Ilam province (Case study: Sirvan county). *Iran J For Range Protect Res*. 2017;15:1–16.
14. Eskandari S, Moradi A. Mapping the land uses and analysing the landscape elements in South-Western Iran: application of Landsat-7, field data, and landscape metrics. *Int J Conserv Sci*. 2020;11(2):557–564.
15. Topaloglu RH, Sertel E, Musaoglu N. Assessment of classification accuracies of Sentinel-2 and Landsat8 data for land cover/use mapping. In: *Proc. Int. Arch. of the Photogrammetry, Remote Sensing and Spatial Information Sciences, Vol. XLI-B8, 2016 XXIII ISPRS Congress; Prague*; 2016:1055–1059.
16. Phan TN, Kappas M, Degener J. Land cover classification using Sentinel-2 image data and random forest algorithm. In: *Proc. 19th Int. Conf. on Geoscience and Remote Sensing; Rome*; 2017:613–617.
17. Mustafa A, Rienow A, Saadi I, Cools M, Teller J. Comparing support vector machines with logistic regression for calibrating cellular automata land use change models. *Europ J Remote Sens*. 2018;51:391–401.
18. Inglada J, Arias M, Tardy B, et al. Assessment of an operational system for crop type map production using high temporal and spatial resolution satellite optical imagery. *Remote Sens*. 2015;7:12356–12379.
19. Immitzer M, Vuolo F, Atzberger C. First experience with Sentinel-2 data for crop and tree species classifications in Central Europe. *Remote Sens*. 2016;8:1–27.
20. Jedrych M, Bogdan Zagajewski B, Marcinkowska-Ochtyra A. Application of Sentinel-2 and EnMAP new satellite data to the mapping of Alpine vegetation of the Karkonosze Mountains. *Pol Cartogr Rev*. 2017;49:107–119.
21. Puletti N, Chianucci F, Castaldi C. Use of Sentinel-2 for forest classification in Mediterranean environments. *Ann Silvicult Res*. 2017. https://doi.org/10.12899/ASR-1463.
22. Szostak M, Hawryło P, Piela D. Using of Sentinel-2 images for automation of the forest succession detection. *Europ J Remote Sens*. 2018;51:142–149.
23. Hawryło P, Bednarz B, Wężyk P, Szostak M. Estimating defoliation of Scots pine stands using machine learning methods and vegetation indices of Sentinel-2. *Europ J Remote Sens*. 2018;51:194–204.
24. ESA (European Space Agency). *User Guide of Sentinel-2 Level-1C*; 2015. https://sentinel.esa.int/web/sentinel/user-guides/sentinel-2-msi/processing-levels/level-1.
25. Lillesand T, Kiefer R, Chipman J. *Remote Sensing and Image Interpretation*. New York: John Wiley and Sons; 2004.

26. SEOS. *Introduction to Remote Sensing*; 2018. http://seos-project.eu/modules/remotesensing/remotesensing-c06-p03.html.
27. Mustapha MR, Lim HS, Mat Jafri MZ. Comparison of neural network and maximum likelihood approaches in image classification. *J Appl Sci*. 2010;10:2847–2854.
28. Chuvieco E. *Fundamentals of Satellite Remote Sensing: An Environmental Approach*. 2rd ed. Boca Raton, FL: CRC Press, Taylor & Francis Group; 2016. 468p.
29. Richards J. *Remote Sensing Digital Image Analysis*. 5th ed. Springer; 2012. 502p.
30. Park B. *Computer vision Technology for Food Quality Evaluation*. Academic Press; Elsevier Inc.; 2008. 658p.
31. Huang C, Davis LS, Townshend JRG. An assessment of support vector machines for land cover classification. *Int J Remote Sens*. 2002;23 (4):725–749.
32. Mountrakis G, Im J, Ogole C. Support vector machines in remote sensing: a review. *ISPRS J Photogramm Remote Sens*. 2011;66(3):247–259.
33. Vapnik VN. *The Nature of Statistical Learning Theory*. New York: Springer-Verlag; 1995.
34. Vapnik VN, Vapnik V. *Statistical Learning Theory*. New York: Wiley; 1998.
35. Burges CJ. A tutorial on support vector machines for pattern recognition. *Data Min Knowl Disc*. 1998;2:121–167.
36. Huang B, Xie C, Tay R. Support vector machines for urban growth modeling. *GeoInformatica*. 2010;14:83–99.
37. Zobeiri M. *Forest Biometry*. Tehran: Tehran University Press; 2008.
38. Fallah A, Zobeiri M, Rahimipour Sisakht A, Naghavi H. Investigation on four sampling methods for canopy cover estimation in Zagros Oak Forests (Case study: Mehrian Forests of Yasuj City). *Iran J For Poplar Res*. 2012;20:194–203.
39. Karlson M, Ostwald M, Reese H, Sanou J, Tankoano B, Mattsson E. Mapping tree canopy cover and aboveground biomass in Sudano-Sahelian woodlands using Landsat 8 and random forest. *Remote Sens*. 2015;7:10017–10041.
40. Fleiss JL, Cohen J, Everitt BS. Large sample standard errors of kappa and weighted kappa. *Psychol Bull J*. 1969;72(5):323–327.
41. Congalton RG, Green K. *Assessing the Accuracy of Remotely Sensed Data: Principles and Practices*. 2rd ed. Boca Raton: CRC Press; 2008.
42. Cohen J. A coefficient of agreement for nominal scales. *Educ Psychol Meas*. 1960;20:37–46.
43. Jenness J, Wynne JJ. *Cohen's Kappa and Classification Table Metrics 2.1a*; 2007. http://www.jennessent.com/arcview/kappa_stats.htm.
44. Yang B, Hawthorne TL, Torres H, Feinman M. Using object-oriented classification for coastal management in the East Central Coast of Florida: a quantitative comparison between UAV, satellite, and aerial data. *Drones*. 2019;3:60. https://doi.org/10.3390/drones3030060.
45. Davoudi Monazam Z, Hajinejad A, Abbasnia M, Pourhashemi S. Detecting of land use change with remote sensing technique (Case study: Shahriar province). *RS GIS J Nat Resour*. 2014;5:1–13.
46. Yousefi S, Tazeh M, Mirzaee S, Moradi HR, Tavangar S. Comparison of different classification algorithms in satellite imagery to produce land use maps (Case study: Noor city). *RS GIS J Nat Resour*. 2014;5:67–76.
47. Mirzayizadeh V, Niknejad M, Oladi QJ. Evaluating non-parametric supervised classification algorithms in land cover map using Landsat-8 images. *RS GIS J Nat Resour*. 2015;6:29–44.
48. Pal M. Random forest classifier for remote sensing classification. *Int J Remote Sens*. 2005;26:217–222.
49. Stefanov WL, Ramsey MS, Christensen PR. Monitoring urban land cover change; an expert system approach to land cover classification of semi-arid to arid urban centers. *Remote Sens Environ*. 2001;77(2):173–185.
50. Erbek FS, Ozkan C, Taberner M. Comparison of maximum likelihood classification method with supervised artificial neural network algorithms for land use activities. *Int J Remote Sens*. 2004;25:1733–1748.
51. Heydarian P, Rangzan K, Maleki S, Taghizadeh A. Land use change detection using post classification comparison Landsat satellite images (Case study: land of Tehran). *RS GIS J Nat Resour*. 2014;4:1–10.
52. Borzafkan AA, Pirbavaghar M, Fatehi P. Capability of Liss III data for forest canopy density mapping in Zagros forests (Case study: Marivan Forests). *Iran J For*. 2015;4:387–401.

13

Identify the important driving forces on gully erosion, Chaharmahal and Bakhtiari province, Iran

Mohammad Nekooeimehr, Saleh Yousefi, and Sayed Naeim Emami

Soil Conservation and Watershed Management Research Department, Chaharmahal and Bakhtiari Agricultural and Natural Resources Research and Education Center, AREEO, Shahrekord, Iran

1 Introduction

Soil is one of the most important natural resources in any country.[1] Soil erosion is a serious threat to human well-being and life. In areas where erosion is not controlled, soils gradually erode and lose their fertility.[2] Erosion not only impoverishes the soil and abandons farms, gardens, and natural areas, but also reduces the useful life of canals, irrigation, and drainage networks, dam reservoirs, port areas, etc.[3, 4]

Among the various forms of water erosion, gully erosion is one of the factors threatening the balance of environmental resources and their sustainability.[5, 6] This type of erosion, especially in arid and semiarid regions of the world, where the use of soil, water, and plant resources is not based on correct principles and in accordance with the natural potential of the region and environmental conditions, causes significant changes in the land landscape and harmful economic and social consequences.[7, 8] The economic effects of gullies on agricultural lands can be much greater than rangelands because the creation of gullies in pastures does not stop grazing, but in agricultural lands, it stops production due to land fragmentation. In the past, experts thought that gullies arise from the development and enlargement of furrows, but studies conducted in different countries of the world have shown that the emergence of gullies is a very complex process; thus, this type of soil erosion can be found in different areas with different geological formations, topographic conditions, soil properties, land use and climatic characteristics.[9–12]

In general, because various aspects of gully erosion, despite extensive research worldwide, have not yet been properly identified and introduced. Therefore, it is absolutely necessary to carry out extensive studies on this type of erosion in different places with different ecological conditions to determine practical measures to control it. The present study was conducted in this direction to determine the most important causes of the creation and expansion of gullies in Chaharmahal and Bakhtiari provinces, Iran; in the latter stages, efforts can be made to control and prevent their expansion.

Gully erosion is an important soil degradation process that causes significant soil loss in different climates. Many researchers believe that the reason for paying attention to gully erosion is the high volume of this type of erosion compared to other types of erosion[12–15]; thus, gully erosion produces up to 50 times more sediment.[15] Soil that is eroded by the gully causes great economic losses by sedimentation in waterways, dam reservoirs, and port areas by reducing their water intake capacity.[8, 11]

Many scientists consider factors such as degradation of natural ecosystems, misuse of lands, degradation of vegetation, overgrazing, climate change, geological conditions, and human interference in natural areas as the most important causes of the creation and expansion of gullies.[7–9, 11–13, 16, 17]

In Iran, several studies have been conducted on the causes of gully erosion and its spread. Among the studies in the field of studying the effects of soil characteristics on the development of gully erosion in the Zanjan-Rud watershed, it was concluded that the formation of gullies is a function of soil properties, including soil texture. The risk of gully erosion in soils with silty and clay texture is much higher than the risk of erosion in light soils.[9]

Studies on the causes of gully erosion in the northern region of Golestan province, Iran, have shown that two factors, the percentage of soluble salts in the soil and the area of the gully watershed, have a great impact on the creation and development of this type of erosion in the region. The variables studied in this study included the percentage of soluble salts, percentage of vegetation, percentage of clay and sand, area of gully watershed, percentage of floor slope, and slope of gully watershed.[7]

The study of the causes of gully erosion in Gilan Province, Iran, has shown that lithological units and land-use changes have been the most important causes of gully erosion in this region.[6]

The effects of soil characteristics such as salinity, acidity, soluble salts, texture, aggregate stability, bulk density, sodium uptake ratio, and exchangeable sodium content of the soil, as well as slope and vegetation factors, on the longitudinal expansion of gorges in Fars Province, and factors such as salinity, acidity, vegetation, and slope have a significant relationship with the longitudinal progression of the abyss.[5]

The results of research conducted on gully erosion in the Chaharmahal and Bakhtiari provinces, Iran, showed that this type of erosion, with an area of 32,520 ha, is more common in the southern and western parts of the province. The high percentage of silt particles in the soil samples of the forehead and the body of the studied gullies indicates the sensitivity of the soil in these areas to erosion. Although many pieces of researches have been published on gully erosion studies in different parts of the world, in Chaharmahal and Bakhtiari province we can't find a comprehensive study on gully erosion and its driving factors. The main aims of the present study are: (i) to identify the spatial distribution of gully erosion in province scale, (ii) to prepare a comprehensive data bank for all gully features in province scale, (iii) to identify the most important driving factors on gully erosion for different counties in province scale.

2 Materials and methods

This research was conducted in the Chaharmahal and Bakhtiari provinces in Southwest Iran. This province, with an area of 16,533 km^2 is located in the central part of the Zagros Mountains. The rainiest part of the province is northwestern, with an average annual rainfall of 1600 mm, and the least rainy part is the northeastern part of the province, with an average annual rainfall of 250 mm. The average annual rainfall for the entire province is approximately 707 mm. The average annual temperature in the province varies from 3.5°C in the northwestern regions to 18.5°C in the southern lowlands.[18, 19] Based on the modified Dumarten climate classification, this province has seven macroclimates, including desert arid, semiarid, Mediterranean, semihumid, humid, and very humid (A and B). Fig. 1 shows the province's position.[20]

From a geological point of view, the Chaharmahal and Bakhtiari provinces are located in the tectonic zones of high Zagros, folded Zagros, and Sanandaj-Sirjan zones. The most important exposed formations in the province are Hormoz, Neyriz, Sarvak-Gurpi, Pabdeh, Asmari-Jahrom-Shahbazan, Gachsaran-Razak, Aghajari-Bakhtiari, and Quaternary alluviums.[18]

3 Methodology

After conducting field visits and recording the geographical location of the gullies, the spatial distribution map of the areas affected by gully erosion in the province was prepared with a scale of 1: 250,000, and among the identified gullies, 24 active and representative gullies were selected for further studies. Gully volume (eroded soil) as a dependent variable and factors such as sodium uptake ratio (SUR), percentage of soluble salts (PSS), percentages of clay, silt and sand in gully soil, gully floor slope percentage (FS), gully upstream slope percentage (US), mean area precipitation (MP) and density percentage vegetation (DPV) were considered as independent variables.[8, 9, 21]

To calculate the volume of gullies, several cross-sections (depending on the changes in section and length of gullies) were considered for each gully and the dimensions of gullies, including their width, width, and depth, were carefully measured in each section. By determining the average area of these sections and the length of each gully, the volume was calculated. Soil sampling was performed from the surface layer (0–30 cm) and the bottom (30–60 cm) of the head and body of each gully and characteristics such as salinity, acidity, sodium adsorption ratio, and percentage of clay, silt, and sand particles of each sample were measured in the laboratory. The slope of the gully floor was calculated using a surveying camera and the slope of the upstream front of the gully was calculated using a slope gauge. The

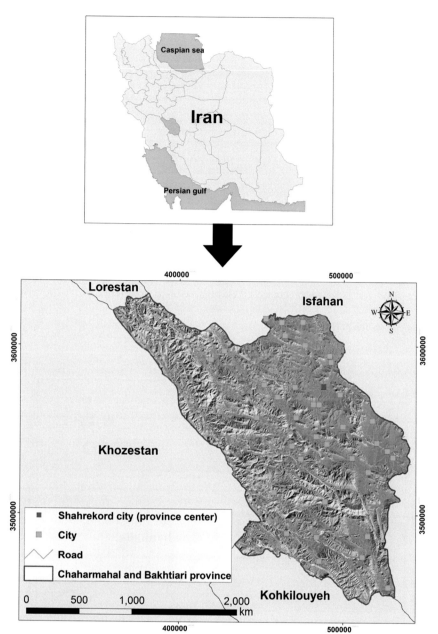

FIG. 1 Location of Chaharmahal and Bakhtiari province.

Vegetation percentage was determined randomly using $1\,m^2$ plot. The long-term mean rainfall in each region was considered a climatic factor. The measured parameters were statistically analyzed using SPSS software with a multivariate regression relationship (Fig. 2).

4 Results and discussion

A map of the spatial distribution of areas affected by gully erosion in the Chaharmahal and Bakhtiari provinces is shown in Fig. 3. The results show that although in most areas of this province, the phenomenon of gully erosion is evident, this phenomenon is more common in the southern and western parts of the province.

A Comparison of the percentage of different soil particles in gullies shows that clay varies between 13.7% and 37.7%, silt between 45.1% and 65.1%, and sand between 11.2% and 40.2%. The high percentage of silt particles in the soil samples of gullies indicates the sensitivity of the soil in these areas to erosion. The sodium uptake ratio in the soil

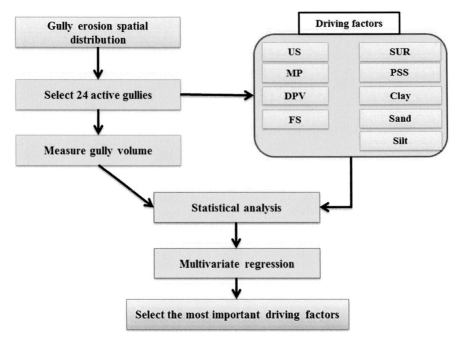

FIG. 2 Flowchart of methodology for the present study.

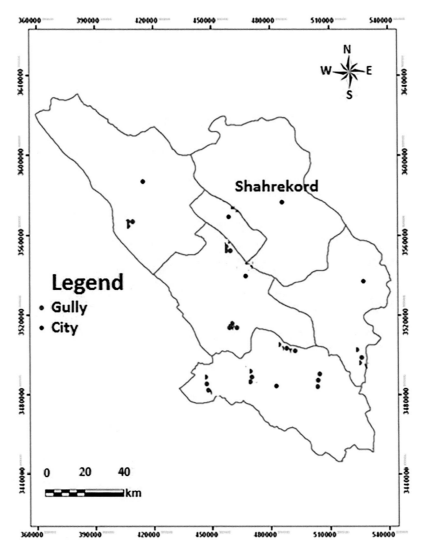

FIG. 3 Spatial distribution map of the studied gullies.

of gully areas varies between 2.13 and 5.3 and their soluble salts vary between 0.38% and 0.67%. The slope of the gully floor is between 3.5% and 16.5% and the slope of the upstream front of the gully is between 4% and 47%. In addition, the percentage of vegetation in the watershed of gullies varies between 10% and 30%.

To determine the most important factors affecting the volumetric expansion of gullies, the volume of gully erosion as a dependent variable, the percentage of clay, silt, and sand particles, percentage of soluble salts, sodium uptake ratio, percentage of gully floor slope, percentage of the slope of the catchment above the gully head, precipitation and vegetation percentage were studied as independent variables. In the first stage, when all the studied variables are included in the model, only the slope of the catchment above the gully head and the percentage of vegetation becomes significant at the level of 0.1% and 5%, respectively, and the remaining factors are insignificant. The model explanation coefficient, in this case, has reached 0.95. In the next steps, step by step, the next variables are entered into the model until finally, the equation is presented as follows:

$$Y = 80558X1 + 3547.02X2 + 1825.49X3 - 102886 \tag{1}$$

where:

$X1 =$ Percentage of soluble salts
$X2 =$ Percentage of the slope of the catchment above the gully
$X3 =$ percentage of vegetation

Only three factors are used in this equation. These factors include the percentage of the slope of the catchment above the gully head, the percentage of vegetation, and the percentage of soluble salts in the soil, which with the explanation coefficient of 0.94 had the greatest effect on the development of gullies in these areas. It is noteworthy that when all the studied variables are entered into the model, the value of the coefficient of explanation increases to 0.95. In other words, other variables, except for the above-mentioned factors, had only 1% effect on gully erosion in the region. On the other hand, in the proposed model, the maximum coefficient of explanation reached 0.95, which indicates that 5% of the effects of other factors have not yet been included in the model. This value is related to factors that were not considered in this study.

The slope percentage of the catchment above the gully head alone accounted for 86% of the changes in gully erosion volume. This can be attributed to the flooding of this factor and the speed of the surface runoff. Because of the high slope of the catchment above the gully head and the abundance of silt in the soil profile, along with the destruction of vegetation due to uncontrolled grazing, take the opportunity to infiltrate the water and cause severe runoff, and runoff is one of the main factors in creating gullies.[5, 7] Fig. 4 shows a view of the slope of the catchment above the gully head.

Constructed roads in the study area are one of the affective factors on gully erosion. Roads increase the rate of runoff and hydraulic power, which eventually increases the rate of soil erosion and gully erosion.[10–12] It is suggested to construct and increase culverts and bridges based on the hydraulic capacity on roads that cross gullies.

FIG. 4   View of the slope of the catchment above the gully head.

FIG. 5   Destruction of vegetation around a gully.

The role of vegetation in reducing runoff and preventing soil erosion is not hidden. In the area where the cover is less dense, the volume of gullies has increased. Fig. 4 shows the destruction of vegetation around the gully and its conversion into bare land. Another factor that is significant at the 5% level is the percentage of soluble salts in the soil. The percentage of soluble solutes was directly related to the soil salinity. The presence of soluble solutes on the soil surface facilitates drainage erosion. The mechanism of formation of these gullies in the early stages is more dissolution and the action of water in the later stages plays an important role in the development of the gullies. In the study area, the density and quantity of vegetation cover around the gullies body have decreased (Fig. 5). Vegetation cover plays a great role in soil erosion controlling and reduces the rate of runoff during rainfall events.[7, 9] Overgrazing in the study area is one of the most important factors in reducing the density of vegetation cover and needs to be controlled.

Based on field trips, increasing the area of agricultural lands (dry farming) is one of the effective factors in increasing the area of gully erosion in the study area.[5, 22] The runoff and soil erosion in natural lands (rangeland and forest) is less than in agriculture lands in the same physical and hydrological conditions.[1, 23] Forest and rangelands control the gully head-cut by using the deep roots and conservative role on soil erosion.[7]

5 Conclusion

The results of this study showed that although many variables can affect the development of gully erosion in the region, the three factors of vegetation density, the slope of the catchment above the gully head, and the percentage of soluble salts in the soil had a greater effect than others. In other words, the destruction of vegetation, the high slope of lands in upstream areas of gullies, and the presence of soluble salts in the soil have caused the expansion of gullies in the area. It is noteworthy that these three factors together account for 94% of the changes in the volume of gully erosion in the region. This can be attributed to the flooding of the region and the speed of the surface runoff. In addition, grazing more than the capacity of pastures and the movement of livestock on the soil has caused the loss of vegetation and the destruction of soil structure, ultimately leading to the production of runoff and the creation and expansion of gullies in the region. The best way to prevent the volume and speed of runoff is to regenerate the vegetation in the area. Therefore, it is suggested that the reduction of surface runoff upstream of the gullies by restoring vegetation has a special priority for watershed management operations. Excessive use of pastures (excessive grazing) is an effective factor in the destruction of vegetation and, as a result, the creation of gullies in different parts of the province, so it is suggested to reduce the number of livestock in rangelands based on rangeland capacity.

Acknowledgment

I would like to express my sincere thanks and gratitude to all my dear colleagues, especially Mr. Raeisian, who helped us in carrying out the various stages of this research project, and I wish them to succeed.

References

1. Martinez G, Weltz M, Pierson FB, Spaeth KE, Pachepsky Y. Scale effects on runoff and soil erosion in rangelands: observations and estimations with predictors of different availability. *Catena.* 2017;151:161–173.
2. Behtari B, Jafarian Z, Alikhani H. Temperature sensitivity of soil organic matter decomposition in response to land management in semi-arid rangelands of Iran. *Catena.* 2019;179:210–219.
3. Pardini G, Gispert M, Dunjó G. Runoff erosion and nutrient depletion in five Mediterranean soils of NE Spain under different land use. *Sci Total Environ.* 2003;309:213–224.
4. Gayen A, Pourghasemi HR, Saha S, Keesstra S, Bai S. Gully erosion susceptibility assessment and management of hazard-prone areas in India using different machine learning algorithms. *Sci Total Environ.* 2019;668:124–138.
5. Soufi M, Bayat R, Charkhabi AH. Gully Erosion in I. R. Iran: characteristics, Processes, causes, and land use. In: *Gully Erosion Studies from India and Surrounding Regions.* Springer; 2020:357–368.
6. Zabihi M, Pourghasemi HR, Motevalli A, Zakeri MA. Gully erosion modeling using gis-based data mining techniques in Northern Iran: a comparison between boosted regression tree and multivariate adaptive regression spline. In: *Advances in Natural and Technological Hazards Research.* 48. Springer; 2019:1–26.
7. Hosseinalizadeh M, Kariminejad N, Chen W, et al. Gully headcut susceptibility modeling using functional trees, naïve Bayes tree, and random forest models. *Geoderma.* 2019;342:1–11.
8. Pourghasemi HR, Yousefi S, Kornejady A, Cerdà A. Performance assessment of individual and ensemble data-mining techniques for gully erosion modeling. *Sci Total Environ.* 2017;609:764–775.
9. Arabameri A, Chen W, Loche M, et al. Comparison of machine learning models for gully erosion susceptibility mapping. *Geosci Front.* 2019.
10. Arabameri A, Pradhan B, Rezaei K. Spatial prediction of gully erosion using ALOS PALSAR data and ensemble bivariate and data mining models. *Geosci J.* 2019;23:669–686.
11. Rahmati O, Tahmasebipour N, Haghizadeh A, Pourghasemi HR, Feizizadeh B. Evaluating the influence of geo-environmental factors on gully erosion in a semi-arid region of Iran: an integrated framework. *Sci Total Environ.* 2017;579:913–927.
12. Amiri M, Pourghasemi HR, Ghanbarian GA, Afzali SF. Assessment of the importance of gully erosion effective factors using Boruta algorithm and its spatial modeling and mapping using three machine learning algorithms. *Geoderma.* 2019;340:55–69.
13. Avand M, Janizadeh S, Naghibi SA, Pourghasemi HR, Bozchaloei SK, Blaschke T. A comparative assessment of Random Forest and k-Nearest Neighbor classifiers for gully erosion susceptibility mapping. *Water.* 2019;11:2076.
14. Shahabi H, Jarihani B, Tavakkoli Piralilou S, Chittleborough D, Avand M, Ghorbanzadeh O. A semi-automated object-based gully networks detection using different machine learning models. *Sensors.* 2019;19:1–21.
15. Nhu VH, Janizadeh S, Avand M, et al. GIS-based gully erosion susceptibility mapping: a comparison of computational ensemble data mining models. *Appl Sci.* 2020;10:2039.
16. Saygın SD, Basaran M, Ozcan AU, et al. Land degradation assessment by geo-spatially modeling different soil erodibility equations in a semi-arid catchment. *Environ Monit Assess.* 2011;180:201–215.
17. Clarke MA, Walsh RPD. Long-term erosion and surface roughness change of rain-forest terrain following selective logging, Danum Valley, Sabah, Malaysia. *Catena.* 2006;68:109–123.
18. Emami SN, Yousefi S, Pourghasemi HR, Tavangar S, Santosh M. A comparative study on machine learning modeling for mass movement susceptibility mapping (a case study of Iran). *Bull Eng Geol Environ.* 2020;1–18.
19. Yousefi S, Pourghasemi HR, Emami SN, et al. Assessing the susceptibility of schools to flood events in Iran. *Sci Rep.* 2020;10.
20. Haghighian F, Yousefi S, Keesstra S. Identifying tree health using sentinel-2 images: a case study on Tortrix viridana L. infected oak trees in Western Iran. *Geocarto Int.* 2020.
21. Azareh A, Rahmati O, Rafiei-Sardooi E, et al. Bin modelling gully-erosion susceptibility in a semi-arid region, Iran: investigation of applicability of certainty factor and maximum entropy models. *Sci Total Environ.* 2019;655:684–696.
22. Sternberg M, Gutman M, Perevolotsky A, Ungar ED, Kigel J. Vegetation response to grazing management in a Mediterranean herbaceous community: a functional group approach. *J Appl Ecol.* 2000;37:224–237.
23. Sun G, McNulty SG. Modeling soil erosion and transport on forest landscape. In: *Proceedings of the Steamboat Springs, CO: International Erosion Control Association.* Miscellaneous Publication: Reno; 1998:189–198.

14

Analysis of social resilience of villagers in the face of drought using LPCIEA indicator case study: Downstream of Dorodzan dam[☆]

Payam Ebrahimi

Forests, Rangelands, and Watershed Research Department, Sistan Agricultural and Natural Resources Research and Education Center AREEO, Zabol, Iran

1 Introduction

The relationship between man and the environment is highly complicated,[1] and as a result, it is difficult to understand the relationship between the two.[2] The social relations between people and their relation to the natural environment are of great importance when extreme events such as floods and droughts occur. Research shows that critical natural events have destroyed past governments[3] or weakened social power.[4]

Floods and droughts have become more frequent[5] in recent decades and extensive damage[6] to the livelihoods of the villagers.[7] To cope with this damage and crisis, villagers have increased their resilience and sustainability by using appropriate and preventative[8] measures or by reducing the effects of the crisis.[9] Resilience is the ability of an endangered system or community to deal with, absorb, adapt, and recover promptly in the event of an effective disaster or type of disaster involves maintaining or restoring essential functions.[10] Farmers play an important role in the economy, food production, and job creation[11] and changing the quality of life, especially livelihood due to natural disasters, is very destructive for the economies of countries.

To maintain and improve the present state of the environment, it is necessary to increase the resilience of villages to make the environment of farmers viable. Viability is another notion of how a system survives. The viability of a system is the capacity of the environment to survive independently in a shared environment.[12] However, a village's resistance to natural disasters increases, making life and survival easier. Farmers lack access to water resources during periods of drought, the negative effects of climate change, and the loss of livestock and crops that play an important role[13] in the household economy[14] increase vulnerability. If there is a drought, social conflict increases,[14] and the household economy continues to weaken. Research and measurements have been conducted[4] to measure the magnitude[15] and frequency[3] of critical events such as drought[16]; however, studies on the drivers of resilience and viability and related indicators are very limited. Droughts and epidemics are interdependent and impact the food supply chain.[17] The impact of these damages increases when drought stress accompanies economic tensions, especially in developing countries. Results from various studies indicate that climate change and its destructive effects[18] are increasing worldwide and locally.[19] Climate change devastatingly impacts poor and developing countries.[20] In recent decades, social conflicts between villagers in arid and semi-arid areas have led to a decline in resilience and viability due to the drought crisis.[21] This matter has several economic and social effects.

At the local community level, particularly farmers, resilience, and viability can be evaluated based on drought.[22] Examining changes in resilience and sustainability helps to identify sustainable communities and strengthen vulnerable and weak communities. The analysis of the resilience and viability of the village facilitates the management of

[☆] Located in Iran.

communities and the selection of vulnerable villages.[23] Effective criteria play a major role in determining the resilience levels. By selecting less resilient and less livable villages and strengthening them in the event of a drought crisis, the social structure of the villages can be maintained and stabilized. This occurs more frequently in developing countries due to the fragility of the ecosystem. One of the determinants of the drought crisis is the existence of controlled irrigation systems. This water supply network will help provide sustainable water during the drought crisis and maintain villagers' livelihoods.

Although the flow of dams is reduced, limited water supplies with specified timings enhance resilience. Hybrid indicators with human and environmental criteria have priority to create a drought resilience and sustainability index. Changes in population and agricultural products are important factors based on the results of the Bayesian network[24] and influence the sustainability of farmers. Land cover and water availability are other important factors that determine resilience.[25] Two key criteria by which drought resilience and viability can be measured: access to water for agricultural production[26] and land use for the production of various crops.[27] A strong link exists between age groups and droughts,[28] and people respond differently to drought and resilience at different ages. Iran is an arid, semi-arid country that has experienced drought over the past several decades.[29]

Climate change has been effective in terms of flooding and drought in Iran,[30] and villagers' lives have been affected by these changes. In areas where access to surface water, scientific knowledge, and financial resources are inadequate, drought crises are intensifying. The arid region's ecosystem is sensitive and vulnerable to climate change and has devastating social consequences. A large part of Iran is covered with arid land with long droughts.[31] People in the villages of Fars province in the Dorodzan section have limited access to water, and many of them use controlled water to farms. In each village, the average literacy level, age of the persons, and access to water regulated by the dam are different. During the dry years, villages are confronted by water crises, and the extraction of water from wells is limited. The distribution of the population is not appropriate, and especially in the case of drought, the use of water is limited by the distance of the dam. Wells are deep in these regions and, during drought, authorized wells are not allowed for further extraction.[32] Increased migration to surrounding cities, marginalization, and criminality are some of the consequences of drought. Considering that identifying resilient and livable villages in the region is essential for addressing the drought crisis in the coming years, based on findings from other researchers, this study uses the criteria of age and education,[33] land use,[28] irrigation, population,[34] and crops.[35] The use of resilient and livable village maps and their consolidation can provide appropriate planning and services to improve village livelihoods. The use of this tool enables policymakers to assist stakeholders that are most affected by drought. Less resilient and less viable villages need more water for their livelihoods. This study focused on areas with irrigation systems supplied by Dorodzan Dam. Through this research, less resilient and less livable communities were identified. Based on the maps obtained, resilience and viability scores of the study area were determined, and decision-makers could modify access to water and allocation for reinforcement in less resilient villages to adapt to drought. The innovation of this research is the invention of a quantitative methodology using combined criteria to evaluate the resilience and viability of a village in response to the crisis. These criteria were suggested in the study. Using the equations of these criteria in the form of an index, it is possible to estimate the drought resilience score.

2 Methodology

2.1 Study area

The study area is in southwest Iran, Dorodzan Dam, Fars Province. The region is located between 35° 42 north latitude and 52° 51 east longitudes. The average rainfall at Dorodzan Dam was 253 mm, and the dam volume was 993 million cubic meters. Dorodzan's dam supplies 112,000 ha of farmland with water.[36] Water from dams is used for drinking and farming purposes. Both Shiraz[a] and Marvdasht[b] use drinking water, and the farmlands of Kamfiroz[c] and Marvdasht use this water. The local villagers' income sources are farming and livestock, where agriculture is the priority, followed by livestock. Major agricultural commodities include wheat, barley, clover, and rice.[37] In addition to agriculture and livestock, several households manage their families' livelihoods through government employment, private workshops, and businesses. In rural communities, most households are dependent on agriculture. Consequently, land ownership is the primary capital and property of the household; it also determines its social

[a] https://en.wikipedia.org/wiki/Shiraz.

[b] https://en.wikipedia.org/wiki/Marvdasht.

[c] https://en.wikipedia.org/wiki/Kamfiruz.

and economic status. In other words, land ownership and social structure are closely interrelated. According to the latest Iranian Census of Population and Housing[d] in 2016, 51,662 people lived in the study area. The area has 78 villages. Field studies of this method have been conducted through observational, library, and face-to-face interviews with farmers. The investigation area is shown in Fig. 1.

Two censuses of population factors and precipitation data were used to determine the year of study.[38] In Iran, a population census was conducted every 10 years. Consequently, changes within two 5-year periods were considered. Between 1986 and 2021, the median precipitation was 435.9 mm. The demographics were consistent with median rainfall. The lower median amount is related to 2006, and the most recent census of the population was completed in 2016. Consequently, from 2006 to 2011, as a post-drought period and from 2012 to 2016 as a pre-drought period were

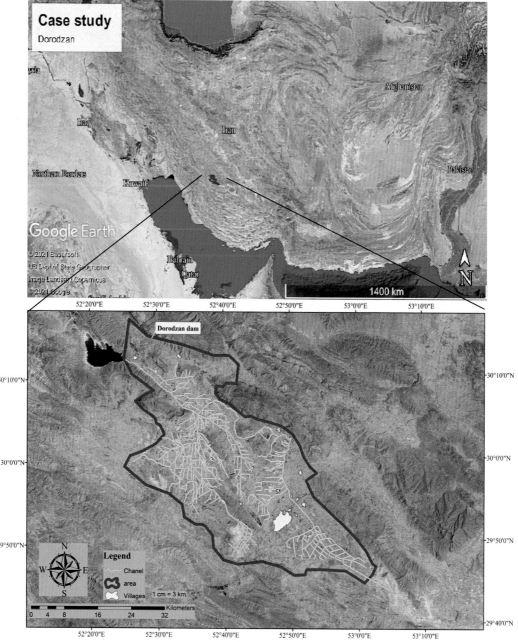

FIG. 1 Study area.

[d] https://www.amar.org.ir/english.

TABLE 1 Dorodzan Dam annual precipitation data from 1986 to 2021 and selected years in the study area.

Year	1986	1988	1989	1990	1991	1992
	130.6	546.6	307.7	548.91	425.82	515.02
Year	1993	1994	1995	1996	1997	1998
	854.51	186.2	712.21	613.21	263.5	750.2
Year	1999	2000	2001	2002	2003	2004
	537.4	259.4	336.8	641	519.5	620.41
Year	2005	2006	2007	2008	2009	2010
	619.61	436.9	423.2	338	205.5	435.9
Year	2011	2012	2013	2014	2015	2016
	379.5	478.5	511.01	332.2	338.6	362.9
Year	2017	2018	2019	2020	2021	Average
	367.8	179.6	506.3	526.2	410.8	446.33

selected (Table 1). Other precipitation values were greater than the median of 35 years and were inconsistent with the population data. Approximately 80% of the selected years had less than the average annual precipitation over a long period. The Dorodzan Dam irrigation network information was obtained from the Regional Water Company of Fars.[e] Using access to water resources, villages that had access to water wells and a surface irrigation system were selected (Table 2). The methodology employed in this study is presented in Fig. 2.

Groundwater utilization in dry years is limited by the Regional Water Company of Fars, and the volume of harvest is reduced. There is no clear border between rural farmland and the rural district border, which is called the official border. In some cases, people from other villages in adjacent villages have farmland in the village under investigation. Many people who live in these villages have higher education, and the rate of population growth in the area is negative.

2.2 LPCIEA indicator

This index uses six criteria: land use, population, crop level, education, irrigation level, and age of village residents. The diversity of land uses, the rural population, and the youth age pyramid increase resilience. Increasing crop cultivation and its diversity, higher levels of education, villages with lower average age, and employed people with standard employment age make rural people more adaptable to drought.[39–52]

Various researchers have recommended using these criteria in their research. Someone from another village may have a piece of land in another village. Therefore, rural districts were used in this study. Using a combination of criteria, a classification can be established based on the impact on resilience and sustainability. The scores for this index were between 0 and 1, indicating resilience to drought. For each criterion, a rating of 1 indicated the highest resilience, and 0 indicated the lowest resilience and drought adaptation. All of the criteria studied were divided into six groups. The final score for each village was determined by the resilience and viability scores between 0 and 100. The

TABLE 2 Rural district area and population (2016).

Rural district	Area (hectare)	Population (person)
Ramjerd 1	15,187.15	10,216
Ramjerd 2	27,468.18	9124
Rodbal	17,846.97	10,638
Majd abad	15,482.7	8447
Mohammad abad	33,803.62	13,237

[e] https://www.frrw.ir/?l=EN.

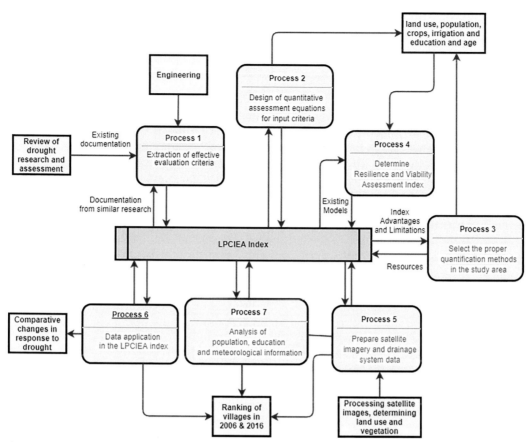

FIG. 2 Flow chart of research methodology.

advantages of this index are its ease of use and fast evaluation according to information for specialists, change according to the characteristics of each country, and data availability. The strength of the index is that the higher the spatial accuracy of satellite imagery, the more accurate the resilience results. Statistics for this index are usually collected regularly in all countries, and there are also free satellite images of Landsat or Sentinel. Google Earth images can also be used to increase image detail. The disadvantage of this indicator is that it requires knowledge of satellite image processing and slows the speed at which field operations are evaluated to control the ground points. The resilience and viability equation for a rural district is based on the following Eq. (1):

$$\text{Resilience} = \frac{(L+P+C+I+E+A)}{6} \times 100 \tag{1}$$

where L is rural land use and diversity per hectare, P is the population of each age group of the total rural population based on the number of people, C is the agricultural products and their variety per hectare, I is the irrigation network of the total agricultural area per hectare, and E is the educational level of the rural population. A refers to the number of persons of working age who have a job. The age of employment in each country is different and must comply with national standards. Four categories of the forest, agricultural, rangeland, and residential land can be used to identify land use.[53] Perennials are more resistant to drought than are annuals. The forested lands of plantations or natural areas have a supporting role in terms of resilience and viability because of these effects. By raising this criterion, other animals and human ecosystems have improved quality in response to drought. Forested areas with a score of one had the highest impact, and residential areas with a score of 0.25, had the lowest impact on resilience and viability. With the increase in the level of agriculture in the region, the livelihoods of its people have improved. In dry years, an increase in farming indicates the water supply of villagers.

In arid and semi-arid areas, Rangeland, after farmland, is a place to meet household economic needs. Processed pastures produce more when the precipitation increases. The increase in crop volume declines in dry years, and the increase in rangelands is not directly related to the increase in production volume. Therefore, it is after forestry and agriculture. As the population grows, residential zones increase. Jobs in the region or a decent livelihood to meet people's needs prevent people from migrating.[54] People tend to adapt to drought, but the pressure is more severe. Eq. (2) is used to compute this criterion, and the sub-criteria of the land use criterion with specific coefficients divided by the total area of the village. The formula is as follows:

$$L = \frac{((Fo \times 1) + (Ag \times 0.75) + (Ra \times 0.5) + (Re \times 0.25))}{A_{ha}} \qquad (2)$$

where Fo is defined as the level of forested land, and Ag is defined as the level of agriculture, Ra is the level of rangeland, Re is the area of rural residential land, and A is the overall area of rural areas. These criteria were set per hectare. In communities where young age groups have little migration, the rate of adaptation to existing conditions and resilience to natural disasters is increasing.[43] The 30–64 age groups with low migration rates were the most resilient and adaptive. If the population increases per unit area of the village, the resilience and viability of the village are higher, but there is more pressure on resources. Based on the above conditions, the population criterion is calculated using Eq. (3).

$$P = \frac{(((P_1) \times 1) + (P_2 \times 0.75) + (P_3 \times 0.5) + (P_4 \times 0.25))}{A_{ha}} \qquad (3)$$

where P_1 is the number of people between 30 and 64 years old, and P_2 is the number of people between 15 and 29 years old. P_3 is the number of persons aged 64 years and older. P_4 is the number of persons aged between 0 and 14 years, and A is the area of the village. If the diversity of agricultural products grows, the livelihoods of populations living in drought conditions will become more stable. Products with different water needs and markets enhance the resilience of the household economy.[55] The main source of household economics is the dominant culture with a larger area; hence, it has a higher score. The increase or stability of the cultivated area against drought indicates the strength of the household. In the study area, dominant agriculture must be identified using suitable scores. The impact on resilience and viability must be determined at each level. The total area of the village was determined to determine the score of this criterion, and the farmland was extracted. Then, depending on the distribution of the products in the area, the dominant products are identified. Dispersed products and less than 30% of all farmlands within the study area were classified as other products. In other words, if the crop variety and area increase, the village will have more resilience and sustainability in the face of drought. Eq. (4) was used to identify the levels of agricultural products.

$$C = \frac{(Ce) + (Fa) + (Ga) + (OP)}{A_{ha}} \qquad (4)$$

where Ce is the level of cereals and Fa is the level of Fabaceae. Ga is the level of the garden, OP is the level of the other products, and A is the area of the village per hectare. Table 3 lists the best scores for the farm products.

The ratings for villages with 4, 3, 2, and a product will be 1, 0.75, 0.5, and 0.25, respectively. The irrigation system is supplied with water from the dam, depending on the volume of water in the dam and the demand for agricultural products. The system releases water by scheduling at specified times. It is easier to manage drought if irrigation is developed using a specific plan and volume.[56] The area of the irrigation network affects the resiliency and viability. This criterion is divided into four categories based on land cover: 0%–25%, 26%–50%, 51%–75%, and 75%–100%. This equation has coefficients ranging from 0.25 to 1. The equations for these criteria are as follows.

$$I = \frac{(Ir_{area} \times X)}{A_{ha}} \qquad (5)$$

where Ir is the irrigated area of a village divided by the area of the village and per hectare. X is the irrigated level coefficient and Table 4 is used.

Education is another measure of drought resilience.[57] people were classified into four groups. The first group was the illiteracy group. They are children or people who have not been educated before. The second group included people with

TABLE 3   Crop diversity coefficients.

Product	Score
If 1 product	0.25
If 2 products	0.5
If 3 products	0.75
If 4 products	1

TABLE 4   Irrigation level score.

Cover	Score
If Ir cover 0–25 Percentage of Rural district	0.25
If Ir cover 26–50 Percentage of Rural district	0.5
If Ir cover 51–75 Percentage of Rural district	0.75
If Ir cover 75–100 Percentage of Rural district	1

first or second high school certificates who had not gone on to further education. The third group consisted of people with a diploma, certificates of competence, internships, or diplomas equivalent to experience in a profession. The fourth group consists of people who have bachelor's, master's, and doctoral degrees. As population knowledge increases, drought resilience methods are implemented more accurately and provide scientific support. It should be noted that the fourth group lives in rural areas, is old, and agriculture is the second job for many of these people. The supply of plant inputs and the use of technologies were funded by the first job in the fourth group. In times of drought, the profits of agricultural mechanization return to the agricultural business cycle. Eq. (6) was used to determine this criterion.

$$E = \frac{(\text{BMD} \times 1) + (\text{CDP} \times 0.75) + (\text{POS} \times 0.5) + (\text{NFE} \times 0.25)}{A_{ha}} \qquad (6)$$

where BMD is the number of persons with a bachelor's, master's, or doctoral degree. CDP, number of individuals with certificates/diplomas/professional qualifications POS are persons with primary or secondary education, and NFE includes persons without formal education. A_{ha} is the total size of the rural area per hectare. Employment has a strong relationship with resiliency and habitability.[58] The minimum legal working age in Iran is 15 years. The age of 15–64 years is considered to be the effective age in the event of drought. While people have good resilience and sustainability between the ages of 15 and 29, they play a lower role because of their willingness to migrate. The jobs of people aged 0–14 and 64 were temporary. Those aged 10–14 study or train and work part-time after graduation or during their studies. People under that age do not have permanent employment or jobs. Beyond the age of 64, the risk level is very low, and employment provides a livelihood. They are highly susceptible to drought. The equation for this criterion is derived from Eq. (7). The number of people having a job in the village at this age was divided according to the total number of people at this age.

$$A = \frac{\sum 15 - 64 \, \text{years} \, E}{\sum 15 - 64 \, \text{years} \, P} \qquad (7)$$

where $15 - 64$ years E is the number of people aged 15–64 who are employed in agriculture, workshops, or public and private employment. $15 - 64$ years P is the total population aged from 15 to 64 years. For land use, population, educational level, irrigation level, and crop level, LPCIEA scores are given in Table 5. The age of employment criterion is used directly and without coefficient because of the elimination of the 0–14 and over 64 age groups.

Finally, to classify villages using quantitative and qualitative methods, the results obtained from the criteria are listed in Table 6.

2.3 Data extraction methods

The LPCIEA index was based on two sets of data. The first group relates to land use, crops, and the level of irrigation, and the second group relates to information on population, employment age, and education. In the first group, three combined field operations, remote sensing, and informational methods from the Regional Water Company of Fars were used. In the second group, the database of the Statistical Center of Iran[f] and the Fars Educational Organization[g] was used. The region was visited during the summer of 2016. Images were processed in spring 2006 and 2016 in 2021. Landsat 7 satellite imagery for spring 2006 was downloaded from the USGS website.[h] The satellite imagery of the IRS was then taken at the Iran National Cartographic Center[i] in April 2016. Landsat 7 images showed striping errors

[f] https://www.amar.org.ir/english.

[g] https://medu.ir/fa?ocode=90230005#.

[h] https://earthexplorer.usgs.gov/.

[i] https://www.ncc.gov.ir/en/.

TABLE 5 Land use, population, level of education, level of irrigation, and agriculture production coefficients.

Score	1	0.75	0.5	0.25
Land use	Forest	Agriculture	Rangeland	Residential
Population	30–64	15–29	Over 64	0–14
Education	Bachelor's / master's / doctorate degree	Certificates / diplomas /professional qualifications	Primary or secondary	No formal education
Irrigation	If cover 0–25 Percentage of Rural district			0.25
	If cover 26–50 Percentage of Rural district			0.5
	If cover 51–75 Percentage of Rural district			0.75
	If cover 75–100 Percentage of Rural district			1
Crop	If 1 product			0.25
	If 2 products			0.5
	If 3 products			0.75
	If 4 products			1

TABLE 6 Quantitative and qualitative classification for the LPCIEA index.

Qualitative	Quantitative
Very low	10
	20
Low	30
	40
Medium	50
	60
High	70
	80
Very high	90
	100

that required correction.[59] Next, the use of 213 GPS points, land use, and types of agricultural products in 2016 was determined through a supervised classification in ENVI 4.5. Land use in the region was categorized into four groups based on their sizes. Forestry, rangelands, agricultural, and residential areas were visited in the summer of 2016 (Figs. 3–8). Others had smaller areas.

The next step was to identify the cultivated areas of the four crops with a higher frequency in the region. These products include cereal, Fabaceae, Garden, and other agricultural products. Products listed as other agricultural products have limited cropland and include eggplant, vegetables, cucumbers, lettuce, and other agricultural products. Field visits and satellite imagery indicated that some regions had a greater variety of products. This diversity enhances risks. The increased risk has increased the resilience of individuals and the viability of the village.[60] Each year, part of the land is not planted for rest after a period of cultivation and rehabilitation to allow the soil to regenerate. According to field studies and visits, approximately one-third of the land in each village during the same cropping season is not used and is out of reach. Therefore, these lands will not contribute to increasing resilience over the same period. According to the irrigation network of the Dorodzan Dam, the area of farmland that uses wells and dam water was extracted from each village. Groundwater wells are generally limited to dry years through the Regional Water Company of Fars. These lands are supplied with water once or twice annually using an irrigation system. The degree of drought resilience varies depending on the area of the rural district, with an irrigation system and the cultivated area.

FIG. 3 Irrigation canal for agriculture.

FIG. 4 Face to face interviews with farmers.

The website of the Statistical Center of Iran was used to obtain information on the population and age groups in the rural areas of the Marvdasht in Fars province. The population was divided into four age groups: 0–14, 15–29, 30–64, and over 64. This statistic is based on two periods: the 2006 and 2016 census of population and housing. The number of people in the study area declined by 7194 in 10 periods. According to observations on the ground, different age groups had different levels of resilience.[61]

3 Results

3.1 Statistical data and image processing

The rural district population surveyed, employment age and education level for 2006 and 2016 were extracted from the census statistical tables. The scores for each group were calculated using the LPCIEA index. This information is provided in Tables 7–9.

FIG. 5   Farmland and irrigation channel.

FIG. 6   Areas without tillage and not cultivated.

A supervised image classification method is used for image classification. Using the ENVI software, the dispersion of the samples, their average, and their variation in the two-dimensional space were studied, and the necessary corrections were made. The number of classes needed for the software was defined, and then, depending on the spectral characteristics of the phenomena, they were placed in specific and defined classes. The supervised method proceeded with classification by selecting educational samples for each of the particular spectral classes using the maximum likelihood algorithm. Under this method, the classes were separated step by step, and in each step, a special class was separated from the other classes.[62] The post-processing of the 2006 and 2016 Landsat and IRS images are presented in Table 10. Farmland and rangeland were larger in the study area than in the other groups. According to field visits, forestland is very limited or manually planted. Based on the research carried out[63] in this research, the resulting information shows that the image of the IRS has a higher kappa coefficient. The kappa coefficient and overall precision for Landsat 2006 and IRS 2016 are listed in Table 10. The overall accuracy of the IRS image was 95.011%.

FIG. 7   Irrigating fields using water from the dam.

FIG. 8   Cereal lands.

The accuracy of the user[j] and the accuracy of the images[k] was used to examine the results.[64] In 2006, the highest image resolution and user precision were linked to agricultural lands and then to rangelands. The user accuracy and image accuracy for agricultural land and rangelands were 94.58, 92.28, 91.25, and 89.25, respectively. In 2016, user and image precision values of 96.75 and 92.54, respectively, refer to rangelands. As a result, forested land is more accurate than residential and agricultural land. Details of this section can be found in Table 11.

The area for each land use in each rural district was extracted using statistical table images processed using the Geographic Information System. The largest part of the forest is in Ramjerd 2, and the smallest is in Rodbal. The agricultural area of the study area is the largest in Ramjerd 2 and the smallest in Ramjerd 1. The largest rangeland area is at Mohammad Abad, and the lowest is at Majdabad. According to the population of residential areas, Ramjerd 2 and

[j] Commission.

[k] Omission.

TABLE 7 Population of rural districts in the study area and their scores for 2006 and 2016 age groups.

Score Rural district	1 30–64	0.75 15–29	0.5 Over 64	0.25 0–14
2006				
Ramjerd 1	3386	4179	1907	2310
Ramjerd 2	3553	3945	2063	2520
Rodbal	3527	3921	2037	2154
Majd abad	3217	3599	2154	1878
Mohammad abad	3409	3852	2212	3033
2016				
Ramjerd 1	3212	3980	1701	1323
Ramjerd 2	2624	3007	1572	1921
Rodbal	3367	3650	1520	2101
Majd abad	3658	1020	2201	1568
Mohammad abad	3550	3870	2407	3310

TABLE 8 Education level of the villages studied and their scores 2006 and 2016.

Score Rural district	1 Bachelor's/master's/ doctorate degree	0.75 Certificates/diplomas /professional qualifications	0.5 Primary or secondary	0.25 No formal education
2006				
Ramjerd 1	3225	1980	2735	3086
Ramjerd 2	3475	2020	2804	3122
Rodbal	2890	1955	2605	3049
Majd abad	2920	1745	2518	2841
Mohammad abad	3110	2102	2902	3276
2016				
Ramjerd 1	3255	1824	2519	2842
Ramjerd 2	3100	1870	2583	2915
Rodbal	3110	1801	2488	2808
Majd abad	2954	1679	2319	2617
Mohammad abad	3455	1936	2673	3017

Mohammad abad have the highest and lowest residential areas, respectively. Overall, in the rural regions studied, the rate of land change in 2016 was lower than that in 2006. This can be explained by the use of land capacity in agriculture. Natural ranges are protected by the forests, range, and watershed management organization, or are steep, rocky, and waterless, which are restricted to cultivation. However, the development of residential areas is limited. This development is the result of building villa gardens. There was no significant growth in residential development as a result of negative population growth (Fig. 3). There are very few forests, and the trees are very far from each other. Forest land is limited to gardens, plantation forests, parks, and a small number of natural forests. The details are listed in Table 12.

Within the agricultural sub-group, the number of products from each village was extracted. Majd abad and Mohammad abad have less than half a hectare of gardens. Due to drought and declining well water tables, surface water has

TABLE 9 Number of people working in rural areas (2006, 2016).

Rural district	30–64	15–29
2006		
Ramjerd 1	1131	723
Ramjerd 2	1851	1059
Rodbal	1292	704
Majd abad	656	451
Mohammad abad	920	582
2016		
Ramjerd 1	992	634
Ramjerd 2	1624	929
Rodbal	1134	617
Majd abad	575	395
Mohammad abad	807	510

TABLE 10 Kappa coefficient, overlay accuracy, and percent image coverage by class in 2006 and 2016.

Landsat 2006		IRS 2016	
Kappa coefficient	0.858	Kappa coefficient	0.910
Overall accuracy	92.22%	Overall accuracy	95.011%
Class	Percentage	Class	Percentage
Forest	0.0007	Forest	0.0006
Agriculture	76.146	Agriculture	76.688
Rangeland	23.136	Rangeland	22.596
Residential	0.716	Residential	0.714
Total	100	Total	100

TABLE 11 Omission and commission for agriculture, forests, rangelands, and residential land in 2006 and 2016.

	2006			2016	
Class	Omission (%)	Commission (%)	Class	Omission (%)	Commission (%)
Forest	85.15	82.54	Forest	91.17	90.29
Agriculture	94.58	92.28	Agriculture	89.15	88.58
Rangeland	91.25	89.25	Rangeland	96.75	92.54
Residential	89.15	85.24	Residential	91.01	90.27

been used to grow cereals and Fabaceae. The purchase of cereals was guaranteed by the Ministry of Jihad. In 2006, there were 102.91 ha of other crops in Mohammadabad, but in 2016, they reached fewer than 0.5 ha. The reason for the change in cultivation is the lack of water discharged from the dam once or twice a year. Table 13 presents information on these groups.

The percentages of irrigation and rural district scores are presented in Table 14. Each coefficient was determined based on the LPCIEA index. Due to a decrease in the volume of water released between 2011 and 2016, the level of irrigation in 2016 decreased. The water dam was released at a specified time, but the reduction in the amount of

TABLE 12 Land use level of the studied areas in 2006 and 2016 (hectare).

Rural district	Area (ha)			
	Forest	Agriculture	Rangeland	Residential
2006				
Ramjerd 1	0.120	11,993.001	2997.357	196.668
Ramjerd 2	0.210	23,201.261	3965.230	301.475
Rodbal	0.018	13,816.956	3916.050	113.950
Majd abad	0.100	12,650.524	2702.084	129.994
Mohammad abad	0.360	21,938.668	11,820.328	44.266
2016				
Ramjerd 1	0.110	12,008.339	2983.468	195.229
Ramjerd 2	0.190	23,201.595	3964.032	302.358
Rodbal	0.020	13,823.731	3910.345	112.878
Majd abad	0.110	12,715.030	2637.752	129.810
Mohammad abad	0.330	22,446.601	11,312.431	44.260

TABLE 13 Cultivation levels of cereals, Fabaceae, orchards, and other products in 2006 and 2016.

Area (ha)		2006		
Rural district	Cereal	Fabaceae	Garden	Other product
Ramjerd 1	8498.871	84.553	2.354	64.836
Ramjerd 2	15,935.88	310.44	0.84	255.17
Rodbal	9204.83	22.01	8.97	225.44
Majd abad	6593.62	68.02	*	42.09
Mohammad abad	6457.12	304.77	*	102.91
		2016		
Rural district	Cereal	Fabaceae	Garden	Other product
Ramjerd 1	8330.348	64.647	2.354	41.312
Ramjerd 2	14,860.60	68.49	0.84	266.27
Rodbal	9568.60	21.28	8.97	227.62
Majd abad	6101.55	68.02	*	39.79
Mohammad abad	6314.93	86.05	*	*

*It is less than 0.5 ha.

water available forced farmers to adapt and reduce their farmland. This issue is in line with the findings of research on reducing the production level with the construction of a dam.[56] The highest rates of irrigation reduction and the lowest were for Ramjerd 2 and Ramjerd 1. The cultivated area of the rural district of Rodbal has increased over the years. This problem is caused by authorized and unauthorized wells and higher water levels than in other villages (Table 15).

The results of all the processes carried out in 2006 and 2016 through satellite images and field visits are presented in Figs. 9 and 10.

3.2 LPCIEA index

Each rural district was ranked according to the resilience and viability index criteria and equations (Table 16). In 2006, the highest and lowest levels of resilience and livability in land use were 71 and 66 Ramjerd 2 and Mohammad

TABLE 14 Level and percent of irrigation based on 2006 and 2016 satellite imagery.

Rural district	Area (ha)		
	Irrigation area	Percentage	Score
2006			
Ramjerd 1	8650.599	57	0.75
Ramjerd 2	16,502.318	60	0.75
Rodbal	9461.239	53	0.75
Majd abad	6703.721	43	0.50
Mohammad abad	6864.714	20	0.25
2016			
Ramjerd 1	8438.66	56	0.75
Ramjerd 2	15,196.209	55	0.75
Rodbal	9826.461	55	0.75
Majd abad	6209.354	40	0.50
Mohammad abad	6400.982	19	0.25

TABLE 15 Percentage change, irrigation levels of villages surveyed in 2016 (hectare).

Rural district	Irrigation area
Ramjerd 1	−211.939
Ramjerd 2	−1306.109
Rodbal	+365.222
Majd abad	−494.367
Mohammad abad	−463.732

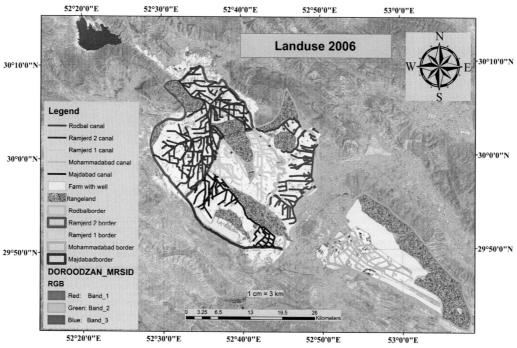

FIG. 9 Satellite image processing (2006).

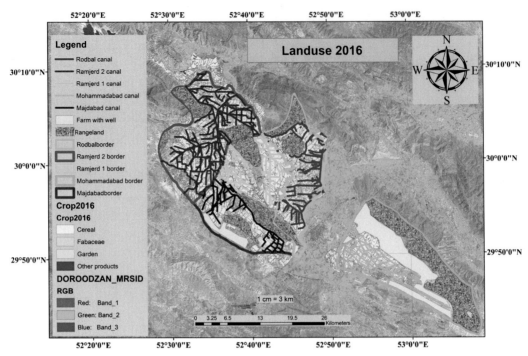

FIG. 10   Satellite image processing (2016).

TABLE 16   LPCIEA scores in 2006 and 2016.

		Ramjerd 1	Ramjerd 2	Rodbal	Majd abad	Mohammad abad
		2006				
	L	69	71	69	70	66
	P	68	68	69	69	65
Resiliance	C	57	60	53	32	15
	E	58	59	55	57	56
	I	57	60	53	43	20
	A	26	42	32	20	24
		56	60	55	49	41
		2016				
	L	69	71	61	70	67
	P	72	67	69	70	64
Resiliance	C	56	55	55	30	3
	E	64	70	57	83	53
	I	56	55	55	40	19
	A	24	49	30	26	21
		57	61	55	53	38

abad, respectively. Regarding the population, the resilience and viability of rural areas are close to each other, with a slight difference. In terms of crop diversity, Ramjard 2 has the greatest diversity and the least cultivated area and is related to the Mohammada bad. The products of Mohammad are cereals and a small quantity of Fabaceae. The decline in water levels in the Dorodzan Dam and wells during the dry years due to the level of cereal production decreased significantly. Consequently, Ramjard 2 has high resilience and viability, and Mohammada bad has the weakest resilience and viability. These findings support research conducted on the construction of dams and their relationship to the cultivated area.[55] Ramjerd 2 has more educated people than Rodbal, and his number is 4 points higher than Rodbal in terms of resilience and viability. Rodbal resilience and viability were 59. The findings of this criterion confirm the relationship between education and energy use.[51]

In 2016, the level of resilience and viability of land use in Majd abad and Ramjard 2 and 1 did not change from 2006. The Mohammad abad has been upgraded to one level. This increase is due to increased farmland and the use of range-lands for agriculture. The 2016 population declined from 2006, and the population pressure on resources decreased, but the situation was still not positive. Ramjerd 1 and Majd abad have higher resilience and viability within the population and lower pressure on resources. The diversity of rodent crops and land has increased their resilience and sustainability. Other villages lack resilience and viability compared to 2006. The educational level of people except Mohammad abad has increased significantly, and resilience and viability have improved in that criterion. The irrigated area increased in Rodbal, but decreased in other rural districts. This is consistent with the findings of other studies on reducing irrigation levels to cope with the increasing stress of drought.[34] employment of young people increased in Ramjerd 2 and Majd abad, but declined in Ramjerd 1 and Mohammad abad. The cause is agricultural growth in Ramjards 1 and 2.

Table 17 provides a comparison of rural districts with changes in the LPCIEA index in 2006 and 2016. This table shows positive changes to improve resilience and viability in green, no change in yellow, and negative changes in red. In this table, villages are subdivided into six groups based on each criterion, and their changes are presented. The LPCIEA suggests different categories of resilience and viability. The villages were divided into three groups based on their resilience and viability scores. The scores ranged from 38 to 49, with low viability and resilience. The second group of rural people had moderate viability and resilience, with a score of 53–57. The third group is rural, with greater resiliency and sustainability than the previous two groups, with scores of 60 and 61. The rural district map was compiled for 2006 and 2016 (Figs. 11 and 12). This division can be viewed in three ways.

This study focused on the resilience of villages to drought. Using the criteria set out in this study as an index, it is possible to classify villages. The trend of change can be assessed using the criteria of cropland, population, age, land use, and educational level incidence before and after the crisis. The study of areas of change shows that after drought, water supply plays a significant role in increasing resilience. In other words, planning before and after the drought crisis can be achieved through this indicator. The primary outcome of this study was the selection of effective criteria for studying the trend of drought changes. These criteria were selected based on the findings of the research in this field. It is possible to determine environmental conditions and social characteristics before and after the crisis. The results show that the effects of drought on the villages studied are different. Distance from the water supply, educational level, and age are important criteria for changing people's livelihoods after the crisis. Second, change assessment is provided quantitatively and qualitatively to decision-making and policy development. Third, change modeling can be achieved by increasing the number of courses studied. Fourth, the process of changing individual criteria was considered separately. Fifth, environmental and social criteria interact with each other. In this study, they were mutually exclusive. In other words, there is no arrangement of the criteria or the weight of one criterion over another criterion. This allows for the use of indexes in other areas. Weighting one criterion causes radical changes in the index. Consequently, to develop and simplify the index, they should be evaluated separately. However, it does not ignore the relationship between humans and the environment. The effects of the environment and humans are well monitored by changes in the criteria. The results show that after the crisis (Table 17), 15 variables experienced negative changes, four variables remained unchanged, and 11 variables performed positively. These results show the efficacy of the selected

TABLE 17   - Results of resilience and viability changes in 2006 and 2016.

	Ramjerd 1	Ramjerd 2	Rodbal	Majd abad	Mohammad abad
L	0	0	−8	0	+1
P	+4	−1	0	+1	−1
C	−1	−5	+2	−2	−12
E	+6	+11	+2	+26	−3
I	−1	−5	+2	3	−1
A	−2	+7	−2	+6	−3

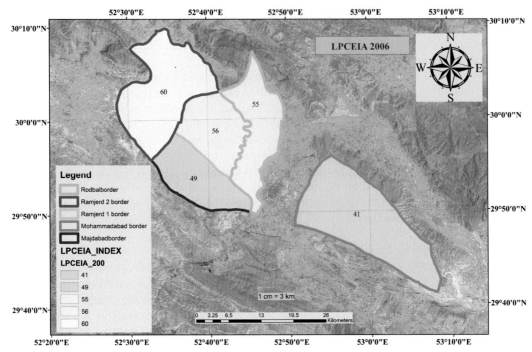

FIG. 11   LPCIEA scores in the study area (2006).

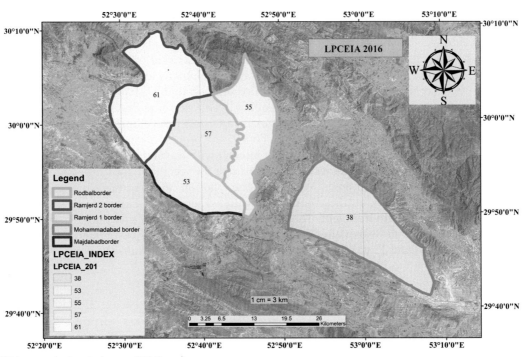

FIG. 12   LPCIEA scores in the study area (2016).

criteria and confirm the search results of Adzawla et al.[58] As the climate changes and the drought crisis, the livelihoods of the villagers change. The process of this change is clearly visible in the results of the LPCIEA index and satellite imagery. The decline in resilience indicates that there have been changes in the level of farmland and the type of land use to address the adverse impacts of drought. These changes are quantified within the research findings and are consistent with the findings of Bowle et al.[55]

4 Discussion

A study of five rural districts showed that the rate of resilience and viability declined in the last 10 years. However, because of the distribution of agricultural water via the Dorodzan dam, the changes are almost similar to each other. The volume of agricultural water supply in villages below Dorodzan Dam decreased from 2006 to 2016. Particularly from 2011 to 2016, the resilience and viability of villages in the face of drought decreased. Decreasing resiliency reduces sustainability, and population growth is negative in regions impacted by various factors that reduce resilience and viability. The comparison shows that areas closer to the dam have greater resilience. Increasing the distance from the dam led to a lower score. Maps and tables of the evolution of resilience and viability over a decade illustrate the effects of drought on the reduction of social cohesion in rural areas. Population growth is negative, but in areas where water has increased during the decade of drought, population pressure has increased resources. People engage in a variety of activities, using water from the dam to live together. The variety of crops in the areas nearest the dam was greater than in the other areas, and the rates of resilience and viability changed smoothly. The range of changes can be larger in other fields. This index can be used in other areas to obtain more accurate results through comparison. This study provides an indicator of slow changes in important resilience and sustainability criteria. This index may be used for planning, economic, and social systems and areas requiring periodic monitoring.

5 Conclusion

The LPCIEA indicator shows that villages with better access to water are more resistant to crises. As the land study shows, there are no changes in the villages of Ramjerd 1 and 2 and the Majd Abad. In other words, by changing the amount of access to water, the strategy to cope with water shortages is adapted to the conditions. It also prevents agricultural land development, but land use is the same as in the past. In Rodbal, due to groundwater exploitation and conditions conducive to water extraction, small changes occur in the area of the farms. In Mohammad Abad, water resources have decreased considerably. This issue is part of the farmland reduction strategy and has increased resilience. The growth of Ramjerd's youth population has led to the use of new tactics to deal with water shortages. Population growth puts more strain on resources, but it is easier to improve performance per unit area and to accept technology by young people than by older people. No significant changes were observed in other villages. Production and crop diversity decreased significantly, with the except for Rodbal villages (due to access to groundwater resources). Because of the distance of the dam and lack of access to water resources, Mohammad abad has less variety of crops and less cultivated land. As educational levels improve and individual knowledge increases, methods of using less water per unit area increase. The educational level in the villages of Majd Abad has increased considerably, and this criterion has been effective in improving other criteria. The educational effect is consistent with research by Wang et al. and Malinowski et al.[51, 57] The effect of drought alters the cultivated area, and this factor can be observed in the irrigated area in all villages except the villages of Rodbal. The level of education in the villages of Majd Abad and Ramjerd 2 increased. The average age in these areas has decreased, and the number of talented people with alternative livelihoods has increased. In other words, small and start-up jobs, most of which are linked to young people, have grown. This issue is closely linked to the speed of learning.

Acknowledgments

This study was funded by the Soil Conservation and Watershed Management Research Institute (SCWMRI) and the Sistan Agricultural and Natural Resources Research and Education Center. We also thank the Fars Agricultural and Natural Resources Research and Education Center, the Regional Water Company of Fars, the Fars Education Organization (Mr. Abbas Salimi Kouchi), and the Statistical Center of Iran for their information. Satellite imagery was obtained from the website of the US Geological Survey and the Iran National Cartographic Center. Dr. Jamileh Salimi Kochi participated in the field visit, and the researcher thanks them for providing this information.

References

1. Ohl C, Johst K, Meyerhoff J, Beckenkamp M, Grüsgen V, Drechsler M. Long-term socio-ecological research (LTSER) for biodiversity protection – a complex systems approach for the study of dynamic human–nature interactions. *Ecol Complex.* 2010;7(2):170–178. https://doi.org/10.1016/j.ecocom.2009.10.002.
2. SchlÜTer M, McAllister RRJ, Arlinghaus R, et al. New horizons for managing the environment: a review of coupled social-ecological systems modeling. *Nat Resour Model.* 2012;25(1):219–272. https://doi.org/10.1111/j.1939-7445.2011.00108.x.
3. Kaniewski D, Van Campo E, Guiot J, Le Burel S, Otto T, Baeteman C. Environmental roots of the late bronze age crisis. *PLoS One.* 2013;8(8): e71004. https://doi.org/10.1371/journal.pone.0071004.
4. Drake BL. The influence of climatic change on the late bronze age collapse and the Greek dark ages. *J Archaeol Sci.* 2012;39(6):1862–1870. https://doi.org/10.1016/j.jas.2012.01.029.

5. Almutairi A, Mourshed M, Ameen RFM. Coastal community resilience frameworks for disaster risk management. *Nat Hazards*. 2020;101 (2):595–630. https://doi.org/10.1007/s11069-020-03875-3.

6. Bukvic A, Rohat G, Apotsos A, de Sherbinin A. A systematic review of coastal vulnerability mapping. *Sustainability*. 2020;12(7). https://doi.org/10.3390/su12072822.

7. Chuang W-C, Eason T, Garmestani A, Roberts C. Impact of Hurricane Katrina on the coastal systems of southern Louisiana. Original research. *Front Environ Sci*. 2019;7(68). https://doi.org/10.3389/fenvs.2019.00068.

8. Sajjad M, Chan JCL. Risk assessment for the sustainability of coastal communities: a preliminary study. *Sci Total Environ*. 2019;671:339–350. https://doi.org/10.1016/j.scitotenv.2019.03.326.

9. Sweeney B, Mordue G, Carey J. Resilient or resistant? Critical reflections on resilience in an old industrial region. *Geoforum*. 2020;110:125–135. https://doi.org/10.1016/j.geoforum.2020.02.005.

10. Roostaie S, Nawari N, Kibert CJ. Sustainability and resilience: a review of definitions, relationships, and their integration into a combined building assessment framework. *Build Environ*. 2019;154:132–144. https://doi.org/10.1016/j.buildenv.2019.02.042.

11. Keshavarz M, Karami E. Drought and agricultural ecosystem services in developing countries. In: Gaba S, Smith B, Lichtfouse E, eds. *Sustainable Agriculture Reviews 28: Ecology for Agriculture*. Springer International Publishing; 2018:309–359.

12. Preece G, Shaw D, Hayashi H. Using the viable system model (VSM) to structure information processing complexity in disaster response. *Eur J Oper Res*. 2013;224(1):209–218. https://doi.org/10.1016/j.ejor.2012.06.032.

13. Karimi V, Karami E, Keshavarz M. Vulnerability and adaptation of livestock producers to climate variability and change. *Rangel Ecol Manag*. 2018;71(2):175–184. https://doi.org/10.1016/j.rama.2017.09.006.

14. Keshavarz M, Karami E, Vanclay F. The social experience of drought in rural Iran. *Land Use Policy*. 2013;30(1):120–129. https://doi.org/10.1016/j.landusepol.2012.03.003.

15. Finné M, Holmgren K, Shen CC, Hu HM, Boyd M, Stocker S. Late bronze age climate change and the destruction of the Mycenaean palace of Nestor at Pylos. *PLoS One*. 2017;12(12). https://doi.org/10.1371/journal.pone.0189447, e0189447.

16. Knapp AB, Sturt WM. Crisis in context: the end of the late bronze age in the eastern Mediterranean. *Am J Archaeol*. 2016;120(1):99–149. https://doi.org/10.3764/aja.120.1.0099.

17. Mishra A, Bruno E, Zilberman D. Compound natural and human disasters: managing drought and COVID-19 to sustain global agriculture and food sectors. *Sci Total Environ*. 2021;754:142210. https://doi.org/10.1016/j.scitotenv.2020.142210.

18. Ayantobo OO, Li Y, Song S, Yao N. Spatial comparability of drought characteristics and related return periods in mainland China over 1961–2013. *J Hydrol*. 2017;550:549–567. https://doi.org/10.1016/j.jhydrol.2017.05.019.

19. Hao Z, Singh VP, Xia Y. Seasonal drought prediction: advances, challenges, and future prospects. *Rev Geophys*. 2018;56(1):108–141. https://doi.org/10.1002/2016RG000549.

20. Morton JF. The impact of climate change on smallholder and subsistence agriculture. *Proc Natl Acad Sci U S A*. 2007;104(50):19680. https://doi.org/10.1073/pnas.0701855104.

21. Udall B, Overpeck J. The twenty-first century Colorado River hot drought and implications for the future. *Water Resour Res*. 2017;53(3):2404–2418. https://doi.org/10.1002/2016WR019638.

22. Wilhite DA, Sivakumar MVK, Pulwarty R. Managing drought risk in a changing climate: the role of national drought policy. *Weather Clim Extremes*. 2014;3:4–13. https://doi.org/10.1016/j.wace.2014.01.002.

23. Levin S, Xepapadeas T, Crépin A-S, et al. Social-ecological systems as complex adaptive systems: modeling and policy implications. *Environ Dev Econ*. 2012;18(2):111–132. https://doi.org/10.1017/S1355770X12000460.

24. Mihunov VV, Lam NSN. Modeling the dynamics of drought resilience in south-central United States using a Bayesian network. *Appl Geogr*. 2020;120:102224. https://doi.org/10.1016/j.apgeog.2020.102224.

25. Liu Y, Chen J. Future global socioeconomic risk to droughts based on estimates of hazard, exposure, and vulnerability in a changing climate. *Sci Total Environ*. 2021;751:142159. https://doi.org/10.1016/j.scitotenv.2020.142159.

26. Lopez-Nicolas A, Pulido-Velazquez M, Macian-Sorribes H. Economic risk assessment of drought impacts on irrigated agriculture. *J Hydrol*. 2017;550:580–589. https://doi.org/10.1016/j.jhydrol.2017.05.004.

27. Conrad C, Usman M, Morper-Busch L, Schönbrodt-Stitt S. Remote sensing-based assessments of land use, soil and vegetation status, crop production and water use in irrigation systems of the Aral Sea basin. A review. *Water Secur*. 2020;11:100078. https://doi.org/10.1016/j.wasec.2020.100078.

28. Salvador C, Nieto R, Linares C, Díaz J, Alves CA, Gimeno L. Drought effects on specific-cause mortality in Lisbon from 1983 to 2016: risks assessment by gender and age groups. *Sci Total Environ*. 2021;751:142332. https://doi.org/10.1016/j.scitotenv.2020.142332.

29. Neisi M, Bijani M, Abbasi E, Mahmoudi H, Azadi H. Analyzing farmers' drought risk management behavior: evidence from Iran. *J Hydrol*. 2020;590:125243. https://doi.org/10.1016/j.jhydrol.2020.125243.

30. Lei Y, Liu C, Zhang L, Luo S. How smallholder farmers adapt to agricultural drought in a changing climate: a case study in southern China. *Land Use Policy*. 2016;55:300–308. https://doi.org/10.1016/j.landusepol.2016.04.012.

31. Pour SH, Wahab AKA, Shahid S. Spatiotemporal changes in aridity and the shift of drylands in Iran. *Atmos Res*. 2020;233:104704. https://doi.org/10.1016/j.atmosres.2019.104704.

32. Khayyati M, Aazami M. Drought impact assessment on rural livelihood systems in Iran. *Ecol Indic*. 2016;69:850–858. https://doi.org/10.1016/j.ecolind.2016.05.039.

33. Singh P, Javed S, Shashtri S, Singh RP, Vishwakarma CA, Mukherjee S. Influence of changes in watershed landuse pattern on the wetland of Sultanpur National Park, Haryana using remote sensing techniques and hydrochemical analysis. *Remote Sens Appl: Soc Environ*. 2017;7:84–92. https://doi.org/10.1016/j.rsase.2017.07.002.

34. Lu J, Carbone GJ, Huang X, Lackstrom K, Gao P. Mapping the sensitivity of agriculture to drought and estimating the effect of irrigation in the United States, 1950–2016. *Agric For Meteorol*. 2020;292–293:108124. https://doi.org/10.1016/j.agrformet.2020.108124.

35. Ciardo F, De Angelis J, Marino BFM, Actis-Grosso R, Ricciardelli P. Social categorization and joint attention: interacting effects of age, sex, and social status. *Acta Psychol*. 2021;212:103223. https://doi.org/10.1016/j.actpsy.2020.103223.

36. Azizi Khalkheili T, Zamani GH. Farmer participation in irrigation management: the case of Doroodzan dam irrigation network, Iran. *Agric Water Manag*. 2009;96(5):859–865. https://doi.org/10.1016/j.agwat.2008.11.008.

37. Afrasiabikia P, Parvaresh Rizi A, Javan M. Scenarios for improvement of water distribution in Doroodzan irrigation network based on hydraulic simulation. *Comput Electron Agric.* 2017;135:312–320. https://doi.org/10.1016/j.compag.2017.02.011.

38. Tashayo B, Honarbakhsh A, Akbari M, Eftekhari M. Land suitability assessment for maize farming using a GIS-AHP method for a semi- arid region, Iran. *J Saudi Soc Agric Sci.* 2020;19(5):332–338. https://doi.org/10.1016/j.jssas.2020.03.003.

39. Xu Y, Zhang X, Wang X, Hao Z, Singh VP, Hao F. Propagation from meteorological drought to hydrological drought under the impact of human activities: a case study in northern China. *J Hydrol.* 2019;579:124147. https://doi.org/10.1016/j.jhydrol.2019.124147.

40. Ayanlade A, Howard MT. Understanding changes in a Tropical Delta: a multi-method narrative of landuse/landcover change in the Niger Delta. *Ecol Model.* 2017;364:53–65. https://doi.org/10.1016/j.ecolmodel.2017.09.012.

41. Sivakumar VL, Radha Krishnappa R, Nallanathel M. Drought vulnerability assessment and mapping using Multi-Criteria decision making (MCDM) and application of Analytic Hierarchy process (AHP) for Namakkal District, Tamilnadu, India. *Mater Today: Proc.* 2020. https://doi.org/10.1016/j.matpr.2020.09.657.

42. Williamson D, Majule A, Delalande M, et al. A potential feedback between landuse and climate in the Rungwe tropical highland stresses a critical environmental research challenge. *Curr Opin Environ Sustain.* 2014;6:116–122. https://doi.org/10.1016/j.cosust.2013.11.014.

43. Muzychko CG. Climatic, land use, human population, and three categories of endemic plants relate drought and biodiversity: an application for ecological management. *Environ Sustain Indicators.* 2021;9:100090. https://doi.org/10.1016/j.indic.2020.100090.

44. Wang M, Jiang S, Ren L, et al. An approach for identification and quantification of hydrological drought termination characteristics of natural and human-influenced series. *J Hydrol.* 2020;590:125384. https://doi.org/10.1016/j.jhydrol.2020.125384.

45. de Carvalho AM, de Carvalho LG, Barbosa HA, et al. Human progress and drought sensitivity behavior. *Sci Total Environ.* 2020;702:134966. https://doi.org/10.1016/j.scitotenv.2019.134966.

46. Jehanzaib M, Shah SA, Yoo J, Kim T-W. Investigating the impacts of climate change and human activities on hydrological drought using non-stationary approaches. *J Hydrol.* 2020;588:125052. https://doi.org/10.1016/j.jhydrol.2020.125052.

47. Dardonville M, Urruty N, Bockstaller C, Therond O. Influence of diversity and intensification level on vulnerability, resilience and robustness of agricultural systems. *Agric Syst.* 2020;184:102913. https://doi.org/10.1016/j.agsy.2020.102913.

48. Degani E, Leigh SG, Barber HM, et al. Crop rotations in a climate change scenario: short-term effects of crop diversity on resilience and ecosystem service provision under drought. *Agric Ecosyst Environ.* 2019;285:106625. https://doi.org/10.1016/j.agee.2019.106625.

49. Sanford GR, Jackson RD, Booth EG, Hedtcke JL, Picasso V. Perenniality and diversity drive output stability and resilience in a 26-year cropping systems experiment. *Field Crop Res.* 2021;263:108071. https://doi.org/10.1016/j.fcr.2021.108071.

50. Sutcliffe C, Knox J, Hess T. Managing irrigation under pressure: how supply chain demands and environmental objectives drive imbalance in agricultural resilience to water shortages. *Agric Water Manag.* 2021;243:106484. https://doi.org/10.1016/j.agwat.2020.106484.

51. Wang S, Xie Z, Wu R. Examining the effects of education level inequality on energy consumption: evidence from Guangdong Province. *J Environ Manag.* 2020;269:110761. https://doi.org/10.1016/j.jenvman.2020.110761.

52. Bartkowiak G, Krugiełka A, Dachowski R, Gałek K, Kostrzewa-Demczuk P. Attitudes of polish entrepreneurs towards knowledge workers aged 65 plus in the context of their good employment practices. *J Clean Prod.* 2021;280:124366. https://doi.org/10.1016/j.jclepro.2020.124366.

53. Lund J, Medellin-Azuara J, Durand J, Stone K. Lessons from California’s 2012–2016 drought. *J Water Resour Plan Manag.* 2018;144(10). https://doi.org/10.1061/(ASCE)WR.1943-5452.0000984, 04018067.

54. Palchaudhuri M, Biswas S. Application of LISS III and MODIS-derived vegetation indices for assessment of micro-level agricultural drought. *Egypt J Remote Sens Space Sci.* 2020;23(2):221–229. https://doi.org/10.1016/j.ejrs.2019.12.004.

55. Bowles TM, Mooshammer M, Socolar Y, et al. Long-term evidence shows that crop-rotation diversification increases agricultural resilience to adverse growing conditions in North America. *One Earth.* 2020;2(3):284–293. https://doi.org/10.1016/j.oneear.2020.02.007.

56. Annys S, Van Passel S, Dessein J, Adgo E, Nyssen J. From fast-track implementation to livelihood deterioration: the dam-based Ribb irrigation and drainage project in Northwest Ethiopia. *Agric Syst.* 2020;184:102909. https://doi.org/10.1016/j.agsy.2020.102909.

57. Malinowski M, Jabłońska-Porzuczek L. Female activity and education levels in selected European Union countries. *Res Econ.* 2020;74(2):153–173. https://doi.org/10.1016/j.rie.2020.04.002.

58. Adzawla W, Azumah SB, Anani PY, Donkoh SA. Analysis of farm households' perceived climate change impacts, vulnerability and resilience in Ghana. *Sci Afr.* 2020;8. https://doi.org/10.1016/j.sciaf.2020.e00397, e00397.

59. Aghamohamadnia M, Abedini A. A morphology-stitching method to improve Landsat SLC-off images with stripes. *Geodesy Geodyn.* 2014;5(1):27–33. https://doi.org/10.3724/SP.J.1246.2014.01027.

60. Matsushita K, Yamane F, Asano K. Linkage between crop diversity and agro-ecosystem resilience: nonmonotonic agricultural response under alternate regimes. *Ecol Econ.* 2016;126:23–31. https://doi.org/10.1016/j.ecolecon.2016.03.006.

61. Koutsou S, Partalidou M, Ragkos A. Young farmers' social capital in Greece: trust levels and collective actions. *J Rural Stud.* 2014;34:204–211. https://doi.org/10.1016/j.jrurstud.2014.02.002.

62. Li N, Martin A, Estival R. Heterogeneous information fusion: combination of multiple supervised and unsupervised classification methods based on belief functions. *Inf Sci.* 2021;544:238–265. https://doi.org/10.1016/j.ins.2020.07.039.

63. Hasim S, Bhar KK. Seasonal cropping pattern extraction using NDVI from IRS LISS-III image of Kangsabati commanded area. *Proc Comp Sci.* 2020;167:900–906. https://doi.org/10.1016/j.procs.2020.03.389.

64. Bastarrika A, Chuvieco E, Martín MP. Mapping burned areas from Landsat TM/ETM+ data with a two-phase algorithm: balancing omission and commission errors. *Remote Sens Environ.* 2011;115(4):1003–1012. https://doi.org/10.1016/j.rse.2010.12.005.

15

Spatial and seasonal modeling of the land surface temperature using random forest

Soheila Pouyan[a], Soroor Rahmanian[b], Atiyeh Amindin[a], and Hamid Reza Pourghasemi[a]

[a]Department of Natural Resources and Environmental Engineering, College of Agriculture, Shiraz University, Shiraz, Iran
[b]Quantitative Plant Ecology and Biodiversity Research Lab, Department of Biology, Faculty of Science, Ferdowsi University of Mashhad, Mashhad, Iran

1 Introduction

LST is used as an important component in various studies, such as climate, hydrological and agricultural processes, land use, and vegetation, to indicate the physical changes on the land surface.[1,2] LST is a crucial indicator in investigations of energy balance models on the land surface, atmosphere, and Earth interactions at the regional and global scales. In addition, LST is known to be a key factor for physical, chemical, and biological control of land processes as well as urban climate studies.[3] Currently, numerous challenges have been created in urban environmental management due to increased urbanization; for instance, global warming is one of the most important environmental problems and pollution.[4,5] Therefore, LST can provide useful information about the physical features of the land surface and climate.[6] LST is of crucial importance in a wide range of geosciences, which is important in urban climatology, global environmental changes, and human-environment interactions[7] and is an important parameter in many environmental models.[8,9]

Measurement of LTS is very difficult in a wide range of spatial and temporal scales because of its time-consuming aspects and expensiveness.[10] Remote sensing data are currently used extensively in studies on land surfaces and their components. Estimation of LST is among the most important practical aspects of remote sensing in climatological studies.[11] The Spatial variations in LST are influenced by numerous factors. Many studies have been performed on the relationship between vegetation and LST, water, land use, and topography. These studies increased the mapping of land vegetation, aiming to achieve stable exploitation and natural source management with high reliability of performance.[12] It has been identified that the type of land use can affect LST and may be used as a possible indicator of the LST trend.[13,14]

The use of machine learning techniques is of particular importance. Over the last two decades, these techniques have been applied to distinguish features and extract more accurate information on land surface studies. In addition, the RF algorithm has the potential to be used as a spatial modeling tool for evaluating environmental problems.[15] Numerous researchers have used the RF algorithm to map vegetation and evaluate the importance of properties.[16–22] RF has high precision for determining the effective factors of LST.[23–27] Therefore, it has been widely used as a machine learning algorithm for remote sensing in different fields over the last few years.[28] This study aimed to develop an LST map of 2019 in two different seasons in the study region using Landsat satellite images and to evaluate and determine the most effective factors used in LST modeling. Random forest algorithm has been used for downscaling land surface temperature in most studies, while in this study, the land surface temperature is modeled using a random forest algorithm. Simultaneously, the random forest algorithm is used to model land use. In addition, the importance of variables effective on LST by this method in other studies has not been done. The results of this study can provide a comprehensive view of the spatial and temporal changes in land surface temperature with regard to spatial changes in urban regions.

2 Methodology

This study was carried out in Shiraz City, aiming to spatially and seasonally model LST using the RF technique. Shiraz is the fifth-largest city in Iran and the capital of Fars Province in southern Iran. In a semi-arid climate, Shiraz is located between 52° 16′ 00″ and 52° 40′ 54″ E longitudes and 29° 29′ 52″ and 29° 53′ 49″N latitudes, and 1540 m altitude, and has an average annual rainfall of 306 mm and the average temperature in July (the hottest month of the year) is 30°C, in December (the coldest month of the year) is 5°C.[29] (Fig. 1). A flowchart of this study is shown in Fig. 2.

2.1 LST mapping using Landsat satellite images

To investigate the LST change trend during the winter and summer of 2019, all images of Landsat 8 in 2019 were downloaded to map the LST in these seasons.

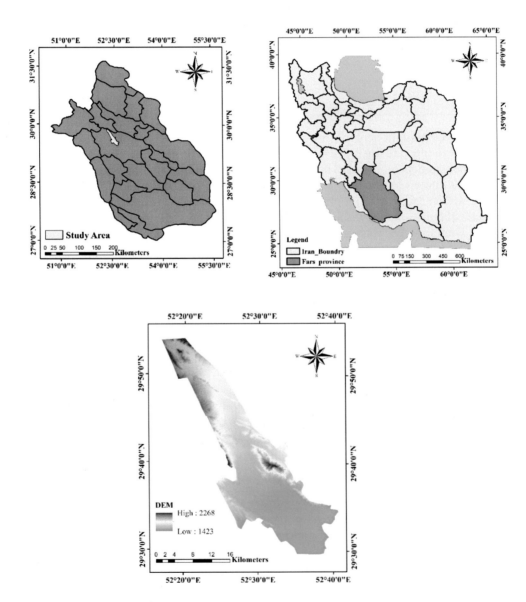

FIG. 1   Location of the study area (Shiraz) in Iran and Fars Province.

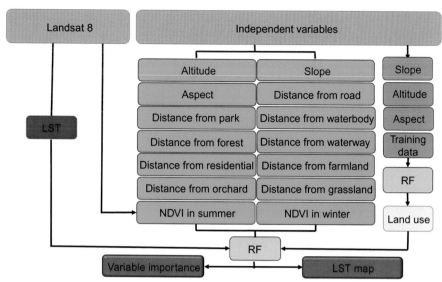

FIG. 2 Flowchart of various steps of this research.

2.2 Preparing effective factors on LST

2.2.1 Land use mapping through the RF model

Land use is one of the most important factors affecting the LST. The Preparation of training samples is necessary for supervised classification. The extensive use of free, high-resolution, and spatially precise images of Google Earth has significantly decreased the costs of field studies so that experts who are familiar with the tools can be utilized for sampling in classifications.[30] To this end, in the present study, we randomly prepared some training points for four land uses, including residential areas, orchards, industrial areas, and bare lands, using Google Earth according to the extent of the land use classes. The accuracy of some samples and their land use were confirmed through field surveys. In addition, a set of different thematic layers, including the digital elevation model (DEM), slope, and aspect, are used for mapping land use.[31–34]

2.2.2 Environmental and topographical features

Geographical and environmental properties are important factors that can affect the spatial changes of LST, such as environmental features such as buildings, roads, orchards, parks in the urban environment, and grasslands, as well as geographical features such as topographic changes, can affect the surrounding land surface temperature.[35–40] In this study, maps of independent variables such as altitude, slope, aspect, distance to main roads, distance to parks, distance to a waterbody, distance to waterways, distance to farmlands, distance to grasslands, land use, and NDVI were prepared (Fig. 3). The spatial resolutions of these factors are listed in Table 1.

2.3 RF algorithm

RF is a simple machine learning algorithm developed by Breiman[41] that can be used for both classification and regression.[42] The RF algorithm uses several decision trees. In fact, an ensemble of decision trees forms a forest that can make better decisions in comparison to a single tree. The RF approach is based on a new method of mixed information, in which many decision trees are developed, and all trees are then mixed for prediction.[43] The key parameters of the RF model are the number of trees and the number of predicting variables. According to these two parameters, the decision tree is grown to the highest possible size and left without pruning.[44,45] The RF prediction model is based on averaging the results of all relevant decision trees and can accurately classify most data sets.[46] The random trees obtain the input vector and classify it with each tree in the forest; the output is the class label received from the majority of votes. To classify a new object, the input vector is located at the end of each RF tree; each tree results in classification, and it is said that this tree has voted to that class.[41]

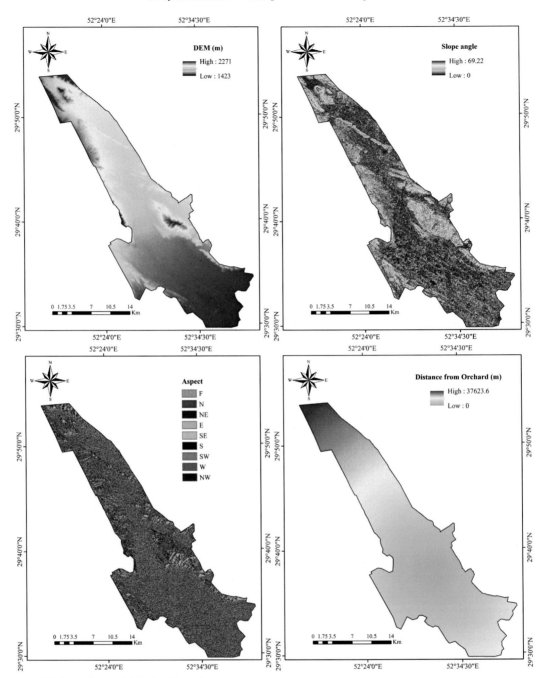

FIG. 3   The maps of independent variables in this study.

(Continued)

2.4 Evaluation of the random forest model

To evaluate the validity of the regression prediction model, the mean root mean squared error (RMSE), and R^2 criteria were used.[47]

2.5 Importance of variables

The variables' importance using the RF model is evaluated through two indices such as, "mean decrease in accuracy (MDA) and mean decrease in Gini coefficient (MDG)." There are two indicators regarding the regression, including IncMSE and IncNodePurity, which are similar to MDA and MDG. These two methods have been used in various studies to evaluate the importance of variables; in general, IncNodePurity is better and more appropriate than IncMSE for

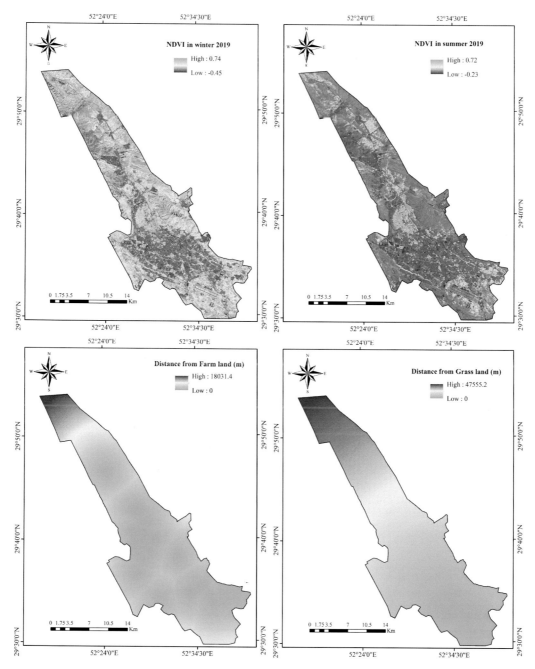

FIG. 3, CONT'D

(Continued)

determining the priority and importance of continuous variables.[48–50] Therefore, IncNodePurity was used in this study to investigate the importance of the variables.

3 Results and discussion

3.1 LST mapping using Landsat images

The LST changes in Shiraz City were mapped in the summer and winter of 2019 using Landsat images 8 (Fig. 4). The thermal information showed that the mean LST was 12.83°C in winter and 40.02°C in summer (Table 2). To investigate the LST map, the variations of LST in the case study were classified into four classes using the "natural break" method

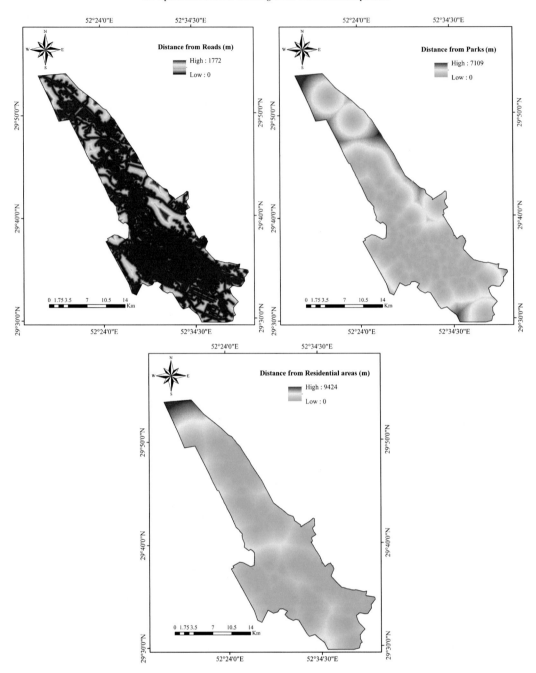

FIG. 3, CONT'D

in the two seasons. In the winter, the highest (39.19%) and the lowest (9%) percent of the area belonged to the fourth (14.57–24.67°C) and the first thermal classes (−6.86°C to 7.76°C), respectively. In the summer, the highest (31.03%) and the lowest (16.63%) percent of an area associated with the third (39.48–42.68°C) and the first thermal classes (27.93°–35.73°C), respectively (Table 3).

3.2 LST map modeling

To model LST, the land use maps of 2019 in four classes of residential areas, orchards, industrial areas, and bare lands were prepared using the RF method (Fig. 5). The highest percentage of land use pertained to bare lands (48.52%), residential areas (43.93%), orchards (7.20%), and industrial areas (0.34). Ghosh et al.[19] indicated that the use of altitude for mapping land use can play a crucial role in the classification of regions with lower topographic

TABLE 1 The spatial resolution of factors.

Factors		Spatial resolution
Distance to	Parks	Open Street Map (1:100,000)
	Farmlands	
	Grasslands	
	Orchards	
	Waterbody	
	Waterways	
	Roads	
DEM	Slope	Shuttle Radar Topography Mission (12.5 m)
	Aspect	
	Land use	
NDVI		Landsat 8 (30 m)

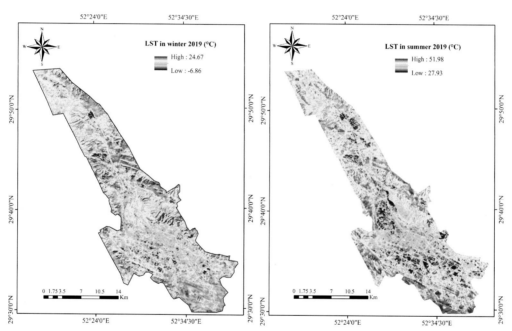

FIG. 4 LST map of Shiraz in 2019 using the Landsat images 8.

TABLE 2 The thermal information of seasonal LST.

Season	Min	Max	Standard error	Mean
Winter (2019)	−6.86	24.67	3.39	12.83
Summer (2019)	27.93	51.98	3.78	40.02

diversity. In this study, altitude, slope, and aspect were used to differentiate the various land uses. In addition, a successful classification method requires training data.[51] Studies have shown that RF-based methods have higher accuracy than maximum likelihood classifier (MLC)-based methods for the classification of land use in urban areas.[52,53] High accuracies have been reported in studies that utilized the RF method for the precise classification of land use.[19,54–58]

Then, LST spatial modeling was performed for two different seasons in 2019 using the maps of independent variables (Fig. 6). The LST map of the winter and summer seasons showed that the mean LST was 12.83°C and 40.02°C, respectively (Table 4). According to the four thermal classes, the highest percent of the area (35.06%) pertained to the

TABLE 3 Classification of seasonal LST.

Season	LST classification (°C)	Percentage of area (%)
Winter (2019)	6.86–7.76	9.00
	7.76–11.49	19.07
	11.49–14.57	32.74
	14.57–24.67	39.19
Summer (2019)	27.93–35.73	16.63
	35.73–39.48	22.09
	39.48–42.68	31.03
	42.68–51.98	30.25

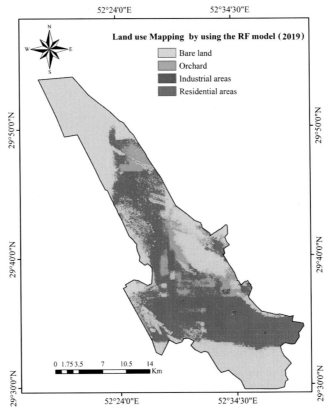

FIG. 5 Land use classes of the study area in 2019.

fourth thermal class (14.14–16.72°C) and the lowest (16.53%) to the second thermal class (10.21–12.36°C). The first and second thermal classes (<12.36) in the winter indicate the coolest fields and are located in the central part about orchards and residential areas with extensive vegetation of parks and scattered patches of mountainous parts. The first and second thermal classes are widely located in the northern, northeastern, and southern parts, with the highest bare land.

In the summer, the highest percentage of the area (46.63%) pertained to the fourth thermal class (41.16–44.81°C) and the lowest percentage of the area (15.07%) to the first thermal class (32.36–32.69°C) (Table 5). In the summer, similar to the winter, the spatial pattern of the land surface temperature of the fourth class in the case study pertained to bare lands. As mentioned, LST modeling for the two seasons showed that the highest thermal class occurred in the bare land of the region, *that is*, bare lands. Bare lands absorb large amounts of energy and warm rapidly because they lack the

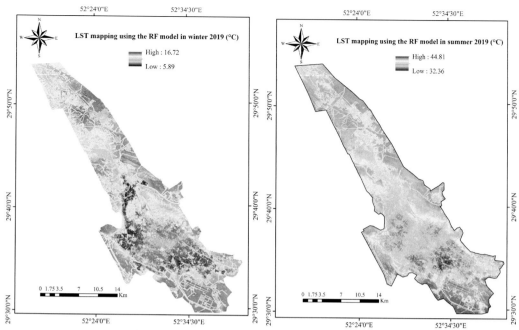

FIG. 6 The seasonal LST map of Shiraz in 2019 using RF.

TABLE 4 The thermal information of seasonal LST using RF.

Season	Min	Max	Standard error	Mean
Winter (2019)	5.89	16.72	2.55	12.83
Summer (2019)	32.36	44.81	2.20	40.02

TABLE 5 Classification of seasonal LST using RF.

Season	LST classification (°C)	Percentage of area (%)
Winter (2019)	5.89–10.21	18.14
	10.21–12.36	16.53
	12.36–14.14	30.27
	14.14–16.72	35.06
Summer (2019)	32.36–36.69	15.07
	36.39–39.97	19.38
	39.97–41.16	18.92
	41.16–44.81	46.63

evapotranspiration cooling mechanism and are exposed to high absorbance and low conductance capacity. Previous studies have also shown that the type of land use is significantly associated with surface temperature and pointed out the impact of NDVI as a cooling factor.[6,13,59] Spatial changes in surface temperature in urban areas are correlated with vegetation.[6,13]

Any changes in the land use of urban areas can affect LST, especially changing green areas to buildings and impermeable surfaces can increase LST.[60] Studies in which the use of green spaces, residential areas, bare lands, and water

TABLE 6 The modeling validation criteria of LST.

Season	R^2	RMSE
Winter (2019)	0.48	2.58
Summer (2019)	0.53	2.61

bodies have shown that higher temperatures in bare lands and residential areas can be seen at any time.[12,61,62] Although the expansion of residential areas over time can increase the land surface temperature, vegetation in urban regions helps decrease LST.[63] Orchards and parks in the urban areas of the study region had a cooling effect and reduced LST. The finding of Wang et al.[64] is indicated that the cooling efficacy of increased vegetation is stronger than the heating efficacy because of urban development and extension. Lower LST is typically in areas where higher vegetation is observed,[65] and increasing vegetation covers are considered an effective reduction procedure for temperature adjustment.[66–69] The major reason for the spatial change in temperature patterns in the urban area is the built-up environment.[70] Urban environments, due to their wide impermeable surface and low vegetation, typically have a high heat capacity and high thermal conductivity.[71–74] This causes less solar radiation energy to be reflected and less energy to be converted into latent heat related to transpiration.[75] Moreover, one of the major factors affecting the urban climate is anthropogenic activities such as fuel burning, automobiles, and machinery.[76,77] The findings of a research in Shiraz have shown that bare lands, industrial and green spaces have a significant impact on LST. In addition, LST is mitigated in residential areas with temperate building density. Besides that, condensed gardens, and lower road density, have decreased LST in Shiraz, a semi-arid urban environment in Iran which is consistent with our findings.[78] The modeling validation criteria for LST are presented in Table 6. The results of this study indicated that the RF technique is an appropriate method with acceptable accuracy and can provide an appropriate estimation of LST.

3.3 Importance of variables

Evaluation of the importance of variables effective on LST showed that in the winter, the NDVI, and distance to roads and in the summer (Fig. 7), NDVI and DEM had the highest effect on LST (Fig. 8). Previous research has shown a negative relationship between NDVI and LST, which indicates the high performance of NDVI for LST assessment.[79,80] In addition, NDVI is considered an important indicator for the measurement of local environmental conditions and the identification of environmental changes.[81] The results of Shafizadeh-Moghadam et al.[82] showed that LST is affected by NDVI, latitude, road, and LULC land use/land cover. Analysis of the relationship between LST and its effective variables can be used for the proper management of urban areas.

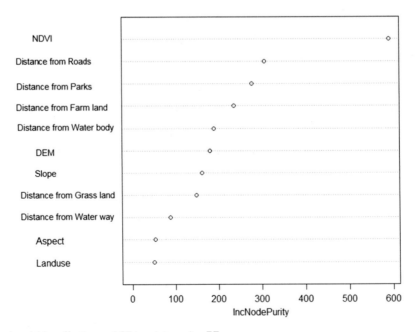

FIG. 7 The importance of variables effective on LST in winter using RF.

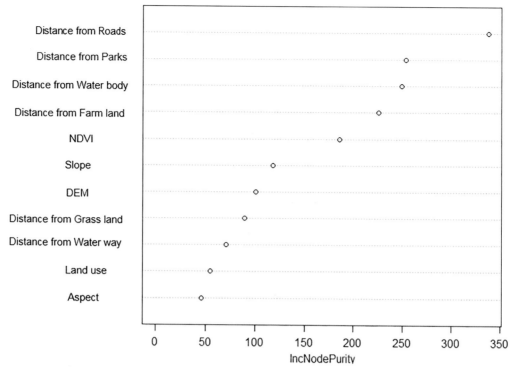

FIG. 8 The importance of variables effective on LST in summer using RF.

4 Conclusion

LST has crucial importance in urban climate studies; as a result, researchers intend to find a relationship between LST and other parameters to precisely estimate it. In this study, the RF algorithm was used to model the spatial and seasonal changes in the land surface temperature of Shiraz City in 2019. The results of this study showed that LST changes according to the different land use and spatial changes in LST arose from changes in land use; in particular, the extent of bare land increased the land surface temperature of the study region and vegetation cover decreased LST. Prioritization of the factors affecting LST in winter and summer showed that factors such as NDVI, distance to roads, and parks play a significant role. In this regard, it seems necessary and important for urban management authorities to preserve vegetation and green space in urban areas as an important variable for moderating weather conditions. Based on the results of this study, using classification algorithms based on machine learning, such as the RF algorithm, is an appropriate method for estimating LST spatial and seasonal data. Therefore, the results of this study can be used by planners and experts in urban areas to obtain information on LST and its relationship with land use.

References

1. Mannstein H. Surface energy budget, surface temperature and thermal inertia. In: *Remote Sensing Applications in Meteorology and Climatology*. Springer; 1987:391–410.
2. Dickinson RE, Henderson-Sellers A. Modelling tropical deforestation: a study of GCM land-surface parametrizations. *Q J Roy Meteorol Soc*. 1988;114(480):439–462.
3. Khandelwal S, Goyal R, Kaul N, Mathew A. Assessment of land surface temperature variation due to change in elevation of area surrounding Jaipur, India. *Egypt J Remote Sens Space Sci*. 2018;21(1):87–94.
4. Zheng Z, Ren G, Wang H, Dou J, Gao Z, Duan C, et al. Relationship between fine-particle pollution and the urban heat island in Beijing, China: Observational evidence. *Boundary-Layer Meteorol*. 2018;169(1):93–113.
5. Feigenwinter C, Vogt R, Parlow E, Lindberg F, Marconcini M, Del Frate F, et al. Spatial distribution of sensible and latent heat flux in the city of Basel (Switzerland). *IEEE J Sel Top Appl Earth Obs Remote Sens*. 2018;11(8):2717–2723.
6. Lu D, Weng Q. Use of impervious surface in urban land-use classification. *Remote Sens Environ*. 2006;102(1–2):146–160.
7. Mallick J, Kant Y, Bharath B. Estimation of land surface temperature over Delhi using Landsat-7 ETM+. *J Ind Geophys Union*. 2008;12(3):131–140.
8. Becker F, Li Z-L. Towards a local split window method over land surfaces. *Remote Sens (Basel)*. 1990;11(3):369–393.
9. Srivastava PK, Majumdar T, Bhattacharya AK. Surface temperature estimation in Singhbhum shear zone of India using Landsat-7 ETM+ thermal infrared data. *Adv Space Res*. 2009;43(10):1563–1574.

10. Hong Y, Nix HA, Hutchinson MF, Booth TH. Spatial interpolation of monthly mean climate data for China. *Int J Climatol*. 2005;25(10):1369–1379.

11. Feizizadeh B, Blaschke T, eds. *Thermal remote sensing for land surface temperature monitoring: Maraqeh County, Iran. 2012 IEEE International Geoscience and Remote Sensing Symposium*, IEEE; 2012.

12. Mustafa EK, Liu G, Abd El-Hamid HT, Kaloop MR. Simulation of land use dynamics and impact on land surface temperature using satellite data. *GeoJournal*. 2019;1–19.

13. Yuan Q, Lu D. A sub-pixel analysis of urbanization effect on land surface temperature and its interplay with impervious surface and vegetation coverage in Indianapolis, United States. *Int J Appl Earth Observ Geoinf*. 2008;10(1):68–83.

14. Dontree S, ed. *Relation of land surface temperature (LST) and land use/land cover (LULC) from remotely sensed data in Chiang Mai—Lamphun basin*: SEAGA Conference; 2010.

15. Booker D, Snelder T. Comparing methods for estimating flow duration curves at ungauged sites. *J Hydrol*. 2012;434:78–94.

16. Pal M. Random forest classifier for remote sensing classification. *Int J Remote Sens*. 2005;26(1):217–222.

17. Lippitt CD, Rogan J, Li Z, Eastman JR, Jones TG. Mapping selective logging in mixed deciduous forest. *Photogramm Eng Remote Sens*. 2008;74(10):1201–1211.

18. Rogan J, Franklin J, Stow D, Miller J, Woodcock C, Roberts D. Mapping land-cover modifications over large areas: a comparison of machine learning algorithms. *Remote Sens Environ*. 2008;112(5):2272–2283.

19. Ghosh A, Sharma R, Joshi P. Random forest classification of urban landscape using Landsat archive and ancillary data: combining seasonal maps with decision level fusion. *Appl Geogr*. 2014;48:31–41.

20. Van Beijma S, Comber A, Lamb A. Random forest classification of salt marsh vegetation habitats using quad-polarimetric airborne SAR, elevation and optical RS data. *Remote Sens Environ*. 2014;149:118–129.

21. Nguyen U, Glenn EP, Dang TD, Pham LT. Mapping vegetation types in semi-arid riparian regions using random forest and object-based image approach: a case study of the Colorado River ecosystem, Grand Canyon, Arizona. *Ecol Inform*. 2019;50:43–50.

22. Phan D, Nguyen N, Pathirana PN, Horne M, Power L, Szmulewicz D. A random forest approach for quantifying gait ataxia with truncal and peripheral measurements using multiple wearable sensors. *IEEE Sens J*. 2019;20(2):723–734.

23. Gualtieri J, Chettri S, eds. *Support vector machines for classification of hyperspectral data. IGARSS 2000 IEEE 2000 International Geoscience and Remote Sensing Symposium Taking the Pulse of the Planet; The Role of Remote Sensing in Managing the Environment Proceedings (Cat No 00CH37120)*, IEEE; 2000.

24. Mpelasoka F, Mullan A, Heerdegen R. New Zealand climate change information derived by multivariate statistical and artificial neural networks approaches. *Int J Climatol*. 2001;21(11):1415–1433.

25. Fasbender D, Tuia D, Bogaert P, Kanevski M. Support-based implementation of Bayesian data fusion for spatial enhancement: applications to ASTER thermal images. *IEEE Geosci Remote Sens Lett*. 2008;5(4):598–602.

26. Hutengs C, Vohland M. Downscaling land surface temperatures at regional scales with random forest regression. *Remote Sens Environ*. 2016;178:127–141.

27. Zhao W, Duan S-B, Li A, Yin G. A practical method for reducing terrain effect on land surface temperature using random forest regression. *Remote Sens Environ*. 2019;221:635–649.

28. Yang Y, Cao C, Pan X, Li X, Zhu X. Downscaling land surface temperature in an arid area by using multiple remote sensing indices with random forest regression. *Remote Sens (Basel)*. 2017;9(8):789.

29. Pasalari H, Nodehi RN, Mahvi AH, Yaghmaeian K, Charrahi Z. Landfill site selection using a hybrid system of AHP-fuzzy in GIS environment: a case study in Shiraz city, Iran. *MethodsX*. 2019;6:1454–1466.

30. Ozdogan M, Thenkabail P. Image classification methods in land cover and land use. In: *Remotely Sensed Data Characterization, Classification, and Accuracies*: 2015:231–258.

31. Samanta S, Pal DK, Lohar D, Pal B. Preparation of digital data sets on land use/land cover, soil and digital elevation model for temperature modelling using remote sensing and GIS techniques. *Indian J Sci Technol*. 2011;4(6):636–642.

32. Rawat J, Kumar M. Monitoring land use/cover change using remote sensing and GIS techniques: a case study of Hawalbagh block, district Almora, Uttarakhand, India. *Egypt J Remote Sens Space Sci*. 2015;18(1):77–84.

33. Nath B, Niu Z, Singh RP. Land use and land cover changes, and environment and risk evaluation of Dujiangyan city (SW China) using remote sensing and GIS techniques. *Sustainability*. 2018;10(12):4631.

34. Nath B, Wang Z, Ge Y, Islam K, Singh RP, Niu Z. Land use and land cover change modeling and future potential landscape risk assessment using Markov-Ca model and analytical hierarchy process. *ISPRS Int J Geo-Inf*. 2020;9(2):134.

35. Oke TR. The energetic basis of the urban heat island. *Q J Roy Meteorol Soc*. 1982;108(455):1–24.

36. Yang JS, Wang YQ, August PV. Estimation of land surface temperature using spatial interpolation and satellite-derived surface emissivity. *J Environ Inf*. 2004;4(1):37–44.

37. Buyantuyev A, Wu J. Urban heat islands and landscape heterogeneity: linking spatiotemporal variations in surface temperatures to land-cover and socioeconomic patterns. *Landsc Ecol*. 2010;25(1):17–33.

38. Tayyebi A, Shafizadeh-Moghadam H, Tayyebi AH. Analyzing long-term spatio-temporal patterns of land surface temperature in response to rapid urbanization in the mega-city of Tehran. *Land Use Policy*. 2018;71:459–469.

39. Zhang Y, Sun L. Spatial-temporal impacts of urban land use land cover on land surface temperature: case studies of two Canadian urban areas. *Int J Appl Earth Observ Geoinform*. 2019;75:171–181.

40. Wu C, Li J, Wang C, Song C, Chen Y, Finka M, La Rosa D. Understanding the relationship between urban blue infrastructure and land surface temperature. *Sci Total Environ*. 2019;694:133742.

41. Breiman L. Random forests. *Mach Learn*. 2001;45(1):5–32.

42. Svetnik V, Liaw A, Tong C, Culberson JC, Sheridan RP, Feuston BP. Random forest: a classification and regression tool for compound classification and QSAR modeling. *J Chem Inf Comput Sci*. 2003;43(6):1947–1958.

43. Cutler DR, Edwards Jr. TC, Beard KH, Cutler A, Hess KT, Gibson J, et al. Random forests for classification in ecology. *Ecology*. 2007;88(11):2783–2792.

44. Allouche O, Tsoar A, Kadmon R. Assessing the accuracy of species distribution models: prevalence, kappa and the true skill statistic (TSS). *J Appl Ecol.* 2006;43(6):1223–1232.

45. Rodriguez-Galiano VF, Ghimire B, Rogan J, Chica-Olmo M, Rigol-Sanchez JP. An assessment of the effectiveness of a random forest classifier for land-cover classification. *ISPRS J Photogramm Remote Sens.* 2012;67:93–104.

46. Lee C-C, Mower E, Busso C, Lee S, Narayanan S. Emotion recognition using a hierarchical binary decision tree approach. *Speech Commun.* 2011;53 (9–10):1162–1171.

47. Mäkelä H, Pekkarinen A. Estimation of forest stand volumes by Landsat TM imagery and stand-level field-inventory data. *For Ecol Manage.* 2004;196(2–3):245–255.

48. Liaw A, Wiener M. Classification and regression by randomForest. *R News.* 2002;2(3):18–22.

49. Kuhn S, Egert B, Neumann S, Steinbeck C. Building blocks for automated elucidation of metabolites: machine learning methods for NMR prediction. *BMC Bioinform.* 2008;9(1):1–9.

50. Hasanuzzaman M, Bhuyan MH, Zulfiqar F, Raza A, Mohsin SM, Mahmud JA, Fujita M, Fotopoulos V. Reactive oxygen species and antioxidant defense in plants under abiotic stress: revisiting the crucial role of a universal defense regulator. *Antioxidants.* 2020;9(8):681.

51. Foody GM, Mathur A. Toward intelligent training of supervised image classifications: directing training data acquisition for SVM classification. *Remote Sens Environ.* 2004;93(1–2):107–117.

52. Mantero P, Moser G, Serpico SB. Partially supervised classification of remote sensing images through SVM-based probability density estimation. *IEEE Trans Geosci Remote Sens.* 2005;43(3):559–570.

53. Belgiu M, Drăguţ L. Random forest in remote sensing: a review of applications and future directions. *ISPRS J Photogramm Remote Sens.* 2016;114:24–31.

54. Wu Z, Deng XY, Zeng RF, Su Y, Gu MF, Zhang Y, Xie CM, Zheng L. Analysis of risk factors for retropharyngeal lymph node metastasis in carcinoma of the hypopharynx. *Head Neck.* 2013;35(9):1274–1277.

55. Millard K, Richardson M. On the importance of training data sample selection in random forest image classification: a case study in peatland ecosystem mapping. *Remote Sens (Basel).* 2015;7(7):8489–8515.

56. Du Z, Ge L, Ng AH, Zhu Q, Yang X, Li L. Correlating the subsidence pattern and land use in Bandung, Indonesia with both Sentinel-1/2 and ALOS-2 satellite images. *Int J Appl Earth Observ Geoinf.* 2018;67:54–68.

57. Ruiz Hernandez IE, Shi W. A random forests classification method for urban land-use mapping integrating spatial metrics and texture analysis. *Int J Remote Sens.* 2018;39(4):1175–1198.

58. Chang S, Wang Z, Mao D, Guan K, Jia M, Chen C. Mapping the essential urban land use in Changchun by applying random Forest and multi-source geospatial data. *Remote Sens (Basel).* 2020;12(15):2488.

59. Gallo KP, Owen TW, Easterling DR, Jamason PF. Temperature trends of the US historical climatology network based on satellite-designated land use/land cover. *J Climate.* 1999;12(5):1344–1348.

60. Ranagalage M, Estoque RC, Zhang X, Murayama Y. Spatial changes of urban heat island formation in the Colombo District, Sri Lanka: implications for sustainability planning. *Sustainability.* 2018;10(5):1367.

61. Kamusoko C, Gamba J, Murakami H. Monitoring urban spatial growth in Harare metropolitan province, Zimbabwe. *Adv Remote Sens.* 2013;5:2013.

62. Peng J, Jia J, Liu Y, Li H, Wu J. Seasonal contrast of the dominant factors for spatial distribution of land surface temperature in urban areas. *Remote Sens Environ.* 2018;215:255–267.

63. Estoque RC, Murayama Y, Myint SW. Effects of landscape composition and pattern on land surface temperature: an urban heat island study in the megacities of Southeast Asia. *Sci Total Environ.* 2017;577:349–359.

64. Wang C, Myint SW, Wang Z, Song J. Spatio-temporal modeling of the urban heat island in the Phoenix metropolitan area: land use change implications. *Remote Sens (Basel).* 2016;8(3):185.

65. Yuan F, Bauer ME. Comparison of impervious surface area and normalized difference vegetation index as indicators of surface urban heat island effects in Landsat imagery. *Remote Sens Environ.* 2007;106(3):375–386.

66. Rosenfeld AH, Akbari H, Bretz S, et al. Mitigation of urban heat islands: materials, utility programs, updates. *Energ Buildings.* 1995;22(3):255–265.

67. Ashie Y, Ca VT, Asaeda T. Building canopy model for the analysis of urban climate. *J Wind Eng Ind Aerodyn.* 1999;81(1–3):237–248.

68. Tong H, Walton A, Sang J, Chan JC. Numerical simulation of the urban boundary layer over the complex terrain of Hong Kong. *Atmos Environ.* 2005;39(19):3549–3563.

69. Yu C, Hien WN. Thermal benefits of city parks. *Energ Buildings.* 2006;38(2):105–120.

70. Schatz J, Kucharik CJ. Seasonality of the urban heat island effect in Madison, Wisconsin. *J Appl Meteorol Climatol.* 2014;53(10):2371–2386.

71. Oke T. The heat island of the urban boundary layer: characteristics, causes and effects. In: *Wind Climate in Cities.* Springer; 1995:81–107.

72. Weng Q. A remote sensing? GIS evaluation of urban expansion and its impact on surface temperature in the Zhujiang Delta, China. *Int J Remote Sens.* 2001;22(10):1999–2014.

73. Bouyer J, Musy M, Huang Y, Athamena K. *Mitigating Urban Heat Island Effect by Urban Design: Forms and Materials.* 28–30.

74. Song J, Wang Z-H. Interfacing the urban land–atmosphere system through coupled urban canopy and atmospheric models. *Bound-Lay Meteorol.* 2015;154(3):427–448.

75. Golden JS. The built environment induced urban heat island effect in rapidly urbanizing arid regions – a sustainable urban engineering complexity. *Environ Sci.* 2004;1(4):321–349.

76. Bohnenstengel S, Evans S, Clark PA, Belcher S. Simulations of the London urban heat island. *Q J Roy Meteorol Soc.* 2011;137(659):1625–1640.

77. Pal S, Xueref-Remy I, Ammoura L, et al. Spatio-temporal variability of the atmospheric boundary layer depth over the Paris agglomeration: an assessment of the impact of the urban heat island intensity. *Atmos Environ.* 2012;63:261–275.

78. Azhdari A, Soltani A, Alidadi M. Urban morphology and landscape structure effect on land surface temperature: evidence from Shiraz, a semi-arid city. *Sustain Cities Soc.* 2018;41:853–864.

79. PANDA S, JAIN MK. Effects of green space spatial distribution on land surface temperature: implications for land cover change as environmental indices. *Int J Earth Sci Eng.* 2017;10(2):180–184.

80. Aggarwal S, Misra M. Comparison of NDVI, NDBI as indicators of surface heat island effects for Bangalore and New Delhi: case study. In: *Remote Sensing Technologies and Applications in Urban Environments III, 2018 October 9*, Vol. 10793:. International Society for Optics and Photonics; 2018:1079314.

81. Yue W, Xu J, Xu LH. An analysis on eco-environmental effect of urban land use based on remote sensing images: a case study of urban thermal environment and NDVI. *Acta Ecol Sin*. 2006;26(5):1450–1460.

82. Shafizadeh-Moghadam H, Weng Q, Liu H, Valavi R. Modeling the spatial variation of urban land surface temperature in relation to environmental and anthropogenic factors: a case study of Tehran, Iran. *GIScience Remote Sens*. 2020;57(4):483–496.

16

Municipal landfill site selection and environmental impacts assessment using spatial multicriteria decision analysis: A case study

Hazhir Karimi[a], Hooshyar Hossini[b], and Abdulfattah Ahmad Amin[c]

[a]Department of Environmental Science, Faculty of Science, University of Zakho, Duhok, Kurdistan Region, Iraq
[b]Department of Environmental Health Engineering, Faculty of Health, Kermanshah University of Medical Sciences, Kermanshah, Iran [c]Department of Road Construction, Erbil Technology College, Erbil Polytechnic University, Erbil, Kurdistan Region, Iraq

1 Introduction

Currently, the disposal of municipal solid waste (MSW) is a major problem in urban areas.[1] Regardless of the extensive efforts to use efficient disposal methods such as composting, reuse, and recycling, landfilling has become an integral part of the municipal solid waste hierarchy, especially in developing countries.[2] Modern landfills should be well-located and well-designed to meet the requirements and standards of national and international regulations.[3] Landfill sites should not be located in environmentally sensitive areas and should be designed using on-site environmental monitoring systems to avoid possible contamination.[4] Additionally, ideal landfill sites should not only focus on environmental requirements but also on social and economic benefits should be considered to minimize financial expenses and social conflicts.[5]

Despite the adverse environmental and health effects of municipal solid waste, waste is still dumped in open areas or as unsanitary landfills. Some regions also burn waste to reduce its volume in disposal sites. The unsuitable dumping sites and inappropriate processes have numerous effects on the natural environment and public health.[4, 6] For instance, studies have highlighted the negative impacts of landfills on groundwater sources and how the leachate can substantially pollute groundwater.[7, 8] In addition, many studies have emphasized the health effects on residents living in proximity to landfills with no proper disposal and treatment.[9]

Locating optimal landfill sites requires a comprehensive set of environmental, social, technical, and economic parameters,[10] which makes the process of site selection complex.[11, 12] The geographic information system (GIS), in combination with multicriteria decision analysis (MCDA) or multicriteria decision making (MCDM), addresses this limitation and it has been widely used for suitability analysis and site selection.[13] GIS combines spatial data such as maps, aerial photographs, and satellite images with quantitative and qualitative information databases, and clearly shows the analyzed results.[14] Many studies have used GIS and MCDA for landfill site selection (e.g., Refs. 10, 13, 15–18).

The present study aimed to employ GIS and multicriteria decision analysis for municipal landfill site selection, considering the city of Kermanshah in the west of Iran as a study area. Although many studies have used GIS and MCDA for landfill site selection, few studies have assessed the environmental impacts of the selected landfill sites during the construction and operational phases. Thus, in this study, an environmental impact assessment (EIA) framework was also developed to predict and identify the negative impacts and to propose appropriate mitigation actions to minimize risks and impacts.

2 Study area

Kermanshah City in the central Kermanshah province, situated in Iran with an area of 564,710 ha, was the case study located in geographical coordinates: 46°24–47°31 longitude and 33°47–34°48 latitude (Fig. 1). According to the 2016 census, the population of Kermanshah City and the total province was 1,046,000 and 1,952,340 inhabitants, respectively. The province has a moderate and mountainous climate. The average annual rainfall is approximately 470 mm, and the average annual temperature is 14°C.[19] The province has various physiographical features ranging from flat in the east to very steep in the west and northwest. The available data demonstrate that the total municipal solid waste generation in the city of Kermanshah accounts for 700 tons per day, and the waste generation rate is 0.67 kg/day. A regular gathering of the wastes is done by human forces and trucks and a part of the waste is composted in a treatment site while the other is buried.

3 Data and methods

3.1 Criteria definition and map preparation

This study aimed to assess and select suitable landfills for the disposal of municipal solid waste in Kermanshah, a major city in western Iran. First, based on the national regulations, literature, and availability of the data, the key criteria affecting the selection of the landfill were selected. In this stage, 11 criteria or parameters were identified: distance from villages, distance from Kermanshah city (point generation), distance from rivers, distance from the aquifer, distance from wells and springs, distance from the fault lines, distance from protected areas, distance from the road networks, slope, land use/land cover, and geology. Fig. 2 shows the framework of the methodology and the selected parameters. The required data were obtained and/or extracted from several sources, including topographical maps, geological maps, satellite images, fieldwork and provincial and national organizations such as Kermanshah Province Management and Planning Organization, Iranian Geological Organization, Kermanshah Meteorology Organization,

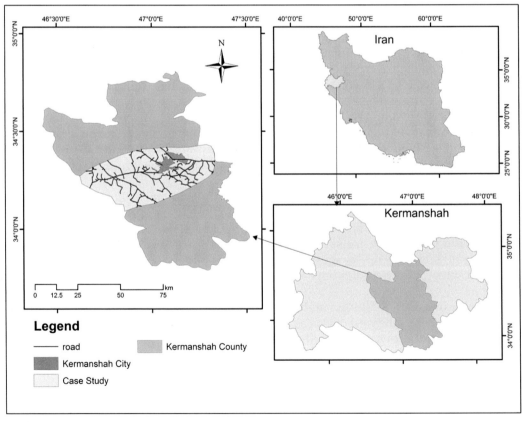

FIG. 1 Location of the study area.

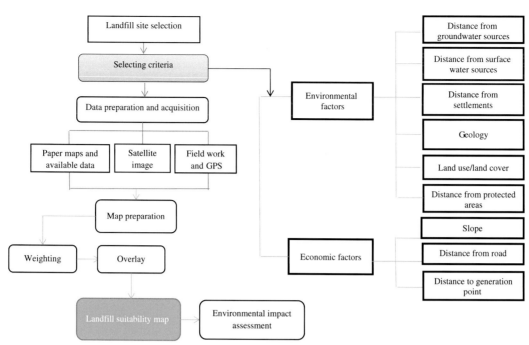

FIG. 2 Methodology framework of the study.

Regional Water Company of Kermanshah, Kermanshah Department of Environment. To prepare the maps of criteria, the data were processed in ArcGIS 10.5 (Figs. 3 and 4). The suitability range for each parameter was determined based on the ranges proposed in the literature and similar studies. In addition, Iran's national regulations and laws were considered to finalize these ranges (Table 1). In the following, the preparation of each parameter is explained in detail.

Surface water sources. Landfill sites are potential threats to surface and Groundwater water sources[7] and the landfill sites should be located far away from water resources. The data and location of the rivers were obtained from the Regional Water Company of Kermanshah at a scale of 1:100,000 in vector format. According to national regulations, landfill sites should not be located within 700 m of the permanent and seasonal rivers, or 1500 m from lakes. There were no ponds or lakes within the study area; thus, only the locations of the rivers were mapped. A 700 m buffer zone was considered around the rivers as unsuitable areas.

Groundwater sources. The data on groundwater sources, including aquifers, wells, springs, and aqueducts, were obtained from the Regional Water Company of Kermanshah with a scale of 1:100,000 in vector format. A buffer of 400 m was considered to be unsuitable for groundwater sources.

Land use/land cover (LULC). Landfills must not be located within or near such land uses as residences, forests, first-class farmlands, municipal parks, and water bodies. The LULC map can be prepared from Landsat data with acceptable accuracy.[20, 21] The LULC map of the study area was obtained from the Kermanshah Natural Resources Organization. Five classes, namely built-up, farmland, rangeland, forest, and bare land, were categorized in the study area and classified into four categories.

Distance from residential areas. Landfill sites near residential areas could annoyed people because of odors, health concerns, and atheistic aspects.[9] This layer was extracted from the LULC map and a distance of 1000 m from the villages, 2000 m from the cities, and 3000 m from the province capital were considered unsuitable.

Distance from roads. The landfill site should be close to the road networks because of transportation costs; however, national laws specify the minimum distances of the landfill sites to the roads. To avoid possible risks, landfills should not be located within 100 m of roads. The road layer was extracted from 1:100000 scale topographical maps.

Geology and fault lines. The geology texture and fault lines play an important role in landfill location. Limestones and clay lands are unsuitable for landfill siting because of their high permeability, whereas metamorphic rocks are more suitable.[17] The geological maps of the 1:100000 scale of the study area was digitized, and the vector layers of the geology and faults were prepared. A distance of 400 m was defined as an unsuitable area around the fault lines.

Distance from protected area. Protected areas, which are environmentally valuable regions, should be sufficiently preserved from any development. In particular, landfilling has adverse effects on protected areas; therefore, the maximum

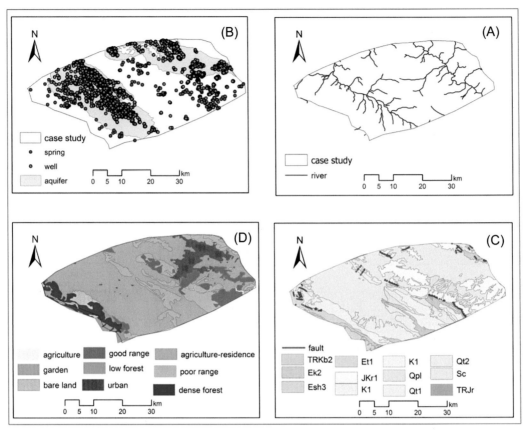

FIG. 3 Maps of rivers (A), groundwater resources (B), geology (C), and land use/cover (D) in the study area.

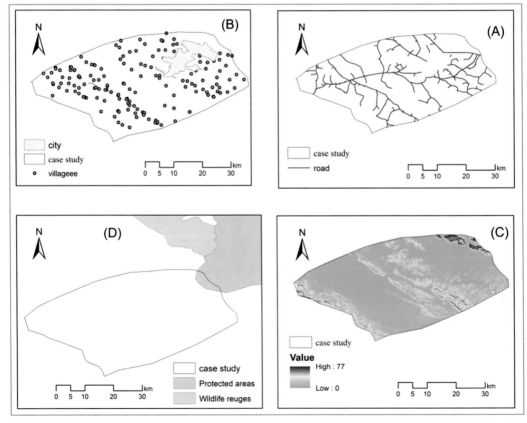

FIG. 4 Maps of roads (A), residential areas (B), slope (C), and protected habitats (D) in the study area.

TABLE 1 The selected factors (criteria) and defined ranges for each factor.

Criteria	Unsuitable	Less suitable	Moderate suitable	High suitable
Distance from rivers (m)	<700	700–1500	1500–2000	>2000
Distance from aquifer (m)	<400	400–800	800–1000	>1000
Distance from well (m)	<400	400–800	800–1000	>1000
Distance from villages (m)	<1000	1000–1500	1500–2500	>2500
Distance from Kermanshah city (km)	<3 <20	3–4	4–6	6–20
Distance from road (m)	<100 ≥2000	1000–2000	500–1000	100–500
Distance from protected areas (m)	<1000	1000–1500	1500–2000	>2000
Distance from fault (m)	<300	300–500	500–700	>700
Slope (%)	≥60	40–60	25–40	<25
Geology	–	Esh3, K1–2, Et1, Uf, EV1	Jkr1.Ek2,Qt1, QPL, k1–1, k1–2-2, k4–2, Sc	TRKb2, Mr., TRJr
Land use/cover	Residential and industrial areas, rocks, water, dense forest	Agriculture lands, good pastures, low-density forest	Poor pastures	Bare land

concern should be considered in landfill site selection. The location of the protected areas within the case study was acquired from the Kermanshah Department of Environment, and a distance of 1500 m was considered unsuitable.

Slope. Slope is a critical factor in landfill site selection. Construction in highly steep areas is not economically ideal, and the cost of construction is maximized.[12] The slope map was prepared from the digital elevation model (DEM) with a 30 m resolution, and the map was classified. A slope of 40% was considered unsuitable, and areas with a lower range were evaluated.

3.2 Landfill suitability analysis

After preparing and gathering all the data, the suitability maps were prepared. Based on Table 1, unsuitable and suitable ranges for each criterion were mapped in ArcGIS 10.5. The maps were resampled to a spatial resolution of 30 m using the nearest neighborhood algorithm. All maps were finally integrated using the raster calculated function in ArcGIS 10.5, and the suitability landfill map was acquired. The final map represents four classes: unsuitable, low-suitability, moderately suitable, and highly suitable.

3.3 Environmental impact assessment

Finally, an environmental impact assessment (EIA) framework was developed to identify, predict, and mitigate the environmental, social, and other related impacts associated with the construction and operational stages of landfilling. The EIA provides useful insights for minimizing negative impacts and is important in the process of municipal solid waste management and landfilling. In this study, the assessment of environmental impacts was conducted for both the construction and operational stages in terms of impact features such as intensity, duration, domain, and possibility.

4 Results and discussion

In this section, based on the methodology explained in Section 3, the findings are explained in two parts: landfill suitability analysis and environmental impact assessment. The suitability analysis for the municipal landfill in the study area was evaluated, and the final map was prepared in four classes. Fig. 5 shows the final suitability map of the study area, and the percentages and areas of the suitability ranges are listed in Table 2. Approximately 7% of

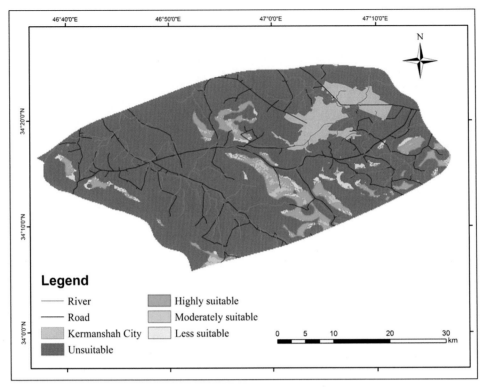

FIG. 5 Final suitability map indicating the suitability for landfilling.

the study area is highly suitable for landfills, while nearly 88% of the total area is unsuitable. Less and moderately suitable classes accounted for 4% of the total. The suitable areas are mostly located in the center, south, and southeast of the study area, whereas the northern and western parts are unsuitable for landfills. A field check was also performed for those regions that fell under high suitability to confirm the results. Acceptable accuracy was observed after field checks.

Water resources can be endangered by landfill leachate, and therefore, they should be seriously protected when landfilling. In our study, maximum concerns were considered. Most of the western and eastern parts were found to be unsuitable areas for rivers. In addition, the aquifer and wells restricted the northern and western parts of the study area for landfills. The lands in these parts are used for irrigation farming because of sufficient underground water resources, which limits landfilling. The fault lines and geology did not act as limitation factors, and it can be stated that most parts of the study area fell under moderate to high suitable ranges. Although Kermanshah is the only city within the study area, the villages were found in most parts of the study area, leading to the unsuitability of land-fills in the northeastern part. Therefore, we considered the highest distance for residences.

The only protected area in our case study is Bistoon Wildlife Refuge and was counted as one of the important pro-tected habitats in western Iran. A very small part of the northeast of the study area was unsuitable for this factor. How-ever, agricultural land restricted the case study for landfilling. This province is known as the agricultural pole in the country, and the lands around the city of Kermanshah are fertile for agriculture. In terms of economic aspects, such as

TABLE 2 Area and percentage of suitability classes.

No.	Suitability class	Percentage
1	Unsuitable	88
2	Less suitable	3
3	Moderately suitable	2
4	Highly suitable	7

road access and land price, the study area has an acceptable network system. Access to roads is not just critical for transportation costs, but it could also increase the risk of waste spread and higher fuel consumption.

An environmental impact assessment was also developed for the landfilling project in two phases of construction and operational. Environmental components such as air quality, water resources, soil, vegetation cover, and wildlife were evaluated, and the impacts were identified. In addition, proper mitigation actions were proposed for the identified impacts. The list of the environmental impacts and mitigation actions for the construction and operation phases are shown in Tables 2 and 3, respectively. During the construction phase, such activities as cut and fill, land leveling, making cells, and transportation are the main activities that cause the degradation of environmental components. In this phase, most of the impacts are temporary and short-term, and the surrounding environment returns to normal states after the completion of activities. However, the impacts on vegetation cover and physiography are permanent, and proper reclamation measures should be considered. The impacts in the operational phase are long-term and permanent, meaning that impacts arise as long as activities continue. The main activities that lead to environmental destruction are the transport of waste from the city to the landfill site, landfilling of wastes using heavy equipment and machinery. Considering the long-term impacts in this phase, mitigation measures such as planting trees (creating a green belt) around the landfill site, utilizing geomembranes and geonets, gravel drainage systems, and on-site monitoring are essential to minimize possible risks (Table 4).

The selected parameters in landfill site selection play an important role in the suitability analysis and in obtaining the final results. In this study, we selected a comprehensive set of criteria, both environmental and economic, and therefore, few areas were evaluated as suitable areas after the maximum considerations. Previous studies have also been conducted based on various comprehensive criteria in the environmental, social, and economic groups; however, the environmental aspects were considered as the most commonly preferred parameter. This point has been highlighted in studies by Karimi et al.,[10] Karimzadeh Motlagh and Sayadi,[22] and Wang et al.[23] It is strongly recommended that future studies consider further economic criteria such as future land-use planning, land price, and cultural aspects.

TABLE 3   Environmental impacts and mitigation actions in construction phase of landfilling.

Environmental components (pollutants or hazards)	Activities	Important mitigation measures
Air quality (CO, CO_2, SOx, NOx, PM, dust)	• Heavy and light vehicles activities • Cut and fill • Making cells • Land grading	• Minimizing activities during wind blow • Using standard vehicles • Using standard fuels (Euro 4) • Planting trees and green space • water spraying of heavy vehicles route • Suitable ventilation system for indoor places
Noise quality	• Construction equipment (e.g., compressors, generators, etc.) • Heavy vehicles activities • Cut and fill	• Creating sound barriers around noisy devices • Regular lubrication of machines and wheels of devices • Shut down activity operation at nights • Use of PPE (e.g., ear muffs)
Soil quality (heavy metals, soil erosion)	• Cut and fill • Land grading • Making cells • Heavy and light vehicles	• Installing worksites and fuel tanks at specific points • Allocating the smallest area to collect contaminated soils • Limiting fill and cut to the determined site • Preventing mixture of vegetable soils with the rest • Using additional soils and waste generated in other construction matters
Water quality (heavy metals, leachate)	• Cut and fill • Land grading • Heavy and light vehicles	• Preventing the discharge of construction debris near rivers • Installing surface water collection system inside landfill site • Creating an impermeable bed (lining)
Fauna and Flora (plant and wildlife)	• Cut and fill • Land grading • Heavy and light vehicles • Personnel activities	• Limiting vegetation clearance within the landfill site • Limiting cut and fill within the landfill site • Educating workers to prevent disturbance or hunting wildlife • Temporary cover for domestic animals crossing

TABLE 4 Environmental impacts and mitigation actions in operational phase of landfilling.

Environmental components (pollutants or hazards)	Activities	Control or preventive measures
Air quality (CO, Ox, methane, dust, etc.)	• Waste transport from city to the landfill site • Landfilling • Fire	• Planting trees and green space around the landfill site (green belt) • Periodic and regular monitoring of air pollutants • Using standards machines and vehicles • Using covers on wastes during waste transport • Using gas collection system for methane
Noise quality	• Waste transport from city to the landfill site • Landfilling	• Planting trees and green space around the landfill site (green belt) • Creating sound barriers around noisy machines • Regular lubrication of machines and wheels of devices • Shut down activity operation at nights • Use of PPE (e.g., ear muffs)
Soil and water pollution (leachate and Pollutants)	• Waste transport from city to the landfill site • Landfilling	• Preventing leakage by proper treatment • Installing monitoring site • Disinfection of wastes • Planting trees and green spaces • Using proper lining • Utilizing geomembranes and geonets instead of daily and intermediate soil covers and gravel drainage systems
Fauna and Flora (plant and wildlife)	• Waste transport from city to the landfill site • Landfilling	• Creating corridor (if needed) • Planting trees and green spaces

The application of GIS-based multi-criteria decision analysis for landfill site selection has been widely used in various regions and countries in recent decades. The results of previous studies were entirely consistent with our findings regarding the efficiency of GIS and MCDA for landfill site selection studies. The high potential of GIS-MCDA in landfill site selection has been reported by Karimi et al.,[10] Bahrani et al.,[15] Aksoy et al.,[16] Kharat et al.,[24] and Demesouka et al.[25] In addition, some studies compared different multi-criteria analysis methods aiming at the selection of the best methods for suitability analysis of landfill site selection. For instance, Lokhande et al.[26] conducted a review of various MCDA methods for locating landfill sites and stated that most of the methods are reliable for landfill site selection. In a similar review by Ozakan et al.,[27] the analytical hierarchy process (AHP) and weighted linear combination (WLC) were found to be the most widely used multi-criteria analysis methods for suitability analysis and ranking the alternatives in recent studies.

The main aim of EIA was to understand the current situation of landfills and enforce appropriate mitigation measures to improve the quality of the degraded environment. Conducting EIA along with site selection has rarely been reported in the literature. In a study by Karimi et al.,[18] site selection and risk assessment for medical landfills were conducted in Iran. Five criteria and two risk aspects, likelihood, and consequence, were considered to calculate the risk of each criterion and total relative risk scores. The results of their study showed that the selected sites for landfilling had a low level of environmental risk because, through the site selection process, environmental considerations were considered. This study strengthens this hypothesis that landfill site selection can significantly decrease the environmental impact of landfills.

Some studies used EIA for comparing the options of solid waste management. Taheri et al.[28] carried out EIA for four municipal solid waste disposal options, including open dumping, sanitary landfill, recycling, and composting in Tabriz. They found that based on the existing situation composting with the sanitary landfill is the best-recommended option. Yang et al.[29] investigated the environmental impact of a landfill in China through life cycle assessment (LCA). The consumption inventory of material and energy during construction and operation was calculated. They found that the diesel consumption used for the daily operation was the main reason for the non-toxic impacts, while the toxic effects were attributed mainly to the use of mineral materials. In this study, they found that utilizing geomembranes and geonets instead of daily and intermediate soil covers and gravel drainage systems could mitigate the environmental burdens by 2% and 27%, respectively.

5 Conclusion

The site selection of municipal landfills requires consideration of various environmental, economic, and social parameters to minimize environmental and social effects as well as economic costs. The multi-criteria decision analysis simplifies the selection of various criteria and decision-making, and the GIS effectively analyzes information layers. Therefore, a combination of GIS and multi-criteria decision analysis reduces time and errors and enhances the accuracy of the analysis. In this study, we selected various parameters for the process of landfill site selection; however, the availability of data was considered in the selection criteria because the criteria should be appropriately mapped with an acceptable scale. The results show that a limited percentage of the study area is highly suitable for landfilling. The locations of the villages, groundwater resources, and rivers were the prominent constraints in the study area. We also developed an EIA approach to identify and mitigate the environmental and socially relevant impacts of landfills. However, geotechnical studies and landfill engineering designs are necessary to avoid possible economic and environmental risks.

Acknowledgments

The authors thank the Kermanshah Department of Environment and Regional Water Company of Kermanshah for providing some data and information.

Conflict of interests

The authors have no conflicts of interest.

References

1. Abdel-Shafy H, Mansour M. Solid waste issue: sources, composition, disposal, recycling, and valorization. *Egypt J Pet.* 2018;27:1275–1290. https://doi.org/10.1016/j.ejpe.2018.07.003.
2. Ferronato N, Torretta V. Waste mismanagement in developing countries: a review of global issues. *Int J Environ Res Public Health.* 2019;16(6):1060. https://doi.org/10.3390/ijerph16061060.
3. Liu A, Ren F, Lin W, Wang J. A review of municipal solid waste environmental standards with a focus on incinerator residues. *Int J Sustain Built Environ.* 2005;4:165–188.
4. Swati TI, Vijay VK, Ghosh P. Scenario of landfilling in India: problems, challenges, and recommendations. In: Hussain C, ed. *Handbook of Environmental Materials Management.* Cham: Springer; 2018. https://doi.org/10.1007/978-3-319-58538-3_167–1.
5. Santhosh LG, Sivakumar BG. Landfill site selection based on reliability concepts using the DRASTIC method and AHP integrated with GIS – a case study of Bengaluru city, India. *Georisk.* 2018;12(3):234–252. https://doi.org/10.1080/17499518.2018.1434548.
6. Khan MH, Vaezi M, Kuma A. Optimal siting of solid waste-to-value-added facilities through a GIS-based assessment. *Sci Total Environ.* 2018;610:1065–1075. https://doi.org/10.1016/j.scitotenv.2017.08.169.
7. Han Z, Ma H, Shi G, He L, Wei L, Shi Q. A review of groundwater contamination near municipal solid waste landfill sites in China. *Sci Total Environ.* 2016;1:1255–1264. https://doi.org/10.1016/j.scitotenv.2016.06.201.
8. Przydatek G, Kanownik W. Impact of small municipal solid waste landfill on groundwater quality. *Environ Monit Assess.* 2019;191:169. https://doi.org/10.1007/s10661-019-7279-5.
9. De Feo G, De Gisi S, Williams ID. Public perception of odour and environmental pollution attributed to MSW treatment and disposal facilities: a case study. *Waste Manag.* 2013;33:974–987. https://doi.org/10.1016/j.wasman.2012.12.016.
10. Karimi H, Amiri S, Huang J, Karimi A. Integrating GIS and multi-criteria decision analysis for landfill site selection, case study: Javanrood County in Iran. *Int J Environ Sci Technol.* 2019;16:7305–7318. https://doi.org/10.1007/s13762-018-2151-7.
11. Siefi S, Karimi H, Soffianian A, Pourmanafi S. GIS-based multi criteria evaluation for thermal power plant site selection in Kahnuj County, SE Iran. *Civil Eng Infrastruct J.* 2017;50(1):179–189. https://doi.org/10.7508/ceij.2017.01.011.
12. Karimi H, Soffianian A, Seifi S, Pourmanafi S, Ramin H. Evaluating optimal sites for combined-cycle power plants using GIS: comparison of two aggregation methods in Iran. *Int J Sustainable Energy.* 2020;39(2):101–112. https://doi.org/10.1080/14786451.2019.1659271.
13. Rahmat ZG, Niri MV, Alavi N, et al. Landfill site selection using GIS and AHP: a case study: Behbahan, Iran. *KSCE J Civ Eng.* 2017;21:111–118. https://doi.org/10.1007/s12205-016-0296-9.
14. Malczewski J. GIS-based multicriteria decision analysis: a survey of the literature. *Int J Geogr Inf Sci.* 2006;20(7):703–726. https://doi.org/10.1080/13658810600661508.
15. Bahrani S, Ebadi T, Ehsani H, Yousefi H, Maknoon R. Modeling landfill site selection by multi-criteria decision making and fuzzy functions in GIS, case study: Shabestar, Iran. *Environ Earth Sci.* 2016;75:337. https://doi.org/10.1007/s12665-015-5146-4.
16. Aksoy E, San BT. Geographical information systems (GIS) and multi-criteria decision analysis (MCDA) integration for sustainable landfill site selection considering dynamic data source. *Bull Eng Geol Environ.* 2017. https://doi.org/10.1007/s10064-017-1135-z.
17. Barakat A, Hilali A, Baghdadi M, Touhami F. Landfill site selection with GIS-based multi-criteria evaluation technique. A case study in Be' ni Mellal-Khouribga Region, Morocco. *Environ Earth Sci.* 2017;76:413. https://doi.org/10.1007/s12665-017-6757-8.
18. Karimi H, Herki BMA, Gardi SQ, et al. Site selection and environmental risks assessment of medical solid waste landfill for the City of Kermanshah-Iran. *Int J Environ Health Res.* 2020. https://doi.org/10.1080/09603123.2020.1742876.
19. Iran Meteorological Organization. Data of meteorological stations; 2020.
20. Karimi H, Jafarnezhad J, Khaledi J, Ahmadi P. Monitoring and prediction of land use/land cover changes using CA-Markov model: a case study of Ravansar County in Iran. *Arab J Geosci.* 2018;11:592. https://doi.org/10.1007/s12517-018-3940-5.
21. Kourosh Niya A, Huang J, Karimi H, Keshtkar H, Naimi B. Use of intensity analysis to characterize land use/cover change in the biggest island of Persian Gulf, Qeshm Island, Iran. *Sustainability.* 2019;11:4396. https://doi.org/10.3390/su11164396.

22. Karimzadeh Motlagh Z, Sayadi M. Siting MSW landfills using MCE methodology in GIS environment (case study: Birjand plain, Iran). *Waste Manag.* 2015;46:322–337. https://doi.org/10.1016/j.wasman.2015.08.013.

23. Wang Y, Li J, An D, Xi B, Tang Y, et al. Site selection for municipal solid waste landfill considering environmental health risks. *Resour Conserv Recycl.* 2018;138:40–46. https://doi.org/10.1016/j.resconrec.2018.07.008.

24. Kharat MG, Kamble SJ, Raut RD, et al. Modeling landfill site selection using an integrated fuzzy MCDM approach. *Model Earth Syst Environ.* 2016;2:53. https://doi.org/10.1007/s40808-016-0106-x.

25. Demesouka OE, Vavatsikos AP, Anagnostopoulosa KP, Eleftheriosb S. Suitability analysis for siting MSW landfills and its multi-criteria spatial decision support system: method, implementation and case study. *Waste Manag.* 2013;33:1190–1206.

26. Lokhande T, Mane S, Mali S. Landfill site selection using GIS and MCDA methods: a review. *Int J Eng Sci Technol.* 2017;2395–6453.

27. Özkan B, Özceylan E, Sarıçiçek İ. GIS-based MCDM modeling for landfill site suitability analysis: a comprehensive review of the literature. *Environ Sci Pollut Res Int.* 2019;26(30):30711–30730. https://doi.org/10.1007/s11356-019-06298-1.

28. Taheri M, Gholamalifard M, Ghazizade Jalili M, Rahimoghli S. Environmental impact assessment of municipal solid waste disposal site in Tabriz, Iran using rapid impact assessment matrix. *Impact Assess Project Apprais.* 2014;32(2):162–169. https://doi.org/10.1080/14615517.2014.896082.

29. Yang N, Damgaard A, Lü F, Shao LM, Brogaard LK, He PJ. Environmental impact assessment on the construction and operation of municipal solid waste sanitary landfills in developing countries: China case study. *Waste Manag.* 2014;34(5):929–937. https://doi.org/10.1016/j.wasman.2014.02.017. Epub 2014 Mar 20 24656422.

17

Predictive habitat suitability models for *Teucrium polium* L. using boosted regression trees

Soroor Rahmanian[a], Soheila Pouyan[b], Sahar Karami[a], and Hamid Reza Pourghasemi[b]

[a]Quantitative Plant Ecology and Biodiversity Research Lab, Department of Biology, Faculty of Science, Ferdowsi University of Mashhad, Mashhad, Iran [b]Department of Natural Resources and Environmental Engineering, College of Agriculture, Shiraz University, Shiraz, Iran

1 Introduction

Urbanization, with consequent impacts on environmental resources and human well-being, has a strong influence on plant distribution.[1] This rising influence of human activities on nature[2] necessitates the consideration of the condition that decides the plant distribution in order to increase plant protection.[3] One of the most crucial components of biodiversity across the globe is medicinal species,[4] with roles in medical attention and cultural importance, economies, and human quality of life, particularly in poor regions.[2, 5] It is crucial to preserve and protect these types of species, containing an enhanced insight into the crucial ecosystem characteristics of medicinal herbs, and to raise awareness among all decision-makers concerning the conservation of this resource.[2]

In ecological sciences, restoration, and environmental policy, SDMs are crucial for evaluating species, ecosystem interactions, and forecasting the distribution of species.[6, 7] SDMs are useful for forecasting the possibility of a plant species being found in a particular area or to measure habitat suitability by the perception of the features of several independent predictors reflecting the main environmental factors relevant to habitat utilization and survival of species.[8] Regression-based models (e.g., general/generalized linear models) have been widely used in the early stages of SDM usage and growth.[6, 9] However, more complicated mathematical methods have been applied to SDMs based on enhanced techniques and ecological knowledge, enhancing the precision of model forecasting.[10] Especially, during the previous years, machine learning methods (MLMs) applied in SDMs significantly, as BRT is particularly one of the most commonly models.[11] The Advantages of BRT include its capability for managing nonlinearity relations, choosing response factors, and computing the relative importance of variables. In many types of research, BRT has a better performance compared to regression-based models in the evaluation of the relations of habitat and species, such as "generalized additive models-GAMs" or "generalized linear models-GLMs".[12, 13]

Due to the importance of machine learning models in predicting the suitable habitats of species in ecological research, however, few studies have focused on medicinal plant SDMs,[14, 15] especially no studies have been done on *Teucrium polium*'s SDMs so far. In this study, a BRT MLM was used to identify the habitat suitability of *T. polium* species. We evaluated this method by using different evaluation techniques to understand its accuracy. In addition, this study compared the effects of various environmental predictors to identify the most important variables affecting the distribution of *T. polium* in its natural habitats.

2 Material and method

2.1 Study area and sampling

Sample data of *T. polium* were recorded from Dakal-kooh, which is located in the Noorabad Region, in the northwest of Fars Province, Iran, with an area of 1756 ha (51.30–55.57°E, and 33.3233.37°N; Fig. 1). The altitude varies between 920 and 1391 m above sea level (Fig. 1). The area has a semiarid climate (based on the de Martonne aridity index). The annual mean rainfall (11-year mean) is 489 mm, with an average temperature of 21°C.[16] We recorded *T. polium* species at 113 points and identified their location using GPS. Among these, 79 locations representing 70% of samples used for training, and the remaining 30% were used as independent samples for validation[17] the BRT model.

2.2 *Teucrium polium* species

Teucrium, a medicinal plant belonging to the genus Labiatae, includes approximately 340 species,[18] which is commonly found in the temperate regions of Africa, northern Europe, and southwest Asia, including Iran. The genus *Teucrium* includes 12 species in Iran, among which *T. polium* is present in many parts of the world, including Europe, Southwest Asia, and northern Africa.[19] Details on its features, especially as traditional medicine, are given in Bahramikia and Yazdanparast,[20] Gharaibeh et al.,[21] and Belmekki and Bendimerad.[22]

2.3 Selection of thematic variables

We picked and categorized 15 conditioning variables based on the current and applicable literature,[23–25] as well as the availability of data, including, elevation, aspect, slope degree, "topographic wetness index (TWI)," "curvatures of plan & profile," annual mean rainfall, and silt, clay, and sand percentages, electrical conductivity-EC, acidity-pH, and nitrogen percentage.

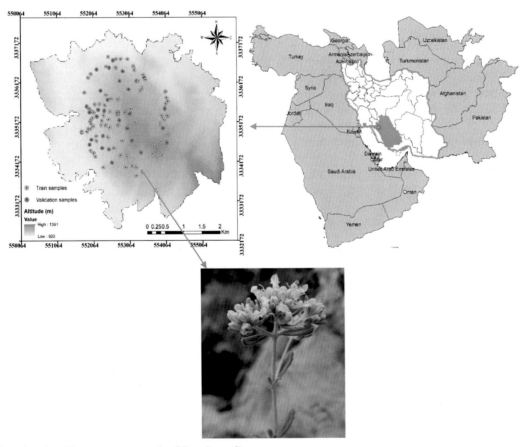

FIG. 1   Region of study with occurrence records of *Teucrium polium*.

2.3.1 Geomorphic factors

The ArcGIS 10.8 program has provided the altitude, slope, aspect, plan, and profile curvature maps as derivatives of the graphical elevation model (DEM).[26]

2.3.2 Elevation/altitude/DEM

In DSM forecasts, elevation is very important.[27, 28] The elevation is taken from the "DEM" with a resolution of 12.5 m × 12.5 m, varying from 920 to 1391 m.

2.3.3 Slope and aspect

These factors were created from the DEM of the study area. The slope was between 0 and 58.5 degrees for the analysis. Aspect reflects the horizontal orientation of the slope features of mountain[29] and was classified into nine distinct classes including, "flat, north, northwest, northeast, south, southeast, west, southwest, and east."

2.3.4 TWI

In the SAGA-GIS program, the TWI map was created from the DEM and calculated according to Eq. (1):

$$TWI = \ln \alpha / \tan\beta \tag{1}$$

Where α is the accumulated region of upslope drainage into a point and $\tan\beta$ is the angle of the slope at the point.[30, 31]

2.3.5 Curvatures

The curvature of the plan is the downward flow acceleration (erosion/deposition rate), whereas the curvature of the profile is the variance of the flow velocity of a slope.[27] The curvature of the profile and plan was classified into three groups: flat, concave, and convex.[32, 33]

2.3.6 Soil factors

One of the important factors affecting habitat suitability is soil features. Therefore, 50 soil samples were collected from depths of 0–30 cm in the T. polium environments. To calculate the various soil characteristics of pH, EC, presence of clay, silt and sand, nitrogen, OC, and OM, soil samples were transmitted to the Shiraz University Laboratory, College of Agriculture. The samples obtained were air-dried and filtered through a 0.2 mm sieve to prepare soil features, eliminating the apparent roots and remaining plants. The hydrometer approach was used to describe the texture of the soil.[34] The EC and pH variables were measured in saturated mud using a conductivity meter and a pH meter.[35] The methods of "Walkley-Black"[36, 37] and "Kjeldahl"[38] were used to calculate the OC, OM, and N. Finally, soil characteristics were acquired from laboratory analyses and then mapped with interpolation techniques such as "inverse distance weight-IDW."

2.3.7 Annual mean rainfall

The key climatic knowledge commonly utilized in SDM is given by a rainfall map.[39] For this research, a thematic map of annual mean rainfall was created using the IDW method based on 15 years of rainfall data in and near the research region acquired from nine rain gauges.

2.4 Correlation consideration of among 15 variables

In this analysis, the Pearson test was used to check the correlation coefficient among various variables. It is important to exclude variables with r correlation values equal to or greater than 0.7.[40]

2.5 BRT model

The BRT model is a nonparametric, strong, and simple ML model used in ecological research, especially for forecasting the distribution of species.[12] This model can be used for various types of data (i.e., categorical and continuous); thus, it provides an accurate estimation of the distribution of species.[11, 12] BRT is a strategy that seeks to boost a single model's efficiency by integrating and incorporating multiple models for prediction.[11] The BRT model is a linear combination of many trees that are consistently boosted and strengthened by the weak models.[11] The advantage of this method is that before the modeling, there is no requirement for the transformation of data or removal of outliers.

Furthermore, nonlinear interactions between variables are well described by the BRT's strong mathematical operation.[11, 41] The "gbm" package of R software[42] version 4.03 has been utilized to apply the BRT model in the current research.[11]

2.6 Model's evaluation

To assess the fitness of the model, we used the training data, while test data were used to show the ability of model estimation and are regarded as an independent means of testing model performance.[43] Three precision metrics, AUC value, TSS, and Cohen's kappa, were calculated for each iteration of each SDM to determine the predictive accuracy of the models. In general, the AUC values range between 0 and 1; AUC < 0.5, indicating that the model estimation is random and AUC > 0.7 indicates strong performance.[44, 45]

For TSS index, the findings varied from −1 to +1. When TSS = +1, a complete consensus is achieved, whereas TSS 0 reveals an output that is not higher than chance.[46]

The Kappa indicators have a score between 0 and 1, whereas a score of 1 indicates a complete predictor with no chance.[47]

3 Results

3.1 Pearson correlation analysis

The Results of the Pearson test are shown in Fig. 2. It was observed that there were two strongly correlated pairs of variables: silt vs sand ($r = -0.97$) and OC and OM ($r = 0.99$; Fig. 2). Therefore, sand and OC were eliminated. The matrix of correlation was prepared based on "mlbench" package in R software version 4.03.[48]

3.2 BRT model's performance

Generally, the system of BRT system relies on the number of trees.[49] In this study, for example, when the BRT technique was fitted with a total fixed number of 700 trees, it was observed that the BRT method could predict with high precision the suitability of the habitat (Fig. 3).

3.3 Explanatory factor importance analysis

There are variations in the weight importance of each element, as shown in Fig. 4. Corresponding to the findings, elevation, silt, OM, pH, and EC had the largest impact on the existence of *T. polium*, with potential contributions of 17%, 11.8%, 10.6%, 9.8%, and 9.3%, respectively (see more details in Supplementary Table S1 in the online version at https://doi.org/10.1016/B978-0-323-89861-4.00029-4). However, the lowest influences on the frequency of the examined plant species were factors such as the curvature of the plan and profile, aspect, and TWI (Figs. 4 and 5).

3.4 Perspective map of suitable habitat distribution

"Habitat suitability mapping-HSM" by BRT model showed that 32%, 27%, 24%, and 17% of the overall research area correlated with "low, moderate, high and very high" classes of suitability for *T. polium* medicinal plant, respectively. The natural breaks method was used for the classification of the final model (Figs. 6 and 7).

3.5 Validation of BRT model

Using AUC, TSS, and Kappa, we evaluated the performance of the BRT model for *T. polium* (Table 1). The AUC, Kappa, and TSS, respectively, were 0.94, 0.78, and 0.8, respectively, illustrating the strong success of the model in the estimation of suitable *T. polium* habitats (Table 1). Thus, the BRT model worked well, and the three separate validation parameters for this model were satisfactory.

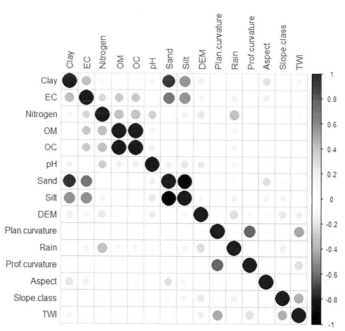

FIG. 2   Pearson's correlations test among various independent variables.

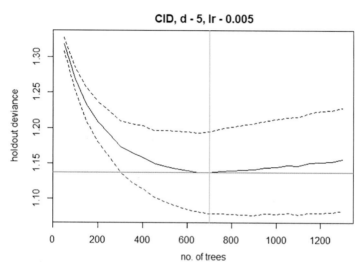

FIG. 3   The relationship between the number of trees for *T. polium* and predictive deviance for BRT model.

4 Discussions

4.1 BRT model

Currently, ML models based on modest mathematics such as GAM and GLM have underperformance in the produced SDMs; therefore, their simulation results would overestimate the future distribution of species in their habitats.[50, 51] However, complex ML models such as BRT, RF, and MaxEnt perform better in spatial distribution compared to simple models.[50] Based on the literature, the BRT model has several advantages, including automatic regulation between predictors, replacing missing data, and changing predictor factors.[11, 52] The BRT model has many benefits based on previous studies, including the automated control of variables, replacement of missing data, and shift in variables.[11] It has also been proposed that this method be implemented in the natural sciences because of its high success in identifying vulnerable areas.[25, 53] In addition, the BRT model is a useful technique or key method for making a decision in a global investigation, as well as in the study of determining the habitat suitability of species.[12]

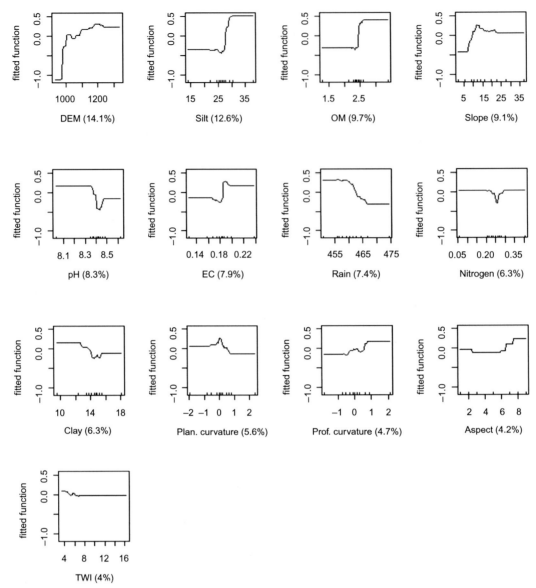

FIG. 4 Partial dependence plots of the predictor variables in the BRT model for predicting the potential map of *T. polium* distribution. The relative contribution of each predictor is shown between *brackets*.

4.2 Evaluation indices

The performance of the model was strong in terms of the AUC value, according to the results of this study. Due to the disadvantage of the AUC consequence of a narrow range of organisms,[2, 54] several reports have questioned the use of this metric.[54, 55] Moreover, if one method is applied to evaluate the ability of SDMs, skewed model performance assumptions are likely to lead.[56] Therefore, it is not appropriate to calculate model efficiency using a specific test parameter.[56] The Kappa and TSS indicators were ranked among the good evaluation metrics that have already been used to provide accurate predictions in SDMs.[57, 58] The TSS and Kappa validation metrics also confirmed that the BRT model satisfactorily predicted habitat suitability for *T. polium*.

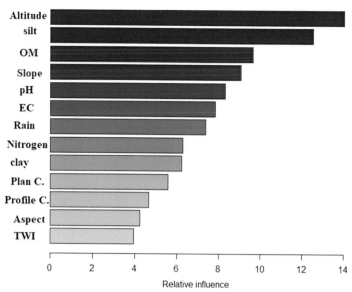

FIG. 5   Analysis of the relative importance of effective factors.

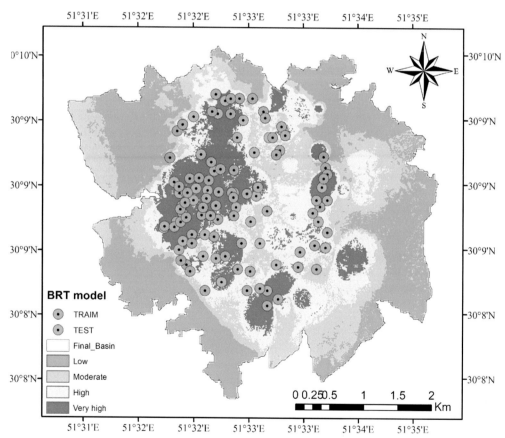

FIG. 6   HSM using BRT model.

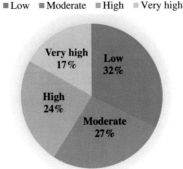

FIG. 7   Percentages in multiple classes of HSM using BRT model.

TABLE 1   Validation outcomes for 13 variables in the BRT model.

Model	AUC	Kappa	TSS
BRT	0.94	0.78	0.8

4.3 Environmental variable

The findings have shown that several effective factors such as soil, climate, and topographic predictors are the most significant variables affecting the distribution of species,[59–61] which is in accordance with the findings of the previous studies.[62] Among these factors, altitude was ranked as the first factor in the spread of *T. polium*. This finding is in line with the findings of Mousazade et al.[63] The other factors influencing the distribution of *T. polium* in the research area were soil physical and chemical data, which are good predictors of species distribution modeling.[64, 65] The percentage of soil silt has a substantial influence on effective plant regeneration, development, and fertility.[66] For instance, the amount of silt is among the most successful elements influencing *Agropyron* sp.[67] OM is another important factor affecting habitat suitability for *T. polium* distribution. Plant productivity is strongly connected to the OM.[68] In addition to supplying nutrients to plant organisms, OM also connects soil particles and increases the soil's ability to maintain water.[69] Therefore, it would be a key factor in the distribution of the current species.

5 Conclusion

This study investigated the complete distribution of *T. polium* in its natural habitat and evaluated the effectiveness of using the BRT model to forecast the future geographical distribution of this species. Elevation (altitude) and silt percentage are bioclimatic variables with a higher effect on *T. polium* distribution. Such information explains the habitat suitability and distribution of species in those areas and may offer options for the survival and management of these habitats in areas under relentless threat from human activity.

Knowledge of the environmental characteristics of the habitat of valuable species is essential because there is a correlation between these factors and the distribution of these species. It should be noted that by identifying the main factors affecting the distribution of species and studying these factors instead of studying all the environmental factors in the region, time and cost will be saved and cost-effective studies will be conducted.

Funding

This work was supported by Shiraz University and grant number of 99GRC1M271143.

Conflict of interest

The authors declare no potential conflicts of interest with respect to the research, authorship, and/or publication of this article.

References

1. Chapin III FS, Zavaleta ES, Eviner VT, et al. Consequences of changing biodiversity. *Nature.* 2000;405:234–242.
2. Kaky E, Gilbert F. Using species distribution models to assess the importance of Egypt's protected areas for the conservation of medicinal plants. *J Arid Environ.* 2016;135:140–146.
3. Dubuis A, Pottier J, Rion V, Pellissier L, Theurillat JP, Guisan A. Predicting spatial patterns of plant species richness: a comparison of direct macroecological and species stacking modelling approaches. *Divers Distrib.* 2011;17(6):1122–1131.
4. Okigbo RN, Eme UE, Ogbogu S. Biodiversity and conservation of medicinal and aromatic plants in Africa. *Biotechnol Mol Biol Rev.* 2008;3(6):127–134.
5. Klein JA, Harte J, Zhao XQ. Decline in medicinal and forage species with warming is mediated by plant traits on the Tibetan Plateau. *Ecosystems.* 2008;11(5):775–789.
6. Guisan A, Zimmermann NE. Predictive habitat distribution models in ecology. *Ecol Model.* 2000;135(2–3):147–186.
7. Robinson NM, Nelson WA, Costello MJ, Sutherland JE, Lundquist CJ. A systematic review of marine-based species distribution models (SDMs) with recommendations for best practice. *Front Mar Sci.* 2017;4:421.
8. Elith J, Leathwick JR. Species distribution models: ecological explanation and prediction across space and time. *Annu Rev Ecol Evol Syst.* 2009;40:677–697.
9. Guisan A, Thuiller W. Predicting species distribution: offering more than simple habitat models. *Ecol Lett.* 2005;8(9):993–1009.
10. Pandurangan AP, Blundell TL. Prediction of impacts of mutations on protein structure and interactions: SDM, a statistical approach, and mCSM, using machine learning. *Protein Sci.* 2020;29(1):247–257.
11. Elith J, Leathwick JR, Hastie T. A working guide to boosted regression trees. *J Anim Ecol.* 2008;77(4):802–813.
12. Hutchinson R, Liu LP, Dietterich T. Incorporating boosted regression trees into ecological latent variable models. In: *Proceedings of the AAAI Conference on Artificial Intelligence, August.* vol. 25; 2011:1.
13. Norberg A, Abrego N, Blanchet FG, et al. A comprehensive evaluation of predictive performance of 33 species distribution models at species and community levels. *Ecol Monogr.* 2019;89(3), e01370.
14. Yi YJ, Cheng X, Yang ZF, Zhang SH. Maxent modeling for predicting the potential distribution of endangered medicinal plant (*H. riparia* Lour) in Yunnan, China. *Ecol Eng.* 2016;92:260–269.
15. Kaky E, Nolan V, Alatawi A, Gilbert F. A comparison between ensemble and MaxEnt species distribution modelling approaches for conservation: a case study with Egyptian medicinal plants. *Ecol Inform.* 2020;60:101150.
16. Karami S, Khosravi AR. Floristic Study of Kuh-e Dakal in Mamasani County, Fars Province. *Tax Biosyst.* 2019;11(39).
17. Liu H, Cocea M. Semi-random partitioning of data into training and test sets in granular computing context. *Granular Comput.* 2017;2(4):357–386.
18. Bonnier G. *La Grande Flore en Couleur.* Paris: Librairie Belin; 1990:943–948.
19. Feinbrun-Dothan N. Flora Palaestina/3. Ericaceae to compositae/by Naomi Feinbrun-Dothan Plates Plates. In: *Flora Palaestina;* 1977.
20. Bahramikia S, Yazdanparast R. Phytochemistry and medicinal properties of *Teucrium polium* L. (Lamiaceae). *Phytother Res.* 2012;26(11):1581–1593.
21. Gharaibeh MN, Elayan HH, Salhab AS. Hypoglycemic effects of *Teucrium polium*. *J Ethnopharmacol.* 1988;24(1):93–99.
22. Belmekki N, Bendimerad N, Bekhechi C. Chemical analysis and antimicrobial activity of *Teucrium polium* L. essential oil from Western Algeria. *J Med Plant Res.* 2013;7:897–902.
23. Dubuis A, Giovanettina S, Pellissier L, Pottier J, Vittoz P, Guisan A. Improving the prediction of plant species distribution and community composition by adding edaphic to topo-climatic variables. *J Veg Sci.* 2013;24(4):593–606.
24. Duque-Lazo J, Navarro-Cerrillo RM, Ruíz-Gómez FJ. Assessment of the future stability of cork oak (*Quercus suber* L.) afforestation under climate change scenarios in Southwest Spain. *For Ecol Manag.* 2018;409:444–456.
25. Bałazy R, Kamińska A, Ciesielski M, Socha J, Pierzchalski M. Modeling the effect of environmental and topographic variables affecting the height increment of Norway spruce stands in mountainous conditions with the use of LiDAR data. *Remote Sens.* 2019;11(20):2407.
26. Ayalew L, Yamagishi H, Ugawa N. Landslide susceptibility mapping using GIS-based weighted linear combination, the case in Tsugawa area of Agano River, Niigata Prefecture, Japan. *Landslides.* 2004;1(1):73–81.
27. Kalantar B, Pradhan B, Naghibi SA, Motevalli A, Mansor S. Assessment of the effects of training data selection on the landslide susceptibility mapping: a comparison between support vector machine (SVM), logistic regression (LR) and artificial neural networks (ANN). *Geomat Nat Haz Risk.* 2018;9(1):49–69.
28. Sezer EA, Pradhan B, Gokceoglu C. Manifestation of an adaptive neuro-fuzzy model on landslide susceptibility mapping: Klang valley, Malaysia. *Expert Syst Appl.* 2011;38(7):8208–8219.
29. Zhang S, Zhang X, Huffman T, Liu X, Yang J. Influence of topography and land management on soil nutrients variability in Northeast China. *Nutr Cycl Agroecosyst.* 2011;89(3):427–438.
30. Sörensen R, Zinko U, Seibert J. On the calculation of the topographic wetness index: evaluation of different methods based on field observations. *Hydrol Earth Syst Sci.* 2006;10(1):101–112.
31. Kopecký M, Čížková Š. Using topographic wetness index in vegetation ecology: does the algorithm matter? *Appl Veg Sci.* 2010;13(4):450–459.
32. Ehsani AH, Malekian A. Landforms identification using neural network-self organizing map and SRTM data. *Desert.* 2012;16(2):111–122.
33. Kalantar B, Ueda N, Saeidi V, Ahmadi K, Halin AA, Shabani F. Landslide susceptibility mapping: machine and ensemble learning based on remote sensing big data. *Remote Sens.* 2020;12(11):1737.
34. Huluka G, Miller R. Particle size determination by hydrometer method. *South Cooperat Ser Bull.* 2014;419:180–184.
35. Carter CR, Rogers DS. A framework of sustainable supply chain management: moving toward new theory. *Int J Phys Distr Log.* 2008.
36. Nelson DW, Sommers LE. *Methods of Soil Analysis.* vol. 2; 1982:539–579.
37. Nosetto MD, Jobbágy EG, Paruelo JM. Carbon sequestration in semi-arid rangelands: comparison of Pinus ponderosa plantations and grazing exclusion in NW Patagonia. *J Arid Environ.* 2006;67(1):142–156.
38. Bremmer J, Mulvaney C, Total N. *Methods of Soil Analysis.* vol. 2; 1982:895–926.
39. Khanum R, Mumtaz AS, Kumar S. Predicting impacts of climate change on medicinal asclepiads of Pakistan using Maxent modeling. *Acta Oecol.* 2013;49:23–31.

40. Senaviratna NA, Cooray TM. Diagnosing multicollinearity of logistic regression model. *Asian J Probab Stat.* 2019;1:1–9.

41. Döpke J, Fritsche U, Pierdzioch C. Predicting recessions with boosted regression trees. *Int J Forecast.* 2017;33(4):745–759.

42. Ridgeway G. *Generalized Boosted Models: A Guide to the GBM Package. Update.* 1; 2007.

43. Wunderlich RF, Lin YP, Anthony J, Petway JR. Two alternative evaluation metrics to replace the true skill statistic in the assessment of species distribution models. *Nat Conserv.* 2019;35:97.

44. Zurell D, Jeltsch F, Dormann CF, Schröder B. Static species distribution models in dynamically changing systems: how good can predictions really be? *Ecography.* 2009;32(5):733–744.

45. van Proosdij AS, Sosef MS, Wieringa JJ, Raes N. Minimum required number of specimen records to develop accurate species distribution models. *Ecography.* 2016;39(6):542–552.

46. Allouche O, Tsoar A, Kadmon R. Assessing the accuracy of species distribution models: prevalence, kappa and the true skill statistic (TSS). *J Appl Ecol.* 2006;43(6):1223–1232.

47. Johnson W. *Effectiveness of California's Child Welfare Structured Decision Making (SDM) Model: A Prospective Study of the Validity of the California Family Risk Assessment.* Madison, WI, USA: Children's Research Center; 2004.

48. Leisch F, Dimitriadou E. *Machine Learning Benchmark Problems.* R Package, mlbench; 2010.

49. Elith J, Leathwick J. *Boosted Regression Trees for ecological modeling;* 2017. R Documentation. Available from: https://cran.r-project.org/web/packages/dismo/vignettes/brt.pdf. Accessed 12 June 2011.

50. Elith J, Kearney M, Phillips S. The art of modelling range-shifting species. *Methods Ecol Evol.* 2010;1(4):330–342.

51. Liu J, Yang Y, Wei H, et al. Assessing habitat suitability of parasitic plant *Cistanche deserticola* in Northwest China under future climate scenarios. *Forests.* 2019;10(9):823.

52. Yang RM, Zhang GL, Liu F, et al. Comparison of boosted regression tree and random forest models for mapping topsoil organic carbon concentration in an alpine ecosystem. *Ecol Indic.* 2016;60:870–878.

53. Leathwick JR, Elith J, Francis MP, Hastie T, Taylor P. Variation in demersal fish species richness in the oceans surrounding New Zealand: an analysis using boosted regression trees. *Mar Ecol Prog Ser.* 2006;321:267–281.

54. Lobo JM, Jiménez-Valverde A, Real R. AUC: a misleading measure of the performance of predictive distribution models. *Glob Ecol Biogeogr.* 2008;17(2):145–151.

55. Austin M. Species distribution models and ecological theory: a critical assessment and some possible new approaches. *Ecol Model.* 2007;200 (1–2):1–9.

56. Yu T, Quillen D, He Z, et al. Meta-world: a benchmark and evaluation for multi-task and meta reinforcement learning. In: *Conference on Robot Learning;* 2020:1094–1100.

57. Freeman EA, Moisen GG. A comparison of the performance of threshold criteria for binary classification in terms of predicted prevalence and kappa. *Ecol Model.* 2008;217(1–2):48–58.

58. Lauzeral C, Grenouillet G, Brosse S. Spatial range shape drives the grain size effects in species distribution models. *Ecography.* 2013;36(7):778–787.

59. Pearson GA, Lago-Leston A, Mota C. Frayed at the edges: selective pressure and adaptive response to abiotic stressors are mismatched in low diversity edge populations. *J Ecol.* 2009;97(3):450–462.

60. Crowther MS, Lunney D, Lemon J, et al. Climate-mediated habitat selection in an arboreal folivore. *Ecography.* 2014;37(4):336–343.

61. Lee DS, Bae YS, Byun BK, Lee S, Park JK, Park YS. Occurrence prediction of the citrus flatid planthopper (Metcalfa pruinosa (Say, 1830)) in South Korea using a random forest model. *Forests.* 2019;10(7):583.

62. Jafarian Z, Kargar M, Bahreini Z. Which spatial distribution model best predicts the occurrence of dominant species in semi-arid rangeland of northern Iran? *Ecol Inform.* 2019;50:33–42.

63. Mousazade M, Ghanbarian G, Pourghasemi HR, Safaeian R, Cerdà A. Maxent data mining technique and its comparison with a bivariate statistical model for predicting the potential distribution of Astragalus Fasciculifolius Boiss. in Fars, Iran. *Sustainability.* 2019;11(12):3452.

64. Coudun C, Gégout JC, Piedallu C, Rameau JC. Soil nutritional factors improve models of plant species distribution: an illustration with *Acer campestre* (L.) in France. *J Biogeogr.* 2006;33(10):1750–1763.

65. Hageer Y, Esperón-Rodríguez M, Baumgartner JB, Beaumont LJ. Climate, soil or both? Which variables are better predictors of the distributions of Australian shrub species? *PeerJ.* 2017;5, e3446.

66. Jensen MT. Robust and flexible scheduling with evolutionary computation. *BRICS.* 2001.

67. Kargar J, Tamratash A, Jalil S. Comparison of parametric and non-parametric species distribution (SDM) models in determining the habitat of dominant rangeland species. *Iran Range Desert Res.* 2018;25(3):512–523.

68. Bauer A, Black AL. Quantification of the effect of soil organic matter content on soil productivity. *Soil Sci Soc Am J.* 1994;58(1):185–193.

69. Grigal DF, Vance ED. Influence of soil organic matter on forest productivity. *N Z J For Sci.* 2000;30(1/2):169–205.

CHAPTER

18

Ecoengineering practices for soil degradation protection of vulnerable hill slopes

R. Gobinath[a], G.P. Ganapathy[b], E. Gayathiri[c], Ashwini Arun Salunkhe[d], and Hamid Reza Pourghasemi[e]

[a]SR University, Warangal, Telangana, India [b]VIT University, Vellore, Tamil Nadu, India [c]Guru Nanak College (Autonomous), Chennai, Tamil Nadu, India [d]Dr. D. Y. Patil Institute of Technology, Pimpri, Pune, India [e]Department of Natural Resources and Environmental Engineering, College of Agriculture, Shiraz University, Shiraz, Iran

1 Introduction

Landslides were recognized as a process of contributing to an outward and backward change in hilly slope resources made of native soil, sand, synthetic cover, and part combination[1,2] the primary causes of landslides are gravitational force and groundwater flow inside the soil. In many parts of the world, land depletion in a sensitive ecosystem of mountainous terrain is becoming a major issue. Because of its effect on the potential loss of production, economic influence, and social losses, land depletion is a serious concern.[3,4] The principal causes of land loss are steep hills with poor vegetation cover and increased human interference. This adds to a number of negative consequences, such as soil deterioration, reduced land quality, degradation of local hydrology networks, and limited capacity to store water with hauled sedimentary rocks and contaminants.[5] Therefore, the current rate of degradation processes may be halted through the implementation of innovative management strategies.[6] Soil has a leading role in influencing or intensifying the loss of productivity that comes from the soil, including agricultural products. The integrated behavior of microclimate variables (aridity, weather events, rainfall erosion) and human mismanagement (intermittent drought, thin soil horizons, poor soil organic matter, scarce vegetation cover), and human pressure (overgrazing, deforestation, agricultural intensification, tourism growth).[6–10] Soil depletion is the depletion of soil ecological resources and functions.[11] Almost 30% of the world's soils has been exhausted, and they need to be analyzed and documented to improve their management and use.[12] The deterioration of land in India is reported to continue on 147 million hectares (MHM or Mha·m), including 94 Mha·m of river erosion, 16 Mha·m of acidification, 14 Mha·m of floods, 9 Mha·m of wind erosion, 6 Mha·m of salinity, and 7 Mha·m of mixing factors. This is incredibly serious since India promotes 18% of the world's population and 15% of the world's farm production, but only 2.4% of the world's population. The cause of soil degradation is often massively complex, with varying dimensions, including low and reduced land fertility, vegetation, ecosystems, and the decline in ecosystem-derived economic and social resources. Six soil depletion processes, such as water, wind and tilling, degradation, topsoil depletion, compressibility, saline intrusion, alkyl nature, pollution, and loss of biodiversity, have been defined as induced or aggravated by poor agricultural practices. Soil erodibility is an indication of land loss sensitivity and is defined by the physical properties[13] of soil. Soil degradation can occur over a short period of time owing to ineffective land management, although soil restoration can take a long time or decades to form. The design and management of land use studies are also needed to support choice-making and public policy, ensure real-time soil quality and lead to sustainability.[14] This chapter delegates knowledge to help better-informed decision-making on the conservation of fragile hillsides and the management of ecosystems, focusing on controlling soil erosion.

2 Materials and method

2.1 Soil erosion and land degradation

Soil degradation, which contributes to 36–75 billion tons of land depletion every year and freshwater shortages, threatens the global food supply.[15–17] Soil is a basic element that must be good to remain diverse and sustainable for the remainder of the ecosystem.[18] Potential challenges to the survival of soil fertility include environmental development and anthropogenic impacts. In coping with negligible, vulnerable, and environmentally responsive habitats, together with that of the semiarid Mediterranean region, for which unsustainable challenging work has contributed to irreversible soil conditions, erosion and desertification, and soil preservation and conservation are especially relevant.[19, 20] More study is required to demarcate the essential borders of land resources below where ecological efficiency is seriously and undoubtedly impaired.[21] Soil erosion is known around the world as a significant issue that needs soil management steps to be taken. Erosion triggers many attributes, which are partly irreversible and cause multiple social or economic issues, flood hazards, or environmental problems, such as arid land desertification.[22] A broad variety of problems along riverbanks are often caused by surface degradation. Here, soil degradation allows water runoff to be decreased and water storage ability to be reduced due to rapid water runoff, as well as decreasing the variety of plants, animals, and microbes. Many herbs, shrubs, and indigenous grass plants play a crucial role in the management of land degradation and have the added advantage of simple filtering through the landscape to control soil erosion. They quickly transplant and take situations that resemble their natural environments. These plants send out root nets that help maintain the topsoil and minimize soil erosion. Such facilities often need less upkeep because they are tailored to the area in which they work and receive much of their specifications in the current location.[23] Climate causes, such as rainstorms, famine, warming, and human reasons, such as unnecessary operations, lack of security, severe land erosion, and ecological destruction also exist or occur in several places.[24]

In addition, an issue causing environmental and economic harm is a worldwide shallow slope collapse, concerned with slope stability, and vegetation plays an important role. Therefore, by integrating knowledge of ecology and engineering, environmental slope engineering aims to enhance the retention of the effect of plants. Slope stability processes using woody plants have been discussed in most experimental studies and associated techniques. Herbaceous plants may also stabilize cliffs, although the same mechanical and ecological concepts are only partially practiced. This information gap adds to the broad belief that woody vegetation is typically stronger than herbaceous vegetation for slope stabilization.[25] Land degradation is commonly determined by three behaviors: soil weakening, transfer, and settlement. Such methods typically lead to the migration of the topsoil that is rich elsewhere on-site in organics, nutrients, and soil life, in which it builds up over the years or is being transported off as far as everything collects in drainage systems. It is normally high in unsecured hilly regions.[26] Soil degradation negatively impacts plant growth, agricultural yields, water quality, and recreation, because it exists naturally in all soils and is a primary cause of soil depletion.[27–30]

2.2 Soil degradation and the quality of water

Soil erosion in hilly regions can have an immense effect on soil quality and water quality. In lower surface water bodies, sediment resulting from soil erosion is a major water quality pollutant. Increased nitrogen (N) and phosphorus (P) levels in surface waters are also trying to contribute to a reduction in water quality. By reducing soil erosion by adopting management strategies, such as conservation tillage, no-till, buffer strips, terracing, and other management techniques, conservation tillage begins with an efficient residue management program that protects the surface of the soil during the nongrowing season and minimizes water erosion. The most effective ways of improving surface water quality, being effective, environmentally conscious, and sustaining innate fertility, and increasing profitability are to control soil erosion and nutrient management. On the one side, for example, the loss of biodiversity caused by changes in land use can affect soil water management by modifying functional features including a clear and indirect path.[31, 32] Water safety problems pose a challenge to the control of water systems. The geographical period difference in water quality may provide a relevant understanding of water environment decision-makers.[33–35] It is also essential that proper patterns in surface water quality are carefully examined and impacted by these factors. Comprehensive quality assurance methods and several factors exist, and many fuzzy occurrences and models are included in the evaluation. Fuzzy testing is frequently utilized in surface water sources, including rivers.[28] The challenges of land degradation, sediment transport, and their subsequent accumulation in water sources have continued for years on the entire surface of the planet.[36] Nevertheless, this condition has recently been exacerbated by increasing interference between citizens and the atmosphere.[37] Land with a high slope promotes rainwater movement or streamflow overload in the region, particularly because the water flows downhill more quickly.[38, 39] The impact of practical recognition and biodiversity on soil degradation was measured at various periods of the period with various climatic conditions in semiarid grasslands; *Artemisia giraldii, Artemisia sacrorum, Carex korshinskii, Cleistogenes hancei, Cynanchum thesioides, edtia verna, Heteropappus altaicus, Leontopodium leontopodioides, Patrinia scabiosaefolia, Phragmites australis, Potentilla bifurca, Potentilla tanacetifolia, Setaria viridis, Stipa grandis, Taraxacum mongolicum, Vicia cracca, Viola verecunda, Viola dissecta* and were suggested as the important aspect of soil degradation during this time, depending on the severity of the runoff and the study shows that FEve. The functional evenness index plays a major role in minimizing degradation when the rate of rainfall is poor at the end of July.[40] The challenges of water erosion on the ground covered by the biomass of plants was effectively reduced because the rainfall is drained away through the biomass layer (i.e., tree leaves and branches intercept and diminish the erosive energy). Losses through efficient heterogeneity in tree height induced by the intensification of land use intensify the impact of water shortage in the rainforest and highlight the significance of preserving the efficient heterogeneity in tree height in the use of tropical land to enhance soil water-related ecological resources, particularly in the light of increased drought occurrence and projected severity.[41] Indeed, there is a need for an hour to protect our land from degradation and erosion to have fertile soil for enhanced agricultural productivity and for better sustainability.

2.3 Need for the implementation of techniques of bioengineering

Numerous slopes across the globe are affected yearly due to degradation and erosion due to several factors, the main being anthropogenic impact. Sliding or landslides are major concerns for mass wastage, through which a large amount of soil is wasted. There are two forms of sliding or mass movements that impact the stabilization of slopes. There were tall slides and shallow-seated slides. High slip issues (the surface depth of failure is more than 3 m) are geotechnical or geological. It could only be approached by considering the effects of slope geometry, land intensity, climatic quality, groundwater conditions, etc., which can be calculated by studying slope stability. In the case of a shallow slip (the depth of the surface of the failure is less than 3 m), the issue is very challenging to measure; moreover, multiple methods of failure occur along the slopes, plains, and valleys throughout the world, engineering methods of protection seem not only costly but also result in poor viability of implementation (both short-term and long-term). Hence, there is a need to provide a sustainable engineering solution that reduces the impact of anthropogenic activities and makes the area sustainable support for long-term soil bioengineering provides solution for the same.

2.4 Soil bioengineering and its slope protection capabilities

Using the unique characteristics of vegetative elements, soil bioengineering combines the specific characteristics of systems with vegetation. There are advantages and disadvantages to the resulting systems and their elements that need to be weighed prior to choosing them for operation. In general, soil bioengineering systems require limited access to machinery and staff and trigger relatively minor site disruptions during construction. On an ecological basis, areas that are sensitive, along with parks, wetlands, riparian areas, and scenery pathways in which architectural effectiveness, natural habitats, and equal values can make significant, are typically priority considerations. When suitable plants are mounted throughout the dormancy period, typically late autumn, snowfall, and early autumn, soil bioengineering systems will be the most successful. The utility of soil bioengineering approaches can be restricted by limitations on planting time or accessibility of the necessary amounts of sufficient flora at the permissible period for planting. Soil bioengineering with the use of live or dead plant material to mitigate climate impacts, including some deep, fast landslides and soil erosion degradation of hills as well as river banks. Plant species are the central systemic elements of bioengineering systems. Each strategy for slope reinforcement is necessary for the true cooperation of many experts, including soil biologists, environmental scientists, botanists, geologists of materials science, construction workers, structural designers, and landscaping professionals. By utilizing locally accessible resources and a minimum of heavy machinery, soil bioengineering more frequently mimics nature, and it may provide roadside administrators an affordable way to address local environmental concerns. In tandem with conventional engineering approaches, such as rock or concrete frames, these strategies may often be used (Figs. 1 and 2).[42]

The field investigation results of numerous research works conducted by researchers across the globe show instances where mixed slope defense schemes have proved to be more cost-effective than the use of either comparable vegetative therapies or structural solutions. Integrated systems (bioengineering with civil engineering structures) can be extremely cost-effective, where building techniques are labor-intensive and labor rates are rational. However,

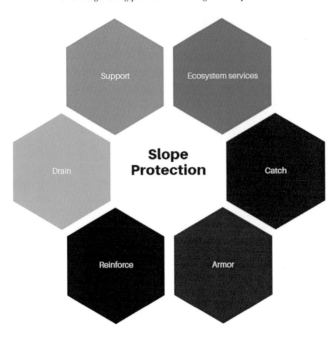

Slope Protection Capability of Plants

FIG. 1 Slope protection capabilities of plants.

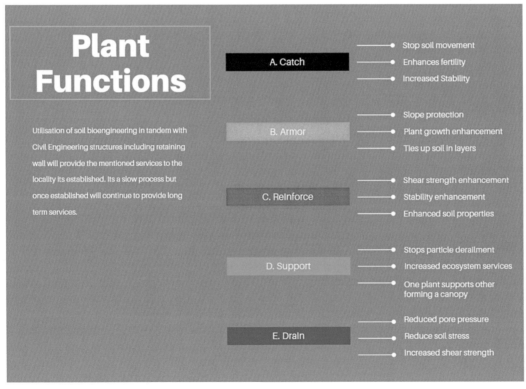

FIG. 2 Functions of plants as bioengineers.

where labor is either insufficient or highly costly, soil bioengineering systems may be less feasible than systemic steps. When other building work is sluggish, this can be overcome by the time of year (fall and winter). Any cost-effectiveness is compensated for by the use of indigenous resources, but expenses for plants are restricted to labor in processing, storage, and specific expenses for shipping the plant species in the field.

3 Role of vegetation in slope stability

Ecoengineering, if used meticulously, can prevent minor problems from escalating into more complex ones. Plant life considerably and significantly impacts both the surface and mass stability of slopes. Vegetation stabilization or protective advantages rely heavily on both plant species and the type of process of soil erosion process. Therefore, in instances of extreme stabilization, the preventive attributes of woody plants vary from mechanical strengthening and root and steam retention to hydrological slope modification as a result of the extraction of soil moisture via evapotranspiration. Plant life benefits the prevention of surface erosion. Catch particles that move down the slope. Vegetation interception and plant residues collect and retain rainfall stability and minimize soil compression. Restrict root systems to physically bind or constrain soil particles while at the same time, filtering sediment out of runoff from above-ground residues. Retardation over ground residues increases the surface roughness and speed of the slow runoff. The exhaustion of surface humidity transpiration by plant species delays the start of saturation and runoff. Plants, when performing as eco-engineers, support the slope in the following ways:

Engineering roles:
(i) Catchment
(ii) Armoring
(iii) Reinforcement
(iv) Anchoring
(v) Supporting
(vi) Drainage
Hydrological effects:
(i) Interception
(ii) Evaporation
(iii) Preservation
(iv) Leaf drip
(v) Lake forming
(vi) Water uptake
(vii) Infiltration

3.1 Benefits of vegetation in slope stabilization

Roots structurally strengthen the soil by transferring soil shear stress to the root tensile strength; therefore, the degree of slope failure is constrained by the vertical and lateral binding of the soil. Modification of soil humidity, evapotranspiration, and the capture of soil moisture stress in the vegetation limit buildup Vegetation also impacts the snowmelt rate, which in turn influences the moisture regime of the soil. Attached and anchored roots counteracting shear pressures can serve as superstructure piles or slope arch abutments. By strengthening and arching, broad trees are planted at the toe of the slope anchor into solid strata to protect the soil mantle of the upslope (Fig. 3).

Although native species may be used to make a significant contribution to reducing inflation in the hills and making land use sustainable, it is essential to replace pioneering plants with later succession communities to restore the full ecological function. Mostly short-lived and unwilling to grow in the shade from their own pioneering shrub and tree populations may only increase stability and growth for a finite time. However, during heavy winds, large trees can be uprooted later.[43] Consequently, pruning or felling will be needed if the plants growing very tall for a vulnerable sloppy area must guarantee slope stability (Bioengineering structure) and ensure that the fall of the trees is not harmed. The weakened slope area may increase in size during succession, making it more difficult and expensive to maintain the slope. New technologies or strategies may also be important in preventing corrosion. The establishment of native vegetation and rapid succession systems will minimize the need for action and will be a long-term regeneration option (requiring stability and much more), offering the best balance between artificially created slopes and naturally occurring slope stabilization.[2] If plant biodiversity is naturally covered, increases on stable slopes using soil bioengineering techniques, for example, the security gained over time should increase. The mechanisms and the inhibiting variables involved in these processes should be analyzed. The capacity of adjacent vegetation (i.e.) the practitioners did not nurture (i) to colonize the target site by spreading seeds or forming a local seed bank by the practitioner; (ii) if the type of soil, in particular the availability of water, adversely affects germinating seeds, growth of seedlings, and eventual development of plants.[44] The task of the eco-engineers is, therefore, to decide how the environmental shift will be affected by the conditions of the site, the relationships of the present organisms, and more probabilistic variables such as the supply of colonizers or seeds or the conditions of the atmosphere at the site. Long-term human-made behavior

SOIL BIOENGINEERING

Soil Bio engineering is widely adopted in many countries and the usage of indigenous plants is promoted for this purpose. They not only provide slope stability but also adds more economic value through the products and services it offers

FIG. 3 Benefits of soil bioengineering.

can be predicted as quickly as possible if these dynamics are not conceivable. Plant species play essential functions in stabilizing soils in many landscapes worldwide, including grasslands, streams, and coastal wetlands. Across these diverse ecosystems, overland shoots, lateral rhizomes, stolons, and root clusters combine to secure soil degradation by physically sheltering and fixing soils, providing protection from rain, runoff, waves, and winds.[45, 46] Grasses, legumes, and small shrubs may have a major reinforcement effect at distances of 0.75 and 1.5 m. Many investigators have attempted to calculate the evident cohesion values due to the existence of roots in the soil by developing a prototype and implementing in situ strength tests for numerous root system permutations.[47–50]

3.2 Hedgerows role in soil conservation

The use of broad native plant species, i.e., trees, shrubs, or grasses cultivated on the contour, also referred to as "contour hedgerows" is an adaptive strategic tool with solutions for reducing land degradation/deterioration on sloppy soils in humid tropical regions.[51] Growing hedgerows along the edge of steep land is a promising technique for improving soil quality in hilly regions. Research has shown that contour hedgerows can minimize runoff and soil erosion.[52, 53] Nonpoint source environmental impacts and improved plant nutrient loss were monitored under control condition.[54, 55] To minimize soil erosion, hedgerow plants may reduce both of these causes. The first factor is that, to avoid splash floods, plants can protect the surface well enough from rain, while the second factor is runoff.[56] In the last 20 years, many studies have also focused on hedgerows and their effect on the reshaping of the micro-topographic characteristics of slopes[56–58] and their effect on the distribution of soil nutrients on slopes effectively reduce soil degradation and drainage.[55] Very few experiments have shown that hedgerows struggle with water, sunshine, nutrients, and crop space.[59–61] Owing to the presence of hedgerows, the A-horizon in the lower alley was compressed and rendered thicker. In this case, the hedgerow treatments were distributed more favorably in the upper alley than in the lower alley. The soil in the top location had less accessible water for plants and did not encourage sufficient infiltration.[62] It is essential to mention that the majority of studies are mainly focused on forest/semiarid and undershrub/grass systems, and relatively fewer records are recorded in the agricultural environment, particularly with regard to the management of tillage.[63] Herbaceous plants have been successfully used to improve the soil in hilly areas. The possibility of dominant herbaceous species in soil reinforcement is a problem.[64]

4 Soil reinforcement

Among the numerous techniques, the soil reinforcement is deemed a more inexpensive and quicker technique. Soil engineers are more involved in natural materials as an option owing to the environmental concerns associated with the different synthetic materials used for soil strengthening. Many studies have investigated the action of various natural fiber/material-enhanced soils in recent years and documented their positive impact on strength behavior.[65, 66] As an infrastructure strategy for slope safety, the latest movement for biodiversity has increasingly encouraged policymakers and engineers to rediscover vegetation. Special Briefing on Global Change from the Intergovernmental Panel,[67] it is well known in ecological slope engineering that vegetation influences stability by two processes: first, by enhancing the hydrogeological requirements of slopes and mechanical reinforcement of soil with plant roots.[68–70]

4.1 Role of roots in soil reinforcement

Plant roots change their natural environment in a number of respects, from altering biophysical, chemical, and mechanical soil properties to fostering microbiota abundance and diversity, and adapted strategies for effective ecosystem regeneration and soil conservation may be formulated by recognizing these main processes. Plant roots can be successfully used on hillsides, riverbanks, and artificial slopes to mechanically strengthen and "fix" soil and it is also an ecological solution to civil and structural strategies to guard against shallow landslides and soil erosion. Root coherence is closely related to the pattern of root distribution patterns. For example, fibrous root systems have a large number of root branching divisions, whereas tap roots have a single main root axis and few lateral roots. Adventitious roots are typically found in soils shallower than tap-root systems.[71] The hydraulic, mechanical, and hydrological characteristics of the soil materials are altered by the roots of living plants, such that the phase of soil erosion is intensified.[72] Both the strength of the soil shear and the aggregate stability of the soil are positively related to each other as they have, to some degree, identical bonding structures.[73] Over time, the use of plant species against shallow landslides and degradation has gained major interest because vegetation provides soil with mechanical and hydrological reinforcement.[74] Lant roots can be effectively used on hillsides, riverbanks, and artificial slopes to physically stabilize and "fix" soil and are also an evolutionary option in contrast to civil engineering methods while providing protection to deep avalanches and avoiding soil erosion. Rootstocks can be cultivated up to 2.0 m below the surface of the soil to physically stabilize the slope against a shallow landslide.[75] Roots that are thicker act as soil holders on hills, strengthening the soil in many ways, and steel rods support the concrete. During slope collapse, thin and fine roots function under stress and, when any reach the slippery rim, stabilize the soil by the incorporation of coherence.[76] Plant roots must cross a shear surface that can be up to 2.0 m below the soil surface to mechanically stabilize the slope against a shallow landslide.[75] Thick roots act as soil nails on hills, strengthening the soil in the same way that steel rods support the concrete. During slope collapse, thin and fine roots function under stress and, if they cross the slippery surface, stabilize the soil by adding cohesion.[76] Root, and root exudates are produced when the roots of plants scatter the soil. Upon meeting the dirt, these exudates lubricate the tip of the root.[77] Root and soil, through diverse interactions, have the ability to develop and influence each other.[71, 78, 79] Large mass density soil (e.g., compacted soil) induces root penetration resistance, evoking a reaction that affects the shape of the root system (growing root diameter),[80] as well as the level at which roots are able to penetrate.[81] From one plant species to another, the reaction to soil pressure at the root-soil boundary can vary (Fig. 4).

4.2 Plant interaction and soil stability

In the early season, plant root wears resistance has a significant influence on land degradation under intermediate and high rainfall intensities. However, the aboveground portion of the plant has a greater impact on soil degradation for intermediate and large rainfall intensities during the middle season. The largest determinant of soil degradation from low to high rainfall levels is planting niche distinction toward the end of the growing season.[40] It is well established and naturally occurring plants in agroecosystems play an important role in the stabilization of habitats and have a major effect on the cycles of drainage and erosion. Many indigenous plants, however, are regarded as weeds that decrease the growth and yield of cultivated crops, and weed control contributes to environmental emissions through complete destruction, decreases plant biodiversity, and invites land degradation.[45]

4.3 Soil instability

Slope disruption occurs mainly through natural causes such as geomorphic and geo-dynamic development, strong seismicity, extreme climatic variations, the loss of rivers and lakes and avalanches, or through human behaviors such

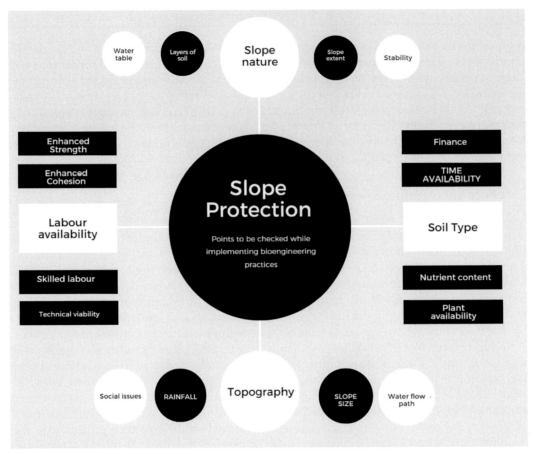

FIG. 4   Slope protection—What to check?

as transfer of land usage, forest degradation, road development, canals, building, and various facilities and infrastructure production.[82–85]

4.4 Root reinforcement

Roots with smaller diameters have greater resilience when they contain a better core[86, 87] between root tensile strength and root diameter, and there is a favorable normal distribution relationship.[76] Many plant species, root-age, rate of growth, inclination to growth,[88] and soil moisture conditions[89] as well as root chemical internal compounds may generate high root tensile strength variability.[86, 87]

4.5 Role of mycorrhiza in soil shear

Mycorrhizal fungi, usually associated with fine roots, are involved in the water-stable aggregate formation, and can increase the intensity of soil shear. As a greater soil volume is influenced by a larger degree of soil exploration by those fungi relative to the root system itself, endo/ecto-mycorrhiza fungi are also important for land degradation control.[73, 90–92]

4.6 Framework and implementation process of bioengineering practice

Eco-bioengineering is sometimes initially seen as merely reinforcing the surface layers, with productivity restricted to the depth of the root. If understanding was at a high degree, trust in eco-engineering systems and plant cover would be strengthened, and funding organizations or consumers might be approached for solutions. To build mechanically stable vegetated hills, soil mechanics engineers collaborate with landscape architects and horticulturists. In various areas of the world, there is a substantial inadequacy of public information and a scarcity of educational and training services for soil bioengineers. Professionals and students should be encouraged to view and refresh the database and should be encouraged to maintain and monitor green slopes with valuable natural vegetation at the end of the day for each resident. In most areas of the world, there is a severe lack of public knowledge and a scarcity of educational and training services for soil bio and eco-engineers. To solve this challenge, the latest ecological and engineering modules can provide ecological engineering solutions for slope stabilization. Qualifications in bioengineering as part of ongoing professional growth, together with the appropriate functional experience of existing bioengineers, must be promoted. Undoubtedly, the most relevant subject for researchers and practitioners to tackle during the next decade is to strengthen teamwork and understand the effects of soft engineering and the use of vegetation to preserve slopes and prevent erosion (Fig. 5).

Land degradation and avalanche disasters may be prevented or reduced by implementing effective remedial steps or by the initial stage of the planning scheme setting or remedial intervention. Ecoengineering is an effective tool for shielding slopes from surface erosion, reducing the chance of slippage, and improving the drainage of surfaces. It is a methodology that can be used virtually everywhere in the world, providing appropriate plants and auxiliary materials on site. At present, mountains must be introduced as a matter of urgency. The effectiveness of this system depends on planting native/indigenous plants at sites of soil depletion. We also observed that precise results can be gathered using fundamental and applied analysis in similar fields for the use of practitioners seeking adapted solutions for the web (e.g., soil structure and biogeochemistry, hydrology, and microbial ecology). Knowledge and recognition of the use of plant materials in slope reconstruction projects are projected to increase dramatically through field studies, open datasets, modeling, and joint projects, particularly in civil and geotechnical societies (Fig. 6).

Data on the availability of planting material and the establishment and maintenance of a particular species at one position cannot be easily moved to other locations or species. Therefore, quick repair solutions for a single plant species can be favored if there is a low probability of immediate slope instability. Alternative species selection involves

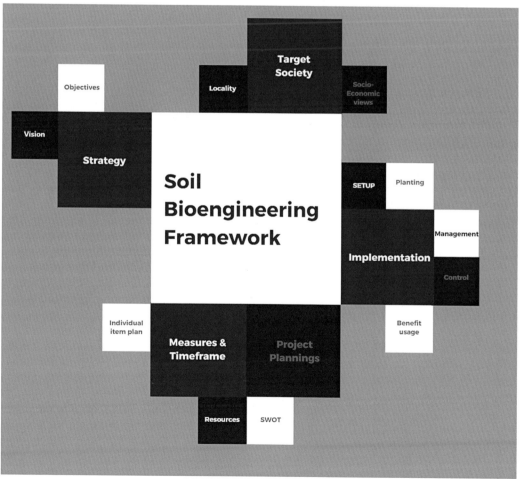

FIG. 5 Soil bioengineering framework decision model.

Implementation Process

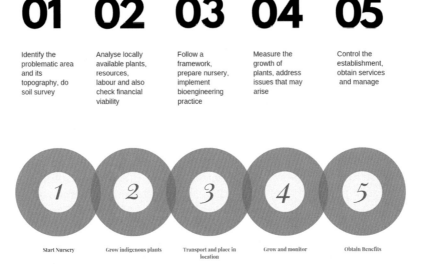

01 Identify the problematic area and its topography, do soil survey

02 Analyse locally available plants, resources, labour and also check financial viability

03 Follow a framework, prepare nursery, implement bioengineering practice

04 Measure the growth of plants, address issues that may arise

05 Control the establishment, obtain services and manage

1 — Start Nursery 2 — Grow indigenous plants 3 — Transport and place in location 4 — Grow and monitor 5 — Obtain Benefits

FIG. 6   Implementation process of slopes.

awareness of the plant characteristics that are attractive, which can entail screening of various species. Slope stabilization plant species need to be studied for their potential to develop and extend their specific physical, chemical, ecological, and biological characteristics in the target environment. Screening for plant traits of particular significance for the stabilization of slopes or the prevention of erosion utilizing established frameworks can supplement plant identification based on environmental suitability; in most cases, slope sustainability can only be achieved by developing succession mechanisms that can minimize interference and provide a long-period remedy for regeneration and safety. Although the conservation of native habitats and the availability of a wide range of ecological resources may be beneficial in some cases, local land use on slopes depends on the longevity of one or two organisms for the stabilization of slopes rather than on the succession of natural vegetation. The relationship between the types of slope hydrologic processes and plant habitat types requires further study, as well as the impact of plant and soil microbiota/fauna on soil structure and physical, chemical, and ecological processes. A better understanding, other than its stabilizing properties, of the services rendered by vegetation on the slopes, is required. Accurate space and time simulation studies may provide useful tools for groups of civil and geotechnical engineers who are still skeptical regarding the use of soft engineering systems and associated vegetation. Through cooperative collaborations, coordination, training, and awareness-raising, knowledge of soil bio-and eco-engineering techniques needs to be significantly enhanced (Fig. 7).

Creating more accurate vegetation roles would provide a clearer view of stand-by patterns and possible growth scenarios. To improve management judgment on species collection for conservation purposes, we advise colleagues to perform flora investigations for other soil bioengineering programs. Multidisciplinary interaction/collaboration between researchers, on the other hand, is considered a significant challenge for the future and is important for a better understanding of our ecosystems and for managing engineering issues.

Hedgerow species may be used as a pioneering succession of vegetation or as a final form of vegetation to stabilize slopes, allowing for various land uses, such as grazing and increasing for food or fuel, while all the plants have been seen to be extremely appropriate for soil reinforcement, there are a variety of risks on slopes connected by relying on a single native or indigenous species for immediate slope stabilization and early succession of species for sustained slope defense. These threats include the potential for widespread disruption or decreased "efficiency" due to pest and disease incursions and the restricted capacity to respond to environmental changes. Genetically distinct planting may

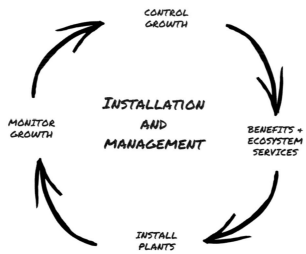

FIG. 7   Plant installation and management process.

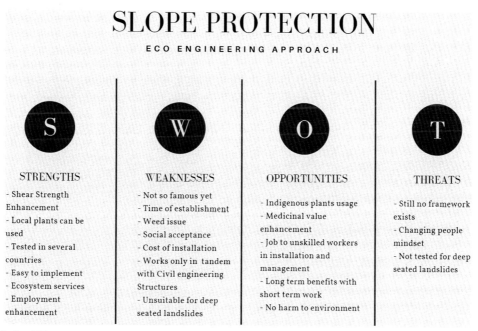

FIG. 8   SWOT analysis of Slope protection measures using soil bioengineering.

result in an invasive plant species, particularly if it is foreign to the place where it is planted. In the same manner, such life forms can convict the unification efficiency and eradicate the colonization of native species by, for example, the creation of dense thickets, the capture of accessible resources, and the elimination of predators from the home range. In addition, the possibility that only early succession organisms can be utilized, including in their natural habitat, is that they may be short-lived. Therefore, understanding that native species adapt to different physical properties of the soil is a critical focus for long-term studies (Fig. 8).

4.7 Requirements and impediments for the collection of alternate organisms

Biophysical and environmental studies are required to test the most acceptable selected plant variety (e.g., development rate, expenses, success rate, criteria of colonization, life type, durability, and plant community successional dynamics). These features are essential for the collection of seeds, trees, under-shrubs, shrubs, and herbaceous plants.

Grasses and soil covering prostrate plants can lessen soil erodibility and the spread of soil gaps, thus inhibiting the creation of advantageous paths along with gaps, resulting in mass failure. Deep-rooted perennials, which are mostly woody, may increase the mechanical strength of the soil. Plant species that are ecologically suitable will display ecological health due to their intended position, are compatible with other plant species representatives, resolve procession, and display no invading trends.[93] Regulations for selecting and selecting suitable plant species could therefore be defined on the basis of environmental factors and geographical characteristics.[94] Public recognition and ease of implementation and usage are still priorities, although[95] recommended this strategy of choosing the most appropriate indigenous native plants to rehabilitate the eroded soil in northern Ethiopia. These studies mainly focused on socio-economic positions, socio-cultural values, and environmental capital.

5 Discussion

Implementing a soil bioengineering strategy will significantly influence ecological regeneration and stabilization with additional benefits such as aiding in natural recovery, ecological replenishment, and slope stabilization. Considering that most roads are built by digging along slopes, the use of plants is a cost-effective approach, and conservation costs can be significantly decreased once the phase of succession occurs, bringing a high influx of new species into the population since each plant species has a unique rooting morphology and root mechanism for combating soil erosion. This approach is useful for slopes that do not exceed a certain degree of risk, which means using conventional engineering approaches such as analytical techniques instead of designing slides and crossings are more applicable to steep and perilous terrain. To investigate and research the relationship between plants, soil resources, biodiversity, and climate, a more sensible system must be used.[96] However, the use of soil bioengineering is restricted to low to moderate risk paths, while the use of standard technical measures is considered more suitable for high-risk slopes and still remains an option for slope stabilization. To accurately track the goal slope instability or erosive process, the profiling of native and indigenous plant species has to be carried out by collecting the main parameters that all plant species have to obey. Plants are known to be suitable for a specific environment (i.e., specifications for temperature, light, nutrients, and water), it is necessary to consider the characteristics of a plant above and below ground level. Characteristic factors to be considered include stem density, soil, and organic debris trap capability, stem rigidity, density of root density, root area ratio (RAR), the morphology of the root structure, and strength of root tensile strength.[76, 88, 97–99] For potentially useful species, a scoring method will suggest the most and least attractive plants in each landscape and contribute to even more growth.[88]

Further, we recommend that intelligent and real-time vegetation resource tracking be investigated through thermal imaging, computer learning software, and aerial unmanned vehicles. The use of natural changes and adequate plantations will facilitate the development of future and productive bioengineered slopes, taking into consideration the infrastructural needs. This leads to further surfing of remote sensing, which plays a clear role in tracking land degradation, as it can provide persistent coverage of global Earth surfaces in the visible, near-infrared, short-wave, and thermal infrared regions with high-resolution details. These are also suitable for tracking ground cover shifts due to the use of land, which are the core factors of land erosion. However, there is currently no specific technique for tracking land degradation/enhancement dependent on remote sensing that has been able to go well beyond the parameters of land cover.[100] Some of the measures implemented are spatial and analytical,[101, 102] but the widespread use of remote sensing techniques allows for easier exploration of large tracts utilizing satellite images and numerous response variable mapping.[103] Remote sensing often offers depictions of time series to facilitate tracking of environmental transformations on a defined time scale.[4] The development of such indicators plays a major role in managing policies, seeking to better manage land use in order to maintain soil quality in the long run and prevent irreversible soil depletion conditions.[104] These models can also be effectively used along with geoprocessing strategies for broad areas of environmental control, characterization, and classification.

This chapter focuses on a potential research topic suitable for designing bioengineered slopes in the future. The ability to track and maintain the biologically developed pathways in real-time redefines the wilting points and the "plant sensors" principle is also highlighted. On the other hand, terrain analysis has traditionally been left to geotechnical engineers and geologists. These last two must explore and verify, but do not generally understand the many of the organic mechanisms of interaction among land, plants, and further, agricultural scientists, foresters, and hydrologists will know how to use the methodology of bioengineering analysis better.[105] In order to protect the soil against erosion, bioengineering, which is usually performed in a simultaneous team, needs to depend on both geoscience and plant science methods.

6 Conclusion

Throughout the world, many slopes are vulnerable due to soil erosion, slope failure, and degradation to eradicate these numerous measures were being taken at the national and international levels, but very few techniques have been developed. The main bottlenecks in implementing the solutions obtained by researchers include social acceptance, relevance, field implementation issues, financial viability, and technical feasibility. Soil bioengineering is a concept obtained through a combination of decades of research and centuries-old practices adopted by humans. This is not only feasible in all locations but also provides bountiful services that are required for the benefit of society, as they protect the slope, stop soil degradation, and increase the fertility of the soil through ecosystem services. However, more location-specific assessments are required to evaluate the entire process and to implement the system effectively, and once implemented in any area, this not only protects the slope but also provides multiple services; hence, the utilization of soil bioengineering practices in tandem with civil engineering structures is strongly suggested for mountainous regions.

References

1. Sidle RC, Ochiai H. *Landslides: Processes, Prediction, and Land Use.* Washington, DC: Water Resources Monogr American Geophysical Union; 2006:18.
2. Walker LR, Shiels AB. *Landslide Ecology.* Cambridge: Cambridge University Press; 2013.
3. Bezerra FGS, Aguiar APD, Alvalá RCS, et al. Analysis of areas undergoing desertification, using EVI2 multi-temporal data based on MODIS imagery as indicator. *Ecol Indic.* 2020;117. https://doi.org/10.1016/j.ecolind.2020.106579, 106579.
4. Rayegani B, Barati S, Sohrabi TA, Sonboli B. Remotely sensed data capacities to assess soil degradation. *Egypt J Remote Sens Space Sci.* 2016;19:207–222. https://doi.org/10.1016/j.ejrs.2015.12.001.
5. Cerdan O, Govers G, Le Bissonnais Y, et al. Rates and spatial variations of soil erosion in Europe: a study based on erosion plot data. *Geomorphology.* 2010;122(1–2):167–177. https://doi.org/10.1016/j.geomorph.2010.06.011.
6. Jish Prakash P, Stenchikov G, Kalenderski S, Osipov S, Bangalath H. The impact of 38 dust storms on the Arabian Peninsula and the Red Sea. *Atmos Chem Phys.* 2015;15(1):199–222. https://doi.org/10.5194/acp-15-199-2015.
7. Puigdefábregas J. Ecological impacts of global change on drylands and their implications for desertification. *Land Degrad Dev.* 1998;9(5):393–406.
8. Villamil BV, Amiotti NM, Peinemann N. Soil degradation related to overgrazing in the semi-arid southern caldenal area of Argentina. *Soil Sci.* 2001;166:441–452. https://doi.org/10.1097/00010694-200107000-00002.
9. Baude M, Meyer BC, Schindewolf M. Use change in an agricultural landscape causing degradation of soil based ecosystem services. *Sci Total Environ.* 2018;659:1526–1536. https://doi.org/10.1016/j.scitotenv.2018.12.455.
10. Yu GH, Chen CM, He XH, Zhang XZ, Li LN. Unexpected bulk density and microstructures response to long-term pig manure application in a Ferralic Cambisol Soil: implications for rebuilding a healthy soil. *Soil Tillage Res.* 2020;203:104668. https://doi.org/10.1016/j.still.2020.104668.
11. Ma X, Asano M, Tamura K, Zhao R, Nakatsuka H, Wuyunna WT. Physicochemical properties and micromorphology of degraded alpine meadow soils in the Eastern Qinghai-Tibet Plateau. *Catena.* 2020;194:104649. https://doi.org/10.1016/j.catena.2020.104649.
12. Nascimento CM, de Sousa Mendes W, Silvero NEQ, et al. Soil degradation index developed by multitemporal remote sensing images, climate variables, terrain and soil attributes. *J Environ Manag.* 2021;277. https://doi.org/10.1016/j.jenvman.2020.111316, 111316.
13. Tejada M, Gonzalez JL. The relationships between erodibility and erosion in a soil treated with two organic amendments. *Soil Tillage Res.* 2006;91(1–2):186–198.
14. Baude, et al. Land use change in an agricultural landscape causing degradation of soil based ecosystem services. *Sci Total Environ.* 2019;659:1526–1536. https://doi.org/10.1016/j.scitotenv.2018.12.455.
15. Pimentel D, Burgess M. Soil Erosion threatens food production. *Agriculture.* 2013;3:443–463.
16. Borrelli P, Robinson DA, Fleischer LR, et al. An assessment of the global impact of 21st century land use change on soil erosion. *Nat Commun.* 2017;8:2013.
17. Dinar A, Tieu A, Huynh H. Water scarcity impacts on global food production. *Glob Food Sec.* 2019;23:212.
18. Whitford WG. Effects of climate change on soil biotic communities and soil processes. In: Peters RL, Lovejoy TE, eds. *Global Warming and Ecological Diversity.* New Haven, CT: Yale University Press; 1992:126136.
19. Perez Trejo E. *Desertification and Land Degradation in the European Mediterranean-EUR 14850.* Luxembourg: Office for Official Publications of the European Communities; 1994.
20. Lopez Bermudez F. La erosion delsueloen el rirsgo de desertificacionde EspaAa. In: *Medio Ambientey Desarrolloantesydcvpues de Ria-92.* Fundacion Marcelino Botin, Santander; 1993:119–147.
21. Lal R, Stewart BA. Soil degradation. Need for action: research and development priorities. *Adv Soil Sci.* 1990;11:331–336.
22. Nill D. Bodenschutz probleme in Entwicklungsländern. In: Richter G, ed. *Bodenerosion, Analyse und Bilanzeines Umwelt problems.* Wissenschaftliche Buchgesellschaft; 1998:222–231.
23. Dey D. Riverine plants of Kansai Basin at Purulia District West Bengal and how they prevent soil erosion and corrosion. *IOSR J Environ Sci Toxicol Food Technol.* 2020;14(5):01–13.
24. Stokes A, Sotir R, Chen W, Chestem M. Soil bio- and eco-engineering in China: past experience and future priorities preface. *Ecol Eng.* 2010;36:247–257.
25. Löbmann MT, Geitner C, Wellstein C, Zerbea S. The influence of herbaceous vegetation on slope stability—a review. *Earth Sci Rev.* 2020;209. https://doi.org/10.1016/j.earscirev.2020.103328, 103328.

26. Shi ZH, Fang NF, Wu FZ, Wang L, Yue BJ, Wu GL. Soil erosion processes and sediment sorting associated with transport mechanisms on steep slopes. *J Hydrol.* 2012;454–455:123–130. https://doi.org/10.1016/j.jhydrol.2012.06.004.

27. Bai ZG, Wu YJ, Dent DL, et al. Land degradation and improvement in China 2. Accounting for soils, terrain and land use change. *ISRIC Rep.* 2010;05.

28. Li LF, Zeng XB, Li GX, Mei XR. Water quality assessment in Chaohe River by fuzzy synthetic evaluation method. *J Agro-Environ Sci.* 2006;25:471–476.

29. Ding L, Chen KL, Cheng SG, Wang X. Water ecological carrying capacity of urban lakes in the context of rapid urbanization: a case study of East Lake in Wuhan. *Phys Chem Earth A/B/C.* 2015;89–90:104–113. https://doi.org/10.1016/j.pce.2015.08.004.

30. Posthumus H, Deeks KL, Rickson RJ, Quinton JN. Costs and benefits of erosion control measures in the UK. *Soil Use Manag.* 2015;31:16–33. https://doi.org/10.1111/sum.12057.

31. Chillo V, Vazquez DP, Amoroso MM, Bennett EM. Land-use intensity indirectly affects ecosystem services mainly through plant functional identity in a temperate forest. *Funct Ecol.* 2018;32:1390–1399. https://doi.org/10.1111/1365-2435.13064.

32. Wen Z, Zheng H, Smith JR, Zhao H, Liu L, Ouyang Z. Functional diversity overrides community-weighted mean traits in linking land-use intensity to hydrological ecosystem services. *Sci Total Environ.* 2019;682:583–590. https://doi.org/10.1016/j.scitotenv.2019.05.160.

33. Wang LQ, Yu WD. Water resources and water environment problems and counter measures in the Zhangweinan river basin. *Haihe River Conserv.* 2008;5:6–8.

34. Varol M, Sen B. Assessment of surface water quality using multivariate statistical techniques: a case study of Behrimaz Stream, Turkey. *Environ Monit Assess.* 2009;159:543–553. https://doi.org/10.1007/s10661-008-0650-6.

35. Wang YY, Zhang RB, Zhao YW, Sun Y. Application of fuzzy mathematical method in lake water quality evaluation. *Jiangsu Agric Sci.* 2010;1:326–328.

36. Rahman MR, Shi ZH, Chongfa C. Soil erosion hazard evaluation-an integrated use of remote sensing, GIS and statistical approaches with biophysical parameters toward. *Ecol Model.* 2009;220(13–14):1724–1734.

37. Shao M, Tang X, Zhang Y, Li W. City clusters in China: air and surface water pollution. *Front Ecol Environ.* 2006;4:353–361. https://doi.org/10.1890/1540-9295 (2006)004[0353:CCICAA]2.0.CO;2.

38. Chen SH, Su HB, Tian J, Zhang RH, Xia J. Estimating soil erosion using MODIS and TM images based on support vector machine and à trous wavelet. *Int J Appl Earth Obs Geoinf.* 2011;13:626–635. https://doi.org/10.1016/j.jag.2011.03.001.

39. Nenadovi S, Kljajevi L, Nenadovi M, Milanovi M, Markovi S. Physico-chemical soil analysis of Rudovci region. *Geonauka.* 2013;1(2):1–8. https://doi.org/10.14438/gn.2013.07.

40. Hou J, Zhu H, Fu B, Lu Y, Zhou J. Functional traits explain the seasonal variation effects of plant communities on soil erosion in semiarid grasslands in the Loess Plateau of China. *Catena.* 2020;194. https://doi.org/10.1016/j.catena.2020.104743, 104743.

41. Wen Z, Zheng H, Smith JR, Ouyang Z. Plant functional diversity mediates indirect effects of land-use intensity on soil water conservation in the dry season of tropical areas. *For Ecol Manag.* 2021;480. https://doi.org/10.1016/j.foreco.2020.118646, 118646.

42. Lewis L. *Rp. Wa-Rd 491.1 Soil Bioengineering for Slopes Soil Bioengineering for Upland Slope Stabilization.* Washington State Department of Transportation Technical Monitor Mark Maurer, L.A., Roadside and Site Development Manager; 2001.

43. Mitchell SJ. Wind as a natural disturbance agent in forests: a synthesis. *Forestry.* 2013;86:147–157.

44. Rey F, Isselin-Nondedeu F, Bédécarrats A. Vegetation dynamics on sediment deposits upstream of bioengineering works in mountainous marly gullies in a Mediterranean climate (Southern Alps, France). *Plant Soil.* 2005;278:149–158.

45. Duran Zuazoc VH, Rodriguez Pleguezuelol R, Panaderoa A, Rayaj M, Francia Martinezb R, Rodriguez C. Soil conservation measures in rainfed olive orchards in south-eastern Spain: impacts of plant strips on soil water dynamics. *Pedosphere.* 2009;19(4):453–464.

46. Gyssels G, Poesen J, Bochet E, Li Y. Impact of plant roots on the resistance of soils to erosion by water: a review. *Prog Phys Geogr.* 2005;29:189–217.

47. Norris JE, Greenwood JR. Assessing the role of vegetation on soil slopes in urban areas. In: *Proc. 10th Congress of the International Association for Engineering Geology and the Environment (IAEG), Nottingham, UK*; 2006.

48. Faisal HA, Normaniza O. Shear strength of soil containing vegetation roots. *Soils Found.* 2008;48(4):587–596.

49. Loughlin CLO, Ziemer RR. The importance of root strength and deterioration rates upon edaphic stability in steepland forests. In: *Proc. I.U.F.R.O. Workshop P.1.07-00 Ecology of Subalpine Ecosystems as a Key to Management, Oregon, USA*; 1982:70–78.

50. Van Beek LPH, Wint J, Cammeraat LH, Edwards JP. Observation and simulation of root reinforcement on abandoned Mediterranean slopes. *Plant Soil.* 2005;278:55–74.

51. Garrity DP. Sustainable land use systems for sloping uplands in Southeast Asia. In: Regland I, Lal R, eds. *Technologies for Sustainable Agriculture.* vol. 56. ASA Spec. Publ.; 1993:41–66.

52. Cullum RF, Wilson GV, McGregor KC, Johnson JR. Runoff and soil loss from ultra-narrow row cotton plots with and without stiff-grass hedges. *Soil Tillage Res.* 2007;93(1):56–63.

53. Salvador-Blanes S, Cornu S, Couturier A, King D, Macaire JJ. Morphological and geochemical properties of soil accumulated in hedge-induced terraces in the massif central, France. *Soil Tillage Res.* 2006;85(1–2):62–77.

54. Wu DM, Yu YC, Xia L-Z, Yin SX, Yang LZ. Soil fertility indices of citrus orchard land along topographic gradients in the three gorges area of China. *Pedosphere.* 2011;21(6):782–792.

55. Fan J, Yan L, Zhang P, Zhang G. Effects of grass contour hedgerow systems on controlling soil erosion in red soil hilly areas, Southeast China. *Int J Sediment Res.* 2015;30(2):107–116.

56. Dabney SM, Liu Z, Lane M, Douglas J, Zhu J, Flanagan DC. Landscape benching from tillage erosion between grass hedges. *Soil Tillage Res.* 1999;51(3–4):219–231.

57. Zheng FL. Effect of vegetation changes on soil erosion on the Loess Plateau. *Pedosphere.* 2006;16(4):420–427.

58. Lin CW, Tu SH, Huang JJ, Chen Y. The effect of plant hedgerows on the spatial distribution of soil erosion and soil fertility on sloping farmland in the purple-soil area of China. *Soil Tillage Res.* 2009;105(2):307–312.

59. Agus F, Garrity DP, Cassel DK. Soil fertility in contour hedgerow systems on sloping oxisols in Mindanao. *Philipp Soil Tillage Res.* 1999;50(2):159–167.

60. Dercon G, Deckers J, Poesen J, et al. Spatial variability in crop response under contour hedgerow systems in the Andes region of Ecuador. *Soil Tillage Res.* 2006;86(1):15–26.

61. Oshunsanya SO. Spacing effects of vetiver grass (*Vetiveri anigritana* Stapf) hedgerows on soil accumulation and yields of maize–cassava inter-cropping system in Southwest Nigeria. *Catena.* 2013;104:120–126.

62. Agus F, Cassel DK, Garrity DP. Soil-water and soil physical properties under contour hedgerow systems on sloping oxisols. *Soil Tillage Res.* 1997;40(3–4):185–199.

63. Wang J, Bowden RD, Lajtha K, Washko SE, Wurzbacher SJ, Simpson MJ. Long-term nitrogen addition suppresses microbial degradation, enhances soil carbon storage, and alters the molecular composition of soil organic matter. *Biogeochemistry.* 2019;142:299–313.

64. Zhang C, Li D, Jiang J, et al. Evaluating the potential slope plants using new method for soil reinforcement program. *Catena.* 2019;180:346–354.

65. Abedin Z, Hassan M, Dewan AS. Bearing capacity of a jute cloth reinforced composite sand bed. In: *Proceedings of the 14th International Conference on Soil Mechanics and Foundation Engineering*; 1997:1553–1556.

66. Vinod P, Ajitha B, Sreehari S. Behavior of a square model footing on loose sand reinforced with braided coir rope. *Geotext Geomembr.* 2009;27:464–474.

67. IPCC. *I.P.O.C.C. Special report on global warming of 1.5 C (SR15)*; 2019.

68. Caviezel C, Hunziker M, Schaffner M, Kuhn NJ. Adapting slope stability measurements to shifting process domains. *Earth Surf Process Landf.* 2014;39:509–521. https://doi.org/10.1002/esp.3513.

69. Guo J, Hasan I, Graeber P. Application of the program PCSiWaPro for the stability analysis in earth dams and dikes considering the influence from vegetation and precipitation—a case study in China. In: Wu W, ed. *Recent Advances in Modeling Landslides and Debris Flows.* Cham: Springer; 2015:195–209.

70. McGuire LA, Rengers FK, Kean JW, et al. Elucidating the role of vegetation in the initiation of rainfall-induced shallow landslides: insights from an extreme rainfall event in the Colorado Front Range. *Geophys Res Lett.* 2016;43:9084–9092.

71. Loades KW, Bengough AG, Bransby MF, Hallett PD. Planting density influence on fibrous root reinforcement of soils. *Ecol Eng.* 2010;36:276–284.

72. Vannoppen W, Vanmaercke M, De Baets S, Poesen J. A review of the mechanical effects of plant roots on concentrated flow erosion rates. *Earth Sci Rev.* 2015;150:666–678.

73. Fattet M, Fu Y, Ghestem M, et al. Effects of vegetation type on soil resistance to erosion: relationship between aggregate stability and shear strength. *Catena.* 2011;87:60–69.

74. Gonzalez-Ollauri A, Slobodan B, Mickovski S. Plant-soil reinforcement response under different soil hydrological regimes. *Geoderma.* 2017;285:141–150.

75. Norris JE, Stokes A, Mickovski SB, et al., eds. *Slope Stability and Erosion Control: Ecotechnological Solutions.* The Netherlands: Springer; 2018.

76. Stokes A, Atger C, Bengough AG, Fourcaud T, Sidle RC. Desirable plant root traits for protecting natural and engineered slopes against landslides. *Plant Soil.* 2009;324:1–30.

77. Bais HP, Weir TL, Perry LG, Gilroy S, Vivanco JM. The role of root exudates in rhizosphere interactions with plants and other organisms. *Annu Rev Plant Biol.* 2006;57:233–266.

78. Preti F, Giadrossich F. Root reinforcement and slope bioengineering stabilization by Spanish Broom (*Spartium junceum* L.). *Hydrol Earth Syst Sci.* 2009;13:1713–1726.

79. Preti F, Dani A, Laio F. Root profile assessment by means of hydrological, pedological and above-ground vegetation information for bio-engineering purposes. *Ecol Eng.* 2010;36:305–316.

80. Materechera SA, Alston AM, Kirby JM, Dexter AR. Influence of root diameter on the penetration of seminal roots into a compacted subsoil. *Plant Soil.* 1992;144:297–303.

81. Pietola L, Smucker AJM. Fibrous carrot root responses to irrigation and compaction of sandy and organic soils. *Plant Soil.* 1998;200:95–105.

82. Roy SC. *A Theoretical Model of the Effects of Timber Harvesting on Slope Stability*; 1992.

83. Shroder J. Slope failure and denudation in the western Himalaya. *Geomorphology.* 1998;26:81–105.

84. Anonymous. *Mountain Risks and Hazards.* Nepal: International Center for Integrated Mountain Development; 2001:32–34. News Letter No. 40.

85. Gupta V, Sah MP. The relationship between Main Central Thrust (MCT) and the spatial distribution of mass movement in the Satluj Valley, Northwestern Higher Himalaya. *Indian J Geomorphol.* 2008;52(2):169–179.

86. Genet M, Stokes A, Salin F, et al. The influence of cellulose content on tensile strength in tree roots. *Plant Soil.* 2005;278:1–9.

87. Zhang CB, Chen LH, Jiang J. Why fine tree roots are stronger than thicker roots: the role of cellulose and lignin in relation to slope stability. *Geomorphology.* 2014;206:196–202.

88. De Baets S, Poesen J, Reubens B, Muys B, De Baerdemaeker J, Meersmans J. Methodological framework to select plant species for controlling rill and gully erosion: application to a Mediterranean ecosystem. *Earth Surf Process Landf.* 2009;34:1374–1392.

89. Zhang CB, Zhou X, Jiang J, Wei Y, Ma JJ, Hallett PD. Root moisture content influence on root tensile tests of herbaceous plants. *Catena.* 2019;172:140–147.

90. Jastrow JD, Miller RM, Lussenhop J. Contributions of interacting biological mechanisms to soil aggregate stabilization in restored prairie. *Soil Biol Biochem.* 1998;30(7):905–916.

91. Leifheit EF, Veresoglou SD, Lehmann A, Morris EK, Rillig MC. Multiple factors influence the role of arbuscular mycorrhizal fungi in soil aggregation—a meta-analysis. *Plant Soil.* 2014;374:523–537.

92. Johnson NC, Gehring CA. Mycorrhiza: symbiotic mediators of rhizosphere and ecosystem processes. In: Cardon ZG, Whitbeck JL, eds. *The Rhizosphere. An Ecological Perspective.* New York, USA: Elsevier Academic Press; 2007:212.

93. Jones HG. *Plants and Microclimate: A Quantitative Approach to Environmental Plant Physiology*; 2013. https://doi.org/10.1017/CBO9780511845727.

94. Evette A, Balique C, Lavaine C, Rey F, Prunier P. Using ecological and biogeographical features to produce a typology of the plant species used in bioengineering for riverbank protection in Europe. *River Res Appl.* 2011;28(10):1830–1842. https://doi.org/10.1002/Rra.1560.

95. Reubens B, Achten WMJ, Maes WH, Danjon F, Aerts R, et al. More than biofuel? Jatropha curcas root system symmetry and potential for soil erosion control. *J Arid Environ.* 2011;75:201–205.

96. Dorairaj D, Osman N. Present practices and emerging opportunities in bioengineering for slope stabilization in Malaysia: an overview. *PeerJ.* 2021;9. https://doi.org/10.7717/peerj.10477, e10477.

97. Giadrossich F, Schwarz M, Cohen D, Preti F, Or D. Mechanical interactions between neighboring roots during pullout tests. *Plant Soil.* 2012;367:391–406.

98. Bischetti GB, Vergani C, Chiaradia EA, Bassanelli C. Root strength and density decay after felling in a Silver Fir-Norway Spruce stand in the Italian Alps. *Plant Soil.* 2014. this issue.

99. Ghestem M, Cao K, Ma W, et al. A framework for identifying plant species to be used as 'ecological engineers' for fixing soil on unstable slopes. *PLoS ONE.* 2014;9(8). https://doi.org/10.1371/journal.pone.0095876, e95876.

100. Bai ZG, Dent DL, Olsson L, Schaepman ME. Proxy global assessment of land degradation. *Soil Use Manag.* 2008;24:223–234. http://onlinelibrary.wiley.com/doi/10.1111/j.1475-2743.2008.00169.x/full.

101. Ladisa MSE, Italy G, Todorovic G, Trisorio Liuzzi A. GIS-based approach for desertification risk assessment in Apulia region. *Phys Chem Earth.* 2012;49:103–113. https://doi.org/10.1016/j.pce.2011.05.007.

102. Kosmas C, Karamesouti M, Kounalaki K, Detsis V, Vassiliou P, Salvati L. Land degradation and long-term changes in agro-pastoral systems: an empirical analysis of ecological resilience in Asteroussia—crete (Greece). *Catena.* 2016;147:196–204. https://doi.org/10.1016/j.catena.2016.07.018.

103. Chikhaoui M, Bonn F, Bokoye AI, Merzouk A. A spectral index for land degradation mapping using ASTER data: application to a semi-arid Mediterranean catchment. *Int J Appl Earth Obs Geoinf.* 2005;7:140–153. https://doi.org/10.1016/j.jag.2005.01.002.

104. Bedoui C. Study of desertification sensitivity in Talh region (Central Tunisia) using remote sensing, G.I.S. and the M.E.D.A.L.U.S. approach. *Geoenviron Disasters.* 2020;16. https://doi.org/10.1186/s40677-020-00148-w.

105. Greenway DR. Vegetation and slope stability. In: Anderson MG, Richards KS, eds. *Slope Stability—Geotechnical Engineering and Geomorphology.* Chichester, U.K: John Wiley & Sons; 1986:187–230.

19

Soft computing applications in rainfall-induced landslide analysis and protection—Recent trends, techniques, and opportunities

Ashwini Arun Salunkhe[a,b], *R. Gobinath*[a], *and Sandhya Makkar*[c]

[a]SR University, Warangal, Telangana, India [b]Dr. D. Y. Patil Institute of Technology, Pimpri, Pune, India [c]Lal Bahadur Shastri Institute of Management, Delhi, India

1 Introduction

The human race is facing unprecedented impacts due to hazards in the recent decade that are aggravated by severe and unplanned anthropogenic activities. Among the hazards; a landslide is the one that causes large devastation. Land movement poses a significant threat and is a very common phenomenon on all continents. In certain areas, it contributes to the evolution or growth of the landscape. Land movement due to the instability of slope instability is termed a landslide. Various factors contribute to instability, such as topography, geology, land utilization, and improper management. The most severe parameters, rainfall, and earthquakes, are responsible for landslides. Natural disasters such as floods, hurricanes, tornadoes, landslides, etc. are severe events that result in major collateral damage or loss of life and property. Recently, the severely high frequency of occurrence of adverse events has attracted the attention of both industrialized and developing countries. The global changes of a growing population, expansion over hazardous areas is a challenging nature to take its course. Hence, it is becoming difficult for 3rd world countries to meet the excessive cost of controlling or mitigating natural hazards.[46] Many developed nations are ready to finance disaster management systems to mitigate the chances of disaster. Reducing or controlling the impact of natural disasters is only possible with early warning systems. These systems are possible with new technologies such as artificial intelligence, the internet of things, cloud computing, etc. Machine learning is another powerful technique that can resolve the issue by triggering an alarm before a landslide occurs or provides a probability of the occurrence of landslides. Worldwide, several attempts have been made for landslide disasters and their distribution over maps. Landslides are the most recurring and disastrous during rainfall, which severely affects hilly areas.[1,29] Such landslides have caused at least 17% of fatalities worldwide.[2] Global compilation by Froude and Petley of the worldwide land disaster database (2004 to 2017) showed that out of 5318 non-seismic landslides, 3285 landslides were triggered by rainfall.[3] Rainfall-runoff relations are unstructured, which diverts the attention of researchers towards soft computing tools.[4] These are multidisciplinary tools that use diversified areas of statistics, optimization, and linear algebra that complement each other and produce three main branches: fuzzy logic, artificial neural networks,[45] and genetic algorithms.[5] The global scenario of the last decade showed that global climatic changes induced sudden heavy rainfall, triggering flash floods which affected so many countries (e.g., China, Nepal, etc.), landslides in hilly areas, debris flows, etc. The irregularity of landslides provides various scenarios for research in anticipating landslide locations and occurrence, the massiveness of landslides, and designing procedures to minimize risk to the structures and life.

2 Classification factor

For every natural phenomenon, classification plays an important role in better understanding and interpretation for further analysis. The classification of landslides is vast and not perfectly repeatable. The prior classification starts with the type of movement (fall, rotational slide, spread, etc.) and type of material (debris, rock, mud, etc.). Further classification is based on causes such as geological (weathered/ fissured material, fault, etc.), morphological (erosion, deposition, etc.), and human (deforestation, land use, etc.). The whole classification narrows down to the most catastrophic causes around the globe: water-induced landslides, seismic-induced landslides, and volcanic lahars.

2.1 Rainfall-induced landslides

Currently, the news has given a dedicated section to climatic change. Climate change in the 19th century was focused on global warming, greenhouse gases, drought, etc., but in the 20th century, more parameters such as cloud bursts and floods were included. The frequency of sudden constant heavy rainfall and cloud bursts increased the slope instability. These are the reasons that trigger the possibility of landslides and mudslides in hilly regions. These landslide prime reasons are water level changes in the ground due to intense rainfall or snowmelt. In this type of landslide, heavy precipitation and water infiltration influence the groundwater pore pressure and results in water level rise. This further leads to the land movement in an outward and downward direction, leading to landslides.

2.2 Seismic induced landslide and volcanic lahars

The seismic waves radiate from the epicenter and travel rapidly within the earth with a shaking effect at the surface. This seismic movement can cause liquefaction in the saturated soil strata. The liquefaction effects caused a decrease in the strength of soil below the level of stability. Hence, further land movement in hilly terrain is caused by static gravitational forces. Volcanic cones surrounded by any of the strata have inherent weaknesses due to the presence of magma and volcanic gases (as they dissolve and form acidic matter). Furthermore, the major impact of acidic matter with the presence of lava leads to internal faults, volcanic debris flow (called "lahars") and produces landslides. In Most cases, intense rainfall near volcanic fields triggers lahars. Landslide classification helps researchers to narrow down the focus on the most devastating triggers of landslides. Moreover, the purpose of landslide analysis will open the options for landslide mitigation.

3 Landslide analysis

This section provides a wide picture of existing research analyses in the field of landslides. The Fourth Industrial Revolution (Industry 4.0) deals with modern smart technology. Smart tools and applications come with terms such as data mining,[64,65] data analysis, and machine learning. It deals with large, collected data with valid sources to discover reliable patterns and make rational interpretations.[34,63] Water has been the center of attention for landslide analysis. Hence, the analysis revolves around the rainfall data. Past rainfall data are available from meteorological services in different countries, which open numerous smart resources. There are two approaches for rainfall-induced landslide analysis: rainfall threshold estimation and smart or machine learning algorithms (Fig. 1).

Rainfall threshold estimation is usually based on rainfall time series extracted from previous records. The threshold is usually obtained by drawing lower-bound lines to the rainfall conditions that result in landslides plotted in Cartesian, semi-logarithmic, or logarithmic coordinates.[6] The distribution of cumulative rainfall values is used by statistical models as input and the deviation gives the rainfall threshold.[38] The rainfall threshold for landslide prediction has been successful in various countries such as Italy,[7] Malaysia,[8] China,[9] the Himalayan Region India,[10] and Bhuta.[11] Rainfall forecasting for the occurrence of landslides can be carried out by calculating rainfall scenarios, subsurface conditions, or analysis of slope analysis. The threshold for minimum rainfall can be categorized as empirical-base,[12] physical-based, and statistically based model.[13] The primary focus of the landslide's statistical approach is to control and forecast different deformation trajectories and complex data segments based on the time series land deformation and to provide mathematical models for landslide prediction. Machine learning tools construct various algorithms by examining data and proposing a decision-making strategy. An algorithm uses mathematical modeling for various functionality approaches that differ from one problem to another. Commonly used machine learning algorithms are categorized into three categories: supervised learning, unsupervised learning, and reinforcement learning. Supervised

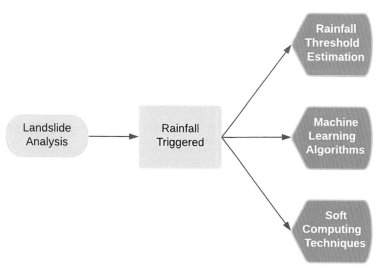

FIG. 1 Landslide analysis methods.

learning builds learning algorithms by pairing input variables and output (dependent) variables. Most past data are used for training the algorithm and it continues until the model reaches the accuracy level. Unsupervised algorithms learn hidden patterns using cluster analysis. The algorithm structure does not involve target variables to predict the interpretation. Reinforced learning trains itself in the field using a trial-and-error strategy. This algorithm makes a sequence of decisions based on previous experience. The popular machine learning algorithms are linear regression, decision tree, support vector machine, and K-means.

Soft computing is an intelligent branch, as its basic principle of working (decision making) is the pursuit of earth's intelligent species, that is, humans. Compared to traditional computing techniques that rely on precise solutions, soft computing seeks to optimize an acceptable solution to a problem in a rapid time. Soft computing methods, such as statistics, probability, and optimization, complement each other or function on their own. They can model complex or unknown interactions that are either nonlinear or noisy.[5] Neural networks and genetic algorithms are prominent members of soft computing and are widely used in real-life problem applications.[4]

4 Landslide monitoring

Monitoring landslide phenomena is important for the identification and recollection of site data, which will be useful for future land movement analyses. The common technique of monitoring technique used by geologists is observing the changes in topography during the site inspection and fractures in the soil. A previous study showed the post-landslide inspection and processing of locations on similar terrain. This study was specifically used to identify and map historic landslides. After the introduction of machine learning, the identification evolved in predicting the approach of future landslides and their mapping. In this approach, the study area was surveyed with ground feature detailing, and machine learning models were trained and tested on a spatial database.[36] Validation of trained and tested models was performed with accuracy measuring parameters to judge the prediction success of the models, and an overview of the methodological approaches for mapping landslide susceptibility is shown in Fig. 2.

5 Machine learning

Machine learning (ML) algorithms are popular techniques that fall under the umbrella of artificial intelligence and have applications in interdisciplinary fields such as fraud detection, sales predictions, medical diagnosis, speech recognition, image recognition, predicting employee job satisfaction,[14] sentiment analysis,[15,30] creation of recommendation systems, chatbots, detecting failures in machines before failure[16] and many more. The applications of ML algorithms are endless because the algorithms can handle abundant data, no human intervention is required for predictions or classifications and there is a continuous improvement of solutions as the machine learns and trains on its own. ML supervised algorithms like random forest (RF),[57] decision tree (DT), logistic regression (LR), etc., helps the decision-maker to

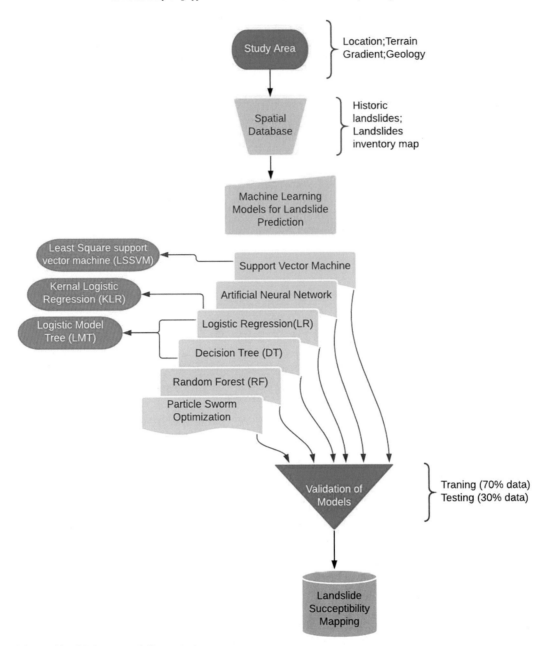

FIG. 2 Methodology of landslide susceptibility mapping.

predict the relation in the occurrence of landslide and causes surrounding it. The algorithms are also useful for predicting the probability of occurrence of land movements in areas before they occur, thereby helping to prepare and evacuate areas beforehand. In addition, ML algorithms manage a variety of data inputs of various types and scales (e.g., ratios, intervals) without the need for a predefined database (e.g., normally distributed or transforming variables).[17]

5.1 Logistic regression

This is a statistical method that is also considered one of the most popular supervised machine learning algorithms.[50] There are many applications of this machine-learning algorithm in different domains. In engineering and computer science, a decision-maker can predict whether a certain email is spam or not spam, prediction of advanced machine failure (Yes/No). It is used by researchers to detect malfunctioning or disease, such as a person suffering from heart disease, cancer, or respiratory disease by assigning symptoms as independent variables such as nausea, chest pain, age, gender, etc., or is used to predict trauma and injury severity score (TRISS), which forecasts the mortality rate in injured patients. The method is used to predict customer churn in the retail industry based on their buying behavior. Logistic regression is generally used to classify various categories of predictor variables. There are many kinds of Logistic Regression viz., Binary Logistic Regression, Multinomial Logistic Regression, Ordinal Logistics Regression to name a few. In binary logistics, the predictor variable that is categorical in nature has two levels: spam or no-spam emails. If the number of levels is increased as in the case of the type of diseases (cardiovascular/Respiratory/Leukemia, etc.), it is always recommended to use multinomial logistic regression. In addition, if the level of categories is ordered in sequence, such as Lecturer/Assistant professor/Sr. Assistant professor/Associate Professor or Professor, in such cases, ordinal logistic regression is used. In landslide problems, where the objective is to predict the probability of landslides in a particular area, binary logistic regression is very useful. A sigmoid function is used in the Binary logistics regression. A binary logistic regression can be used when the categories can be linearly separable, which indicates that the two categories can be divided with the help of a plane, as shown in Fig. 1. The objective of the logistic regression is to maximize the optimizer.

$$\text{Max} \sum y_i w^T x_i \tag{1}$$

The parameter to update is w_i to max expression (1) or the best fit.

Logistic regression can be considered a generalized linear model that uses the Sigmoid function. If Y is a binary variable and X is a set of independent variables, the sigmoid function in Eq. (2) provides the probability of occurrence of an event.

$$P(Y_i = 1 X_i = x_i) = \exp(\beta_0 + \beta_1 x_i) \backslash (1 + \exp(\beta_0 + \beta_1 x_i)) \tag{2}$$

The Distribution of Y_i is Bin (n_i, π_i), that is, the binary logistic regression model assumes a binomial distribution of the response. The relationship between landslide occurrence and its dependency on the predictors is expressed in Eq. (2).

5.2 Decision tree (DT)

These are the utmost popular supervised machine learning algorithms. This technique was used for both classification and regression.[58,59,61] If the target variable or predictor variable is binary, discrete, or categorical, the method is termed as a classification decision tree. If the target or predictor variable is continuous or on an interval scale, the technique is termed a regression decision tree. Response variables or input variables can be categorical or continuous in nature. The technique sets Boolean rules to split nodes. The method partitions the datasets into subsets and continues partitioning until the final output is achieved; thus, an inverted tree is formed with roots at the top and leaves at the bottom. Each leaf or node represents the splitting of the attribute in consideration and the path of each leaf can be expressed as a Boolean rule.

The impurities in the leaves were obtained by tree pruning. All the above steps are performed using recursive partitioning based on the greedy algorithm, which starts from the root node itself. Achieving the split point for the root node is the very first step of the greedy algorithm, which chooses a locally optimal point. The splitting criterion reduces impoverishment and variability in the child nodes. The steps were repeated for each child node. The impurity was measured by the Gini index and entropy. For a categorical variable, the measure used is the Gini index, the basis of which is high diversity, low impurity, and vice-versa. For a pure node, the Gini index was zero. The Gini index is computed as follows:

$$\text{Gini Index} = p_i(1 - p_i) \, for \, i = 1 \, to \, k$$

where k is the number of classes of the response variable and p_i is the probability of class i at a node. The feature that gives the least value of the Gini index is chosen for splitting the dataset at a node. Entropy is another measure of impurity or disorder in nodes, the value of which lies between 0 and 1. Information gain is another measure that is based on entropy and measures which feature provides the maximum information. Overall, decision trees are widely used by data scientists and are applied in detecting landslide susceptible zones, landslide prediction, etc. A DT measures the probability of belonging to a certain class and thus the probability of predicting the frequency of landslide pixels can be used.[18] The decision tree is a hierarchical model that, in terms of probability, recursively separates landslide conditioning variables into two groups: landslide and non-landslide.[35]

5.3 Random forest (RF)

RF is regarded as a non-parametric machine learning algorithm and is famous for spatial prediction problems, including landslide susceptibility mapping.[19] This is a supervised ML ensemble model. It is used for both classification and regression problems. It was introduced by Breimen,[20,31] where he first discussed classification and regression trees in 1983 and again about random forests in 2001.[21,32] The method consists of many decision trees from the sample data and chooses the best solution from the samples. Voting was performed for each solution from random trees to obtain the final prediction result. The most voted solution is prediction. The technique is widely used owing to its simplicity and ability to obtain the best solutions. There are many advantages of this technique, some of which are as follows:

• The technique overcomes the problem of overfitting
• This technique is highly flexible and produce results with high accuracy
• Since it uses multiple decision trees, hence each decision tree brings less variance
• This method is best suited for problems with big data or large datasets. One such application is predicting landslides that require a large dataset.
• Scaling of data is not required.
• It uses different combinations of rows and variables for predictions, which results in training a high variety and diversity of data sets.

The data that are withheld from training are called out-of-bag samples. Observations in the training sample are called bagged observations, and the training data for a specific decision tree are called bagged data. Many researchers have used the RF to predict landslide susceptibility mapping. Some are.[22-24]

5.4 Support vector machines (SVM)

This is another machine learning algorithm that has applications in various fields such as face detection, text and hypertext categorization, image classification, and bioinformatics.[55] Since SVM is also a supervised machine-learning algorithm that predicts the output based on historical data. It is primarily used for classification problems but also has applications in regression problems, where the data are split with the help of a hyperplane line so that the distance between the points and the line (hyperplane in this case) is as far as possible or technically the distance between the hyperplane and support vectors should be as far as possible for classification. The Support vectors are the exterior points of the dataset. A line is called a hyperplane, and the classification is performed on the multidimensional plane and not only on 2D. The distance between the two support vectors is called the Distance margin (Fig. 3).

After determining the largest distance margin, an optimal hyperplane can be evaluated. Once an optimal hyperplane is created, one can easily locate the side at which the new data fits into and belongs to which class. If the hyperplane is not optimal, there is a high chance of misclassification; in particular, the support vector machine is now called the Lagrangian support vector machine (LSVM). For such applications, a hyperplane cannot be used, instead, a kernel function[33,49] is used for transformations that transform the plane into a new higher dimension (Fig. 4). Therefore, there are many advantages of SVM, such as (i) SVM supports high dimensional (HD input space, (ii) SVM can work on sparse document vectors, and (iii) SVM also supports regularization parameters that help determine the overfitting of data or in biasedness.

For landslide modeling, SVM was used to report that the prediction power of the derived models outperformed those obtained by conventional methods. Owing to their ability to generate complex curved boundaries, SVMs can be used for more complex, nonlinear contexts or linear separable issues.[52]

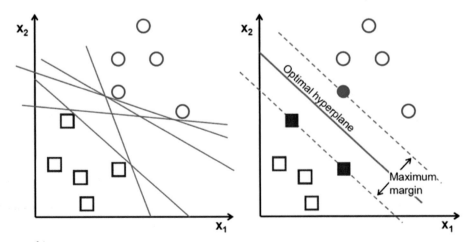

FIG. 3 Support vector machine.

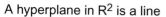

A hyperplane in R^2 is a line

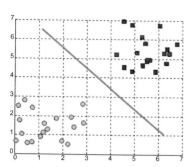

A hyperplane in R^3 is a plane

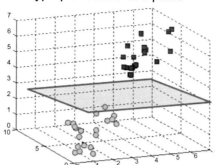

FIG. 4 Hyperplanes and R^2.

5.5 Artificial neural networks

An ANN is a powerful machine learning algorithm that overcomes the challenges of many other techniques, such as parametric and non-parametric non-linear regression models. Nonlinear regression models are difficult to estimate in the presence of a large number of input variables, that is, why the term is coined as "the curse of dimensionality." Neural networks play an important role as they do not require the functional form to be specified and they perform well in high-dimensional spaces. Neural networks form a base of deep learning, where the algorithms are inspired by the cells of the human brain called neurons. A Neural network takes into account data in its network, trains them, and generates the output. It consists of three layers: input layer, hidden layer, and output layer. The hidden layer lies between the input layer and output layer (Fig. 5). In a network, necessary computations are mostly done by hidden layers and each neuron in the input layer is an input variable. The channels provide a connection between neurons present in consecutive layers. The weights allotted to channels represent the numerical value and play an important role as every input is multiplied by these weights. All these inputs in layers are calculated, added, and then sent to the hidden next layer. Each neuron has its bias value which is added to the input sum. This sum is processed through a

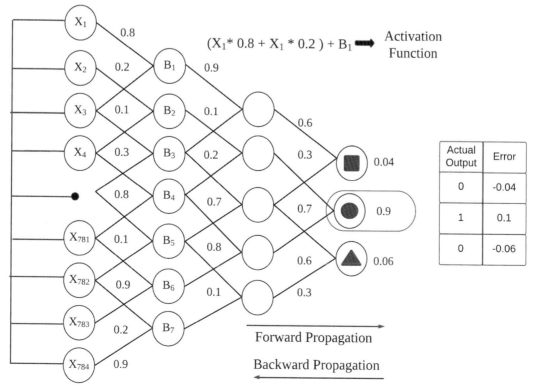

FIG. 5 Neural network process.

threshold function, termed as an activation function. This function determines whether a particular neuron will be activated and the activated neuron transmits data to next layer neurons over the channel. In this fashion, the data propagated and called forward propagation. From the output layer, the greatest valued neuron finds the network output. These are mostly probabilities.

The network is yet to be trained, as predictions using forward propagation may provide incorrect results. The predicted output was compared with the actual output to determine the error in the prediction. The magnitude of an error indicates how incorrect the predictions are and suggests if the predicted value is higher or lower than expected. The errors provide the path and extent of adjustment to minimize the error. These details are reverted back to the network, and this process is known as backpropagation. From this backpropagation, weights are adjusted.[44] The forward and backward propagation process is performed in iteration with multiple inputs. The process continued till the assigned weights in the network were predicted in precision. Mostly this is the end of the training process.

There have been studies that predicted landslides and used this powerful technique. Landslide hazard map of Penang Island was formed with two ANN models namely, cascade forward NN (CFNN) and multilayer perceptron (MLP) by Al-Batah et al.[66] Another study (Roy and Islam)[67] tried the landslide prediction in Bangladesh at Chittagong City.

6 Recent trends

Data analysis, in particular, artificial intelligence is aiding many engineering solutions in the recent decade, let it be industrial production or market analysis, it playing a key role in prediction to control of various attributes. Urbanization and rapid conversion of land into constructed areas create a lot of ecological stress to the particular region which affects the landform and triggers hazards. Abatement of hazard is being studied in multiple levels including administration and on-field level, one of the prime hazards that affects huge areas is landslides which will be triggered mostly during heavy rainfall. Recent inventions in the scientific domain, in particular to the soft computing applications in engineering problems provide a perfect solution for the analysis and protection from landslides. Algorithms, protocols, and functions that are suitable for landslide prediction are being introduced at all levels, with increasing computing power and software, this is getting established rapidly.

7 Opportunities

Landslides are perennial issues around the globe and it affects huge land areas in particular to developing countries. Controlling landslides requires a lot of understanding about soil profile, topography, climate, vegetation, and several other factors. Available data is not sufficient to predict the landslides in many areas, research works are going on to develop models by using several techniques. Also, due to anthropological interaction with hilly terrain, a lot of new landslides are being triggered across many countries. Development vs ecological impact is being discussed at various levels, the outcome of which is developed should be there in a balanced way. One of the ways to develop hilly slopes which are vulnerable to landslides if disturbed is to predict, analyze and propose solutions that avoid future landslides. To predict and propose better landslide abatement methods are currently available. Some methods are not suitable which is proved beyond doubt in many cases which ended up in huge landslides beyond prediction. Uncertainties that exist with the data and multiple attributes create serious issues in the prediction, hence novel methods that can overcome this is the need of the hour. Artificial intelligence-based methods are upcoming in many domains; it's not only revolutionizing the way the data is handled but also providing solutions that are not possible in many other methods. Landslide analysis and prediction including slope stability analysis is gaining paradigm shift in the recent decade, thanks to the innovations in computing methods, tools, and techniques. Researchers and administrators are seeing this as a good opportunity to predict unpredictable landslides in real-time level and with more accuracy, hope this trend will continue with inventions in the AI domain.

8 Importance and future

Land movement or landslides are serious geo-hazards that pose a great threat to life and property. Although physical causes cannot be controlled to reduce their impact, landslide analysis and prediction mapping with smart tools are becoming world-famous. Predicting the location and timing of mudslides with sufficient lead time is a technically demanding issue that is also crucial for measuring landslide impact.[25] The data of land movement and displacement in real-time are useful in smart technologies such as the Internet of Things (IoT) and artificial intelligence (AI) to track and forecast the deformation pattern. Moreover, the same effectively worked for landslide mitigation for early warning and minimizing the impact.[26] Landslide mapping shows the vulnerable areas and the geographical extent of possible landslides. This can limit the town growth and engineering infrastructure for proper planning of land use, which is either surrounded or exposed to landslide hazards. By the late 19th century, most countries launched Earth observation satellites. Presently, satellites have a massive amount of data with a variety of sensors that can retrieve deep data using satellite images. In addition, these images are compatible with GIS, which can extract a variety of geological and meteorological data for analyzing and forecasting events. Mostly, landslides occur in hilly regions[27]; hence, a significant disadvantage to follow a common technique is that they do not have variations over a brief time interval, and it is difficult to precisely predict the time and location of possible landslide occurrences in the future. Landslides are a common disaster and it is difficult for conventional methods to accurately monitor and provide early warnings. In this context, the development of the landslide system was done through remote monitoring and forecasting.[28] Remote sensing techniques include multiple time series of SAR imagery analysis, GPS, synthetic aperture radar interferometry, and light detection and ranging (LiDAR). All these methods are not affected by weather conditions and hence very helpful for measuring slope displacement in large terrain.

9 Historical outline of rainfall-induced landslide monitoring and mapping

The history of landslide mapping began with the application of machine-learning algorithms. The last 20 years of data were collected for rainfall-induced landslides and are presented in Table A.1 (Appendix A). A country with mountainous terrain faces landslides every year. The past landslide data provide the pattern or trigger points for a particular region's landslide. These are termed landslide conditioning factors. The researchers studied the landslide hazard sites and determined the conditioning factors for that area, the details of which are given in Table A.1. These factors play important roles in determining the accuracy of future landslide predictions. The summary shows various machine-learning algorithms utilized for landslide prediction and mapping. A comparative study provides the best algorithm for prediction. It can be seen that region-wise, the algorithms provide variation in performances with one another as the conditioning factors change. Few researchers have achieved advancements in algorithms by combining two machine-learning methods: 1. for the selection of landslide conditioning factors, 2. landslide mapping. Most researchers have utilized the area under the Receiver Operating characteristic (ROC) curve (AUC) for judge model accuracy. The summarized applications of this study cover landslide susceptibility, monitoring, and land displacement.

9.1 Advantages

Landslides occur in many inaccessible locations that are becoming a tedious task to monitor and control, this is being studied at various levels but an amicable solution is not yet obtained by any research work. The major problem of analyzing landslide data is the quantum of data available which makes normal processes cumbersome. The most important advantage of this study is making alerts before the occurrence of a disaster through proper data analysis and prediction. The time of landslide and according to that plan of action will benefit in early mitigation can surely lead to saving lives and property. The cost for collecting data is saved. A lot of prehistoric satellite imagery data can be used to work out the algorithms. One of the advantages of machine learning algorithms is that the precision and reliable performance are achieved. Most landslides repeatedly occur in the same area due to several reasons, if the data is obtained through proper methods and develop models, it's easy to predict the landslides using the methods given in this work. Newer tools and techniques are being introduced in soft computing techniques which makes them suitable to handle any amount of data in any shape, this is one needed feature for landslide analysis that has multi-domain attributes.

9.2 Disadvantages

Despite all the above-mentioned advantages, there are few disadvantages also in the study which includes the following

- Availability of data for a particular landslide is not yet established on a regional scale

TABLE A.1
Summary of rainfall-induced landslide prediction.

Year	Author details	Country or study area	Algorithm used	Landslide conditioning factors	Results obtained	Application
2004	Neaupane et al.[43]	Nepal	Artificial Neural Network-backpropagation (ANN-BP)	Groundwater table; Antecedenta rainfall; Rainfall intensity; infiltration coefficient; Shear strength; slope gradient; slope movement	BPNN gives promising result and predicts slope movement considering the rainfall relation.	Landslide monitoring
2009	Yilmaz[42]	Turkey	Conditional probability (CP) Logistic regression (LR) Artificial neural networks (ANN) Support vector machine (SVM)	Geological (lithology and distance from faults), topographical (slope angle and aspect, topographical elevation; distance from drainage; topographic wetness index, TWI; stream power index, SPI) and environmental (the normalized difference vegetation index, NDVI; distance from roads and settlements	The area under curve shows better performance in the map from ANN model. However, simplicity of CP and its high compatibility with GIS takes the attention of the new researchers.	Landslide susceptibility map
2010	Ren et al.[25]	China	Scalable and Extensible Geo-fluid Model of the Environment (SEGMENT)	Real-time rainfall data, digital elevation data, Hydrological Data, soil parameters, land classification map	The RMS error, is only a $0.42\,mm\,yr^{-1}$ within the instrumental error range	Predicting storm-triggered landslides
	Nefeslioglu et al.[62]	Turkey	Decision Tree (DT)	Geological formations, slope gradient, slope aspect, altitude, plan and profile curvatures, and stream power index	The map has 89.6% AUC which gives a good enough performance	Landslide susceptibility map
	Nefeslioglu et al.[62]	Tokyo	Decision Tree (DT) Random forest (RF)	Altitude, slope angle, slope aspect, curvature, lithology, distance to streams, drainage density, compound topographic index (CTI), stream power index (SPI), and rainfall	Research concludes that rainfall and lithology factors are significant in prediction of landslide. The RF model works better in land susceptibility as compare to the DT model.	Rainfall-induced landslide Susceptibility
2011	Marjanović et al.[54]	Serbia	Support vector machine (SVM) Decision Tree (DT) Logistic regression (LR)	Elevation, Slope, Aspect, Slope Length, Topographic Wetness Index, Plan Curvature, Profile Curvature, Distance from Stream, Lithology, Distance from Fault, Distance from geo-boundary, NDVI	SVM is termed to be model of choice as it outperforms other models	Landslide susceptibility map
	Ballabio and Sterlacchini[51]	Italy	Support vector machines (SVM) Logistic regression (LR) Linear discriminant analysis (LDA) Naive Bayes (NB)	Channel base level; Convergence index; CTI; Distance from faults; Distance from thrust; Downslope distance gradient; Elevation; Insolation; Internal relief; Morphological protection Idx; Slope; Stream power index	The model performances are addressed through ROC curves, classification performances in the testing set, and cross-validation set. For both set SVM provide more performance and as compare to other techniques gives more accuracy.	Landslide Susceptibility Mapping; The Staffora River Basin

Year	Author	Country	Methods	Factors/Features	Results	Objective
2012	Dieu Tien Bui et al.	Vietnam	Radial basis function (RBF) Kernel-Support vector machines (SVM); Polynomial kernel (PL)-Support vector machines (SVM); Decision Tree (DT); Naïve Bayes (NB)	Slope angle; Slope aspect; Relief amplitude; Lithology; Land use; Soil type; Rainfall; Distance to roads; Distance to rivers; Distance to faults	Prediction accuracy based on maximum area under success rate curve AUC is for RBF-SVM (0.961) followed by PL-SVM, AUC =0.957	Landslide Susceptibility Assessment
2013	Pradhan[66]	Malaysia	Decision Tree (DT); Support vector machines (SVM); An adaptive neuro-fuzzy inference system (ANFIS)	Landslide inventory; Topographic map; Geological map; Drainage; land cover; Soil map; Normalized difference vegetation index (NDVI)	The higher prediction performance was measured by AUC for DT (83.07) then ANFIS (82.80) and SVM (81.46)	Landslide susceptibility mapping
2014	Lian et al.[70]	China	Least squares support vector machine (LSSVM); Extreme learning machine (ELM)	Rainfall; reservoir level elevation; landslide accumulative displacement	Selection of conditional factors can lead to various prediction performances.	Prediction for landslide displacement
2015	Jie Dou et al.	Japan	Support vector machines (SVM) Radial basis function (RBF)	Elevation, slope, aspect, curvature, lithology, distance from the nearest geologic boundary, density of geologic boundaries, distance from drainage network, the compound topographic index (CTI) and the stream power index (SPI), Lithology, Distance from the nearest geological boundary, Density of geological boundaries	The training and testing accuracy of landslide prediction were 89.2% and 77.8%.	Probable future landslides differentiated into shallow and deep-seated landslides.
	Bui et al.[41]	Vietnam	Support vector machine (SVM); MLP Neural network; RBF Neural network; Kernel logistic regression (KLR); Logistic Model tree (LMT)	Slope, aspect, altitude, relief amplitude, topographic wetness index (TWI), stream power index (SPI), sediment transport index (STI), lithology, fault density, land use, and rainfall	Results show positive land susceptibility mapping	For decision making and policy planning in areas prone to landslides.
2016	Bui[39]	Vietnam	Least squares support vector machine bee colony (LSSVM-BC); Support vector machine (SVM)	Historical landslide dataset: DEM; Slope; Aspect; Releif amplitude; Valley Depth; stream power index (SPI); sediment transport capacity index (STC); topographic wetness index(TWI); Landuse;	The prediction power of LSSVM-BC model from (AUC)=0.900 pointed the better prediction performance than SVM.	Spatial prediction of rainfall-induced landslides

Continued

TABLE A.1 Summary of rainfall-induced landslide prediction—cont'd

Year	Author details	Country or study area	Algorithm used	Landslide conditioning factors	Results obtained	Application
2017	Pourghasemi and Rahmati[53]	Iran	Artificial neural networks (ANN) Boosted regression tree (BRT) Classification and regression trees (CART) Generalized linear model (GLM) Generalized additive model (GAM) Multivariate adaptive regression splines (MARS) Naïve Bayes (NB). Quadratic discriminant analysis (QDA) Random forest (RF) Support vector machines (SVM)	Altitude, slope angle, slope aspect, slope-length, plan curvature, profile curvature, drainage density, distance from river, distance from faults, land-use, lithology, and distance from roads	AUC values for MLTs vary from 62.4% to 83.7%. It has been found that the RF (AUC = 83.7%) and BRT (AUC = 80.7%) have the best performances comparison to other MLT	Landslide spatial modeling
	Binh Thai Pham and Indra Prakash	India	Logit Boost Ensemble (LBE) Support Vector Machines (SVM) Logistic Regression (LR) Fisher's Linear Discriminant Analysis (FLDA)	Slope angle, elevation, curvature, slope length, soil type, slope aspect, land use, valley depth, lithology, Normalized Difference Vegetation Index (NDVI), Stream Power Index (SPI), Sediment Transport Index (STI), Topographic Ruggedness Index (TRI), Topographic Wetness Index (TWI), distance to lineaments, distance to rivers, distance to roads, and rainfall	The hybrid intelligent approach LBE gives the best performance with (AUC = 0.962), next is SVM (0.945), LR (0.873), and last FLDA (0.870).	Landslide Susceptibility Mapping
	Wei Chen et al.	China	Kernel logistic regression (KLR) Naïve-Bayes tree model (NB) Alternating decision tree model (ADT)	Slope aspect, slope angle, altitude, profile curvature, plan curvature, NDVI, land-use, lithological unit, distance to rivers, distance to roads, distance to faults, and mean annual precipitation	The model performance shows better performance KLR model because of its good classification ability in training and validation datasets.	Spatial prediction of landslide susceptibility

Year	Author	Country	Methods	Data/Parameters	Findings	Application
2018	Wei Chen et al.	China	WoE-LMT; EBF-LMT	Elevation, profile curvature, plan curvature, slope aspect, slope angle, stream power index (SPI), sediment transport index (STI), topographic wetness index (TWI), distance from roads, distance from river networks, distance from faults, lithology, the Normalized Difference Vegetation Index (NDVI), and land use	The comparison of ensemble for predictability gives better performance to EBF-LMT (86.21% and 85.23%) as compare to WoE-LMT (85.11% and 83.98%).	Spatial prediction of landslide susceptibility
	Gariano[11]	Bhutan	Calculation of Thresholds for Rainfall-Induced Landslides-Tool (CTRL-T)	Spatial and temporal information on 269 landslides that occurred along the Phuentsholing-Thimphu highway from July 1998 to July 2015; continuous Rainfall series	The threshold has a validity period of around 1–13days, hence not preferred for sub-daily landslide prediction.	Rainfall thresholds for landslide occurrence
	Dikshit and Satyam[10]	India	The threshold for landslide occurrences by power law equation	Rainfall data and landslide information during 2010–2016	Antecedent rainfall of 10 to 20- days with intensity 88.37 and 133.5mm can cause landslides in the region	Rainfall thresholds for landslide occurrences
2019	Wei Chen et al.	China	Evidential belief function (EBF); Logistic regression (LR); Logistic model tree (LMT)	Slope aspect, elevation, slope angle, profile curvature, plan curvature, topographic wetness index (TWI), stream sediment transport index (STI), stream power index (SPI), distance to rivers, distance to faults, distance to roads, lithology, normalized difference vegetation index (NDVI), and land use.	To overcome the deficiency of the single-based model researchers assed effectiveness by a combination approach. This shows decision tree-logistic regression-based algorithm (EBF-LMT) provides better landslide prediction.	Spatial prediction of landslide susceptibility
	Sahin and Colkesen[19]	Turkey	Random Forest (RF); Rotation Forest (RotFor); Canonical Correlation Forest (CCFA); AdaBoost; Bagging	Aspect, drainage density, elevation, lithology, LULC, NDVI, slope, slope length, Sediment transport index (STI), Stream Power Index (SPI), Topographic roughness index (TRI) and TWI	Performance measured based on AUC, chi-squared based McNemar's test, and ROC curve. The CCFA method gives better performance followed by RF, RotFor, Bagging, and AdaBoost.	Landslide susceptibility mapping (LSM)
2020	Abraham et al.[37]	India	Sistema integrato gestione monitoraggio allerta (SIGMA) - integrated system for management, monitoring, and alerting	Spatial and temporal distribution of rainfall-induced landslide events during 2010–2017	The model prediction is satisfactory having 92% efficiency and a likelihood ratio of 11.28%	Rainfall Threshold Estimation and Landslide Forecasting
	He et al.[9]	China	Original rainfall event-duration (E-D) thresholds and normalized and (EMAP-D) rainfall thresholds	Landslide events occurred in China during 1998–2017; rainfall thresholds for landslide occurrence;	This model improved the thresholds for landslide and debris flow occurrences by 16%–20%, 10%–17%, and 20%–38% in the whole year, rainy season, and non-rainy season, respectively	Thresholds for Landslide Occurrences

Continued

TABLE A.1 Summary of rainfall-induced landslide prediction—cont'd

Year	Author details	Country or study area	Algorithm used	Landslide conditioning factors	Results obtained	Application
	Yuqiu Lin et al.	China	Extended ascendant strategy PSO Elman neural network (EESPSO-ENN) Artificial Neural Network-backpropagation (ANN-BP) Support vector machines (SVM)	Time-series of landslide cumulative displacement; data of rainfall and reservoir water level	EESPSO-ENN gives high precision as compared to other models providing less error in training and validating datasets.	Landslide Displacement Prediction

•Cooperation between authorities to support data collection is not good in many areas.
•This particular work is interdisciplinary in nature and hence a team of Geologists, Data scientists, Geotechnical engineering should work in coherence to develop a better prediction model
•Cost of collection of data is too high in certain areas
•Reliability of data is not so accurate concerned with several soil types and hence careful consideration is required when collecting soil samples for analysis.
•Data collection is related to multiple divisions and departments including geology, meteorology, etc. which creates a lot of uncertainties in the data collection

10 Conclusion

As soft computing reaches its success in various phenomena civil engineering is also adopting this tool in various forms. In recent decades, a large number of landslides have occurred in all urban and peri-urban areas, which is aggravated by anthropogenic activities. However, solutions available to predict, monitor, and abate landslides are not implemented in full swing. Many governments and research institutions are giving attention to focusing on pre-mitigation, that is, before the loss of events. This leads to the use of modern smart tools that can perform satisfactorily and evolve in better versions. Machine learning methods are receiving significant attention owing to their ability to interlink the variables. Some of the classic algorithms assume linearity in variables to predict the outcome; hence, some advanced algorithms such as EBF-LMT, EESPSO-ENN, SVM-RBF, and AdaBoost, perform better.[47,48,56] Although most of these methods provide good classification and prediction ability, some of these methods are gaining attention owing to their ease of interpretation and understanding. Hence careful utilization of a specific method based upon various attributes including available variables, need, etc. is strongly suggested.

Appendix A

See Table A.1.

References

1. Dou J, et al. Assessment of advanced random forest and decision tree algorithms for modeling rainfall-induced landslide susceptibility in the Izu-Oshima Volcanic Island Japan. *Sci Total Environ*. 2019;662:332–346. https://doi.org/10.1016/j.scitotenv.2019.01.221.
2. Chae B-G, Park H-J, Catani F, Simoni A, Berti M. Landslide prediction monitoring and early warning: a concise review of state-of-the-art. *Geosci J*. 2017;21(6):1033–1070. https://doi.org/10.1007/s12303-017-0034-4.
3. Froude MJ, Petley DN. Global fatal landslide occurrence from 2004 to 2016. *Nat Hazards Earth Syst Sci*. 2018;18(8):2161–2181. https://doi.org/10.5194/nhess-18-2161-2018.
4. Chandwani V, Vyas SK, Agrawal V, Sharma G. Soft computing approach for rainfall-runoff modelling: a review. *Aquat Proc*. 2015;4:1054–1061. https://doi.org/10.1016/j.aqpro.2015.02.133.
5. Suykens JAK, Vandewalle J. Least squares support vector machine classifiers. *Neural Process Lett*. 1999;9(3):293–300. https://doi.org/10.1023/a:1018628609742.
6. Kanungo DP, Sharma S. Rainfall thresholds for prediction of shallow landslides around Chamoli-Joshimath region Garhwal Himalayas, India. *Landslides*. 2013;11(4):629–638. https://doi.org/10.1007/s10346-013-0438-9.
7. Segoni S, Lagomarsino D, Fanti R, Moretti S, Casagli N. Integration of rainfall thresholds and susceptibility maps in the Emilia Romagna (Italy) regional-scale landslide warning system. *Landslides*. 2014;12(4):773–785. https://doi.org/10.1007/s10346-014-0502-0.
8. Althuwaynee OF, Pradhan B, Ahmad N. Estimation of rainfall threshold and its use in landslide hazard mapping of Kuala Lumpur metropolitan and surrounding areas. *Landslides*. 2014;12(5):861–875. https://doi.org/10.1007/s10346-014-0512-y.
9. He S, Wang J, Liu S. Rainfall event duration thresholds for landslide occurrences in China. *Water*. 2020;12(2):494. https://doi.org/10.3390/w12020494.
10. Dikshit A, Satyam DN. Estimation of rainfall thresholds for landslide occurrences in Kalimpong India. *Innovat Infrastruct Solut*. 2018;3(1). https://doi.org/10.1007/s41062-018-0132-9.
11. Gariano SL, et al. Automatic calculation of rainfall thresholds for landslide occurrence in Chukha Dzongkhag Bhutan. *Bull Eng Geol Environ*. 2018;78(6):4325–4332. https://doi.org/10.1007/s10064-018-1415-2.
12. Wu Y-M, Lan H-X, Gao X, Li L-P, Yang Z-H. A simplified physically based coupled rainfall threshold model for triggering landslides. *Eng Geol*. 2015;195:63–69. https://doi.org/10.1016/j.enggeo.2015.05.022.
13. Zhang SJ, et al. A physics-based model to derive rainfall intensity-duration threshold for debris flow. *Geomorphology*. 2020;351:106930. https://doi.org/10.1016/j.geomorph.2019.106930.
14. Jain D, Makkar S, Jindal L, Gupta M. Uncovering employee job satisfaction using machine learning: a case study of Om Logistics Ltd. In: *Advances in Intelligent Systems and Computing*. Singapore: Springer; 2020:365–376.

15. Makkar S, Singhal M, Gulati N, Agarwal S. Detecting medical reviews using sentiment analysis. In: *Privacy Vulnerabilities and Data Security Challenges in the IoT*; 2020:199–216.

16. Nangia S, Makkar S, Hassan R. IoT based predictive maintenance in manufacturing sector. *SSRN Electron J*. 2020. https://doi.org/10.2139/ssrn.3563559.

17. Merghadi A, et al. Machine learning methods for landslide susceptibility studies: a comparative overview of algorithm performance. *Earth Sci Rev*. 2020;207:103225. https://doi.org/10.1016/j.earscirev.2020.103225.

18. Bui DT, Pradhan B, Lofman O, Revhaug I. Landslide susceptibility assessment in Vietnam using support vector machines decision tree, and Naïve Bayes models. *Math Probl Eng*. 2012;2012:1–26. https://doi.org/10.1155/2012/974638.

19. Sahin EK, Colkesen I. Performance analysis of advanced decision tree-based ensemble learning algorithms for landslide susceptibility mapping. *Geocarto Int*. 2019;36:1253–1275. https://doi.org/10.1080/10106049.2019.1641560.

20. *Classification and Regression Trees*. https://www.routledge.com/Classification-and-Regression-Trees/Breiman-Friedman-Stone-Olshen/p/book/9780412048418.

21. Breiman L. Random forests. *Mach Learn*. 2001;45(1):5–32. https://doi.org/10.1023/a:1010933404324.

22. Kim J-C, Lee S, Jung H-S, Lee S. Landslide susceptibility mapping using random forest and boosted tree models in Pyeong-Chang Korea. *Geocarto Int*. 2017;33(9):1000–1015. https://doi.org/10.1080/10106049.2017.1323964.

23. Park S, Kim J. Landslide susceptibility mapping based on random forest and boosted regression tree models and a comparison of their performance. *Appl Sci*. 2019;9(5):942. https://doi.org/10.3390/app9050942.

24. Wang Y, Sun D, Wen H, Zhang H, Zhang F. Comparison of random forest model and frequency ratio model for landslide susceptibility mapping (LSM) in Yunyang county (Chongqing China). *Int J Environ Res Public Health*. 2020;17(12):4206. https://doi.org/10.3390/ijerph17124206.

25. Ren D, Leslie LM, Fu R, Dickinson RE, Xin X. A storm-triggered landslide monitoring and prediction system: formulation and case study. *Earth Interact*. Oct. 2010;14(12):1–24. https://doi.org/10.1175/2010ei337.1.

26. Dong M, et al. Deformation prediction of unstable slopes based on real-time monitoring and DeepAR model. *Sensors*. 2020;21(1):14. https://doi.org/10.3390/s21010014.

27. Zhao F, Mallorqui J, Iglesias R, Gili J, Corominas J. Landslide monitoring using multi-temporal SAR interferometry with advanced persistent scatterers identification methods and super high-spatial resolution TerraSAR-X images. *Remote Sens*. 2018;10(6):921. https://doi.org/10.3390/rs10060921.

28. Tao Z, Wang Y, Zhu C, Xu H, Li G, He M. Mechanical evolution of constant resistance and large deformation anchor cables and their application in landslide monitoring. *Bull Eng Geol Environ*. 2019;78(7):4787–4803. https://doi.org/10.1007/s10064-018-01446-2.

29. Dou J, Paudel U, Oguchi T, Uchiyama S, Hayakawa YS. Shallow and deep-seated landslide differentiation using support vector machines: a case study of the Chuetsu Area Japan Terrest. *Atmos Ocean Sci*. 2015;227.

30. Agarwal S, Makkar S, Tran D-T. In: *Privacy Vulnerabilities and Data Security Challenges in the IoT*. Taylor & Francis; 2020.

31. *Classification and Regression Trees*; 1984.

32. Breiman L, Last M, Rice J. Random forests: finding quasars. *Statistical Challenges in Astronomy*. Springer-Verlag; 2003:243–254.

33. Cawley GC, Talbot NLC. Efficient approximate leave-one-out cross-validation for kernel logistic regression. *Mach Learn*. 2008;71(2–3):243–264. https://doi.org/10.1007/s10994-008-5055-9.

34. Mining D. *Practical Machine Learning Tools and Techniques*. 3rd ed. Burlington, USA: Morgan Kaufmann; 2011.

35. Doetsch P, Buck C, Golik P, et al. Logistic model trees with AUC split criterion for the KDD cup 2009 small challenge. In: *KDD-CUP'09 Proceedings of the 2009 International Conference on KDD-Cup 2009*. vol. 7; 2009:77–88.

36. Panahi M, Gayen A, Pourghasemi HR, Rezaie F, Lee S. Spatial prediction of landslide susceptibility using hybrid support vector regression (SVR) and the adaptive neuro-fuzzy inference system (ANFIS) with various metaheuristic algorithms. *Sci Total Environ*. 2020;741:139937. https://doi.org/10.1016/j.scitotenv.2020.139937.

37. Abraham MT, Satyam N, Kushal S, Rosi A, Pradhan B, Segoni S. Rainfall threshold estimation and landslide forecasting for Kalimpong India using SIGMA model. *Water*. 2020;12(4):1195. https://doi.org/10.3390/w12041195.

38. Frattini P, Crosta G, Sosio R. Approaches for defining thresholds and return periods for rainfall-triggered shallow landslides. *Hydrol Process*. 2009;23(10):1444–1460. https://doi.org/10.1002/hyp.7269.

39. Bui DT, et al. Spatial prediction of rainfall-induced landslides for the Lao Cai area (Vietnam) using a hybrid intelligent approach of least squares support vector machines inference model and artificial bee colony optimization. *Landslides*. 2016;14(2):447–458. https://doi.org/10.1007/s10346-016-0711-9.

40. Lian C, Zeng Z, Yao W, Tang H. Multiple neural networks switched prediction for landslide displacement. *Eng Geol*. 2015;186:91–99. https://doi.org/10.1016/j.enggeo.2014.11.014.

41. Bui DT, Tuan TA, Klempe H, Pradhan B, Revhaug I. Spatial prediction models for shallow landslide hazards: a comparative assessment of the efficacy of support vector machines artificial neural networks, kernel logistic regression, and logistic model tree. *Landslides*. 2015;13(2):361–378. https://doi.org/10.1007/s10346-015-0557-6.

42. Yilmaz I. Comparison of landslide susceptibility mapping methodologies for Koyulhisar Turkey: conditional probability, logistic regression, artificial neural networks, and support vector machine. *Environ Earth Sci*. 2009;61(4):821–836. https://doi.org/10.1007/s12665-009-0394-9.

43. Neaupane KM, Achet SH. Use of backpropagation neural network for landslide monitoring: a case study in the higher Himalaya. *Eng Geol*. 2004;74(3–4):213–226. https://doi.org/10.1016/j.enggeo.2004.03.010.

44. Kavzoglu T, Mather PM. The use of backpropagating artificial neural networks in land cover classification. *Int J Remote Sens*. 2003;24(23):4907–4938. https://doi.org/10.1080/0143116031000114851.

45. Lippmann R. Book review: neural networks a comprehensive foundation, by Simon Haykin. *Int J Neural Syst*. 1994;05(04):363–364. https://doi.org/10.1142/s0129065794000372.

46. Guzzetti F, Carrara A, Cardinali M, Reichenbach P. Landslide hazard evaluation: a review of current techniques and their application in a multi-scale study, Central Italy. *Geomorphology*. 1999;31(1):181–216.

47. Lv Y, et al. A comparative study of different machine learning algorithms in predicting the content of ilmenite in titanium placer. *Appl Sci*. 2020;10(2):635. https://doi.org/10.3390/app10020635.

48. Shariati M, et al. Application of a hybrid artificial neural network-particle swarm optimization (ANN-PSO) model in behavior prediction of channel shear connectors embedded in normal and high-strength concrete. *Appl Sci.* 2019;9(24):5534. https://doi.org/10.3390/app9245534.

49. Kavzoglu T, Colkesen I. A kernel functions analysis for support vector machines for land cover classification. *Int J Appl Earth Observat Geoinf.* 2009;11(5):352–359. https://doi.org/10.1016/j.jag.2009.06.002.

50. de Mello RF, Ponti MA. Statistical learning theory. In: *Machine learning.* Springer International Publishing; 2018:75–128.

51. Ballabio C, Sterlacchini S. Support vector machines for landslide susceptibility mapping: the Staffora river basin case study Italy. *Math Geosci.* 2012;44(1):47–70. https://doi.org/10.1007/s11004-011-9379-9.

52. Broséus J, Vallat M, Esseiva P. Multi-class differentiation of cannabis seedlings in a forensic context. *Chemom Intel Lab Syst.* 2011;107(2):343–350. https://doi.org/10.1016/j.chemolab.2011.05.004.

53. Pourghasemi HR, Rahmati O. Prediction of the landslide susceptibility: which algorithm which precision? *Catena.* 2018;162:177–192. https://doi.org/10.1016/j.catena.2017.11.022.

54. Marjanović M, Kovačević M, Bajat B, Voženílek V. Landslide susceptibility assessment using SVM machine learning algorithm. *Eng Geol.* 2011;123(3):225–234. https://doi.org/10.1016/j.enggeo.2011.09.006.

55. Guo Q, Kelly M, Graham CH. Support vector machines for predicting distribution of sudden oak death in California. *Ecol Model.* 2005;182(1):75–90. https://doi.org/10.1016/j.ecolmodel.2004.07.012.

56. Breiman L. Population theory for boosting ensembles. *Ann Stat.* 2003;32(1):1–11. https://doi.org/10.1214/aos/1079120126.

57. Prasad AM, Iverson LR, Liaw A. Newer classification and regression tree techniques: bagging and random forests for ecological prediction. *Ecosystems.* 2006;9(2):181–199. https://doi.org/10.1007/s10021-005-0054-1.

58. Cho JH, Kurup PU. Decision tree approach for classification and dimensionality reduction of electronic nose data. *Sens Actuators B.* 2011;160(1):542–548. https://doi.org/10.1016/j.snb.2011.08.027.

59. Brown SD, Myles AJ. Decision tree modeling. In: *Comprehensive Chemometrics.* Elsevier; 2020:625–659.

60. Pradhan B. A comparative study on the predictive ability of the decision tree support vector machine and neuro-fuzzy models in landslide susceptibility mapping using GIS. *Comput Geosci.* 2013;51:350–365. https://doi.org/10.1016/j.cageo.2012.08.023.

61. Myles AJ, Feudale RN, Liu Y, Woody NA, Brown SD. An introduction to decision tree modeling. *J Chemometr.* 2004;18(6):275–285. https://doi.org/10.1002/cem.873.

62. Nefeslioglu HA, Sezer E, Gokceoglu C, Bozkir AS, Duman TY. Assessment of landslide susceptibility by decision trees in the metropolitan area of Istanbul Turkey. *Math Probl Eng.* 2010;2010:1–15. https://doi.org/10.1155/2010/901095.

63. Chien C-F, Chen L-F. Data mining to improve personnel selection and enhance human capital: a case study in high-technology industry. *Expert Syst Appl.* 2008;34(1):280–290. https://doi.org/10.1016/j.eswa.2006.09.003.

64. Microsoft Academic. https://academic.microsoft.com/paper/1540461373/reference?showAllAuthors=1Server.

65. Langit L. Introduction to data mining. In: *Foundations of SQL Server 2005 Business Intelligence.* Apress; 2007:243–276.

66. Al-Batah MS, Alkhasawneh MS, Tay LT, Ngah UK, Lateh HH, Isa NM. Landslide occurrence prediction using trainable cascade forward network and multilayer perceptron. *Math Problem Eng.* 2015;2015. https://doi.org/10.1155/2015/512158, 512158.

67. Roy AC, Islam MM. Predicting the probability of landslide using artificial neural network. In: *2019 5th International Conference on Advances in Electrical Engineering (ICAEE), Dhaka, Bangladesh;* 2019:874–879. https://doi.org/10.1109/ICAEE48663.2019.8975696.

20

Remote sensing and machine learning techniques to monitor fluvial corridor evolution: The Aras River between Iran and Azerbaijan

Khosro Fazelpoor[a], Vanesa Martínez-Fernández[b], Saleh Yousefi[c], and Diego García de Jalón[a]

[a]Department of Systems and Natural Resources, ETSI of Mountains, Forest and the Natural Environment, Polytechnic University of Madrid, Madrid, Spain [b]National Museum of Natural Sciences, CSIC, Madrid, Spain [c]Soil Conservation and Watershed Management Research Department, Chaharmahal and Bakhtiari Agricultural and Natural Resources Research and Education Center, AREEO, Shahrekord, Iran

1 Introduction

One of the most politicized natural resources all around the world is frontier rivers. Dynamism, due to spatial displacement in fluvial zone and change in its character, are important for human life, whether happen by the influence of artificial drivers or natural processes, can conducive political issues, conflict, and struggle between countries in this kind of rivers.[1] Natural events and human impacts are changing channel conditions in riverine borders, which affect morphological adjustments (e.g., channel changes, river dynamism reduction, and simplification of channel planforms)[2, 3] together with vegetation responses such as vegetation encroachment[4] and decrease of pioneer recruitment as well as the increase in cover of late-seral species.[5] Therefore, the evolution of the river geomorphology can be determined as a tool to investigate the impact of the changed drivers of the system.

Anthropogenic, man-made structures and urbanization pressures are driving fluvial corridor evolution limiting channel dynamism, and favoring riparian vegetation encroachment.[6] Residential has flourished along the major rivers of the world[7] as fluvial areas are crucial for human development[8] that can lead the river to changes in channels over short timescales[9] River morphology have been altered by interplaying several inputs (e.g., reservoir dam, sediment extraction from rivers, organic material, boundary condition, and artificialization).[10, 11] Also, significant morphological changes can occur over a longer period of time due to land-use changes in the basin and the river floodplain (e.g., incision, channel narrowing, and channel widening).[12] Specifically, land use activities such as agriculture and urbanization are imposing relevant pressure on riparian vegetation and floodplain geomorphology which its evolution and characteristics depend on river channel profile.[13] It is worth mentioning that the elimination of riparian vegetation for agriculture (human alteration) is the most evident that also can alter channel banks, increase runoff, and introducing fine-grained sediments causing channel aggradation.[11, 14] Additionally, human alterations could show opposite or negative effects and lead the system to changes in the riverine processes that should be considered for artificialization, erosion, and sedimentation control.[11]

Nowadays many types of research have contributed to the interactive role between anthropogenic and fluvial river geomorphology.[3, 6, 9] From a geomorphological point of view, a research gap has been reported on investigating riverine borders that just focused on hydro-political aspects without considering the morphological changes of a frontier river.[15, 16]

In the last decades, analyzing river geomorphology by using modern technologies such as remote sensing has been prevailed.[17] In the recent year, some methodology that has been flourished (e.g., adaptive neuro-fuzzy inference system (ANFIS), artificial neural networks (ANNs) and support vector machine (SVM)).[3, 11, 18] which have shown merits over conventional modeling due to their ability to resolve and handle the noisy data from the process.[18] In this regard, SVM has been considered as an acceptable and accurate tool to classify different categories and is gaining recognition in ecology.[9] The successful monitoring of the fluvial system could greatly contribute to quantifying the affections caused by human pressures. Together with river elements characterization in a GIS environment, machine learning offers the potential for efficient classification of remotely sensed imagery. Accordingly, in this study SVM is used to detect the geomorphic changes and also investigate how river dynamism has affected the land cover in the floodplain. In this context, Landsat satellite images were used to analyze land cover changes and some indexes were offered to assess geomorphological changes.

2 Methodology

2.1 Study area

The study area is located in the Aras basin between Iran and Azerbaijan. The Aras River which is a common boundary between Iran, Turkey, Armenia, and Azerbaijan, has 1072 km and lies within a water catchment of around 100,220 km$^{2.19, 20}$ The annual average temperature in the catchment ranges from 0 to 2°C in the mountains and from 14 to 14.5°C in plains and lowland areas. The annual rainfall varies between 200 and 1600 mm. However, fog, as hidden precipitation, provides an effective additional water supply particularly at elevations between 1000 and 2000 m.

The main species of riparian vegetation in the study area are *Tamarix* (82%), *Salix* (4.5%), and *Populus* (4.4%) (Author communication from the Iranian Ministry of Energy). At the catchment scale, the vegetation composition of this zone is made up of both the Irano-Turanian and Euro-Siberian flora and forms a transitional zone where both elements are intermingled.[21] Three different areas of vegetation have been reported,[22] considering elevation, slope, and impact of artificialization:

I.   The high elevation zone, mainly from 1000 to 1800 m.a.s.l., is forest, and the alpine zone (1800–2700 m.a.s.l.) can be split up into short shrub grasslands and pure grasslands and forbs and grasses.

II. In the mid-elevation zone (600–1250 m.a.s.l.), dense stands of thorny shrubs dominate the landscape and are slowly replaced by later-successional hardwoods.[21–23]

III. The lower elevation zone (265–600 m.a.s.l.) consists mainly of abandoned agricultural lands with secondary vegetation types of mostly Irano-Turanian origin.[23, 24]

The study area is located after Mil-Moghan Dam just close to the Pars-Abad city in the northwest of Iran (Fig. 1). The Mil-Moghan Reservoir has been operated since 1972.

2.2 Analysis of the floodplain buffer zone

The analysis of the floodplain buffer zone has been divided into two phases. First, the geomorphic calculus based on the extraction of variables (e.g., extracting training sample points, active channel area, active channel width, RNCI, and CMI) in a GIS environment by using the Google earth images; and second, the application of machine learning by using Support Vector Machine methods to assess the evolution of the land covers in the river reaches along the considered period (e.g., land cover and transition matrix) (Fig. 2). The combination of both approaches results is extremely useful in fluvial corridor mapping and morphological changes assessment. The approach here exemplified uses a sequence of Landsat and Google Earth images to assess land cover changes in a floodplain buffer by Support Vector Machine algorithms that served as one of the tools for interpretation of the change detection.

2.2.1 Geomorphic analysis

In this study, to assess the morphological and land cover changes in the segment, a database was created by using ArcGIS 10.2. By using the Google earth images, aerial ortophotography from the years 1984, 2000, and 2020 were investigated. It should be noted that the orthophotographs were provided correctly georeferenced. After identifying the floodplain buffer zone within a radius of 2 km from the river centerline, the active channel was digitized and then the centerline and active channel width were determined at 10-m intervals using Fluvial Corridor 10.1.[9] It is noteworthy to say that channel displacement was calculated systematically by drawing cross-sections at 1-km intervals and

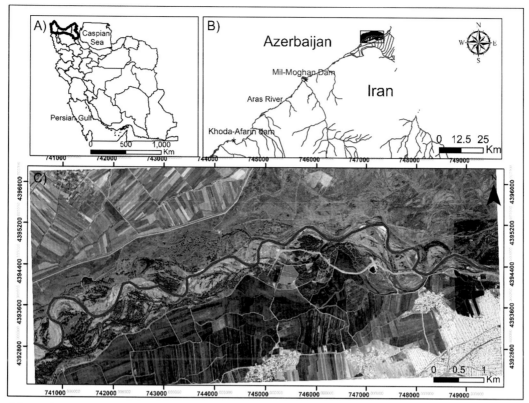

FIG. 1 Map of the study area (A); Iran (B); Aras Basin and (C); with focus on the study segment that is 13.3 km long.

calculating the distance between the intersection point and the centerline in 1984 and the centerline in 2019. Channel movements to the detriment of Iranian territory, namely to the north, take negative values in the presentation of results.

Regarding the interpretation of the geomorphological evolution, indexes such as River Network Change Index (RNCI)[9] and Channel Mobility Indexes (CMI)[25] have been calculated (Table 1).

2.2.2 Support vector machine

In this study, a classification method including SVM was applied to extract land cover categories.

Land cover maps were grouped into five classes: (1) Agriculture, (2) no-farming, (3) residential area; any developed area and other human construction, (4) water; the area that is covered by water, including irrigation pools and rivers, and (5) natural cover; the riparian area is the interface between land and a river or stream, this area includes natural straw, grass, and lawn.[9] Continually, a transition matrix was done in order to assess the floodplain buffer zone

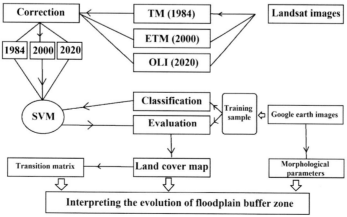

FIG. 2 Flow chart of the used methodology based on remote sensing analysis.

TABLE 1 Geomorphic indexes with the equation.

Index	Equation	Interpretation
RNCI[a]	$\left(\frac{\sum EA - \sum DA}{L}\right)/Y$	If it's negative, the main process is sedimentation, while if it is positive, the main process is river erosion
CMI	$\left(\frac{\text{New Channel} + \text{Abandoned Channel}}{\text{Preserved Channel}}\right)$	A high value means that the channel has experimented with high mobility

[a] *Where EA is erosion area, DA is deposition area, L is the segment length and Y is the number of years between the beginning and end of the period.*

dynamism. The different classes of habitat changes have been quantified.[26] These transitions are as follows: (1) stationary: areas that maintained the same land cover type along the period were considered as "no-change," (2) succession: development changes toward a more mature state or to an aged riparian forest, (3) rejuvenation: when regression changes are directed toward pioneer and early stages of vegetation and bare gravel bars in the channel; and finally, (4) artificialization, when the anthropization process transforms habitats into roads, buildings, and agriculture land.[11]

3 Result

3.1 Morphological evolution of floodplain buffer zone

One of the objectives of this study was to analyze channel evolution in the reach of the case study that forms the border between Iran and Azerbaijan. The result shows that the channel has been narrowed through passing 1984 to the 2020 (−88.9%) which is close to a decrease in width ratio of 16.6 m/year for the entire period (Fig. 3). As it can be seen in Table 2, the active channel area has been notably decreased over the last two decades (−91.5%).

The sedimentation and erosion area was calculated for the study period along the reach. The results show that the main process was sedimentation with an average of −38.9 (Table 3). The outcomes of RNCI demonstrate its maximum value in the first period (−60.1) while it decreased dramatically in the second period (−17.6). It should be said that although the erosion rate in the whole study period was so similar, the sedimentation process decreased from 9.5 km² in 1984 to 3.86 km² in 2020. The Channel Mobility Index show an increase through passing the first period to the second one which means it freely moved over the river corridor.

Channel displacement for the entire period has been shown in Fig. 4. The highest movement toward Iranian territory was −770 m, while the displacement toward Azerbaijan is +470 m. On average, it has been moved toward Azerbaijan (133 m) along the whole study reach.

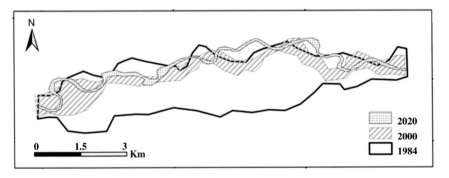

FIG. 3 Spatial change of the active channel in the study reach for 1984, 2000, and 2020.

TABLE 2 Morphological parameters calculated during the study period in the active channel.

Year	Active channel area (km^2)	Active channel width (m)
1984	13.47	673
2000	4.45	302
2020	1.15	75

TABLE 3 Erosion, sedimentation, and preserved channel areas (km^2), River Change Network Index (RNCI, m/year), and Channel Mobility Index (CMI, unitless) for consecutive study periods.

Period	Erosion area (km^2)	Sedimentation area (km^2)	Preserved channel area (km^2)	RCNI (m/year)	CMI
1984–2000	0.48	9.50	4.45	−60.11	2.24
2000–2020	0.57	3.86	0.59	−17.59	7.52

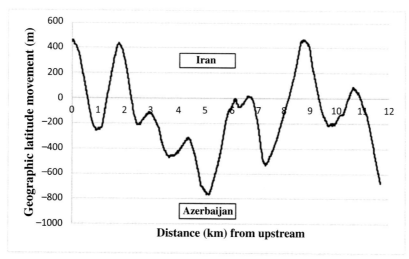

FIG. 4 Aras channel centreline movement along the international border for the period 1984–2020.

3.2 Buffer zone analysis based on SVM

3.2.1 Land cover changes

The SVM has been used to produce the land cover maps (Fig. 6). The result of the evaluation shows that the map of the year 2020 has an overall accuracy of 92.4%, while the least accurate is for 1984, presenting values of 86.0%. The generated land cover map demonstrated that during the study period land cover has been altered dramatically (Figs. 5 and 6). Vegetation has been removed almost by half continuously over the study period to the benefit of residential and agriculture. Indeed, urbanization and farming lands have been incremented significantly. On the contrary, no farming like vegetation and water body observed a rapid reduction (−68.9%).

3.2.2 Geomorphologic evolution based on transition analysis

As it is seen in Fig. 7, through passing 1984 to 2020, only 24% of the surface remained in the same land cover type (13.4 km^2) and 42.4 km^2 of the total area has been changed. Looking specifically to the change category, degradation

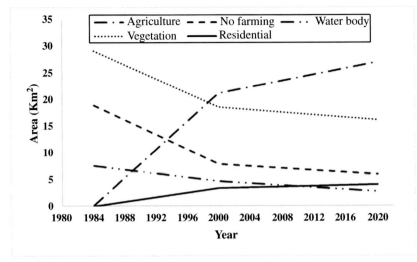

FIG. 5　Evolution of land cover types in the study area along the period (1984–2020).

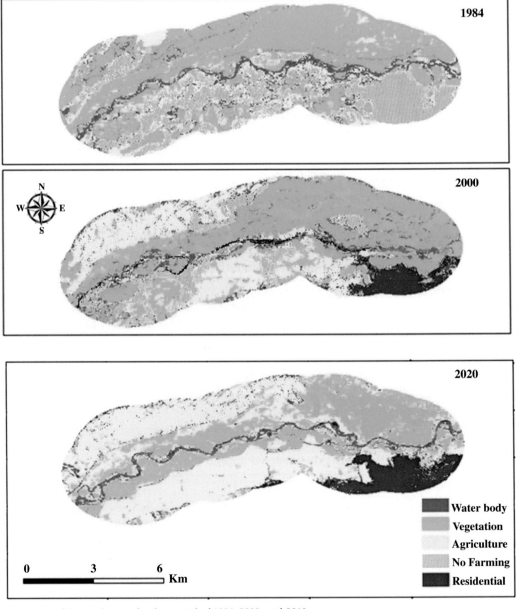

FIG. 6　Land cover maps of the study area for the period of 1984, 2000, and 2019.

Percentage changes of different land cover type

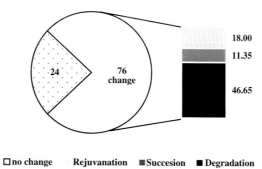

□ no change   Rejuvanation   ■ Succesion   ■ Degradation

FIG. 7   Frequency of land cover types in floodplain buffer zone (1984–2020).

(e.g., agriculture, urbanization, anthropogenetic) was prevailed (46.7%) while dynamism (rejuvenation and succession) was 29.4%.

4 Discussion

In general, for the entire period, the increasing urbanization occupied the fluvial area, particularly, causing channel decreasing due to human pressure and economic activity. The extensive artificialization in the buffer zone has had multiple effects on the study reach. Indeed, as a consequence of human interruption, degradation was higher than dynamism in the whole study period.[9] Channel narrowing is a phenomenon that has been reported by other researchers to be a result of human activity such as urbanization and agricultural uses.[11, 27, 28] With the need for irrigation and economic activity, urban was growing close to the river which caused tension and pressure to the river dynamism.[29] According to previous authors,[3, 13, 30] riparian vegetation has been removed due to urban developing and growth in the population. The population density in the north of Iran is higher than in the rest of the country[31] and as a result of an increasing working population and tourism, artificialization processes increased dramatically in the floodplain buffer zone as they became a significant process. Channel narrowing and vegetation reduction due to human activity have been reported previously in Iran.[3, 9] It should be said human activity in the farming land (i.e., plowing) is conducive realizing much sediment to the buffer zone which can erode the banks. Also, the economic situation can impose people to converting natural riparian vegetation zone to agriculture. In the present study, the result shows a significant change in morphological indexes and according to the subsequent observation of land use maps produced, the study area, specifically, in the floodplain, has been interrupted by human activities. The outcomes of RNCI in the Talar River showed that sedimentation has prevailed and river narrowing was the main process that express spatial variations of the river channel during the last 36 years.[32] During the past decades, the RNCI in the study river changed and the average of this index was about −38.8. The SVM result showed a high accuracy to classify floodplain buffer zone which can be considered having perfect estimation.[17, 18, 28, 33] The fluvial corridor tool was used to analyze channel movement between the country with common boundaries. The investigation demonstrated that the Hirmand River in Iran moved toward a non-Iranian part for 60 years.[34] This result can highlight the outcomes of this study which demonstrate a high movement toward Azerbaijan. This kind of movement can cause conflict and struggle between neighbor countries.

The results show that the study reach, and specifically the floodplain buffer zone, has been strongly affected by human pressure, with farmland activities, urbanization, and damming being the most important types.

5 Conclusion

Dynamics in frontier borders can be considered an essential tool in political relationships between countries. Under the influence of natural processes and external drivers (e.g., land-use change in river catchments), Rivers move and evolve. In this study, the geomorphological indexes were analyzed to understand the river's evolution. Regarding this, SVM was used to analyzed the Landsat images and extracting the land cover map, and continually calculating transition processes in the buffer zone. As it can be expected, channel narrowing has happened during the last 36 years and active channel width has been decreased dramatically. RNCI afforded a negative result which means deposition was the prominent process. Due to human activity (e.g., agriculture, urbanization) riparian vegetation, no-farming, and water body have been reduced through passing 1984 to 2020 which can highlight that fluvial zone are being suffered by artificialization.

It should be described that the most challenging result of this study is channel displacement. As the channel has been moved to Azerbaijan territory (133 m on average), the result must be extended to the country's treaty and political situation between the country with common boundary. Further investigation on the rate of alteration and evaluation in the other rivers in northern Iran should be done to better assess anthropogenic in river geomorphology.

References

1. Engel FL, Rhoads BL. Interaction among mean flow, turbulence, bed morphology, bank failures and channel planform in an evolving compound meander loop. *Geomorphology*. 2012;163–164:70–83. https://doi.org/10.1016/j.geomorph.2011.05.026.
2. Ibisate A, Díaz E, Ollero A, Acin V, Granado D. Channel response to multiple damming in a meandering river, middle and lower Aragón River (Spain). *Hydrobiologia*. 2013;712(1):5–23.
3. Mirzaee S, Yousefi S, Keesstra S, Pourghasemi HR, Cerdà A, Fuller IC. Effects of hydrological events on morphological evolution of a fluvial system. *J Hydrol*. 2018;563:33–42.
4. García de Jalón D, Martínez-Fernández V, Fazelpoor K, González del Tánago M. Vegetation encroachment ratios in regulated and non-regulated Mediterranean rivers (Spain): an exploratory overview. *J Hydro Environ Res*. 2020;30:35–44. https://doi.org/10.1016/j.jher.2019.11.006.
5. Martínez-Fernández V, González E, López-Almansa JC, González SM, García de Jalón D. Dismantling artificial levees and channel revetments promotes channel widening and regeneration of riparian vegetation over long river segments. *Ecol Eng*. 2017;108:132–142.
6. Zhu L, Meng J, Zhu L. Applying Geodetector to disentangle the contributions of natural and anthropogenic factors to NDVI variations in the middle reaches of the Heihe River basin. *Ecol Indic*. 2020;117:106545.
7. Rhoads BL, Lewis QW, Andresen W. Historical changes in channel network extent and channel planform in an intensively managed landscape: natural versus human-induced effects. *Geomorphology*. 2016;252:17–31. https://doi.org/10.1016/j.geomorph.2015.04.021.
8. Overeem I, Kettner AJ, Syvitski JPM. Impacts of humans on river fluxes and morphology. *Treatise Geomorphol*. 2013;9:828–842. https://doi.org/10.1016/B978-0-12-374739-6.00267-0.
9. Yousefi S, Mirzaee S, Keesstra S, et al. Effects of an extreme flood on river morphology (case study: Karoon River, Iran). *Geomorphology*. 2018;304:30–39.
10. Imwangana FM, Dewitte O, Ntombi M, Moeyersons J. Topographic and road control of megagullies in Kinshasa (DR Congo). *Geomorphology*. 2014;217:131–139.
11. Fazelpoor K, Martínez-Fernández V, García de Jalón D. Exploring the hydromorphological response to human pressure in Tagus River (1946–2014) by complementary diagnosis. *Catena*. 2021;105052.
12. Keesstra SD, van Huissteden J, Vandenberghe J, Van Dam O, de Gier J, Pleizier ID. Evolution of the morphology of the river Dragonja (SW Slovenia) due to land-use changes. *Geomorphology*. 2005;69:191–207. https://doi.org/10.1016/j.geomorph.2005.01.004.
13. Esfandiary F, Rahimi M. Analysis of river lateral channel movement using quantitative geomorphometric indicators: Qara-Sou River, Iran. *Environ Earth Sci*. 2019;78(15):469.
14. González del Tánago M, Bejarano MD, García de Jalón D, Schmidt JC. Biogeomorphic responses to flow regulation and fine sediment supply in Mediterranean streams (the Guadalete River, southern Spain). *J Hydrol*. 2015;528:751–762.
15. Güneralp I, Abad JD, Zolezzi G, Hooke J. Advances and challenges in meandering channels research. *Geomorphology*. 2012;163-164:1–9.
16. Nelson NC, Erwin SO, Schmidt JC. Spatial and temporal patterns in channel change on the Snake River downstream from Jackson Lake dam, Wyoming. *Geomorphology*. 2013;200:132–142.
17. Choubin B, Moradi E, Golshan M, Adamowski J, Sajedi-Hosseini F, Mosavi A. An ensemble prediction of flood susceptibility using multivariate discriminant analysis, classification and regression trees, and support vector machines. *Sci Total Environ*. 2019;651:2087–2096.
18. He Z, Wen X, Liu H, Du J. A comparative study of artificial neural network, adaptive neuro fuzzy inference system and support vector machine for forecasting river flow in the semiarid mountain region. *J Hydrol*. 2014;509:379–386.
19. Fateai E, Mosavi S, Imani AA. Identification of anthropogenic influences on water quality of Aras River by multivariate statistical techniques. In: *Proceedings of the 2nd International Conference on Biotechnology and Environment Management IPCBEE (Vol. 42, No. 1)*; 2012.
20. Nasehi F, Hassani AH, Monavvari M, Karbassi AR, Khorasani N. Evaluating the metallic pollution of riverine water and sediments: a case study of Aras River. *Environ Monit Assess*. 2013;185(1):197–203.
21. Ramezani E, Talebi T, Alizadeh K, Shirvany A, Hamzeh'ee B, Behling H. Long-term persistence of steppe vegetation in the highlands of Arasbaran protected area, northwestern Iran, as inferred from a pollen record. *Palynology*. 2020;1–12.
22. Hamzeh'ee B, Safavi SR, Asri Y, Jalili ADEL. Floristic analysis and a preliminary vegetation description of Arasbaran biosphere reserve, NW Iran. *Rostaniha*. 2010;11(1):1–16.
23. Talebi KS, Sajedi T, Pourhashemi M. *Forests of Iran: A Treasure from the Past, a Hope for the Future*. Vol. 10. Springer Science and Business Media; 2013.
24. Jalili A, Hamzeh'ee B, Asri Y, et al. Soil seed banks in the Arasbaran protected area of Iran and their significance for conservation management. *Biol Conserv*. 2003;109(3):425–431.
25. Sanchis-Ibor C, Segura-Beltrán F, Navarro-Gómez A. Channel forms and vegetation adjustment to damming in a Mediterranean gravel-bed river (Serpis River, Spain). *River Res Appl*. 2019;35(1):37–47.
26. Díaz-Redondo M, Egger G, Marchamalo M, Hohensinner S, Dister E. Benchmarking fluvial dynamics for process-based river restoration: the upper Rhine River (1816–2014). *River Res Appl*. 2017;33:03–414.

27. Alayande AC, Ogunwamba JC. The impacts of urbanisation on Kaduna River flooding. *J Am Sci.* 2010;6:28–35.
28. Yousefi S, Pourghasemi HR, Hooke J, Navratil O, Kidova A. Changes in morphometric meander parameters identified on the Karoon River, Iran, using remote sensing data. *Geomorphology.* 2016;271:55–64. https://doi.org/10.1016/j.geomorph.2016.07.034.
29. Das TK, Haldar SK, Das Gupta I, Sen S. River bank erosion induced human displacement and its consequences. *Living Rev Landsc Res.* 2014;8. https://doi.org/10.12942/lrlr-2014-3.
30. Yanan L, Yuliang Q, Yue Z. Dynamic monitoring and driving force analysis on rivers and lakes in Zhuhai City using remote sensing technologies. *Procedia Environ Sci.* 2011;10:2677–2683. https://doi.org/10.1016/j.proenv.2011.09.416.
31. Yousefi S, Moradi HR, Tevari A, Vafakhah M. Monitoring of fluvial systems using RS and GIS (case study: Talar River, Iran). *J Selçuk Univ Nat Appl Sci.* 2015;4:60–72.
32. Yousefi S, Moradi HR, Keesstra S, Pourghasemi HR, Navratil O, Hooke J. Effects of urbanization on river morphology of the Talar River, Mazandarn Province, Iran. *Geocarto Int.* 2019;34(3):276–292.
33. Shu C, Ouarda TBMJ. Regional flood frequency analysis at ungauged sites using the adaptive neuro-fuzzy inference system. *J Hydrol.* 2008;349:31–43.
34. Yousefi S, Keesstra S, Pourghasemi HR, Surian N, Mirzaee S. Interplay between river dynamics and international borders: the Hirmand River between Iran and Afghanistan. *Sci Total Environ.* 2017;586:492–501.

21

Studies on plant selection framework for soil bioengineering application

E. Gayathiri[a], R. Gobinath[b], G.P. Ganapathy[c], Ashwini Arun Salunkhe[b,d], J. Jayanthi[a], M.G. Ragunathan[a], and Hamid Reza Pourghasemi[e]

[a]Guru Nanak College (Autonomous), Chennai, Tamil Nadu, India [b]SR University, Warangal, Telangana, India [c]VIT University, Vellore, Tamil Nadu, India [d]Dr. D. Y. Patil Institute of Technology, Pimpri, Pune, India [e]Department of Natural Resources and Environmental Engineering, College of Agriculture, Shiraz University, Shiraz, Iran

1 Introduction

The most important environmental challenges faced by humans are soil degradation and desertification. Land degradation is one of the biggest challenges faced by us and has been a challenge for the sustainable development of regional communities and the economy. The challenges are complicated on a state, regional, and global scale.[1] Soil loss, erosion, and biodiversity loss are interconnected processes with complex consequences.[2] Land degradation indicates a substantial decline in the land's productive potential. It is a dynamic mechanism entailing several explanatory dimensions that play a dominant role in climate change, land use/land shifts, and human-controlled land use.[3–6] In addition, over the last few decades, several inaccurate methods have been introduced by humans to control and make use of land guided by a single-sided focus on societal gains that have exacerbated a variety of soil depletion problems, such as climate hazards and river erosion.[7, 8] Soil erosion has resulted from the continuous destruction of dryland habitats induced by both climate conditions and human activity, including improper land use/land conservation.[9] Land destruction has had a major impact on the sustainable protection of land and has caused a significant risk to food safety and has had an impact on sustainable social and economic development. Soil loss is a socio-economic problem and is not just an environmental concern affecting humanity.[10] The impact of climate change has been slowly increasing; indeed, those caused by human activities are rapidly increasing and becoming more serious.[11] In the twenty-first century, the study of soil degradation, in particular its mechanism, adaptive development, temporal and spatial spread, as well as defensive measures for soil ecosystem regeneration and reconstruction, has been a high priority in many disciplines, such as ecology and geography.[12] Land degradation studies currently focus largely on the guiding factors of land degradation, land degradation reduction and management steps, and methodologies for measuring land degradation.[13] The land degradation mechanism includes two interconnected dynamic processes, that is, the natural ecosystem and social system of humans, and all shifts in the physical and socio-economic factors can influence the process of soil deterioration.[14] Multiple factors impact soil erosion through recurrent pressures induced by severe and prolonged climate change, where human actions contribute to soil deterioration.

(1) Deforestation, destruction of biodiversity, and suburban development lead to shifts in terms of ground usage and cover.
(2) Inappropriate methods of agricultural land management, such as fertilizer usage and violence, chemical pesticides, and metals.
(3) Rotation of the field incorrectly and devastating drainage techniques, as well as overgrazing.[5]

1.1 Innovation strategies

- Land degradation is causing severe damage to humanity and the economy, and monitoring its occurrence is a need of the hour in many places, and it is necessary to introduce practices that abate land degradation with novelty. This work is a novel attempt to use indigenous plants in a properly formed framework to protect land degradation.
- Fewer practices have been carried out concerning the knowledge of researchers concerned with soil degradation protection using soil bioengineering, and we are sure that; if this work is implemented on a small scale or large scale, will be of immense benefit to society.
- This chapter may suggest a newer approach, such as landslide mitigation through bioengineering tool/methodology, which is being applied/practiced in more than 50 countries worldwide. This practice combines novel plant features in tandem with civil engineering solutions which will support both ecosystem functions and also in engineering prepositions.
- Bioengineering is a cost-effective practice that will establish the area permanently and will ensure that a particular locality is free from landslips and smaller landslides which is not possible via any civil engineering solutions.
- Important arguments that support soil bioengineering practice as an extremely desirable solution for a developing world is the use of indigenous plant resources, along with civil engineering systems as well as the likelihood of including native communities in management and maintenance.
- In this study, we propose a few potential plants and improved native plant communities and exotic species, which have been designed to improve soil workability, which demonstrates positive mechanical and architectural characteristics to fix the soil in place. Additionally, we present an optimization technique that explains how and where to properly plant the plants, based on root system traits, on an unstable site, both of which have not been attempted by any researchers at the regional level.

2 Materials and methods

Finally, the implementation of the Seventeen Sustainable Development Goals (SDGs) by the United Nations in the 2030 framework for Sustainable Growth persuaded, the research groups to produce knowledge for the planning and monitoring of socio-economic progress and the environmental compartments underlying it. Objectives in SDGs 1, 2, 3, 6, 8, 11, 12, 13, 14, and 15 directly consider soil wealth. Five SDG categories and designated SDG measures were in positions where land was one of the key components. It proposes several implementations for sustainable development targets relevant to soil and related metrics that can be tracked in existing tracking schemes. SDGs 2, 3, 6, 11, 13, 14, and 15 apply to goals that require explicit consideration of soil properties. For example, food safety (SDGs 2 and 6), good health and well-being (SDG 3), land-based ocean nutrient contamination (SDG 14), semi-urban and urban growth (SDG 11), and land-based ecosystem service sustainability (SDG 15) rely on the provision of environmental resources where the properties of soil play a key role. Specifically, SDG 15.3 analyses the neutrality of land depletion, desertification by 2030, preserving damaged soil, including land impacted by floods, desertification, and drought, aiming to create an environment that is neutral in terms of land degradation. In addition, SDG 13 relates how soils serve as a major factor in mitigation and adjustment to rising temperatures due to global warming. In addition, SDGs 7 and 12 implicitly check the equilibrium of the existence of land products. Concerning the rest of the SDGs, it is likely to find ties to sustainable soil management to some degree[15] (Figs. 1 and 2).

An overview of the quality of the SDGs and their metrics explains the degree to which soils are incorporated into them. Five SDG classes and allocated indicators are accessible where land plays a prime function, namely:

(1) including efficiency directly (2.3, 2.4),
(2) soil depletion is specifically included (15.3),
(3) while no soil-based marker has indeed been suggested to name soil in the SDG (3.9),
(4) there is an obvious reference to land resources, but there is no connection to soil, which has a clear significance to land resources (11.3), and
(5) by naming the soil in SDG or using the soil-related SDG predictor, the soil has clear relevance to SDG (6.4, 6.5, 13.2, 14.1, 15. 5).

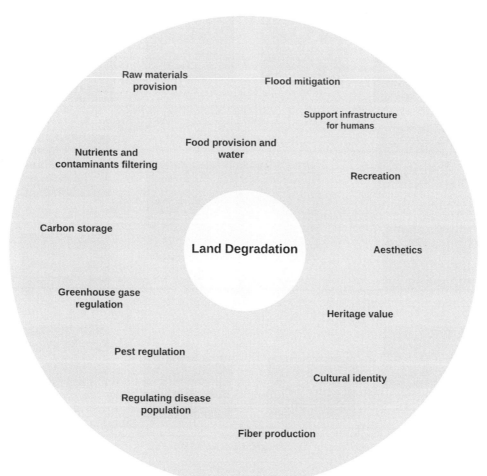

FIG. 1   Relation of different domains within the SDGs: earth, Socioeconomic. One of the four objectives of the biosphere ring is SDG15. *Azote Images for Stockholm Resilience Centre, Stockholm University.*

2.1 Soil bioengineering and its importance

Many methods are available for slope stability enhancement and soil degradation protection such as soil nailing, retaining structure, geosynthetic reinforcement, geotextile reinforcement, shotcrete, Vibro compaction, and one of the recent techniques being used is Soil Bioengineering. Among them, bioengineering is the use of plant species that are locally available to protect the slopes from failure and erosion. Plant species are carefully chosen for their function and suitability for the stabilization of road slopes. It is commonly used in conjunction with civil engineering systems. Bioengineering offers a different range of methods to the developer but does not usually substitute the usage of constructs in civil engineering. Typically, integrating bioengineering strategies provides a more powerful approach to this problem. Soil bioengineering is a modern technique that underlines the significant role of root properties in slope stabilization. It is a science that reduces small-scale landslides by using plant/tree roots and understanding how the root contributes to slope stability. Soil bioengineering is a strategy to use plants to support engineering functions such as catch, armor, reinforcement, drain, and support, and is found to be an effective tool for converting the potential failure surface to a stable and non-eroding surface. This practice is widely adopted worldwide as a potential solution for soil erosion and the prevention of unstable slopes from failure.

2.2 Role of bioengineering systems in slope protection and degradation control

Bioengineering technology functions in the same way as civil engineering systems, which also prevent shallow slips and control slopes from erosion. They were successful at a distance of up to 500 mm below the surface. They do not perform on deep-seated slope faults. Bioengineering involves the use of grasses, shrubs, and some varieties of trees to protect the slopes from the mechanical and hydrological effects of plants. Erosion protection has been proven to

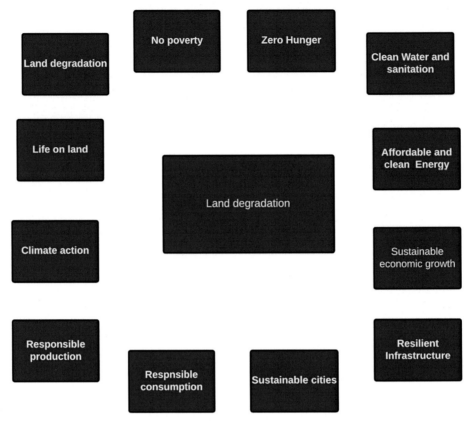

FIG. 2   Showing Land degradation and its attributes related to ecosystem services. *Redrawn as per Keesstra S, Soil-related sustainable development goals: four concepts to make land degradation neutrality and restoration work. Landarzt 2018;7(4):133.*

increase due to the presence of vegetation; this parameter is well recognized by researchers. Vegetation cannot be used alone but in tandem with engineering structures, and many regions have factors that support the use of this technique, such as abundant labor, most of the population working on agriculture-based works, and an enormous variety of indigenous plants appropriate for cultivation in the area. Bioengineering is limited because it cannot impact soil stability below the depth of the rooting depth. This is effectively restricted to 1 m. The results of vegetation on a slope often apply to particular and it is unfair to generalize what will produce false hope. The use of bioengineering methods clearly costs more in the near run than zero. However, the long-term advantages of decreased operating expenses may be extra.

2.3 Bioengineering systems and soil erosion control (Physical land degradation control measure)

Slope soil erosion is a multi-step mechanism that includes landslides, soil detachment, breakup or wear, as well as movement and sedimentation induced by erosional agents such as water, wind, and gravity.[16] Human actions aggravate this mechanism by cutting mature trees, affecting cultivation and overloaded pastures, and altering ecological paths through transportation cuts. This increased erosion has significant environmental, social, and economic implications worldwide, soil fertility reduction, and encouraging sedimentation of soil in stream flow.[17] As a result, erosion management has gained prominence in recent times, with approaches varying from hard engineering to more natural technologies such as bioengineering. Bioengineering is a cutting-edge technique that has recently gained attention and experience in the field of erosion management. This blends living plants with inert elements such as rocks, wood, metal, and geosynthesis, which enhance the earth's environment, prevent and stabilize the process of erosion, increase water infiltration, and lower surface runoff. Land degradation in many countries, covering not only agriculture but also pasture and forestland, is the foremost problem. LDN was established by the United Nations Convention to Combat Desertification (UNCCD) in 1994 as "a system in which the quantity as well as the quality of resources on the land needed to maintain the ecological balance and to enhance stable food security or maintain the spatial and temporal scales of the environment".[18] The world status of LDN is assessed by SDG's predictor 15.3 "Proportion of land depleted over a total land area" and its three sub-indicators (land cover changes, soil fertility, and carbon footprint), both of which are based on the "a one-and-part" theory.[19]

The use of vegetation for engineering purposes differs from theory. On gentle slopes, simple planting patterns on gentle slopes are often sufficient. However, on steep and often intrinsically unstable roadside slopes, experience has shown that only a relatively small number of robust techniques serve the range of engineering functions required. The slopes addressed by the techniques in this manual are extremely long and steep, the disturbance and weakness of the materials of which they are composed, and the intensity of periodic monsoon rainfall. At this stage, it is necessary to consider only the engineering functions required to stabilize them. To date, the principles of vegetation in engineering have been considered, and their contributions are made by individual plants. Plants used in combination can provide significantly more effects than single plants. For example, a single grass plant can catch a small number of debris and reinforce a small volume of soil with its roots. However, grasses can be planted across a slope to form a continuous line to catch debris, thus providing a line rather than a point of reinforcement. However, in the process of serving these functions, the contour line of grass also increases the infiltration extent of the soil's potential. Although bioengineering techniques are simple, the size of most sites necessitates the use of a variety of techniques, much like civil engineering projects typically involve a variety of distinct yet complementary frameworks.

Soil erosion control may be achieved by employing mechanical and biological methods; the former, which will have an immediate effect, will work to the fullest efficiency planned during the design but proved to be costly for construction and operation. To avoid these, biological methods have been introduced, such as utilizing plants grown from seeds and cuttings in soil erosion protection techniques; they are generally cheaper but will not have any immediate effect in the shorter span of implementation. However, once they are established, they provide permanent control and grow on their own, increasing the overall efficiency.[20] Several researchers have found that vegetation controls erosion, and it has been found that the plant stem, canopy, and roots interact and reduce soil detachment and transportation. The canopy of the plant reduces the raindrop size and controls its distribution, which also reduces the fall velocities, which in turn reduces the kinetic energy available in the water to induce soil erosion through impact. Plant stems disturb the flow paths of water over the land and induce roughness to the flow, which in turn reduces the flow velocity and energy available for soil detachment and transportation. The physical nature and characteristics of the soil are altered by plant roots, which promote infiltration and thereby reduce the incidence of overland flow. Plant roots also have the potential to enhance the shearing soil resistance at the surface, which acts against detachment due to the water force in overland flow. In large plants, the roots may go deep and penetrate a considerable depth, thereby increasing the overall strength of the soil, which decreases the mass movement. Plants also help to fix nitrogen in the soil through roots, which may act as a barrier or filter against sediment-laden runoff.[21] The plant canopy effect in reducing soil erosion through the interception of rain droplets has also been studied by several researchers. The anchorage or stabilization amount strengthened by a plant may be correlated with the individual root's tensile strength, which may be based on the plant type, width, maturity level, moisture levels, and chemical composition of the roots.[22–24]

Among the several techniques adopted in slope soil erosion protection, soil bioengineering is an emerging technique that focuses on utilizing plant root traits to enhance slope stability, and the presence of roots in the soil enhances its shearing resistance along the shear distortion plane. The frictional strength of a slope decreases with the presence of water, and further presence of pore water pressure reduces the geomechanical properties, which is mitigated by the plant roots, which drain the water through evapotranspiration. Tension cracks that will be formed due to seepage may augment the failure, but they will be reduced owing to the holding capacity of the root by the soil.[25, 26] However, there is a need to ascertain the plant characteristics that are suitable for land degradation protection, not all plants can offer all the services that are required, but there are plants that can offer the majority of the services required for slope stability and land degradation protection. An engineer or a botanist deciding the plant that can be used needs to analyze multiple parameters focusing on the utility, usage pattern, growth characteristics, and cost of establishing the plants in the locality. In this work, we provide information about the plant selection criteria that should be looked upon by an engineer, planner, or any authority who wishes to decide a suitable plant for ecoengineering purposes in their locality.

2.4 The roots' relation towards slope stability

The root behaves in two different ways: the first is to act as a mechanical reinforcement and the second is to provide efficient hydrological properties. The soil has high strength in compression but less resistance to tension. The presence of cellulose in the root provides resistance to tension and bending stress. The size of the root is inversely proportional to the presence of cellulose in the cell wall. However, older trees with roughly equal strain values at the stem under both tension and compression. Owing to the wind and gravity forces, variations in tension, compression stress, and strain are produced.[27] This prompts the tree to make asymmetrical root growth adjustments to the tree for better anchorage. Many experiments have been conducted to quantify the root's soil shear in stabilizing the slope so that stability relies on the root system of the plant. In soil bioengineering, the selection of suitable plants plays a key role in the local application of this technique. Certain bioengineering practices such as cutting, brush mattresses and live fascines in tandem with geogrids that are geo-gabions and vegetated have been besieged in many rivers' edge constructions.[28, 29] The capacity of root anchorage has been studied both by in situ and ex situ. Through pull-out tests, it is found that soil-root interaction creates more shear strength.[29–34] However, the soil-root system should perform the duty of a reinforcing agent and absorb the tensile stress acting on the soil by increasing the shear strength, which depends on the length of the root, penetration depth, lateral spread, thickness, and root hair development. Several researchers have suggested that the deeper the penetration of the root, the greater the slope strength against failure.[35] Root growth is defined by the type of soil, and roots are confined to cracks in clay or clayey loam texture, but they expand through large pores in the granulated soil structure, and the hydraulic characteristics of the root rely on the type of plan, characteristics of the soil and installation method of plants.[36]

The root system of the oldest stand was the most tolerant in tension to roots of the same diameter as those of the middle and young stands. The disparity in tensile strength between plantations can be clarified by variations in the root structure, as older trees contain higher cellulose levels. Therefore, it is interesting to study the quality of cellulose in the roots of vegetation.[37]

3 Results and discussion

3.1 Plant selection process

To analyze the needs, demand, and usage pattern of plants that are proposed for ecoengineering purposes, one should first understand the nature of the soil, slope pattern (topography), water availability, and hazard zonation in that area to have a clear understanding of what exactly is the present scenario and how new plants can be utilized. Fig. 3A–C shows that the analysis was conducted for one study area in Nilgiris district in Tamil Nadu, India, where this practice is suggested as an outcome of the current research work.

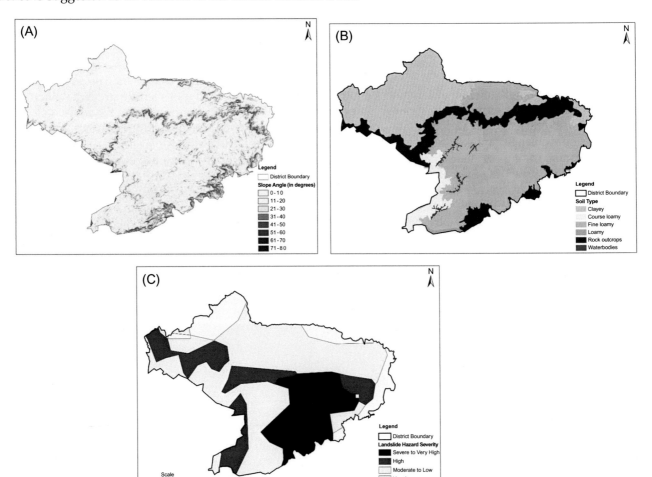

FIG. 3   Map of the study area where the research work is conducted. (A) Slope angle; (B) soil type; and (C) landslide hazard severity.

Fig. 3A shows the slope angle in the study area, which varies drastically in certain regions that are prone to soil erosion where specific varieties of plants that can grow fast should be used where the soil slope is too high. Certain plants whose roots can penetrate to greater depths can also be recommended, but avoiding tall trees in this area is suggested, which if planted will create an overturning moment that will further aggravate the situation.

Fig. 3C shows the hazard mapping of the area, which indicates several areas that fall under highly hazardous concerns with landslides, which can occur as deep-seated or shallow landslides including small-to-medium land-slips. Plants that can cover more areas through leaves and can also produce more added cohesion via clustered roots are prescribed in this zone; however, deep-seated landslides cannot be prevented, which may induce severe land degradation if they occur.

3.2 Functions of plants related to soil degradation protection

Soil erosion is a significant global issue, so the plant may serve several functional tasks, but it is important to choose the best plant for a specific task to ensure the effectiveness of the eco-technological solution. The collection of potential native plants will meet the goal of the plantation program.[38] This leads to too many fatalities, building destruction, and agricultural losses per year. A collapse in land is calculated using a three-dimensional (3D) method in which some of the earth moves downward. Gravity and natural forces, such as precipitation, have an effect.[39] There have been occurring in several countries due to natural disasters, such as heavy rains, drought, temperature increase, and sociological behavior, such as excessive deforestation, lack of protection, major soil degradation, and ecological degradation.[40] In soil protection, plants may have hydraulic and mechanical impacts.[41] The results of soil protection were either optimistic or detrimental. The potential capacity of the vegetation to resist degradation of soil and landslides is partly attributed to the beneficial hydrological effects of the roots.[42] Selecting the most suitable plant species from the population to lead eco-restoration initiatives using root reinforcement is a successful option and remains a great challenge. Furthermore, the effectiveness of regeneration projects relies on the choice of a desired native species.[43] Increasing the successful regeneration rate would not only be economical and ecological but would also improve the soil to the maximum practicable by using the same suitable plants for restoration.

Bioengineering or naturalistic engineering is an area in which living plants, such as log piles and earth frameworks, are used to monitor erosion and soil stabilization steps.[44] Soil bioengineering is a modern technology that can focus on the stabilization of slopes by root traits. It is a science that reduces small-scale landslides by using plant/tree roots and understanding how the root contributes to slope stability. The roots were performed in two distinct forms: first, to serve as a mechanical support and to have appropriate hydrological properties.[45, 46]

The contribution of perennial herbs and shrubs to slope stabilization is often indirect (hydraulically), rather than direct (mechanical). In the postharvest or landslide periods, the occupation of forest and sterile stream bank sites by herbs under natural conditions is a consequence of the loss of tree canopies that provide light to the established pioneer species colonization. This plant life protects the soil from rain splashes and decreases below the canopy of a tree, and roots and rhizomes connect the soil.[47, 48] The idea of utilizing a natural stabilizing mechanism has been popular for improving shear resistance and soil stiffness. Geotechnical engineers often consider the function of roots in reinforcing soil at the surface of the field and the importance of the strength of the root to enhance slope stability. The tensile ability of the roots of native plant species was investigated. Different external influences, such as vegetation type, root diameter, and other environmental factors, have been shown to influence root tensile ability.[49]

Studies have shown that those plant root traits which should be examined when deciding a plant for soil bioengineering are root area ratio, tensile strength, root thickness, shallow rooting, deep rooting, root length density, adventitious root development, forcing, types of root clustering, root resistance to soil stress, root decay rate, sprouting ability, and mycorrhizal engagement. While deciding a plant for inclusion in soil bioengineering practice, the above-mentioned traits will be the defining parameters. Some of the plants that can be used in this process include woody plants such as *B. halimifolia, C. occidentalis* L., *C. amomum, C. drummondii, C. foemina, C. racemose, C. rugose, C. sericea* L., *P. opulifolius, P. balsamifera, P. deltoids*, and *R. palustris*. Moreover, grasses, such as *A. gigantean* and *A. breviligulata, A. gerardi, A. Psathyrostachys junceus, E. trichodes* (Nutt.) Alph. Wood, *F. rubra, H. altissima, L. perenne*, and *P. amarum* Ell. var. amarulum (A.S. Hitchc. & Chase) P.G. Palmer, *P. clandestinum, P. virgatum* L., *P. arundinacea, P. pratensis, S. scoparium, S. nutans, S. pectinata* Bosc ex Link, *Z. miliacea, C. zizanioides, C. citratus* and *Z. miliacea* can also be used. In this work, we propose potential plant selection parameters that are suitable for bioengineering applications and a methodology on how to use them to abate soil degradation and to enhance slope strength in vulnerable slopes.

3.3 Plant root traits

The use of plant traits as screening standards has been increasingly focused on helping soil bioengineers classify appropriate organisms for stabilizing the slopes.[50] Plant characteristics are described in conditions related to plant structure, physiology, or biomechanics as a separate and quantitative attribute of a species.[51] Plant system design allows for the explanation of root and stems morphology, both of which affect earth slope stability. The root traits involved in maintaining soil quality during drought are fine root diameter, long root length, and width capacity, particularly at soil depths where water is available. Root characteristics have no exceptions, as the benefit of designing better-rooting systems has always been neglected.[52] Roots are intense under pressure when the soils are strong in stiffness but poor in pressure; hence, the combined impact of soil and roots strengthens the soil further. The tensile strength of roots mobilizes the tensile stresses and directs them to the root sheath by interfacing tension with the root matrix.[53] The root matrix structure should be able to absorb soil tensile stresses by acting as a root reinforcement, thus increasing the soil shear strength. However, the root matrix system differs in length, penetration depth, lateral distribution, thickness, and root hair growth. The deeper the root penetration depth, the greater the potential sustainability of the soil slope to failure.[35] If the earth has a granular or sandy composition, the root can expand through broad pores. Meanwhile, if the soil is clay or loamy, the root would be limited to cracks.[36]

3.4 Root area ratio

There are many works reported on the technique used to track the distribution of roots in a wide direct shear specimen.[54] The root area ratio (RAR) is the ratio between the shear plane total area and the shear plane crossing the cross-section of the roots. The percentage of the root area corresponds to the proportion of the overall cross-section of the area of the root-occupied area.[55] The root area ratio, which is precisely determined by the cross-section of the region in the shear plane, plays a significant role in the shear force activity of root fibers. Eventually, it is very difficult to calculate the root area ratio perpendicular to the path of root growth because it differs with depth. The parallel plane calculation of the RAR was used in this analysis because it is easy to observe the average contribution of root-fibers in the soil.

3.5 Tensile strength

Storm-induced shallow landslides damage the area throughout the year, which minimizes the root strength in stabilizing steep hill stopes.[33] Shearing forces may cause roots to respond in three different ways: stretching, sliding, and splitting.[56–58] Root tensile strength is a significant soil reinforcement index and its measurement by lab testing will be influenced by root diameter, root species, season, test displacement intensity, root existence or decay, and environmental conditions.[59–61] The function of root rigidity and root-soil adhesion in root reinforcement mobilization was illustrated by.[62, 63] Root tensile strength is generally assumed to depend on diameter, commonly following negative power-law curves through a blend of root populations.[64]

Root fiber reinforcement creates an obvious cohesion that facilitates the stabilization of the slope in shallow soils. Landslide inventories that indicate a rise in landslide occurrence after vegetation removal help to stabilize the reinforcement of roots in the soil. While root cohesion possibly differs with vegetation form or age, the standing density and vegetation composition in dense forests pose significant spatial and temporal variance. The plant species at the landslide site will continue to change, depending on growth conditions, as well as the history of arboreal growth, wildfire, climate change, and pathological conditions. Root cohesion was inversely proportional to the root distribution trends. The adventitious root system has a wide range of root segments, while tap roots have a primary axis with only a few secondary and tertiary roots that emerge from the secondary root system. Fibrous/adventitious roots are generally simpler to develop in soil than in tap roots.[65]

Properly planned slope safety and stabilization must have two components: a vegetational biological component and a mechanical structural component. Until road building, all elements must be integrally designed for full effects. Well-built and planted vegetative covers play a major role in preventing surface erosion and shallow mass collapse. The selected native plant's root should have more affinity towards the soil and the root hairs should have a binding agent/system for individual soil particles or aggregates on shallow soils on steep slopes. To improve slope stabilization, they work in many ways: (1) attaching unstable mantles of soil to secure subsoils or substrates, (2) providing a cover at the top for laterally heavy fine root systems; and (3) providing clustered reinforcement centers near individual trees where implanted stems act on a slope like a pile of buttress or edge.[66]

3.6 Root thickness

Three stabilizing functions are given by the tree roots: (i) soil reinforcement, (ii) leakage current of excess pore stresses, and (iii) maintaining adequate matric suction to improve shear power. The normal soil strengthening induced by an individual root with strong tensile strength improves the peak shear strength of the soil matrix.[42] This reinforcing root impact offers shear resistance before or after intense rainfall on shallow landslides. By affecting the geomechanical composition and hydrological processes of the soil, root thickness increases the shallow slope stability.[67] Accessible pieces of evidences suggest that the structural support offered by the roots has been the subject of most experiments to analyze the potential state of vegetation in the land. In addition, root thickness causes matric suction, which serves as a crucial factor regulating the properties of unsaturated land, which specifically imparts slope stability and resistance to action. The effects of root thickness preserve moisture content and soil settling in the ground.[68]

3.7 Shallow/deep rooting

To be more precise in modeling the mechanical root contribution to the deep rooting habit of plants, slope stability, and their role should be included in the fight against deep-rooted instability.[69] Trees with superficial root systems also lack a taproot, whereas trees with shallow root systems have more deeper and substantial root systems.[70] Deep growing roots helps to absorb nutrient and water, but they also have the potential to affect plant populations via hydraulic raises (HL). Conversely, Schenk asserted that roots expand as shallow as possible or as deeply as required, depending on the available water. The services available in the shallow rhizosphere have a significant impact on underground fauna and microbial organisms, and deep roots may have an impact on soil pedogenesis and carbon storage.[71] Woody plants are more deeply rooted than herbaceous plants.[72] Deep roots facilitate the easy exchange of gases such as O_2 and CO_2 and maintain osmosis in the soil.

3.8 Root length density

It is stated that the soil erosion decreases with increasing root weight and length density.[73] Increased root length, root density, microorganism interaction variability, and percentage coverage improve soil aggregate stability.[74] Arboreal roots in the forest have a $1\,cm/cm_3$ root length density that can establish a very robust and challenging soil composition. Increasing the root length density does not inherently increase the soil permeability.[75] To enhance the consistency of soil aggregates, limited root length density is required, and an increased thickness of the mobile water film correlated with lower root density was observed.[76]

3.9 Specific root length

Root systems have been reported to exhibit considerable variation in growth concerning local soil heterogeneity. Several common criteria are used to define the root structure.[77, 78] The specific root length (SRL, mg − 1) is commonly evaluated as a fine root morphological parameter.[78] The specific root length and root diameter are independent of relative growth rates and the range of the entire plant economy and contribute to environmental change, as well as adaptations to soil fertility at broad spatial scales,[79] and have a significant effect on water and nutrient absorption.[80, 81] The severity of the response to a particular root length depends on the root mechanism inside the fine root structure. Specific root length reactions to all fertilizations varied greatly across groups of root diameters.[78]

3.10 Root angle

Root angle has a substantial effect on the acquisition of mineral nutrients and high levels of phosphate in plants and animals, while low phosphate induces minimal root progress as an adaptive reaction. Therefore, defining genes and root angle regulation mechanisms is of great importance for plant growth.[82]

The root angle and orientation define the complexity of the failure modes. Their cross-sectional area would be greater as the roots cross an area under a certain angle, giving an improved RAR value,[22] which further adds the shear force between the root fibers. The larger lateral root angles lower the amount of force used to lift the roots out of the soil. Between the main lateral and secondary lateral roots, the best branching angle region with the highest anchorage exists up to 20 °C.[83]

The shear test concluded that roots-oriented away from the shear direction broke rather than slipped, while roots-oriented toward and perpendicular to the shear direction slid instead of shattered. As fine roots, roots inclined horizontally slipped rather than split, but the vertical and lateral roots were split instead of slipping, causing a large, completely mobilized resistance to shear stress.[84]

3.11 Production of adventitious roots

The type of root design is determined by a variety of factors: (i) by the degree of root growth, (ii) the rate at which peripheral roots and adventitious roots (ARs) are developed, and (iii) the node at which peripheral roots and adventitious roots extend relative to the gravity matrix. Environmental signs are generated to maintain plant survival and delivery, nutrition, and water supply.[85] Adventitious roots are a special type of root formed from non-root tissues by natural growth and react to stressful situations such as floods and nutritional deficiency, as well as wounding.[86] Shifting adventitious roots plays a vital role in stabilizing changing habitats such as coastal areas, ecological estuaries, and river flood plains..[87] Adventitious roots promote the movement of gas and the processing of water and nutrients through floods. After floods, they help to accumulate nutrients and ensure plant life.[88] The adventitious root plays an important role in hydromineral absorption, together with the lateral roots, by increasing the amount of soil surveyed and utilized and in the anchorage of the plant that is needed for the strong aerial vegetative structure to be created.[89]

3.12 Root clustering

Cluster roots are suited to the acquisition of nutrients from nutrient-poor soils.[90] The change in the sheer power of the soil was proportional to the number of roots in the soil. Root clusters are intangible and are constantly substituted by the primary root axis expansion,[90] the diameter of the roots, distribution of the roots inside the soil, root-soil composite material, the tension between the roots and the soil, the mobilized shear stress resistance between the soil and roots during failure, and the root modules[90, 91]; deepening and wider growth may also occur with roots in the soil to fix the ground from expanding out and reducing the chance of siltation.[35] Tiny, dense, fine roots that closely connect the soil are often typical of species in the Cyperaceae and Restionaceae families.[92, 93] Plants that have clustered roots are normally non-mycorrhizae associated and depend on their special tissues to obtain phosphorus and other scarce nutrients.[90]

3.13 Root response to soil stresses

Because of the stress placed on root cells by soil, root development in the region may be impeded, especially if there is a hard-packed condition, drought, low oxygen levels, or a turbine is present. The operating pressures can differ depending on the root position, prevailing soil water conditions, and degree of compaction.[94] The external tension of the soil can mitigate root elongation if the matrix capacity is too poor, inadequate water supply, when the soil is compacted: for example, water cannot reach the tip of the roots. This causes hypoxia.[95] The limiting water content was specified for each physical stress. Physical stresses of the soil have often been shown to combine to minimize root elongation rather than expected by a mixture of stresses that function separately.[94] The model[96] provide an intensively studied prospect of root length size changes, that occur during soil physical stress

3.14 Root decay rate

Soil water infiltration is linked to root decay, land drainage, soil degradation, groundwater recharge, and farm water storage.[97] Moreover, water movement activity is a key factor in soil water delivery and root decay.[98] Fine roots degrade more rapidly than coarse roots.[99] Tree roots, including both young and old, were found to have live and deteriorating roots, which have a beneficial impact by increasing macroporosity and aggregate stability on soil penetration.[100] Roots can also undergo second-order deterioration owing to other foliage or cambium accidents coinciding with tree demise.[101]

3.15 Mycorrhizal interaction

Fungi promote plant protection and environmental stress tolerance. According to research, introducing native species along with ecto- and endo-mycorrhizal fungi community is an appropriate focus and promotes autogenically degraded ecosystem recovery.[102, 103] Mycorrhizal symbiosis is a relationship between soil fungi and plant roots. Vesicular—arbuscular mycorrhizal fungi (VAM) initiate a mutual and symbiotic union with 80% of the roots of ground crops.[104]

VAM fungi clearly influence the stabilization of the soil matrix via associated plants and foster their growth.[105–107] VAM has a significant impact on the macro-aggregate scale, whereas bacteria and archaea are more specifically involved in micro-aggregate formation and stabilization.[108] Mycorrhizal fungi often influence water conductivity and gas exchange within the roots and foliage. In general, cascade pathways caused by arbuscular mycorrhizal (AM) will improve water transportation and nutrient accessibility in drought soil.

The list of plants that can be used for eco-technological solutions (e.g., soil stabilization, erosion control, soil depletion safety, and the role of the plant) are listed in Table 1. The table includes details on the scientific names, family, common names, type of plant and structural attributes, characteristics of root intensity, and ecological significance.

TABLE 1 List of plants that can be used for eco technological solutions and land degradation protection.

S. no.	Type	Scientific name	Family	Common name	Plant type and structural characteristics	Root strength characteristics	Ecological significance
	Shrub	*Baccharis halimifolia*	Asteraceae	Eastern baccharis	Perennial, usually found in wetlands, salt-tolerant, and often found along salty or brackish shores of marshes and estuaries	An extensive root system can be grown on sand or thin coastal soils to bind the soil	Weed is used in the restoration of coastal barrier inland shores
	Shrub	*Cephalanthus occidentalis* L.	Rubiaceae	Buttonbush	Perennial, growth is moderately fast	Native to stream banks, its fibrous roots hold soil in place	Suitable for stream sides and margins of wetlands, often also developing in standing deeper water. Stabilize to stop the erosion of riparian areas
	Shrub	*Cornus amomum*	Cornaceae	Silky dogwood	Perennial, medium-sized vigorous shrub	Deep rooted soil, even penetrate barren rocks	Streambank stabilization
	Clumping shrub	*Cornus drummondii*	Cornaceae	Rough leaf dogwood	Perennial, grows quite fast and is a very hardy plan	The root system normally consists of a woody branching taproot	Effective in controlling soil erosion
	Woody shrub	*Cornus foemina*	Cornaceae	Stiff dogwood	Perennial, moderate growth	The root system consists of woody branching taproot; vegetative offsets are sometimes produced from underground runners	A promising choice in damp or soggy sites for naturalizing or planting in shrub borders
	Medium shrub	*Cornus racemosa*	Cornaceae	Gray dogwood	Perennial, grows in full sun in wet or dry sites but best in well-drained soil	Well spread shallow fibrous root system	Used on stream banks subject to mechanical disruption, it spreads to form thickets
	Medium shrub	*Cornus rugosa*	Cornaceae	Roundleaf Dogwood	Shrub medium-sized perennial woody plant	Shallow fibrous. The spreading root system readily suckers, creating colonies	Tolerant to weak soils and drought, balance wet soils
	Shrub	*Cornus sericea* L.	Cornaceae	Red Osier Dogwood	Loose, spreading perennial	Moderate fibrous root system	Effective erosion control on banks and slopes

Continued

TABLE 1   List of plants that can be used for eco technological solutions and land degradation protection—cont'd

S. no.	Type	Scientific name	Family	Common name	Plant type and structural characteristics	Root strength characteristics	Ecological significance
	Shrub	*Physocarpus opulifolius*	Rosaceae	Common ninebark	Fast-growing perennial	Fibrous root	On large banks, it can be used for erosion control
	Tree	*Populus balsamifera*	Salicaceae	Balsam poplar	Fast-growing planted as a windbreaker	Shallow-rooting plant, extremely wide, wide-spreading, with a few large roots, penetrating obliquely to the water table or a layer of hardened soil	Prevent soil degradation and abatement of contaminated soils and water
	Tree	*Populus deltoides*	Salicaceae	Eastern cottonwood	Medium sized, short-lived tree	Stabilizing roots, it is tolerant of the stream and river floods	Extensive root structure maintains stream banks in position and is successful in preserving the shoreline and revegetation of degradation in stream channels
	Shrub	*Rosa palustris*	Rosaceae	Swamp Rose	Erect perennial shrub	The root forms a wood taproot with rhizomes which allow the plant to spread vegetatively	Effective erosion control on banks and slopes
	Grass	*Agrostis gigantea*	Poaceae	Redtop	Perennial herb	Fibrous, rhizome roots that can grow to a depth of four feet. Its root system is well suited for holding soils on wetlands, waterways, ditch bank	Suitable for rehabilitating disturbed sites, such as watershed banks and abandoned mines
	Grass	*Ammophila breviligulata*	Poaceae	American beachgrass	Fibrous, Perennial grass arising from long subsurface rhizomes	Dense root structure and ability to grow through accreting sand	Useful as an erosion control plant in non-dune areas where soils are very sandy or inherently dry and the conditions of the site render it very difficult to develop seed species
	Grass	*Andropogon gerardi*	Poaceae	Big bluestem	Perennial warm-season bunchgrass	Dense Roots well established and reach depths of 7–8 ft.	Combat soil erosion and favored shrubs at sites with relatively well-drained to overly well-drained soils
	Grass	*Arundo donax*	Poaceae	Giant reed	Tall, erect, perennial cane. One of the largest herbaceous grasses	Its root structure is very strong with the fibrous roots, penetrating deep into the soil	Planted along ditches for erosion control
	Grass	*Psathyrostachys junceus*	Poaceae	Russian wildrye	Cool season perennial bunchgrass	The roots are fibrous, extended horizontal spread, and may establish to a depth of 6–8 ft.	Excellent restorative species for rangeland

TABLE 1 List of plants that can be used for eco technological solutions and land degradation protection—cont'd

S. no.	Type	Scientific name	Family	Common name	Plant type and structural characteristics	Root strength characteristics	Ecological significance
	Grass	Eragrostis trichodes (Nutt.) Alph. Wood	Poaceae	Sand Lovegrass	Short-lived, perennial, bunch grass	Shallow, widely spreading root system	Provides improved land protection that prevents both wind and stream degradation at these locations
	Grass	Festuca rubra	Poaceae	Red Fescue	Perennial lawn grass	Very deep root system and is very resistant to wear and drought	Used for flood management and road and highway maintenance, and; stream and bank stabilization
	Grass	Hemarthria altissima	Poaceae	Limpograss	warm-season perennial grass	Deep rooting with good soil contact	Have improved capacity to withstand water erosion and improve stabilization of slopes
	Grass	Lolium perenne	Poaceae	Perennial ryegrass	Tall and have tough perennial grass	Fibrous root system, with thick main roots and thinner lateral branches	The quickest growing erosion control grass. These properties render this grass a suitable choice for erosion prevention
	Grass	Panicum amarum Ell. var. amarulum (A.S. Hitchc. & Chase) P.G. Palmer	Poaceae	Coastal panicgrass	Clump-forming, rhizomatous, perennial grass	Deep-rooted robust, long-lived were hardy, and grows to heights of 3–6 ft.	Best erosion mitigation grasses. These characteristics render this grass a strong choice for the prevention of short-term erosion
	Grass	Panicum clandestinum	Poaceae	Deer tongue	Rhizomatous perennial grass	The root system is composed of thickened root stalks	Excellent for flood management and acid mine spoil revegetation and pipelines across wooded areas; a strong part of a stabilization combination of streambanks
	Grass	Panicum virgatum L.	Poaceae	Switchgrass	Hardy, perennial rhizomatous grass	Deep-rooted, root systems are most prominent in surface soil, with roots in the top 30 cm	Excellent erosion protection, such as river spillway channels, or used as filter strips, lawn hedges, or cover
	Grass	Phalaris arundinacea	Poaceae	Reed canary grass	Tall, perennial bunchgrass	Roots are short, stout, scaly, creeping rhizomes. Roots develop from the nodes of the rhizomes to form a thick, fibrous root mass	Used in rivers and along wetlands for flood prevention
	Poaceae	Poa pratensis	Poaceae	Kentucky bluegrass	Herbaceous perennial	Roots develop from the underground nodes of rhizomes and the basal nodes of above-ground shoots, the crown	Because of its thick, vigorous turf formation, it is an excellent plant to control soil erosion

Continued

TABLE 1 List of plants that can be used for eco technological solutions and land degradation protection—cont'd

S. no.	Type	Scientific name	Family	Common name	Plant type and structural characteristics	Root strength characteristics	Ecological significance
	Grasses	*Schizachyrium scoparium*	Poaceae	Little bluestem	Perennial bunchgrass and is prominent in tallgrass prairie	Deep-rooting for anchoring soil. Deep and fibrous, with some roots developing horizontally. clumping grass	Adaptable to multiple stressful factors in the climate, including low productivity and acid soils, freezing temperatures, and drought
	Grass	*Sorghastrum nutans*	Poaceae	Indian grass	Perennial grass warm season bunch grass	Roots are tall, bunching sod-former, with tangled root systems	Drought, Erosion, Dry Soil, Shallow-Rocky Soil
	Grass	*Spartina pectinata Bosc ex Link*	Poaceae	Prairie cordgrass	Tallgrass robust, native grass. Strong rhizomes	Fibrous and strongly rhizomatous	Stiff stem and sturdy rhizomes that enable good shoreline coverage to be provided and lead to dissipation of wave energy. Earthfill dams, spillways, and irrigation channels have proved to be effective in avoiding erosion
	Grass	*Zizaniopsis miliacea*	Poaceae	Giant cutgrass	Large Perennial grass	Spread via its rhizome and above-ground root system	Used near reservoirs and onstream banks for erosion prevention
	Grass	*Chrysopogon zizanioides*	Poaceae	Vetiver grass	Medium-sized perennial grass	Dense fibrous roots penetrate to large depth	Native grass for erosion control and soil improvement
	Grass	*Cymbopogon citratus*	Poaceae	Lemongrass	Medium-sized perennial grass	Dense fibrous roots penetrate to moderate depth	Introduced grass for prevention of erosion and soil enhancement
	Grass	*Zizaniopsis miliacea*	Poaceae	Giant cut grass	Rapidly grow Tall perennial grass	Root spread via rhizome and above-ground root system form stolons	It is sometimes cultivated in wetlands for flood prevention

4 Discussion

In the last 3 decades, steep slopes and heavy rains have triggered numerous landslides worldwide. To address this issue geotechnical engineers are collaborating with landscaping architectures and botanists to build protected vegetated slopes. Researchers are allowed to view the archive and refresh it. However, there is a serious shortage of public awareness in much of the world and little experience, learning tools, and regular training services for soil bio- and eco-engineers are needed. To address this issue, existing ecological and engineering modules should provide ecological engineering solutions for slope stabilization. The biggest challenge for soil bioengineers is to enhance coordination and understanding of the advantages of soft infrastructure solutions and the use of exotic and native species to maintain steep slopes and combat erodibility. Owing to the massive strong erosive forces aided by complex soil systems, installing native/exotic plant species on severely deteriorated slopes is difficult, particularly in semi-arid and drought climates and on weak soils with mineral deficiencies, low organic content, and low water-holding capacity.[38] Hence, improved native plant communities and exotic species (Table 1) have been designed to improve soil workability, which demonstrates a positive approach for physically fixing soil in place. In addition, we suggest a simulation description for plant positioning, based on root system traits, at an unstable site. New techniques (e.g., soil-vegetation interface models, technical systems, better analytical approaches, and guidelines) must be established to increase the implementation of bioengineering strategies by a broader population for use in the construction of bioengineering structures. Similar ideologies have been proposed by many researchers to determine the best vegetation forms or combinations of vegetation and substrate for landslides, taking into account the range of soil slides, geometry, and loading constraints.[17, 109]

4.1 Advantages

Additional savings can arise from the use of native plant materials and seeds that are available in a particular locality. Costs for harvesting, handling, and delivery to the job site were restricted to labor only, which may also be available at a lower rate. Indigenous plant species are commonly accessible and well suited to the local environment and soil conditions. During the dormant season of late fall, winter, and early spring, soil bioengineering projects may be mounted. This is the perfect period to add bioengineered soil work because while other building work is late, it also fits time-wise. For vulnerable or steep sites, where heavy equipment is not possible, soil bioengineering work is also useful. Years of monitoring have also shown that soil bioengineering systems are initially strong and develop stronger as the vegetation becomes formed over time. If plants die, during the resetting of other plants, roots, and surface organic debris continue to play a significant role. If plants are formed, the soil mantel is strengthened by root systems, and excess moisture is extracted by profiling the soil mantle. Finally, I strongly believe that enhanced ecosystem and biodiversity qualities will be created by soil bioengineering techniques.

Civil engineering systems such as retaining walls, lakes, barriers, and conventional design methods for drainage are effective instruments for managing land depletion and degradation, which is a major disadvantage. Such methods have a high carbon footprint, are costly and often hazardous to build, interrupt local and regional environmental systems, require regular upkeep, and ultimately require restoration or replacement. Ecological interventions by soil bioengineering have a lower footprint, promoting a larger variety of environmental resources to facilitate ecological processes. In comparison, environmental techniques, such as intense rainstorms and earthquakes, are more robust to ongoing disruptions.[109] It remains to be elucidated how an ecological solution reacts to biotic and non-biotic disruptions, interacts with physical frameworks, and satisfies the need of an environment (preventing landslides as they improve slope stability and maintain ecological balance), but adapting ecological tools in combination with conventional bio-engineering methods would have to be taken care of.

4.2 Disadvantages

Areas for the implementation of bioengineering techniques require frequent supervision until the plants are established. On highly erosive sites, once the plants have been created, maintenance of the combined system is needed. Developed vegetation may be susceptible to drought, shortages of soil nutrients and sunshine, and side-cast debris for road repair and overgrazing should be considered to ensure the long-term sustainability of the ecosystem. Bioengineering techniques require more than the evaluation and assessment of sites. Design can take into consideration the natural history and evolution of the surrounding landscape, as well as its cultural and social applications. For further progress, an appreciation of these variables and how they form the current and possible future environment is important. Awareness of existing and potential priorities for land management is also important. For example, a potential soil bioengineering project within a forested environment requires an understanding of the geological and glacial background of the area: its tendency for wildland fires, wind storms, and floods; frequency and patterns of natural and management-related erosion; history of methods of road building and present maintenance practices; sequence of removal and re-vegetation of vegetation.

5 Conclusion

The use of suitable technologies for environmental management, protection, and regeneration of the environment to reduce land degradation is an essential part of natural resource management. Optimal management and selection of plant species are key to the success of eco-restoration for soil erosion control. Land degradation is defined as shifting soil, rock, and organic debris under gravity. Soil erosion management, particularly landslide protection, is conventionally regarded as a resource-intensive operation. However, historical vegetation development and nature-based erosion mitigation strategies have developed into a wider sense of bioengineering. The listed species can support professionals and researchers concerned with the preparation and application of soil bioengineering strategies by offering practical knowledge on plant selection and usage in a broad range of circumstances; moreover, the plant species chosen would promote the natural succession phase and serve as a front-line species and would build a win-win situation where bioengineering technologies will protect soil health while contributing adequately to landslide management strategies.

The listed species were found to be highly resistant to every form of soil degradation and erosion caused by mechanical stress. Moreover, the listed plant species adopted the criteria of plant root characteristics to maintain soil properties and promote soil stabilization. It is emphasized that the overall development includes the selection and planting of a proper plant. Both herb and wood plants are desired for lakeshore and stream-bank protection. Perennial herbs or grassland plants at and near the edge of the lake may be required. These plants may grow underwater with their roots. Root development brings tremendous intensity to the soil. Generally, the use of specific wetland plant species increases the probability of effective planting. However, woody plants that are very close to the water or flow depth may not have good structural support and will not thrive. Perennial trees can be used on the upper slopes and upland areas where their roots can rise in the soil above the water level.

References

1. Kosmas C, Kairis O, Karavitis C, Ritsema C, Salvati L, Acikalin SA. Ziogas evaluation and selection of indicators for land degradation and desertification monitoring: methodological approach. *Environ Manag.* 2014;54(5):951–970.
2. Lal R. Climate change and soil degradation mitigation by sustainable management of soils and other natural resources. *Agricult Res.* 2012;1:199–212.
3. Barbier EB. The economic determinants of land degradation in developing countries. *Philos Trans R Soc Lond B Biol Sci.* 1957;352:891–899. https://doi.org/10.1098/rstb.1997.0068.
4. Symeonakis E, Calvo CA, Arnaurosalen E. Land use change and land degradation in south eastern Mediterranean Spain. *Environ Manag.* 2007;40(1):80–94.
5. Sivakumar K, Mannava V, Ndegwa N. *Climate and Land Degradation.* XXVI. Springer-Verlag Berlin Heidelberg; 2007:623. https://doi.org/10.1007/978-3-540-72438-4.
6. Bajocco S, De Angelis PA, Ferrara L, Salvati L. The impact of land use/land cover changes on land degradation dynamics: a Mediterranean case study. *Environ Manag.* 2012;49(5):980–989. https://doi.org/10.1007/s00267-012-9831-8.
7. Liu H. Types and characteristics of land degradation and countermeasures in China. *J Nat Resour.* 1995;4:26–32.
8. Cai Y, Meng J. Ecological reconstruction of degraded land: a social approach. *J Sci Geogr Sin.* 1999;19(3):198–204.
9. Xie S, Qu J, Lai Y, Xu X, Pang Y. Key evidence of the role of desertification in protecting the underlying permafrost in the Qinghai-Tibet Plateau. *Sci Rep.* 2015;5:15152.
10. Zhang T, Wang X. Research progress and trend of soil deterioration. *J Nat Resour.* 2000;15(3):280–284.
11. Wijitkosum S. The impact of land use and spatial changes on desertification risk in degraded areas in Thailand. *Sustain Environ Res.* 2016;26(2):84–92.
12. Liu L, Gong Z. Global land degradation evaluation. *J Nat Resour.* 1995;1:10–15.
13. Zhao X, Dai J, Wang J. GIS-based evaluation and spatial distribution characteristics of land degradation in Bijiang watershed. *Springer Plus.* 2013;2:S8. https://doi.org/10.1186/2193-1801-2-S1-S8.
14. Millennium Ecosystem Assessment. *Ecosystems and Human Well-Being: Synthesis.* Washington, DC: Island Press; 2005.
15. Keesstra SD, Bouma J, Wallinga J, et al. The significance of soils and soil science towards realization of the United Nations sustainable development goals. *Soil.* 2016;2(2):111–128. https://doi.org/10.5194/soil-2-111-2016.
16. Ellison WD. Soil erosion. *Soil Sci Soc Am J.* 1948;12:479–484.
17. Stokes A, Douglas GB, Fourcaud T, et al. Ecological mitigation of hillslope instability: ten key issues facing researchers and practitioners. *Plant and Soil.* 2014;377:1–23.
18. ICCD/COP (12)/4. *Report of the IWG on LDN. Integration of sustainable development goals and targets in the implementation of the UNCCD;* 2015.
19. Akhtar-Schuster M, Stringer LC, Erlewein A, Metternicht G, Minelli U, Sommer SS. Unpacking the concept of land degradation neutrality and addressing its operation through the Rio conventions. *J Environ Manage.* 2017;195(1):4–15.
20. Coppin NJ, Richards IJ. *Use of Vegetation in Civil Engineering.* Butterworths, London: CIRIA; 1990.
21. Greenway DR. Vegetation and slope stability. In: Rich-ards MG, ed. *Slope Stability: Geotechnical Engineering and Geomorphology.* New York: Wiley; 1987:187–230.
22. De Baets S, Poesen J, Reubens B, et al. Root tensile strength and root distribution of typical Mediterranean plant species and their contribution to soil shear strength. *Plant Soil.* 2008;305:207–226. https://doi.org/10.1007/s11104-008-9553-0.
23. Comino E, Marengo P. Root tensile strength of three shrub species: Rosa canina, Cotoneaster dammeri and Juniperus horizontalis soil reinforcement estimation by laboratory test. *Catena.* 2010;82:227–235. https://doi.org/10.1016/j.catena.2010.06.010.
24. Saifuddin M, Osman N. Evaluation of hydro-mechanical properties and root architecture of plants for soil reinforcement. *Curr Sci.* 2014;107(5):845–852.
25. Singh JS, Singh SP. *Forest of Himalaya, Structure, Functioning and impact of Man.* Nainital, India: Gyanodaya Prakashan; 1992.
26. Verma D. Prediction of strength parameters of himalayan rocks: a statistical and ANFIS approach. *Geotech Geol Eng.* 2013;33(5). https://doi.org/10.1007/s10706-015-9899-z.
27. Stokes A. Strain distribution during anchorage failure of Pinus pinaster at different ages and tree growth response to wind-induced root movement. *Plant Soil.* 1999;217:17–27.
28. Schiechtl HM. *Bioengineering for land reclamation and conservation [Buch].* Edmonton: University of Alberta; 1980.

29. Liu Y, Rauch HP, Zhang J, Gao J. Development and soil reinforcement characteristics of five native species planted as cuttings in local area of Beijing. *Ecol Eng.* 2014;71:190–196. https://doi.org/10.1016/j.ecoleng.2014.07.017.

30. Shewbridge NS. Formation of shear zones in reinforced sand. *J Geotech Eng.* 1996;122(11):873. https://doi.org/10.1061/(ASCE)0733-9410.

31. Mickovski SB, Bengough AG, Bransby MF, MCR D, Hallett PD, Sonnenberg R. Material stiffness, branching pattern and soil matric potential affect the pullout resistance of model root systems. *Eur J Soil Sci.* 2007;58:1471–1481. https://doi.org/10.1111/j.1365-2389.2007.00953.x.

32. Dev DJC, Tomar OS, Kumar Y, Bhagwan H, Yadav RK, Tyagi NK. Performance of some under-explored crops under saline irrigation in a semiaridclimate in Northwest India. *Land Degrad.* 2006;17(3):285–299. https://doi.org/10.1002/ldr.712.

33. Tosi M. Root tensile strength relationships and their slope stability implications of three shrub species in Northern Apennines (Italy). *Geomorphology.* 2007;87(4):268–283. https://doi.org/10.1016/j.geomorph.2006.09.019.

34. Burylo M, Rey F, Roumet C, et al. Linking plant morphological traits to uprooting resistance in eroded marly lands (Southern Alps, France). *Plant Soil.* 2009;324:31–42.

35. Noorasyikin MN, Zainab M. Geotechnical engineering analysis on the electrical resistivity model of composite bedrock. *Electr J Geotech Eng.* 2015;20(10):4263–4275.

36. Noorasyikin MN, Zainab M. Assessment of soil-root matrix of vetiver and bermuda grass for mitigation of shallow slope failure. *EJGE.* 2015;20(15).

37. Genet M, Stokes A, Salin F, et al. The influence of cellulose content on tensile strength in tree roots. *Plant Soil.* 2005;278:1–9.

38. Norris JE, Di Iorio A, Stokes A, Nicoll BC, Achim A. Species selection for soil reinforcement and protection. In: Norris JE, et al., eds. *Slope Stability and Erosion Control: Ecotechnological Solutions.* Dordrecht: Springer; 2008. https://doi.org/10.1007/978-1-4020-6676-4_6.

39. Sidle RC, Ochiai H. Landslides: processes, prediction, and land use. *Water Resour Monograph.* 2006;18.

40. Stokes A, Sotir R, Chen W, Chestem M. Soil bio- and eco-engineering in China: past experience and future priorities preface. *Ecol Eng.* 2010;36:247–257.

41. Pollen N, Bankhead SA. Hydrologic and hydraulic effects of riparian root networks on streambank stability: is mechanical root-reinforcement the whole story. *Geomorphology.* 2010;116:353–362.

42. Fan CH, Su CF. Role of roots in the shear strength of root-reinforced soils with high moisture content. *Ecol Eng.* 2008;33:157–166. https://doi.org/10.1016/j.ecoleng.2008.02.013.

43. Withrow-Robinson B, Johnson R. *Selecting Native Plant Materials for Restoration Projects: Ensuring Local Adaptation and Maintaining Genetic Diversity.* Oregon State University Extension Service, Corvallis (Publication EM 8885-E); 2006.

44. Motta E. Earth pressure on reinforced earth walls under general loading. *Soils Found.* 1996;4:113–117.

45. Fatahi B, Khabbaz H, Indraratna B. Bioengineering ground improvement considering root water uptake model. *Ecol Eng.* 2010;36:222–229.

46. Bariteau L, Bouchard D, Gagnon G. A riverbank erosion control method with environmental value. *Ecol Eng.* 2013;58:384–392. https://doi.org/10.1016/j.ecoleng.2013.06.004.

47. Gyssels G, Poesen J, Bochet E, Li Y. Impact of plant roots on the resistance of soils to erosion by water: a review. *Prog Phys Geogr.* 2005;29:189–217.

48. Bochet E, Poesen J, Rubio JL. Runoff and soil loss under individual plants of a semiarid Mediterranean shrub land: influence of plant morphology and rainfall intensity. *Earth Surf Process Landf.* 2006;31:536–549. https://doi.org/10.1002/esp.1351.

49. Capilleri PP, Motta E, Raciti E. Experimental study on native plant root tensile strength for slope stabilization. *Procedia Eng.* 2016;158:116–121. https://doi.org/10.1016/j.proeng.2016.08.415.

50. Stokes A, Atger C, Bengough AG, Fourcaud T, Sidle RC. Desirable plant root traits for protecting natural and engineered slopes against landslides. *Plant Soil.* 2009;324:1–30. https://doi.org/10.1007/s11104-009-0159-y.

51. Pérez HN, Díaz S, Garnier E, Lavorel S, Poorter H. New handbook for standardised measurement of plant functional traits worldwide. *Aust J Bot.* 2013;61:167–234. https://doi.org/10.1071/BT12225.

52. Herder GD, Van Isterdael G, Beeckman T, De Smet I. The roots of a new green revolution. *Trends Plant Sci.* 2010;15:600–607. https://doi.org/10.1016/j.tplants.2010.08.009.

53. Gray DH, Barker D. Root-soil mechanics and interactions. In: Bennett JJ, Simon A, eds. *Riparian Vegetation and Fluvial Geomorphology. Water Science and Application 8.* New York: American Geophysical Union; 2004:113–123.

54. Alsheimer L, Hughes BO. *Ch 4 – Black and White in Photoshop Black and White in Photoshop CS3 and Photoshop Lightroom: Create Stunning Monochromatic Images in Photoshop CS3, Photoshop Lightroom, and Beyond.* Elsevier; 2007:91–139.

55. Gray, Sotir. *Biotechnical and Soil Bioengineering Slope Stabilization: A Practical Guide for Erosion Control.* 1996. New York: John Wiley & Sons; 1996.

56. Ennos RA. The mechanics of anchorage in seedlings of sunflower, Helianthus anus (L.). *New Phytol.* 1989;113:185–192.

57. Abe K, Ziemer RR. Effect of tree roots on a shear zone: modeling reinforced shear stress. *Can J For Res.* 1991;21:1012–1019.

58. Waldron LJ, Dakessian S. Soil reinforcement by roots: calculation of increased soil shear resistance from root properties. *Soil Sci.* 1981;132:427–435.

59. Cofie P, Koolen AJ. Test speed and other factors affecting the measurements of tree root properties used in soil reinforcement models. *Soil Tillage Res.* 2001;63(1–2):51–56.

60. Schmidt KM, Roering JJ, Stock JD, Dietrich WE, Montgomery DR, Schaub T. Root cohesion variability and shallow landslide susceptibility in the Oregon Coast Range. *Can Geotech J.* 2001;38:995–1024.

61. Preti F. Forest protection and protection forest: tree root degradation over hydrological shallow landslides triggering. *Ecol Eng.* 2013;61:633–645.

62. Mickovski S, Hallett P, Bransby M, Davies M, Sonnenberg R, Bengough A. Mechanical reinforcement of soil by Willow roots: impacts of roots properties and root failure mechanism. *Soil Sci Soc Am J.* 2009;73(4):1276–1285. https://doi.org/10.2136/sssaj2008.0172.

63. Schwarz M, Cohen D. Root-soil mechanical interactions during pullout and failure of root bundles. *Case Rep Med.* 2015;115 https://doi.org/10.1029/2009JF001603, F04035.

64. Bischetti GB, Chiaradia EA, Simonato T, et al. Root strength and root area ratio of forest species in lombardy (Northern Italy). *Plant Soil.* 2005;278:11–22. https://doi.org/10.1007/s11104-005-0605-4.

65. Loades KW, Bengough AG, Bransby MF, Hallett PD. Planting density influence on fibrous root reinforcement of soils. *Ecol Eng.* 2010;36:276–284.

66. FAO, Surface and Slope Protective Measures n.d.

67. Cazzuffi D, Cardile G, Gioffrè D. Geosynthetic engineering and vegetation growth in soil reinforcement applications. *Transport Infrastruct Geotech.* 2014;1:262–300. https://doi.org/10.1007/s40515-014-0016-1.

68. Hadi K, Behzad F, Buddhima I. Bioengineering ground improvement considering root water uptake model. *Ecol Eng.* 2010;36(2):222–229.

69. Docker BB, Hubble TCT. Quantifying root-reinforcement of river bank soils by four Australian tree species. *Geomorphology.* 2008;100:401–418. https://doi.org/10.1016/j.geomorph.2008.01.009.

70. Schneemann J. *Rooting Patterns of Tropical trees.* Wageningen, The Netherlands: Department of Silviculture and Forestry Ecology; 1988:23.

71. Maeght J-L, Rewald B, Pierret A. How to study deep roots and why it matters. *Front Plant Sci.* 2013;13. https://doi.org/10.3389/fpls.2013.00299.

72. Schenk HJ, Jackson RB. The global biogeography of roots. *Ecol Monogr.* 2002;72:311–328. https://doi.org/10.1890/0012-9615072[0311:TGBOR]2.0.CO;2.

73. Gyssels G, Poesen J. The importance of plant root characteristics in controlling concentrated flow erosion rates. *Earth Surf Process Landf.* 2003;28:371–384.

74. Rillig MC, Wright SF, Eviner VT. The role of arbuscular mycorrhizal fungi and glomalin in soil aggregation: comparing effects of five plant species. *Plant Soil.* 2002;238:325–333.

75. Vergani C, Graf F. Soil permeability, aggregate stability and root growth: a pot experiment from a soil bioengineering perspective. *Ecohydrology.* 2015;9(5):830–842. https://doi.org/10.1002/eco.1686.

76. Lange B, Lüescher P, Germann PF. Significance of tree roots for preferential infiltration in stagnic soils. *HESS.* 2009;13:1809–1821.

77. Leuschner C, Hertel D, Schmid I, Koch O, Muhs A, Hölscher D. Stand fine root biomass and fine root morphology in old-growth beech forests as a function of precipitation and soil fertility. *Plant Soil.* 2004;258:43–56.

78. Ostonen I, Püttsepp Ü, Biel C, et al. Specific root length as an indicator of environmental change. *Plant Biosyst.* 2007;141(3):426–442. https://doi.org/10.1080/11263500701626069. An International Journal Dealing With All Aspects of Plant Biology.

79. Kramer-Walter K, Kramer-Walter K, Bellingham PJ. Root traits are multidimensional: specific root length is independent from root tissue density and the plant economic spectrum. *J Ecol.* 2016;104(5):1299–1310. https://doi.org/10.1111/1365-2745.12562.

80. Jackson RB, Mooney HA, Schulze ED. A global budget for fine root biomass, surface area and nutrient contents. *Proc Natl Acad Sci U S A.* 1997;94:7362–7366.

81. Comas LH, Becker SR, Cruz VMV, Byrne PF, Dierig DA. Root traits contributing to plant productivity under drought. *Front Plant Sci.* 2013;4:1–16.

82. Huang G, Liang W, Sturrock CJ, et al. Rice actin binding protein RMD controls crown root angle in response to external phosphate. *Nat Commun.* 2018;9:2346.

83. Stokes A, Guitard D. Tree root response to mechanical stress. In: Waisel AA, ed. *Biology of Root Formation and Development.* New York: Plenum Press; 1997:227–236.

84. Ghestem M. *Quelles espèces-outils et quelles architec-tures racinaires pour la stabilisation des points chauds dedégradation en Chine du sud ? Which tool species and which root architectures can best stabilize degradation hot-spots in Southern China?.* PhD thesis, France: AgroParis Tech; 2012.

85. Atkinson JA, Rasmussen A, Traini R, et al. Branching out in roots: uncovering form, function, and regulation. *Plant Physiol.* 2014;166:538–550. https://doi.org/10.1104/pp.114.245423.

86. Steffens B, Rasmussen A. The physiology of adventitious roots. *Plant Physiol.* 2015;170(2):603–617. https://doi.org/10.1104/pp.15.01360.

87. Krauss KW, Allen JA, Cahoon DR. Differential rates of vertical accretion and elevation change among aerial root types in Micronesian mangrove forests. *Estuar Coast Shelf Sci.* 2003;56:251–259.

88. Sauter M. Root responses to flooding. *Curr Opin Plant Biol.* 2013;16:282–286.

89. Coudert Y, Le TVA, Gantet P. Rice: a model plant to decipher the hidden origin of adventitious roots. *Plant Roots Hidden Half.* 2013;4:157–166.

90. Shane MW, Lambers H. Cluster roots: a curiosity in context. *Plant Soil.* 2005;274:101–125. https://doi.org/10.1007/s11104-004-2725-7.

91. Schmidt KM, Roering JJ, Stock JD, Dietrich WF, Montgomery DR, Schaub T. The variability of root cohesion as an influence on shallow landslide susceptibility in the Oregon Coast Range. *Can Geotech J.* 2001;38:995–1024.

92. Dinkelaker B, Hengeler C, Marschner H. Distribution and function of proteoid roots and other root clusters. *Bot Acta.* 1995;108:183–200.

93. McCully ME. Roots in soil: unearthing the complexities of roots and their rhizospheres. *Annu Rev Plant Physiol Plant Mol Biol.* 1999;50:695–718. https://doi.org/10.4236/ajps.2015.61012.

94. Bengough AG, Bransby MF, Hans J, McKenna SJ, Roberts TJ, Valentine TA. Root responses to soil physical conditions; growth dynamics from field to cell. *J Exp Bot.* 2005;57(2):437–447. https://doi.org/10.1093/jxb/erj003.

95. Da Silva AP, Kay BD, Perfect E. Characterization of the least limiting water range of soils. *Soil Sci Soc Am J.* 1994;58:1775–1781. https://doi.org/10.2136/sssaj1994.03615995005800060028x.

96. Chavarria-Krauser A, Schurr U. A cellular growth model for root tips. *J Theor Biol.* 2004;230:21–32. https://doi.org/10.1016/j.jtbi.2004.04.007.

97. Jiang XJ, Liu WJ, Chen CF, et al. Effects of three morphometric features of roots on soil water flow behavior in three sites in China. *Geoderma.* 2018;320:161–171.

98. Jiang XJ, Liu W, Wu J, Wang P, Liu C, Yuan ZQ. Land degradation controlled and mitigated by rubber–based agroforestry systems through optimizing soil physical conditions and water supply mechanisms: a case study in Xishuangbanna, China. *Land Degrad Dev.* 2017;28:2277.

99. Vergani C, Werlen M, Conedera M, Cohen D, Schwarz M. Investigation of root reinforcement decay after a forest fire in a Scots pine (*Pinus sylvestris*) protection forest. *For Ecol Manage.* 2017;400:339–352. https://doi.org/10.1016/j.foreco.2017.06.005.

100. Benegas L, Ilstedt U, Roupsard O, Jones J, Malmer A. Effects of trees on infiltrability and preferential flow in two contrasting agroecosystems in Central America. *Agric Ecosyst Environ.* 2014;183:185–196. https://doi.org/10.1016/j.agee.2013.10.027.

101. Jackson M, Roering JJ. Post fire geomorphic response in steep, forested landscapes: Oregon Coast Range, USA. *Quatern Sci Rev.* 2009;28:131–1146. https://doi.org/10.1016/j.quascirev.2008.05.003.

102. Requena N, Perez-Solis E, Azcon AC, Jeffries P, Barea JM. Management of indigenous plant–microbe symbioses aids restoration of desertified ecosystems. *Appl Environ Microbiol.* 2011;67:495–498.

103. Graf F, Frei M, Böll A. Effects of vegetation on the angleof internal friction of a moraine. *Forest Snow Landsc Res.* 2009;82:61–77.

104. Prasad R, Bhola D, Akdi K, Cruz C, Tuteja N, Varma A. *Introduction to Mycorrhiza: Historical Development Mycorrhiza—Function, Diversity, State of the Art.* Springer International Publishing; 2017:1–7.

105. Budi SW, Van TD, Martinotti G, Ginainazzi S. Isolation from the Sorghum bicolor mycorrhizosphere of a bacterium compatible with arbuscular mycorrhiza development and antagonistic towards soil-borne fungal pathogens. *Appl Environ Microbiol.* 1999;65:5148–5150.

106. Bezzate S, Aymerich S, Chambert R, Czarnes S, Berger O, Heulin T. Disruption of the Paeni bacillus polymyxalevansucare gene impairs its ability to aggregate soil in the wheat rhizosphere. *Environ Micorbiol.* 2000;2:333–342. https://doi.org/10.1046/j.1462-2920.2000.00114. x.

107. Hildebrandt U, Janetta K, Bothe H. Towards growth of arbuscular mycorrhizal fungi independent of a plant host. *Appl Environ Microbiol.* 2002;68:1919–1924.

108. Oades JM. The role of biology in the formation, stabilization and degradation of soil structure. *Geoderma.* 1993;56:377–400.

109. Giupponi L, Borgonovo G, Giorgi A, et al. How to renew soil bioengineering for slope stabilization: some proposals. *Landsc Ecol Eng.* 2019;15:37–50. https://doi.org/10.1007/s11355-018-0359-9.

IoT applications in landslide prediction and abatement—Trends, opportunities, and challenges

U. Sinthuja[a,b], S. Thavamani[b], Sandhya Makkar[c], R. Gobinath[d], and E. Gayathiri[e]

[a]Hindusthan College of Arts and Science, Coimbatore, Tamil Nadu, India [b]Sri Ramakrishna College of Arts and Science, Coimbatore, Tamil Nadu, India [c]Lal Bahadur Shastri Institute of Management, Delhi, India [d]SR University, Warangal, Telangana, India [e]Guru Nanak College, Chennai, Tamil Nadu, India

1 Overview

Landslides are among the most dangerous environments and can cause significant damage to human life and business habits.[1] The drive impact of landslide assortments from the gradual measurement of material to a rapid landslide of a significant volume of debris in the range of millimeters/centimeters per year.[2] There are two primary factors for landslides. First, it is a cause by nature (inevitable) that is away from the control of the human being and order. Second, man-made landslides are generated by rising environmental forces in the form of construction work, illegal mining, strong ground apparatus movements, and sometimes hills-cutting. There are several examples of negative landslide signs in different regions of India. India has experienced harmful landslides in the past few years.

In 2020, states such as Kerala, Karnataka, Tamilnadu, and Manipur have played a vital role in landslides near the TNEB colony, Emerald, Kundah Taluk, which is located in Nilgiris district, Tamil Nadu, after a hex of heavy rainfall of about 34.6 cm (for the period ending 24 h, another one has occurred in Talacauvery on August 6, 2020, around 02:30 am, because of the same heavy rain, near Talacauvery Temple on Brahmagiri hills near Bhagamandala in Kodagu district, Karnataka. The Longmai landslide occurred during the last week of June 2020, along the upslope of the NH-37, the Imphal-Jiribam road approximately 500 m from Noney Bazar towards Imphal, Manipur. In July 2020 in Mao town, which has affected an area of about 0.6 sq.km in the same Manipur. Landslides in Papumpare District, Arunachal Pradesh, two notable landslide occured in a village called Tigdo, which was destroyed on July 10, 2020. Another in Donyi Colony around 8:12 am on June, 25.06.2020. Two landslides occurred in each Dhare and Sitala area, and the other occurred in Kabi Village in North Sikki.[3]

Internet of Things (IoT) or inter-machine communication (M2M) over the World Wide Web is a common platform that enables interaction between machines such as computers and smartphones via the Internet. The number of connected devices is increasing rapidly, and Cisco IBSG anticipates that the number of IoT devices will meet 1.5 trillion by 2020.[4] To have access, analysts predict that the Internet of Things will be composed of 20.4 billion units by 2020.[5]

Internet Phase-I: People to People (P2P)
Internet Phase-II: Device to People (M2P)
Internet Phase-III: Device to Device (M2M)

Inter-machine communication (M2M), which is referred to as IoT, plays a key role in smart city execution such as smart transportation, smart home appliances, and smart parking, military, automobiles, smartphones, and supervisory control and data acquisition (SCADA). Although IoT has footprints in these application areas, it is also possible that landslides are predicted with the most accuracy; hence, different sensors and proficient techniques are used to sense landslides within a specified period.

2 Technology landscape

A novel system that mainly evaluates the prevalence of landslides They introduced an innovative device that includes some devices, such as sensors, to locate landslide movements. The author's ideas are based on a 3-degree set of rules. First, the receptors sense tiny changes in the slippery surface. Shortly after the presence of surface movement was identified, the displacement of the shift sensor was calculated inside the two-dimensional section. Eventually, in addition to the locations of the decided-to-move nodes, the path of the displacements was used to pinpoint the position of the sliding ground. After estimating the sliding floor space, a version of the finite element prefigures, and not if the landslide will take region.[6]

A lightweight earthquake surveillance system[7] used the Wi-Fi sensor community to detect and monitor changes in the landslide frame on a challenging site. This simulation device was developed and constructed to measure the geographical extent of the sensors employed and the video clips of the sensors employed. Through visual interface statistical analysis, the system can generate signals for viable unsafe conditions composed of avalanches or earth's crust disasters. The exploratory result shows the console's early capabilities and proof-of-concept capacities. Based on the literature review, the use of accelerators, rain detectors, temperature sensors, Landsat pictures, sonar, etc. can be identified by avalanches. However, these mechanisms are the simplest to alert the tragedy management government and are no longer unusual.

The predominant issue is the delay in resolving the affected departments and common human beings living in remote areas of landslide warnings during landslides. Transitory data on the alerts consume a significant amount of time manually as statistics are moved from one branch to the alternative, and any loophole will serve as extreme damage. There is a need for an automation process to send an alert message to all concerned at once and at the same time without even any person requiring energy. This computerized slope-failure alert gadget was developed to meet these requirements. The primary aim of every research is to design, extend, or introduce a model of any unique problem for an optimal solution. The main aim of the proposed system is to extend the real-time identification of earthquakes and to expand an internet portal to enable humans to interact with the authorities concerned.[8]

The issue seems to be the time and circumstances that help guide the incidence of operations. G_connect is an investigation of a practical life initial noticing gadget for landslide development that entirely depends on a computer system, detector and cloud storage, and advanced alert framework. G_Connect was performed in Sendang Village near Wonogiri, Indonesia, which is a well-known neighboring disaster. The "G_Connect Landslides Primary Alert Gadget" which is called EWS by authors, this device upgrade is mapped into three subsystems. These modules consist of electric power or power generation, record management structures, and record acquisition.[9]

Landslides remain among the most natural disasters in the Himalayas, encouraging a wonderful dearth of living and valuables. Rainfall is one of the main causes of landslides globally. This paper provides a comparative topographical and hydro-meteorological analysis of the ends of the slope instability and enhancement of IoT and the equipment of study-based, completely wise monitoring structures for the earlier detection of natural disasters. Collected statistics have been used to gain the right output via utilization of returned-to-evolved software systems, which has also been integrated into the IoT-based "social early-warning computer system" to evaluate accumulated extreme weather records in such an effort to measure and forecast landslides in real-time.[10]

Excessive fees for standard landslide surveillance systems enable the discovery of new technologies that may be relatively inexpensive and appear underneath the ground floor. "MEMS-based sensors" have been used in a variety of applications in the automation, healthcare, defense, and verbal communiqué sectors.[11]

However, because the effectiveness of "MEMS-based sensors" has still not been assessed for land damages processes in factual artworks and extra research is required in the development and design of IoT systems which include such sensors utilization. To plan and enhance such workflows, it is essential to first confirm the strengths of these detectors. Laboratory environment-based mockups, after which they can be verified for potential applications. The design and enhancement of the IoT architecture are provided where even the lab grid "MEMS-based sensors" have been utilized.[12]

In a landslide, nearly any position in the world is revealed. Mudslides could not be stopped at times, whether they were shallow or enormous earthquakes. However, an outstanding decline to something like the surrounding environment can actually be introduced by the use of the Landslide Advance warning gadget, which also signifies the surrounding residents and concern supervision preceding the beginning of the landslides. Every Landslide's measurement is usually no matter how much experience is experienced in India, and active alert control is necessary to aid living things from leaving the area under such conditions.[13]

Relevant studies and technological developments in IoT have been utilized in the field of geohazard prevention. It first examines the IoT programs inside the tracking and early warning of seven common geohazard styles, along with landslides, proton trickles, sandstone drops, exterior sedimentation crashes, surface defects, and volcanic eruptions, after which start investigating the interesting invention of geo-danger preventive measures when using IoT and eventually describes the challenging situations in IoT-primarily based tracking and early warning frameworks for geo-risk prevention.[14]

To provide excellent layouts of natural disasters, the predicted length machine is a customizable infrastructure for geographically transferred Wi-Fi and smart sensors. The surveyed proximity is divided into community locations, where each area is observed using a shrewd sensor. Topographic and environment-based records were used to classify the destination, which enabled the installation of the community and the surveillance system. In accordance with the decision being made, a set of rules needs to set the scale and number of zones in the immediate area. The singular system can detect the gliding and progressive movements of the soil, in addition to means of pressure sensors. Every implementation of knowledge transports statistical data to the HSDPA web page. The final records were then examined remotely.[15]

The scheme contains a collection of autonomous detecting gadgets engineered with such a detector match tailored specifically to supervise landslides. The gadgets take perceptual observations at prevalent periods at the same time as functioning at a limited experience switching frequency transfer between them via the Zigbee community to a record database operated by the use of the ELK stack for the time frame and visual analytics. The system was successfully dispatched at a landslide site in Bournemouth, UK, providing the government with a new method of environmentally friendly and far-flung tracking.[16]

3 General view of IoT system for landslide prediction and abatement

By reading the echoes from ultrasonic sound waves that are comparable to the activity of sonar, vibration sensors measure the distance of a target. The HC-SR04 module offers a 2–400 cm non-contact measurement feature, and the range of precision can exceed 3 mm. The modules include transmitters, ultrasonic devices, receiving devices, and control circuits. A switching system triggered by rainfall is a rain sensor or rain switch. Water is sensed by a rain sensor that completes the circuits on the printed leads of the sensor boards.[17] The sensor board functions as a resistor vector that varies from 100 kΩ when wet, when dry to 2 MM. In short, as the board becomes wet, more current is applied. Computers or smartphones can act as nodes to transmit data. IoT hubs can have devices such as Arduino, which are used to make machines to help detect and monitor the physical environment. The alert node is capable of receiving alerts and trying to turn on the alarm when the expected situation has been detected. On the same side, the signals are stored in the IoT server system database to create a log. Currently, Android applications and/or web portals are capable of displaying data to user systems. Fig. 1 depicts the process of IoT sensor-based landslide prediction and abatement systems.

4 Discussion of traditional and trending technologies

By detecting small shifts using few technologies, we can predict when a landslide is imminent, so few of the traditional and latest technologies discussed in this section are as follows.

A. *Machine learning*: supervised learning and unsupervised learning algorithms were used for land slope prediction. Supervised learning involves the creation of a prototype that connects established input data to output data that are unknown. Subsequently, based on the relations between the earlier labeled and trained data, the performance standards for the raw data can be estimated.[18] This methodology has been divided into regression and classification to further differentiate between landslides and non-landslides. Clustering, where samples are clustered based on similarity, is a very popular unsupervised learning algorithm. Dimensionality reduction, which aims to reduce the variance in a dataset and eliminate outliers, is another common process.

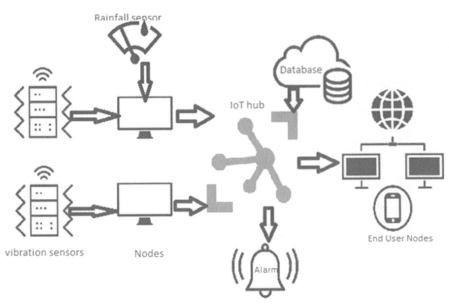

FIG. 1 IoT system for landslide prediction and abatement.

B. *Deep learning:* Deep learning trained with the concept of deep neural networks at multiple levels to progressively extract structures from the input. The broad distinction between deep learning and traditional artificial learning neural networks is the size and sophistication of networks. Popular methods of deep learning can be separated into two types: unsupervised and supervised learning approaches. Dissimilar groups have distinct designs, allowing them to be extremely versatile.

C. *Artificial intelligence (AI):* Methods of AI Multiple Earth Observation datasets were used to generate landslide hazard maps. The global inventory of historical landslides is constructed based on open data sources collected and developed as learning inputs for supervised machine learning algorithms aimed at producing hazardous landslides, including charts.

D. *Wireless sensor networks:* WSNs can have nodes that are wireless, gateway, base radio, and server to relay information from the landslide monitor area to the remote-controlled site via geo-sensors. A wireless system was used for the sensor data downlink and uplink. For contact, the wireless node uses the ZigBee technology.[19] The WSN system contains Wi-Fi-based computers, gateways, radio hubs, server systems, geo-sensors, and solar-based power arrangements. The machine incessantly displays the constraints that can disturb landslides.

E. *Optical fiber sensing technique:* Many fiber optic sensors, called FOS, have been suggested for strain measurement in geotechnical applications over the last decades, including landslide observation. All the sensors projected are single-unit, emulating old-style current devices, such as extensometers and inclinometers.[20]

F. *Persistent scatterer interferometry:* Initially satellite PSI data and morphological data with the current record for landslides. Subsequently, valuing the PSI data velocities, the rate and density of local soil distortions were evaluated during a PSI data post-process phase and the PSI displacements were estimated in the direction of the sheer topography slope in the direction of LOS. The outcome of the same kinetic energy is used as an insight for the third-tier prediction.[21]

G. *Radar technology:* Light Detection of Ranging System (LiDAR) is the most commonly used radar-based technology that is able to take care of landslide zones. In geodesic measurements, the submission of LiDAR procedures is comparatively novel, but it has developed quickly. Two LiDAR technologies were implemented, which were differentiated by those of the area, including its working system. Airborne laser scanning is known as ALS and ground-based global laser scanning (TLS methodologies.[22]

5 IoT device communication technique

IoT device connectivity methods are known as Wired and Wireless, which are also popular in computer networks. The same methods used with IEEE Standardization 802 extension for short-range technologies (WLAN), medium-range technology (WMAN), and long-range technology (WMAN), such as Bluetooth, ultra-wideband, ZigBee, WBAN, WHART, WiMAX, and so on. Once the connection information or message between the devices can be transferred via the famous communication protocol which is called Message Queue Telemetry Transport. MQTT is a widely accepted and emerging protocol for IoT, and Facebook and Messenger use it for the speedy delivery of messages (Fig. 2). The size of the header is 2 bytes in length followed by 256 bytes of message.[23]

Similar to Facebook, the user can subscribe to friends, and they can see a post when their friend posts something on their timeline. The Facebook server manages to send a post from a friend (publisher) to his/her friends (subscribers). This is how the MQTT protocol works, except that MQTT friends are the subjects. Therefore, it can be used in IoT devices, such as publishing and subscribing. Publishers can request data from the device, which is called a subscribe. It will subscribe to the topic first, and then further communication can proceed, including acknowledgments such as SUBACK and PUBACK. In this way, landslides can be monitored and alerted to the user.

5.1 Challenges of using IoT in landslide prediction and abatement

In this section, the reasons for some of the challenges faced by IoT applications already after implementation of the IoT techniques are highly recommended to prevent landslide hazards.

A. *Energy efficiency*: Energy utilization is a key aspect of low-energy networks. Owing to the stable usage of consistent data transmission and power, the life of the network will increase. The detection of unprocessed data from the system was performed by the sensors, and interaction and path computing were carried out by all the relay nodes.[24] Sarwesh et al. achieved this mechanism based on routing. Previously[25] recommended a "power-efficient clustering-based routing algorithm" for IoT aided by a WSN. Already[26] given an EE-IoT report and accomplishes liveliness issues in IoT competently. The planned approach was envisioned to efficiently reduce the maximum utilization of energy with the Internet of Things.

B. *Data interoperability:* This suggests inter-communication transmission between devices without background considerations, such as manufacturing and specifications,[27] proved that non-semantic, semantic foundations of data are interoperable for the IoT environment,[28] suggested a method that enables correlation and uses platform-specific data in another platform,[29] depicted platform-based interoperability for better data communication.

C. *Data administration*: The IoT data management consists of three layers. This model was designed with a data cleaning layer, event processing layer, and data storage and analysis layer.[30] Thirty-one[31] added data management layers for the practical management of useful data mined from the large volume of data produced in numerous Internet of Things applications.

5.2 Opportunities by using more methods in landslide prediction and abatement

This section discusses more methods that are also used in the Internet of Things as addons to detect or predict and save human life from hazards such as landslides.

A. *Generative adversarial network (GAN) unsupervised machine learning method*: Conceptual analysis utilizing deep learning techniques, such as CNN, which can automatically analyze patterns or shapes based on a given set of

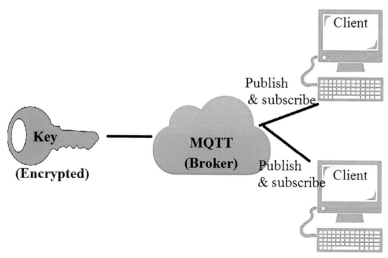

FIG. 2   MQTT communication sample.

inputs. These designs could be regarded as supervised learning techniques for two scenarios: the power station prototype and the discriminator prototype. The intent, including the generator model, captures the distribution of the data. The purpose of the discriminator framework is to evaluate the likelihood that the datasets may simply be used instead of the generation prototype. Incremental antagonistic learning between such designs has been completed repetitively. Popular analyses of earthquake preventive measures (e.g., sensitive earthquake evaluations) mainly gather information from external policies. This immense quantity with remote sensing data could even make the task of recognizing only that information extremely expensive - duration saving and cost-effective. GAN is also an incredible opportunity as it can generate training data from a power source.[32]

B. *Graph neural network (GNN)—deep learning method*: As being one of the risk factors besides earthquakes, spatial relevant documents can sometimes be regarded as chart camshaft information, that can be established as just a chart sensor for both the features within each juncture throughout the target objct. From such a point of view, GNN designs would then give the most viewpoint besides earthquake forecasting. Initially, earthquake activities have been supposed to be objective. In other words, dependency has little or no impact when it comes to preventive measures. To web applications increasing faster throughout density, wide range as well as multiplicity, numerous new types of information channels are much more useful for preventing landslides, including the IoT technology, wireless communications but mostly cloud-based internet services.[33]

C. *Convolutional neural network (CNN, or ConvNet)—deep learning method:* When ConvNets tried to apply to soil erosion, entire landslide inventory levels could be considered a source module in which each point has many landslide-influenced characteristics. Thus, every other phase of the condition factor can become a platform. The findings confirm that ConvNet can efficiently retrieve location data utilizing real connections and help reduce the number of sensor nodes by exchanging loads; individuals can be used to obtain consistent landslide hazard layouts. However, both of these studies were limited. Most research is warranted to validate distinct ConvNet models, in addition to land suitability evaluations. Related to many other deep learning techniques, ConvNet is faced with a classification issue: under-fitting or over-results throughout underperformance with different classifiers. As an optimization technique, gradient-based techniques seem to be justifiable and have been frequently used to tune ConvNet to reduce downtime.[34]

D. *Recurrent neural network (RNN)—deep learning method:* Primarily shown as a supervised learning technique that can be used for handling linear models. Informing this same history, as well as one's choices, is affected by what they have learned and experienced. RNN is composed of clusters, and the process is repeated only after the information is inserted, and the same outcome has been brought back to it. The entire procedure enables the assessment of vibrant changes, where even information is persistent.[35]

6 Conclusion

This work presents a survey of the challenges of applications and technologies of the IoT in landslide/land-slope prevention. The overview section discusses landslides that have recently occurred in 2020 and introduces the Internet of Things as well. Some landslide-oriented research works have been reviewed in detail. The generalized IoT system is depicted to bring better ideas about the role of IoT in landslide detection/abatement. Few traditional and trending technologies, rather than the IoT, have been briefly discussed. Finally, the importance and challenges of using sensors, MQTT protocol-based device communications, and cloud-based methods have been presented. In the future, it may focus on some of the upcoming techniques for landslide prevention.

References

1. Cruden DM, Varnes DJ. Landslide types and processes. In: *Landslides: Investigation and Mitigation*. Transportation Research Board Special Report; 1996:247 [Chapter 3].
2. Pradhan SP, Vishal V, Singh TN. *Landslides: Theory, Practice and Modelling*. Springer; 2019:50.
3. LandSlide Hazard; 2020. https://www.gsi.gov.in/webcenter/portal/OCBIS/pageQuickLinks/pageLandslideIncidents2020.
4. Evans D. Cisco Internet Business Solution Group. *The Internet of Things: How the Next Evolution of the Internet Is Changing Everything*; 2011.
5. Gartner. IoT, Online. Available: http://www.gartner.com/newsroom/id/3598917.
6. Terzis A, Razvan M-E, Cogan J, et al. Wireless sensor networks for soil science. *Int J Sens Netw*. 2010;7(1–2):5370.
7. Chen Y-R, Chen J-W, Hsieh S-C, Ni P-N. The application of remote sensing technology to the interpretation of land use for rainfall-induced landslides based on genetic algorithms and artificial neural networks. *IEEE J Sel Top Appl Earth Observat Remote Sens*. 2009;2:87–95.
8. Kapoor S, Pahuja H, Singh B. *Real Time Monitoring & Alert System for Landslide*. IEEE; 2016. ISBN:978-1-5090-5256-1.

9. Riasetiawan M, Prastowo BN, Putro NAS, Dhewa OA. G-connect: real-time early warning system for landslide data monitoring. In: *6th International Conference on Instrumentation, Control, and Automation (ICA), Bandung, Indonesia*; 2019.

10. Singh VK, Vishal V, Angara KP, Singh TN. Shallow landslides monitoring using the Internet of Things and machine learning technique. In: *Second International Conference on Smart Systems and Inventive Technology (ICSSIT) IEEE Xplore Part Number: CFP19P17-ART*; 2019.

11. Crone WC. A brief introduction to MEMS and NEMS. In: *Springer Handbook of Experimental Solid Mechanics*. Springer, US; 2008:203–228.

12. Manconi A, Giordan D. Landslide failure forecast in near-real-time. *Geomat Nat Haz Risk*. 2016;7(2):639–648.

13. Joshi A, Grover J, Kanungo DP, Panigrahi RK. *Edge Assisted Reliable Landslide Early Warning System*. IEEE; 2019.

14. Mei G, Xu N, Qin J, Wang B, Qi P. A survey of internet of things (IoT) for geo-hazards prevention: applications, technologies, and challenges. *IEEE Internet Things J*. 2020;33:10881–10907.

15. Morello R, De Capua C, Lugarà M. The design of a sensor network based on IoT technology for landslide hazard assessment. In: *4th Imeko TC19 Symposium on Environmental Instrumentation and Measurements Protecting Environment, Climate Changes and Pollution Control June 3–4*; 2013. ISBN: 9788896515204.

16. Butler M, Angelopoulos M, Mahy D. *Efficient IoT-enabled Landslide Monitoring*. IEEE; 2019. 978-1-5386-4980-0/19.

17. Kanchan S, Shaikh JN, Shakeel H, Nachankar M, Suryawanshi SK. *Landslide Detection*; 2007. https://doi.org/10.17148/IJARCCE.2017.6945. ISO 3297.

18. Bzdok D, Krzywinski M, Altman N. Machine learning: supervised methods. *Nat Methods*. 2018;15:5–6. https://doi.org/10.1038/nmeth.4551.

19. Shukla SK, Chaulya SK, Mandal R, et al. Real-time monitoring system for landslide prediction using wireless sensor networks. *Int J Mod Commun Technol Res*. 2014;2:12.

20. Schenato L, Palmieri L, Camporese M, et al. Distributed optical fibre sensing for early detection of shallow landslides triggering. *Sci Rep*. 2017;7. https://doi.org/10.1038/s41598-017-12610-1.

21. Aslan G, Foumelis M, Raucoules D, De Michele M, Bernardie S, Cakir Z. Landslide mapping and monitoring using persistent scatterer interferometry (PSI) technique in the French Alps. *Remote Sens*. 2020;12:1305. https://doi.org/10.3390/rs12081305.

22. Ozdogan MV, Deliormanli AH. Landslide detection and characterization using terrestrial 3d laser scanning (LIDAR). *Acta Geodyn Geomater*. 2019;16:4(196):379–392. https://doi.org/10.13168/AGG.2019.0032.

23. Naik N. *Choice of Effective Messaging Protocols for IoT Systems*. MQTT, CoAP, AMQP and HTTP, IEEE Xplore; 2017.

24. Sarwesh P, Shet NSV, Chandrasekaran K. Energy efficient and reliable network design to improve lifetime of low power IoT networks. In: *International Conference on Wireless Communications, Signal Processing and Networking (WiSPNET), Chennai*; 2017:117–122.

25. Wang Z, Qin X, Liu B. An energy-efficient clustering routing algorithm for WSN-assisted IoT. In: *IEEE Wireless Communications and Networking Conference (WCNC), Barcelona*; 2018:1–6.

26. Suresh K, RajasekharaBabu M, Patan R. EEIoT: energy efficient mechanism to leverage the internet of things (IoT). In: *International Conference on Emerging Technological Trends (ICETT), Kollam*; 2016:1–4.

27. Kibria MG, Ali S, Jarwar MA, Chong I. A framework to support data interoperability in web objects based IoT environments. In: *International Conference on Information and Communication Technology Convergence (ICTC), Jeju*; 2017:29–31.

28. Kiourtis A, Mavrogiorgou A, Kyriazis D, Maglogiannis I, Themistocleous M. Towards data interoperability: turning domain specific knowledge to agnostic across the data lifecycle. In: *30th International Conference on Advanced Information Networking and Applications Workshops (WAINA), Crans-Montana*; 2016:109–114.

29. Schneider M, Hippchen B, Abeck S, Jacoby M, Herzog R. Enabling IoT platform interoperability using a systematic development approach by example. In: *Global Internet of Things Summit (GIoTS), Bilbao*; 2018:1–6.

30. Ma M, Wang P, Chu C. Data management for internet of things: challenges, approaches and opportunities. In: *IEEE International Conference on Green Computing and Communications and IEEE Internet of Things and IEEE Cyber, Physical and Social Computing, Beijing*; 2013:1144–1151.

31. Cerbulescu CC, Cerbulescu CM. Large data management in IOT applications. In: *2016 17th International Carpathian Control Conference (ICCC), Tatranska Lomnica*; 2016:111–115.

32. Tenenbaum J, Kemp C, Griffiths T, Goodman N. How to grow a mind: statistics, structure, and abstraction. *Science*. 2011;331:1279–1285. https://doi.org/10.1126/science.1192788.

33. Mei G, Xu N, Qin J, Wang B, Qi P. A survey of internet of things (IoT) for geo-hazards prevention: applications, technologies, and challenges. *IEEE IoT J*. 2019;1–16. https://doi.org/10.1109/JIOT.2019.2952593.2019.

34. Ma Z, Mei G, Piccialli F. Machine learning for landslides prevention: a survey. *Neural Comput Applic*. 2020. https://doi.org/10.1007/s00521-020-05529-8.

35. Vamathevan J, Clark D, Czodrowski P, et al. Applications of machine learning in drug discovery and development. *Nat Rev Drug Discov*. 2019;18:1. https://doi.org/10.1038/s41573-019-0024-5.

23

Application of WEPP model for runoff and sediment yield simulation from ungauged watershed in *Shivalik* foot-hills

Abrar Yousuf, Anil Bhardwaj, Sukhdeep Singh, and Vishnu Prasad

Department of Soil and Water Engineering, Punjab Agricultural University, Ludhiana, India

1 Introduction

Soil degradation is a severe environmental problem, affecting approximately 1.9 billion ha of land worldwide, and almost 24 billion tons of soil is irrevocably washed or carried away every year.[1] Practically, the loss of 80 percent of the landscape impacted by water erosion is light to moderate. Africa, after Asia, is the continent with the most severe soil erosion among the major continents.[2] Soil erosion occurs in all parts of the world, but the plight of agricultural stakeholders to replace missing soils and nutrients has wreaked havoc in developing countries.[3] In India, 147 million ha (m ha) of land has been degraded out of 329 m ha.[4] The *Shivalik* foot-hills in northwest India are portrayed as one of the eight major fragile ecosystems of India and are highly susceptible to accelerated soil erosion.[5] This region produces considerable bulk of sediments as approximately 35% 45% of rainfall goes as runoff during the monsoon season, resulting in flash floods and downstream sedimentation. In some watersheds, soil erosion is as high as 244 Mg/ha/year.[6] The monsoon downpour entirely influences the *Shivalik* foot-hills; thus, forecasting and estimation of runoff and sediment yield from a particular watershed is a fundamental precursor for the conservation of aquatic resources and boosting agricultural productivity.[7]

Soil erosion is a brunt for food safety in developing countries, which are further perplexed by harsh vagaries of climate and poor socio-economic and political stability.[8] Therefore, it is important to conserve the soil to sustain life on earth and ensure global food security. Soil erosion is a multifaceted process because it is influenced by various parameters, such as vegetation cover, management practices, soil, topography, and climate of a particular place. To implement the best management practices to reduce erosion hazards, it is important to comprehend the erosion process about topographic conditions, soil characteristics, and land use of the watersheds. Therefore, quantifying runoff and sediment yield from disturbed watersheds is necessary to achieve this objective. Because the direct computation of soil loss on each piece of land is unrealistic or impractical because of high expenditure, data procurement complexities, extensive requisite of land area, field staff, and requirements of automated apparatus, there is an urgent need to develop different and effective methods to estimate runoff and erosion from watersheds. The use of numerical hydrological models to determine runoff and sediment yield in a watershed for planning and analyzing alternative land use and best management practices is a useful technique for understanding and quantifying the impact of complex management practices in a diverse environment.[9] To simulate the complete interactions of hydrologic processes and determine sediment and runoff from watersheds, many computer-based simulation models have been developed and implemented on different watersheds.

Several hydrological models are available to approximate runoff and sediment yields from watersheds. Some of these models are "RUSLE, EPIC, ANSWERS, CORINE, ICONA, MIKE SHE, CASC 2D, AGNPS, CREAMS, SWAT and WEPP," etc.[10–12] These models are usually site-specific and hence need to be properly calibrated before applying them to a watershed in a particular region or problematic area. The calibration of any model is cumbersome because it requires considerable time and technical knowledge. Several factors must be considered while calibrating the model, and a large amount of input data is also required. Calibration is a tedious and complex process that often hinders the application of hydrological models. Hence, a study was conducted to ascertain whether the WEPP model must be calibrated every time it is applied to a new watershed in the region.

2 Materials and methods

2.1 Study area

Two watersheds (WS-I and WS-II) having areas of 1.85 and 21.3 ha located at latitude 30°44′N and longitude 76°51′E, and latitude 30°45N and longitude 76°45E, respectively, were selected as the study watersheds. The study region has a subhumid climate.[13] The maximum temperature recorded in May ranges between 41 and 43°C, whereas the minimum temperature recorded in January falls in the range of 6–8°C.[14] The average annual rainfall of both watersheds is 1100 mm, of which 80% is received during the monsoon season. Rainfall is highly erratic, and high-intensity rainfall is very common in this region. Both watersheds were gauged for monitoring of runoff and soil loss by the Indian Institute of Soil and Water Conservation (IISWC), Research Centre, Chandigarh.

2.2 Topography

The average slope of the WS-I and WS-II was 46.7% and 36.1%, respectively. In hydrologic studies, watershed geomorphology is of immense significance. The geomorphic characteristics of the study watersheds illustrate a swift diminution of storm runoff, leading to a high peak rate of runoff (Table 1).

2.3 Soil characteristics

The soil texture of the watersheds (WS-I and WS-II) varied from sandy-to-sandy loam with pH ranging from 6.3 to 8.5. Organic carbon varied from low to medium in both watersheds. The hydraulic conductivity of the soils of the study watersheds varied from 1.4 to 13 mm/h.[13]

TABLE 1   Geomorphometric characteristics of the study watersheds.

Characteristics	WS-I	WS-II
Watershed area (ha)	1.85	21.3
Watershed length (m)	290	560
Relief (m)	73.4	81
Circulatory ratio	0.96	1.31
Form factor	0.22	0.68
Compactness coefficient	1.55	0.76
Elongation ratio	0.53	0.92
Drainage density (m/ha)	449	126
Time of concentration (min)	2.83	4.7
Length of main channel (m)	305	500
Total length of drains (m)	832.5	2580
Infiltration rate (cm/h)	0.14–1.3	0.11–7.44
Average slope (%)	46.7	36.1
Land use	Forest	Forest, scrub, range grass

Source: Yadav RP, Aggarwal RK. Bhattacharyya P, Bansal RC. Infiltration characteristics of different aspects and topographical locations of hilly watershed in Shivalik-lower Himalayan region in India. Indian J Soil Conserv. 2005;33:44–48.

2.4 Land use and management practices

In both watersheds, the landscape is inhabited by shrubs (mostly *Lantana camara*) and trees, such as *Acacia catechu* and *Dalbergia sissoo*. The percentage of grasses increased and that of *Lantana camara* and trees decreased with elevation. Grasses were very low in the lower half of the slope. At the top, the grasses mostly comprise *Eulaliopsis binata* (Bhabbar) and *Chrysopogon* sp. (Dholu grass).

2.5 Data collection

The study watersheds (WS-I and WS-II) were equipped with an automatic water-stage recorder to continuously record the stage hydrograph for each rainfall event. During each storm, runoff samples were collected and analyzed to determine the sediment mass per unit volume of water. Weather data including daily rainfall, maximum and minimum temperatures, relative humidity, wind speed, and wind direction were obtained from the agro-meteorological observatory located within WS-II. The storm-wise runoff and sediment yield data for the period 1982–2004 for WS-I and 2001–4 for WS-II were collected from IISWC, Research Centre, Chandigarh.

2.6 Preparation of input files for WEPP model

The requisite input files of the WEPP model are climate, soil, slope, and land use management files for the simulation of runoff and sediment yield. The watershed is divided into hillslopes and channels. All input files were prepared for each hillslope and channel.

2.7 Watershed hillslopes

The primary step in applying the WEPP model is to discretize the watershed into hillslopes and channels. The WS-I was divided into 29 hillslopes and 18 channels, and WS-II was divided into 35 hillslopes and 25 channels, as shown in Figs. 1 and 2. The soil, slope, land use, and management file were prepared for each hillslope. Similarly, channel characteristics, including slope, soil type, length, and width, were assigned to each channel segment. All input data files in this model have input file builders, except for the climate input file, which has a program called CLIGEN.

2.8 Calibration and validation of a model

The WEPP watershed model was calibrated on WS-I using data from 22 storms and validated on WS-II for the data of 13 storms. The model was calibrated using the trial-and-error method.[15] The model was calibrated using soil parameters such as effective hydraulic conductivity, rill erodibility, interrill erodibility, and critical shear of WS-I.[16–18] The model was validated on WS-II using the calibrated values of WS-I. The overall methodology adopted in this study is shown in Fig. 3.

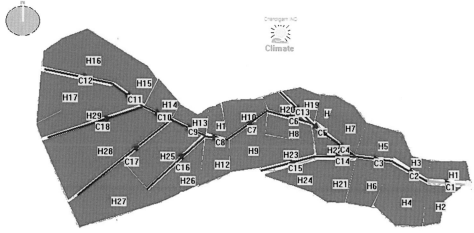

FIG. 1 Hillslopes and channels of WS-I.

FIG. 2 Hillslopes and channels of WS-II.

2.9 Performance evaluation of model

To calibrate and validate the model, statistical parameters such as mean, standard deviation, percent error, coefficient of correlation, root mean square error, and Nash-Sutcliffe model efficiency[19] were calculated and compared with the measured and predicted daily runoff and sediment yield values.

3 Results and discussion

3.1 Model calibration

The WEPP model was calibrated to simulate runoff and sediment yield from WS-I with an area of 1.85 ha using hydrological data from 22 storms. During the calibration process, the WEPP model was found to be sensitive to the soil characteristics, including interrill erodibility, rill erodibility, critical shear, and hydraulic conductivity. The runoff was found to be sensitive to soil hydraulic conductivity, while sediment yield was sensitive to interrill erodibility, rill erodibility, critical shear, and hydraulic conductivity. The sediment yield increased with an increase in rill and interrill erodibility, whereas a negative correlation was observed between sediment yield and critical shear and soil hydraulic conductivity. As soil hydraulic conductivity increases, the infiltration rate increases, and the runoff amount decreases, which might have reduced the number of sediments reaching the watershed outlet, as reported by various researchers.[20,21]

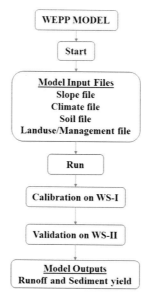

FIG. 3 Flowchart for the overall methodology of the WEPP model application.

The storm-wise observed and predicted values of runoff and sediment yield for WS-I were plotted for the calibration period, as shown in Figs. 4 and 5, respectively. The observed and predicted values are in close approximation for most rainfall events. The model simulated smaller events better than the larger events. The higher values of runoff and sediment yield are under-predicted by the model because, during high rainfall events, splash erosion results in the clogging of soil pores, which reduces the infiltration, resulting in higher surface runoff. This phenomenon of surface sealing is not well considered in the WEPP model.[7] Nearing[22] stated that the under-prediction of larger events is due to the limitation of the WEPP model in representing the random component of the observed data. The under-prediction of larger events by the WEPP model in forest watersheds has also been reported by several studies that have been attributed to the underestimation of subsurface lateral flows.[23,24] To increase the simulation accuracy of the WEPP model in forest watersheds, subsurface flow parameters need to be calibrated more accurately.[7] The total observed and predicted runoff during the calibration period was 119.09 and 118.85 mm respectively with the corresponding values of sediment yield as 14.91 and 13.08 Mg/ha.

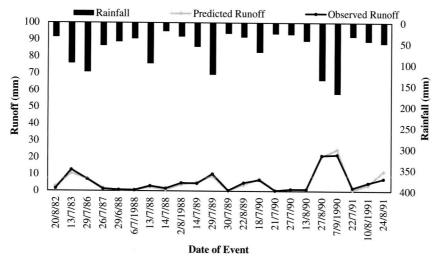

FIG. 4 Observed and predicted runoff during model calibration for WS-I.

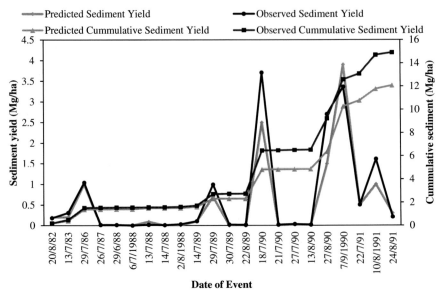

FIG. 5   Observed and predicted sediment yield during model calibration for WS-I.

TABLE 2   Statistics for runoff and sediment yield simulation during calibration for WS-I.

Parameter	Runoff (mm)		Sediment yield (Mg/ha)	
	Observed	Simulated	Observed	Simulated
Mean	5.40	5.42	0.67	0.59
Std. Dev.	6.02	6.45	1.11	1.01
Maximum	21.65	24.75	3.71	4.20
Total	119.09	118.85	14.91	13.08
RMSE	1.40		0.41	
Percent error (%)	7.01		5.01	
Correlation coefficient	0.97		0.93	
Model efficiency (E_{NS})	94.0%		86.31%	

The summary statistics of the observed and predicted storm-wise runoff and sediment yield during the calibration period for WS-I are quite close to each other, as shown in Table 2. The RMSE of 1.40 mm, a correlation coefficient of 0.97, percent error of 7.01, and model efficiency (94.0%) for the calibration period indicate reasonably accurate simulation of surface runoff by the model. Similarly, the RMSE of 0.41 Mg/ha, a correlation coefficient of 0.93, percent error of 5.01, and model efficiency of 86.31% indicate reasonably accurate simulation of sediment yield by the model.

3.2 Model validation

The WEPP model calibrated on a smaller watershed (WS-I) has been validated on a comparatively larger watershed (WS-II) with an area of 21.3 ha to ascertain whether the model must be calibrated every time it will be applied to a new watershed in the region. The storm-wise observed and predicted values of runoff and sediment yield are shown in Figs. 6 and 7. The scatter plot for the storm-wise observed and predicted runoff illustrates that the model slightly over-predicted the runoff values for WS-II (Fig. 8). The scatter plot between the observed and predicted values of runoff and sediment yield shows that the model accuracy in the simulation of comparatively larger events is lower than that of smaller events (Fig. 8). The simulation error for the largest event on March, 31.07.2003, was 40%.

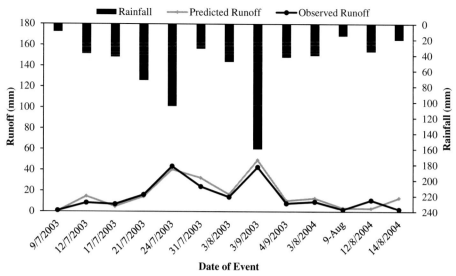

FIG. 6   Observed and predicted storm wise runoff for WS-II.

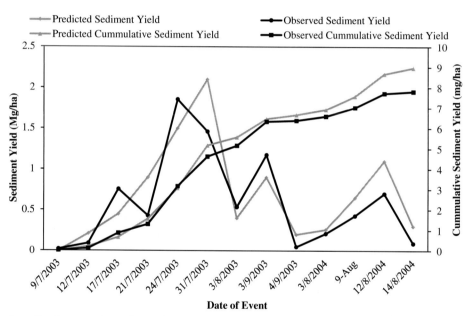

FIG. 7   Observed and predicted storm wise and cumulative sediment yield for WS-II.

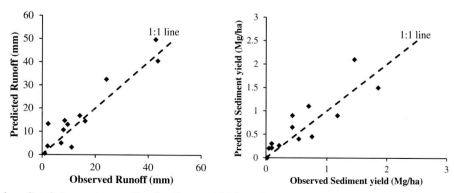

FIG. 8   Observed and predicted storm wise runoff and sediment yield for WS-II.

TABLE 3 Statistics for runoff and sediment yield simulation for WS-II by calibrated model through WS-I.

	Runoff (mm)		Sediment yield (Mg/ha)	
Parameter	Observed	Simulated	Observed	Simulated
Mean	15.07	16.75	0.63	0.69
Std. Dev.	14.04	14.48	0.569	0.58
Maximum	43.57	49.65	1.86	1.50
Total	180.93	217.7	7.59	8.97
RMSE	5.43		0.31	
Storm wise error	Maximum = 10.97 Minimum = 0.49 Average = 2.09		Maximum = 0.64 Minimum = 0.02 Average = 0.09	
Correlation coefficient	0.93		0.86	
Model efficiency (E_{NS})	83.90%		69.50%	

The results of the statistical analysis of runoff and sediment yield for WS-II are presented in Table 3. The RMSE value of 5.43 mm and 0.31 Mg/ha, correlation coefficient of 0.93 and 0.86, and model efficiency of 83.90 % and 69.50% indicate reasonably accurate simulation of runoff and sediment yield, respectively. The satisfactory results of the WEPP model during the validation period show that the WEPP model if once calibrated on a representative watershed can be successfully applied to uncalibrated/ungauged watersheds with similar characteristics in *Shivalik* foot-hills.

4 Conclusions

In the present study, the WEPP model was calibrated and validated for simulating runoff and sediment yield. The model was calibrated on watershed WS-I and validated on watershed WS-II to ascertain whether the model calibrated on one watershed can be applied to another similar watershed uncalibrated. The model performance was measured based on statistical parameters, including RMSE, percent error, correlation coefficient, and model efficiency. The lower values of RMSE and percent error and higher values of correlation coefficient and model efficiency during the validation period indicate a reasonably accurate simulation of runoff and sediment yield. The model performance was found satisfactorily when the WS-I calibrated model was applied to WS-II without calibration. Hence, it can be concluded that the WEPP model if once calibrated on a representative watershed can be successfully applied uncalibrated on similar ungauged watersheds in *Shivalik* foot-hills.

References

1. Young R, Orsino S. *Soil Degradation: A Major Threat to Humanity*. Bristol, UK: Sustainable Food Trust; 2015.
2. Oldeman LR. *Global Extent of Soil Degradation*. ISRIC Biennial Annual Report, Wagningen, The Netherlands; 1992:19–36.
3. Mohamed HH. *Cause and Effect of Soil Erosion in Boqol-Jire Hargeisa, Somaliland*. Ph.D. Thesis, Somalia: University of Hargeisa; 2015.
4. Bhattacharyya B, Ghosh BN, Mishra PN, et al. Soil degradation in India: challenges and potential solutions. *Sustainability*. 2015;7:3528–3570. https://doi.org/10.3390/su7043528.
5. Sidhu GS, Walia CS, Sachdev CB, et al. Soil resource of NW Shivaliks for prospective land use planning. In: Mittal SP, Aggarwal RK, Samra JS, eds. *Fifty Years of Research on Sustainable Resource Management in Shivaliks*. India: Central Soil & Water Conservation Research and Training Institute, Research Centre Chandigarh; 2000:3–34.
6. Bhardwaj A, Kaushal MP. Two-dimensional physically based finite element runoff model for small agricultural watershed. I. Model development. *Hydrol Process*. 2011;23:397–407.
7. Singh RK, Panda RK, Satapathy KK, Ngachan SV. Simulation of runoff and sediment yield from a hilly watershed in the eastern Himalaya, India using the WEPP model. *J Hydrol*. 2011;405:261–276.
8. Blanco-Canqui H, Lal R. Soil erosion and food security. In: Blanco-Canqui H, Lal R, eds. *Principles of Soil Conservation and Management*. Dordrecht: Springer; 2010:493–512.
9. Xevi E, Christiaens K, Espinao A, et al. Calibration, validation and sensitivity analysis of the MIKE-SHE model using the Neuenkirchen catchment as a case study. *Water Resour Manage*. 1997;11:219–242.

10. Yousuf A, Bhardwaj A, Tiwari AK, Bhatt VK. Simulation of runoff and sediment yield from a forest micro-watershed in *Shivalik* foot-hills using WEPP model. *Indian J Soil Conserv.* 2017;45:21–27.

11. Yousuf A, Singh MJ. Runoff and soil loss estimation using hydrological models, remote sensing and GIS in Shivalik foothills: a review. *J Soil Water Conserv.* 2016;15:205–210.

12. Yousuf A, Bhardwaj A, Prasad V. Simulating impact of conservation interventions on runoff and sediment yield in a degraded watershed using WEPP model. *ECOPERSIA.* 2021; [Accepted; ID: 47895].

13. Yadav RP, Aggarwal RK, Bhattacharyya P, Bansal RC. Infiltration characteristics of different aspects and topographical locations of hilly watershed in *Shivalik*-lower Himalayan region in India. *Indian J Soil Conserv.* 2005;33:44–48.

14. Yousuf A, Bhardwaj A, Tiwari AK, Bhatt VK. Modeling runoff and sediment yield from a small forest micro-watershed in *Shivalik* foot-hills using WEPP model. *Int J Agric Sci Res.* 2015;5:67–78.

15. Sorooshian S, Gupta VK. Model calibration. In: Singh VP, ed. *Computer Models of Watershed Hydrology.* Highlands Ranch, CO, USA: Water Resour Pub; 1995:23–68.

16. Nearing MA, Deer-Ascough L, Laflen JM. Sensitivity analysis of the WEPP hillslope profile erosion model. *Trans ASAE.* 1990;33:839–849.

17. Bhuyan SJ, Kalita PK, Janssenc KA, Barnesa PL. Soil loss predictions with three erosion simulation models. *Environ Model Softw.* 2002;17:135–144.

18. Pandey A, Chowdary VM, Mal BC, Billib M. Runoff and sediment yield modelling from a small agricultural watershed in India using the WEPP model. *J Hydrol.* 2008;348:305–319.

19. Nash JE, Sutcliffe JV. River flow forecasting through conceptual models. Part 1. A discussion of principles. *J Hydrol.* 1970;10:282–290.

20. Fu B, Wang Y, Xu P, Yan K. Assessment of the performance of WEPP in purple soil area with simulated rainfall experiments. *J Mount Sci.* 2012;9:570–579.

21. Han F, Ren L, Zhang X, Li Z. The WEPP model application in a small watershed in the Loess Plateau. *PLoS One.* 2016;11, e0148445.

22. Nearing MA. Why soil erosion models over-predict small soil losses and under-predict large soil losses. *Catena.* 1998;32:15–22.

23. Dun S, Wu JQ, Elliot WJ, et al. Adapting the Water Erosion Prediction Project (WEPP) model for forest applications. *J Hydrol.* 2009;336:45–54.

24. Saghafian B, Meghdadi AR, Sima S. Application of the WEPP model to determine sources of run-off and sediment in a forested watershed. *Hydrol Process.* 2015;29:481–497.

24

Parameter estimation of a new four-parameter Muskingum flood routing model

Majid Niazkar[a] and Mohammad Zakwan[b,c]

[a]Department of Civil and Environmental Engineering, Shiraz University, Shiraz, Iran [b]Civil Engineering Department, IIT Roorkee, Roorkee, India [c]Civil Engineering Department, MANUU, Hyderabad, India

1 Introduction

One of the natural hazards is floods that threaten hydraulic structures and critical infrastructures and, in some cases, human lives. Vulnerability assessment and risk analysis against this natural-based threat require prediction methods, such as the flood routing approach. Flood routing is a technique used to determine the time and magnitude of a flood at a section on a watercourse from either known or assumed hydrographs at one or more upstream sections. Flood routing is an integral part of flood protection and management.[1-3] In this regard, flood routing techniques can be broadly divided into hydraulic (distributed) and hydrologic (lumped) routing approaches.[1,4] In the former, the flow passing through the channel reach is commonly computed as a function of time and space, while the latter assumes that the flow is dependent on time alone. As a result, the hydrologic routing approach is applicable even in the absence of cross-sectional and roughness data of a river reach. On the other hand, not only does the hydraulic routing method require hydraulic geometries and reach-average bed roughness of the channel under consideration, but also the variation of the grain roughness coefficient can be implemented in a one-dimensional hydraulic routing model.[5] Depending on the amount of data available, hydraulic flood routing models may require more input data and consequently, may be more complicated than hydrologic models. Hence, there is inevitably a trade-off between accuracy and complexity in selecting an appropriate flood routing model.

Owing to its simplicity, the Muskingum flood routing method is one of the most popular approaches for channel routing.[6,7] In essence, it bifurcates storage into two parts: prism storage and wedge storage. The former is proportional to outflow, while the latter is proportional to the difference between inflow and outflow.[1,4]

The Application of the Muskingum model is a two-step procedure that includes calibration (optimization step) and simulation.[8,9] Over the years, researchers have used various optimization techniques to estimate the parameters of the Muskingum model.[1,10] Gill[11] suggested the Least-squares method for estimating the parameters of the nonlinear Muskingum model. Tung[12] utilized hybrids of the Hooke-Jeeves algorithm and found that Hooke-Jeeves and Davidon-Fletcher-Powell to be most reliable. Mohan[13] proposed the use of a genetic algorithm to estimate the parameters of the nonlinear Muskingum model, while Kim[14] applied Harmony search to estimate its parameters. Das[8] transformed the Muskingum parameter estimation problem into an unconstrained problem by applying a Lagrange multiplier. Karahan[15] compared the performances of the linear and nonlinear forms of the Muskingum model. Similarly, Luo and Xie[16] applied immune clonal selection, while Xu[10] applied differential evolution to estimate the parameters of the Muskingum equation. Hirpurkar and Ghare[17] compared three different forms of the nonlinear Muskingum model; however, in their analysis, they considered only two data sets. To improve the performance of nonlinear Muskingum models, several modifications have also been proposed.[18] A four-parameter version of the Muskingum model was proposed by Easa[19] and a five-parameter version was proposed by Vatankhah.[18] However, most of these models result in dimensional nonhomogeneity.[20] Moreover, increasing the number of parameters in the model enhances the complexity of the model.[20] Akbari[21] utilized a hybrid particle swarm optimization-genetic algorithm to improve the estimates of a four-parameter Muskingum flood routing equation. Gęsiorowski and Szymkiewicz[22] identified the parameters defining the accuracy of the outflow of the Muskingum flood routing procedure. Similarly, a few other researchers have compared one or more forms of the Muskingum routing equation, but a comprehensive analysis of various forms of the Muskingum equation considering a wide range of data sets, which could provide some decisive evidence, is still lacking. In this regard, Nawaz[23] presented a comprehensive comparative analysis of variants of the Muskingum equation by considering six data sets.

Recently, there have been many discussions on the application of different numerical methods to obtain accurate results for the Muskingum flood routing procedure.[24] Vatankhah[18] concluded that an explicit numerical scheme is more accurate than an implicit scheme. However, Karahan[24] objected to it by referring to how comparative analysis was carried out by Vatankhah[18] and Karahan[24] stated that if the objective function is changed, different optimal parameters are obtained for the same data set. Further, Karahan[24] insisted that in most practical cases, implicit schemes would result in more appropriate flood routing. Although numerous studies have been conducted on improving the Muskingum flood routing method, the results of parameter estimations of available nonlinear versions of the Muskingum method are significantly dependent on the optimization algorithm used. To be more accurate, different values associated with differently routed outflows may be achieved using different techniques applied to the parameter estimation of the common three-parameter Muskingum model. This variation in the routed results, which is evident based on the current literature, brings about a major concern about implementing flood routing. Therefore, more efforts are required not only to develop new versions of Muskingum models but also to increase the accuracy of the simulation process of this flood routing model.

In the present chapter, an attempt was made to improve the estimate of outflows by proposing a new Muskingum flood routing model with four parameters. In a bid to test the performance of the new Muskingum model, different optimization techniques were applied for different sets of flood data considering two simulation processes. The comparison of flood routing results achieved by different simulation processes indicates the importance of the simulation process used in nonlinear Muskingum models.

2 Materials and methods

2.1 Muskingum method and its variants

The Muskingum flood routing method is one of the most common hydrological approaches. Its popularity among different channel routing methods is due not only to its relatively simple physical background but also to its ease of use in practical applications. The linear Muskingum model was proposed for developing flood protection schemes in the Muskingum River Basin, Ohio, in the late 1930s. Using the mass-conservation equation (Eq. 1), this method defines the channel storage in each time interval as a function of the inflow entering the channel and the outflow exiting the channel. This relationship is called the storage function. Depending on whether the storage function is either linear or nonlinear, based on channel historical flood records, the Muskingum model is linear or nonlinear. Finally, employing the Muskingum model requires parameter estimation of the storage function using the measured flood data.

$$\frac{\Delta S_t}{\Delta t} = I_t - Q_t, \tag{1}$$

where, S_t, I_t, and Q_t are the storage, inflow, and outflow at time t, respectively.

The literature is filled up with numerous efforts on improving the Muskingum model and its parameter estimation. In this regard, various variants of the nonlinear Muskingum model have been developed. To review the linear and nonlinear versions of the Muskingum model, the symbolic storage function shown in Eq. (2) is adopted here. Depending on the substitution instead of the parameters in Eq. (2), different versions of the Muskingum flood routing model can be obtained.

$$S_t = p_1 \left[p_2 I_t^{p_3} + p_4 Q_t^{p_5} \right]^{p_6}, \tag{2}$$

where, $p_1, p_2, \ldots, p_6$ are dummy parameters.

Table 1 lists various substitutions recommended in the literature[25–27] for six parameters (p_i for $i = 1, 2, \ldots, 6$) of the symbolic storage relation shown in Eq. (2), which are basically different variants of Muskingum models. In Table 1, K (or K_i) is a storage parameter, x (or x_i) is a weighting parameter, α (or α_i) is flow exponent, m (or m_i) is a power exponent, β is a lateral inflow factor so that βI_t is a lateral inflow at time t, $K(u_i)$, $x(u_i)$, $m(u_i)$ and $\alpha(u_i)$ are variable Muskingum parameters, u_i is a parameter for dividing inflow hydrograph into an arbitrary number of parts. Among different

TABLE 1 Various substitutions suggested for the parameters of the symbolic storage function.

No.	Authors	Year	p_1	p_2	p_3	p_4	p_5	p_6
1	McCarthy	1938	K	x	1	$1-x$	1	1
2	Chow	1959	K	x	α	$1-x$	α	1
3	Gill	1978	K	x	1	$1-x$	1	m
4	Gavilan and Houck	1985	K	x	α_1	$1-x$	α_2	1
5	O'Donne	1985	K	$x(1+\beta)$	1	$1-x$	1	1
6	Easa	2013	K	x	1	$1-x$	1	$m(u_i)$
7	Easa	2013	K	x	α	$1-x$	α	m
8	Vatankhah	2014	K	x	α_1	$1-x$	α_2	m
9	Karahan	2014	K	$x(1+\beta)$	1	$1-x$	1	m
10	Easa et al.	2014	1	xK_1	α_1	$(1-x)K_2$	α_2	m
11	Easa et al.	2014	K	x	α	$1-x$	α	$m(u_i)$
12	Easa	2014	$K(u_i)$	$x(u_i)$	$\alpha(u_i)$	$[1-x(u_i)]$	$\alpha(u_i)$	$m(u_i)$
13	Easa	2014	K	$x(u_i)(1+\beta)^\alpha$	α	$[1-x(u_i)]$	1	m
14	Karahan et al.	2015	K	$x(1+\beta)$	1	$1-x$	1	m
15	Easa	2015	K	x	1	$1-x$	1	$m(u_i)$
16	Haddad et al.	2015	1	$xK_1^{\alpha 1}$	α_1	$(1-x)K_2^{\alpha 2}$	α_2	m
17	Zhang et al.	2016	K	$x(1+\beta)$	1	$1-x$	1	$m(u_i)$
18	Niazkar and Afzali	2016	K	$x_1(1+x_2 I_t^{\alpha 2})$	α_1	x_3	α_3	m
19	Niazkar and Afzali	2017	K	x_1	α_1	x_2	α_2	m

versions of the Muskingum model in Table 1, McCarthy's version is the only linear one, while others are nonlinear. Unlike the linear version, nonlinear Muskingum models theoretically have more degree of freedom to capture the channel storage relationship. This advantage probably brings about more attention towards nonlinear Muskingum models, which yielded to the development of many nonlinear versions in the literature. Among nonlinear Muskingum models, Gill's model, which has three Muskingum parameters, is the most common version. Furthermore, some Muskingum versions presented in Table 1 consider variable Muskingum models, which may even provide more flexibility in this flood routing model.

For a flood routing problem with fixed flood data and the calibration method, Muskingum models with variable parameters gave better routing results than those with constant parameters in most of the studies conducted in the literature. Nonetheless, there is generally a trade-off between improving the routing results and the complexity of flood routing methods. Obviously, the calibration of the Muskingum model with variable parameters demands more computational effort in comparison with that with constant parameters. Likewise, the more the number of parameters a version of the Muskingum model, the more complicated the parameter estimation becomes. Therefore, the number of Muskingum parameters and taking into account the variability of Muskingum parameters during a flood period are two factors that may affect the complexity of the parameter estimation process. These two factors are related to the type of Muskingum flood routing method. In addition to these factors, the simulation process and optimization algorithm utilized for estimating the Muskingum parameters may play considerable roles in flood routing results. For this purpose, numerous optimization techniques have been adopted for parameter estimation of Muskingum models in the literature.

2.2 Proposed Muskingum model

In this study, a new nonlinear version of the Muskingum flood routing model has four constant parameters. The storage function of the new version is given by Eq. (3). As shown, it will be achieved by putting $p_1 = K$, $p_2 = x_1$, $p_3 = m_1$, $p_4 = x_2$, $p_5 = m_2$, and $p_6 = 1$ into the symbolic storage function.

$$S_t = K[x_1 I_t^{m_1} + x_2 Q_t^{m_2}] = K_1 I_t^{m_1} + K_2 Q_t^{m_2}, \qquad (3)$$

where, K_1, m_1, K_2, and m_2 are the Muskingum parameters of the new version.

2.3 Simulation process of the proposed Muskingum model

The simulation process is basically the calibration process of the Muskingum flood routing model. This process utilizes historical flood records to determine Muskingum parameters so that they can be used for future routing applications for the same river reach. To apply the proposed Muskingum model, two straightforward simulation processes were considered. These are described as follows:

First simulation process (Simulation 1): This calibration process comprises the following six steps. For the first time interval, all six steps needed to proceed, while the third to sixth steps should be conducted successively for the remaining time intervals of the flood period.

(1) First, an initial guess for the four constant parameters (K_1, m_1, K_2, and m_2) of the new Muskingum model is required to start the simulation process. These values are randomly selected in a zero-order search-based optimization algorithm (such as the genetic algorithm or MHBMO algorithm), whereas first-order (like GRG) or second-order (such as Broyden-Fletcher-Goldfarb-Shanno) optimization techniques require the user to suggest a set of arbitrary values. In hybrid algorithms (such as MHBMO-GRG), the very first initial in the first stage is selected randomly, while the outcome of the first stage can be used as the initial guess of the second optimization algorithm in the second stage.

(2) For the first time interval, S_2 was computed using Eq. (4) by assuming $Q_1 = I_1$. This assumption, which is associated with uniform flow passing throughout the channel reach under investigation, is common in the calibration process of the Muskingum models.

$$S_2 = K_1 I_1^{m_1} + K_2 Q_1^{m_2}, \qquad (4)$$

(3) After computing the storage volume in each time step, the following criterion should be checked for each set of values of the Muskingum parameters to avoid producing complex values for the outflow at any time interval. The inequality shown in Eq. (5) was obtained by rearranging and manipulating the storage function of Eq. (3) so that a feasible outflow can be calculated using the corresponding equation. If the values chosen for the Muskingum parameters fail to satisfy the criterion introduced in Eq. (5), they were excluded from the simulation process.

$$m_1 < \frac{\log\left(\dfrac{S_t}{K_1}\right)}{\log\left(I_t\right)}. \qquad (5)$$

(4) The expression for calculating the change in channel storage at each time step is shown in Eq. (6), which can be determined by substituting Eq. (4) into Eq. (1):

$$\frac{\Delta S_t}{\Delta t} = I_t - Q_t = I_t - \left(\frac{S_t}{K_2} - \frac{K_1 I_t^{m_1}}{K_2}\right)^{\frac{1}{m_2}}, \qquad (6)$$

(5) The storage volume at the next time interval can be obtained by the first-order Euler scheme,[28] as shown in Eq. (7), whereas negative values for S_{t+1} are not acceptable.

$$S_{t+1} = S_t + \Delta S_t, \qquad (7)$$

(6) The outflow of the next time interval (Q_{t+1}) in terms of S_{t+1} and I_{t+1} can be obtained by rearranging Eq. (3). This equation is given by Eq. (8):

$$Q_{t+1} = \left(\frac{S_{t+1}}{K_2} - \frac{K_1 I_{t+1}^{m_1}}{K_2} \right)^{\frac{1}{m_2}}, \tag{8}$$

For better clarification, Fig. 1 depicts the flowchart of the first simulation process.

Second simulation process (Simulation 2): The major difference between the first and second calibration processes is the inflow values considered in the storage function. To clarify this issue, the second simulation process is described in the following steps.

(1) The simulation process requires an initial set of values for the Muskingum parameters (K_1, m_1, K_2, and m_2).
(2) By assuming $Q_1 = I_1$, S_2 can be computed using Eq. (4).
(3) To improve the routing results, it was suggested in several studies that adopted the average inflow at time t and $t-1$ instead of the inflow at time t in the Muskingum models.[7, 29] This modification was considered in the simulation process. Consequently, the criteria introduced in Eq. (5) must be modified in Eq. (9).

$$m_1 < \frac{\log \left(\frac{S_t}{K_1} \right)}{\log (I_{ave})}, \tag{9}$$

where, $I_{ave} = \frac{I_{t-1} + I_t}{2}$.

(4) By considering the average inflow, the change in the storage volume is modified in Eq. (10):

$$\frac{\Delta S_t}{\Delta t} = I_{ave} - Q_t = I_{ave} - \left(\frac{S_t}{K_2} - \frac{K_1 I_{ave}^{m_1}}{K_2} \right)^{\frac{1}{m_2}}, \tag{10}$$

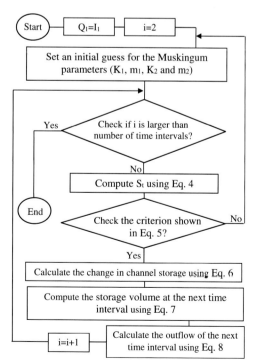

FIG. 1   Flowchart of the first simulation process of the new four-parameter Muskingum model.

(5) The storage volume at the next time interval can be obtained using Eq. (7).

(6) Eq. (11) can be utilized to compute the outflow at the next time interval:

$$Q_t = \left(\frac{S_t}{K_2} - \frac{K_1 I_{ave}^{m_1}}{K_2} \right)^{\frac{1}{m_2}}, \tag{11}$$

2.4 Techniques for parameter estimation of the new Muskingum model

In this study, three techniques were adopted to estimate the four parameters of the new Muskingum model. These algorithms are briefly described as follows.

(1) The first technique used for parameter estimation of the new Muskingum model is the MHBMO algorithm. This is a zero-order metaheuristic optimization algorithm, which was previously applied to the common three-parameter nonlinear Muskingum models in the literature.[29] As the title of the MHBMO algorithm indicates, it was originally inspired by the natural process of honey bee mating. In other words, this algorithm was developed by simulating a similar multi-step process and assigning a functional parameter to each character (drones, a queen, workers, and broods) involved. Therefore, it is a search-based optimization algorithm with several controlling parameters, while the most dominant ones include the size of the initial population, number of workers, queen's speed at the start and end of the mating flight, and speed reduction factor. A sensitivity analysis was previously conducted for these controlling parameters for parameter estimation of the Muskingum model,[29] and the recommended values for them in this study.

(2) The GRG algorithm, which was previously used for parameter estimation of the three-parameter Muskingum model,[4, 9] is a first-order optimization algorithm that is embedded in an MS Excel spreadsheet. It requires a set of initial guesses for the Muskingum parameters, while the outcome depends on the initial values. Hence, various initial guesses were attempted to ensure that the results were optimum. In this study, the default properties were selected for this technique.

(3) The hybrid MHBMO-GRG is a two-stage optimization algorithm that exploits the MHBMO and GRG algorithms in its first and second stages, respectively. This hybrid technique was suggested for parameter estimations of the six-parameter[25] and eight-parameter[26] Muskingum models. Because of its successful performance in estimating parameters of other versions of the Muskingum model, it was also utilized in this study to calibrate the new four-parameter Muskingum model.

2.5 Flood routing data

The present analysis was accomplished by analyzing three data sets from the literature of Muskingum flood routing models to arrive at a rational conclusion on the reliability of the new version of the Muskingum model in practice to route floods. The inflow and outflow hydrographs of the three datasets (Wilson,[30] Viessman and Lewis[31] and Wye) used in this study are shown in Fig. 2. Usually, flood hydrographs have a single peak, but in certain cases, multiple peaks may be observed. Given this, the double-peaked data set of Viessman and Lewis was also included in the analysis.

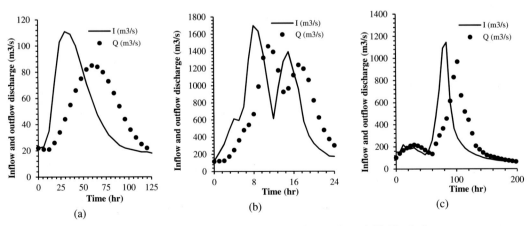

(a)   (b)   (c)

FIG. 2 The inflow and outflow hydrographs of (A) Wilson's, (B) Viessman and Lewis's, and (C) Wye's data.

2.6 Performance evaluation criteria

The performances of various equations were assessed based on the sum of squares of deviation in outflow (SSQ), Nash-Sutcliffe efficiency (NE), deviation of peak outflow (DPO), and deviation of peak outflow time (DPOT). SSQ represents the disparity between the observed and routed outflows, while NE shows the degree of agreement between the observed and routed flows. However, these two criteria alone may not justify the reliability of routed outflows as outflow peaks and the time at which the peak discharge would occur at the downstream station are important components of flood routing. Therefore, the deviation of peak outflow (DPO) and deviation of peak outflow time (DPOT) were also considered to analyze the reliability of the new Muskingum model. The simulation process that results in the lowest SSQ, DPO, DPOT, and highest NE would be considered the most reliable. The outflows estimated by different techniques using the four-parameter Muskingum model were quantitatively compared based on the performance evaluation criteria[4,29,32,33] that may be presented as follows:

$$SSQ = \sum_{t=1}^{N} \left(Q_t^{obs} - Q_t^{cal} \right)^2 \tag{12}$$

$$NE = 1 - \frac{\sum_{t=1}^{N} \left(Q_t^{obs} - Q_t^{cal} \right)^2}{\sum_{i=1}^{N} \left(Q_t^{obs} - \overline{Q} \right)^2} \tag{13}$$

$$DPO = \left| Q_{peak}^{cal} - Q_{peak}^{obs} \right| \tag{14}$$

$$DPOT = \left| t_{peak}^{cal} - t_{peak}^{obs} \right| \tag{15}$$

where, Q_t^{obs} and Q_t^{cal} are observed and routed outflows at time t, respectively; $\overline{Q}$ is the average outflow; Q_{peak}^{cal} and Q_{peak}^{obs} are peak calculated and observed outflows at the downstream station, respectively; t_{peak}^{cal} and t_{peak}^{obs} are the time of peak calculated and observed outflows at the downstream station, and N is the number of time intervals in each flood record.

3 Results

In the present chapter, an attempt was made to improve the estimate of outflows from Muskingum flood routing equations by applying different optimization techniques to a new form of Muskingum flood routing equations. The optimum values of the Muskingum parameters obtained by MHBMO, GRG, and hybrid MHBMO-GRG for the first and second simulations are presented in Tables 2 and 3, respectively. As can be observed from Tables 2 and 3, no significant difference was observed in the parameters obtained by the MHBMO, GRG, and hybrid MHBMO-GRG techniques; however, the parameters significantly differed based on the simulation. Unlike the variation in the routed results obtained by various optimization algorithms for the common three-parameter Muskingum model, Tables 2 and 3 demonstrate that the optimum values of the parameters of the new Muskingum model can be easily obtained by adopting any type of optimization algorithm used in this study. This may indicate that the proposed four-parameter Muskingum model has a more reliable structure, while further investigations may be required by utilizing other optimization algorithms for the parameter estimation.

Tables 4 and 5 present the performance evaluation criteria for the first and second simulation processes, respectively. The outflow estimates from different optimization techniques were similar as could be seen from Tables 4 and 5. Only in the case of Viessman and Lewis[31] data in simulation application of GRG and the hybrid MHBMO-GRG reduced SSQ and DPO in comparison to the MHBMO algorithm, otherwise, for all other cases, performance evaluation remained the same for all the optimization techniques. Additionally, the MHBMO algorithm achieved 36.242 for SSQ of Wilson's data using Gill's model,[28] while the proposed Muskingum model reaches 149.76 for SSQ in Simulation 2. However, the new version improves the SSQ of Viessman and Lewis's data by 28.63% as it reduces it from 7751.80 to 5524.96 using the MHBMO algorithm.[28] Furthermore, significant improvement was observed in the estimates of outflow was observed when weighted inflow (Simulation 2) was used. Also, the application of Simulation 2 to Wilson's data led to a reduction of over 50% and 70% in SSQ and DPO, respectively. Hence, it may be observed from Tables 4 and 5 that SSQ and DPO significantly reduced on the application of Simulation 2.

TABLE 2 Muskingum parameters obtained for Simulation 1.

Methods	Parameters	Wilson	Viessman and Lewis	Wye
MHBMO	K_1	8.80E−09	0.00065	0.002777
	K_2	0.0667	0.8234	0.0151
	m_1	5.2167	1.9035	2.027
	m_2	2.2830	1.1615	2.0549
GRG	K_1	8.80E−09	0.00075	0.002777
	K_2	0.0667	0.8229	0.0151
	m_1	5.2167	1.8840	2.027
	m_2	2.2830	1.1612	2.0549
MHBMO-GRG	K_1	8.80E−09	0.00065	0.002777
	K_2	0.0696	0.8234	0.0151
	m_1	5.2168	1.9035	2.027
	m_2	2.2830	1.1615	2.0549

TABLE 3 Muskingum parameters obtained for Simulation 2.

Methods	Parameters	Wilson	Viessman and Lewis	Wye
MHBMO	K_1	3.66E−09	0.02539	4.99E−05
	K_2	0.0801	1.3631	0.0311
	m_1	5.3549	1.3493	2.5880
	m_2	2.2197	1.0657	1.9148
GRG	K_1	3.66E−09	0.02185	5.55E−05
	K_2	0.0801	1.3796	2.5742
	m_1	5.3549	1.36872	0.02319
	m_2	2.2197	1.161	2.0549
MHBMO-GRG	K_1	3.66E−09	0.02185	4.99E−05
	K_2	0.0801	1.3796	0.0311
	m_1	5.3549	1.36872	2.5880
	m_2	2.2197	1.161	1.9148

TABLE 4 Performance evaluation criteria for Simulation 1.

Data	Method	SSQ (m^6/s^2)	NE	DPO (m^3/s)	DPOT (hr)
Wilson	MHBMO	334.09	0.97	2.03	6.0
	GRG	334.15	0.97	2.04	6.0
	MHBMO-GRG	334.09	0.97	2.03	6.0
Viessman and Lewis	MHBMO	63892.56	0.99	37.54	0.0
	GRG	63897.26	0.99	37.70	0.0
	MHBMO-GRG	63892.56	0.99	37.54	0.0
Wye	MHBMO	135166.77	0.92	196.69	12.0
	GRG	135393.83	0.92	196.02	12.0
	MHBMO-GRG	135166.77	0.92	196.69	12.0

TABLE 5 Performance evaluation criteria for Simulation 2.

Data	Method	SSQ (m⁶/s²)	NE	DPO (m³/s)	DPOT (h)
Wilson	MHBMO	149.76	0.99	0.54	6.0
	GRG	149.76	0.99	0.54	6.0
	MHBMO-GRG	149.76	0.99	0.54	6.0
Viessman and Lewis	MHBMO	5532.11	0.99	3.22	0.0
	GRG	5524.96	0.99	3.07	0.0
	MHBMO-GRG	5524.96	0.99	3.07	0.0
Wye	MHBMO	76134.74	0.95	208.34	6.0
	GRG	76489.24	0.95	208.83	6.0
	MHBMO-GRG	76134.74	0.992	208.34	6.0

4 Discussion

Figs. 3–5 depict the relative-error plots for the three datasets for different optimization techniques and both the simulation processes. As shown, the general pattern of relative errors of this new model is dependent on not only the dataset but also the simulation process, whereas the optimization algorithms exploited for the parameter estimation have much less influence on the pattern. To be more specific, different techniques led to quite similar temporal variations of relative errors for each dataset considered, whereas changes of relative errors over time are unique for each dataset. Moreover, the bounds (maximum and minimum values) of relative errors are different from one dataset to another. To be more precise, the relative errors achieved for Wye's data have a wider range than two other datasets. Regardless of the technique adopted for the parameter estimation, the maximum relative errors for Wilson's data were observed in the sixth time interval for Simulation 1, whereas the maximum relative errors were obtained by Simulation 1 in the third- and first-time intervals of Viessman and Lewis's data and Wye's data, respectively.

Finally, Figs. 3–5 indicate that irrespective of the optimization technique used, Simulation 2 overall results in the reduction of relative errors and improve the estimates of outflows for all three data sets. Basically, Simulation 2 utilizes the average inflow rather than the inflow of the current time step alone. Hence, the improvement in the estimates in the routed outflows obtained by the use of the average inflows indicates that the outflow at any time step is also influenced by the inflow at the previous time step.

5 Conclusions

This chapter proposes a new four-parameter Muskingum flood routing model. The applicability of the new Muskingum model was assessed by routing three flood data points using two simulation processes. Three types of optimization algorithms were utilized to calibrate the four parameters of the new Muskingum model. They include the MHBMO, GRG, and MHBMO-GRG algorithms, which are zero-order, first-order, and hybrid optimization techniques, respectively. The results indicated that the new Muskingum model yielded a higher SSQ for the single-peak data, whereas it improved SSQ by 28.63% of double-peak flood data using the MHBMO algorithm. When the parameters of the new Muskingum model were calibrated by the MHBMO algorithm for different datasets, the average inflow simulation process decreased SSQ values between 43.52% and 91.35% compared to the common simulation process. Furthermore, investigating the relative errors of routed outflows during the flood period demonstrates that the general pattern of relative errors depends not only on the dataset under consideration but also on the simulation process applied, whereas the optimization technique used for parameter estimation negligibly influences the patterns for the new Muskingum model. Since different types of optimization techniques yielded quite the same results for the parameter estimation of the new Muskingum model, it may be a more reliable version of the Muskingum model than the common three-parameter nonlinear Muskingum model, while further studies are indeed required to investigate this issue.

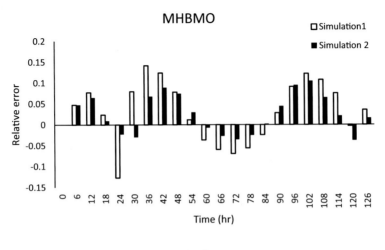

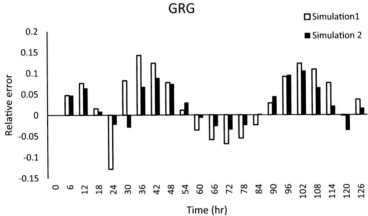

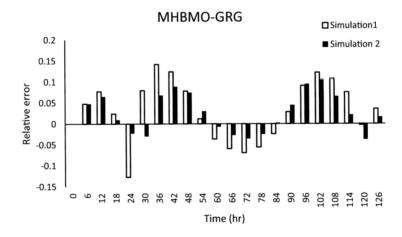

FIG. 3   Relative error plot for Wilson's data obtained by the MHBMO, GRG, and MHBMO-GRG algorithms.

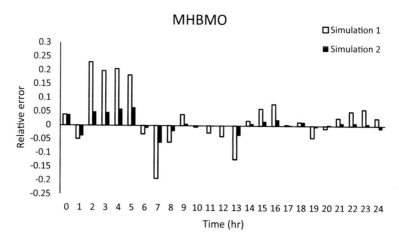

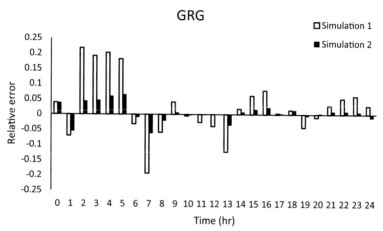

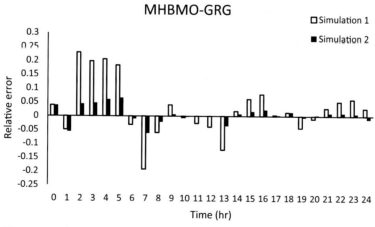

FIG. 4 Relative error plot for Viessman and Lewis's data obtained by the MHBMO, GRG, and MHBMO-GRG algorithms.

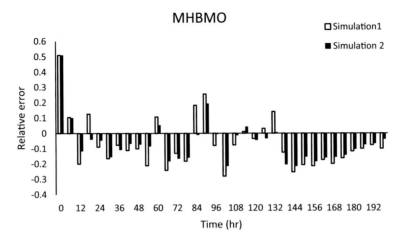

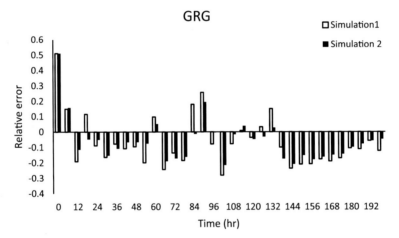

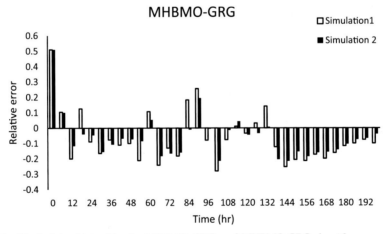

FIG. 5   Relative error plot for Wye's data obtained by the MHBMO, GRG, and MHBMO-GRG algorithms.

References

1. Geem ZW. Issues in optimal parameter estimation of non-linear Muskingum flood routing model. *Eng Optim.* 2014;46(3):328–339.
2. Zakwan M, Ahmad Z, Sharief SMV. Magnitude-frequency analysis for suspended sediment transport in the Ganga river. *J Hydrol Eng.* 2018;23 (7), 05018013.
3. Pandey M, Zakwan M, Khan MA, Bhave S. Development of scour around a circular pier and its modelling using genetic algorithm. *Water Supply.* 2020;20(8):3358–3367.
4. Zakwan M, Muzzammil M. Optimization approach for hydrologic channel routing. *Water Energy Int.* 2016;59(3):66–69.
5. Niazkar M, Talebbeydokhti N, Afzali SH. One dimensional hydraulic flow routing incorporating a variable grain roughness coefficient. *Water Resour Manag.* 2019;33(13):4599–4620. https://doi.org/10.1007/s11269-019-02384-8.
6. Niazkar M, Afzali SH. New nonlinear variable-parameter Muskingum models. *KSCE J Civ Eng.* 2017;21(7):2958–2967. https://doi.org/10.1007/s12205-017-0652-4.
7. Chu HJ, Chang LC. Applying particle swarm optimization to parameter estimation of the nonlinear Muskingum model. *J Hydrol Eng.* 2009;14 (9):1024–1027.
8. Das A. Parameter Estimation for Muskingum Models. *J Irrig Drain Eng.* 2004;130(2):140–147.
9. Barati R. Application of Excel solver for parameter estimation of the nonlinear Muskingum models. *KSCE J Civ Eng.* 2013;17(5):1139–1148.
10. Xu D, Qui L, Chen S. Estimation of nonlinear muskingum model parameter using differential evolution. *J Hydrol Eng.* 2012;17(2):348–353.
11. Gill MA. Flood routing by the Muskingum method. *J Hydrol Eng.* 1978;36:353–363.
12. Tung YK. River flood routing by nonlinear Muskingum method. *J Hydraul Eng.* 1985;111(12):1447–1460.
13. Mohan S. Parameter estimation of nonlinear Muskingum models using genetic algorithm. *J Hydraul Eng.* 1997;123(2):137–142.
14. Kim JH, Geem ZW, Kim ES. Parameter estimation of the nonlinear Muskingum model using harmony search. *J Am Water Resour ASSQc.* 2001;37 (5):1131–1138.
15. Karahan H. *Predicting Muskingum Flood Routing Parameters Using Spreadsheet.* Wiley Periodicals Inc; 2009:280–286.
16. Luo J, Xie J. Parameter estimation for nonlinear Muskingum model based on immune clonal selection algorithm. *J Hydrol Eng.* 2010;15 (10):844–851.
17. Hirpurkar P, Ghare AD. Parameter estimation for the nonlinear forms of the Muskingum model. *J Hydrol Eng.* 2014;20(8), 04014085.
18. Vatankhah AR. Evaluation of explicit numerical solution methods of the Muskingum model. *J Hydrol Eng.* 2014;19(8), 06014001.
19. Easa SM. New and improved four-parameter non-linear Muskingum model. *Proc Inst Civil Eng.* 2013;167(5):288.
20. Vatankhah AR. Non-linear Muskingum model with inflow-based exponent. *Proc Inst Civil Eng–Water Manag.* 2017;170(2):66–80.
21. Akbari R, Hessami-Kermani MR, Shojaee S. Flood routing: improving outflow using a new non-linear Muskingum model with four variable parameters coupled with PSO-GA algorithm. *Water Resour Manag.* 2020;34(10):3291–3316.
22. Gąsiorowski D, Szymkiewicz R. Identification of parameters influencing the accuracy of the solution of the nonlinear Muskingum equation. *Water Resour Manag.* 2020;34(10):3147–3164.
23. Nawaz AR, Zakwan M, Khan I, Rahim ZA. Comparative analysis of variants of Muskingum model. *Water Energy Int.* 2020;63(7):64–73.
24. Karahan H. Discussion of evaluation of explicit numerical solution methods of the Muskingum model by Ali R. Vatankhah. *J Hydrol Eng.* 2015;20 (8):07015005.
25. Niazkar M, Afzali SH. Application of new hybrid optimization technique for parameter estimation of new improved version of Muskingum model. *Water Resour Manag.* 2016;30(13):4713–4730. https://doi.org/10.1007/s11269-016-1449-9.
26. Niazkar M, Afzali SH. Parameter estimation of an improved nonlinear Muskingum model using a new hybrid method. *Hydrol Res.* 2017;48 (4):1253–1267. https://doi.org/10.2166/nh.2016.089.
27. Easa SM. Channel flood routing: review of recent hydrologic Muskingum models. In: Yurish S, ed. *Advances in Measurements and Instrumentation.* Vol. 1. 1st ed. IFSA Publishing; 2019. ISBN-13: 978-8409073214 (Chapter 7).
28. Niazkar M, Afzali SH. Closure to assessment of Modified Honey Bee Mating Optimization for parameter estimation of nonlinear Muskingum models. *J Hydrol Eng.* 2018;23(4). https://doi.org/10.1061/(ASCE)HE.1943-5584.0001602, 07018003.
29. Niazkar M, Afzali SH. Assessment of Modified Honey Bee Mating Optimization for parameter estimation of nonlinear Muskingum models. *J Hydrol Eng.* 2015;20(4). https://doi.org/10.1061/(ASCE)HE.1943-5584.0001028, 04014055.
30. Wilson EM. *Engineering Hydrology.* Hampshire, UK: MacMillan Education Ltd.; 1974.
31. Viessman W, Lewis GL. *Introduction to Hydrology.* 5th ed. Prentice Hall; 2003. New Delhi, India.
32. Niazkar M, Zakwan M. Assessment of artificial intelligence models for developing single-value and loop rating curves. *Complexity.* 2021;2021:1–21. https://doi.org/10.1155/2021/6627011. 6627011.
33. Zakwan M, Niazkar M. A comparative analysis of data-driven empirical and artificial intelligence models for estimating infiltration rates. *Complexity.* 2021;2021:1–13. https://doi.org/10.1155/2021/9945218.

25

Predicting areas affected by forest fire based on a machine learning algorithm

Mahdis Amiri[a] and Hamid Reza Pourghasemi[b]

[a]Department of Watershed and Arid Zone Management, Gorgan University of Agricultural Sciences and Natural Resources, Gorgan, Iran [b]Department of Natural Resources and Environmental Engineering, College of Agriculture, Shiraz University, Shiraz, Iran

1 Introduction

In nature, favorable places for Ff$_s$ are completely undeniable, and they act as a serious way to displace land-cover and prepare beautiful landscapes.[1] On the other hand, the sudden destruction of forests may cause many problems for the environment and societies in the world. Destruction of forests, deforestation, and forest fires cause serious damage to communities and humans, as well as endangers the sustainability of the biological community. Therefore, it is important to conserve the world's forests.[2] Today, the risk of fire is a serious and fundamental problem in nature. As the crowd/population has grown and the mystery/industry has expanded in recent years, the use of natural resources, particularly forests, for obtaining food and nutrition is augmenting.[3,4] Many factors, including human and natural factors, can lead to Ff$_s$. For example, the conversion of natural to dummy lands by farmers and foresters, pollution from urban activities such as noise pollution, air, groundwater, and land degradation increases the likelihood of fires.[5] In Iran, the notable factors known to cause forest fires includes lighting fires in camps and forests which becomes uncontrollable due to carelessness of tourists that destroys five thousand to six thousand hectares of forest lands annually.[6] Ff$_s$ have many negative and detrimental effects on vegetation and forest ecosystems because they cause the destruction of vegetation and carbon reserves (CR), change the atmosphere (CA), and change vegetation and biodiversity (CVB) of forests.[7] Given the significant effects that forest fires have on ecosystems and the special economic and social conditions of communities to prevent forest fires, this is one of the issues that should be given special attention by governments and researchers around the world.[8] Therefore, creating a fire prevention plan is to build a fire susceptibility map (FSM) on local, regional, and even global scales. The FSM can make effective factors as well as fire-prone areas available based on past recorded events to prevent future fires and are also very useful for planning land-use changes.[9] To reduce the negative effects and damage caused by a fire in hazardous areas, it is necessary to investigate the fire occurrence and to consider various effective factors for further planning and protection. Rapid advances in geographic information systems (GIS) and remote sensing techniques (RS) can help researchers to zone forest fires owing to the capabilities of rapid analysis.[6,8,10] In recent decades, machine learning algorithms have been used to spatially predict fire hazards. Among the algorithms used in various studies, the following can be mentioned: Support vector machines, artificial neural networks, random forests, logistic regression classifier with kernel function, neural fuzzy, linear and quadratic discriminant analysis, multivariate adaptive regression splines, fuzzy logic, the generalized additive model, radial-basis function neural network (RBFANN), logit boost models, dynamic Bayesian network, adaptive neuro-fuzzy inference system, boosted regression tree, and decision tree.[11–17]

Therefore, one of the main features of the present study is to prepare a reliable map of forest fires using SVM$_A$ at Zohreh Watershed in Fars Province, Iran, and the important goal is utilization of the results of the present study in management measures to prevent damage. Another positive point of the present study is the prioritization of influencing variables using MDAm from a random (RF) forest technique. Since most of the study area is forest cover, thus, it is necessary to predict the possible occurrence of future forest fires based on the perception of past fire occurrence points and some effective factors. Therefore, spatial forecasting greatly reduces the sensitivity of forest and pasture fires to reduce land degradation and the damage caused by this sensitivity to local people and natural areas.

2 Materials and methods

2.1 The case study and the GIS database

Details of the spatial modeling of fire forest susceptibility at the Zohreh watershed are presented in the form of a flow diagram/chart and Fig. 1.

2.2 A general explanation of Zohreh watershed

The Zohreh watershed is an open basin in Iran and is a subset of the Persian Gulf and the Sea of Oman with an area of 40,790 km^2, of which 52.3% are mountains and 47.7% are plains and foothills. The main rivers are Jarahi and Zohreh. In general, the watershed is located on the southern slopes of the Middle Zagros and is enclosed between the geographical coordinates of 48° 18' to 52° 19' east longitude and 30° 00' to 31° 42' N (Fig. 2).

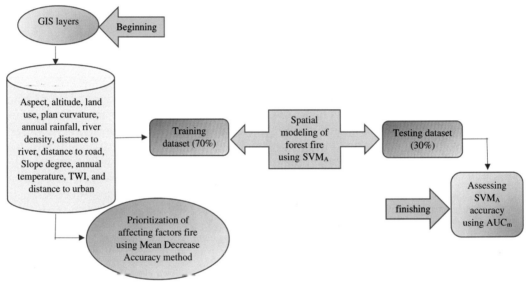

FIG. 1   Demonstration of the current research method.

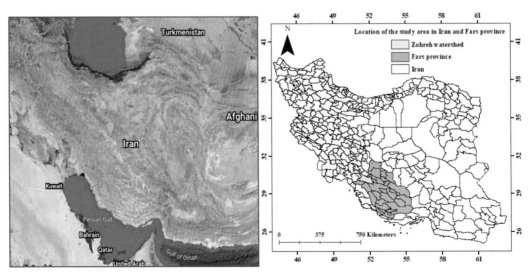

FIG. 2   Study area in Iran and Fars province.

2.3 Historical forest and pasture fire events

Methods of collecting forest fire events at the Zohreh watershed in Fars Province were discussed in two ways: in this study, the data on 53 forest fire locations that occurred from May 2008 to May 2018 was used as a database of historical forest fires. These fire places have been detected in the national forest fire database created by the Fars Province Natural Resources Organization (FNRO). In addition, to confirm the recorded incidents of fires in the forests of the study area, extensive inspections were carried out and forest fire locations were recorded by the Global Position System "GPS." Of the 53 epochal/historic spots, 37 (70%) were fortuitously elected, it was used to build/train the model and the remaining 16 spots (30%) were applied to validate the model.[15,18,19]

2.4 Forest fire and pasture influencing factors

One of the most basic steps in spatial modeling in natural hazard research, such as forest fires, is the stage of selecting the influencing variables on the occurrence of fire in an area. The selection of influential factors in the present study is based on a review of past research as well as the availability of basic layers.[16,20–22] In general, variables can be segregated into four groups: climate, vegetation, topography, and humans.[23–26] In the present study, 12 significant variables (aspect, altitude, land use, plan curvature, annual rainfall, river density, distance to river, distance to road, slope degree, annual temperature, TWI, and distance to urban) were selected, which are explained in detail below.

2.4.1 Altitude/elevation

A digital elevation model (DEM) map with a pixel size of 30 m was prepared in ArcGIS.10.6.1, using a topographic map obtained from the Department of Natural Resources of Fars Province (Fig. 3). The altitude variable has a significant effect on temperature and vegetation because areas with high altitude have lower temperatures and higher humidity; therefore, the probability of forest fires is lower and consequently, it can be stated that the probability of fires is inversely related to the altitude variable.[27] On the other hand, altitude also affects the speed and direction of the wind, and in this way, it can affect the expansion of the fire.[28]

2.4.2 Slope degree

Another influential topographic factor affecting the occurrence of forest fires is the slope degree. The angle of inclination can affect the direction and vastness of the fire. In general, fires are more frequent on steeper and sharp slopes.[29] Therefore, this important and influential variable was prepared from the DEM in ArcGIS.10.6.1, using the extension "Spatial Analyze" (Fig. 3).

2.4.3 Slope aspect

The Slope direction is another important topographical variable for the occurrence of fire in any region. Slope direction has a significant effect on soil moisture, sunlight, wind direction, and rainfall, and since this variable affects many environmental factors such as humidity, sunlight, wind, and rainfall, all of which occur in the event of fire and play a significant role in forest and range fires.[30] In general, the highest incidence of fires occurs in the geographical directions of south and east, because in the mentioned geographical directions, they are more exposed to sunlight and consequently more exposed to fire. The aspect map was prepared from a 30-m DEM map in ArcGIS.10.6.1 software and using the extension "Spatial Analyze" (Fig. 3) and divided into nine main classes.

2.4.4 Topographic wetness index (TWI)

The Next most significant topographic factor is the TWI. This variable can be applied to distinguish topographic verifications over hydrological processes. TWI illustrates the impact of geodesy on a specific point or location that is relevant to the soil situation of a region. The input of the TWI map is a 30-m DEM map prepared in SAGA-GIS software (Fig. 3). The *TWI* map calculation formula is as follow[21,31,32]:

$$TWI - \ln (H/M) \qquad (1)$$

where H is the specific area of the watershed (m^2/m), and M is the steepness (°).

2.4.5 Plan curvature

Plan curvature is defined as the vertical to the highest value of the slope. It also shows the convergence or divergence of water downstream of a watershed.[3] Therefore, it is a well-known variable for predicting fire susceptibility.

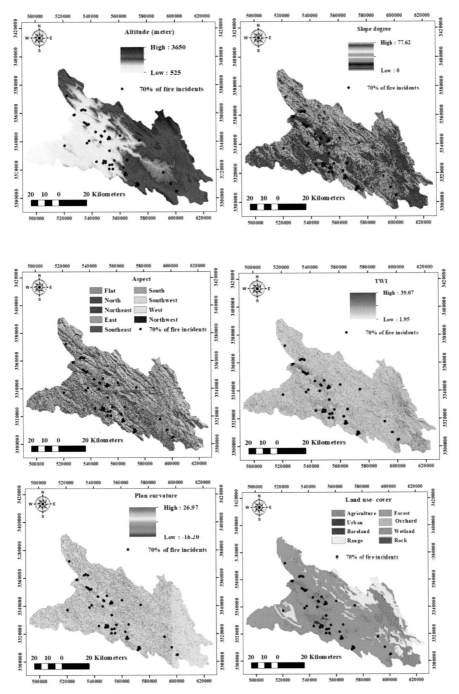

FIG. 3   Output layers taken from ArcGIS software.

(Continued)

The "plan curvature" variable, like other topographic parameters, was created from a 30-m DEM map in ArcGIS.10.6.1 and classified into three classes including, concave, flat, and convex (Fig. 3).

2.4.6 *Land use/land cover (LULC)*

Land use/land cover" is typically used to identify and target fire-sensitive areas. It also identifies locations that increase the likelihood of fires due to land-use changes.[33] The effective land use/land cover map was taken from the Fars Province Natural Resources Organization with a pixel size of 30 m × 30 m and Google Earth software was used to update and improve the desired layer (Fig. 3). The Land use map of the study area in the present study was

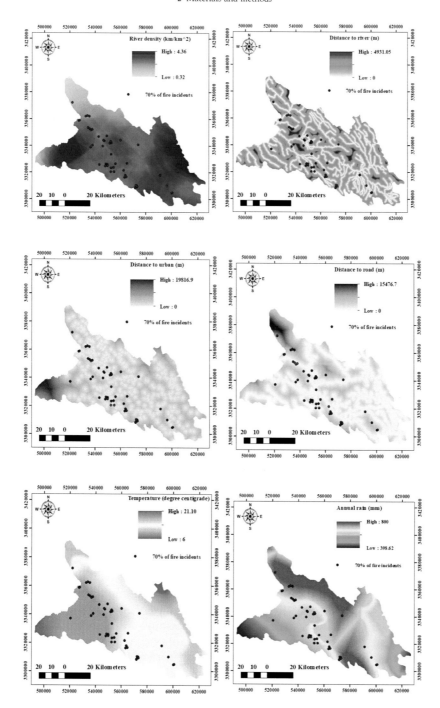

FIG. 3, CONT'D

categorized into eight important and main categories: agriculture, urban, bare land, rangeland, forest areas, orchards, wetlands, and rock lands (Fig. 3).

2.4.7 River density and distance to the river

Natural sites near streams and rivers have more groundwater and soil moisture. Therefore, there is an inverse relationship between water resources and rivers and the occurrence of fires. In general, water resources, including drainage density/river density and distance to rivers, affect the occurrence of fire susceptibility. Thus, we can see fires in an environment where there are these water resources.[34] Both maps were prepared from the waterway layer of the study area with a pixel size of 30 m in ArcGIS.10.6.1 software (Fig. 3).

2.4.8 Distance to urban and road

Generally, natural areas that are close to cities, villages, and residential areas are known to be more prone to fire because human activities often lead to the destruction of natural lands.[35] In addition, man-made operations such as road construction and the movement of vehicles, especially on forest roads, increase the likelihood of land being prone to fire. It is noteworthy that forest areas, which are often located near road networks and there is a short distance between the road and the forest, are very vulnerable.[36] The inputs of these variables are road and urban. The layers of distance to urban areas and roads were created using the "Distance Function" in ArcGIS.10.6.1 (Fig. 3).

2.4.9 Annual temperature and rainfall

Temperature is one of the most important climatic factors affecting the occurrence of forest fires. This means that at higher temperatures, forest fires are more likely to occur. Hence, an increase in ET results in combustion. On the other hand, rising temperatures can affect forest moisture as increasing temperatures reduce the humidity of the vegetation and increase heating, resulting in more fires.[37] Annual mean rainfall is another climatic parameter. Rainfall can determine characteristics such as period and intensity, penetration and runoff, or periodic droughts.[38] The annual mean of both climatic maps during the last 10 years (2010–2020) was prepared by Fars Regional Water Organization and was made using the "inverse-distance weighting (IDW)" method in ArcGIS.10.6.1[39–41] (Fig. 3).

- The output of all the layers affecting the occurrence of fire, as described, is shown in Fig. 3.

2.5 Random forest and determination of accuracy using MDAm

Random forest (RF), a non-parametric method of machine learning, belongs to the family of group methods (ensemble methods). This algorithm is an extended type of tree classification and regression model (CRm).[42] The RF model is based on an information synthesis method in which a large number of decision trees are created and then the results of all the trees are combined to predict. One of the advantages of using multiple decision trees is that it reduces the instability and sensitivity of the model and increases the predictive power of the model.[43] The RF model examines the importance of variables in two ways: mean decrease in accuracy and mean decrease in Gini coefficients.[44] In general, using the mean decreasing accuracy is better and more appropriate than the Gini index in determining the priority and importance of variables.[45–47]

2.6 Spatial modeling

2.6.1 SVM$_A$

One efficient method for continuous and discrete data modeling based on statistical training is SVM$_A$. The SVM model is a "Supervised Classification System" (SCS) based on learning theory and statistical dimensional theory.[48] In fact, the SVM is a strong learning system based on constrained optimization theory that uses the inductive principle of structural error minimization and leads to an overall optimal solution.[49] In this way, a set of classification functions that can evaluate the error and generalize the information properly and reduce the complexity of forest fire behavior using the information contained in the layers of effective factors and high repetition of modeling.[50] To minimize the rate of generalized errors, the training error rate, and the complexity of the classification should be reduced, which is done using the "Separating hyper-plane," which is responsible for increasing the margin between classes. The larger the "Separating hyper-plane," the greater the stability of the model to distortion and noise, and thus the higher the power to generalize.[51]

2.7 Evaluating the SVM$_A$ using AUC value

To evaluate the accuracy of SVM$_A$, the "receiver operating characteristic (ROC)" and the "Area Under the ROC curves" were used. It should be noted here that accurate validation or evaluation in spatial modeling is necessary to determine the importance/emphasis of a particular model or to evaluate the results of a study.[52] In fact, ROC is a graphical diagram that defines the efficiency of models through diagnostic testing. In other words, the ROC method is a graphical representation of the balance between sensitivity, true positivity rate, and specificity, and false-negative rate". The AUC$_m$ value, which indicates the quality of an algorithm for predicting fire susceptibility in an area, varies between 0.5 and 1.[53,54] The closer this value is to the number one, the model has high prediction accuracy and the SVM$_A$ is known as a reliable method; conversely, the closer it is to the number 0.5, the less accurate.[55]

3 Results

3.1 Results of prioritizing the influencing factors

The importance of the 12 variables affecting forest fire susceptibility was based on MDAm (Fig. 4). According to the MDAm method, the "land use/land cover" variable is the most important factor in the occurrence of fire at Zohreh watershed, followed by "aspect," "annual temperature," "distance to road," "river density," "altitude," "slope degree," "plan curvature," "distance to urban," "annual rain," "distance to the river" and "TWI." Therefore, "land use" and "TWI" variables were recognized as the most and the least effective factors on the occurrence of fire, respectively. Other research results on prioritizing factors affecting fire susceptibility confirmed the results of the present study. For example, the researchers showed that human activities such as land-use changes are highly correlated with fire susceptibility.[3,56] In another study, it was proved that among the various topographic factors, the slope direction has a strong and positive relationship with forest fire sensitivity because it is more affected by other factors.[57,58] On the other hand, among climatic factors, the annual temperature has a significant relationship with fire susceptibility, which has been proven in some studies.[59,60]

3.2 Forest fire susceptibility zonation using SVM$_A$

The Final Forest fire susceptibility map using SVMA) at the Zohreh watershed, Fars Province is shown in Fig. 5. The forest fire map was prepared using the "SVM" in R.3.5.3 software and was classified into 4 susceptibility classes ("Low, Moderate, High and Very high") based on the "Natural break" method in ArcGIS.10.6.1.[2,22,54] Table 1 shows the percentages of forest fire susceptibility areas in the study area. As it is known, the highest percentage of fire sensitivity area in the "Moderate" class is 39.11%. After that, the highest percentage of forest fire susceptibility was in the high

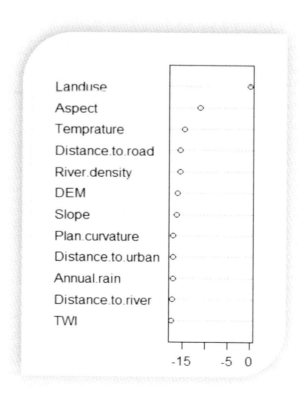

FIG. 4 Importance of effective factors using MDAm.

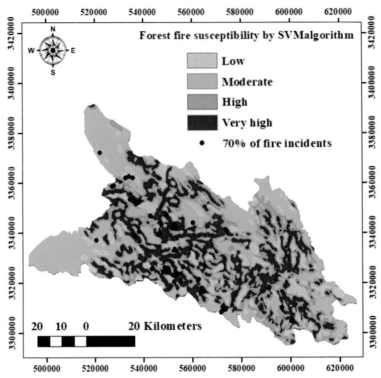

FIG. 5   The final map of forest fire susceptibility.

TABLE 1   Percentage of forest fire susceptibility area in the study area.

Forest fire classes	Area (%) SVM$_A$	The value of pixels in each class of sensitivity
Low (<0.24)	7.78	464,088
Moderate (0.24–0.50)	39.11	2,331,757
High (0.50–0.76)	26.96	1,607,584
Very high (0.76–1.19)	26.13	1,558,051
	100	

(26.96%), very high (26.13%), and low (7.78%) classes, which shows that the Zohreh watershed in Fars Province is prone to fire occurrence.

3.3 Checking the accuracy of SVM$_A$ using the ROC curve

Evaluating the algorithm used in research can be a fundamental step in creating susceptibility and determining the quality of the model in predicting the occurrence of disasters such as forest fires.[61] In the present study, the ROC curve was used to evaluate the accuracy and efficiency of the SVM model. Fig. 6 and Table 2 show that the SVM algorithm with an AUC value of 0.619 and acceptable accuracy at the Zohreh watershed. Various studies have shown that the SVM algorithm has excellent accuracy in predicting forest fires, which is consistent with the present study. For example, Ngoc Thach et al.[24] showed that the SVM algorithm, with an AUC value of 0.867, has good accuracy in predicting forest fire susceptibility. Details of the accuracy of Algorithm SVM in predicting fire sensitivity are given in Table 2.

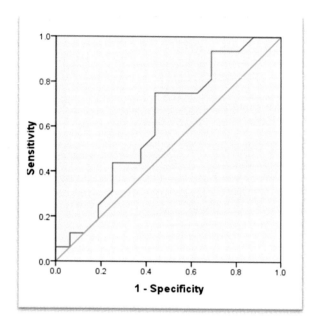

FIG. 6   ROC curve for SVM_A.

TABLE 2   Assessing the accuracy of the model used for zoning forest fire sensitivity.

Area	Standard error	Asymptotic significant	Lower bound	Upper bound
0.619	0.101	0.250	0.421	0.817

Asymptotic 95% Confidence Interval.

4 Discussion

Forest fires have many consequences, including reducing biodiversity and the production or biomass of forest eco-systems. It also affects forest carbon reserves, which in turn reduces soil fertility and increases air pollution.[62] The comparative significance of the effective factors using the MDAm method indicated that "land use/land cover," "aspect," "annual temperature" are the first three factors in terms of prioritization and impact, respectively. The damage caused by recent fires in Iran has shown that land-use changes play a significant role in fire sensitivity. In the present study, the effect of human activities, especially land use, intensified the fire in the study area. Therefore, vegetation density can be considered as the main criterion. This is consistent with previous research.[10,26] The second important factor of the topographic parameters is the direction of the slope. The aspect classes and categories in the area also caused greater fire sensitivity. Especially in the plains and southern directions, this can be due to the continuity and greater impact of sunlight, increasing the temperature and reducing the relative humidity of the surface in the directions, which subsequently leads to a forest fire that in previous studies has emphasized the importance of this issue.[3,63–65] This is because it strongly affects other factors such as wind direction, temperature, and humidity, which are important factors affecting fire sensitivity.[66,67]

On the other hand, the third most important factor is the annual temperature, which indicates that the climate is strongly related to the patterns of fire combustion.[68,69] The annual temperature is one of the most effective factors in forest and rangeland fire susceptibility and the frequency of fires in areas that annual temperature increases.[59,70] The Humidity of fuels in areas that experience higher temperatures throughout the year always has higher flammability due to low humidity.[71] The RF algorithm is widely used as a common machine learning algorithm.[72] An important and vital policy in spatial modeling and predicting the sensitivity of the study area to fire is that the results obtained are reliable; thus, a meaningful and reliable interpretation can be provided according to the fire sensitivity in the forest.[73] So far, statistical methods and machine learning have been used to map forest fires, which has provided very good and reliable results.[74,75]

It can be said that different models of machine learning have a high accuracy in predicting the sensitivity of natural disasters such as fire. Machine learning algorithms are extremely powerful, as they can be easily classified. It also uses a large amount of information and data with a large number of effective factors as dependent and independent variables without deleting even one variable.[76]

5 Conclusion

As mentioned, in the present study, 53 forest fire sites in the Zohreh watershed and 12 effective factors were used to map forest fire susceptibility. Spatial modeling and variable importance were performed using the SVM_A and random forest techniques. Despite the limitations created during a spatial modeling study, such as time, machine learning models provide a very useful and credible tool for planners, local authorities, and environmental managers in fire control. Thus, the sensitivity of fire sites can be predicted, and the resulting damage can be reduced.

References

1. Wang X, Wotton BM, Cantin AS, et al. cffdrs: an R package for the Canadian Forest fire danger rating system. *Ecol Process*. 2017;6(1):5. https://doi.org/10.1186/s13717-017-0070-z.
2. Razavi-Termeh SV, Sadeghi-Niaraki A, Choi S-M. Ubiquitous GIS-based forest fire susceptibility mapping using artificial intelligence methods. *Remote Sens (Basel)*. 2020;12(10):1689. Retrieved from https://www.mdpi.com/2072-4292/12/10/1689.
3. Su Z, Tigabu M, Cao Q, Wang G, Hu H, Guo F. Comparative analysis of spatial variation in forest fire drivers between boreal and subtropical ecosystems in China. *For Ecol Manage*. 2019;454:117669. https://doi.org/10.1016/j.foreco.2019.117669.
4. Xiong Q, Luo X, Liang P, et al. Fire from policy, human interventions, or biophysical factors? Temporal–spatial patterns of forest fire in southwestern China. *For Ecol Manage*. 2020;474:118381. https://doi.org/10.1016/j.foreco.2020.118381.
5. Yin H-W, Kong F-H, Li X-Z. RS and GIS-based forest fire risk zone mapping in da hinggan mountains. *Chin Geogr Sci*. 2004;14(3):251–257. https://doi.org/10.1007/s11769-003-0055-y.
6. Adab H, Kanniah KD, Solaimani K. Modeling forest fire risk in the northeast of Iran using remote sensing and GIS techniques. *Nat Hazards*. 2013;65(3):1723–1743. https://doi.org/10.1007/s11069-012-0450-8.
7. Alexander ME, Cruz MG. Fireline intensity. In: Manzello SL, ed. *Encyclopedia of Wildfires and Wildland-Urban Interface (WUI) Fires*. Cham: Springer International Publishing; 2018:1–8.
8. Nami MH, Jaafari A, Fallah M, Nabiuni S. Spatial prediction of wildfire probability in the Hyrcanian ecoregion using evidential belief function model and GIS. *Int J Environ Sci Technol*. 2018;15(2):373–384. https://doi.org/10.1007/s13762-017-1371-6.
9. Bax V, Francesconi W. Environmental predictors of forest change: an analysis of natural predisposition to deforestation in the tropical Andes region, Peru. *Appl Geogr*. 2018;91:99–110.
10. Parajuli A, Gautam AP, Sharma SP, et al. Forest fire risk mapping using GIS and remote sensing in two major landscapes of Nepal. *Geomat Nat Haz Risk*. 2020;11(1):2569–2586. https://doi.org/10.1080/19475705.2020.1853251.
11. Chuvieco E, Aguado I, Yebra M, et al. Development of a framework for fire risk assessment using remote sensing and geographic information system technologies. *Ecol Model*. 2010;221(1):46–58 https://doi.org/10.1016/j.ecolmodel.2008.11.017.
12. Eskandari S. Investigation on the relationship between climate change and fire in the forests of Golestan Province. *Iran J For Range Protect Res*. 2015;13(1):Pe1–Pe10.
13. Gigović L, Pourghasemi HR, Drobnjak S, Bai S. Testing a new ensemble model based on SVM and random forest in forest fire susceptibility assessment and its mapping in Serbia's Tara National Park. *Forests*. 2019;10(5):408. Retrieved from https://www.mdpi.com/1999-4907/10/5/408.
14. Hong H, Naghibi SA, Moradi Dashtpagerdi M, Pourghasemi HR, Chen W. A comparative assessment between linear and quadratic discriminant analyses (LDA-QDA) with frequency ratio and weights-of-evidence models for forest fire susceptibility mapping in China. *Arab J Geosci*. 2017;10(7):167. https://doi.org/10.1007/s12517-017-2905-4.
15. Oliveira S, Oehler F, San-Miguel-Ayanz J, Camia A, Pereira JMC. Modeling spatial patterns of fire occurrence in Mediterranean Europe using multiple regression and random Forest. *For Ecol Manage*. 2012;275:117–129. https://doi.org/10.1016/j.foreco.2012.03.003.
16. Tien Bui D, Hoang N-D, Samui P. Spatial pattern analysis and prediction of forest fire using new machine learning approach of multivariate adaptive regression splines and differential flower pollination optimization: a case study at Lao Cai province (Viet Nam). *J Environ Manage*. 2019;237:476–487. https://doi.org/10.1016/j.jenvman.2019.01.108.
17. Tien Bui D, Le K-TT, Nguyen VC, Le HD, Revhaug I. Tropical Forest fire susceptibility mapping at the cat Ba National Park Area, Hai Phong City, Vietnam, using GIS-based kernel logistic regression. *Remote Sens (Basel)*. 2016;8(4):347. Retrieved from https://www.mdpi.com/2072-4292/8/4/347.
18. Ljubomir G, Pamučar D, Drobnjak S, Pourghasemi HR. 15—Modeling the spatial variability of forest fire susceptibility using geographical information systems and the analytical hierarchy process. In: Pourghasemi HR, Gokceoglu C, eds. *Spatial Modeling in GIS and R for Earth and Environmental Sciences*. Elsevier; 2019:337–369.
19. Pourghasemi HR, Gayen A, Lasaponara R, Tiefenbacher JP. Application of learning vector quantization and different machine learning techniques to assessing forest fire influence factors and spatial modelling. *Environ Res*. 2020;184:109321. https://doi.org/10.1016/j.envres.2020.109321.
20. Boychuk D, Perera AH, Ter-Mikaelian MT, Martell DL, Li C. Modelling the effect of spatial scale and correlated fire disturbances on forest age distribution. *Ecol Model*. 1997;95(2):145–164. https://doi.org/10.1016/S0304-3800(96)00042-7.

21. Eskandari S, Amiri M, Sādhasivam N, Pourghasemi HR. Comparison of new individual and hybrid machine learning algorithms for modeling and mapping fire hazard: a supplementary analysis of fire hazard in different counties of Golestan Province in Iran. *Nat Hazards.* 2020;104(1):305–327. Retrieved from https://EconPapers.repec.org/RePEc:spr:nathaz:v:104:y:2020:i:1:d:10.1007_s11069-020-04169-4.

22. Moayedi H, Mehrabi M, Bui DT, Pradhan B, Foong LK. Fuzzy-metaheuristic ensembles for spatial assessment of forest fire susceptibility. *J Environ Manage.* 2020;260:109867. https://doi.org/10.1016/j.jenvman.2019.109867.

23. Li X-Y, Jin H-J, Wang H-W, et al. Influences of forest fires on the permafrost environment: a review. *Adv Clim Chang Res.* 2021. https://doi.org/10.1016/j.accre.2021.01.001.

24. Ngoc Thach N, Bao-Toan Ngo D, Xuan-Canh P, et al. Spatial pattern assessment of tropical forest fire danger at Thuan Chau area (Vietnam) using GIS-based advanced machine learning algorithms: a comparative study. *Eco Inform.* 2018;46:74–85. https://doi.org/10.1016/j.ecoinf.2018.05.009.

25. Wang S, Hu Y. A forest fire rescue strategy based on variable extinguishing rate. *Alex Eng J.* 2021;60(1):1271–1289. https://doi.org/10.1016/j.aej.2020.10.050.

26. Ying L, Han J, Du Y, Shen Z. Forest fire characteristics in China: spatial patterns and determinants with thresholds. *For Ecol Manage.* 2018;424:345–354. https://doi.org/10.1016/j.foreco.2018.05.020.

27. Chuvieco E, Congalton RG. Application of remote sensing and geographic information systems to forest fire hazard mapping. *Remote Sens Environ.* 1989;29(2):147–159. https://doi.org/10.1016/0034-4257(89)90023-0.

28. López-Vicente M, González-Romero J, Lucas-Borja ME. Forest fire effects on sediment connectivity in headwater sub-catchments: evaluation of indices performance. *Sci Total Environ.* 2020;732:139206. https://doi.org/10.1016/j.scitotenv.2020.139206.

29. Vadrevu KP, Eaturu A, Badarinath KVS. Fire risk evaluation using multicriteria analysis—a case study. *Environ Monit Assess.* 2010;166(1):223–239. https://doi.org/10.1007/s10661-009-0997-3.

30. Eskandari S, Chuvieco E. Fire danger assessment in Iran based on geospatial information. *Int J Appl Earth Obs Geoinf.* 2015;42:57–64. https://doi.org/10.1016/j.jag.2015.05.006.

31. Beven KJ, Kirkby MJ. A physically based, variable contributing area model of basin hydrology/Un modèle à base physique de zone d'appel variable de l'hydrologie du bassin versant. *Hydrolog Sci Bull.* 1979;24(1):43–69. https://doi.org/10.1080/02626667909491834.

32. Moore ID, Grayson RB, Ladson AR. Digital terrain modelling: a review of hydrological, geomorphological, and biological applications. *Hydrol Process.* 1991;5:3. https://doi.org/10.1002/hyp.3360050103.

33. Butsic V, Kelly M, Moritz MA. Land use and wildfire: a review of local interactions and teleconnections. *Landarzt.* 2015;4(1):140–156. Retrieved from https://www.mdpi.com/2073-445X/4/1/140.

34. Feurdean A, Florescu G, Tanţău I, et al. Recent fire regime in the southern boreal forests of western Siberia is unprecedented in the last five millennia. *Quat Sci Rev.* 2020;244:106495. https://doi.org/10.1016/j.quascirev.2020.106495.

35. Monjarás-Vega NA, Briones-Herrera CI, Vega-Nieva DJ, et al. Predicting forest fire kernel density at multiple scales with geographically weighted regression in Mexico. *Sci Total Environ.* 2020;718:137313. https://doi.org/10.1016/j.scitotenv.2020.137313.

36. Stevens JT, Collins BM, Miller JD, North MP, Stephens SL. Changing spatial patterns of stand-replacing fire in California conifer forests. *For Ecol Manage.* 2017;406:28–36. https://doi.org/10.1016/j.foreco.2017.08.051.

37. Couto FT, Iakunin M, Salgado R, Pinto P, Viegas T, Pinty J-P. Lightning modelling for the research of forest fire ignition in Portugal. *Atmos Res.* 2020;242:104993. https://doi.org/10.1016/j.atmosres.2020.104993.

38. Drobyshev I, Niklasson M, Ryzhkova N, Götmark F, Pinto G, Lindbladh M. Did forest fires maintain mixed oak forests in southern Scandinavia? A dendrochronological speculation. *For Ecol Manage.* 2021;482:118853. https://doi.org/10.1016/j.foreco.2020.118853.

39. Shukla K, Kumar P, Mann GS, Khare M. Mapping spatial distribution of particulate matter using kriging and inverse distance weighting at supersites of megacity Delhi. *Sustain Cities Soc.* 2020;54:101997. https://doi.org/10.1016/j.scs.2019.101997.

40. van Mierlo C, Faes MGR, Moens D. Inhomogeneous interval fields based on scaled inverse distance weighting interpolation. *Comput Methods Appl Mech Eng.* 2021;373:113542. https://doi.org/10.1016/j.cma.2020.113542.

41. Yao X, Yu K, Deng Y, Zeng Q, Lai Z, Liu J. Spatial distribution of soil organic carbon stocks in Masson pine (*Pinus massoniana*) forests in subtropical China. *Catena.* 2019;178:189–198. https://doi.org/10.1016/j.catena.2019.03.004.

42. Rohmer J, Idier D, Paris F, Pedreros R, Louisor J. Casting light on forcing and breaching scenarios that lead to marine inundation: combining numerical simulations with a random-forest classification approach. *Environ Model Software.* 2018;104:64–80. https://doi.org/10.1016/j.envsoft.2018.03.003.

43. Fouedjio F. Exact conditioning of regression random forest for spatial prediction. *Artif Intell Geosci.* 2021. https://doi.org/10.1016/j.aiig.2021.01.001.

44. Masroor M, Rehman S, Sajjad H, et al. Assessing the impact of drought conditions on groundwater potential in Godavari Middle Sub-Basin, India using analytical hierarchy process and random forest machine learning algorithm. *Groundw Sustain Dev.* 2021;100554. https://doi.org/10.1016/j.gsd.2021.100554.

45. Carranza C, Nolet C, Pezij M, van der Ploeg M. Root zone soil moisture estimation with random forest. *J Hydrol.* 2021;593:125840. https://doi.org/10.1016/j.jhydrol.2020.125840.

46. Khammas BM. Ransomware detection using random forest technique. *ICT Express.* 2020;6(4):325–331. https://doi.org/10.1016/j.icte.2020.11.001.

47. Zhang S, Liu G, Chen S, Rasmussen C, Liu B. Assessing soil thickness in a black soil watershed in Northeast China using random forest and field observations. *Int Soil Water Conserv Res.* 2020. https://doi.org/10.1016/j.iswcr.2020.09.004.

48. Yu L, Yao X, Zhang X, Yin H, Liu J. A novel dual-weighted fuzzy proximal support vector machine with application to credit risk analysis. *Int Rev Financ Anal.* 2020;71:101577. https://doi.org/10.1016/j.irfa.2020.101577.

49. Jiménez-Cordero A, Morales JM, Pineda S. A novel embedded min-max approach for feature selection in nonlinear support vector machine classification. *Eur J Oper Res.* 2020;293:24–35. https://doi.org/10.1016/j.ejor.2020.12.009.

50. Moeini R, Babaei M. Hybrid SVM-CIPSO methods for optimal operation of reservoir considering unknown future condition. *Appl Soft Comput.* 2020;95:106572. https://doi.org/10.1016/j.asoc.2020.106572.

51. Liu M-Z, Shao Y-H, Li C-N, Chen W-J. Smooth pinball loss nonparallel support vector machine for robust classification. *Appl Soft Comput.* 2021;98:106840. https://doi.org/10.1016/j.asoc.2020.106840.

52. Bradley AP. The use of the area under the ROC curve in the evaluation of machine learning algorithms. *Pattern Recognit.* 1997;30(7):1145–1159. https://doi.org/10.1016/S0031-3203(96)00142-2.

53. Azareh A, Rahmati O, Rafiei-Sardooi E, et al. Modelling gully-erosion susceptibility in a semi-arid region, Iran: investigation of applicability of certainty factor and maximum entropy models. *Sci Total Environ.* 2019;655:684–696. https://doi.org/10.1016/j.scitotenv.2018.11.235.

54. Roy J, Saha S. Integration of artificial intelligence with meta classifiers for the gully erosion susceptibility assessment in Hinglo river basin, Eastern India. *Adv Space Res.* 2021;67(1):316–333. https://doi.org/10.1016/j.asr.2020.10.013.

55. Tripepi G, Jager KJ, Dekker FW, Zoccali C. Diagnostic methods 2: receiver operating characteristic (ROC) curves. *Kidney Int.* 2009;76(3):252–256. https://doi.org/10.1038/ki.2009.171.

56. Gale MG, Cary GJ, Van Dijk AIJM, Yebra M. Forest fire fuel through the lens of remote sensing: review of approaches, challenges and future directions in the remote sensing of biotic determinants of fire behaviour. *Remote Sens Environ.* 2021;255:112282. https://doi.org/10.1016/j.rse.2020.112282.

57. Guo F, Su Z, Wang G, et al. Understanding fire drivers and relative impacts in different Chinese forest ecosystems. *Sci Total Environ.* 2017;605-606:411–425. https://doi.org/10.1016/j.scitotenv.2017.06.219.

58. Sá ACL, Pereira JMC, Charlton ME, Mota B, Barbosa PM, Stewart Fotheringham A. The pyrogeography of sub-Saharan Africa: a study of the spatial non-stationarity of fire–environment relationships using GWR. *J Geogr Syst.* 2011;13(3):227–248. https://doi.org/10.1007/s10109-010-0123-7.

59. Luo C, Shen Z, Li Y, et al. Determinants of post–fire regeneration demography in a subtropical monsoon–climate forest in Southwest China. *Sci Total Environ.* 2020;142605. https://doi.org/10.1016/j.scitotenv.2020.142605.

60. Oliveira S, Pereira JMC, San-Miguel-Ayanz J, Lourenço L. Exploring the spatial patterns of fire density in southern Europe using geographically weighted regression. *Appl Geogr.* 2014;51:143–157. https://doi.org/10.1016/j.apgeog.2014.04.002.

61. Chung C-JF, Fabbri AG. Validation of spatial prediction models for landslide hazard mapping. *Nat Hazards.* 2003;30(3):451–472. https://doi.org/10.1023/B:NHAZ.0000007172.62651.2b.

62. Amiro BD, MacPherson JI, Desjardins RL. BOREAS flight measurements of forest-fire effects on carbon dioxide and energy fluxes. *Agric For Meteorol.* 1999;96(4):199–208. https://doi.org/10.1016/S0168-1923(99)00050-7.

63. Çolak E, Sunar F. The importance of ground-truth and crowdsourcing data for the statistical and spatial analyses of the NASA FIRMS active fires in the Mediterranean Turkish forests. *Remote Sens Appl Soc Environ.* 2020;19:100327. https://doi.org/10.1016/j.rsase.2020.100327.

64. Dhall A, Dhasade A, Nalwade A, Raj VKM, Kulkarni V. A survey on systematic approaches in managing forest fires. *Appl Geogr.* 2020;121:102266. https://doi.org/10.1016/j.apgeog.2020.102266.

65. Ng J, North MP, Arditti AJ, Cooper MR, Lutz JA. Topographic variation in tree group and gap structure in Sierra Nevada mixed-conifer forests with active fire regimes. *For Ecol Manage.* 2020;472:118220. https://doi.org/10.1016/j.foreco.2020.118220.

66. Kherchouche D, Slimani S, Touchan R, Touati D, Malki H, Baisan CH. Fire human-climate interaction in Atlas cedar forests of Aurès, Northern Algeria. *Dendrochronologia.* 2019;55:125–134. https://doi.org/10.1016/j.dendro.2019.04.005.

67. Viccaro M, Cozzi M, Fanelli L, Romano S. Spatial modelling approach to evaluate the economic impacts of climate change on forests at a local scale. *Ecol Indic.* 2019;106:105523. https://doi.org/10.1016/j.ecolind.2019.105523.

68. Ai J, Sun X, Feng L, Li Y, Zhu X. Analyzing the spatial patterns and drivers of ecosystem services in rapidly urbanizing Taihu Lake Basin of China. *Front Earth Sci.* 2015;9(3):531–545. https://doi.org/10.1007/s11707-014-0484-1.

69. Guo F, Su Z, Tigabu M, et al. Spatial modelling of fire drivers in urban-forest ecosystems in China. *Forests.* 2017;8:180.

70. Davis R, Yang Z, Yost A, Belongie C, Cohen W. The normal fire environment—modeling environmental suitability for large forest wildfires using past, present, and future climate normals. *For Ecol Manage.* 2017;390:173–186. https://doi.org/10.1016/j.foreco.2017.01.027.

71. Hamilton R, Stevenson J, Li B, Bijaksana S. A 16,000-year record of climate, vegetation and fire from Wallacean lowland tropical forests. *Quat Sci Rev.* 2019;224:105929. https://doi.org/10.1016/j.quascirev.2019.105929.

72. Nhu V-H, Mohammadi A, Shahabi H, et al. Landslide susceptibility mapping using machine learning algorithms and remote sensing data in a tropical environment. *Int J Environ Res Public Health.* 2020;17(14):4933. https://doi.org/10.3390/ijerph17144933.

73. Rossi J-L, Chetehouna K, Collin A, Moretti B, Balbi J-H. Simplified flame models and prediction of the thermal radiation emitted by a flame front in an outdoor fire. *Combust Sci Technol.* 2010;182(10):1457–1477. https://doi.org/10.1080/00102202.2010.489914.

74. Arunakranthi G, Rajkumar B, Chandra Shekhar Rao V, Harshavardhan A. Advanced patterns of predictions and cavernous data analytics using quantum machine learning. *Mater Today Proc.* 2021. https://doi.org/10.1016/j.matpr.2020.11.062.

75. Raju B, Bonagiri R. A cavernous analytics using advanced machine learning for real world datasets in research implementations. *Mater Today Proc.* 2020. https://doi.org/10.1016/j.matpr.2020.11.089.

76. Yin J, Medellín-Azuara J, Escriva-Bou A, Liu Z. Bayesian machine learning ensemble approach to quantify model uncertainty in predicting groundwater storage change. *Sci Total Environ.* 2021;144715. https://doi.org/10.1016/j.scitotenv.2020.144715.

26

Pest-infected oak trees identify using remote sensing-based classification algorithms

Saleh Yousefi[a], Farshad Haghighian[b], Mojtaba Naghdyzadegan Jahromi[c], and Hamid Reza Pourghasemi[d]

[a]Soil Conservation and Watershed Management Research Department, Chaharmahal and Bakhtiari Agricultural and Natural Resources Research and Education Center, AREEO, Shahrekord, Iran [b]Research Division of Natural Resources, Chaharmahal and Bakhtiari Agricultural and Natural Resources Research and Education Center, AREEO, Shahrekord, Iran [c]Department of Water Engineering, School of Agriculture, Shiraz University, Shiraz, Iran [d]Department of Natural Resources and Environmental Engineering, College of Agriculture, Shiraz University, Shiraz, Iran

1 Introduction

Forest with an area of 12.4 million hectares covering 7.4% of Iran.[1,2] The two main forest ecosystems in Iran are the Hyrcanian and Zagros Forest habitats. The Zagros forests spread across 11 provinces from the northwest to the southwest through a mountainous region and comprise 6 million hectares, accounting for 40% of the Iranian forests.[3] Oak species account for more than 70% of Zagros forests. Despite deforestation by direct (i.e., land-use change, civil forest projects, wood harvesting, and leaf grazing by domesticated animals) and indirect (i.e., pollution, climate change, transportation development, and electrical power generation and transmission) disturbances, forests are also destroyed by a variety of pests. These direct and indirect human activities may also increase or decrease the distribution and intensity of pests at several spatial and temporal scales.[4-8]

Tortrix viridana and *Tibicina plebejus* are the two most destructive pests in oak forests and together cause a high decline in oaks across thousands of hectares throughout northern and western Iran. *T. viridana* belongs to the family Tortricidae and subfamily Tortricinae. It is an oligophagous species (feeding primarily on different Oak species) in oak forests in western Iran.[9] *T. viridana* also feeds on the shrubs *Vaccinium* and *Urtica*.[10-12] Currently, it is flourishing throughout the western Palearctic region (including continental Europe, North Africa, Cyprus, Iran, and other Mediterranean countries). The pest is commonly found in oaks during the early defoliation season.[12] In the worst case, it is capable of completely defoliating an oak tree. Unfortunately, it is widely distributed throughout the Zagros oak forest and has caused severe damage, especially in the Chaharmahal-e-Bakhtiari Province. *T. viridana* has a life cycle of only 1 year; moths are present during the late spring and early summer seasons. Females lay eggs under a scale on twigs.[13] The larvae feed on buds and the host tree leaves and as the larvae age, the damages will increase. During mass outbreaks, larvae can cause complete defoliation.[4] Defoliated trees become more susceptible to secondary pests and other pathogens[10] and after 2 or 3 years of damage and defoliation, the trees begin to decline and may die within 4 years.[14] On-site monitoring and detection of pests are time-consuming and costly.[14-16]

T. plebejus is a cicada found throughout Europe and the Middle East (or southwest Asia). This cicada belongs to the family Cicadidae, subfamily Tibicininae, and the order Hemiptera.[17] The males call females to attract them to their habitat for pairing.[18] The females oviposit eggs into the bark of twigs. After hatching, the nymphs drop to the soil and burrow to the tree roots. Nymphs live underground for most of their lives deep in the soil.[19] Nymphs, with strong front legs, burrow and dig through the soil, feed on xylem sap from tree roots. Trees attacked by *T. plebejus* nymphs were weakened. They become fragile and photosynthesis is hampered, as nourishment is parasitized from the roots. The maturing nymphs exit the soil and emerge as adults by molting their exoskeletons.[17,19]

Knowledge of the distribution, severity, and extent of forest pests and the damage they cause is a vital need for ecologists and managers of resources and the environment to understand the implications of infestations to assess mortality and to prioritize rehabilitation and restoration activities to mitigate further damage.[20] One of the most costly and time-consuming activities in forest management is the on-site monitoring and detection of pests.[14,20] Therefore, the development of methods that decrease the time and funding needed for these activities can be very helpful to forest managers. During the last two decades, remote sensing (RS) has been widely used in natural science research.[14,21-27] Imagery from remotely sensed data has great potential to provide detailed information about pest infestation. The processes of data collection and analysis are easily repeated, data can be collected over extensive areas quickly, and data collection can be accomplished at a much lower cost than field-work. RS has been used to map many pest species in crops and agricultural lands.[28-31] Studies of RS applications for agricultural pest monitoring have examined numerous types of satellite data.[32-36] Forests and rangelands are also extensively affected by pests, yet only a few studies have examined pest monitoring in forests using RS.

Ismail et al.[37] used high-resolution (50 cm) imagery to study damage to tree crowns by *Sirex noctilio* (Lep. Siricidae). Their study showed that high-resolution airborne imagery vegetation indices can be used to discriminate between infected and healthy trees. Goodwin et al.[15] used Landsat data and the normalized difference moisture index (NDMI) to monitor insect infestation in British Colombia. They found that mountain pine beetle attacks increased from 0% to 15% in the study region between 1992 and 2006. Vogelmann et al.[38] studied the changes in forest health in the southern United States using the Landsat data over a period of 18 years and observed that the vegetation SWIR (short-wave infrared) /NIR (near-infrared) index revealed a decrease in crown greenness over the studied period. Furthermore, their study showed that Landsat satellite data can be used to monitor long-term forest health. Meddens et al.[39] examined the use of single- and multi-date Landsat data to locate dying trees. They reported that Landsat single- and multi-date imagery aided in the detection of tree mortality caused by bark beetles. Hawryło et al.[40] also studied the defoliation of Scot pine stands in western Poland. Spruce et al.[20] used MODIS NDVI to map tree mortality levels in forest lands affected by beetle outbreaks in Colorado, USA. These results indicate that NDVI products can be successfully used to map tree mortality. In another study in northern Iran, Abdi[16] studied insect defoliators using multispectral Landsat data, showing that NDVI and GEMI derived from OLI Landsat images have great promise for insect-defoliation detection.

Apparent from the literature is that image-classification methods almost have even been used and compared to land use/cover mapping processes[41-43] and there is no specific image-classification algorithm that is deemed most effective for pest infestation detection. In contrast to the aforementioned studies, it might be useful to test the use of classification methods to map infestation areas instead of using only vegetation indices. This study attempts to fill this gap by comparing the predictions of oak tree infestations by two insect species made with seven classification algorithms.

Sentinel-2 is a high spatial-resolution satellite data series (10-60m) and its mission is to provide for a wide range of uses, including land cover mapping, vegetation and forest monitoring (vegetation indices, chlorophyll, and greenness concentrations), water quality, and natural disasters mapping, and others.[40,44] Pest-infestation detection by RS analysis is based on spectral changes occurring in the green parts of trees and on canopy structures.[14,40,44]

The Zagros Forest is a Mediterranean forest that is important for soil conservation and ecosystem health in Iran. Pest-infestation mapping with Sentinel-2 data and RS techniques in these forest ecosystems could be valuable for managing and protecting forests. Therefore, the purpose of this study was to determine an image-classification algorithm that can accurately predict infestation at two sites in the Zagros Forest using Sentinel-2 imagery.

2 Materials and methods

2.1 Study area and field data collections

The study area comprises two forest sites (A and B) forest sites located in southwestern Chaharmahal-e-Bakhtiari Province in western Iran (Fig. 1). The elevations of the study sites were 1630 (Site A) and 1820 m. a. s. l. (Site B). The daily average temperature is 17°C, and annual precipitation is approximately 530 mm at both sites. The average slope of Site A was 30% and that of Site B was 36%. Site A was infested by *T. plebejus* and Site B was infested with *T. viridana*. The sites were approximately 4 km apart. The main forest species at both sites is *Quercus branti*, an important host for both pests, and is known as a vital species in the Zagros Forest and plays a significant role in maintaining ecosystem health. Areas within the sites affected by *T. viridana* and *T. plebejus* were mapped using GPS during the spring of 2016. Inputting 501 field locations from the two study sites that reference the conditions caused by the two pests (comprising 316 locations that were not affected and 185 that were affected; Table 1) into the seven classification algorithms, predictive maps of

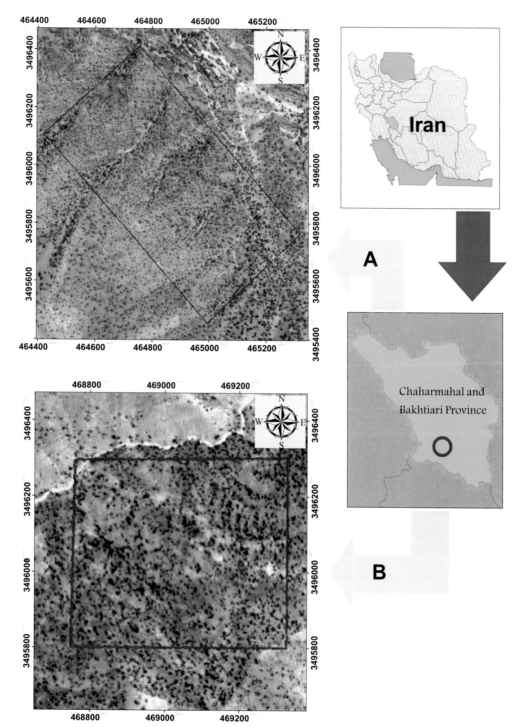

FIG. 1   The geographical location of two sites (A & B) infested by pests.

TABLE 1   Distribution of reference sites in this study.

Pest	Training	Assessment	Total
T. plebejus	142	61	203
T. viridana	208	90	298
Both	350	151	501

T. viridana and *T. plebejus* were produced. The field points were located at least 10 m apart to match the spatial resolution of the Sentinel-2 satellite images. In the dataset, 87 sites were affected by *T. plebejus* and 98 were affected by *T. viridana*. Additionally, 116 locations were not affected by *T. plebejus* and 200 were not affected by *T. viridana*.

3 Methodology

Fig. 2 shows the flowchart of methodology in the present study.

3.1 Data source

Sentinel-2 satellite images of the study area, taken on, September 25/09/2016 were downloaded from the USGS website (https://earthexplorer.usgs.gov/). To correct the images for atmospheric influences, the fast line-of-sight atmospheric analysis of spectral hypercubes (FLAASH) correction method was used in the Sentinel Application Platform (SNAP) software.[45–47]

4 Image classification algorithms

4.1 Maximum likelihood (ML)

ML is one of the most common methods used for image classification.[48–50] ML is a highly accurate algorithm.[51–53] It calculates the probability that each pixel is in a defined class and assigns it to the most likely class. Classes should follow a Gaussian normal distribution.[41]

4.2 Neural network (NN)

The NN is a nonparametric algorithm that does not assume a normal distribution of sample sites.[43,54,55] Compared to ML, this algorithm can detect classes that do not necessarily have a spectral linear distribution. NN can discretely detail the process and incorporate input data in the output[56]; it is usually referred to as the "black box" method.[48] In

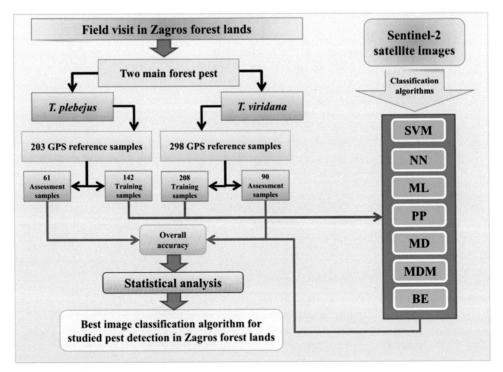

FIG. 2 Flow chart of the methodology.

this study, NN weights were initialized using a uniform distribution and the learning rate (LR) was defined as 100 for the hidden layer and 0.01 for the output layer; thus, the stopping criterion (SC) was fixed at 0.001.

4.3 Support vector machine (SVM)

The SVM algorithm is based on the theory introduced by Vapnik and Chervonenkis[57] and was later improved by others.[58–60] SVM is a non-parametric method that uses a set of learning algorithms (LA) for classification and regression.[60–64] It is based on statistical learning theory, which decreases uncertainty in the structure of the model.[65] This algorithm can classify images with greater accuracy by using small training sites.[66] SVM has been used in land use/land cover (LULC) mapping and natural hazards modeling. Studies have demonstrated that SVM is a more accurate classification method than other algorithms.[21,53,66,67]

4.4 Minimum distance to mean (MDM)

The average of each class defined by the training sample is calculated and the Euclidean distance of each pixel reflection from the average of all classes is calculated. Each pixel is assigned to the class based on the nearness of the pixel value of the classes' averages; the pixel is assigned to the class that has an average that is closest to the pixel's value.[41,68,69]

4.5 Mahalanobis distance (MD)

MD is similar to the MDM classification method, however, the covariance matrix can be used for the classification process.[41] MD is defined as a value between two points in a space that is considered most similar and appropriate based on its features.[70] Because it counts correlations between features for unequal variances, it also determines distance by calculating the weights or magnitudes of the variables for each data point.[70] The distance under a Mahalanobis distance metric is identical to the Euclidean distance metric when the features are not correlated.[21] Geometrically, a Mahalanobis distance metric can adjust the geometrical distribution of data when the distance between similar data points is low.[70,71] It is assumed that the histogram bands are normally distributed.[68]

4.6 Parallel piped (PP)

PP is a supervised classification method based on the variance in the spectral values of a training sample[72] that has been widely used to produce LULC maps from multispectral data.[42,72,73] The maximum and minimum pixel values were determined from the training samples for each class.[72] Pixel values are checked and classified sequentially to determine whether they are located inside any defined parallelepiped. If the pixel value is not within one of the defined classes, then the pixel is left unclassified.[73] The PP classification can be completed quickly, but the accuracy is quite variable, as parallelepipeds are formed using only the maximum and minimum pixel values.[42]

4.7 Binary encoding (BE)

BE is a simple and quick classification method.[74] The structure of this algorithm is based on spectral end-number coding with 0 and 1 values.[75] The values 0 and 1 are assigned if the band value is below or above the spectral average, respectively.[74–76]

5 Accuracy assessment

To evaluate the effectiveness of the above techniques, an accuracy assessment was performed. Using random selection, the sample sets were divided into two groups: training (70% of reference sites and validation (or assessment) (30% of reference sites). The results of the image classification procedures of the seven algorithms based on the training samples were analyzed to assess their accuracy by testing them using ground-truthed samples. The ground-truthed samples, 61 and 90 (for *T. plebejus* and *T. viridana*, respectively) sets were used to develop the confusion matrix with ENVI 5.3.1. The overall accuracy (OA) was used as an index for the algorithm-produced pest infection maps. OA is usually presented as a percentage. If the classification is perfect and shows that all reference sites are classified correctly, then OA = 100%. The calculation of OA is the ratio of the number of correctly classified sites (for all classes) to the total number of reference sites.

6 Statistical analysis

After determining the OA coefficient for each classification algorithm at both study sites, a one-way comparison of means test was used for the seven classification algorithms using IBM SPSS Statistics 22.[41] Further, Tukey's (honestly significant difference (HSD) was used to test the homogenous subset of the algorithms using the OA coefficient.

7 Results

Field visits to the study area showed that the situation was critical at both forest sites during the autumn of 2016 with both suffering from extreme infestations of *T. viridana* and *T. plebejus* (Fig. 3). The classification algorithms indicated that Sentinel-2 images performed well in the procedures to map these two pests in the Zagros Forest.

The results of the classifications of affected areas by the seven algorithms are shown in Fig. 4 (*T. plebejus*) and Fig. 5 (*T. viridana*) for both studied sites (A and B). The accuracy assessment revealed that SVM (with OAs of 86.73% and 70.37%) was the most accurate classification algorithm or mapped the pest-infested areas at sites A and B (Table 2). The one-way comparison of the means test shows that there are statistical differences among the products of the seven classification algorithms (based on OAs at 99% confidence levels (Table 3).

In addition, the results of Tukey's homogenous subset test indicate that in the case of OA, the seven studied classification algorithms fall into four categories (Fig. 6).

According to the parametric and nonparametric tests, the results show that pure parametric classifiers produce the most accurate pest-infection maps. The results of a one-way ANOVA test showed that there was no significant difference between the parametric and nonparametric classifiers in terms of the overall accuracy of the pest-detection maps (Table 4).

The results of classification of the affected and non-affected areas using classification algorithms for *T. plebejus* (Fig. 7) showed that 40% of the study area was classified as areas affected by *T. plebejus* using the SVM classification method (Fig. 7).

Similarly, the percentages of the study area affected by *T. viridana* varied between 2% and 56% at the study site (B) among the seven classification methods (Fig. 8). SVM is the most accurate classification method and only 6% of the study area was affected by *T. viridana*.

FIG. 3   (A) Defoliation of oak tree by *T. viridana* at study site B; (B) Adult moth of *T. viridana* after eclose; (C) Fifth instar *T. viridana* larva on oak leaf; (D) Nymphal skin of *T. plebejus* at study site A; and (E) Adult *T. plebejus* resting on an oak tree.

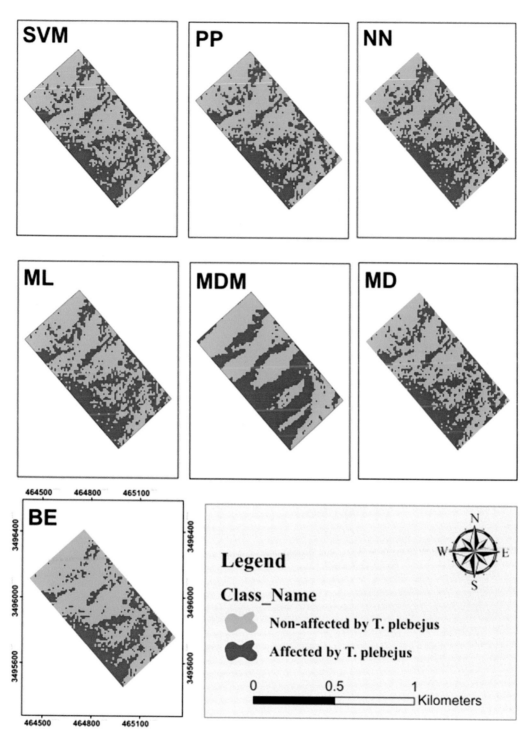

FIG. 4 *T. plebejus*-infestation maps predicted by the seven classification algorithms at study site A.

8 Discussion

Globally, *T. viridana* is one of the most defoliating forest pests.[11] This pest leaves visual evidence of its impacts on trees in spring, but most trees recover from the damage and grow new leaves after infestation.[14] New leaves are of poorer quality in terms of greenness and chlorophyll and are fewer in number.[12] The trees showed thinning canopies and lower vigor after 3 or 4 months.

T. plebejus has two main effects on attacked trees. First, females slice twigs for oviposition. When they are sliced, twigs dry out and tree growth stops in the infested parts of the tree. Second, nymphs feeding on roots interrupt

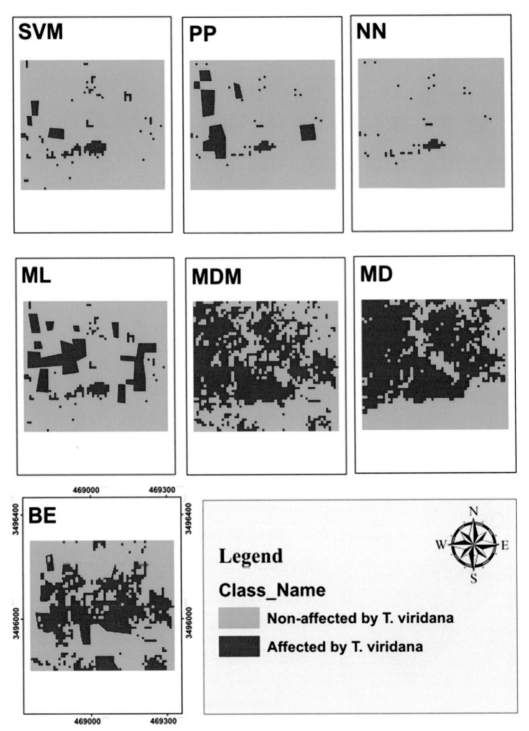

FIG. 5 *T. viridana*-infestation maps predicted by the seven classification algorithms at study site B.

the transfer of nutrients to the green parts of the tree. Eventually, the host trees loses their greenness and canopy cover. Tree decline in an area will occur if the pest-attacks' continue for several years or if the infestations are particularly severe.

The classification algorithms classified areas as affected and non-affected areas using multispectral images. The visual results indicated that oak decline occurred after 2 or 3 years of repeated defoliation. Our study found that at both sites A and B and for both pests, SVM is a more accurate classification procedure. On the other hand, BE was the least accurate classification method. ML, NN, and PP, with OAs greater than 60%, are acceptable algorithms for mapping infestations by *T. plebejus* and *T. viridana* (Fig. 6). All algorithms used in this study employed

TABLE 2 Accuracies (%) of the predicted distributions of *T. viridana* and *T. plebejus* infestations by the seven algorithms.

Algorithm	*T. viridana*	*T. plebejus*
BE	24.29	27.98
MD	52.5	55.74
MDM	53.29	55.25
ML	68.49	68.78
PP	66.92	65.74
SVM	70.37	86.73
NN	68.65	71.22

TABLE 3 ANOVA analyses of accuracies of the seven algorithms at two study sites.

	Sum of squares	df	Mean square	F	Significance
Between group	3542.19	6	590.36	27.21	.000[*]
Within group	151.84	7	21.69		
Total	3694.04	13			

[*] *The mean difference is significant at the 0.01 level or P value <.01.*

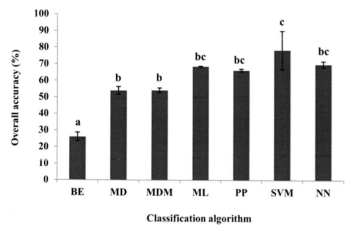

FIG. 6 Tukey's homogenous subset for seven image-classification algorithms.

TABLE 4 ANOVA analyses of the accuracies of parametric and nonparametric classifiers.

Pest	Groups	Sum of squares	df	Mean square	F	Significance
T. viridana and *T. plebejus*	Between group	326.62	1	326.62	1.164	.302
	Within group	3367.41	12	280.61		
	Total	3694.04	13			
T. viridana	Between group	106.33	1	106.33	0.346	.582
	Within group	1538.19	5	307.63		
	Total	1644.52	6			
T. plebejus	Between group	232.46	1	232.46	0.658	.454
	Within group	1765.24	5	353.04		
	Total	1997.71	6			

FIG. 7 The distributions of affected *(red)* and non-affected *(green)* areas by *T. plebejus* were predicted by the seven classification algorithms.

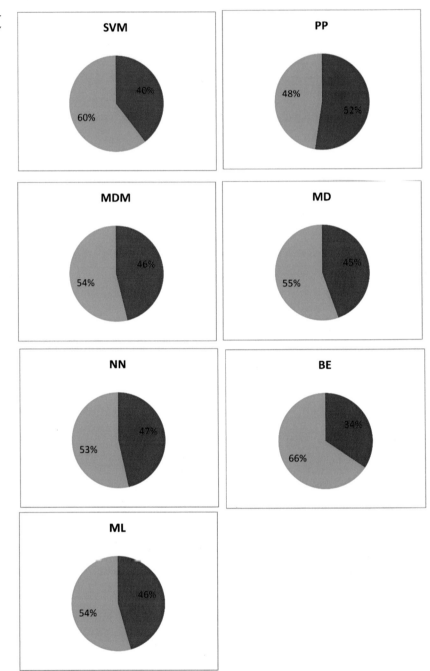

probability-based supervised classification. However, these algorithms can be grouped into parametric and non-parametric approaches. ML, SVM, MDM, and MD are parametric classification methods, and NN, PP, and BE are nonparametric. NN and SVM had the highest accuracies in the nonparametric and parametric groups, respectively. Parametric classifiers are based on the statistical probability distributions of the defined classes in the training samples. According to the results (Table 4), there is not enough statistical evidence to say that parametric classifiers are better than nonparametric classifiers for pest infestation mapping with Sentinel-2 images. However, the average OA for the parametric (63.89) classifiers is greater than the average OA of the nonparametric (54.13) algorithms.

One of the most important advantages of SVM classification is that it can solve the imbalance problems between training sites.[77,78] It is very important to have highly accurate classified images for both affected and non-affected

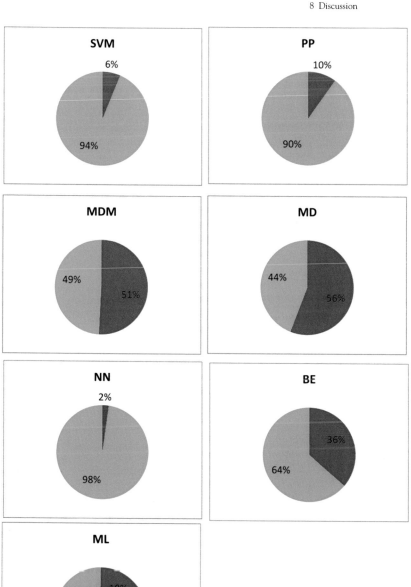

FIG. 8 The distributions of affected *(red)* and non-affected *(green)* areas by *T. viridana* predicted with each of the seven classification algorithms.

reference sites that can rely on RS for pest mapping. Both SVM and NN were acceptably accurate classifiers of Sentinel-2 images for both the affected and non-affected classes, and for both *T. plebejus* and *T. viridana*.

The greenness of aboveground biomass and canopy cover of stand trees are very important factors when using multispectral images. The attacks of *T. plebejus and T. viridana* in the forest stand affect the health and greenness of trees. Thus, canopy cover and spectral reflection are different and this is the basis for RS detection of pest infestations.

Most of the satellite image bands are basically and functionally related to vegetation canopy covers, such as leaf-area index, above ground biomass, and chlorophyll.[79–82] Therefore, they could be key indices for the detection of pests in surface vegetation. Gooshbor et al. stated that the NDVI values in areas affected by *T. viridana* decreased from 2007 to 2014. *T. viridana* destroys tree leaves, decreasing greenness levels, and chlorophyll in host trees. Therefore, greenness is the key factor to be used in our study to classify and detect affected and non-affected areas.[14] Many researchers have stated that healthy forest stands almost always have higher values of NDVI infested stands with pests because the NIR band decreases.[83–85] Forest canopy is the component most often used to estimate vegetation health, which is related to biotic and abiotic stress factors. The results confirm the study of[16] who stated that OLI images have great potential for the identification of forests infested by *T. viriadan* in northern Iran. Our results also support the conclusions of,[15] who state that Landsat images are capable of detecting insect infestation dynamics in forests.

Globally, pests and diseases can have significant negative impacts on natural ecosystems. One of the greatest concerns for forest management is pest control, which ensures forest survival. Monitoring pests in large, inaccessible forests is a significant challenge for managers. RS analysis alleviates this challenge. The results demonstrate the potential for integrated pest (microbial control, biological control, physical control, behavior control) management, and forest conservation. Over the past few decades, RS has successfully overcome important management challenges, but there are still unmet needs. RS is valuable for pest management because the data have wide spectral ranges (beyond the visible range) that can be used to assess the health of vegetation; satellites can cover a range of spatiotemporal scales, inaccessible areas can be monitored at large scales, spatial monitoring (RS combined with ground-truthing) can be very precise, and because RS is efficient and cost-effective.

9 Conclusion

Historically, forests have played very important roles in the evolution of ecosystems. Some forests such as the Zagros Mountain Forest are the heritage of past eras of Earth's history, therefore protection and integrated management of these valuable resources are key for sustainable development of the region. Identification of forest pests and their distribution is the first step toward an integrated forest pest management strategy. Identification is a costly endeavor when conducted in the field. RS techniques can provide valuable replacements, saving both time and money. This study sought to determine the best classification algorithm for mapping the areas affected by *T. plebejus* and *T. viridana* in the Zagros Forest in western Iran. The results indicate that SVM and NN have great potential for classification that can identify areas infested by *T. plebejus* and *T. viridana*. This study shows that there is immense potential for these techniques to be used for the detection of other pests, using other satellite imagery sources, and deriving indices from other imagery.

Acknowledgments

We would like to thank Shiraz University for supporting us with this study. This study was supported by the College of Agriculture, Shiraz University (Grant No. 99GRC1M271143).

References

1. Yousefi S, Moradi H, Boll J, Schönbrodt-Stitt S. Effects of road construction on soil degradation and nutrient transport in Caspian Hyrcanian mixed forests. *Geoderma*. 2016;284:103–112.
2. Jahani J, Ardajani SR, Poursheykhian AR. Watershed and land use management in the Hyrcanian forests, north of Iran. *Int J Agric Crop Sci*. 2013;6:1068.
3. Damavandi AA, Rahimi M, Yazdani MR, Noroozi AA. Assessment of drought severity using vegetation temperature condition index (VTCI) and Terra/MODIS satellite data in rangelands of Markazi Province. *Iran J Rangel Sci*. 2016;6:33–42.
4. Rubtsov VV, Utkina IA. Interrelations of green oak leaf roller population and common oak: results of 30-year monitoring and mathematical modeling. In: McManus ML, Liebhold AM, eds. *Proc. Ecol. Surv. Manag. For. Insects; 2002 Sept. 1–5; Krakow, Poland. Gen. Tech. Rep. NE-311. Newt. Square, PA US Dept. Agric. For. Serv. Northeast. Researc*; 2003:311.
5. Bahrami A, Emadodin I, Atashi MR, Bork HR. *Land-Use Change and Soil Degradation: A Case Study*. Tehran, Iran: North of Iran Department of Soil Science, Tarbiat Modares University; 2010.
6. Sanjuán Y, Gómez-Villar A, Nadal-Romero E, et al. Linking land cover changes in the sub-alpine and montane belts to changes in a torrential river. *L Degrad Dev*. 2016;27:179–189.
7. Yousefi S, Moradi HR, Pourghasemi HR, Khatami R. Assessment of floodplain landuse and channel morphology within meandering reach of the Talar River in Iran using GIS and aerial photographs. *Geocarto Int*. 2018;33:1367–1380.
8. Barber CP, Cochrane MA, Souza CM, Laurance WF. Roads, deforestation, and the mitigating effect of protected areas in the Amazon. *Biol Conserv*. 2014;177:203–209.
9. Fazeli MJ, Abai M. Green oak leaf-roller moth in Kohkiluyeh and Boyer-Ahmad province (*Tortrix viridana* L., Lep.: Tortricidae). *Appl Entomol Phytopathol*. 1990;57:1–2.
10. Rubtsova NN. *Fortrix viridana* L. in stands of late leafing oak [RSFSR-in-Europe] [Russian]. *Lesovedenie*. 1981;1:83–88.
11. Kalapanida-Kantartzi M, Glavendekic M. Observation on the appearance and the development *Tortrix viridana* L. (Lepidoptera, Tortricidae). *Acta Entomol Serbica*. 2002;7:59–65.
12. Schroeder H, Degen B. Genetic structure of the green oak leaf roller (*Tortrix viridana* L.) and one of its hosts, *Quercus robur* L. *For Ecol Manage*. 2008;256:1270–1279.

13. Hunter MD. Differential susceptibility to variable plant phenology and its role in competition between two insect herbivores on oak. *Ecol Entomol.* 1990;15:401–408.

14. Gooshbor L, Pir Bavaghar M, Amanollahi J, Ghobari H. Monitoring infestations of oak forests by *Tortrix viridana* (Lepidoptera: Tortricidae) using remote sensing. *Plant Prot Sci.* 2016;52:270–276.

15. Goodwin NR, Coops NC, Wulder MA, Gillanders S, Schroeder TA, Nelson T. Estimation of insect infestation dynamics using a temporal sequence of Landsat data. *Remote Sens Environ.* 2008;112:3680–3689.

16. Abdi O. Climate-triggered insect defoliators and forest fires using multitemporal landsat and TerraClimate data in NE Iran: an application of GEOBIA TreeNet and panel data analysis. *Sensors.* 2019;19:3965.

17. Mehdipour M, Sendi JJ, Zamanian H. External morphology and calling song characteristics in *Tibicen plebejus* (Hemiptera: Cicadidae). *C R Biol.* 2015;338:103–111.

18. Baker E, Price BW, Rycroft S, Villet MH. Global Cicada Sound Collection I: Recordings from South Africa and Malawi by BW Price & MH Villet and harvesting of BioAcoustica data by GBIF. *Biodivers Data J.* 2015;3:e5792.

19. Mozaffarian F, Sanborn AF. Two species of the genus Tibicen Latreille, 1825 (Hemiptera: Cicadidae) in Iran, with an identification key to the genera of the family Cicadidae (Hemiptera) in the country. *Acta Zool Bulg.* 2016;68:469–476.

20. Spruce JP, Hicke JA, Hargrove WW, Grulke NE, Meddens AJH. Use of MODIS NDVI products to map tree mortality levels in forests affected by mountain pine beetle outbreaks. *Forests.* 2019;10:811.

21. Yousefi S, Mirzaee S, Tazeh M, Pourghasemi H, Karimi H. Comparison of different algorithms for land use mapping in dry climate using satellite images: a case study of the Central regions of Iran. *Desert.* 2015;20:1–10.

22. Beck PSA, Atzberger C, Høgda KA, Johansen B, Skidmore AK. Improved monitoring of vegetation dynamics at very high latitudes: a new method using MODIS NDVI. *Remote Sens Environ.* 2006;100:321–334.

23. Zhu X, Liu D, Chen J. Remote sensing of environment a new geostatistical approach for filling gaps in Landsat ETM + SLC-off images. *Remote Sens Environ.* 2012;124:49–60.

24. Perbet P, Fortin M, Ville A, Béland M. *Near Real-Time Deforestation Detection in Malaysia and Indonesia Using Change Vector Analysis with Three Sensors.* Vol. 40. Taylor & Francis; 2019.

25. Zirlewagen D, von Wilpert K. Modeling water and ion fluxes in a highly structured, mixed-species stand. *For Ecol Manage.* 2001;143:27–37.

26. Vadrevu K, Heinimann A, Gutman G, Justice C. Remote sensing of land use/cover changes in south and southeast Asian countries. *Int J Digit Earth.* 2019;12:1099–1102.

27. Van Eck CM, Nunes JP, Vieira DCS, Keesstra S, Keizer JJ. Physically-based modelling of the post-fire runoff response of a Forest catchment in Central Portugal: using field versus remote sensing based estimates of vegetation recovery. *L Degrad Dev.* 2016;27:1535–1544.

28. Alves DS. Space-time dynamics of deforestation in Brazilian Amazonia. *Int J Remote Sens.* 2002;23:2903–2908.

29. Qin Z, Zhang M, Christensen T, Li W, Tang H. Remote sensing analysis of rice disease stresses for farm pest management using wide-band airborne data. In: *Proceedings of the IGARSS 2003. 2003 IEEE International Geoscience and Remote Sensing Symposium. Proceedings (IEEE Cat No 03CH37477).* Vol. 4. IEEE; 2003:2215–2217.

30. Du Q, Chang N-B, Yang C, Srilakshmi KR. Combination of multispectral remote sensing, variable rate technology and environmental modeling for citrus pest management. *J Environ Manage.* 2008;86:14–26.

31. Hall RJ, Castilla G, White JC, Cooke BJ, Skakun RS. Remote sensing of forest pest damage: a review and lessons learned from a Canadian perspective. *Can Entomol.* 2016;148:S296–S356.

32. Liaghat S, Balasundram SK. A review: the role of remote sensing in precision agriculture. *Am J Agric Biol Sci.* 2010;5:50–55.

33. Nansen C. The potential and prospects of proximal remote sensing of arthropod pests. *Pest Manag Sci.* 2016;72:653–659.

34. Nansen C, Elliott N. Remote sensing and reflectance profiling in entomology. *Annu Rev.* 2016;61:139–158.

35. Nansen C, Stewart AN, Gutierrez TAM, Wintermantel WM, McRoberts N, Gilbertson RL. Proximal remote sensing to differentiate nonviruliferous and viruliferous insect vectors—proof of concept and importance of input data robustness. *Plant Pathol.* 2019;68:746–754.

36. Nibali A, Ross R, Parsons L. Remote monitoring of rodenticide depletion. *IEEE Internet Things J.* 2019;6:7116–7121.

37. Ismail R, Mutanga O, Bob U. *The Use of High Resolution Airborne Imagery for the Detection of Forest Canopy Damage Caused by Sirex Noctilio;* 2006.

38. Vogelmann JE, Tolk B, Zhu Z. *Monitoring Forest Changes in the Southwestern United States Using Multitemporal Landsat Data.* Vol. 113. Elsevier; 2009.

39. Meddens AJH, Hicke JA, Vierling LA, Hudak AT. *Evaluating Methods to Detect Bark Beetle-Caused Tree Mortality Using Single-Date and Multi-Date Landsat Imagery.* Vol. 132. Elsevier; 2013.

40. Hawryło P, Bednarz B, Wężyk P, Szostak M. *Estimating Defoliation of Scots Pine Stands Using Machine Learning Methods and Vegetation Indices of Sentinel-2.* Vol. 51. Taylor & Francis; 2018.

41. Yousefi S, Khatami R, Mountrakis G, Mirzaee S, Pourghasemi HR, Tazeh M. Accuracy assessment of land cover/land use classifiers in dry and humid areas of Iran. *Environ Monit Assess.* 2015;187:641.

42. Abburu S, Golla SB. Satellite image classification methods and techniques: a review. *Int J Comput Appl.* 2015;119:20–25.

43. Lu D, Weng Q. A survey of image classification methods and techniques for improving classification performance. *Int J Remote Sens.* 2007;28:823–870.

44. Clevers JGPW, Gitelson AA. Remote estimation of crop and grass chlorophyll and nitrogen content using red-edge bands on sentinel-2 and-3. *Int J Appl Earth Obs Geoinf.* 2013;23:344–351.

45. Chavez Jr PS. An improved dark-object subtraction technique for atmospheric scattering correction of multispectral data. *Remote Sens Environ.* 1988;24:459–479.

46. Xu B, Gong P. Land-use/land-cover classification with multispectral and hyperspectral EO-1 data. 2007;73:955–965.

47. Campbell J, Dougherty M. *Digital Image Classification Geography 4354 – Remote Sensing;* 2001.

48. Qiu F, Jensen JR. Opening the black box of neural networks for remote sensing image classification. *Int J Remote Sens.* 2004;25:1749–1768.

49. Bargiel D. Capabilities of high resolution satellite radar for the detection of semi-natural habitat structures and grasslands in agricultural landscapes. *Ecol Inform.* 2013;13:9–16.

50. Unger Holtz TS. *Introductory Digital Image Processing: A Remote Sensing Perspective;* 2007.

51. Hopkins PF, Maclean AL, Lillesand TM. Assessment of thematic mapper imagery for forestry applications under Lake states conditions. *Photogramm Eng Remote Sens.* 1988;54:61–68.

52. Jia X, Richards JA. Cluster-space representation for hyperspectral data classification. *IEEE Trans Geosci Remote Sens.* 2002;40:593–598.
53. Halder A, Ghosh A, Ghosh S. Supervised and unsupervised landuse map generation from remotely sensed images using ant based systems. *Appl Soft Comput J.* 2011;11:5770–5781.
54. Dixon B, Candade N. Multispectral landuse classification using neural networks and support vector machines: one or the other, or both? *Int J Remote Sens.* 2008;29:1185–1206.
55. Foody GM, Mathur A. A relative evaluation of multiclass image classification by support vector machines. *IEEE Trans Geosci Remote Sens.* 2004;42:1335–1343.
56. Kavzoglu T, Mather PM. The use of backpropagating artificial neural networks in land cover classification. *Int J Remote Sens.* 2003;24:4907–4938.
57. Vapnik VN, Chervonenkis AY. On the uniform convergence of relative frequencies of events to their probabilities. *Dokl Akad Nauk.* 1968;181:781–783.
58. Huang C, Davis LS, Townshend JRG. An assessment of support vector machines for land cover classification. *Int J Remote Sens.* 2002;23:725–749.
59. Salberg A-B, Jenssen R. Land-cover classification of partly missing data using support vector machines. *Int J Remote Sens.* 2012;33:4471–4481.
60. Zhang H, Jiang Q, Xu J. Coastline extraction using support vector machine from remote sensing image. *J Multimed.* 2013;8:175–182.
61. Bray M, Han D. Identification of support vector machines for runoff modelling. *J Hydroinformatics.* 2004;6:265–280.
62. Han D, Chan L, Zhu N. Flood forecasting using support vector machines. *J Hydroinformatics.* 2007;9:267–276.
63. Remesan R, Bray M, Shamim MA, Han D. Rainfall-runoff modelling using a wavelet-based hybrid SVM scheme. In: *Proceedings of the IAHS-AISH Publication.* Vol. 331. IAHS Press; 2009:41–50.
64. Abyaneh HZ, Nia AM, Varkeshi MB, Marofi S, Kisi O. Performance evaluation of ANN and ANFIS models for estimating garlic crop evapotranspiration. *J Irrig Drain Eng.* 2011;137:280–286.
65. Oommen T, Misra D, Twarakavi NKC, Prakash A, Sahoo B, Bandopadhyay S. An objective analysis of support vector machine based classification for remote sensing. *Math Geosci.* 2008;40:409–424.
66. Gualtieri JA, Cromp RF. Support vector machines for hyperspectral remote sensing classification. In: *Proceedings of the 27th AIPR Workshop: Advances in Computer-Assisted Recognition; International Society for Optics and Photonics.* Vol. 3584; 1999:221–232.
67. Yousefi S, Keesstra S, Pourghasemi HR, Surian N, Mirzaee S. Interplay between river dynamics and international borders: the Hirmand River between Iran and Afghanistan. *Sci Total Environ.* 2017;586:492–501.
68. Richards JA. *Remote Sensing Digital Image Analysis: An Introduction.* 9783642300. Springer; 2013. ISBN:9783642300622.
69. Ghimire S, Wang H. Classification of image pixels based on minimum distance and hypothesis testing. *Comput Stat Data Anal.* 2012;56:2273–2287.
70. Zhang Y, Huang D, Ji M, Xie F. Image segmentation using PSO and PCM with Mahalanobis distance. *Expert Syst Appl.* 2011;38:9036–9040.
71. Xing EP, Ng AY, Jordan MI, Russell S. Distance metric learning, with application to clustering with side-information. In: *Proceedings of the Advances in Neural Information Processing Systems*; 2003:521–528.
72. Xiang M, Hung C-C, Pham M, Kuo B-C, Coleman T. A parallelepiped multispectral image classifier using genetic algorithms. In: *Proceedings of the Proceedings. 2005 IEEE International Geoscience and Remote Sensing Symposium, 2005. IGARSS'05.* Vol. 1. IEEE; 2005:4.
73. Lü Q, Tang M. Detection of hidden bruise on kiwi fruit using hyperspectral imaging and parallelepiped classification. *Procedia Environ Sci.* 2012;12:1172–1179.
74. Xie H, Tong X. A probability-based improved binary encoding algorithm for classification of hyperspectral images. *IEEE J Sel Top Appl Earth Obs Remote Sens.* 2013;7:2108–2118.
75. Bandyopadhyay S, Maulik U. Genetic clustering for automatic evolution of clusters and application to image classification. *Pattern Recogn.* 2002;35:1197–1208.
76. Akata Z, Perronnin F, Harchaoui Z, Schmid C. Label-embedding for image classification. *IEEE Trans Pattern Anal Mach Intell.* 2015;38:1425–1438.
77. Srivastava PK, Han D, Rico-Ramirez MA, Bray M, Islam T. Selection of classification techniques for land use/land cover change investigation. *Adv Sp Res.* 2012;50:1250–1265.
78. Yousefi S, Pourghasemi HR, Hooke J, Navratil O, Kidová A. Changes in morphometric meander parameters identified on the Karoon River, Iran, using remote sensing data. *Geomorphology.* 2016;271:55–64.
79. Kong W, Sun OJ, Chen Y, Yu Y, Tian Z. Patch-level based vegetation change and environmental drivers in Tarim River drainage area of West China. *Landsc Ecol.* 2010;25:1447–1455.
80. Wan Z, Wang P, Li X. Using MODIS land surface temperature and normalized difference vegetation index products for monitoring drought in the southern Great Plains, USA. *Int J Remote Sens.* 2004;25:61–72.
81. Dutta R. Drought monitoring in the dry zone of Myanmar using MODIS derived NDVI and satellite derived CHIRPS precipitation data. *Sustain Agric Res.* 2018;7:46–55.
82. Wang G, Wang J, Zou X, Chai G, Wu M, Wang Z. Estimating the fractional cover of photosynthetic vegetation, non-photosynthetic vegetation and bare soil from MODIS data: assessing the applicability of the NDVI-DFI model in the typical Xilingol grasslands. *Int J Appl Earth Obs Geoinf.* 2019;76:154–166.
83. Jepsen JU, Hagen SB, Høgda KA, et al. Monitoring the spatio-temporal dynamics of geometrid moth outbreaks in birch forest using MODIS-NDVI data. *Remote Sens Environ.* 2009;113:1939.
84. Hoagland SJ, Beier P, Lee D. Using MODIS NDVI phenoclasses and phenoclusters to characterize wildlife habitat: Mexican spotted owl as a case study. *For Ecol Manage.* 2018;412:80–93.
85. Mkhabela MS, Bullock P, Raj S, Wang S, Yang Y. Crop yield forecasting on the Canadian Prairies using MODIS NDVI data. *Agric For Meteorol.* 2011;151:385–393.

27

The COVID-19 crisis and its consequences for global warming and climate change

Abdullah Kaviani Rad[a], Mehdi Zarei[a,b], Hamid Reza Pourghasemi[c], and John P. Tiefenbacher[d]

[a]Department of Soil Science, School of Agriculture, Shiraz University, Shiraz, Iran [b]Department of Agriculture and Natural Resources, Higher Education Center of Eghlid, Eghlid, Iran [c]Department of Natural Resources and Environmental Engineering, College of Agriculture, Shiraz University, Shiraz, Iran [d]Department of Geography, Texas State University, San Marcos, TX, United States

1 Introduction

In late 2019, Chinese doctors and health researchers announced that they were facing a new virus, the dimensions of which were mysteries to them. The first cases of the virus were recognized on November 17th and people became infected with the virus every day. By December 17, the number of people infected with the virus reached 27. Five days later, 60 people fell ill. On December 27, Zhang Jixian, a doctor from Hubei Provincial Hospital of Integrated Chinese and Western Medicine, announced that the cause of these illnesses was a new coronavirus.[1] Coronaviruses are a family of viruses that, according to research, can cause diseases ranging from as mild as a common cold to more dangerous conditions such as mild Middle East Respiratory Syndrome (MERS) or even the deadlier severe acute respiratory syndrome (SARS).[2] In 2019 and by early 2020, the pandemic rapidly spread to people in all world regions.[3]

Governments have imposed shelter-in-place orders to counteract the spread of the disease. Most industries stopped operating, and movement, transportation systems, and traffic declined. The demand for petroleum and several other fossil fuels has decreased. Economic challenges have emerged in most countries. Global lockdowns have drastically changed the energy demand pattern, leading to a recession.[4]

According to the Asian Development Bank, the coronavirus pandemic could cost the global economy as much as $8.8 trillion (£7 trillion) in losses or 9.7% of global gross domestic product.[5] COVID-19 may have had positive, indirect effects on our environment.[6] Global warming is driven by the concentrations of CO_2 and other greenhouse gases in the atmosphere, which is a consequence of anthropogenic emissions of these chemicals. Human activity has accelerated global warming and has led to changing climates around the world. The reduction of air pollution is a good thing, and reducing greenhouse gas emissions can help mitigate warming and slow climate change. Although the epidemic has helped to reduce greenhouse gas emissions, for the time being, the consequent death of hundreds of thousands of people was a steep and unacceptable cost for achieving the desired goal. Climate researchers and activists see no success in reducing environmental impacts at the expense of human lives and health; it is an anathema to their destinations. The threat of this disease may be only temporary, but global warming, climate change, and extreme weather events will challenge societies for years to come. Human health cannot be separated from the health of global natural systems and the environment.[7] Although the consequences of the coronavirus crisis are enormous, the climate change crisis cannot be ignored.[8] There have been many studies on the impact of the coronavirus crisis on the environment, global warming, and climate change. Helm[9] studied the environmental effects of the coronavirus crisis, and Lidskog et al.[10] showed the Corona as an opportunity to re-implement climate change policies, and Ching and Kajino[11] led the battle against COVID-19 to bring short-lived, long-lasting, and positive and negative impacts on the warming climate. Fischedick and Schneidewind[12] simultaneously examined the coronavirus crisis and climate protection.

If governments have been able to undertake measures to control COVID-19 transmission, they are undoubtedly capable of changing the sources of energy production and energy consumption habits.[13] For decades, scientists have been calling for severe commitments to activities that will reduce the rate of global warming. As oil prices fall, governments have opportunities to implement carbon-emission reduction policies, make direct investments in renewable energy production, and promote a transition to cleaner energy. Coronavirus pandemic's impact on the oil industry have reduced greenhouse gases to some extent. However, reducing employment by destroying industries has never been the preferred approach for reducing CO_2 emissions and global warming. The most cost-effective ways to deal with carbon emissions are to maintain vegetation and soil biological resources, use conservation, improved efficiencies, and renewable energy, and reduce energy demand. Global warming has caused a great deal of concern worldwide, and clean energy must be used to solve this problem.[14] If investments in the low-carbon economy increased after the corona crisis, economic progress could be put back on track.[8] According to previous studies, the cost of countermeasures will be lower if the principle of prevention against climate change is implemented, such as the prevention of coronavirus.[15]

The purpose of this study is to review the evidence and effects of the COVID-19 crisis on air pollution, global warming, climate change, and a transition to a low-carbon economy. Numerous reports, analytical notes, and news sites on global warming, climate change, and Coronavirus are reviewed.

2 Results and discussion

2.1 Reduction of air pollution and greenhouse gases

The global response to the COVID-19 pandemic has led to a sudden reduction in greenhouse gas (GHG) emissions and air pollutants[16], the most vexing global environmental issue faced by the planet along with the global build-up of GHGs. GHGs warm the surface and the atmosphere with significant implications.[17]

Air pollution is usually the emission of pollutants from internal combustion engines, factories, and other natural and human activities. According to the WHO, air pollution kills 7 million people annually worldwide. A typical engine combustion reaction (fuel + air (nitrogen (N_2) + oxygen (O_2))) may emit the following gases: carbon monoxide (CO), ozone (O_3), nitrogen oxides (NO_x) (including nitrogen dioxide (NO_2), nitrous oxide (N_2O) and nitric oxide (NO)), and volatile organic compounds, which may be toxic to human respiratory systems or serve as precursors to toxins (such as O_3). Carbon dioxide (CO_2), emitted from engines and all other combustion processes, and N_2O and methane (CH_4) are gaseous pollutants that contribute to global warming. Also emitted from engines are respirable particulate matter.[18] Sulfur dioxide (SO_2), another widespread air pollution problem that, along with NO_2, contributes to acid rain, is usually released from processes that require combustion of coal combustion.

The life of every child born today is affected by changes in contemporary climate induced by global warming. The climate crisis is not a virus, but manifests itself in other extreme atmospheric phenomena such as floods and droughts, and intensifies and complicates other issues, such as air pollution. A consequence of the coronavirus pandemic was a sharp drop in atmospheric emissions due to the cessation of industrial activities and reduced use of transportation modes (automobiles, trucks, buses, airplanes, ships, and trains). The data indicate that the coronavirus lockdown may have saved the lives of 4000 children under the age of 5 and 73,000 adults over 70 in China alone by reducing pollution.[19]

In Iran, following the COVID-19 outbreak, social distancing was implemented and people were informed not to go on Nowruz (Iranian New Year) trips or to visit the National Headquarters of the Coronavirus Administration. The governors-general were told that the facilities available in past years should not be used this year. Many tourists, leisure, and religious centers have been closed. Schools and universities were closed. The government's guidance was well received by the public. Train travel was reduced by 94%, bus travel by 75%, and air travel by 70%. About 80% of Nowruz's trips were forgone in the first week of the New Year.

The Air quality index in Iran decreased during sequestration (March 5 to April 3, 2020), as indicated by data provided by the Iranian National Air Quality Monitoring System.[20] A similar observation was reported by Asna-ashary et al.[21] Data describing the concentrations of several air pollutants in the air quality index archive of INAQMS[20] from February 2 to April 8, 2020, were analyzed (Fig.1), revealing that the trend curve changed whenever social rules and quarantine requirements were not met. Limiting the outbreak of Corona in Canada through lockdown led to a reduction in NO_2 emissions in southern Ontario and an average reduction of 40% in Toronto.[22]

One of the highest rates of decline in pollutant concentration was observed in Wuhan, China, which began a city-wide quarantine in late January 2020. Nitrogen dioxide (NO_2) concentrations dropped significantly in China as

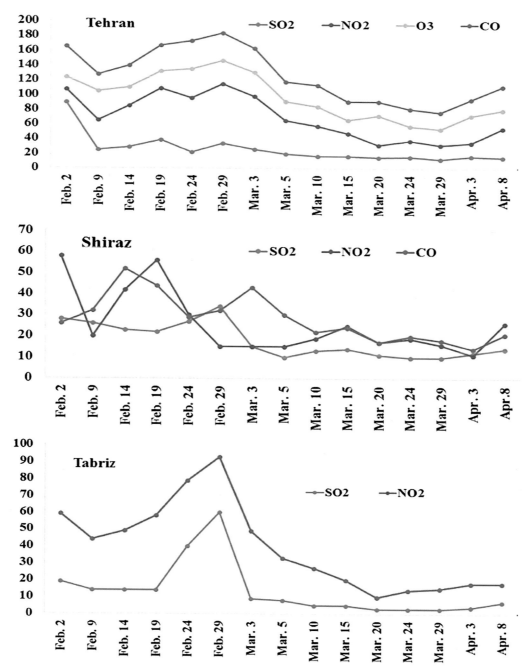

FIG. 1 Air pollutants (NO_2, CO: (ppm), O_3, SO_2: (ppb)) in three Iranian cities during the period of February 2 to April 8, 2020. Social distancing was from March 5 to 3 April 2020. *Data source: INAQMS (Iran National Air Quality Monitoring System). 2020, http://aqms.doe.ir/.*

quarantine continued (10–February 25, 2020). The decline began in Wuhan and spread to other cities with active responses to squelch the spread of the virus.[23]

In a study of 44 cities in northern China, mobility was reduced by 69.85% after the government banned travel, air quality index and five air pollutants (i.e., SO_2, $PM_{2.5}$, PM_{10}, NO_2, and CO) decreased.[24] According to the Copernicus Atmosphere Monitoring Service, air pollution concentrations in China fell by approximately 20% to 30% in February 2020.[25] A study in eastern China based on National Aeronautics and Space Administration (NASA) data found that NO_2 decreased during lock-down (February 1 to February 14, 2020). At the start of 2020 were between 1/3 and 1/6 of the normal pollution levels in Wuhan, Shanghai, Beijing, and other cities.[26] Closures of highways, factories, and commercial centers in many countries (China, Italy, Great Britain, Germany, and others) have generated temporary 40%

reductions in CO_2 and NO_2.[27] Since Italy shut down on March 9, NO_2 levels in Milan and other northern regions have dropped by approximately 40%.[28] In 2020, NO_2 concentrations were significantly lower (from 10% to 30% lower) than in the previous year.[23] Sentinel-5 satellite data show how air quality improved in Wuhan, China, compared to a similar period in 2019. Recent data released by NASA and European Space Agency (ESA) indicate that pollution in some of the epicenters of COVID-19, such as Wuhan, Italy, Spain, and the USA, has been reduced by up to 30%.[29] Compared to the same period in 2017–2019, the daily O_3 mean concentrations increased at urban stations by 24% in Nice, 14% in Rome, 27% in Turin, and 2.4% in Valencia during the lockdown in 2020.[30]

Mahato et al.[31] indicated that NO_2, CO, PM_{10}, and $PM_{2.5}$, were lower than pre-lockdown levels in Delhi, India. Compared to the five-year monthly mean concentrations, the three pollutant concentrations decreased in the urban areas of the state of São Paulo, Brazil. NO was reduced by 3.73%, NO_2 diminished by 4.54%, and CO decreased by 64.8%).[32] The average concentrations of atmospheric $PM_{2.5}$, PM_{10}, and SO_2 in the three cities (Central China) in 2020 were 46.1, 50.8, and 2.56, 30.1%, 40.5%, and 33.4% lower than the levels in February 2017–2019.[33] Agriculture is one of the sources of GHG emissions. Research has shown that agriculture is a significant source of NO_x pollution in California.[34] Approximately 60% of global NO emissions, 39% of methane, and 1% of global NO_2 emissions are related to agriculture.[35] The results of Sadeghi et al.[36] showed that carbon, methane, and nitrous oxide footprints of economic sectors in Iran were 646 million tons, 51,000 tons, and 12,000 tons, respectively. Meanwhile, the share of the agricultural industry was 10.2%, 10.5%, and 17%.

Although the outbreak of the disease has caused significant economic damage to agriculture,[37] the result could be a reduction in greenhouse gases reduced consumption, and production of chemical fertilizers.

Coronavirus has improved air quality in many regions and has reduced the risk of lung and heart disease. The World Health Organization has described NO_2 as being toxic to humans. NO_2 concentrations above $200\,mg\,m^{-3}$ cause severe inflammation in the respiratory tract.[38] Ogen[39] found a correlation between death rates and NO_2 concentrations with death rates from coronavirus been higher in areas that have experienced prolonged exposure to high NO_2 concentrations.

If industrial and transportation activities resume sooner or at an accelerated rate, air pollution will quickly return. Forster et al.[16] estimate that the pandemic-driven response's direct effect will be negligible, with cooling off around 0.01 ± 0.005°C by 2030 compared to a baseline scenario that follows current national policies. In contrast, with green economic and fossil fuel investments reduction, it is possible to avoid future warming of 0.3°C by 2050.

2.2 Reduction of carbon emissions

The Keeling Curve, which charts the systematic measurement of CO_2 concentration in the atmosphere at a Hawaii site since 1958, shows that CO_2 concentrations have risen steadily for 61 years with no apparent interruption (Fig. 2). Also, a similar trend is observed in the global carbon dioxide emissions (Fig. 3).

FIG. 2 The trend of increasing carbon dioxide emissions from 1980 to 2020 at the Mona Loa station. *Data source: Tans P. Trends in Atmospheric Carbon Dioxide, NOAA/GML and Dr. Ralph Keeling, Scripps Institution of Oceanography 2020. https://scrippsco2.ucsd.edu/; https://www.esrl.noaa.gov/gmd/ccgg/trends/data.html.*

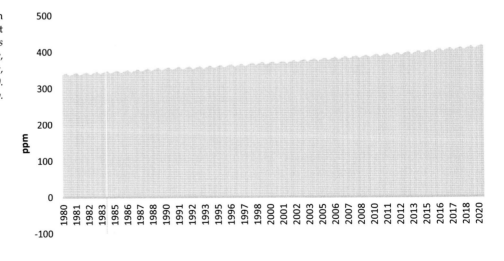

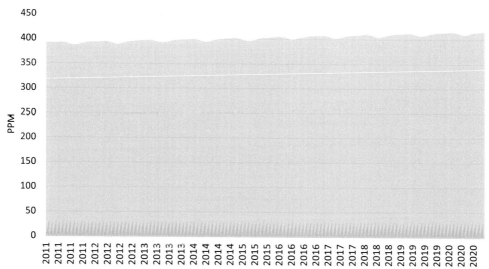

FIG. 3 The trend of daily global carbon dioxide emissions from 2011 to 2020. *Data source: Dlugokencky E, Tans P. Trends in Atmospheric Carbon Dioxide. NOAA/GML 2020. https://www.esrl.noaa.gov/gmd/ccgg/trends/gl_data.html.*

With the outbreak of COVID-19 and reduced fossil fuel use, for the first time, we see that a crisis has been manifested in CO_2 measurements. Recent data show a 10% reduction in CO_2 emissions from fossil fuels, which will undoubtedly bend the curve, which is the highest rate of emissions reduction since World War II. Carbon emissions fell 1.4% for the first time during the 2008 financial crisis,[40] but emissions in 2020 are expected to decrease by approximately 5% (2.5 billion). The trends of changes in global CO_2 emissions growth in 2000–2009, 2010–2018, and 2019–2020 were +3, +0.9, and +0.6%, respectively. The reasons for this slow increase in the demand for coal and, most recently, the coronavirus pandemic consequences.[40]

In 2018, the United States was the second-largest producer of fossil fuels, according to data from The Global Carbon Atlas 2018, but CO_2 emissions are projected to fall by 7.5%. China's combustion of less coal in February 2020, a reduction equivalent to the amount used over an entire year in a small European country, is sufficient to prevent growth in this year's emissions.[26] It is estimated that continental Europe will produce 390 million tons of CO_2 by 2020.[27] The Breakthrough Institute in California also predicted that global greenhouse gas emissions would be 0.5 to 2.2% lower.[40]

The First-quarter 2020 global daily CO_2 emissions from surface transport, power, and industry fell by 36%, 7.4%, and 19%, respectively, relative to the annual mean daily emissions in 2019. CO_2 emissions from surface transport, power, and industry accounted for 86% of the total global reductions in CO_2 emissions. The sum of daily CO_2 emissions in 2020 corresponded to the levels of emissions in 2006.[41]

Much of the oil demand comes from the aviation industry. Typically, 99,700 flights flies per day worldwide, but this has been reduced to only one-fourth of the number to prevent COVID-19 outbreaks. Air travel is responsible for 5.2% of the carbon emissions and 2.5% of the greenhouse gas emissions. Analysis of Flightradar24 data showed that CO_2 emissions from aircraft decreased by 31% (equivalent to 28 million tons). In the UK, Hong Kong, and Switzerland, air traffic has fallen by more than 90% since last year.

The demand for gasoline and diesel has fallen by an average of 9.4% per year with fewer cars on the road. As a result, oil demand in road transport lessened by an average of 2.6 million barrels per day in 2020. Gasoline sales at U.S. gas stations fell 46.6% in the week to March 28, according to IHS Marketing,[42] and the government's weekly figures show that gasoline supplies have fallen to their lowest level since 1994. Gasoline demand in Spain fell by 83% yearly from the week to March 29. The British Gasoline Dealers' Union also said that gasoline sales lessened by 66% in late March compared to an average of 2 months earlier. Road traffic in the UK has dropped by 70%[27] and data from the US Traffic Transportation Center show that traffic volume has fallen by 4%.[42]

In Iran, gasoline consumption at 75 million liters per day in February 2020 dropped to 65 million liters/day in the first half of March 2020. Daily consumption is now below 50 million liters, less than half 5 months ago. During social distancing (5 March to 3 April 2020), fuel consumption decreased by 30%–72%, and traffic on different provinces' roads decreased by 41%–80% (Table 1).

The International Energy Agency estimates that the coronavirus crisis's shock will significantly reduce total global oil consumption by 2020. The COVID-19 crisis has affected several energy markets, including coal, natural gas, and renewable energy, but its impact on the oil market has been severe. This was most apparent in China because it

TABLE 1   Reduce fuel consumption and traffic in different provinces of Iran during social distancing (5 March to 3 April 2020).

Province	Reduce traffic (%)	Reference	Reduce fuel consumption (%)	Reference
East Azarbaijan	48	www.irna.ir/news/83761577	44	www.irna.ir/news/83739833
West Azarbaijan	63	www.irna.ir/news/83740167	47	www.mehrnews.com/news/4891691
Ardabil	72	www.irna.ir/news/83743250	48	www.mehrnews.com/news/4890425
Esfahan	66	www.isna.ir/news/99011607552	54	www.irna.ir/news/83736243
Alborz	50	www.irna.ir/news/83729736	31	www.irna.ir/news/83737879
Ilam	41	www.mehrnews.com/news/4904784	50	www.yjc.ir/fa/news/7308523
Bushehr	68	www.irna.ir/news/83741809	55	www.irna.ir/news/83739210
Tehran	73	www.tasnimnews.com/fa/news/2232032	50	www.donya-e-eqtesad.com/fa/tiny/news-3640864
Chaharmahal va Bakhtiari	52	www.irna.ir/news/83738591	50	www.yjc.ir/fa/news/7296689
South Khorasan	70	www.irna.ir/news/83735441	39	www.iribnews.ir/fa/news/2678936
Khorasan Razavi	–	–	50	www.yjc.ir/fa/news/7278560
North Khorasan	69	www.qudsonline.ir/news/699706	49	www.irna.ir/news/83732859
Khuzestan	48	www.iribnews.ir/fa/news/2674242	40	www.irna.ir/news/83731166
Zanjan	50	www.isna.ir/news/99012413237	57	www.isna.ir/news/990119102225cg7'0t
Semnan	46	www.iribnews.ir/fa/news/2687016	72	www.irna.ir/news/83727911
Sistan and Baluchestan	47	www.irna.ir/news/83731598	45	www.irna.ir/news/83739562
Fars	65	www.irna.ir/news/83738250	32	www.iribnews.ir/fa/news/2677616
Qazvin	64	www.irna.ir/news/83731844	41	www.yjc.ir/fa/news/7293497
Qom	70	www.mehrnews.com/news/4889434	65	www.yjc.ir/fa/news/7299833
Kurdistan	67	www.isna.ir/news/99011406760	43	www.isna.ir/news/99011206025
Kerman	52	www.isna.ir/news/99011809477	30	www.irna.ir/news/83732058
Kermanshah	40	www.iribnews.ir/fa/news/2679676	50	www.mehrnews.com/news/4888872
Kohgiluyeh and Boyer-Ahmad	50	www.tasnimnews.com/fa/news/2250241	51	www.isna.ir/news/99011708955
Golestan	62	www.yjc.ir/fa/news/7310679	–	–
Gilan	80	www.iribnews.ir/fa/news/2680051	34	www.isna.ir/news/98122619792
Lorestan	49	www.isna.ir/news/99011809907	50	www.isna.ir/news/99010803969
Mazandaran	70	www.iribnews.ir/fa/news/2686259	40	www.irna.ir/news/83721000
Markazi	68	www.irna.ir/news/83737059	55	www.irna.ir/news/83734939
Hormozgan	–	–	–	–
Hamedan	58	www.irna.ir/news/83734339	51	www.mehrnews.com/news/4895832
Yazd	26	www.iribnews.ir/fa/news/2683922	30	www.yazdeemrooz.ir

accounted for 80% of the world's energy demand growth in 2019.[43] The EIA estimates that COVID-19 will reduce the demand for oil and other liquid fuels in China by an average of 190,000 barrels per day. China bought about 200,000 barrels of oil per day from the United States, while US oil exports were 8.5 million barrels per day. A few weeks after the outbreak in China, oil consumption fell by 20%, impacting a considerable number of oil companies. Fifty million people in China's Hubei Province no longer consume oil. The corresponding drop in oil prices will reduce the profits of producers such as Saudi Arabia and Iraq by 10%.[44]

Most automakers and parts manufacturers have suspended production and orders due to declining demand and non-favoring government policies. Eighty percent of automakers quotes that the coronavirus will have a direct impact on 2020 revenues. In January, the sales of Chinese cars fell by 18%. At least 1.2 million jobs in Europe are directly dependent on car production, which are now in jeopardy in the wake of factory closures. The sale of approximately 729,400 units of light trucks in the United States in March 2020 was more than one million units less than sales in March 2019.

The International Energy Agency estimates that the coronavirus crisis's shock will significantly reduce the total global oil consumption by 2020. However, it is unlikely that plummeting oil prices will stimulate demand and increase consumption.[42] Falling oil prices have given countries a chance to cut or eliminate subsidies for fossil fuel consumption. These subsidies can either be directed to promote renewable energy or be forgone to allow the free market to determine the prices of conventional and alternative forms of life.

The effects of these emission reductions are not apparent for several months because they have been released for years. Therefore, government measures and post-crisis economic stimulus will have important influences on future emissions,[41] and countries must adhere to the commitments of Paris. If we return to the previous rate of greenhouse gas emissions after the COVID-19 pause, a small but favorable effect on global temperature will be eliminated.

2.3 Reduction of deforestation

Forests cover 31% of the world's land area (4.06 billion). It is estimated that about 420 million hectares of forests have been destroyed for human consumption since 1990 with annual destruction of 10 million hectares of forests from 2015 to 2020.[45]

With deforestation, biodiversity is declining and the risk of animal-to-human transmission is increasing. Viruses such as Ebola, SARS, and COVID-19 are thought to have been transmitted from wild animals to humans in tropical forest areas. A 2019 study found that with a 10% increase in deforestation, the number of malaria cases increased by 3.3%.[46]

Forests play an essential role in reducing global warming by absorbing carbon from the atmosphere and absorbing air pollutants. In Iran, due to the outbreak of the disease and the limited cessation of tourism activities, forests have been preserved to some extent, and other incidents such as fires have not occurred. Forest fires are significant sources of pollution, so reducing fires may improve air quality in the southeastern part of the United States. Fewer fires may have occurred due to economic constraints on the use of forests and the requirements for social distancing in March 2020.[47]

Due to the negative economic consequences of the disease, there is a possibility that investments will be reduced to destroy forests and pastures turning them into residential and agricultural areas. Indirect coronavirus disease has led to a slight reduction in global deforestation, and if this situation continues, it will be useful in reducing global warming.

2.4 Reduction of noise pollution

The issue of noise pollution in cities is another pervasive global problem.[48] In addition to damaging the human body and its function, loud noise has adverse psychological effects.[49] There are a variety of sources of noise pollution, but the most important are industries and transportation. A study on noise pollution determined that more than 30% of Europeans were exposed to more than 55 dB during the day and 20% were exposed to more than 65 dB and suffered from sleep disorders.[50] According to the Iranian NAQMS, from March 5, 2020, the average sound level was reduced by 1–2 dB compared to 2 years ago. Due to the Corona situation, The New York Times reported that noise levels in some parts of New York City had dropped by about five decibels.[51]

3 Conclusions

Global challenges do not have national borders. Some think that climate changes occasionally occur on Earth and have nothing to do with them. Currently, practically, no part of the world is immune to COVID-19, and this is also true of global warming and the prospects for changing climates. COVID-19 has created an opportunity for solidarity among all people that have never been seen before. This is precisely the tool that must be used to deal with climate change. The principle that prevention is better than treatment, a fixed rule in health and medicine, should be adopted to slow global warming and reduce climate change. It will be much easier to challenge the momentum of global warming now, than, it will be to deal with the enormous consequences we face with changing climates. In the post-COVID-19 world, economic redevelopment has become a priority for environmental concerns. The people and governments of the world, by staying at home, forgoing demonstrations, and caring for and protecting each other, have shown that they can unite to prevent disasters and save humanity. The same decisive spirit is needed to deal with climate crises. However, there is a risk that if people abandon public transportation because of the perceived health threat, it poses and returns to their cars, greenhouse gas emissions will be higher than before. Plans should be drawn to encourage people to invest in renewable and low-carbon energy sources. Today's reduced carbon emissions are temporary because there is insufficient infrastructure to migrate to primary reliance on renewable energy; we remain dependent on fossil fuels. We find ourselves (inadvertently) in the world's most massive experiment to reduce, if not halt, the combustion of carbon. However, the oil industry has been the primary engine of technological progress and innovation for decades, and it will be challenging to dispense with it, if not overcome its political power, to achieve clean energy. Clean air and stable climates will only return when we move to make concerted commitments and changes to develop low-carbon economies in the future.

Conflict of interest

The authors declare that they have no known competing financial interests or personal relationships that could have influenced the work reported in this paper.

References

1. Economic Times. *First COVID-19 case can be traced back to November 17 in China's Hubei province: Report*; 2020. https://economictimes.indiatimes.com/news/international/world-news/first-covid-19-case-can-be-traced-back-to-november-2017-in-chinas-hubei-province-report/articleshow/74608199.cms. Accessed 13 March 2020.
2. WHO. *Middle East respiratory syndrome coronavirus (MERS-CoV)*; 2018. https://apps.who.int/mediacentre/factsheets/mers-cov/en/index.html. Accessed January 2020.
3. WHO. *WHO coronavirus disease (COVID-19) dashboard*; 2020. https://covid19.who.int/. Accessed 15 May 2020.
4. Kumar A, Malla MA, Dubey A. With corona outbreak: nature started hitting the reset button globally. *Front Public Health*. 2020;8:569353. https://doi.org/10.3389/fpubh.2020.569353.
5. Brenchley D. *Report from the Asian Development Bank: Coronavirus could cost global economy $8.8trn*; 2020. https://www.investmentweek.co.uk/news/4015244/asian-development-bank-coronavirus-cost-global-economy-usd-8trn. Accessed 15 May 2020.
6. Zambrano-Monserrate MA, Ruano MA, Sanchez-Alcalde L. Indirect effects of COVID-19 on the environment. *Sci Total Environ*. 2020;728:138813. https://doi.org/10.1016/j.scitotenv.2020.138813.
7. de Paula N, Mar KA. *Is the Coronavirus "good" for climate change? This question misses the point. The blog of the Institute for Advanced Sustainability Studies (IASS)*; 2020. www.iass-potsdam.de/en/blog/2020/03/coronavirus-climate-change. Accessed 30 March 2020.
8. Elkerbout M, Egenhofer C, Ferrer JN, et al. *The European Green Deal after Corona: Implications for EU climate policy, Policy Insights*; 2020. https://www.ceps.eu/wp-content/uploads/2020/03/PI2020-06_European-Green-Deal-after-Corona.pdf.
9. Helm D. The environmental impacts of the coronavirus. *Environ Resource Econ*. 2020;76:21–38. https://doi.org/10.1007/s10640-020-00426-z.
10. Lidskog R, Elander I, Standring A. COVID-19, the climate, and transformative change: comparing the social anatomies of crises and their regulatory responses. *Sustainability*. 2020;12(16):6337. https://doi.org/10.3390/su12166337.
11. Ching J, Kajino M. Rethinking air quality and climate change after COVID-19. *Int J Environ Res Public Health*. 2020;17(14):5167. https://doi.org/10.3390/ijerph17145167.
12. Fischedick M, Schneidewind U. The Corona crisis and climate protection keeping long term goals in mind. *Sustain Manag Forum*. 2020;28:71–88. https://doi.org/10.1007/s00550-020-00494-1.
13. Bordoff J. *Sorry, but the virus shows why there won't be global action on climate change*; 2020. www.foreignpolicy.com/2020/03/27/coronavirus-pandemic-shows-why-no-global-progress-on-climate-change. Accessed 27 March 2020.
14. Jianu O, Rosen M, Naterer G. Noise pollution prevention in wind turbines: status and recent advances. *Sustainability*. 2012;4:1104–1117. https://doi.org/10.3390/su4061104.
15. Heyd T. COVID-19 and climate change in the times of the Anthropocene. *Anthropocene Rev*. 2020;8:21–36. https://doi.org/10.1177/2053019620961799.
16. Forster PM, Forster HI, Evans MJ, et al. Current and future global climate impacts resulting from COVID-19. *Nat Clim Change*. 2020;10:913–919. https://doi.org/10.1038/s41558-020-0883-0.
17. Ramanathan V. Global dimming by air pollution and global warming by greenhouse gases: Global and regional perspectives. In: O'Dowd CD, Wagner PE, eds. *Nucleation and Atmospheric Aerosols*. Dordrecht: Springer; 2007. https://doi.org/10.1007/978-1-4020-6475-3_94.
18. Demirbas A. *Biodiesel: A Realistic Fuel Alternative for Diesel Engines*. 1st ed. Springer; 2008. 208 p https://www.springer.com/gp/book/9781846289941.

19. Forbes. *Coronavirus lockdown likely saved 77,000 lives in China just by reducing pollution*; 2020. https://www.forbes.com/sites/jeffmcmahon/2020/03/16/coronavirus-lockdown-may-have-saved-77000-lives-in-china-just-from-pollution-reduction/#48a1c15634fe. Accessed 16 March 2020.

20. INAQMS (Iran National Air Quality Monitoring System); 2020. http://aqms.doe.ir/.

21. Asna-ashary M, Farzanegan MR, Feizi M, Malek Sadati S. *COVID-19 Outbreak and Air Pollution in Iran: A Panel VAR Analysis*; 2020. No. 16 https://www.uni-marburg.de/fb02/makro/forschung/magkspapers/paper_2020/16-2020_asna.pdf.

22. Griffin D, McLinden CA, Racine J, et al. Assessing the impact of Corona-virus-19 on nitrogen dioxide levels over southern Ontario, Canada. *Geophys Res Lett.* 2020. https://doi.org/10.1002/essoar.10503538.2.

23. NASA Earth Observatory. *Airborne nitrogen dioxide plummets over China*; 2020. https://earthobservatory.nasa.gov/images/146362/airborne-nitrogen-dioxide-plummets-over-china.

24. Bao R, Zhang A. Does lockdown reduce air pollution? Evidence from 44 cities in northern China. *Sci Total Environ.* 2020;731:139052. https://doi.org/10.1016/j.scitotenv.2020.139052.

25. CNN. *The world is coming together to fight coronavirus. It can do the same for the climate crisis*; 2020. https://edition.cnn.com/2020/03/18/world/coronavirus-and-climate-crisis-response-intl-hnk/index.html. Accessed 8 April 2020.

26. Financial Times. *How coronavirus stalled climate change momentum*; 2020. https://www.ft.com/content/052923d2-78c2-11ea-af44-daa3def9ae03. Accessed 14 April 2020.

27. Guardian. *Climate crisis: in coronavirus lockdown, nature bounces back – but for how long?*; 2020. https://www.theguardian.com/world/2020/apr/09/climate-crisis-amid-coronavirus-lockdown-nature-bounces-back-but-for-how-long. Accessed 9 April 2020.

28. WIRED. *The pandemic has led to a huge, global drop in air pollution*; 2020. https://www.wired.com/story/the-pandemic-has-led-to-a-huge-global-drop-in-air-pollution/. Accessed 28 March 2020.

29. Muhammad S, Long X, Salman M. COVID-19 pandemic and environmental pollution: a blessing in disguise? *Sci Total Environ.* 2020;728:138820. https://doi.org/10.1016/j.scitotenv.2020.138820.

30. Sicard P, De Marco A, Agathokleous E, et al. Amplified ozone pollution in cities during the COVID-19 lockdown. *Sci Total Environ.* 2020;735:139542. https://doi.org/10.1016/j.scitotenv.2020.139542.

31. Mahato S, Pal S, Ghosh KG. Effect of lockdown amid COVID-19 pandemic on air quality of the megacity Delhi, India. *Sci Total Environ.* 2020;730:139086. https://doi.org/10.1016/j.scitotenv.2020.139086.

32. Nakada LYK, Urban RC. COVID-19 pandemic: impacts on the air quality during the partial lockdown in São Paulo state, Brazil. *Sci Total Environ.* 2020;730:139087. https://doi.org/10.1016/j.scitotenv.2020.139087.

33. Xu K, Cui K, Young LH, et al. Impact of the COVID-19 event on air quality in Central China. *Aerosol Air Qual Res.* 2020;20:915–929. https://doi.org/10.4209/aaqr.2020.04.0150.

34. Almaraz M, Bai E, Wang C, et al. Agriculture is a major source of NO_x pollution in California. *Sci Adv.* 2018;4(1). https://doi.org/10.1126/sciadv.aao3477.

35. Moradi R, Pourghasemian N. Investigation of greenhouse gas emissions and global warming potential due to the use of chemical inputs in the cultivation of essential crops in Kerman province: Cereals. *J Agric Ecol.* 2017;9(2):389–405 [in Persian] 10.22067/jag.v9i2.42033.

36. Sadeghi SK, Karimi Takanlou Z, Motefakker Azad MA. Study of the carbon, methane, and nitrous oxide footprint in Iran's agricultural sub-sectors compared to other economic sectors: the social accounting matrix (SAM) approach. *J Econ Growth Dev Res.* 2016;5(20):13–30. http://egdr.journals.pnu.ac.ir/article_4665.html. [in Persian].

37. Zarei M, Kaviani Rad A. COVID-19. Challenges and recommendations in agriculture. *J Bot Res.* 2020;2(1):12–15. https://doi.org/10.30564/jrb.v2i1.1841.

38. WHO. *Ambient (outdoor) air pollution*; 2018. https://www.who.int/news-room/fact-sheets/detail/ambient-(outdoor)-air-quality-and-health. Accessed 2 May 2020.

39. Ogen Y. Assessing nitrogen dioxide (NO_2) levels as a contributing factor to coronavirus (COVID-19) fatality. *Sci Total Environ.* 2020;726:138605. https://doi.org/10.1016/j.scitotenv.2020.138605.

40. Nasralla S, Volcovici V, Green M. *Coronavirus could trigger biggest fall in carbon emissions since World War II*. Reuters; 2020. https://www.reuters.com/article/us-health-coronavirus-emissions/coronavirus-could-trigger-biggest-fall-in-carbon-emissions-since-world-war-two-idUSKBN21L0HL. Accessed 3 April 2020.

41. Le Quéré C, Jackson RB, Jones MW, et al. Temporary reduction in daily global CO_2 emissions during the COVID-19 forced confinement. *Nat Clim Change.* 2020;10:647–653. https://doi.org/10.1038/s41558-020-0797-x.

42. Financial Times. *Coronavirus puts the brake on America's gas-guzzling ways*; 2020. https://www.ft.com/content/45c9829a-6573-11ea-b3f3-fe4680ea68b5. Accessed 14 March 2020.

43. IEA. *Global oil demand to decline in 2020 as coronavirus weighs heavily on markets*; 2020. https://www.iea.org/news/global-oil-demand-to-decline-in-2020-as-coronavirus-weighs-heavily-on-markets. Accessed 9 March 2020.

44. U.S. EIA. *EIA revises global liquid fuels demand growth down because of the coronavirus*; 2020. https://www.eia.gov/todayinenergy/detail.php?id=42855. Accessed 18 February 2020.

45. FAO. *The state of the world's Forests*; 2020. http://www.fao.org/state-of-forests/en.

46. Chaves LSM, Fry J, Malik A, et al. Global consumption and international trade in deforestation-associated commodities could influence malaria risk. *Nat Commun.* 2020;11:1258. https://doi.org/10.1038/s41467-020-14954-1.

47. NASA Earth Observatory. *Satellites show a decline in fire in the U.S. Southeast*; 2020. https://earthobservatory.nasa.gov/images/146714/satellites-show-a-decline-in-fire-in-the-us-southeast.

48. Barbosa A, Cardoso M. Hearing loss among workers exposed to road traffic noise in Sao Paulo in Brazil. *Auris Nasus Larynx.* 2005;32:17–21. https://doi.org/10.1016/j.anl.2004.11.012.

49. Smith A. A review of the non-auditory effects of noise on health. *Work Stress.* 1991;5(1):49–62. https://doi.org/10.1080/02678379108257002.

50. Muzet A. Environmental noise, sleep and health. *Sleep Med Rev.* 2007;11:135–142. https://doi.org/10.1016/j.smrv.2006.09.001.

51. Bui Q, Badger E. *The Coronavirus Quieted City Noise. Listen to What's Left*; 2020. https://nytimes.com/interactive/2020/05/22/upshot/coronavirus-quiet-city-noise.html. Accessed 22 May 2020.

28

Earthquake ionospheric and atmospheric anomalies from GNSS TEC and other satellites

Munawar Shah

Space Education and GNSS Lab, National Center of GIS and Space Application,
Institute of Space Technology, Islamabad, Pakistan

1 Introduction

The EQ precursors have been reported for global events in the ionosphere and atmosphere over the epicenter under the Lithosphere-Atmosphere-Ionosphere Coupling (LAIC) hypothesis, which assisted in the demonstration of the world's most severe seismogenic regions.[1–6] The main focus of precursory studies is on pre-, co-, and post-EQ anomalies consequent to large-magnitude and shallow-depth events in long-and short-term estimations of Total Electron Content (TEC), f_oF2 (ionospheric F2 layer critical frequency), and other atmospheric parameters from ground and other satellite measurements.[6–9] Similarly, there are many different reports on physical coupling of LAIC to delineate the relationship between seismo-ionospheric anomalies (SIAs) and their coupling in multiple model calculations.[7, 10–12] In addition to ionospheric EQ anomalies, there are some reports of anomalous ionospheric perturbations caused by geomagnetic storms, where different satellite-based storm disturbances have been reported with more morphological characteristics.[13, 14] However, some reports contradict the LAIC coupling of SIAs and even denied the propagation of seismic energy in the atmosphere to ionosphere over seismogenic zones during the preparation period.[15, 16]

Moreover, there are different reports on different methods for SIA detection and monitoring over the seismogenic zone before and after the mainshock day from different satellites during the EQ breeding period.[17, 18] To highlight the ionospheric anomalies associated with EQs, a statistical relationship between co-located and synchronized Chi-Chi EQ anomalies was established in Taiwan, where the Chi-Chi event occurred in 1999.[19] Similarly, many other characteristics of SIAs from GPS TEC were discussed before and after the 1339 global EQ events of $M_w > 6.0$ during 2003–2014, and further morphological characteristics were studied for SIAs coupling.[20] Additional information about the occurrence and propagation of ionospheric anomalies in a statistical analysis was provided by Shah and Jin.[21] They also differentiate the different categories of seismo-ionospheric anomalies associated with global EQs based on different fault systems, magnitudes, and hypocentral depths in different regions of the world. More analyses include SIAs from the Detection of Electro-Magnetic Emissions Transmitted from Earthquake Regions (DEMETER) satellite in the form of case studies[8, 22, 23] and statistically ionospheric anomalies associated with several sea quakes worldwide.[23] These satellite-based atmospheric and ionospheric anomalies provide more stringent evidence for LIAC coupling over the seismogenic zones with augmented monitoring properties. Furthermore, these previous reports emphasized the need to perform more analyses of SIAs with more enhanced clusters of ionospheric and atmospheric satellites. Similarly, SIAs have also been reported during the mainshock preparation period in the long-term data of the Swarm satellite for global EQs, which supports the LAIC coupling from multiple satellites to strengthen the phenomena.[24, 25]

There are two main hypotheses regarding the relationship between EQ anomalies in the atmosphere and ionosphere: SIA due to the emanation of an increasing number of radon gas emissions from the seismogenic zone during the EQ preparation period,[26] and the existence of positive holes (p holes) from EQ breed zones before and after the main shock day.[27] According to Pulinets and Ouzounov,[26] radon and other gases rise from the epicentral and associated fault lineaments regions before the main shock day during the EQ breeding period, which caused different reactions in the air at the Earth atmosphere interface in the form of ion-molecular reactions, followed by air ionization.

These processes further initiate ion cluster generation and increase or decrease the vertical electric field around the epicenter. This vertical electric field rises to the atmosphere and lower edge of ionosphere to complete the LAIC connection over the epicenter of the impending EQ, which later rises to high altitudes in the ionosphere. On the other hand, the concept of p-hole emission from stressed rock in the seismogenic zones concerns the mutual coupling between the lithosphere and ionosphere, resulting in the ionization of the atmosphere near the lithosphere.[27] Positive ions mix near the atmosphere due to adiabatic changes and move upward to the ionosphere. Ultimately, these variations occur in the form of complex reactions that result in the deviation of electromagnetic signals travel up to the ionosphere, which can be sensed as ionospheric anomalies. Furthermore, the SIA anomalies above the epicenter of the impending EQ rise to different altitudes in the ionosphere owing to the phenomenon of acoustic gravity waves (AGWs).[10, 28, 29] More reports on LAIC coupling via AGWs demonstrated short-term perturbations in the ionosphere immediately after the EQs in the form of co/postseismic responses.[28]

We aim to study the integrated analysis of LAIC from multiple satellites associated with global EQs of large magnitude and low depth to determine the disturbances before and after the mainshock. Similarly, another objective of this chapter is to study the different morphological characteristics of SIAs in the ionosphere and atmosphere associated with global EQs. The main objective is to correlate the atmospheric and ionospheric indices with future EQ using the LAIC phenomena from ground and satellite measurements.

2 Data and method

2.1 Data sources

This section focuses on the description of different datasets and their sources to possibly complete LAIC coupling associated with different global EQs. For this study, geographical locations of different EQs and their epicentral information were obtained from the United States Geological Survey (USGS) via their website (https://earthquake.usgs.gov/earthquakes/search). Moreover, the study of space weather conditions is also important for monitoring before and after the EQs to clearly correlate the seismic anomalies using the LAIC hypothesis. If the geomagnetic activity is abnormal before and after the EQ, anomalies in the ionosphere are most likely due to storms ($Kp > 3$). Geomagnetic activity indices used to be retrieve from the International Services of Geomagnetic Indices (ISGI) (http://isgi.unistra.fr/index.php) and Kyoto University (http://wdc.kugi.kyoto-u.ac.jp/dstdir/) to differentiate SIA from storm anomalies.

To investigate the ionospheric anomalies over the epicenter of future EQ, TEC data from the International GNSS Service (IGS) Continuous Operating Stations (CORS) network are available at http://www.ionolab.org/index.php?page=index&language=en.[30] Moreover, ionospheric observational data are also available from in situ satellites, such as the French satellite DEMETER and European Space Agency, Swarm satellites. Similarly, atmospheric anomalies over the seismogenic zones and associated fault lineaments regions can be investigated using different atmospheric parameters (e.g., OLR, geopotential height, humidity, etc.), which are available at https://psl.noaa.gov/about/ and https://giovanni.sci.gsfc.nasa.gov/. Similarly, satellite-based thermal images are also useful in identifying the EQ-induced abnormal surface temperature over the breeding zone of an impending mainshock to integrate the LAIC hypothesis, which is available from the land surface temperature (LST) of MODIS (MODIS is onboard the Aqua and Terra satellites).

2.2 Methodology

It is also important to organize the ionosphere and atmosphere indices of LAIC coupling for probable SIA connections from ground and satellite measurements within the EQ preparation zone. In this study, SIAs were estimated within an EQ preparation zone. The EQ preparation zone of the impending main shock was calculated using the following relation[31]:

$$R = 10^{0.43 M_w} \tag{1}$$

where, M_w is the magnitude of the EQ, and R denotes the radius of the critical region within the mainshock preparation zone of any event. Eq. (1) shows that the larger the magnitude of EQ, the larger the critical zone, and vice versa.[32, 33] Moreover, the spatial atmospheric anomalies associated with impending EQ in the LAIC hypothesis over the epicenter of the impending main shock can be calculated from the logarithmic relation of the EQ magnitude and radius of the preparation zone,[34] as a statistical formula of the logarithm of $R \propto 0.5 M$; where M and R are the magnitudes of EQ and radius of the mainshock preparation zone, respectively. The TECs from GNSS stations were estimated from IGS stations via IONOLAB (http://www.ionolab.org/index.php?page=iriplasopt&language=en) within the critical zone calculated from Eq. (1), and the station must operate within the seismogenic zone.

The slant TEC (STEC) from IGS permanent stations around the epicenter within the critical regions is converted into vertical TEC (VTEC) along the line of sight from the satellite to the receiver to study the ionospheric anomalies. STEC is estimated by the total number of electrons in a 1 m^2 tube along the signal ray path, and it is expressed in TEC units (1 TECU $= 10^{16}$ electrons/m^2). The STEC from different satellites of the GNSS cluster is formulated by the following equations.[35, 36]

$$STEC = \frac{f_1^2 f_2^2}{40.28\left(f_1^2 - f_2^2\right)}\left(L_1 - L_2 + \lambda_1(N_1 - N_2) - \lambda_2(N_1 - N_2) + \varepsilon\right) \qquad (2)$$

$$STEC = \frac{f_1^2 f_2^2}{40.28\left(f_1^2 - f_2^2\right)}\left(P_1 - P_2 - (d_1 - d_2) + \varepsilon\right) \qquad (3)$$

In Eqs. (1)–(2), f_1 and f_2 are the carrier phase frequencies. P and L are the pseudorange and carrier phase observations of the ionospheric delay in the ray path, whereas λ, N, d, and ε are the wavelength, ambiguity, pseudorange, and random residual of the GNSS signal, respectively. Moreover, the VTEC from STEC is estimated using the following equation[37]:

$$VTEC = STEC.\left(\cos\left(\sin^{-1}\left(\frac{R.\sin Z}{R + H}\right)\right)\right) \qquad (4)$$

Here, R is the Earth's radius, and H is the height of the top ionospheric layer at atmospheric altitude ($H = 350$ km for the Klobuchar model in this case). Furthermore, Z is the elevation angle of the satellite at the ionospheric pierce point (IPP).[10, 38]

To identify the ionospheric anomalies for the monitoring of the LAIC hypothesis, median ($\tilde{X}$) and associated interquartile range (IQR)-based confidence bounds were applied to the daily TEC. The abnormal TEC above either the upper or lower bound can be the SIA associated with the EQ. Furthermore, the bounds for a single day are determined by the median and IQR of 15 days before and 5 days after the observed day (the day for which bounds can be calculated). The upper bound (UB) and lower bound (LB) were calculated using the following equations:

$$UB = \tilde{X} + 1.5 \times IQR \qquad (5)$$

$$LB = \tilde{X} - 1.5 \times IQR \qquad (6)$$

Eqs. (5) and (6) show the upper and lower bounds of the normal TEC distribution from the GNSS stations within the critical zone of the mainshock. The median and IQR-based bounds statistically show the abnormal TEC as SIA with a 95% confidence interval.[21]

Another method is the differential TEC (ΔTEC) to quantify the SIA above the UB or below the LB. The calculated TEC above or below the bounds can be expressed in the form of deviations, as shown below:

$$\Delta TEC = \begin{cases} TEC - X_{LB}; & \text{while } TEC < X_{LB} \\ 0; & \text{while } X_{LB} \leq TEC \leq X_{UB} \\ TEC - X_{UB}; & \text{while } TEC > X_{UB} \end{cases} \qquad (7)$$

In Eq. (7), X_{LB} and X_{UB} are the lower and upper confidence bounds of the TEC, respectively. Moreover, the deviation from the normal TEC distribution for every GNSS station was estimated as a percentage deviation. After implementing these methods, we can easily discriminate anomalous TEC between the impending seismic activity and any other geomagnetic activity.

We provide more evidence of LAIC coupling before and after the mainshock day with a spatial and temporal investigation of different atmospheric indices, including geopotential height, air temperature, and Outgoing Longwave Radiation (OLR) over the epicenter. This aids in further underlining the EQ variations in the near atmosphere after it moved upward beyond the Earth-atmosphere interface, specifically in the cross-section of the pressure level between 300 and 600 mb. These maps can be viewed on the webpage http://www.esrl.noaa.gov/psd/data/composite of NOAA/NCEP. Furthermore, all atmospheric index maps were retrieved over the epicentral region estimated by.[34] The daily and monthly values of atmospheric indices were monitored at temporal and spatial resolutions to delineate the epicentral region in the atmosphere. In this study, the LAIC mechanism is calculated over the epicentral region and associated fault lineaments between 300mb and 600mb pressure levels using the following calculation.[1, 26]

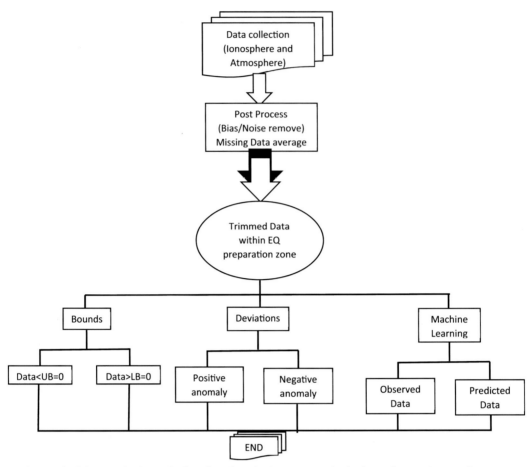

FIG. 1   The complete methodology in the form of a flowchart for seismic precursors in the ionosphere and atmosphere.

$$\Delta AP_{max} = (AP_{current} - AP_{median} | AP_{median}) \qquad (8)$$

where, ΔAP_{max} is the difference between the calculated atmospheric parameters (e.g., geopotential height, OLR). Moreover, in Eq. (8), APmedian and APcurrent are the median and current values of any index of the atmosphere for EQ associated study, respectively. All these datasets validate the coupling of the EQ, atmosphere, and ionosphere over the epicenter of the mainshock. The complete flowchart of the seismic precursor is shown in Fig. 1.

3   Results

There are different reports about possible EQ precursors from ground- and space-based atmospheric and ionospheric measurements to demonstrate the LAIC coupling over the epicenter of impending EQ. The enhancement/depletion in the ionosphere and atmosphere are correlated with different EQs from ground and space measurements, which may provide insights about the variations of future EQs.[39] In addition to these studies, the EQ precursors in the ionosphere from GNSS provides more ways to connet the LAIC phenomena, which are discussed in detail.

The M_w 6.2 EQ hits Japan in October 2016 with a shallow hypo central depth of 5.6 km, and the EQ preparation zone was 460 ± 45 km.[34] The ionospheric anomalies in GNSS TEC and f_oF2 values in Fig. 2 manifested significant perturbations on the day of EQ by the mutual enhancement of nearby station data; however, the EQ variation in f_oF2 is not as clear as the TEC from the nearby GNSS stations. Additionally, the ionospheric precursor on the mainshock day may be due to the EQ, as the geomagnetic activity in the top panels of Fig. 2 is quiet. Because of quiet geomagnetic activity, the ionospheric anomalies in TEC and f_oF2 are bonafide EQ precursors. The mutual enhancement in TEC in nearby stations on the EQ day indicated enormous energy emanation during the seismic preparation period that triggered the ionospheric variations. However, one can see no exceptional variations in f_oF2 of Okinawa station in 2019 before and after the M_w 6.2 EQ, and no anomaly in Okinawa station may be due to its operation outside the EQ critical region. These results emphasize on a unanimous aim to detect the ionospheric anomalies from only those GNSS and other ground stations that operate within the vicinity of the epicentral region. On the other hand, differential TEC beyond the

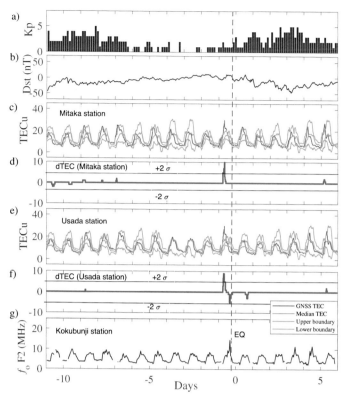

FIG. 2 Daily VTEC variations from the nearby GNSS stations and f_oF2 frequency analysis for the 2016 Japan EQ (M_w 6.2) in Shah et al.[1] VTEC and differential TEC with associated confidence bounds for every station are presented in different colors. The EQ is indicated by a *red dashed line*.

confidence bounds further clarified the ionospheric perturbations and a significant 5–10 TECU variation associated with EQ occurred in the daily values from the normal distribution. The ionospheric anomalies are specifically related to the mainshock, as the geomagnetic storm ($Kp < 3$) is quiet during 5 days before the EQ breeding day. Moreover, positive EQ anomalies (anomalies beyond the upper bound are positive) were observed during the seismogenic period, which links the lithosphere and ionosphere under the hypothesis of the LIAC hypothesis.

Additionally, the criteria in Fig. 2 are implemented on shallow hypocentral depth and large magnitude ($M_w > 6.0$) EQs in a statistical analysis for seismo-ionospheric anomalies associated with global EQs during 1998–2019 (Fig. 3). The index number in Fig. 3 is calculated by the following equation[40–42] and is discussed in detail in Ref. 43.

$$\text{Index No} = \frac{P_{SD}}{P_{Total}}. \tag{9}$$

In Eq. (9), P_{SD} and P_{Total} are the ionospheric anomalies associated with a single EQ and the total seismo-ionospheric anomalies of all events during the study period, respectively. The index number of Zone B shows a significant rise for EQs of $6.1 < M_w < 7$ in the range of 8%–10% and approximately 15% for $M_w > 7.0$ (Fig. 3). Moreover, the index No for the global EQ precursor dropped to 5% within 5 days after the mainshock during the EQ preparation period. On the other hand, one can see prominent ionospheric anomalies before the EQ for most events within 5 days before the mainshock day than postseismic variations, followed by the EQ day. Moreover, seismo-ionospheric precursors become spiky in the case of EQs of $M_w > 7.0$, in the zone of high latitudes (i.e., Zone A and Zone C). In Fig. 2, the number of seismo-ionospheric anomalies before the mainshocks in Zone A and Zone C for $M_w > 7.0$ are 7% and 10%, respectively. The EQ-associated ionospheric precursors in Zone B are more significant and frequent than those in Zone A and Zone C. However, Index No also shows that EQs in Zone B produced more seismic anomalies than Zone A and Zone C by the calculation of ionospheric anomalies from EQs in all the zones. It is also evident that Zone-B has more EQs than the other zones. Ionospheric enhancements in equatorial regions have also been reported for different case studies and statistical analyses.[21,43]

The spatial ionospheric maps from the GIM provide more evidence to support the hypothesis of LIAC coupling over the epicenter during the seismic preparation period. The spatial maps of TEC (dTEC maps) show a clear enhancement of electron density over M_w 6.5 Padilla, Bolivia EQ epicenter, which hits the region on February 21, 2017. Moreover, abnormal dTEC clouds are clear at UT = 10, and they occur until UT = 15h with the same high intensity (Fig. 4).

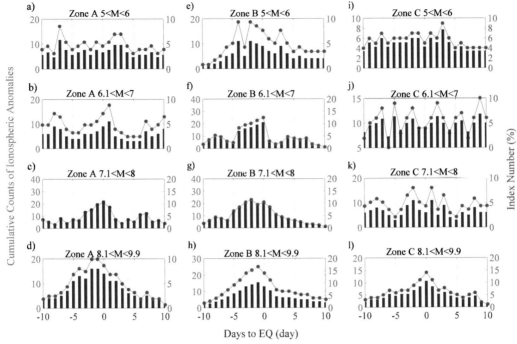

FIG. 3 The cumulative counts of ionospheric anomalies associated with global EQs within 10 days before and after the mainshocks in a statistical analysis during 1998–2019, as in Ref. 43. For this study, 1182 ($M_w > 5.5$) EQs are divided into three different zones of the world: Zone A (80°N–25°N), Zone B (25°N–25°S), and Zone C (25°S–80°S) to delineate significant seismic anomalies from long term GNSS TEC data.

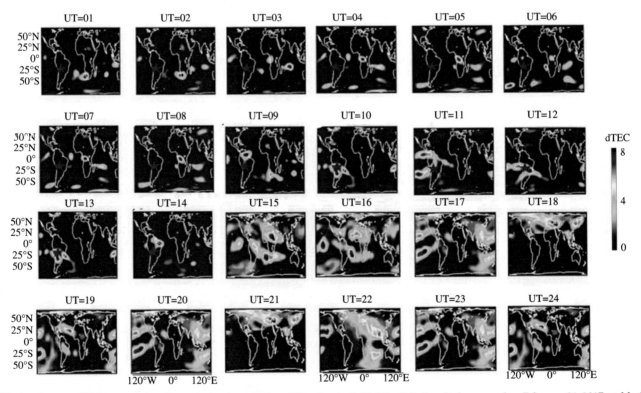

FIG. 4 The spatial TEC maps in hourly resolution from GIMs on EQ epicenter in Padilla, Bolivia, which occurred on February 21, 2017, as M_w 6.5. The epicenter is clearly shown as a *red-filled star* and differential TEC values in the legend. The EQ-induced electron clouds are clearly visible during UT = 10–15 h with more than 5 TECU intensity from normal GIMs over the epicenter, as shown in Ref. 43. The rest of the maps UT = 01–09 h and UT = 16–24 h have no clear variations on the epicenter but variations are clear on equatorial regions due to equatorial ionization anomaly, which is another phenomenon of ionospheric TEC enhancement.

The differential TEC map at a specific UT hour was calculated by subtracting the previous one-month TEC maps from the background of the same UT hours of the observed map. As discussed in Ref. 43, spatial maps show enhancement during the same UT hours as anomalous temporal daily TEC values from nearby GNSS stations of the epicenter. Therefore, it is important to confirm the ionospheric enhancement over the epicenter from the GIM maps to detect anomalies in specific UT hours associated with the mainshock. It also provides more evidence to further develop the monitoring mechanism of the LAIC system. It should be noted that TEC value is larger than 10 TECU in GIMs during the hourly TEC in 10–15 h of UT over the EQ epicenter (Fig. 4). Moreover, the anomalous map shows that the dTEC values over the epicenter of the impending EQ were greater than 30% from the normal distribution at the epicenter during these hours. Additionally, the deviation in the anomalous maps was larger than that in the other analyzed maps. The dTEC shows that electron clouds linger over the epicenter away from the equatorial ionization anomaly to form seismo-ionospheric anomalies over the epicenter and associated fault lineaments. The ionospheric anomalies in the form of GIM maps validate the coupling of LAIC over the epicenter associated with EQ, which is produced by stress rocks inside the Earth during the tectonic activity period and rises to the ionosphere by different propagation paths at the interaction of Earth's atmosphere interface.

For a complete understanding of the LAIC process associated with the epicenter of the future EQ, the analysis of atmospheric anomalies can barely be omitted. This study also includes the analysis of different atmospheric indices from multiple satellites over the epicenter of the M_w 7.2 Pakistan event to complete the LAIC process, which can provide deep insights into the emanation of seismic associated atmospheric and ionospheric anomalies. Moreover, these analyses will also assist in the design of future studies in this field to map EQ energy from the epicenter in the atmosphere and ionosphere. The investigation of atmospheric indices includes OLR, surface temperature (ST), Aerosol Optical Depth (AOD), and NO_2 from satellites in the context of the M_w 7.2 Pakistan EQ. These datasets were studied for 2 months before and 1 month after the mainshock by implementing statistical bounds on it. The upper and lower confidence bounds were generated from the median and standard deviation of the entire dataset (Fig. 5). The shallow hypocenter EQ occurred at 68 km depth as a result of normal faulting on January 18, 2011, in the southern region of Pakistan. The results demonstrate anomalous atmospheric indices within 1 month before the impending EQ,

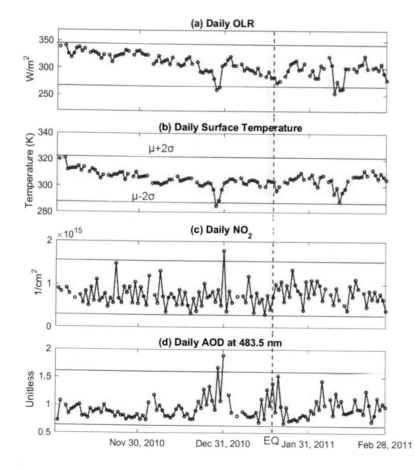

FIG. 5 The analysis of atmospheric anomalies in daily, (A) OLR, (B) Surface Temperature, (C) NO_2, and (D) AOD related to M_w 7.2 southwestern Pakistan EQ, where the variations occur within a month before the seismic event. More discussion about the generation and evolution of different observables in the atmosphere is in Shah et al.[2]

indicating a huge emanation of energy from the epicenter to the Earth's atmosphere interface. Moreover, the enhancement on December 31, 2010, in the time series of NO_2 and AOD deviation of nearly $13e14\,cm^2$ and $483.5\,nm$, respectively, was noticed within 15–21 days before the seismic event (Fig. 5). Similarly, the other atmospheric indices also show anomalous perturbations triggered by the M_w 7.2 EQ of Pakistan, which confirms the formation of complex air ionization and aerosol motion over the epicentral region. Furthermore, the enhancements in multiple satellite observations associated with M_w 7.2 EQ might be the pre-EQ variations and cannot be the usual semiannual cycle during November-December. These atmospheric anomalies are generated at epicentral regions due to stress rock at hypocentral depth and rise to different atmospheric-ionospheric altitudes to formulate the LAIC mechanism, which is discussed in detail in Ref. 2.

Atmospheric anomalies were further observed in the Dobrovolsky et al.[31] region over the epicenter of M_w 7.2, Pakistan EQ, to support the anomalous variations in the time-series analysis. In Shah et al.,[2] the generation mechanism and its travel over the epicenter within 15–30 days before the mainshock are discussed in correlation with previous reports on seismic hazards in this region. The OLR maps monitored anomalies of atmospheric clouds over the seismogenic zone and associated fault lineaments in Fig. 6, which is clear evidence of LAIC in Fig. 5. The mean composite anomaly maps of OLR observed spatially distributed perturbations in a daily map on December 28, 2010, which occurred 21 days before the mainshock, resulting in abnormal energy exhalation to the atmosphere by the epicenter (Fig. 6). Based on the analysis of OLR maps, the deviations suggest a bonafide EQ anomaly on December 28, 2010, which led to the development of atmospheric anomalies in LAIC coupling. More analyses are needed to study the different characteristics of atmospheric anomalies and further develop the hypothesis of lithospheric-atmosphere-coupling. The study of atmospheric indices must be included in the LAIC phenomenon to properly monitor epicentral energy in the lower and upper atmosphere.

The anomalous ionospheric perturbations were further investigated for M_w 6.4 from the DEMETER satellite for the Taiwan EQ on December 19, 2009 (Fig. 7), and a detailed explanation of the generation from the source region (i.e., epicenter) is discussed in Abbasi et al.[44] The analysis of different ionospheric indices from DEMETER onboard ISL and IAP payloads showed significant variations within 2 days before and after the M_w 6.4, Taiwan EQ from the rest of the measurement, which justified the existence of ionospheric variations within 5–10 days' window of the previous findings (Fig. 7; Panels, B–C). Moreover, the electron density, ion density, and electron temperature from DEMETER onboard ISL vary by more than 5% from the normal distribution due to the main shock of the Taiwan EQ. Similarly, H+, O+, and ion density from the IAP sensor of the DEMETER satellite have variations of more than 10% associated with EQ, which leads to the enormous energy emanation from the epicentral region due to stress rocks. Panel (A) of Fig. 7 shows significant TEC perturbations from the nearby operating GNSS stations of M_w 6.4, Taiwan EQ, which

FIG. 6 Spatial maps of OLR over the epicenter of M_w 7.2 southwestern Pakistan EQ completely correlate with the anomalous days in temporal plots in Fig. 5, as in Shah et al.[2] The variations in daily anomaly maps of a mean composite of OLR are obtained from National Oceanic and Atmospheric Administration/ National Centers for Environmental Prediction (NOAA/NCEP) before the M_w 7.2 EQ, where clear anomalous clouds were observed on the epicentral region.

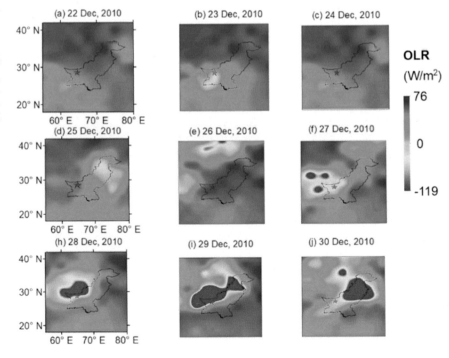

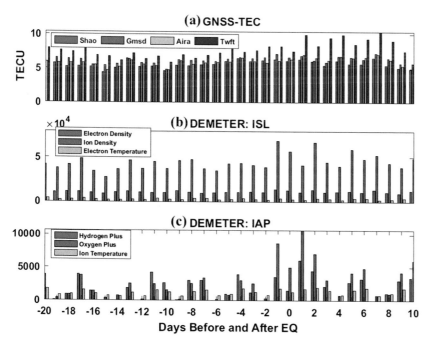

FIG. 7 The analysis of GNSS and DEMETER satellite with clear anomalies within 5 days before and after the M_w 6.4, Taiwan EQ on December 19, 2009,[44] which aims to develop the LAIC coupling for portraying EQ anomalies in the ionosphere.

culminated as a result of LAIC coupling over the seismogenic zone above the epicenter. Additionally, ISL onboard DEMETER demonstrated anomalies in electron and ion density in 2 days before the mainshock, followed by the mainshock.[44] The electron temperature of DEMETER's ISL has no explicit variations, which means that the EQ-induced ionospheric electron and ion densities varied owing to the enormous energy exhalation over the seismic breeding zone. The data over the epicenter from the IAP sensor revealed abnormal daytime H+ and O+ variations as EQ perturbations immediately before and after the EQ, which endorsed the mechanism of LAIC during the EQ preparation period. In contrast, the ion temperature has no anomaly due to the EQ. The LAIC coupling from multiple satellite measurements (i.e., DEMETER, GNSS, and other atmospheric satellites) provides a strong monitoring system of EQ energy in the upper and lower atmosphere and ionosphere.

4 Discussions

The reported ionospheric and atmospheric anomalies from multiple satellite measurements reveal significantly different EQ-induced perturbations but no significant geomagnetic storm anomalies. These analyses require to study for the monitoring of EQ precursors in the ionosphere, which rise from epicentral regions to different altitudes in the atmosphere and ionosphere.[25, 45] Moreover, we further emphasized the necessity of more work in this field to develop the hypothesis of LAIC by integrating multiple satellite measurements. The integration of ground and satellite measurements for portraying the lithospheric attributes in the atmosphere and ionosphere is clearly explained in the description of the LAIC model.[26] They presented Radon as the source of the formation of atmospheric and ionospheric anomalies from a normal distribution. An increasing amount of Radon emission from seismogenic zones of the epicenter of impending main shock transformed into different atmospheric and ionospheric anomalies after crossing a threshold point. The abnormal Radon amount over the epicenter reached up to the ionosphere in different ways, which further deviate the radio signal propagation from their normal travel path. Radon-based atmospheric and ionospheric anomalies under the LAIC mechanism have been presented in different reports.[2, 46–48]

The p-hole emission at hypo central depth from squeezed rocks as a result of tectonic stress also varies the morphology of the air column at different atmospheric altitudes over the seismogenic zone.[49, 50] The overabundance of p holes after crossing the threshold amount reached the ionosphere in the form of Acoustic Gravity Waves (AGWs) over the epicenter and associated fault lineaments.[51] Moreover, Shah et al.[8] reported EQ-induced ionospheric anomalies over the Chile EQ by comparing the electron density from the DEMETER satellite as a ratio over the epicenter latitude and its conjugate axis. Similarly, Kuo et al.[51] demonstrated the coupling of the lithosphere and ionosphere over the epicenter due to squeezing rocks from tectonic stress activation through numerical simulation by assuming p-hole emissions from seismogenic zones. Furthermore, they correlated the variations in the ionosphere with potential differences in epicentral electric field variations in the mainshock region. They further highlighted the initial source and ultimate position of the LAIC mechanism in the ionosphere above the seismic breeding region. Later, Kuo et al.[11] modified their LAIC coupling model by introducing arbitrary paths of electric fields produced around the epicenter of the future EQ. Additionally, they showed the intensification of mid-latitude ionospheric anomalies on the evening side in their updated model. The existence of observable seismic precursors that have been reported in the literature appears plausible; but, it should be noted that they are by no means universally accepted. The various components of the proposed LAIC coupling between the lithosphere and atmosphere, followed by ionospheric anomalies throughout this study, were confirmed by multiple satellite observations (Fig. 8). In Fig. 8, the complete caption of the LAIC mechanism is shown for correlating the EQ energy with the atmosphere and ionospheric layers. On the other hand, many reports have suggested that LAIC coupling is valid for shallow depth and high magnitude terrestrial EQs throughout the world and other ionosphere anomaly cases.

The anomalous lithospheric variations were quantified in the atmosphere for different EQs distributed worldwide using the LAIC mechanism. For example, Daneshvar and Freund[52] studied the EQ-associated atmospheric anomalies in the long-term data of different satellites. Furthermore, they showed that EQ enhancement and depletion initiated within a few months before the EQ, which validates the statistical relation of the EQ preparation period from the law of Rikitake.[53] In most cases, atmospheric anomalies occurred owing to an increasing number of p-holes added to the outer atmosphere after the commencement of a threshold. Moreover, Awais et al.[54] also studied the EQ perturbation in the form of thermal anomalies from MODIS satellite images. They implemented a statistical analysis based on the median and standard deviation of satellite-based land surface temperature to measure abnormal variations associated with the mainshock. Furthermore, the spatial images revealed abnormal clouds in the EQ epicenter during the mainshock preparation period. Ganguly[55] also demonstrated short-and long-term EQ-induced atmospheric clouds over the epicenter of the Nepal EQ, where the abnormal atmospheric clouds can be triggered by the p-hole emissions from the hypocentral zone of the impending mainshock. Nepal EQ caused a lot of demolition of buildings and other infrastructure after the mainshock day, and Ganguly[55] correlated some abnormal atmospheric clouds with dust emissions from buildings and other infrastructure. These clouds may have been formed by both abnormal gas emissions from the epicentral regions and dust from the demolished buildings. Moreover, these atmospheric anomalies produced at the epicenter regions further reached up to high altitudes in the ionosphere in the seismogenic zone.

The ionospheric anomalies associated with different EQs are shown in the case studies and statistical analyses to delineate more characteristics in the epicentral region. For example, Shah et al.[56] studied two different EQs in Turkey from ground GNSS stations and space-based GIM TEC maps to detect abnormal ionospheric delays before the

FIG. 8 The proposed LAIC coupling model for the propagation of atmosphere and ionospheric anomalies over the epicenter during the seismic preparation period for different EQs before and after the mainshock.[46] Shah et al.[46] presented a complete description of the generation mechanism of different anomalies during the preparation period.

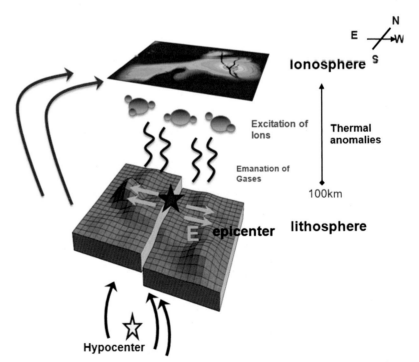

mainshock day. Furthermore, they measured the TEC anomalies associated with geomagnetic storms to differentiate between seismic and storm anomalies in the ionosphere. More analyses include day and nighttime ionospheric perturbations before and after the mainshock day, as well as the detection of suspected local time ionospheric clouds in the GIM TEC maps. Moreover, abrupt TEC enhancement and depletion associated with the worldwide EQs were also calculated from the GPS network for global events.[21,43] They showed that the mainshock energy became significant within 5 or 10 days before the mainshock, and similar anomalies were found within 5–10 days after the EQ day. It is noted in their studies that EQ-induced ionospheric anomalies are more prominent for high magnitude and low hypocentral depth EQ events. Furthermore, the statistical analysis showed that positive seismo-ionospheric anomalies were observed more than negative anomalies for global $M_w > 6.0$, during the long-term data of 1998–2020. Similarly, there are other reports about seismo-ionospheric anomalies, which discussed the different morphological characteristics of EQ-induced variations.[57–59] Recently, more analyses were done on the ionosphere, and particularly, seismo-ionospheric variation to develop the hypothesis of LAIC for different worldwide EQs.[60–63] More results are published by the research community with new findings on LIAC coupling to enhance our knowledge.[64–69] All these analyses suggest that EQ precursors in the atmosphere and ionosphere are possible with improved clusters of multiple sensors onboard GNSS and other satellite missions. In addition, there is a need for more scientific groups to work dedicatedly on LAIC phenomena to search for adequate EQ precursors.

5 Conclusion

The reported ionospheric and atmospheric anomalies from different space-based observations consequent to major EQs in the world under LAIC phenomena considerably enhance the knowledge of seismic event monitoring. All these analyses provide meaningful insights of the lithospheric attributes in the atmosphere and ionosphere, which have led to the attraction of early-career scientists in this field. The main points of all the observations are listed below.

- Ionospheric anomalies from TEC of the GNSS network provided significant evidence for the monitoring and detection of global $M_w > 5.0$ EQs with a more adequate description. The different characteristics of ionospheric anomalies associated with global EQs are presented in the form of case studies and statistical analysis to discuss the LAIC mechanism over the epicenter.
- The atmospheric anomalies over the epicenter of future EQs support the LAIC mechanism to quantify the path of propagation of ionospheric precursors during the interaction at the Earth-atmosphere interface. Moreover, it assists in completing the monitoring system by calculating anomalies in the atmosphere and ionosphere. Similarly, ionospheric heating due to the lower atmosphere over the epicentral region of the EQ can be distinguished from other weathering forces by the incorporation of multiple satellite-based atmospheric index measurements.
- The study of ionospheric anomalies from in situ satellites, such as DEMETER, Swarm, and GNSS, enhances our ability to identify probable EQ precursors to complete the LAIC phenomenon. Similarly, the analysis of different ionospheric constituents from these multiple satellites provides more knowledge about the confirmation of SIAs associated with impending mainshocks, which may help in future forecasting systems for EQs.
- The incorporation of different ionospheric and atmospheric models (i.e., SAMI2, SAMI3, IRI, etc.) further facilitates the study of LAIC by studying different ions and particles at different altitudes above the epicenter and associated fault lineament regions. These models also help to study the different characteristics and morphologies of seismo-ionospheric anomalies. However, more studies with multiple satellites and models are encouraged to forecast EQs in the future with the LAIC phenomenon to assist humanity in eliminating the EQ catastrophe.

References

1. Shah M, Aibar AC, Tariq MA, Ahmed J, Ahmed A. Possible ionosphere and atmosphere precursory analysis related to Mw > 6.0 earthquakes in Japan. *Remote Sens Environ.* 2020;239:111620.
2. Shah M, Tariq MA, Naqvi NA. Atmospheric anomalies associated with Mw > 6.0 earthquakes in Pakistan and Iran during 2010–2017. *J Atmos Sol Terr Phys.* 2019;191:105056.
3. Liu X, Zhang Q, Shah M, Hong Z. Atmospheric-ionospheric disturbances following the April 2015 Calbuco volcano from GPS and OMI observations. *Adv Space Res.* 2017;60(12):2836–2846.
4. Liu JY, Chen YI, Huang CC, et al. A spatial analysis on seismo-ionospheric anomalies observed by DEMETER during the 2008 M8.0 Wenchuan earthquake. *J Asia Earth Sci.* 2015;114(Part 2):414–419.
5. Fujiwara H, Kamogawa M, Ikeda M, et al. Atmospheric anomalies observed during earthquake occurrences. *Geophys Res Lett.* 2004;31(17).
6. Hayakawa M, Tomizawa I, Ohta K, et al. Direction finding of precursory radio emissions associated with earthquakes: a proposal. *Phys Earth Planet In.* 1993;77(1):127–135.

7. Oyama KI, Kakinami Y, Liu JY, Abdu MA, Cheng CZ. Latitudinal distribution of anomalous ion density as a precursor of a large earthquake. *J Geophys Res Space Phys.* 2011;116(A4).
8. Shah M, Tariq MA, Ahmad J, Naqvi NA, Jin S. Seismo ionospheric anomalies before the 2007 M7.7 Chile earthquake from GPS TEC and DEME-TER. *J Geodyn.* 2019;127:42–51.
9. Shah M, Qureshi RU, Khan NG, Ehsan M, Yan J. Artificial neural network based thermal anomalies associated with earthquakes in Pakistan from MODIS LST. *J Atmos Sol Terr Phys.* 2021;215:105568.
10. He L, Heki K. Ionospheric anomalies immediately before Mw7.0–8.0 earthquakes. 2017;122(8):8659–8678.
11. Kuo CL, Lee LC, Huba JD. An improved coupling model for the lithosphere-atmosphere-ionosphere system. *J Geophys Res Space Physics.* 2014;119 (4):3189–3205.
12. Namgaladze AA, Klimenko MV, Klimenko VV, Zakharenkova IE. Physical mechanism and mathematical modeling of earthquake ionospheric precursors registered in total electron content. *Geomagn Aeron.* 2009;49(2):252–262.
13. Afraimovich EL, Astafyeva EI, Zhivetiev IV, Oinats AV, Yasyukevich YV. Global electron content during solar cycle 23. *Geomagn Aeron.* 2008;48 (2):187 200.
14. Thomas EG, Baker JBH, Ruohoniemi JM, Coster AJ, Zhang S-R. The geomagnetic storm time response of GPS total electron content in the north American sector. *J Geophys Res Space Physics.* 2016;121(2):1744–1759.
15. Geller RJ, Jackson DD, Kagan YY, Mulargia F. Earthquakes cannot be predicted. *Science.* 1997;275(5306):1616.
16. Rishbeth H. How the thermospheric circulation affects the ionospheric F2-layer. *J Atmos Sol Terr Phys.* 1998;60(14):1385–1402.
17. De Santis A, Balasis G, Pavón-Carrasco FJ, Cianchini G, Mandea M. Potential earthquake precursory pattern from space: the 2015 Nepal event as seen by magnetic swarm satellites. *Earth Planet Sci Lett.* 2017;461:119–126.
18. Jin S, Jin R, Li JH. Pattern and evolution of seismo-ionospheric disturbances following the 2011 Tohoku earthquakes from GPS observations. *J Geophys Res Space Physics.* 2014;119(9):7914–7927.
19. Liu J, Chen Y, Chuo Y, Tsai H. Variations of ionospheric total electron content during the chi-chi earthquake. *Geophys Res Lett.* 2001;28 (7):1383–1386.
20. Zhu, F., F. Su, J. Lin, Statistical analysis of TEC anomalies prior to M6.0+ earthquakes during 2003–2014. Pure Appl Geophys, 2018.
21. Shah M, Jin S. Statistical characteristics of seismo-ionospheric GPS TEC disturbances prior to global mw ≥5.0 earthquakes (1998–2014). *J Geodyn.* 2015;92:42–49.
22. Ryu K, Parrot M, Kim SG, et al. Suspected seismo-ionospheric coupling observed by satellite measurements and GPS TEC related to the M7.9 Wenchuan earthquake of 12 May 2008. *J Geophys Res Space Phys.* 2014;119(12):10,305–10,323.
23. Parrot M, Tramutoli V, Liu TJY, et al. Atmospheric and ionospheric coupling phenomena related to large earthquakes. *Nat Hazards Earth Syst Sci Discuss.* 2016;2016:1–30.
24. Marchetti D, De Santis A, D'Arcangelo S, et al. Pre-earthquake chain processes detected from ground to satellite altitude in preparation of the 2016–2017 seismic sequence in Central Italy. *Remote Sens Environ.* 2019;229:93–99.
25. De Santis A, Marchetti D, Pavón-Carrasco FJ, et al. Precursory worldwide signatures of earthquake occurrences on swarm satellite data. *Sci Rep.* 2019;9(1):20287.
26. Pulinets S, Ouzounov D. Lithosphere–atmosphere–ionosphere coupling (LAIC) model – an unified concept for earthquake precursors validation. *J Asian Earth Sci.* 2011;41(4–5):371–382.
27. Freund F. Stress-activated positive hole charge carriers in rocks and the generation of pre-earthquake signals. In: *Electromagnetic Phenomena Associated with Earthquakes;* 2009:41–96.
28. Jin S, Su K. Co-seismic displacement and waveforms of the 2018 Alaska earthquake from high-rate GPS PPP velocity estimation. *J Geod.* 2019;93 (9):1559–1569.
29. Yang S-S, Asano T, Hayakawa M. Abnormal gravity wave activity in the stratosphere prior to the 2016 Kumamoto earthquakes. 2019;124 (2).1410–1425.
30. Sezen U, Arikan F, Arikan O, Ugurlu O, Sadeghimorad A. Online, automatic, near-real time estimation of GPS-TEC: IONOLAB-TEC. 2013;11 (5):297–305.
31. Dobrovolsky IP, Zubkov SI, Miachkin VI. Estimation of the size of earthquake preparation zones. pure and applied geophysics. 1979;117 (5):1025–1044.
32. Barkat A, Ali A, Rehman K, et al. Multi-precursor analysis of Phalla earthquake (July 2015; mw 5.1) near Islamabad, Pakistan. *Pure Appl Geophys.* 2018;175(12):4289–4304.
33. Ahmad N, Barkat A, Ali A, et al. Investigation of spatio-temporal satellite thermal IR anomalies associated with the Awaran earthquake (Sep 24, 2013; M 7.7), Pakistan. *Pure Appl Geophys.* 2019;176(8):3533–3544.
34. Bowman DD, Ouillon G, Sammis CG, Sornette A, Sornette D. An observational test of the critical earthquake concept. 1998;103 (B10):24359–24372.
35. Roma-Dollase D, Hernández-Pajares M, Krankowski A, et al. Consistency of seven different GNSS global ionospheric mapping techniques during one solar cycle. *J Geod.* 2018;92(6):691–706.
36. Li Z, Wang N, Wang L, Liu A, Yuan H, Zhang K. Regional ionospheric TEC modeling based on a two-layer spherical harmonic approximation for real-time single-frequency PPP. *J Geod.* 2019;93(9):1659–1671.
37. Klobuchar JA. Ionospheric time-delay algorithm for single-frequency GPS users. *IEEE Trans Aerosp Electron Syst.* 1987;AES-23(3):325–331.
38. Okazaki I, Heki K. Atmospheric temperature changes by volcanic eruptions: GPS radio occultation observations in the 2010 Icelandic and 2011 Chilean cases. *J Volcanol Geotherm Res.* 2012;245–246:123–127.
39. Nĕmec F, Santolík O, Parrot M. Decrease of intensity of ELF/VLF waves observed in the upper ionosphere close to earthquakes: a statistical study. *J Geophys Res Space Phys.* 2009;114(A4).
40. Tariq MA, Shah M, Inyurt S, Shah MA, Liu L. Comparison of TEC from IRI-2016 and GPS during the low solar activity over Turkey. *Astrophys Space Sci.* 2020;365(11):179.
41. Tariq MA, Shah M, Ulukavak M, Iqbal T. Comparison of TEC from GPS and IRI-2016 model over different regions of Pakistan during 2015–2017. *Adv Space Res.* 2019;64(3):707–718.

42. Tariq MA, Shah M, Hernández-Pajares M, Iqbal T. Ionospheric VTEC variations over Pakistan in the descending phase of solar activity during 2016–17. *Astrophys Space Sci.* 2019;364(6):99.

43. Shah M. A. Ahmed, M. Ehsan, M. Khan, M.a. Tariq, A. Calabia, and Z.u. Rahman, Total electron content anomalies associated with earthquakes occurred during 1998–2019. *Acta Astronaut.* 2020;175:268–276.

44. Abbasi AR, Shah M, Ahmed A, Naqvi NA. Possible ionospheric anomalies associated with the 2009 Mw 6.4 Taiwan earthquake from DEMETER and GNSS TEC. *Acta Geodaet Geophys.* 2020.

45. Shah M, Jin S. Pre-seismic ionospheric anomalies of the 2013 Mw = 7.7 Pakistan earthquake from GPS and COSMIC observations. *Geod Geodyn.* 2018;9(5):378–387.

46. Shah M, Majid K, Ullah H, Ali S. Thermal anomalies prior to the 2015 Gorkha (Nepal) earthquake from Modis Land Surface Temperature and Outgoing Longwave Radiations. *Geodyn Tectonophys.* 2018;9(1):123–138.

47. Ulukavak M, Inyurt S. Seismo-ionospheric precursors of strong sequential earthquakes in Nepal region. *Acta Astronaut.* 2020;166:123–130.

48. Kiyani A, Shah M, Ahmed A, et al. Seismo ionospheric anomalies possibly associated with the 2018 Mw 8.2 Fiji earthquake detected with GNSS TEC. *J Geodyn.* 2020;140:101782.

49. Freund FT, Kulahci IG, Cyr G, et al. Air ionization at rock surfaces and pre-earthquake signals. *J Atmos Sol Terr Phys.* 2009;71(17–18):1824–1834.

50. Freund F. Toward a unified solid state theory for pre-earthquake signals. *Acta Geophys.* 2010;719.

51. Kuo CL, Huba JD, Joyce G, Lee LC. Ionosphere plasma bubbles and density variations induced by pre-earthquake rock currents and associated surface charges. *J Geophys Res Space Phys.* 2011;116(A10). p. n/a-n/a.

52. Mansouri Daneshvar MR, Freund FT. Remote sensing of atmospheric and ionospheric signals prior to the Mw 8.3 Illapel earthquake, Chile 2015. *Pure Appl Geophys.* 2017;174(1):11–45.

53. Rikitake T. Earthquake precursors in Japan: precursor time and detectability. *Tectonophysics.* 1987;136(3):265–282.

54. Awais M, Barkat A, Ali A, Rehman K, Ali Zafar W, Iqbal T. Satellite thermal IR and atmospheric radon anomalies associated with the Haripur earthquake (Oct 2010; mw 5.2), Pakistan. *Adv Space Res.* 2017;60(11):2333–2344.

55. Ganguly ND. Atmospheric changes observed during April 2015 Nepal earthquake. *J Atmos Sol Terr Phys.* 2016;140:16–22.

56. Shah M, Inyurt S, Ehsan M, et al. Seismo ionospheric anomalies in Turkey associated with Mw ≥ 6.0 earthquakes detected by GPS stations and GIM TEC. *Adv Space Res.* 2020;65(11):2540–2550.

57. Yildirim O, Inyurt S, Mekik C. Review of variations in mw < 7 earthquake motions on position and TEC (mw = 6.5 Aegean Sea earthquake sample). *Nat Hazards Earth Syst Sci.* 2016;16(2):543–557.

58. Melgarejo-Morales A, Vazquez-Becerra GE, Millan-Almaraz JR, Pérez-Enríquez R, Martínez-Félix CA, Gaxiola-Camacho JR. Examination of seismo-ionospheric anomalies before earthquakes of Mw ≥5.1 for the period 2008–2015 In Oaxaca, Mexico using GPS-TEC. *Acta Geophys.* 2020;68(5):1229–1244.

59. Inyurt S, Peker S, Mekik C. Monitoring potential ionospheric changes caused by the Van earthquake (Mw7.2). *Ann Geophys.* 2019;37(2):143–151.

60. Hussain A, Shah M. Comparison of GPS TEC with IRI models of 2007, 2012, & 2016 over Sukkur, Pakistan. *NASIJ.* 2020;1(1):1–10. https://doi.org/10.47264/idea.nasij/1.1.1.

61. Mehmood M, Filjar R, Saleem S, Shah M, Ahmed A. TEC derived from local GPS network in Pakistan and comparison with IRI-2016 and IRI-PLAS 2017. *Acta Geophys.* 2021. https://doi.org/10.1007/s11600-021-00538-0.

62. Adil MA, Shah M, Abbas A, Ehsan M, Naqvi NA. Investigation of ionospheric and atmospheric anomalies associated with three Mw > 6.5 EQs in New Zealand. *J Geodyn.* 2021;101841. https://doi.org/10.1016/j.jog.2021.101841.

63. Shah M, Qureshi RU, Khan NG, Ehsan M, Yan J. Artificial Neural Network based thermal anomalies associated with earthquakes in Pakistan from MODIS LST. *J Atmos Sol-Terr Phys.* 2021;215(105568). https://doi.org/10.1016/j.jastp.2021.105568.

64. Shah M, Ehsan M, Abbas A, Ahmed A, Jamjareegulgarn P. Possible thermal anomalies associated with global terrestrial earthquakes during 2000–2019 based on MODIS-LST. *IEEE Geosci Remote Sens Lett.* 2021;1–5. https://doi.org/10.1109/LGRS.2021.3084930.

65. Tariq MA, Shah M, Li Z, Wang N, Shah MA, Liu L. Lithosphere ionosphere coupling associated with three earthquakes in Pakistan from GPS and GIM TEC. *J Geodyn.* 2021;147(101860). https://doi.org/10.1016/j.jog.2021.101860.

66. Ahmed J, Shah M, Awai M, et al. seismo-ionospheric anomalies before the 2019 Mirpur earthquake from ionosonde measurements. *Adv Space Sci.* 2021. https://doi.org/10.1016/j.asr.2021.07.030.

67. Adil MA, Senturk A, Shah M, Naqvi NA, Saqib M, Abbasi AR. Atmospheric and ionospheric disturbances assocaited with M > 6.0 earthquakes in the East Asian regions: a case study from Taiwan. *J Asian Earth Sci.* 2021;220(104918). https://doi.org/10.1016/j.jseaes.2021.104918.

68. Mehdi S, Shah M, Naqvi NA. Lithosphere atmosphere ionosphere coupling associated with the 2019 Mw 7.1 California earthquake using multiple precursors. *Environ Monit Assess J.* 2021;193(501). https://doi.org/10.1007/s10661-021-09278-6.

69. Rahman ZU. Possible seismo ionospheric anomalies before the 2016 Mw 7.6 Chile Earthquake from GPS TEC, GIM TEC and Swarm Satellites. *Nat Appl Sci Int J.* 2020;1:11–20. https://doi.org/10.47264/idea.nasij/1.1.2.

29

Landslide spatial modeling using a bivariate statistical method in Kermanshah Province, Iran

Mojgan Bordbar[a], Sina Paryani[a], and Hamid Reza Pourghasemi[b]

[a]Department of GIS/RS, Faculty of Natural Resources and Environment, Science and Research Branch, Islamic Azad University, Tehran, Iran [b]Department of Natural Resources and Environmental Engineering, College of Agriculture, Shiraz University, Shiraz, Iran

1 Introduction

In recent decades, population growth, residential areas, and the construction of buildings and roads have significantly impacted on slope failure in various countries worldwide. One of the most dangerous natural disasters in the world is landslides that threaten human life and economic losses.[1] In some cases, damage caused by landslides has been estimated more than other natural disasters. Also, deaths and economic damages caused by landslides have been reported to be higher than the actual number. According to the Center for Disaster Epidemiology Research (CRED), at least 17% of the damage caused by natural disasters worldwide has been reported due to landslides.[2] Every year, many people affected as a result of this disaster worldwide.[3] The occurrence of landslides depends on various factors such as earthquakes, heavy rains, geological characteristics, and vegetation.[3] Iran is also very prone to landslides due to its mountainous topography (Alborz and Zagros Mountains), physiographic conditions, seismicity, climate change, and geological conditions. According to available statistics, 4900 landslides occurred in 2007, killing 187 people and causing the US $ 12,700 million in economic losses.[4] Therefore, the management of areas susceptible to landslides is of great importance to reduce their damages.[5]

Landslide susceptibility refers to the spatial distribution of the probability of landslides. One of the most widely used methods that can be used to reduce the damage caused by natural disasters is landslide susceptibility assessment.[6] The Geographical Information System (GIS) is an efficient tool for assessing landslide susceptibility and has been widely used in recent years.[7–9] Various methods, including quantitative and qualitative methods, have been applied to prepare landslide susceptibility maps. Qualitative methods such as multicriteria decision-making methods are based on the knowledge and opinions of experts.[10, 11] While quantitative methods include mathematical analyses and statistical probabilities for analyzing the relationship between landslide occurrence and parameters affecting it, multivariate and binary statistical methods are used for evaluation in statistical methods.[3]

Bivariate and multivariate probabilistic statistical methods, including weights of evidence method,[12, 13] logistic regression,[14, 15] and statistical index[16, 17] have been the most used among the available methods. The Frequency ratio model is also a common bivariate statistical method for preparing landslide susceptibility maps, which has been used in different regions, including Iran,[18] Turkey[19] Korea,[20] India,[21] and Pakistan.[22] In the FR method, landslide occurrence maps are placed on the maps of each factor affecting landslides, and their relationship is determined, which is considered as one of the advantages of this model.

Kermanshah Province is one of the areas at risk of landslide amplitude movements due to its location in the Zagros Mountains and the presence of loose formations. Therefore, careful study of landslides in this province and identification of landslide areas, as well as their potential to reduce future damage, is of great importance. For this purpose, landslide susceptibility assessment and spatial modeling were performed using a bivariate statistical model of FR. To the best knowledge of the author, this study for the first time has examined landslide hazards in Kermanshah Province.

2 Study area

Kermanshah Province is located in the west of Iran at 45° 20 39″ to 48° 01 58″ longitude and 33° 37 08″ to 35° 17′ 08″ latitudes (Fig. 1). This province is a mountainous region located between the Iranian plateau and the Mesopotamian plain, which is covered by the summits and heights of the Zagros Mountains (https://www.irimo.ir). The Zagros Mountains have emerged within this province as a series of parallel mountain ranges between which high mountainous plains have formed, shaping the bed of important Zagros passages. The area of this province is 24,650 km². The highest altitude of the study area above sea level was 3372 m. This region has a temperate, mountainous climate. Kermanshah Province is exposed to the Mediterranean humid fronts, and its intersection with the Zagros heights causes snow and rain. In this region, four different climates can be distinguished: mild winters and hot and dry

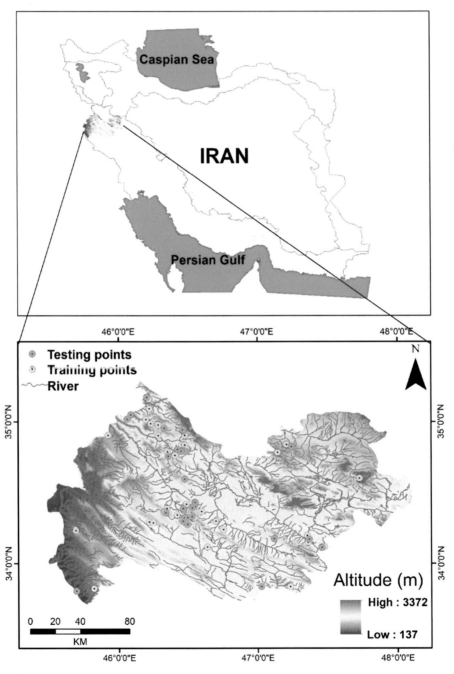

FIG. 1　Location map of the study area.

summers, cool winter and summer, semiarid and cool steppe climate, and semiarid and warm steppe (http://www.yjc.ir/fa/news/4787902). The average annual rainfall in the study area is 434 mm.[23] This area is very important because of its agricultural and economic use. According to the land use map, 43.02% of the region includes agricultural and dry farming. In addition, the range of land in this province is 37.23%, which shows the great importance of this land use. Qft2 and EMas-sb are the most important lithological units in the study area. The Qft2 unit consists of a low-level piedment fan and valley terrace deposits. In addition, the EMas-sb unit consists of undivided Asmari and Shahbazan formations.

3 Methodology

3.1 Data collection and preparation

3.1.1 Inventory map

In studies on landslides, a database of landslide locations is needed to prepare an inventory map, which is considered as the main step in modeling landslide susceptibility and hazard maps.[24] In the present study, the spatial distribution of 115 landslide occurrence locations in Kermanshah Province was used to prepare an inventory map. First, the landslide locations were prepared through the historically recorded data of the study area, and the landslide points were randomly investigated as 81 samples (70%) were randomly selected for modeling and 34 points (30%) were selected for validation of the results (Fig. 1).

3.1.2 Conditioning factors

Various factors, such as topography, lithology, tectonics, climate, human activities, and environment cause landslides.[25] In this study, 10 factors affecting landslides, including altitude, slope aspect, slope degree, profile curvature, distance to faults, distance to roads, distance to rivers, lithology, rainfall, and land use were used spatially to model the landslides. These factors were selected based on available data, landslide occurrence mechanisms, and research literature.[26, 27] Various sources, as shown in Table 1, were used to prepare the data. Maps of each factor were prepared in the environment of ArcGIS 10.4.1 software with raster format and a pixel size of 30×30 m to create the landslide susceptibility map. Each of the factors affecting landslide occurrence is described briefly below.

Altitude

Altitude is one of the most widely used factors for landslide susceptibility.[28] Fig. 2A shows the altitude map, which was prepared using a digital elevation model (DEM) with a resolution of 30 m. This map was classified with distances of 500 was classified into six classes of <500, 500–1000, 1000–1500, 1500–2000, 2000–2500, and >2500 m.

TABLE 1 Data and sources for the landslide spatial modeling.

Data	Sources
DEM (digital elevation model) (m)	ASTER (Global DEM) 30 m
Slope degree (°)	Extracted from DEM
Slope aspect	Extracted from DEM
Profile curvature	Extracted from DEM
Distance to roads (m)	Topographical maps (1:100,000)
Distance to faults (m)	Geological Survey of Iran (1:100,000)
Distance to rivers (m)	National Cartographic Center (1:25,000)
Lithology	Geological Survey of Iran (1:100,000)
Land use	National Geographic Organization (30 m)
Rainfall (mm)	Weather stations located in the Kermanshah Province, Iran

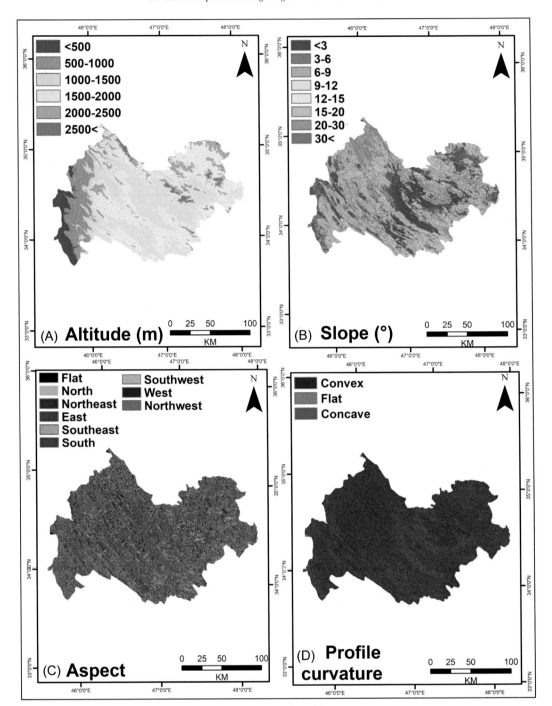

FIG. 2 Conditioning factors: (A) altitude, (B) slope degree, (C) aspect, (D) profile curvature,

(Continued)

Slope degree factor

The slope degree factor is one of the most important and influential factors on landslide susceptibility maps.[29] The slope map was prepared based on the DEM of the area and was classified into eight classes (Fig. 2B). These classes included <3, 3–6, 6–9, 9–12, 12–15, 15–20, 20–30, and more than >30 degrees.

Slope aspect

The maximum slope direction is referred to as the slope aspect.[29] Another factor that needs to be evaluated for preparing a landslide susceptibility map is the factor slope direction. This factor determines the amount of topographic

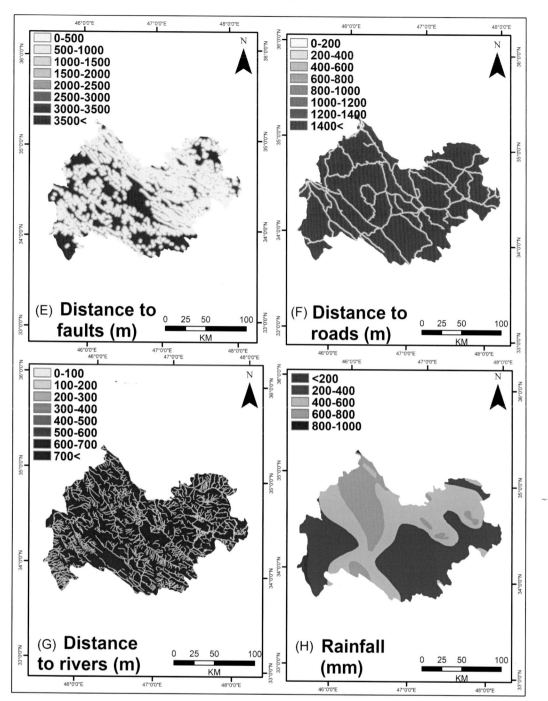

FIG. 2, CONT'D (E) distance to faults, (F) distance to roads, (G) distance to rivers, (H) rainfall,

(Continued)

moisture under the influence of sunlight and rainfall. The slope direction determines the effects of solar heat, air dryness, and air humidity.[30] Fig. 2C shows the slope direction map for the Kermanshah Province.

Profile curve

The profile curve is defined as the surface curvature in the direction of the steepest slope.[31] The profile curve extracted from the DEM of the region is shown in Fig. 2D. If the surface is convex, the curvature is positive and if it is concave, the curvature is negative. A value of zero indicates a smooth surface.

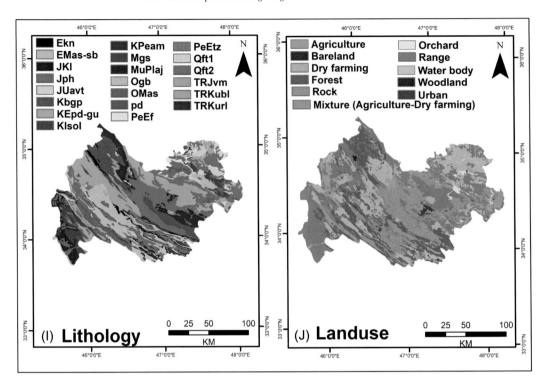

FIG. 2, CONT'D   (I) lithology, (J) land use.

Distance to faults

Faults are crustal fractures that reduce rock resistance and landslide.[3] The greater the distance to faults, the lower the probability of landslides.[32] The map of distance to faults in Fig. 2E shows eight classes with distances of 500 m. This map was classified into classes of 0–500, 500–1000, 1000–1500, 1500–2000, 2000–2500, 2500–3000, 3000–3500, and > 3500 m.

Distance to roads

Distance to roads is another factor affecting landslides, which was investigated in this study. The greater is the distance to roads, the lower is the probability of landslides.[32] The map of distance to roads was divided into eight classes with distances of 200 m. The classes of this map include 0–200, 200–400, 400–600, 600–800, 800–1000, 1000–1200, 1200–1400, and > 1400 m (Fig. 2F).

Distance to rivers

Rivers are also effective in landslides by affecting slope stability by eroding the side of the slope and rising water levels due to the accumulation of downstream sediments.[33] The map of distance to rivers was classified into eight classes and their distances were considered as 100 m. This map was divided into classes of 0–100, 100–200, 200–300, 300–400, 400–500, 500–600, 600–700, and > 700 m (Fig. 2G).

Rainfall

Rainfall is another major factor in assessing landslide susceptibility and has a significant effect on landslide occurrence. The number of landslides caused by rainfall was greater than the number of landslides caused by earthquakes.[34] The rainfall map of the Kermanshah Province rainfall map was classified into five classes with a distance of 200 m. This map includes classes of <200, 200–400, 400–600, 600–800, and >800 mm (Fig. 2H).

Lithology

The susceptibility of different lithologies varied depending on the landslide. Therefore, lithology is considered to be one of the main factors in landslide susceptibility assessment.[35] The lithological units are shown in Fig. 2I. The map was divided into 21 geological units in the study area. Information about the lithology of the area is presented in Table 2.

TABLE 2   Description of lithological units in the study region.

Lithological units	Description
Ekn	Tine-bedded argillaceous limestone and calcareous shale
EMas-sb	Undivided Asmari and Shahbazan Formation
JKl	Crystalized limestone and calc- schist
Jph	Phyllite, slate, and meta-sandstone
JUavt	Andesitic volcanic Tuff
Kbgp	Undivided Bangestan Group, mainly limestone and shale, Albian to Companian, comprising the following formations: Kazhdumi, Sarvak, Surgah, and Ilam
KEpd-gu	Gray and brown, medium-bedded to massive fossiliferous limestone
Klsol	Gray thick-bedded to massive orbitolina limestone
KPeam	Dark olive-brown, low weathered siltstone, and sandstone with local development of chert conglomerates and shelly limestone
Mgs	Anhydrite, salt, gray, and red marl alternating with anhydrite, argillaceous limestone, and limestone
MuPlaj	Brown to gray, calcareous, feature-forming sandstone and low weathering, gypsum-veined, red marl, and siltstone
Ogb	Gabbro
OMas	Cream to brown—weathering, feature-forming, well-jointed limestone with intercalations of shale
pd	Peridotite including harzburgite, dunite, lerzolite and websterite
PeEf	Flysch turbidite, sandstone, and calcareous mudstone
PeEtz	Gray and brown, medium-bedded to massive fossiliferous limestone
Qft1	High-level piedmont fan, and valley terrace deposits
Qft2	Low-level piedment fan, and valley terrace deposits
TRJvm	Meta-volcanic, phyllites, slate, and meta- limestone
TRKubl	Kuhe Bistoon limestone
TRKurl	Purple and red thin-bedded radiolarian chert with intercalations of neritic and pelagic limestone

Land use/land cover

Land use is also an important factor in landslide susceptibility assessments. The map of land use was classified into 11 classes. Land use in the study area includes agriculture, bare land, dry farming, forest, mixture (agriculture-dry farming), orchards, range, rocks, urban areas, water bodies, and woodlands. Table 3 shows the information on the factors affecting the landslides (Fig. 2J).

3.2 Model description

3.2.1 Frequency ratio model

One of the methods used to show the relationship between the position of landslides and factors in the study area is the frequency ratio (FR) method, where, the maps of each factor overlapped with the map of the landslides that occurred individually. The frequency ratio method is the ratio of the areas where the landslide occurred to the entire study area, as well as the ratio of the probability of the landslide occurrence to its nonoccurrence for a factor. Using Eq. (1), the weight of each class of each factor is calculated[36] as follows:

$$\text{Frequency Ratio} = \frac{\frac{A}{B}}{\frac{C}{D}}, \tag{1}$$

TABLE 3 Frequency ration values for landslide spatial modeling.

Conditioning factor	Class	Number of pixels in domain	Number of landslides	% Total of area (a)	% Total of landslide area (b)	FR (b/a)
Altitude (m)	<500	1,115,741	0	4.90	0	0
	500–1000	2,633,798	5	11.56	6.17	0.53
	1000–1500	8,038,917	28	35.30	34.56	0.97
	1500–2000	8,685,492	44	38.14	54.32	1.42
	2000–2500	2,002,334	4	8.79	4.93	0.56
	2500<	292,948	0	1.28	0	0
Slope degree (°)	<3	5,322,641	8	23.37	9.87	0.42
	3_6	4,388,611	16	19.27	19.75	1.02
	6_9	2,902,246	15	12.74	18.51	1.45
	9_12	2,160,330	9	9.48	11.11	1.17
	12_15	1,715,403	8	7.53	9.87	1.31
	15_20	2,162,429	6	9.49	7.40	0.77
	20_30	2,682,110	14	11.77	17.28	1.46
	30<	1,435,460	5	6.30	6.17	0.97
Aspect	Flat	141,298	0	0.62	0	0
	North	2,859,791	9	12.55	11.11	0.88
	Northeast	3,249,068	11	14.26	13.58	0.95
	East	2,342,249	7	10.28	8.64	0.84
	Southeast	2,088,089	6	9.17	7.40	0.80
	South	3,028,063	14	13.29	17.28	1.29
	Southwest	3,777,892	10	16.59	12.34	0.74
	West	2,940,713	15	12.91	18.51	1.43
	Northwest	2,342,067	9	10.28	11.11	1.08
Profile curvature	Convex	10,086,932	35	44.30	43.20	0.97
	Flat	1,675,351	4	7.35	4.93	0.67
	Concave	11,006,947	42	48.34	51.85	1.07
Distance to fault (m)	0–500	7,163,454	16	31.46	19.75	0.62
	500–1000	4,099,084	22	18.00	27.16	1.50
	1000–1500	2,860,606	12	12.56	14.81	1.17
	1500–2000	2,060,015	6	9.04	7.40	0.81
	2000–2500	1,515,708	7	6.65	8.64	1.29
	2500–3000	1,158,875	8	5.08	9.87	1.94
	3000–3500	897,134	3	3.94	3.70	0.93
	3500<	3,014,354	7	13.23	8.64	0.65
Distance to roads (m)	0–200	880,091	11	3.86	13.58	3.51
	200–400	799,192	8	3.50	9.87	2.81
	400–600	773,308	2	3.39	2.46	0.72

TABLE 3 Frequency ration values for landslide spatial modeling—cont'd

Conditioning factor	Class	Number of pixels in domain	Number of landslides	% Total of area (a)	% Total of landslide area (b)	FR (b/a)
	600–800	722,524	2	3.17	2.46	0.77
	800–1000	713,654	1	3.13	1.23	0.39
	1000–1200	706,179	3	3.10	3.70	1.19
	1200–1400	682,901	3	2.99	3.70	1.23
	1400<	17,491,381	51	76.82	62.96	0.81
Distance to rivers (m)	0–100	1,729,927	14	7.59	17.28	2.27
	100–200	1,469,569	13	6.45	16.04	2.48
	200–300	1,474,508	9	6.47	11.11	1.71
	300–400	1,307,163	6	5.74	7.40	1.29
	400–500	1,280,419	4	5.62	4.93	0.87
	500–600	1,250,989	8	5.49	9.87	1.79
	600–700	1,110,886	0	4.87	0	0
	700<	13,145,769	27	57.73	33.33	0.57
Lithology	Ekn	454,324	5	1.99	6.17	3.09
	EMas-sb	2,428,907	5	10.66	6.17	0.57
	JKl	1070	0	0.004	0	0
	Jph	133,755	0	0.58	0	0
	JUavt	357,781	0	1.57	0	0
	Kbgp	2,020,786	12	8.87	14.81	1.66
	KEpd-gu	2,150,441	35	9.44	43.20	4.57
	Klsol	484,695	1	2.12	1.23	0.57
	KPeam	378,398	2	1.66	2.46	1.48
	Mgs	799,309	0	3.51	0	0
	MuPlaj	877,371	3	3.85	3.70	0.96
	Ogb	317,889	0	1.39	0	0
	OMas	758,835	2	3.33	2.46	0.74
	pd	187,664	1	0.82	1.23	1.49
	PeEf	201,196	1	0.88	1.23	1.39
	PeEtz	422,539	0	1.85	0	0
	Qft1	602,368	0	2.64	0	0
	Qft2	6,722,129	5	29.52	6.17	0.20
	TRJvm	315,627	0	1.38	0	0
	TRKubl	1,794,547	0	7.88	0	0
	TRKurl	1,359,599	9	5.97	11.11	1.86
Land use	Agriculture	893,058	1	3.92	1.23	0.31
	Bare land	20,013	0	0.08	0	0

Continued

TABLE 3   Frequency ration values for landslide spatial modeling—cont'd

Conditioning factor	Class	Number of pixels in domain	Number of landslides	% Total of area (*a*)	% Total of landslide area (*b*)	FR (*b/a*)
	Dry farming	3,943,756	32	17.32	39.50	2.28
	Forest	3,806,480	17	16.71	20.98	1.25
	Mixture (agriculture-dry farming)	4,960,376	6	21.78	7.40	0.34
	Orchard	387,279	1	1.70	1.23	0.72
	Range	8,477,053	20	37.23	24.69	0.66
	Rock	62,263	0	0.27	0	0
	Urban	173,020	4	0.75	4.93	6.49
	Water body	17,087	0	0.07	0	0
	Woodland	28,845	0	0.12	0	0
Rainfall (mm)	<200	226,901	0	0.99	0	0
	200–400	10,435,097	15	45.82	18.51	0.40
	400–600	8,338,544	32	36.62	39.50	1.07
	600–800	3,755,723	34	16.49	41.97	2.54
	800–1000	12,965	0	0.056	0	0

where, *A* is the number of landslide points in each class of each factor, *B* is the total number of landslide points in the region, *C* is the area of each class of each factor, and *D* is the total area of the study region. A value of 1 indicates an average value, and if the value is greater than 1, it indicates a high probability of landslide, and a value less than 1 indicates a low probability of landslide.[36]

3.2.2 Landslide susceptibility index (LSI)

The frequency ratio method was first obtained for each class of the parameter, and then the resulting values were added together to calculate the landslide susceptibility index (Eq. 2).[37]

$$\text{LSI} - Fr_1 + Fr_2 + \dots + Fr_n \tag{2}$$

In the above equation, LSI is the landslide susceptibility index and Fr is the frequency ratio of each factor.

4 Result and discussion

4.1 FR model

The FR model is a comprehensible statistical probabilistic model with the capability to provide highly accurate results. In this model, input, calculation, and output processes are easily processable in the ArcGIS software environment; especially when available data is sufficient, this model can be used as a practical tool to assess landslide susceptibility.[38] Moreover, the FR method generates landslide susceptibility maps and examines the relationship between landslides and factors related to landslides.[38,39] In this regard, the FR method was used to weigh each class of each parameter affecting the occurrence of landslides (Table 3). For altitude, the highest amount of FR was obtained in the high-altitude class of 1500–2000 m (1.42). in the class of more than 2500 m, the FR value was the lowest (0).The results of the altitude are consistent with a study by Paryani et al.[15] They concluded that the highest FR is in the high-altitude class of 1500–1700, which has the greatest effect on landslides. In the case of the slope degree, the maximum value of FR belongs to the class of 20–30 degrees (1.46). for slope class >30, the FR value decreased (0.97). A similar result was obtained in the study by Wang and Li .[40] It can be concluded that landslide occurrence increases with an increase in slope and decreases in a certain slope. In the study by Lee and Sambath,[41] the FR value increased on high slopes of 21–25°. Concerning the slope aspect, the highest value (1.43) was obtained on the west slope of FR, followed

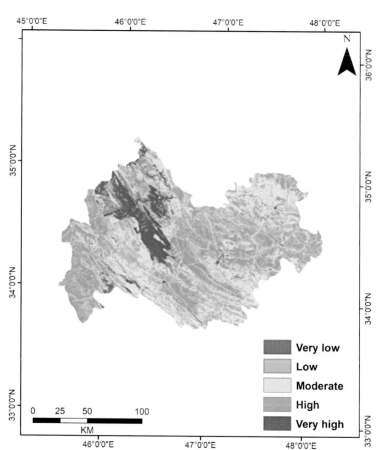

FIG. 3 Landslide susceptibility map using FR model.

by the south slope (1.29). The FR in the profile curvature showed almost the same values for both concave and convex shapes. These results are in accordance with those of Lee and Sambath.[41] They also obtained relatively similar values for the two curvatures. For the factor of the distance to faults, the results of FR showed that at distances of 2500–3000 m, the maximum value of FR was 1.94. At a distance of 0–500 m, the lowest FR value was 0.62. For the parameter of distance to roads at distances of 0–200 m, the maximum value of FR was obtained as 3.51. At distances closer to the road, the probability of landslide occurrence is more. However, at a distance of 800–1000 m, the lowest value of FR was obtained (0.39). For the factor of distance to rivers, the FR value belongs to the class of 100–200 m (2.48) and the lowest value belongs to the class of 600–700 m (0). The value of FR decreased with increasing distance from the rivers. In terms of lithology, the highest value of FR belonged to the classes of KEpd-gu (4.57) and Ekn (3.09). For the factor of land use, it was found that urban land use has the highest FR (6.49). For rainfall, the maximum value of FR belongs to the class of 600–800 mm (2.54). In the study by Wang and Li,[40] the maximum FR belongs to the class of <500 mm/year. They also concluded that the class of higher rainfall with a higher FR value had a greater effect on landslide occurrence. A Landslide susceptibility map was obtained from the sum of the FR-weighted maps. Fig. 3 shows that there is a very high susceptibility to landslides in the northwestern areas, while in the eastern and northeastern regions, susceptibility to landslides is low. The results obtained from a study by Thanh et al.[42] show that the performance of the FR model is acceptable for assessing landslide susceptibility in Vietnam. Haung et al.[43] also showed that the FR model is more efficient than the SVM model for investigating landslide susceptibility in the Nantian region of China.

4.2 Producing landslide susceptibility map

Various classification methods have been used for the classification of LSM in ArcGIS platforms, such as natural breaks, equal intervals, quantiles, and standard deviation.[44–46] The susceptibility index was divided into equal sizes according to equal intervals. For natural break classification, the classification is determined statistically according to the natural grouping inherent in the data. The Quantile method is based on unequal size intervals and is suitable for linear data. The

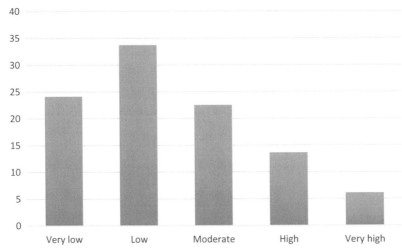

FIG. 4   The percentage of landslide areas.

Standard deviation method applies the mean value of the data for classification.[46] In the current study, the aforementioned methods tested and the natural break classifier yielded the best results. Therefore, the LSM map was classified using the natural break classification method. The results showed that 24.1%, 33.7%, 22.5%, 13.6%, and 6.1% of the regions were in the susceptibility classes of very low, low, moderate, high, and very high susceptibility, respectively (Fig. 4).

4.3   Validation of LSM map

4.3.1   ROC curve

The receiver operating characteristics (ROC) curve is a useful technique for determining the predictive quality of probabilistic and definitive diagnosis.[47] The ROC diagram consists of two axes: the x-axis showing the false-positive value and the y-axis showing the true-positive value. In addition, the area under the curve (AUC) determines the quality of the prediction system by describing the system's ability to accurately estimate the occurrence or nonoccurrence of a predefined event. Because the value of the area under the curve shows the accuracy of the method quantitatively, it is necessary to estimate this value to compare the performance of the models. The values of the area under the curve vary from 0.5 to 1, so that the closer it is to 1, the better the performance, and the closer it is to 0.5, the poorer the performance.[29]

Seventy percent of the data on landslides that occurred in the study area were used for modeling and 30% were used for model validation. In this study, the area under the ROC curve was used to evaluate the performance of the FR landslide susceptibility model. Based on the ROC diagram shown in Fig. 5, the AUC value for this model was 90.3%. Therefore, the results of this model can be trusted.

4.3.2   Frequency ratio plot

The FR plot was also applied to validate the landslide susceptibility map using the percentage of landslide distribution. Therefore, the landslide points were overlaid on the landslide susceptibility classification map and FR values were extracted for each point. An FR plot was drawn based on this plot. The results showed that the FR value in the plot increased when moving from the very low to the very high susceptibility classes (Fig. 6). The very high (57.4%) and very low (3.5%) classes had the highest and lowest landslide densities, respectively.

Landslide as a natural phenomenon has been modeled frequently by various methods. In other words, the goal is to use a method to achieve logical and accurate results. In the present study, we applied the FR probabilistic model to spatial modeling of landslides in Kermanshah Province, Iran.

One of the key steps in developing research is to compare the results with similar studies. The probabilistic FR model has performed well in other studies.[48–50] For example, Ozdemir and Altural[35] used three methods: frequency ratio (FR), logistic regression (LR), and weights of evidence (WOE) to produce landslide susceptibility maps in Turkey. The values of AUC values for the maps generated with FR, LR, and WOE were 0.976, 0.952, and 0.937, respectively. They concluded that the FR model is the best model for assessing landslide susceptibility in this area. In another study by Vakhshoori and Zare,[51] the FR model had the highest prediction accuracy compared to the fuzzy and WoE models. Demir et al. [52] prepared LSM in Turkey using multivariate logistic regression and bivariate frequency ratio models.

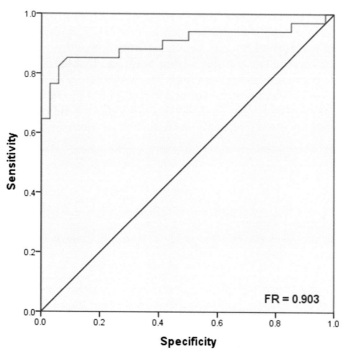

FIG. 5 Prediction rate for FR model.

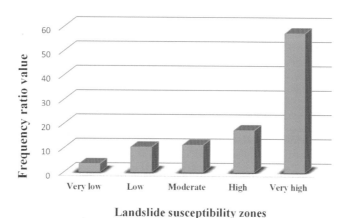

FIG. 6 Frequency Ratio plot.

Their results showed that FR with an AUC of 0.744 performed better than the LR model (AUC=0.708). They concluded that although the results of both methods were successful, the FR model was slightly accurate. For spatial modeling of landslides in India, Sharma and Mahajan[53] applied information value (InV), frequency ratio (FR), and analytical hierarchy process (AHP). According to their results, the FR model with an AUC of 89.61% has a slightly better performance compared to InV and AHP with an AUC of 88.26% and 87.12%, respectively. Similar to the literature review, the results of the present study show that the FR probabilistic model with an accuracy of more than 90% has an excellent performance in geospatial modeling of landslides. This method is suggested for other studies because it is easy to implement and suitable for any type of data.

5 Conclusion

Landslides are one of the most dangerous natural disasters in many parts of the world, therefore, it is of great importance to evaluate and prepare susceptibility maps to reduce human and financial losses. In this study, the FR model was applied to generate the landslide susceptibility map of Kermanshah Province, Iran,. To achieve this aim, first, the correlation between conditioning factors (altitude, slope degree, slope aspect, profile curve, distance to faults, distance to roads, distance to rivers, lithology, rainfall, and land use) and landslide points were evaluated. In the next step, the landslide susceptibility index was calculated and the related map was prepared from the sum of the FR weighted layers. The area under the ROC curve was used to validate the performance of the applied model. The area under the ROC curve indicated that the FR model performed well in the spatial modeling of landslides. (AUC = 0.903). In addition, the results of the FR plot illustrate that the LSM has a standard trend from very low to very high classes. The outcomes of this study can be provided to managers and planners to better manage and reduce damage resulting from landslides.

References

1. Moayedi H, Mehrabi M, Mosallanezhad M, Rashid AS, Pradhan B. Modification of landslide susceptibility mapping using optimized PSO-ANN technique. *Eng Comput.* 2019;35(3):967–984.
2. Lacasse S, Nadim F. Landslide risk assessment and mitigation strategy. In: *Landslides–Disaster Risk Reduction.* Berlin, Heidelberg: Springer; 2009:31–61.
3. Wang Y, Fang Z, Wang M, Peng L, Hong H. Comparative study of landslide susceptibility mapping with different recurrent neural networks. *Comput Geosci.* 2020;138:104445.
4. Party IL. *Iranian Landslides List.* Tehran, Iran: Forest, Rangeland and Watershed Association; 2007:60.
5. Thai Pham B, Shirzadi A, Shahabi H, et al. Landslide susceptibility assessment by novel hybrid machine learning algorithms. *Sustainability.* 2019;11(16):4386.
6. Nohani E, Moharrami M, Sharafi S, et al. Landslide susceptibility mapping using different GIS-based bivariate models. *Water.* 2019;11(7):1402.
7. Wang Q, Guo Y, Li W, He J, Wu Z. Predictive modeling of landslide hazards in Wen County, northwestern China based on information value, weights-of-evidence, and certainty factor. *Geomat Nat Haz Risk.* 2019;10(1):820–835.
8. Ali SS. Towards Maqāṣid al-Sharī'ah-based index of socio-economic development: an introduction to the issues and literature. In: *Towards a Maqāṣid al-Sharī'ah Index of Socio-Economic Development.* Cham: Palgrave Macmillan; 2019:1–20.
9. Juliev M, Mergili M, Mondal I, Nurtaev B, Pulatov A, Hübl J. Comparative analysis of statistical methods for landslide susceptibility mapping in the Bostanlik District, Uzbekistan. *Sci Total Environ.* 2019;653:801–814.
10. Yan F, Zhang Q, Ye S, Ren B. A novel hybrid approach for landslide susceptibility mapping integrating analytical hierarchy process and normalized frequency ratio methods with the cloud model. *Geomorphology.* 2019;15:170–187.
11. Achour Y, Boumezbeur A, Hadji R, Chouabbi A, Cavaleiro V, Bendaoud EA. Landslide susceptibility mapping using analytic hierarchy process and information value methods along a highway road section in Constantine, Algeria. *Arab J Geosci.* 2017;10(8):194.
12. Cui K, Lu D, Li W. Comparison of landslide susceptibility mapping based on statistical index, certainty factors, weights of evidence and evidential belief function models. *Geocarto Int.* 2017;32(9):935–955.
13. Kose DD, Turk T. GIS-based fully automatic landslide susceptibility analysis by weight-of-evidence and frequency ratio methods. *Phys Geogr.* 2019;40(5):481–501.
14. Nourani V, Pradhan B, Ghaffari H, Sharifi SS. Landslide susceptibility mapping at Zonouz plain, Iran using genetic programming and comparison with frequency ratio, logistic regression, and artificial neural network models. *Nat Hazards.* 2014;71(1):523–547.
15. Paryani S, Neshat A, Javadi S, Pradhan B. GIS based comparison of the GA-LR ensemble method and statistical models at Sefiedrood Basin, Iran. *Arab J Geosci.* 2020;13(19):1–7.
16. Razavizadeh S, Solaimani K, Massironi M, Kavian A. Mapping landslide susceptibility with frequency ratio, statistical index, and weights of evidence models: a case study in northern Iran. *Environ Earth Sci.* 2017;76(14):1–6.
17. Aghdam IN, Varzandeh MHM, Pradhan B. Landslide susceptibility mapping using an ensemble statistical index (Wi) and adaptive neuro-fuzzy inference system (ANFIS) model at Alborz Mountains (Iran). *Environ Earth Sci.* 2016;75(7):1–20.
18. Abedini M, Tulabi S. Assessing LNRF, FR, and AHP models in landslide susceptibility mapping index: a comparative study of Nojian watershed in Lorestan province, Iran. *Environ Earth Sci.* 2018;77(11):1–3.
19. Yilmaz I. Landslide susceptibility mapping using frequency ratio, logistic regression, artificial neural networks and their comparison: a case study from Kat landslides (Tokat—Turkey). *Comput Geosci.* 2009;35(6):1125–1138.
20. Park S, Choi C, Kim B, Kim J. Landslide susceptibility mapping using frequency ratio, analytic hierarchy process, logistic regression, and artificial neural network methods at the Inje area, Korea. *Environ Earth Sci.* 2013;68(5):1443–1464.
21. Balamurugan G, Ramesh V, Touthang M. Landslide susceptibility zonation mapping using frequency ratio and fuzzy gamma operator models in part of NH-39, Manipur, India. *Nat Hazards.* 2016;84(1):465–488.
22. Khan H, Shafique M, Khan MA, Bacha MA, Shah SU, Calligaris C. Landslide susceptibility assessment using frequency ratio, a case study of northern Pakistan. *Egypt J Remote Sens Space Sci.* 2019;22(1):11–24.
23. Hashemian AH, Rezaei M, Kashefi H, Pirsaheb M, Kharajpour H. Trend step changes of seasonal and annual precipitation over Kermanshah during a 60-year period using non-parametric methods. *Biosc Biotech Res Comm.* 2017;10(4):662–671.
24. Ercanoglu M, Gokceoglu C. Use of fuzzy relations to produce landslide susceptibility map of a landslide prone area (West Black Sea region, Turkey). *Eng Geol.* 2004;75(3–4):229–250.
25. Peng J, Wang S, Wang Q, et al. Distribution and genetic types of loess landslides in China. *J Asian Earth Sci.* 2019;170:329–350.
26. Paryani S, Neshat A, Javadi S, Pradhan B. Comparative performance of new hybrid ANFIS models in landslide susceptibility mapping. *Nat Hazards.* 2020;103:1961–1988.
27. Wu Y, Ke Y, Chen Z, Liang S, Zhao H, Hong H. Application of alternating decision tree with AdaBoost and bagging ensembles for landslide susceptibility mapping. *Catena.* 2020;187:104396.

28. Raghuvanshi TK, Negassa L, Kala PM. GIS based grid overlay method versus modeling approach–a comparative study for landslide hazard zonation (LHZ) in Meta Robi District of west Showa zone in Ethiopia. *Egypt J Remote Sens Space Sci.* 2015;18(2):235–250.

29. Dehnavi A, Aghdam IN, Pradhan B, Varzandeh MH. A new hybrid model using step-wise weight assessment ratio analysis (SWARA) technique and adaptive neuro-fuzzy inference system (ANFIS) for regional landslide hazard assessment in Iran. *Catena.* 2015;135:122–148.

30. Anbalagan R, Kumar R, Lakshmanan K, Parida S, Neethu S. Landslide hazard zonation mapping using frequency ratio and fuzzy logic approach, a case study of Lachung Valley, Sikkim. *Geoenv Disasters.* 2015;2(1):1–7.

31. Ayalew L, Yamagishi H, Ugawa N. Landslide susceptibility mapping using GIS-based weighted linear combination, the case in Tsugawa area of Agano River, Niigata prefecture, Japan. *Landslides.* 2004;1(1):73–81.

32. Gholami M, Ghachkanlu EN, Khosravi K, Pirasteh S. Landslide prediction capability by comparison of frequency ratio, fuzzy gamma and landslide index method. *J Earth Syst Sci.* 2019;128(2):1–22.

33. Yalcin A. GIS-based landslide susceptibility mapping using analytical hierarchy process and bivariate statistics in Ardesen (Turkey): comparisons of results and confirmations. *Catena.* 2008;72(1):1–2.

34. Conforti M, Pascale S, Robustelli G, Sdao F. Evaluation of prediction capability of the artificial neural networks for mapping landslide susceptibility in the Turbolo River catchment (northern Calabria, Italy). *Catena.* 2014;113:236–250.

35. Ozdemir A, Altural T. A comparative study of frequency ratio, weights of evidence and logistic regression methods for landslide susceptibility mapping: Sultan Mountains, SW Turkey. *J Asian Earth Sci.* 2013;64:180–197.

36. Yalcin A, Reis S, Aydinoglu AC, Yomralioglu T. A GIS-based comparative study of frequency ratio, analytical hierarchy process, bivariate statistics and logistics regression methods for landslide susceptibility mapping in Trabzon, NE Turkey. *Catena.* 2011;85(3):274–287.

37. Lee S, Talib JA. Probabilistic landslide susceptibility and factor effect analysis. *Environ Geol.* 2005;47(7):982–990.

38. Li L, Lan H, Guo C, Zhang Y, Li Q, Wu Y. A modified frequency ratio method for landslide susceptibility assessment. *Landslides.* 2017;14(2):727–741.

39. Guo C, Montgomery DR, Zhang Y, Wang K, Yang Z. Quantitative assessment of landslide susceptibility along the Xianshuihe fault zone, Tibetan Plateau, China. *Geomorphology.* 2015;248:93–110.

40. Wang Q, Li W. A GIS-based comparative evaluation of analytical hierarchy process and frequency ratio models for landslide susceptibility mapping. *Phys Geogr.* 2017;38(4):318–337.

41. Lee S, Sambath T. Landslide susceptibility mapping in the Damrei Romel area, Cambodia using frequency ratio and logistic regression models. *Environ Geol.* 2006;50(6):847–855.

42. Thanh DQ, Nguyen DH, Prakash I, et al. GIS based frequency ratio method for landslide susceptibility mapping at Da Lat City, lam dong province, Vietnam. *Vietnam J Earth Sci.* 2020;42:55–66.

43. Huang F, Yao C, Liu W, Li Y, Liu X. Landslide susceptibility assessment in the Nantian area of China: a comparison of frequency ratio model and support vector machine. *Geomat Nat Haz Risk.* 2018;9(1):919–938.

44. Das I, Sahoo S, van Westen C, Stein A, Hack R. Landslide susceptibility assessment using logistic regression and its comparison with a rock mass classification system, along a road section in the northern Himalayas (India). *Geomorphology.* 2010;114(4):627–637.

45. Kumar R, Anbalagan R. Landslide susceptibility mapping using analytical hierarchy process (AHP) in Tehri reservoir rim region, Uttarakhand. *J Geol Soc India.* 2016;87(3):271–286.

46. Baeza C, Lantada N, Amorim S. Statistical and spatial analysis of landslide susceptibility maps with different classification systems. *Environ Earth Sci.* 2016;75(19):1–7.

47. Swets JA. Measuring the accuracy of diagnostic systems. *Science.* 1988;240(4857):1285–1293.

48. Demir G, Aytekin M, Akgün A, Ikizler SB, Tatar O. A comparison of landslide susceptibility mapping of the eastern part of the north Anatolian fault zone (Turkey) by likelihood-frequency ratio and analytic hierarchy process methods. *Nat Hazards.* 2013;65(3):1481–1506.

49. Shahabi H, Khezri S, Ahmad BB, Hashim M. Landslide susceptibility mapping at central Zab basin, Iran: a comparison between analytical hierarchy process, frequency ratio and logistic regression models. *Catena.* 2014;115:55–70.

50. Ding Q, Chen W, Hong H. Application of frequency ratio, weights of evidence and evidential belief function models in landslide susceptibility mapping. *Geocarto Int.* 2017;32(6):619–639.

51. Vakhshoori V, Zare M. Landslide susceptibility mapping by comparing weight of evidence, fuzzy logic, and frequency ratio methods. *Geomat Nat Haz Risk.* 2016;7(5):1731–1752.

52. Demir G, Aytekin M, Akgun A. Landslide susceptibility mapping by frequency ratio and logistic regression methods: an example from Niksar–Resadiye (Tokat, Turkey). *Arab J Geosci.* 2015;8(3):1801–1812.

53. Sharma S, Mahajan AK. A comparative assessment of information value, frequency ratio and analytical hierarchy process models for landslide susceptibility mapping of a Himalayan watershed, India. *Bull Eng Geol Environ.* 2019;78(4):2431–2448.

30

Normalized difference vegetation index analysis of forest cover change detection in Paro Dzongkhag, Bhutan

Sangey Pasang[a,b], *Rigzin Norbu*[a], *Suren Timsina*[a], *Tshering Wangchuk*[a], and *Petr Kubíček*[b]

[a]Civil Engineering Department, College of Science and Technology, Royal University of Bhutan, Rinchending, Bhutan
[b]Department of Geography, Masaryk University, Brno, Czech Republic

1 Introduction

Land use and land cover (LULC) are the biophysical characteristics of the land, such as how the soil, water, and vegetation are distributed and how these particular lands are being used.[1] One of the most important land cover types in the world is forest cover which is progressively decreasing due to the increasing socio-economic development, with several countries losing most forest coverage in their nation's history. In 2019, the global forest coverage decreased by about 11.9 million hectares, at a rate 2.8% higher than the rate of loss in 2018.[2] The main driving force for the decrease in forest coverage was attributed to the occurrence of wildfires caused by climatic conditions and human intervention. Another important cause is deforestation due to the demand for agricultural land and construction work.[3]

In Bhutan, the landscape is mostly dominated by forests, with total coverage of 70.77%, followed by shrubs (9.74%), snow and glaciers (5.35%), and cultivated agricultural land at 2.75%.[4] Among the different types of land cover, forest cover remains one of the most important as it is a vital renewable resource, providing timber, grazing lands for animals, wildlife habitats, and water resources.[5]

Given the fragile ecosystem and dependence of the population on natural resources, it is mandated by the constitution of the Kingdom of Bhutan to always maintain a forest coverage of more than 60%. However, there is an increasing amount of evidence indicating that the environment and the resources it provides are undergoing degradation over time because of natural and manmade causes. A significant factor contributing to the loss of forest cover is the outbreak of wildfires, with a total of 1043 forest fires in Bhutan during the last decade.[6] Some of the Dzongkhags (districts), such as Wangdue, lost more than 40,923 acres to forest fires. To combat this predicament, one of the main programs the government plans to implement is sustainable natural resource management and utilization. However, the lack of research due to technical and financial constraints affects the development and effective implementation of any policies and regulations undertaken by the forestry sector of the Royal Government of Bhutan. An important supporting tool for decision-making processes is understanding the characteristics, extent, and pattern of land use land cover change (LULCC). A detailed study on LULCC will be useful for prioritizing regions that require more measures for sustainable management of the resources.[7] Further, the factors affecting LULC change in that region can be identified, thus increasing the clarity of how the land and its resources are utilized. This will in turn make it possible to predict the state of the land and the demands of the inhabitants for future decades.[8] The Ministry of Agriculture and Forests (MoAF) of the Royal Government of Bhutan stresses the importance of the use of GIS in carrying out field enumeration work and generating preliminary information to ease fieldwork and help allocate resources more judiciously and assertively.

Hence, in the current study, we aimed to evaluate the pattern and extent of forest cover in Paro Dzongkhag and identify the major contributing factors using GIS and related technology. The study in Paro Dzongkhag is particularly important because the socio-economic growth and the pressures exerted on natural resources are relatively higher than any other place in the country. In the most recent study on the forest coverage of Bhutan conducted by MoAF in 2016, the country had an overall forest coverage of 70.77%,[4] with an average increase in forest cover of 13.38% from 2010. However, eight of the 20 Dzongkhags were reported to have undergone a decrease in the total coverage with Paro Dzongkhag, with a maximum decrease of 8.9%, followed by Gasa and Thimphu Dzongkhags with 6.65% and 2.82%, respectively. The total forest coverage in Paro Dzongkhag was 60.9% in 2010 and 52% in 2016.[4]

2 Study area

Paro Dzongkhag is a historic valley, both in terms of culture and trade, with a population of 46,316 in 2017.[9] It is located to the west of the capital Thimphu at coordinates 27.4287°N and 89.4164°E and has a total area of approximately 1293 km² (Fig. 1). The only international airport in Bhutan is also located in the valley at an altitude of 2200 m above mean sea level, making it the only gateway to the country. The annual average maximum temperature is 19°C and the minimum is 6.4°C, with an annual average rainfall of 472.7 mm.[10] The land cover of Paro was divided into 52% forest, 4.28% alpine scrub, 22.70% shrub, 9.97% meadow, and 4.26% cultivated agricultural fields.[4] The dzongkhag is a popular tourist destination in Bhutan, with its rich landscape and cultural heritage, especially the famous Paro Taktsang.

3 Methodology

3.1 Landsat scenes

The primary data used were multispectral data from Landsat-8, a satellite program launched in July 1972. There have been eight different iterations of the Landsat program, with Landsat-8 being the latest, while Landsat-9 is currently in development. The first Landsat satellite was named Landsat-1 and launched on 23 July 1972. The satellite was able to measure only three types of spectral data: the combined blue-green band, orange-red band, and red-near infrared band. After it was decommissioned at the end of its life cycle, seven more upgraded versions of the satellite were

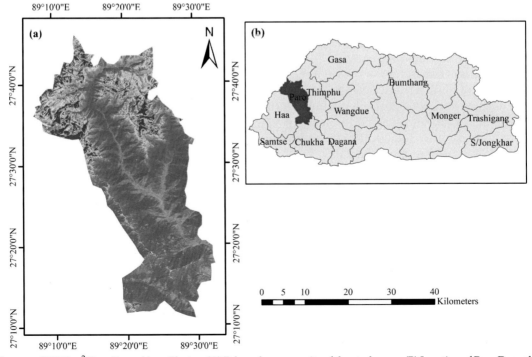

FIG. 1   Study area of 1293 km² Paro Dzongkhag, Bhutan. (A) False color composite of the study area. (B) Location of Paro Dzongkhag in Bhutan.

TABLE 1   Landsat-8 band designations.

Band	Wavelength in μm	Useful for mapping
Band 1—coastal aerosol	0.43–0.45	Coastal and aerosol studies
Band 2—blue	0.45–0.51	Bathymetric map works, distinguishing soil from vegetation
Band 3—green	0.53–0.59	Signifies peak vegetation, which is useful for checking plant vigor
Band 4—red	0.64–0.67	Differentiates vegetation slopes
Band 5—near infrared	0.85–0.88	Signifies biomass content and shore area
Band 6—shortwave Infrared	1.57–1.65	Maps the amount of moisture in soil and vegetation; penetrates clouds
Band 7—shortwave infrared 2	2.11–2.29	Better moisture detection of soil and vegetation
Band 8—panchromatic	0.50–0.68	15 m resolution for supplying sharper image definition
Band 9—cirrus	1.36–1.38	Improved detection of cirrus cloud contamination
Band 10—TIRS 1	10.60–11.19	100 m resolution, thermal mapping, and estimated wetness index of soil
Band 11—TIRS 2	11.5–12.51	100 m resolution, better thermal mapping, and estimated soil wetness

launched by NASA, each having improvements in its build quality and possessing superior technology compared to earlier iterations. The newer satellites could measure more data, with Landsat-8 having the highest capability.

The Landsat-8 satellite comprises of two sensors, the Operational Land Imager (OLI) which is used for the collection of spectral data, while the thermal infrared sensor (TIRS) is a thermal imager for the supplementation of the OLI data.[11] Regions on the earth are divided into scenes, 185 km by 180 km in dimension, identified by their path and row numbers.[12] Landsat-8 collects up to 400 scenes per day. The Landsat satellite provides data in the form of 11 different spectral bands, each providing details on different wavelengths reflected by the objects on the surface of the earth (Table 1).

The data acquired from Landsat have been instrumental in mapping and monitoring the biophysical and geophysical characteristics of earth surfaces for LULC studies.[13] The multispectral data provided by Landsat-8 has been vital for LULC study and monitoring of the earth for more than 40 years after its inception, with a key aspect of it being in the mapping of the present vegetation cover of an area.[14]

3.2 Geographical information system (GIS)

GIS is a system used in the storage, manipulation, and analysis of geospatial data through the use of remote sensing (RS).[15] It has multiple applications in various fields. The use of GIS and data acquired through RS has been used in environmental studies such as natural resource management and environmental modelling.[16] GIS and other related technologies can aid in forestry management through the use of various tools that can be implemented to study the details about the forest resources[5] and the modification of the earth's surface due to human activities.[17] Fig. 2 shows the detailed flow of the method used in this study.

3.3 Data collection

The principal data required for the study were the first eight of the 11 spectral band images recorded by the Landsat-8 satellite (Fig. 3). Bhutan is encompassed by scenes with path and row numbers: 137/41, 138/41, and 139/41.[4]

For the Paro area, the grid with path and row numbers 138/41 was sufficient, as it fully encompassed the entire Dzongkhag. USGS provides the options to download scenes containing different percentages of cloud cover, with the minimum being the option to download scenes with cloud cover less than 10%. As of the time during which this study was conducted, the most recent map with cloud cover less than 10% was dated 16 December 2019. More than 400 high-resolution images of the various regions of Paro Dzongkhag obtained from Google Earth Pro were also used for preliminary observations of the study area and the accuracy assessment of the research.

3.4 Data cleaning

Landsat satellites measure the wavelength of the light reflected by the various features present on the earth's surface, but there can be some inaccuracies in the measurement due to the presence of the atmosphere, which can cause

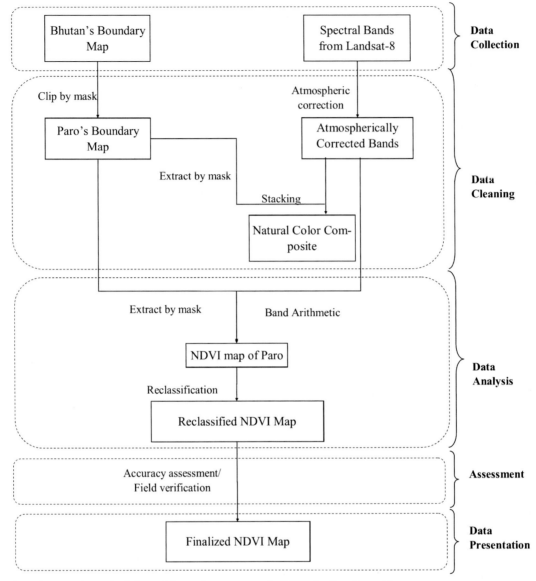

FIG. 2   Flowchart showing the methodology of the study to prepare the NDVI for Paro Dzongkhag.

some light to undergo refraction or reflection.[18] The variables needed for the correction were specified in the *metadata* file which was provided with the rest of the multispectral data obtained from Landsat. The formula used for atmospheric correction of the spectral images is[19]:

$$L_\lambda = M_L \times Q_{cal} + A_L \tag{1}$$

Where:

L_λ = Top of atmospheric spectral radiance
M_L = multiplicative rescaling factor
Q_{cal} = Quantized and calibrated standard data pixels
A_L = additive rescaling factor

3.5 Composite band creation

The images obtained from Landsat are composed of 11 maps of different individual bands in gray-scale color, which do not depict the natural color of the area under observation. To create a natural color image of the study area, the

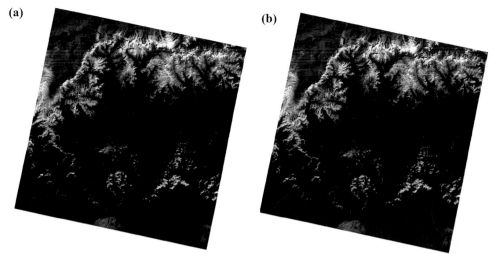

FIG. 3 Data downloaded from the USGS Landsat 8. (A) True color representation of the path and row numbers 138/41 (B) False color composite representing the band 5 of Landsat 8 of the area.

bands were added into ArcGIS, and a composite image was created by combining bands 1–7 into a single map using the tool composite bands (Fig. 3).

A natural color composite of the study area was created, with which it was possible to make preliminary observations and predictions of the land class features before the actual analysis itself.

3.6 Data analysis

Using GIS techniques and different methods of analysis, it is possible to determine the vegetation cover of a region from the spatial data acquired from satellites. The method of analysis used in this study to investigate forest cover change was the calculation of the normalized difference vegetation index (NDVI). NDVI is the most commonly used method for classifying vegetation classes owing to its simple but effective algorithm.[20] It can both quantitatively and qualitatively evaluate not only the class of vegetation but also the vigor and health of the vegetation cover. As such, it is the most widely implemented method for mapping vegetation cover.[21] Among the different vegetation index calculation methods, NDVI has the maximum accuracy in being able correctly identifying the vegetation classes.[22]

The principle of NDVI is the phenomenon by which high vegetation classes reflect more near-infrared light and absorb light of red wavelength.[23] By quantifying the amount of NIR light reflected and red light absorbed, it can be determined whether the region from which the reflection and absorption occur is an area populated by forms of vegetation, such as forests, agricultural fields, shrubs, and grasslands, or whether the area contains nonvegetative features such as barren soil, rocky outcrops, or rivers.[24] NDVI can be calculated from the NIR band (band 5) and the red spectral band (band 4) acquired from Landsat.

The formula used to calculate the NDVI values for a particular region is given by[25]:

$$NDVI = \frac{(\rho_{NIR} - \rho_R)}{(\rho_{NIR} + \rho_R)} \tag{2}$$

Where:

ρ_{NIR} = spectral reflectance of the NIR wavelength.

ρ_R = spectral reflectance of the red wavelength.

The value of NDVI lies in the range of −1 to +1, with values from −1 to 0 indicating areas of nonvegetation such as water and rocks.[16] Values from 0 to 1 indicate the presence of vegetation. The NDVI range that signifies certain vegetation classes in an area must be determined by trial and error.

3.7 Accuracy assessment

To assess the accuracy of the NDVI map created for Paro Dzongkhag, a random point was selected to cross validate the result using Yamane's formula:

$$n = \frac{N}{1 + N(e)^2}$$

(3)

Where:

n = Required number of points

N = Total number of points/pixels in the map

e = Level of precision/sampling error of 5%

The coordinates were then entered into Google Earth Pro, and the type of land feature present in the map was compared to the type of land feature classified in the NDVI map. Using this method, the total number of features correctly and incorrectly identified is noted. The accuracy was then calculated using Eq. (4), where the accuracy was expressed as a percentage:

$$Accuracy = \frac{Correctly\ identified\ points}{Total\ number\ of\ points} \times 100$$

(4)

4 Results

4.1 Vegetation cover by NDVI

For the geospatial analysis of the data procured from Landsat, NDVI analysis was conducted using the NIR and red bands which are represented by bands 5 and 4, respectively. To implement the NDVI algorithm using the two designated bands, the Raster Calculator was used which allows the performance of arithmetic operations on the Landsat bands.

Using the Raster Calculator, a new raster band (Fig. 4) output was created which had NDVI values embedded in each pixel of the band image. The darker regions represent areas with low NDVI values, whereas the lighter regions represent areas with higher NDVI values. To extract the study area from the output, the previously created boundary map of Paro was used along with an Extract by Mask tool, which is used to create a new raster image, from a

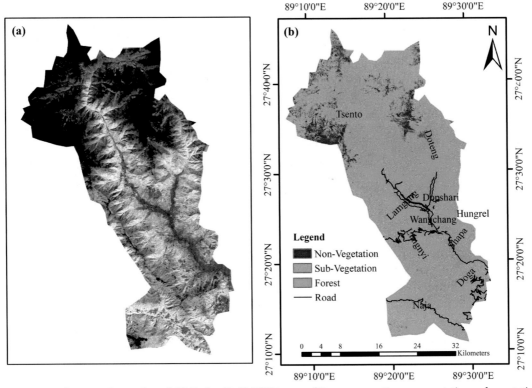

FIG. 4 (A) NDVI image of Paro in Grayscale and (B) Reclassified NDVI image of Paro representing nonvegetation, subvegetation, and forest.

TABLE 2 NDVI range classification determined through observation.

Sl. No	Classification	Features	NDVI range
1	Nonvegetation	Rivers, lakes, barren soil, rocky outcrops, etc.	−1 to 0
2	Subvegetation	Alpine shrubs, agricultural fields grasslands, meadows, etc.	0–0.35
3	Forests	Blue pine forests, broadleaf forests, etc.	0.35–1

TABLE 3 The coverages of different vegetation categories in terms of pixels, area, and percentage for 2019.

Sl. No.	Classification	Number of pixels	Area in hectares	Percentage area
1	Nonvegetation	95,369	8640.53	6.71
2	Subvegetation	635,882	57,611.55	44.71
3	Forests	690,938	62,599.67	48.58

preexisting one, after having specified a boundary map that will serve as the region of interest. After the creation of the NDVI map of Paro Dzongkhag, the next step was to identify and calculate the total quantities of the features present in the map based on their NDVI values. The classification was made as specified in Table 2.

The images were classified to produce NDVI maps with green, orange, and purple colors representing forest, subvegetation, and nonvegetation regions, respectively (Fig. 4). The number of pixels was determined along with the percentage coverage in each category, as shown in Table 3.

The total forest coverage determined using the NDVI analysis was 48.58%, which was a decrease of 3.42% or 3683 ha compared to the total coverage in 2016 (Fig. 5). The total nonvegetation cover was 6.71% which remained roughly the same as that in 2016, whereas the percentage coverage of subvegetation increased by 3.5% from 41.21% in 2016 to 44.71%.

4.2 Vegetation cover assessment

Using Eq. (3), 400 pixels on the map were randomly selected out of the 1,422,189 pixels in the attribute table to assess the accuracy. The accuracy was determined to be 95.75% using Eq. (4), field verification, and Google Earth. Thus, the total percentage cover calculated can have an error of ±4.25%.

5 Discussion

It was determined that the total nonvegetation cover of Paro was 6.71% ± 0.28%, the total subvegetation cover (excluding forests) was 44.71% ± 1.9%, and the total forest coverage was 48.58 ± 2.06%.

The decrease in forest cover could be mainly due to the loss of 3237 ha of forests during the years 2017 and 2018 wildfires.[26] After an area has undergone a forest fire, due to the decrease in the health and amount of vegetation cover, there can be a decrease in NDVI values of the region which can cause the area to possess the properties of

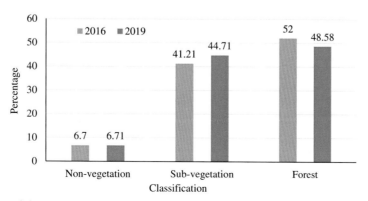

FIG. 5 Graphical representation of change in vegetation cover in Paro dzongkhag from the year 2016 to 2019.

subvegetation, causing them to be classified as such during the NDVI analysis.[27] Hence, this could be the reason why the amount of subvegetation increased by roughly the same amount; the forest cover decreased.

Other contributing factors to the decrease in coverage could be the increase in deforestation of the land with a boom in the construction sector[26] and the construction of institutional and commercial buildings, such as the recently established law school. During the year 2017, 80,077.66 ha of the Paro forest was allotted for major development of construction of transmission lines and road[28] in addition to 70 ha of forest in year 2016.[29]

Another factor contributing to the decrease could be the conversion of forests to agricultural fields, with Paro expected to increase crop production by more than 90% which has detrimental effects on the forest cover.[30]

6 Conclusion

In this study, data from Landsat-8 OLI/TIRS were acquired to determine the total forest cover of Paro Dzongkhag. The Landsat scenes were downloaded and cleaned. NDVI analysis was performed on two of the spectral bands of the Landsat-8 scenes, namely the NIR and red bands. Through the analysis, the different vegetation coverage of Paro Dzongkhag was determined. After making a comparison with the forest coverage determined for 2016 by the government of Bhutan, the decrease in total forest coverage was determined. The result was validated by conducting an accuracy assessment by selecting 400 random points of the NDVI map and comparing the classification to Google Earth Pro images.

With the resolution of Landsat images, the accuracy of the vegetation map is limited when trying to identify smaller features such as roads, rivers, small patches of agricultural land, or subvegetation. Because forests have areas much larger than 900 square meters, the Landsat-8 images were suitable for determining the total forest coverage, but it posed a challenge in trying to correctly identify the smaller features. To overcome this limitation as much as possible, high-resolution images and field data prepared by the National Land Commission (NLC) for the road network of Paro were used for the analysis.

Paro is one of the Dzongkhags where there are relatively more occurrences of forest fires, and as per the comparison made between the 2010 and 2016 reports, it is also one of the eight Dzongkhags which is undergoing a decrease in the total forest coverage.[6] Forest fires are the driving force that initiates the changes which could undermine the country's efforts to conserve natural resources and maintain a minimum forest cover percentage of 60%. Forest fires also do not provide any economic benefits to the country but result in the devastation of wildlife and natural resources. It is evident that for the country to meet its environmental goals, an emphasis must be placed on combating forest fires.

Acknowledgments

We sincerely thank Earth Explorer, USGS, and the National Land Commission Secretariat (NLCS) for providing the satellite images and maps of the study area. We would also like to thank the College of Science and Technology and Masarykova Univerzita for providing the opportunity to conduct this study.

Conflicts of interest

The authors declare no conflicts of interest regarding the publication of this paper.

References

1. Liping C, Yujun S, Saeed S. Monitoring and predicting land use and land cover changes using remote sensing and GIS techniques—a case study of a hilly area, Jiangle, China. *PLoS One*. 2018;1–23.
2. Weisse M, Goldman E. *How Much Forest Did the World Lose in 2019?* World Resources Institute; 2020. https://www.wri.org/insights/we-lost-football-pitch-primary-rainforest-every-6-seconds-2019. Accessed 19 April 2021.
3. Negassa MD, Mallie DT, Gemeda DO. Forest cover change detection using geographic information systems and remote sensing techniques: a spatio-temporal study on Komto protected forest priority area, east Wollega zone, Ethiopia. *Environ Syst Res*. 2020;9(1):1–14. https://doi.org/10.1186/s40068-020-0163-z.
4. Rai A, Phuntsho PS, et al. *Land Use and Land Cover Assessment of Bhutan2016 Technical report*; 2017. Thimphu http://www.dofps.gov.bt/wp-content/uploads/2018/07/LULC2016_Maps-and-Statistics.pdf.
5. Geogr J. Geography & natural disasters application of geographic information system (GIS) in forest management. *J Geogr Nat Disasters*. 2015;5(3). https://doi.org/10.4172/2167-0587.1000145.

6. Wangmo C. *Bhutan's Forest (Coniferous) Prone to Fire—Kuensel Online*; 2020. Kuensel Online https://kuenselonline.com/bhutans-forest-coniferous-prone-to-fire/. Accessed 19 November 2020.
7. Voight C, Hernandez-aguilar K, Gutierrez S, Garcia C. *Utilizing GIS and Remote Sensing to Inform Spatial Conservation Planning : Assessing Vulnerability to Future Tropical Forest Loss in Southern Belize*; 2018. https://doi.org/10.3390/ecrs-2-05150.
8. Han H, Yang C, Song J. Scenario simulation and the prediction of land use and land cover change in Beijing, China. *Sustain*. 2015;7:4260–4279. https://doi.org/10.3390/su7044260.
9. National Statistics Bureau. *Statistical Yearbook of Bhutan 2020*; 2020. https://www.nsb.gov.bt/publications/statistical-yearbook/.
10. Wangmo P, Dorji S. *Report on Bhutan State of the Climate 2020*; 2021. http://nchm.gov.bt/attachment/ckfinder/userfiles/files/Bhutan. State of the Climate 2017.pdf.
11. Roy DP, Wulder MA, Loveland TR, et al. Landsat-8: Science and product vision for terrestrial global change research. *Remote Sens Environ*. 2014;145:154–172. https://doi.org/10.1016/j.rse.2014.02.001.
12. Wachiye SA, Kuria DN, Musiega D. GIS based forest cover change and vulnerability analysis : a case study of the Nandi North forest zone. *J Geogr Reg Plan*. 2013;6(5):159–171.
13. Wulder MA, White JC, Loveland TR, Woodcock CE, Belward AS. The global landsat archive : status, consolidation, and direction. *Remote Sens Environ*. 2015.
14. Sothe C, de Almeida CM, Liesenberg V, Schimalski MB. Evaluating Sentinel-2 and Landsat-8 data to map sucessional forest stages in a subtropical forest in Southern Brazil. *Remote Sens*. 2017;9(8). https://doi.org/10.3390/rs9080838.
15. Yadav KK, Gupta N, Kumar V. *NCEETR 2016 National Conference*. Researchgate; 2016. January 2017.
16. Hussain S, Mubeen M, Ahmad A, et al. Using GIS tools to detect the land use/land cover changes during forty years in Lodhran District of Pakistan. *Environ Sci Pollut Res*. 2019. https://doi.org/10.1007/s11356-019-06072-3. August.
17. Akbar TA, Hassan QK, Ishaq S, Batool M, Butt HJ. Investigate spatial distribution and modelling of existing and future urban land changes and its impact on urbanization and economy. *Remote Sens*. 2019;11(105). https://doi.org/10.3390/rs11020105.
18. Iqbal MF, Khan IA. Spatiotemporal land Use land cover change analysis and erosion risk mapping of Azad Jammu and Kashmir, Pakistan. *Egypt J Remote Sens Sp Sci*. 2014;17(2):209–229. https://doi.org/10.1016/j.ejrs.2014.09.004.
19. Pirotti F, Parraga MA, Stuaro E, Dubbini M, Masiero A, Ramanzin M. NDVI from Landsat 8 vegetation indices to study movement dynamics of Capra ibex in mountain areas. *Int Arch Photogramm Remote Sens Spat Inf Sci—ISPRS Arch*. 2014;40(7):147–153. https://doi.org/10.5194/isprsarchives-XL-7-147-2014.
20. Boqer S, Science O. Use of NDVI and land surface temperature for drought assessment : merits and limitations. *J Climate*. 2009;23:618–633. https://doi.org/10.1175/2009JCLI2900.1.
21. Ghebrezgabher MG, Yang T, Yang X, Wang X, Khan M. Extracting and analyzing forest and woodland cover change in Eritrea based on landsat data using supervised classification. *Egypt J Remote Sens Sp Sci*. 2016;19(1):37–47. https://doi.org/10.1016/j.ejrs.2015.09.002.
22. Towers PC, Strever A, Poblete-echeverr C. *Comparison of Vegetation Indices for Leaf Area Index Estimation in Vertical Shoot Positioned Vine Canopies with and without Grenbiule Hail-Protection Netting*; 2019.
23. Borowik T, Pettorelli N, Sönnichsen L. Availability for ungulates in forest and field habitats normalized difference vegetation index (NDVI) as a predictor of forage availability for ungulates in forest and field habitats. *Eur J Wildl Res*. 2013;59(January 2015):675–682. https://doi.org/10.1007/s10344-013-0720-0.
24. Gandhi GM, Parthiban S, Thummalu N, Christy A. Ndvi: vegetation change detection using remote sensing and Gis—a case study of Vellore District. *Procedia Comput Sci*. 2015;57(March):1199–1210. https://doi.org/10.1016/j.procs.2015.07.415.
25. Panda SS, Ames DP, Panigrahi S. Application of vegetation indices for agricultural crop yield prediction using neural network techniques. *Remote Sens (Basel)*. 2010;2:673–696. https://doi.org/10.3390/rs2030673.
26. Nima. *Fire Destroyed About 16,000 Acres of Forest Reserve in 2017–18—Kuensel Online*. Kuenselonline; 2018. https://kuenselonline.com/fire-destroyed-about-16000-acres-of-forest-reserve-in-2017-18/. Accessed 19 April 2021.
27. Ryu JH, Han KS, Hong S, Park NW, Lee YW, Cho J. Satellite-based evaluation of the post-fire recovery process from the worst forest fire case in South Korea. *Remote Sens*. 2018;10(6). https://doi.org/10.3390/rs10060918.
28. Department of Forests and Park Services/Ministry of Agriculture and Forests. *Facts and Fig.s 2017*; 2017.
29. Department of Forests and Park Services/Ministry of Agriculture and Forests. *Forest Facts & Fig.s 2016*; 2016.
30. Norbu C. *The Changing Landscape of Paro Valley—Kuensel Online*. Kuenselonline; 2019. https://kuenselonline.com/the-changing-landscape-of-paro-valley/. Accessed 21 April 2021.

31

Rate of penetration prediction in drilling wells from the Hassi Messaoud oil field (SE Algeria): Use of artificial intelligence techniques and environmental implications

Ouafi Ameur-Zaimeche[a], Rabah Kechiched[a], and Charaf-Eddine Aouam[b]

[a]Laboratory of Underground Reservoirs: Petroleum, Gas and Aquifers, University of Kasdi Merbah Ouargla, Ouargla, Algeria [b]Faculty of Hydrocarbons, Renewable Energies and Earth Sciences and the Universe, University of Kasdi Merbah Ouargla, Ouargla, Algeria

1 Introduction

Groundwater is the main source of drinking water in semi-arid and arid areas, such as the Algerian Sahara. The North West Sahara Aquifer System (NWSAS) is one of the largest aquifers in the world, with a surface exceeding 10 million km^2. It is shared by three countries: Algeria, Libya, and Tunisia, where aquifers can be found at depths of up to 3000 m.[1,2] Algeria hosts the largest number of aquifers in other countries. Groundwater resources are contained in three principal reservoirs[3–5]: (1) the phreatic (2) complex terminal (CT), and (3) continental intercalaire (CI) aquifers. The CI aquifer hosts confined waters of acceptable quality in a potential reservoir thickness varying between 120 and 1000 m.[6–8] This huge drinking water resource is vulnerable to many problems of contamination and degradation of water quality because the drilling activities near water reservoirs may pose a danger of pollution due to the risk of releasing toxic elements in water, especially when mud is lost.

The Hassi Messaoud oilfield, located approximately 650 km southeast of Algiers, is well known for its immense industrial activity. The city of Hassi Messaoud and its surrounding areas depend only on groundwater as a source for domestic and agricultural uses.[9] As the Hassi Messaoud is a world-class oil field, it has undergone many drilling operations since the 1950s. To reach oil reservoirs, the wells pass through aquifers and can impact water resources. For instance, during 1980, an accident in the drilling well (ONK-32) had fatal consequences in the short and long term.[10] Owing to geological characteristics, a crater was created at 80 m depth and 200 m in diameter caused by salt dissolution as a result of poor cementation of the well. The craters continue to expand inexorably, with major ecological consequences. The prevention of water resources has become a real concern; thus, the development of efficient protection methods is necessary.

During the drilling operation, the monitoring of drilling parameters is key to safe operations, and their importance appears in reacting immediately in case of a problem to avoid catastrophic impacts. Monitoring, which includes parameters measured by sensors and calculated ones, is performed using surface logging units. The rate of penetration (ROP) is one of the most important parameters as it is a significant factor in reducing drilling costs. In addition, ROP is also a signal of various geological phenomena and problems that occur during drilling. For instance, the value of ROP can register indirectly lost zones that imply the loss of drilling mud, and therefore, predicting ROP can help in identifying these sectors.

Several empirical techniques can be found in the literature for the optimization and prediction of ROP,[11–15] such as the Bingham equation as a hybrid semi-analytical model and the linear regression approach as a statistical tool. These physics-based models have certain limitations, such as the necessity for auxiliary data (bit design, bit properties, mud properties, etc.). In the last few years, to predict the drilling rate of penetration, many researchers have used artificial intelligence algorithms including artificial neural networks,[16–18] adaptive neuro-fuzzy inference systems,[19] decision trees,[14,20] random forests,[21,22] and support vector machine algorithm.[23,24] In this study, two artificial intelligence techniques were applied and evaluated to predict ROP in drilling wells from the Hassi Messaoud oilfield, by using offsets to improve the identification of the lost circulation zones, in which the contamination of groundwater resources can occur through the dispersion of hazardous chemical composition in the groundwater of the study area.

2 Material and methods

2.1 Study area description

2.1.1 Geological setting

The Hassi Messaoud city is situated 650 km southeast of Algiers and about 350 km from the Algerian-Tunisian border (Fig. 1). This area is well known in Algeria and throughout the world for its large petroleum industry activities, where the population is concentrated in the city, exceeding 45.000 inhabitants in 2008.[9] The studied area is located between longitudes of 6° 2 11.47″ E to 6° 4 41.30″ E and latitudes 31° 39 16.47″ N and 31°45 12.47″ N. In this area, the dominant climate is hyper-arid with a very hot and dry summer and moderate winter. The average temperatures can exceed 45 °C in summer whereas rainfalls are characterized by their scarcity and extreme variability, with an average monthly: 0.1–7 mm.[9]

The geology of Hassi Messaoud is well recognized because of the wide geological work carried out during exploration by petroleum companies. A synthesis of the petroleum geology in Algeria, including the detailed geological background, was published by Sonatrach-Schlumberger.[25] The area of Hassi Messaoud is considered a part of the intracratonic basin of the Oued Mya.[26] The lithologies consist of a sedimentary succession of ~4393 m thick; the sedimentary strata range from the Cambrian to Quaternary in age and locally display the Hercynian discontinuity effects.[9,27]

FIG. 1 Location map of the study area (Hassi Messaoud oilfield).

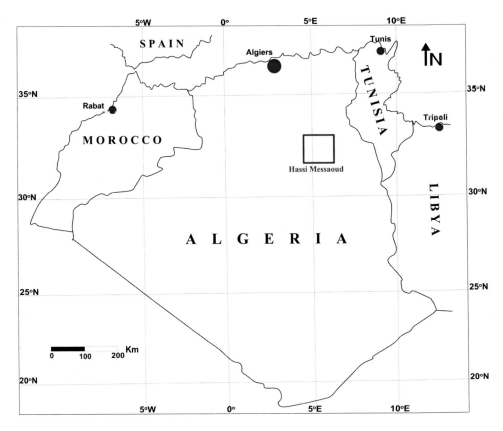

2.1.2 Hydrogeological setting

From a hydrogeological viewpoint, the Hassi Messaoud aquifers belong to the North West Sahara Aquifer System (NWSAS), in which two separated aquifer systems extend to Tunisia and Libya and host considerable water quantities.[28,29] aquifers were exploited from the following reservoirs:

1. *The Complex Terminal (CT)* is represented by a multi-layer aquifer displaying an average thickness of 300 m, in which the Albian aquifer denotes the most exploited aquifer and represents the principal focus for drilling wells for water supply in the entire northern Sahara.
2. *The Continental Intercalaire (CI)* is situated in a complex series of Mesozoic siliciclastic deposits with thicknesses ranging between 200 and 400 m. Laterally, the aquifer exhibits large variation in terms of lithology, which is typically represented by Mio-Pliocene sand and Senonian carbonates (Fig. 2).

2.2 Methods

2.2.1 Artificial intelligence

Artificial intelligence (AI) enables computers to perform tasks involving human intelligence.[30] AI aims to construct an algorithm or model that allows machines to perform tasks that are typically similar to human intelligence. A wider definition of AI includes problem-solving, understanding of language, and processes that are conscious and unconscious.[31,32] Moreover, artificial intelligence is defined as a subfield of computer science that includes the utilization of computers in tasks that typically require skills in reasoning, intelligence, learning, and understanding. The following methods were used in this study:

(a) Genetic Algorithm (GA)

The genetic algorithm (GA) was suggested, for the first time, by Holland (1975)[33,34] and is defined as a technique of stochastic optimization, inspired by the method of biological evolution. This population-based algorithm simulates the present process in a natural system, where only the strongest individual survives. The GA produces and recombines an initial population containing possible solutions to direct the quest for areas with high potential for desirable responses. Each of these potential solutions is encoded as a chromosome, and an objective function assigns them a value called fitness (cost function), the low and high fitness values provide the best solution to problems with minimization and maximization, respectively. Strong candidates of each group are moved to a new population matrix with unique genetic operators, such as crossover, reproduction, and mutation, undergoing some changes.[35] The crossover is recombination between two parents, a generation component that merges to produce new members for the next generation.

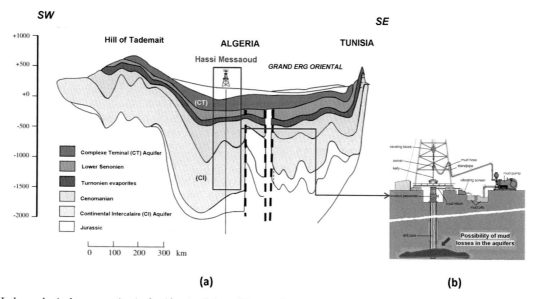

(a)   **(b)**

FIG. 2   (A) Hydrogeological cross-section in the Algerian Sahara/Tunisian basins presenting the three aquifer systems. (B) Hazardous losses of the mud and risk of groundwater contamination. *(A) After Castany G. Bassin sédimentaire du Sahara septentrional (Algérie-Tunisie). Aquifères du continental intercalaire et du complexe terminal. Bull BRGM Sect III 1982;2:127–147. Chkir N, Guendouz A, Zouari K, Hadj Ammar F, Moulla AS. Uranium isotopes in groundwater from the continental intercalaire aquifer in Algerian Tunisian Sahara (Northern Africa). J Environ Radioact 2009;100(8):649–656. Modified.*

The most compatible candidates of each generation are selected for reproduction, and mutation is another GA operator that is identical to a normal biological mutation, modifying genes with new unknown genes.[36] In this study, GA was used as a feature selection in the input parameters for the ROP estimation.

(b) Artificial Neural Networks (ANN)

Nonlinear simulations based on artificial intelligence (AI) approaches have been widely used in many scientific research fields in recent years. Among them, artificial neural networks (ANNs) are the most significant. By definition, ANN is known as a nonlinear mathematical model inspired by the function of the human brain, and it is regarded as an information-processing system composed of many processing elements arranged in several parallel layers.[37,38] An artificial neural network is composed of many different elements, including neurons, hidden layers, training functions, transfer functions, and epoch durations. Neurons are components that have a single input/output that is coupled to form a network of neural network nodes. Weights and biases were used to determine the relationship between neurons and the source to obtain the input parameters. The efficiency of a network in an ANN depends on the selection of weights and biases. The multilayer perceptron neural network (MLPNN) is the most common ANN model and is regarded as a universal approximator as long as adequate training is carried out.[39,40] The MLPNN model used in this study consists of three layers of processing units, one hidden layer, and a sigmoid activation function, which is the best structure of robustness, as listed in the literature.[41,42] The determination of the number of hidden neurons is based on trial-and-error rules.[43,44]

(c) Support Vector Machines (SVM)

Support vector machines (SVMs) were derived from Vapnik's statistical learning theory. Over the years, it has become a common technique for classification and regression tasks among machine learning models.[45] The SVM employs the optimum separation principle for classification problems. This concept selects a hyperplane with the highest margin of linearly separable groups (from an infinite number of linear classifiers). The selected optimum hyperplane of separation is established by minimizing the generalization error or the given upper bound on the error using the minimization of structural risk. The one with the highest distance from the closest point of the two groups is an ideal separation hyperplane.[45,46] Nevertheless, for non-separable groups, the SVM seeks a hyperplane with maximum margin and minimizes the sum with direct proportionality to the number of misclassification errors. For this process, a tradeoff between the number of misclassification errors and the total margin was selected using a predefined positive constant. To create linear decision surfaces with SVM, a finite set of variables can be transformed into a higher-dimensional space where linear classification can then be performed.[45]

2.3 Data pre-processing

2.3.1 Data set collection

In the present study, a data set was obtained from a vertical well drilled in the Hassi Messaoud oilfield. The data set includes11 real-time surface drilling parameters that were used in the prediction of ROP representing the input variables as follows: standpipe pressure, weight on bit, torque, pump flow rate, rotation speed, bit time, bit depth, ROP, D-exponent, and temperature in/out. The data set involved 2837 samples with a sampling rate of 1 m penetration and covered a depth from 24 to 2900 m passing through the aquifers (phreatic, CT, and CI), which are vulnerable to contamination by the loss of mud. Table 1 summarizes the descriptive statistics of the data set, where X_{mean}, X_{max}, X_{min}, and Std dev represent the mean, maximum, minimum, and standard deviation, respectively.

The next step consists of preparing the data set into a format that can be used for the ANN or SVM techniques as well as preprocessing such as normalization of the inputs and target,[47] allowing the ANN and SVM training to be more effective. This ending on either the applied kernel function in designing the SVM or the transfer function in designing the ANN. To boost the network convergence, all datasets were normalized to be within the interval $[-1, 1]$ using the following equation:

$$Z_i = \frac{x_i - \bar{x}}{S} \tag{1}$$

where x_i is a data point $(x_1, x_2 \dots x_n)$, $\bar{x}$ is the sample mean, S is the sample standard deviation.

2.3.2 Input data selection and separation

Understanding the impact of input variables is a primary concern when designing AI models, and thus, a feature selection technique using a genetic algorithm. This was used to select the unrivaled features and reduce the input

TABLE 1 Statistical analysis of input and output parameters.

Drilling parameters	X_{mean}	X_{min}	X_{max}	Std Dev
Weight on bit (klb$_f$)	9.40	0.01	19.94	4.74
Rotation speed (rpm)	133.96	39	750	92.94
Torque (kft.lb$_f$)	9532.60	901	18.443	4886.31
Pump flow rate (l/min)	2389.39	495	3000	748.05
Stand pipe pressure (psi)	1808.62	2	3388	643.32
Bit time (h)	18.61	0.04	93.2	17.65
Bit depth (m)	1481.5	63	2900	819.404
D-exponent	1.21	0.32	15.35	2.00
ROP (mn/m)	8.66	0.24	529.31	21.17
Temperature in (°C)	56.596	28	67.20	7629
Temperature out (°C)	61.710	33	73	8229

variables and vectors. Based on the results of the feature selection approach, seven factors were identified as the most influential parameters in the prediction of ROP ($P < .05$) and are presented in Table 2 and Fig. 3.

The dataset was separated into three subdivisions: training, validation, and testing. Among the 2837 data, a random selection of 2271 input-output pairs (70%) were used in the training set, and 283 input-output-pairs representing 15% of the total data were used for the validation collection, while the remaining data consisting of 283 output pairs (15%) were used for the test for the model. Note that validation is used for the final result to ensure that there is no overfitting by applying the function *dividevec* in MATLAB, where the three subdivisions were randomly taken from the original results. The validation samples were used to confirm that the network parameters were robust, to determine when to stop training, and to track errors. Therefore, when the error increased, the training was stopped. The test subset was used with completely unknown data to test the model.

2.4 Model performance assessment criteria

The efficiency of the models was evaluated using the coefficient of determination (R^2), mean absolute error (MAE), and root mean squared error (RMSE). These parameters are widely used to determine efficiency, especially in petroleum engineering.[48] In other words, the selected performance aligns with the best practices that ensure a rational evaluation of the developed model. In the literature,[49,50] the equations for R^2, MAE, and RMSE are given as follows:

TABLE 2 Best drilling parameters predictors for ROP estimation.

Drilling parameters	F-value	P-value
Stand pipe pressure	18.17688	0.000000
D-exponent	17.30932	0.000000
Rotation speed	16.23956	0.000000
Pump flow rate	13.19263	0.000000
Torque	6.56250	0.000000
Temperature n	3.86273	0.000000
Temperature out	3.25576	0.000000
Weight on bit	0.85560	0.998318
Bit time	0.52951	1.000000

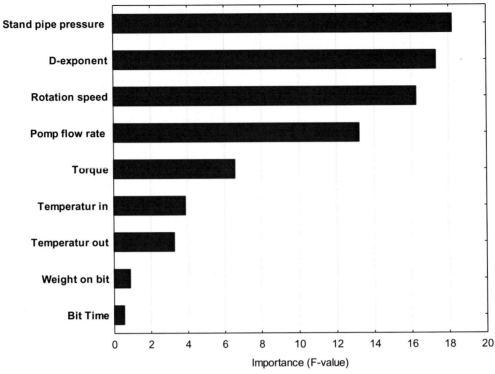

FIG. 3 Sensitivity analysis of each parameter on ROP prediction using GA technique.

$$R^2 = \left[\frac{\frac{1}{N} \sum (O_i - O_m)(P_i - P_m)}{\sqrt{\frac{1}{N} \sum_{i=1}^{n} (O_i - O_m)^2} \sqrt{\frac{1}{N} \sum_{i=1}^{n} (P_i - P_m)^2}} \right] \tag{2}$$

$$MAE = \frac{1}{N} \sum_{i=1}^{N} |O_i - P_i| \tag{3}$$

$$RMSE = \sqrt{\frac{1}{N} \sum_{i=1}^{N} (O_i - P_i)^2} \tag{4}$$

where N is the number of data points, O_i is the real and P_i is the corresponding model estimated. O_m and P_m are the average values of O_i and P_i.

To apply artificial intelligence tools (ANN and SVM) for ROP prediction, MATLAB and R software were used. Fig. 4 illustrates the methodology used in this study.

3 Results and discussion

The penetration rate (ROP) is considered one of the most critical variables determining the overall cost of drilling and is influenced by many parameters that can be classified into controllable and uncontrollable parameters.[51,52] Therefore, the optimization of ROP has a significant effect on reducing the total cost. On the other hand, weak monitoring of this parameter can result in the loss of mud and can damage the environment due to hazardous mud loss through aquifers. The study area of the Hassi Messaoud oilfield is situated within the large aquifer systems used by the population for drinking groundwater.[9] It is worth noting that this study was conducted to improve the prediction of ROP using ANN and SVM and to evaluate these two different approaches in terms of performance.

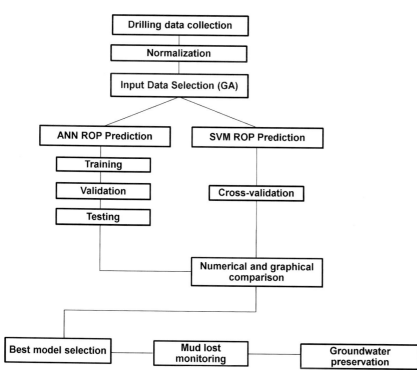

FIG. 4 Methodology flowchart of ROP predicting conducted in this study.

The use of a genetic algorithm to select the best influencing input factors in constructing AI models showed that only seven parameters had a significant influence on the ROP prediction ($P < .05$). The list includes the standpipe pressure, D-exponent, rotation speed, pump flow rate, torque, the temperature in, and temperature out without weight on bit, which was not significant in our case ($P < .998318$) (Table 2).

This was unlike previous studies where some authors have introduced weight on bit and rotary speed as the most influential operational parameters to be controlled in the ROP optimization process.[22,53–58] For example, Abbas et al. (2019) found that the weight on the bit was the first parameter influencing ROP prediction. This may indicate that the ROP can be impacted differently owing to the complexity of its calculation.[59] Therefore, generalization of such a model appears to be difficult, and ROP prediction needs to be performed for each study area.

3.1 ROP prediction using ANN approach

The ANN implementation was conducted in many trials for choosing the optimal parameters: training functions, number of neurons, transfer functions, and the best network function to estimate the ROP. The adopted architecture consists of one layer with different numbers of neurons based on trial-and-error rules[43,44] (Fig. 5). The architecture of the ANN with of 10 neurons, one hidden layer, and the training algorithm of *Levenberg-Marquardt back-propagation* (*LMBP*), the transfer function of *Logsig*, and the network function of *purlin* (Table 3), showed the highest performance, expressed by the highest coefficient of determination (R^2) and the lowest mean absolute error (MAE) and root mean square error (RMSE).

Following the selection of the ideal ANN structure, a detailed statistical and graphical representation of the ANN performance was performed, as shown in Table 4. In addition, Fig. 6 shows a graph illustrating the variation of the estimated versus real ROP for the built model for the training, validation, and testing datasets. The results of the predicted ROP were similar to those of the real model with high accuracy.

According to the performance assessment criteria, the architecture of the ANN, with a three-layer network, of (7-10-1) neurons, transfer functions *Logsig* for hidden layers and *purelin* for the output layer, provides the best performance for ROP prediction ($R^2 = 0.9332, 0.9031, 0.9456$; RMSE $= 0.2593, 0.1511, 0.1101$, and MAE $= 0.1060, 0.0255$, and 0.0223 in training, validation, and testing, respectively).

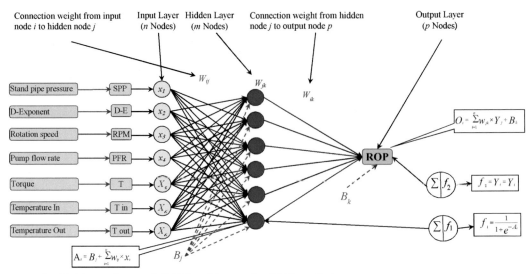

FIG. 5   The proposed artificial neural network (ANN) architecture for (ROP) prediction.

TABLE 3   The optimum ANN parameters/setting for ROP prediction.

Parameter/setting	Value/type
Training function	*Levenberg-Marquardt*
Transfer function (in put-hidden layer)	*Logsig*
Transfer function (hidden layer-output)	*Purlin*
Number of hidden layers	1
Number of neurons in a hidden layer	10

TABLE 4   Prediction performance of the optimum ANN model.

Performance criteria	Training	Valldation	Test
R^2	0.9332	0.9031	0.9456
RMSE	0.2593	0.1511	0.1101
MAE	0.1060	0.0255	0.0223

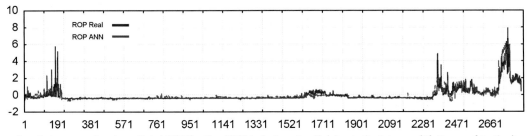

FIG. 6   Comparison between real and predicted ROP by the ANN technique in all datasets (training, validation, and testing).

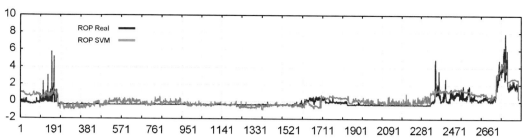

FIG. 7 Comparison between real and predicted ROP by the SVM technique.

3.2 ROP prediction using SVM approach

The SVM input parameters used for ROP prediction in this section are the same significant parameters used in the ANN implementation. The accuracy of the estimated values was compared to the actual values. The results show a good match between the ROP, indicating that modeling with SVM was successful (Fig. 7), with the best kernel type: RBF (radial basis function) and K-folder (10) for the cross-validation step. The assessment performance was $R^2 = 0.8480$, RMSE $= 0.5148$, and MAE $= 0.3815$.

3.3 Comparison between ANN and SVM approaches

A comparison between the two techniques was carried out to select the best technique for predicting ROP with high precision. A graphical representation of the results is provided in Fig. 8, and the performance parameters are listed in Table 5.

The results showed that ANN displayed the highest performance in terms of R^2 (ANN $= 0.9456$; SVM $= 0.8480$), RMSE (ANN $= 0.1101$; SVM $= 0.5148$), and MAE (ANN $= 0.0223$; SVM $= 0.3815$), indicating the success of modeling with ANN compared to SVM (Table 5).

A graphical comparison between the real and predicted ROP values is shown in Fig. 9. The boxplots and violin plots for real and predicted ROP using the ANN model showed the best performance in predicting ROP when compared to the SVM method.

This study further proved that ANN is an effective technique for solving complex problems as well as predicting the ROP, and is commonly applied for the training of ANNs. The log-sigmoid function, which is a widely used activation function for performing ANN, is used together with the linear (*purelin*) function for the output layer transition. Based on the performance assessment criteria and the graphical illustration (boxplots, violin plots), the ANN model with 10 neurons in the hidden layer showed the best performance when associated with a topology is 7-10-1 with the LMBP learning algorithm and sigmoidal function purlin. The selected input variables that proved useful in predicting the ROP include the standpipe pressure, D-exponent, rotation speed, pump flow rate, torque, temperature, and temperature out). Accordingly, this method appears to be a solution for estimating the ROP and can be integrated into surface-logging systems. The highest values of ROP were recorded between 100 m and 300 m, indicating possible losses of the mud, and should be taken into account during careful monitoring to avoid the release of toxic elements in the aquifers.

3.4 Influencing of the ROP on lost circulation prediction and environment implication

Lost circulation is a common challenge in drilling activities. Many negative consequences can occur, including economic loss due to the cost of mud losses and increased well drilling time[60]; thus, it can potentially cause the

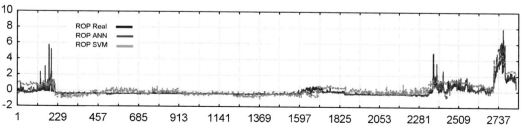

FIG. 8 Comparison between real and predicted ROP using ANN and SVM.

TABLE 5   Performance parameters of ANN versus SVM.

Performance criteria	ANN	SVM
R^2	0.9456	0.8480
RMSE	0.1101	0.5148
MAE	0.0223	0.3815

hydrocarbon industry to spend about US 800 million per year.[61] Loss of circulation has become a dangerous issue because of environmental pollution and contamination of groundwater resources.[62]

Recently, many statistical and intelligent techniques have been developed to predict the loss of circulation.[51,60,62–64] The authors have indicated that among many parameters, the ROP appears to be the most effective and has significant sensitivity in predicting the loss circulation rate. Furthermore, Abbas et al. indicated that the flow rate and ROP are among the important parameters that affect circulation loss. This can be increased by increasing these variables.[65]

In the first step, this study generates an effective model for effective ROP prediction, thus enabling drillers to identify possible lost circulation intervals and, therefore, can be used in the drilling of new wells. A high ROP should be considered to avoid mud loss and allow the preservation of groundwater in the study area. This study suggests the use of artificial intelligence to calculate the ROP by surface units to obtain accurate values based on significant input values.

The introduction of artificial intelligence-based approaches in the drilling of oil wells requires further efforts to be considered in both theory and practice. Although the present study has provided a comprehensive analysis of ROP prediction models using ANN and SVM techniques, the AI techniques are developing at a rapid rate; thus, its application in the oil field should be updated over time to improve the prediction quality. It is important that automated systems based on AI methods will help engineers to develop efficient monitoring in real-time smart systems, allowing the best formula for the ROP calculation (selection of best input parameters) and prediction by the use of offsets and real-time data, thus helping to identify lost circulation zones. It is noteworthy that these lost circulation zones can be detected throughout several parameters where the ROP appears to be one of the best indicators. Therefore, by accurately predicting the ROP using offset data from the Hassi Messaoud oilfield, the lost circulation zones can be successfully detected earlier; thus, engineers can provide an effective remedy and prevent contamination of groundwater resources by dispersing the hazardous chemical composition of the drilling mud. However, AI models depend on the quality of data and the efficiency of sensors in recording different parameters. In addition, these models cannot be generalized to other areas because of the complexity of geology and technical problems; hence, the prediction of ROP must be performed in each area separately.

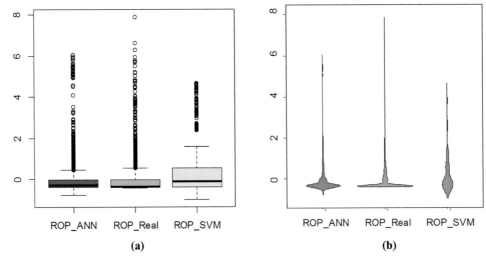

FIG. 9   Graphical performance representation: (A) boxplots of real and predicted ROP (B) violin plots of real and predicted ROP.

4 Conclusion

This study assessed two artificial intelligence techniques (ANN and SVM) to predict the ROP, which is one of the most critical parameters during petroleum well drilling. The Hassi Messaoud world-class oilfield was chosen for the case study. The prediction of ROP was performed using a selection of real-time surface drilling parameters that were chosen using genetic algorithm optimization for the best predictors of ROP in the study area. The list includes the standpipe pressure, D-exponent, rotation speed, pump flow rate, torque, temperature, and temperature. ROP prediction is performed to ensure an efficient method for ROP calculation on the one hand and to prevent environmental damage due to hazardous losses of the mud through aquifers. Based on this comparative study, the main conclusions are as follows.

1. The genetic algorithm (AG) is powerful tool for improving the selection of the best input parameters for ANN and SVM methods.
2. The results indicated that the ANN was the most accurate model with the best performance parameters ($R^2 = 0.9456$, RMSE $= 0.1101$, and MAE $= 0.0223$) compared to the SVM approach ($R^2 = 0.848$; RMSE $= 0.5148$; MAE $= 0.3815$).
3. The study confirms the robustness of AI models in predicting ROP, and the ANN model allows good control of ROP in drilling new wells, which can be used to predict mud losses and reduce the risk of groundwater contamination.
4. High ROP in the zone of aquifers was recorded which represents a risk of mud losses.

References

1. Mamou A, Besbes M, Abdous B, Latrech DJ, Fezzani C. North Western Sahara aquifer system (NWSAS). In: *Non-Renewable Groundwater Resources: A Guidebook on Socially-Sustainable Management for Water-Policy Makers, IHP-VI Series on Groundwater.* vol. 10. Paris, France: UNESCO; 2006:68–74.
2. Foster S, Loucks DP. *Non-Renewable Groundwater Resources: A Guidebook on Socially-Sustainable Management for Water-Policy Makers.* Paris, France: Unesco; 2006.
3. UNESCO. *Projet ERESS: Etude des ressources en eau du Sahara septentrional. Rapport final ERESS Project: study of the northern Sahara water resources* [Final report]. Paris, France: United Nations Educational Scientific and Cultural Organisation (UNESCO); 1972.
4. Eckstein GE, Eckstein Y. A hydrogeological approach to transboundary ground water resources and international law. *Am Univ Int Law Rev.* 2003;19:201–258.
5. Besbes M, Abdous B, Abidi B, et al. Système Aquifère du Sahara septentrional Gestion commune d'un bassin transfrontière. *Houille Blanche.* 2003;5:128–133.
6. Castany G. Bassin sédimentaire du Sahara septentrional (Algérie-Tunisie). Aquifères du continental intercalaire et du complexe terminal. *Bull BRGM Sect III.* 1982;2:127–147.
7. Edmunds WM, Guendouz AH, Mamou A, Moulla A, Shand P, Zouari K. Groundwater evolution in the continental Intercalaire aquifer of southern Algeria and Tunisia: trace element and isotopic indicators. *Appl Geochem.* 2003;18(6):805–822. https://doi.org/10.1016/S0883-2927(02)00189-0.
8. Petersen JO, Deschamps P, Gonçalvès J, et al. Quantifying paleorecharge in the Continental Intercalaire (CI) aquifer by a Monte-Carlo inversion approach of 36 Cl/Cl data. *Appl Geochem.* 2014;50:209–221. https://doi.org/10.1016/j.apgeochem.2014.04.014.
9. Kechiched R, Nezli IE, Foufou A, et al. Fluoride-bearing groundwater in the complex terminal aquifer (a case study in Hassi Messaoud area, southern Algeria): hydrochemical characterization and spatial distribution assessed by indicator kriging. *Sustain Water Resour Manag.* 2020;6 (4). https://doi.org/10.1007/s40899-020-00415-6.
10. Akretche S, Okbi L. History of the Okn-32 incident in the northern part of the Algerian Sahara. *Int J Rock Mech Min Sci.* 1996;7(33):331.
11. Warren TM. Penetration rate performance of roller cone bits. *SPE Drill Eng.* 1987;2(01):9–18. https://doi.org/10.2118/13259-PA.
12. Motahhari HR, Hareland G, James JA. Improved drilling efficiency technique using integrated PDM and PDC bit parameters. *J Can Pet Technol.* 2010;49(10):45–52. https://doi.org/10.2118/141651-PA.
13. Deng Y, Chen M, Jin Y, Zhang Y, Zou D, Lu Y. Theoretical and experimental study on the penetration rate for roller cone bits based on the rock dynamic strength and drilling parameters. *J Nat Gas Sci Eng.* 2016;36:117–123. https://doi.org/10.1016/j.jngse.2016.10.019.
14. Hegde C, Wallace S, Gray K. Using trees, bagging, and random forests to predict rate of penetration during drilling. In: *Day 1 Tue, September 15.* SPE; 2015:2015. https://doi.org/10.2118/176792-MS.
15. Bezminabadi SN, Ramezanzadeh A, Esmaeil Jalali S-M, Tokhmechi B, Roustaei A. Effect of rock properties on ROP modeling using statistical and intelligent methods: a case study of an oil well in southwest of Iran. *Arch Min Sci.* 2017;62(1):131–144. https://doi.org/10.1515/amsc-2017-0010.
16. Bilgesu HI, Tetrick LT, Altmis U, Mohaghegh S, Ameri S. A new approach for the prediction of rate of penetration (ROP) values. In: *All Days SPE;* 1997. https://doi.org/10.2118/39231-MS.
17. Momeni M, Hosseini SJ, Ridha S, Laruccia MB, Liu X. An optimum drill bit selection technique using artificial neural networks and genetic algorithms to increase the rate of penetration. *J Eng Sci Technol.* 2018;13(2):361–372.
18. Elkatatny S. New approach to optimize the rate of penetration using artificial neural network. *Arab J Sci Eng.* 2018;43(11):6297–6304. https://doi.org/10.1007/s13369-017-3022-0.
19. Yavari H, Sabah M, Khosravanian R, Wood DA. Application of an adaptive neuro-fuzzy inference system and mathematical rate of penetration models to predicting drilling rate. *Iran J Oil Gas Sci Technol.* 2018;7(3):73–100. https://doi.org/10.22050/IJOGST.2018.83374.1391.
20. Eskandarian S, Bahrami P, Kazemi P. A comprehensive data mining approach to estimate the rate of penetration: application of neural network, rule based models and feature ranking. *J Petrol Sci Eng.* 2017;156:605–615. https://doi.org/10.1016/j.petrol.2017.06.039.
21. Breiman L. Bagging predictors. *Mach Learn.* 1996;24(2):123–140. https://doi.org/10.1007/BF00058655.

22. Hegde C, Daigle H, Millwater H, Gray K. Analysis of rate of penetration (ROP) prediction in drilling using physics-based and data-driven models. *J Petrol Sci Eng.* 2017;159:295–306. https://doi.org/10.1016/j.petrol.2017.09.020.

23. Smola AJ, Schölkopf B. A tutorial on support vector regression. *Stat Comput.* 2004;14(3):199–222. https://doi.org/10.1023/B:STCO.0000035301.49549.88.

24. Gan C, Cao W-H, Wu M, et al. Prediction of drilling rate of penetration (ROP) using hybrid support vector regression: a case study on the Shennongjia area, Central China. *J Pet Sci Eng.* 2019;181(106200):106200. https://doi.org/10.1016/j.petrol.2019.106200.

25. El Ouahed AK, Tiab D, Mazouzi A. Application of artificial intelligence to characterize naturally fractured zones in Hassi Messaoud oil field, Algeria. *J Pet Sci Eng.* 2005;49(3–4):122–141.

26. Benamrane O, Messaoudi M, Messelles H. Geology and hydrocarbon potential of the Oued Mya Basin, Algeria. In: *55th EAEG Meeting.* vol. 77. EAGE Publications BV; 1993:9.

27. Lerat S. Hassi Messaoud. *Cah O M.* 1971;24(93):16–31.

28. Edmunds WM, Gaye CB. Naturally high nitrate concentrations in groundwaters from the Sahel. *J Environ Qual.* 1997;26(5):1231–1239. https://doi.org/10.2134/jeq1997.00472425002600050006.

29. Edmunds WM, Shand P, Guendouz AH, Moulla AS, Mamou A, Zouari K. Recharge characteristics and groundwater quality of the Grand Erg Oriental basin. Final report. EC (Avicenne) contract CT93AVI0015, BGS Tech Rep WD/97/46R. In: *Hydrogeology Series*; 1997.

30. McCarthy J. Epistemological problems of artificial intelligence. In: Webber BL, Nilsson NJ, eds. *Readings in Artificial Intelligence.* Elsevier; 1981:459–465.

31. Minsky M. Steps toward artificial intelligence. *Proc IRE.* 1961;49(1):8–30.

32. Nilsson N. *Principles of Artificial Intelligence.* 1st ed. Morgan Kaufmann; 2014.

33. Goldberg DE. *Genetic Algorithms in Search, Optimization, and Machine Learning.* 1st ed. Addison Wesley; 1989.

34. Holland JH. Genetic algorithms. *Sci Am.* 1992;267(1):66–72.

35. Ab Wahab MN, Nefti-Meziani S, Atyabi A. A comprehensive review of swarm optimization algorithms. *PLoS One.* 2015;10(5). https://doi.org/10.1371/journal.pone.0122827, e0122827.

36. Tang KS, Man KF, Kwong S, He Q. Genetic algorithms and their applications. *IEEE Signal Process Mag.* 1996;13(6):22–37.

37. Fausett LV. *Fundamentals of Neural Networks: Architectures, Algorithms and Applications.* United States Edition: Pearson; 1993.

38. Haykin S. *Neural Networks: Comprehensive Foundation.* 2nd ed. IEEE Press; 1999.

39. Hornik K. Approximation capabilities of multilayer feedforward networks. *Neural Netw.* 1991;4(2):251–257.

40. Hornik K, Stinchcombe M, White H. Multilayer feedforward networks are universal approximators. *Neural Netw.* 1989;2(5):359–366. https://doi.org/10.1016/0893-6080(89)90020-8.

41. Heddam S. Multilayer perceptron neural network-based approach for modeling phycocyanin pigment concentrations: case study from lower Charles River buoy, USA. *Environ Sci Pollut Res Int.* 2016;23(17):17210–17225. https://doi.org/10.1007/s11356-016-6905-9.

42. Ameur-Zaimeche O, Zeddouri A, Heddam S, Kechiched R. Lithofacies prediction in non-cored wells from the Sif Fatima oil field (Berkine basin, southern Algeria): a comparative study of multilayer perceptron neural network and cluster analysis-based approaches. *J Afr Earth Sci.* 2020;166 (103826):103826. https://doi.org/10.1016/j.jafrearsci.2020.103826.

43. Ramadevi R, Rani S, Prakash V. Role of hidden neurons in an Elman recurrent neural network in classification of cavitation signals. *Int J Comput Appl.* 2012;37:9–13.

44. Sheela KG, Deepa SN. Review on methods to fix number of hidden neurons in neural networks. *Math Probl Eng.* 2013;2013:1–11. https://doi.org/10.1155/2013/425740.

45. Cortes C, Vapnik V. Support-vector networks. *Mach Learn.* 1995;20(3):273–297.

46. Cristianini N, Shawe-Taylor J. *An Introduction to Support Vector Machines and Other Kernel-Based Learning Methods.* Cambridge University Press; 2013. https://doi.org/10.1017/CBO9780511801389.

47. Quackenbush J. Microarray data normalization and transformation. *Nat Genet.* 2002;32(Suppl. 4):496–501.

48. Jahanbakhshi R, Keshavarzi R, Aliyari Shoorehdeli M, Emamzadeh A. Intelligent prediction of differential pipe sticking by support vector machine compared with conventional artificial neural networks: an example of Iranian offshore oil fields. *SPE Drill Complet.* 2012;27 (04):586–595. https://doi.org/10.2118/163062-PA.

49. Willmott CJ, Ackleson SG, Davis RE, et al. Statistics for the evaluation and comparison of models. *J Geophys Res.* 1985;90(C5):8995. https://doi.org/10.1029/JC090iC05p08995.

50. Krause P, Boyle DP, Bäse F. Comparison of different efficiency criteria for hydrological model assessment. *Adv Geosci.* 2005;5:89–97. https://doi.org/10.5194/adgeo-5-89-2005.

51. Ahmed A, Elkatatny S, Ali A, Abughaban M, Abdulraheem A. Application of artificial intelligence techniques in predicting the lost circulation zones using drilling sensors. *J Sens.* 2020;2020:1–18. https://doi.org/10.1155/2020/8851065.

52. Al-AbdulJabbar A, Elkatatny S, Abdulhamid Mahmoud A, et al. Prediction of the rate of penetration while drilling horizontal carbonate reservoirs using the self-adaptive artificial neural networks technique. *Sustainability.* 2020;12(4):1376. https://doi.org/10.3390/su12041376b.

53. Ayoub M, Shien G, Diab D, Ahmed Q. Modeling of drilling rate of penetration using adaptive Neuro-Fuzzy inference system. *Int J Appl Eng Res.* 2017;12(22):12880–12891.

54. Bataee M, Kamyab M, Ashena R. Investigation of various ROP models and optimization of drilling parameters for PDC and roller-cone bits in Shadegan oil field. In: *Proceedings of International Oil and Gas Conference and Exhibition in China.* SPE; 2010.

55. Chapman CD, Flores JL, De Leon PR, Yu H. Automated closed-loop drilling with ROP optimization algorithm significantly reduces drilling time and improves downhole tool reliability. In: *Paper Presented at the IADC/SPE Drilling Conference and Exhibition, San Diego, California, USA.* SPE; 2012.

56. Dunlop J, Isangulov R, Aldred WD, et al. Increased rate of penetration through automation. In: *SPE/IADC Drilling Conference and Exhibition. Society of Petroleum Engineers.* SPE; 2011.

57. Dupriest F, Koederitz W. Maximizing drill rates with real-time surveillance of mechanical specific energy. In: *Proceedings of SPE/IADC Drilling Conference.* SPE; 2005.

58. Eren T, Ozbayoglu M. Real-time drilling rate of penetration performance monitoring. In: *Offshore Mediterranean Conference*; 2011.

59. Abbas AK, Rushdi S, Alsaba M, Al Dushaishi MF. Drilling rate of penetration prediction of high-angled wells using artificial neural networks. *J Energy Resour Technol.* 2019;141(11):112904. https://doi.org/10.1115/1.4043699.

60. Hou X, Yang J, Yin Q, et al. Lost circulation prediction in South China Sea using machine learning and big data technology. In: *Day 4 Thu, May 07.* OTC; 2020:2020. https://doi.org/10.4043/30653-MS.

61. Agin F, Khosravanian R, Karimifard M, Jahanshahi A. Application of adaptive neuro-fuzzy inference system and data mining approach to predict lost circulation using DOE technique (case study: maroon oilfield). *Petroleum.* 2020;6(4):423–437. https://doi.org/10.1016/j.petlm.2018.07.005.

62. Abbas AK, Bashikh AA, Abbas H, Mohammed HQ. Intelligent decisions to stop or mitigate lost circulation based on machine learning. *Energy.* 2019;183(C):1104–1113. https://doi.org/10.1016/j.energy.2019.07.020.

63. Toreifi H, Rostami H, Manshad AK. New method for prediction and solving the problem of drilling fluid loss using modular neural network and particle swarm optimization algorithm. *J Pet Explor Prod Technol.* 2014;4(4):371–379. https://doi.org/10.1007/s13202-014-0102-5.

64. Abbas AK, Hamed HM, Al-Bazzaz W, Abbas H. Predicting the amount of lost circulation while drilling using artificial neural networks: An example of southern Iraq oil fields. In: *SPE Gas & Oil Technology Showcase and Conference.* Society of Petroleum Engineers; 2019.

65. Abbas AK, Alqatrani G, Mohammed HQ, Dahm HH, Alhumairi MA. Determination of significant parameters affecting the risk level of lost circulation while drilling. In: *54th U.S. Rock Mechanics/Geomechanics Symposium.* OnePetro; 2020.

32

Soil erodibility and its influential factors in the Middle East

Yaser Ostovari[a], Ali Akbar Moosavi[b], Hasan Mozaffari[b], Raúl Roberto Poppiel[c], Mahboobeh Tayebi[c], and José A.M. Demattê[c]

[a]Chair of Soil Science, Research Department of Ecology and Ecosystem Management, TUM-School of Life Sciences Weihenstephan, Technical University of Munich, Freising, Germany [b]Department of Soil Science, College of Agriculture, Shiraz University, Shiraz, Iran [c]Department of Soil Science, Luiz de Queiroz College of Agriculture, University of São Paulo, São Paulo, Brazil

1 Introduction

Soil erosion is an important environmental phenomenon that threatens ecosystems, soils, and water.[1,2] It has negative impacts on soil productivity, food production, soil and water quality, and biodiversity.[3,4] The United Nations (UN) reported that "the majority of the world's soil is in poor, fairly poor,"[5] and annually around 12 million hectares of productive lands are lost due to soil erosion.[5] In the Middle East, the climate is generally arid and semi-arid and soils are mainly susceptible to water erosion due to low/or no surface vegetation, little organic matter, and low resistance to erosive factors.[4–9] Soil erosion is not only the main cause of land degradation and productivity loss in the Middle East but also threatens human health and sustainable development of rural areas.[5,7,9] According to the literature,[3,5–9] many regions of the Middle East are drastically faced with soil erosion, which can be considered a serious environmental, economic, and social problem.

According to the Food and Agriculture Organization,[6] soil erosion affected 29% of agricultural land in Egypt and caused the loss of a huge number of fertile soils in Iran due to land-use change and inappropriate soil and water management systems. Approximately $50,070\,km^2$ (11.86%) of Iraq territory is affected by severe water erosion and $46,910\,km^2$ (11.06%) by low to moderate water erosion. In Oman, Syria, Palestine, and Saudi Arabia, soil erosion is considered a primary threat to productivity.[6]

Soil erodibility, described by the K-factor, has been introduced as important information for estimating soil loss (soil erosion). The K-factor is an inherent property that reflects the resistance of soil to detachment by erosive forces.[3,10] According to the universal soil loss equation,[11,12] the K-factor can be determined as the average rate of soil loss per unit plot ($1.83 \times 22.3\,m$ with a uniform 9% slope in continuous clean-tilled fallow) divided by the rainfall erosivity index as follows:

$$K = \frac{A}{R} \tag{1}$$

where K is the soil erodibility ($t\,h\,(MJ\,mm)^{-1}$), A soil loss ($t\,ha^{-1}$) and R is the rainfall erosivity factor ($MJ\,mm\,(ha\,h)^{-1}$). There are several methods for determining the K-factor, such as rainfall simulation, wind tunnel experiments, and erosion plot experiments.[3,13] Because field measurements of the K-factor are resource-intensive, the development of pedotransfer functions (PTFs) for predicting the K-factor using easily measurable soil properties has received considerable attention in the last decades.

The well-known USLE nomograph was initially proposed by Wischmeier et al.[14] to estimate the K-factor, which has been used widely around the world. This model considered some basic and important parameters, including soil particles, organic matter, soil structure, and permeability. The linear regression form of the USLE model was further developed by Wischmeier and Smith[12] as follows:

$$K = \frac{[2.1 \times 10^{-4}(12 - OM)M^{1.14} + 3.25(S - 2) + 2.5(P - 3)]}{100} \tag{2}$$

where K represents the USLE soil erodibility (t h $(MJ\,mm)^{-1}$), M ((100 − %clay) × (%very fine sand + %silt)), OM organic matter (%), S is the soil structure, and P is the profile permeability class. Although USLE (Eq. 2) model has been accepted and widely used to predict the K-factor, particularly in medium-textured soils, and its usefulness in other soil texture categories has often been questioned.[9,15]

Later, Römkens et al.[16,17] investigated the role of soil particle size on the K-factor and, therefore, developed a new model called the RUSLE K-factor, using 10 soil particle size classes from 225 global soils, as follows:

$$K = 7.594 \left\{ 0.0034 + 0.0387 \exp \left[-0.5 \left(\frac{loglog\,(Dg) + 1.533}{0.7671} \right)^2 \right] \right\} \tag{3}$$

$$D_g = \exp \left[0.01 \left(P_{sand} \ln 1.025 + P_{silt} \ln 0.026 + P_{clay} \ln \ln 0.001 \right) \right] \tag{4}$$

where D_g represents the geometric mean diameter of soil particles (mm), and P_{sand}, P_{silt} and P_{clay} are the percentages of sand, silt, and clay, respectively. In the Middle East, the USLE K-factor model was used to estimate the soil erodibility factor (Table 1). However, low prediction performances of the soil erodibility factor for the USLE model have been frequently reported in the literature,[13,18,19] mainly because of the soil and climate datasets from specific regions used for developing the model. This means that the model does not completely reflect the soil and climate conditions of the world.

The USLE K-factor was calibrated using rainfall simulation data for the United States, where soils with an aggregation index < 0.3 and surface rock fragments < 10% are mostly medium-textured and have no or low CaCO3 content due to the high intensity of rainfall (> 63 mm h^{-1}).[3,16,17] Therefore, the USLE and RUSLE equations for the Middle East, with semi-arid and arid climates and most areas covered by calcareous soils, reveal a great uncertainty in predicting the K-factor, and subsequently for soil erosion prediction.[3,9,13]

Some studies[9,15,28] have demonstrated the noticeable effects of soil structure and aggregation, influenced by soil carbonates (e.g., calcium and magnesium carbonates), on the K-factor. In calcareous soils, carbonates increase soil fluctuation and stability of aggregates, soil water infiltration, resulting in a decrease in the K-factor.[13] Vaezi et al.,[19] Shabani et al.,[13] and Ostovari et al.[3] showed the weak applicability of the USLE model in predicting soil loss in southern Iran. Thus, many attempts have been made to modify and improve the USLE model for predicting the K-factor with higher performance in the Middle East (Table 2).

Extensive research has investigated soil erosion and erodibility factors in some regions and countries from the Middle East.[4,23,29–31] However, no study has assessed the soil erodibility factor for the entire extension of the Middle East, which is one of the most critical regions regarding wind and water erosion worldwide. Hence, the main purpose of this chapter is to evaluate the soil erodibility factor calculated by the RUSLE model and by some PTFs developed in

TABLE 1   Studies used the USLE/RUSLE equation for estimating the K-factor in some regions in the Middle-East.

Reference	Country	Soil type	Equation	K-factor range
Bahrawi et al.[20]	Saudi Arabia	–	RUSLE	0.29–0.68
Qaryouti et al.[21]	Jordan	–	RUSLE	0.0–0.09
Abdullah et al.[22]	Kuwait	Inceptisols	RUSLE	09–0.42
Ostovari et al.[23]	Iran	Aridisols, Inceptisols Entisols	USLE/RUSLE	0.02–0.05
Ebrahimi et al.[24]	Iran	Entisols, Inceptisols, Ultisols, Vertisols, Mollisols	USLE	0.02–0.10
Dutal and Reis[25]	Turkey	–	USLE	0.08–0.28
Al-Abadi et al.[26]	Iraq	Entisols	RUSLE	0.04–0.38
Keya[27]	Iraq	–	USLE	0.47–0.79

TABLE 2 Some developed PTFs for predicting the K-factor in the Middle East.

Reference	Location	Developed PTFs	R^2	Unit	Remark
Vaezi et al.[15]	Iran	$K = 0.0185 - 2.1 \times 10^{-5} Sa - 1.3 \times 10^{-5} Cl - 2.1 \times 10^{-5} PE$	0.68	(t h (MJ mm)$^{-1}$)	
Vaezi et al.[19]	Iran	$K = 0.0123 - 5.7 \times 10^{-5} Cl - 5.2 \times 10^{-5} CaCO_3 - 0.00129 PE$	0.84	(t h (MJ mm)$^{-1}$)	
Shabani et al.[13]	Iran	$K = 0.01 \times (71.95 + 0.327 Si + 0.31 Cl - 6.43 OM + 0.19 VFS + 0.02 PE - 2.83\ CaCO_3)$	0.67	(t h (MJ mm)$^{-1}$)	

Sa, sand content (% particle size of 0.05–2.0 mm); Si, silt (% particle size 0.002–0.05 mm); Cl, clay (% particle size less than 0.002 mm); VFS, very fine sand; OM, organic matter; D_g, mean geometric diameter of soil particles (mm), $CaCO_3$, calcium carbonate equivalent (%); PE, permeability (cm h^{-1}).

the Middle East and to investigate the spatial variation of the PTFs. We expect that our outcomes can inform decision-making for the implementation of activities against soil degradation in these large areas.

2 Material and methods

2.1 Study area

The Middle East includes the lands around the southern and eastern Mediterranean Sea, Arabian Peninsula, Iran, and North Africa (Fig. 1A), which covers approximately seven million km^2 of the land's earth and has around 370 million people. The Middle East has diverse topography, geology, and climatic and vegetation cover conditions, which have resulted in the formation of different soils. The Elevation of the Middle East varies from −422 m in Syria-Lebanon to 5434 m in Damavand, Iran (Fig. 1B). The climate is generally arid and semi-arid. The average annual precipitation based on the 1 km resolution WorldClim ranged from 3 mm in the dry land to 855 mm in the coastal of the Mediterranean and Caspian Sea[32] (Fig. 1D).

Land cover varies from complex mountains to deserts. In this region, agriculture is the main source of income. Arable lands cover 6.8% of the total area, while pasture and forest constitute 26% and 7%, respectively.[5] The major land uses/land covers (LULC) for the year 2018 base on the 500 m resolution global University of Maryland land cover classification data[35] were non-vegetated lands (69.51%), grasslands (12.56%), croplands (6.32%), open shrublands (5.37%), cropland/natural vegetation mosaics (2.46%), savannas (1.02%), woody savannas (0.72%), and urban areas (0.60%) (Fig. 1E).

According to the pedological map (Fig. 1F) obtained from SoilGrids,[36] which were obtained from soil grids (Fig. 1F), the soils were classified as Regosols (25.7%), Leptosols (22.1%), Arenosols (17.7%), Calcisols (8.8%), Cambisols (8.1%), Solonchaks (5.3%), Gypsisols (3.8%), Fluvisols (3.8%), Luvisols (2.3%) and Kastanozems (1.1%). The soils are mainly advanced on sedimentary materials, carbonate formations, metamorphic-basic and acid rocks.[37]

2.2 Soil data

In this study, we used topsoil data (average 0–30 cm depth) from 690 sites of the World Soil Information Service (WoSIS) dataset[38] spread across the study site. The WoSIS is a quality-standardized soil profile dataset with a variety of soil types, textures, and other attributes. The WoSIS dataset was collected during the last 50 years (1970–2018). Most soil data had similar measurement methods and units. The location of the soil samples was recorded in the World Geodetic System 1984 (WGS84, EPSG code 4326). The WoSIS data were used to investigate and map the soil erodibility factor and its related soil properties. According to the US Department of Agriculture,[39] the soil samples were classified into all 12 soil texture classes (Fig. 2A). Various soil properties (clay, silt, sand, coarse fragments, soil organic carbon (SOC), bulk density, cation exchange capacity (CEC), and pH), which are mainly linked to soil fertility and plant root growth, were selected. In addition, we used a dataset obtained from 40-unit erosion plots for validating the developed PTFs (Table 2) for the K-factor prediction (more details of the data refer to Ostovari et al.[3]). The flowchart of this study is presented in Fig. 2B.

2.3 Spatial analysis

We acquired raster maps with a spatial resolution of 250 m from the SoilGrids dataset (https://soilgrids.org/) using the Google Earth Engine (GEE)[40] for topsoil attributes. We used the maps of clay, silt, and sand to calculate the soil erodibility for the study area by applying the RUSLE K-factor model, as described in Eqs. (3), (4).

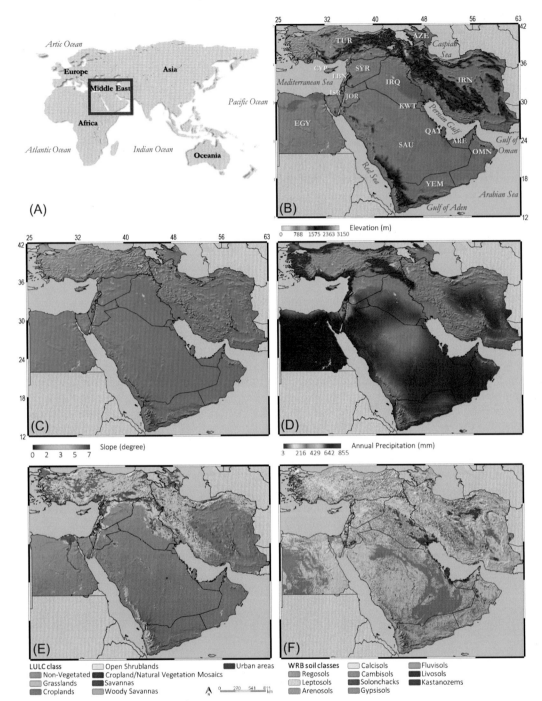

FIG. 1   Physiographical maps of the study region (Middle East) were acquired from different sources and resolutions. (A) Elevation map obtained from the 30 m resolutions ALOS data,[33] employed to calculate the (B) slope map using the Terrain Analysis in GEE package.[34] (C) Map of annual precipitation obtained from the 1 km resolution WorldClim data.[32] (D) Major land use/land cover (LULC) map acquired from the 500 m resolution Global University of Maryland land cover classification data[35] for the year 2018. Major World Reference Base (WRB) soil classes covering the region. *ARE*, United Arab Emirates; *AZE*, Azerbaijan; *CYP*, Cyprus; *EGY*, Egypt; *IRN*, Islamic Republic of Iran; *IRQ*, Iraq; *JOR*, Jordan; *KWT*, Kuwait; *LBN*, Lebanon; *OMN*, Oman; *QAT*, Qatar; *SAU*, Saudi Arabia; *SYR*, Syria; *TUR*, Turkey; *YEM*, Yemen.

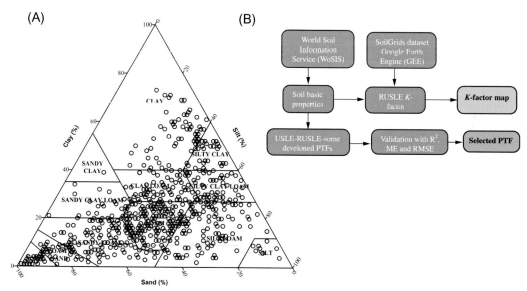

FIG. 2 The USDA textural distribution of the studied soils (A) and flowchart of the study (B). *Filled* and *blank circles* are the calibration and validation datasets, respectively.

2.4 Statistical analysis and validation of the PTFs

Statistical analysis of data was performed using Statistica 8.0 and figures were depicted in Excel 2016. The performance of the designed PTFs was tested using the R^2, root mean squared error (RMSE) and mean error (ME) as follows:

$$R^2 = \frac{\left[\sum_{i=1}^{n} \left(O_i - \overline{O_i}\right)\left(P_i - \overline{P_i}\right)\right]^2}{\sum_{i=1}^{n} \left(O_i - \overline{O_i}\right)^2 \sum_{i=1}^{n} \left(P_i - \overline{P_i}\right)^2} \tag{5}$$

$$RMSE = \sqrt{\left(\frac{\sum_{i=1}^{n} (P_i - O_i)^2}{n}\right)} \tag{6}$$

$$ME = \frac{\sum_{i=1}^{n} (P_i - O_i)}{n} \tag{7}$$

where n is the number of observations, and O and P are the measured and estimated values, respectively.

3 Results and discussion

3.1 Soil description

Table 3 provides a statistical summary of soil properties from the Middle Eastern countries. The pH with a mean of 8.0 varied from 4.8 to 9.2 was slightly variable (CV = 6.5%), while OC with a CV = 119.4% was highly variable. This occurs because, in this area, there are different land use and cover types, which lead to different sources of soil OM in this region.

Among the soil particles, clay with mean and CV values of 23.3% and 66.1%, respectively, had the highest variability (Table 1). Most of the samples presented high silt content with an alkaline reaction (mean pH = 8.0), which is due to the high amount of calcium carbonate in this area. As the Middle East is located on the earth's dry land belt with prolonged hot summer, due to low annual precipitation and high evapotranspiration along with the low solubility of $CaCO_3$, soils are mainly calcareous. Similar amounts of soil properties, particularly $CaCO_3$ content, have been widely documented in the Middle East.[25,31,41] The RUSLE K-factor[16] with a mean of 0.25 t h $(MJ\,mm)^{-1}$ ranged from 0.06 0.33 t h $(MJ\,mm)^{-1}$, which is consistent with the findings of Addis and Klik[42] and Bonilla and Johnson,[43] who reported K-factor values in the range of 0.02 to 0.039 t h $(MJ\,mm)^{-1}$ Chile.

TABLE 3 Descriptive statistics of soil properties ($n=772$) in the study area (Middle East).

Soil properties	Unit	Mean	Minimum	Maximum	Standard deviation, SD	Coefficient of variation, CV (%)
Organic carbon	g kg-1	7.50	0.1	97.0	8.3	119.4
CaCO$_3$	g kg^{-1}	184.5	0.1	750. 0	144.4	89.7
pH H$_2$O		8.1	4.8	9.2	0.5	6.6
Sand	%	38.4	1.0	98.0	23.6	61.6
Silt	%	38.3	1.0	93.0	16.8	43.8
Clay	%	23.3	1.0	73.0	15.4	66.1
K-factor	th (MJ mm)$^{-1}$	0.25	0.07	0.33	0.06	26.7

TABLE 4 Pearson's correlation coefficient (r) between the K-factor and some basic soil properties ($P<.05$).

	Organic carbon	CaCO$_3$	pH H$_2$O	Sand	Silt	Clay	K-factor
Organic carbon	1.00						
CaCO$_3$	−0.06	1.00					
pH H$_2$O	−0.36	−0.04	1.00				
Sand	−0.21*	−0.15*	0.18*	1.00			
Silt	0.07	0.14*	−0.07	−0.78*	1.00		
Clay	0.25*	0.08	−0.21*	−0.71*	0.12*	1.00	
K-factor	−0.24*	0.1	−0.05	−0.68*	0.70*	0.29*	1.00

Pearson's correlation coefficients (r) between the K-factor and some soil properties are summarized in Table 4. CaCO$_3$ has a high amount of Ca^{2}+, which plays an important role in creating large and stable aggregates by binding agents for flocculating soil minerals, resulting in a decrease in the K-factor and an increase in the resistance of soil aggregates against runoff and raindrop detachment. The K-factor had a significant negative relationship with bulk density ($r=-0.46$, $P<.05$, $n=120$), which is supported by Ostovari et al.,[29] who highlighted that bulk density was negatively and positively correlated to the K-factor and soil loss tolerance, respectively.

3.2 Relationship between soil erodibility and soil texture

Soil textural components (sand, silt, and clay content) strongly affected soil erosion. Clay is a binding agent that can improve soil structure and make greater aggregates, resulting in increased resistance against erosive forces.[44] On the other hand, sand is not sticky but is a weighted particle that can resist water flow in transporting soil particles. Silt and fine sand particles are defenseless particles that are highly susceptible to soil detachment and transport because of their low sticky and low weight.[45]

As mentioned above, the soils in this study have all 12 USDA soil texture classes (Fig. 2). The K-factor in clayey texture was similar to the K-factor in sandy classes. Sand content had a strong negative correlation with the K-factor ($r=-0.68$; $P<.05$), which means that with increasing sand particles, the K-factor decreased. Particles >200 µm are difficult to detach and transport because of their high particle mass.[46] With increasing sand content (Fig. 3A) by sand $=40\%$, the K-factor smoothly decreased and then sharply decreased after sand $\geq40\%$. Among the soil particles, silt had the highest positive correlation with K-factor ($r=0.70$, $P<.05$; Table 4). Soils with a high amount of silt, which is the most sensitive soil particle to erosive forces,[43] have weak and small aggregates, resulting in higher K-factor values. The lowest K-factor was found in the classes with silt content $<\sim30\%$ (Fig. 3B), which is similar to the results reported by Bonilla and Johnson.[43] As shown in Fig. 3B, with increasing silt content, the K-factor increased. There was a significant positive correlation between the silt content and K-factor. However, a weak relationship was found between the clay content and the K-factor (Fig. 3C). Di Stefano and Ferro[46] pointed out that the detachment of soil particles decreases out of the range of 20–200 µm (silt content). Consequently, soil particles such as silt and very fine sand are more prone to erosion. Particles <20 µm have strong cohesive forces against

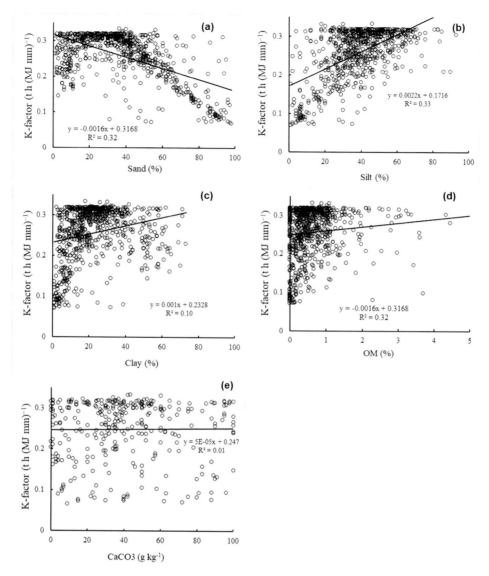

FIG. 3 The relationship between K-factor and soil particle fractions (A–C), organic carbon (D), and $CaCO_3$ (E).

particle detachment. Similar findings were highlighted in England by Ampontuah et al.,[47] who investigated soil particle size distribution in two different cultivated hillslopes and addressed that a field with a high silt content (67%–80%) was very susceptible to water erosion.

Our results showed that the correlation of the K-factor and soil properties increased when soil particles were isolated. This indicates a synergistic influence between soil properties,[45,48] as an increase in the particle fraction reduces another fraction. Thus, it is more reliable to assess the K-factor based on soil texture than on a single particle fraction,[43] as soil texture reflects the relative proportion of sand, silt, and clay. A significant positive relationship was also observed between SOC and clay content ($r = 0.49$, $P < .05$, $n = 120$). Ostovari et al.[9] reported a correlation coefficient of 0.36, between OM and clay. Strong relationships have been reported between soil textural components and the K-factor or erosion rate. For example, Duiker et al.[45] found an inverse relationship between erosion rate and sand content. Vaezi et al.[15] in northwest Iran reported that the K-factor had a significant and positive correlation with silt and silt + very fine sand. They stated that the sand content could increase the infiltration rate and decrease the K factor. Dimoyiannis et al.,[49] Zhang et al.,[50] and Rodríguez et al.[51] also reported a negative correlation and relationship between clay and the K-factor.

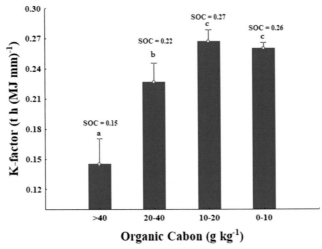

FIG. 4 The relationship between *K*-factor and four different amounts of SOC.

3.3 Relationship between soil erodibility and organic carbon

For our study area, the results showed that there was a weak negative relationship between the *K*-factor and SOC ($r = -0.24$, $P < .05$), which is consistent with previous research in this region. For instance, Vaezi et al.[15] and Ostovari et al.[3] reported a correlation coefficient of -0.48 and -0.6 between the *K*-factor and the SOC, respectively. Soil organic matter acts as a gluing agent between mineral particles to build stable aggregates, thereby decreasing soil erosion. Strong aggregates significantly reduce the *K*-factor because of the increasing resistance of soils to splash erosion.[52]

The important impact of SOC on soil aggregation was reported by Hoyos,[53] Rodríguez et al.,[51] Vaezi et al.[15] and Ostovari et al.[9] Although the relationship between the *K*-factor and SOC was low, a significant difference in the average soil erodibility was found among the four different ranges of SOC (Fig. 4). As can be seen in Fig. 4, the group with SOC > 40 g kg^{-1} had the lowest *K*-factor among other groups, which shows the important influence of SOC on the *K*-factor, with respect to a low SOC content. However, no significant difference was found between the 10–20 and 0–10 g kg^{-1} SOC groups. Generally, SOC is associated with fine particles, particularly clay particles.[43] In our study, SOC was linked to clay content with a significant positive correlation coefficient of 0.25 (Table 4). It is worth mentioning that SOC cannot explain the soil *K*-factor because it has different interactions with soil particles, infiltration, and structure.[54,55] This could be a concern for soil conservation plans where soils have a high amount of SOC and are considered less vulnerable to erosion without attention to soil texture.

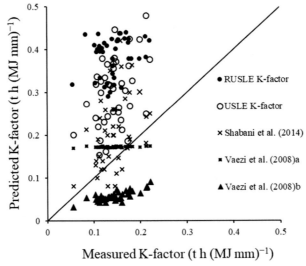

FIG. 5 Measured *K* versus the predicted *K* for RUSLE, USLE, and some developed PTFs.

3.4 Soil structure representative parameters

The Soil structure predominantly affects the K-factor. A good soil structure causes a higher infiltration rate and larger aggregates that decrease water erosion and consequently decrease the K-factor. Some physicochemical properties, such as clay, SOC, and calcium carbonates (due to the Ca^{2+} effect) are the binding agents and positively improve the soil structure by creating large, rebuts and stable aggregates.[15] Charman and Murphy[52] illustrated that surface aggregates can decrease soil erodibility due to their stability against particle detachment. Vaezi et al.[15] reported that the K-factor was noticeably correlated with SOC ($r = -0.478$; $P < .05$), lime content ($r = -0.41$; $P < .05$), MWD ($r = -0.35$; $P < .05$), and permeability ($r = -0.88$; $P < .05$), which supports our findings. Permeable soils can decrease the K factor due to lower water flow on the soil surface.[56–58]

The positive relationship between sand content and organic matter on infiltration rate and consequently decreasing the K-factor also have been stated by Santos et al.,[28] Evrendliek et al.,[59] Tejada and Gonzalez,[55] and Rodríguez et al.[51] The negative effects of MWD and permeability on the K-factor have also been reported by Barthès and Roose,[60] Gupta,[61] and Hoyos.[53]

3.5 Soil carbonates

The USLE K-factor did not consider lime and calcium carbonate contents, which are considered the most important factors in calcareous soils. Lime significantly affects basic properties, which are significantly associated with soil structure. Hassan and Agha[18] in northern Iraq showed that with the removal of the Ca-carbonates, the K-factor significantly decreased due to dynamic changes in physical properties such as texture, structure, and permeability. They also stated that the removal of carbonate from soil significantly decreased the weight of sand content and increased the percentage of clay and silt. This means that compared to the silt and clay fractions, carbonate is highly distributed in the sand fraction. In calcareous soils, calcium in the lime components is an important factor that affects aggregate stability and infiltration rate, and can consequently affect the K-factor. The negative effects of lime on the K-factor have been reported by Castro and Logan,[62] Orts et al.,[63] Charman and Murphy,[52] Duiker et al.,[45] and Vaezi et al.,[15] who stated the importance of Ca^{2+} which affects particles flocculation and stability of soil aggregate and therefore decreases the K-factor.

In addition to these basic physicochemical properties, Pérez-Rodríguez[64] in Spain reported that physiographic units could significantly influence the K-factor. They explained that the values of the K-factor ranked in descending order, as gypsiferous hills > transition reliefs ≈ alluvial plains > paramos and alcarrias. The soils of the gypsiferous hills could be sensitive to erosion owing to the low content of SOC and weak structure.

3.6 Validation of developed PTFs

As mentioned above, we used a dataset obtained from standard erosion plots to test the performance of the RUSLE K-factor, USLE K-factor, and some PTFs (Table 2) that have been developed in recent years in the Middle East. Fig. 5 shows the measured K-factor using standard erosion plots versus the predicted K-factor using the PTFs. As shown in Fig. 5, an important match between the measured and predicted K-factor values was delivered with the developed PTFs compared with the RUSLE and USLE K-factor models. The RUSLE with ME = 0.239 th (MJ mm)$^{-1}$ and USLE K-factor with ME = 0.158 th (MJ mm)$^{-1}$ showed high overestimation in predicting the K-factor among the PTFs (Table 5).

A mean predicted K-factor using the RUSLE K-factor and USLE K-factor was 0.38 th (MJ mm)$^{-1}$ and 0.30 th (MJ mm)$^{-1}$, respectively, 2.71 and 2.14 times higher than the mean measured K-factor (0.14 th (MJ mm)$^{-1}$) using

TABLE 5 Statistical indices for some PTFs in predicting soil erodibility factor.

PTFs	Mean K-factor	R^2	RMSE (th (MJ mm)$^{-1}$)	ME (th (MJ mm)$^{-1}$)
RULSE K-factor	0.38	0.06	0.244	0.239
USLE K-factor	0.30	0.17	0.173	0.158
Shabani et al.[13]	0.21	0.18	0.097	8.22
Vaezi et al.[15]	0.17	0.36	0.028	0.044
Vaezi et al.[19]	0.09	0.61	0.042	−0.086

standard erosion plots, respectively. The mean K-factor predicted by Shabani et al.[13] was 0.21, which is 1.5 times higher than the mean measured K-factor. As mentioned above, the RULSE and USLE K-factor models were developed in US soils with no carbonate content under rainfall higher than 63 mm h^{-1}. Therefore, using these models in the Middle East, where the climate is dry and soils have a high amount of carbonate that makes strong and large soil aggregates, provides poor results for predicting the K-factor. This is in agreement with the reports by Vaezi et al.,[19] Hassan and Agha,[18] and Shabani et al.[13] Among the PTFs, Vaezi et al.[15] with respect to the highest R^2 ($R^2 = 0.61$), acceptable RMSE (RMSE $= 0.042$ t h (MJ mm)$^{-1}$) and low ME (ME $= -0.086$ t h (MJ mm)$^{-1}$) was selected as the best model for predicting the K-factor for the Middle East soils. As shown in Table 2, CaCO$_3$ had a significant coefficient and appeared in the second PTFs by Vaezi et al.,[15] which indicates the important effects of soil carbonates on the K-factor. However, in the

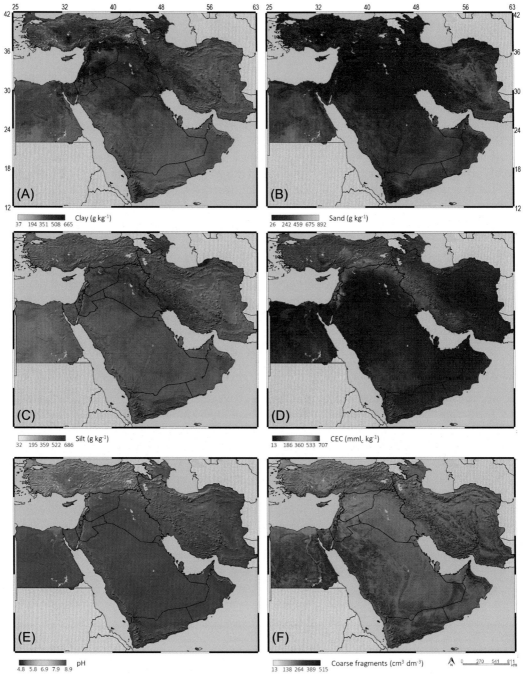

FIG. 6   Soil maps of the study area were acquired from the SoilGrids dataset (https://soilgrids.org/). Maps of soil attribute content (averaged to 0–30 cm depth) for (A) clay, (B) sand, (C) silt, (D) cation exchange capacity (CEC), (E) pH, (F) coarse fragments.

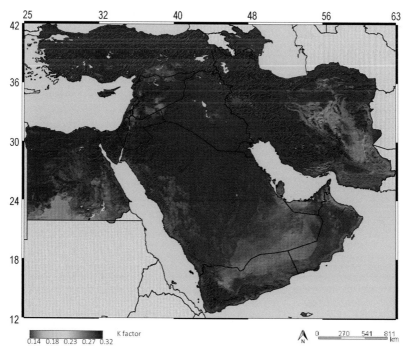

FIG. 7 Map of soil erodibility (*K* factor) calculated using equations.

present study, no significant relationship was found between *K*-factor and carbonate. The reason could be that the Vaezi et al.[19] equation was developed in the north of Iran, where soils are formed on carbonate formation and are classified mainly as Calcisols (Fig. 1F), whereas in the present study, there was a wide range of soil types with different amounts of soil carbonate.

3.7 Spatial analysis

The Spatial distributions of clay, silt, sand, and organic matter are shown in Fig. 6. The Sand content varied from 2.6% (26 g kg^{-1}), mostly in the central areas of the Middle East (Iraq, Jordan, western Iran, and northern Saudi Arabia) to 89.2% (892 g kg^{-1}) in the southern and western Middle East (Egypt, southern Saudi Arabia, and Yemen). In contrast, the maximum silt content was observed in the central areas of the Middle East and the lowest in the western Middle East (mainly in Egypt and western Saudi Arabia). This result was expected because there was a significant negative correlation ($r = -0.78$; $P < .05$) between sand and silt content (Table 4).

Interestingly, clay followed the same trend as that of silt (Fig. 6A and C). As shown in Fig. 6D, the SOC content varied from 0.3 to 95.6 g kg^{-1} and most areas has almost the same content. However, the highest SOC content was found in Turkey and northern Iran (close to the Caspian Sea). These areas are mainly mountainous (Fig. 6D) and have greater land cover, the dominated by forest and grassland, and receive higher annual precipitation (Fig. 1D). In addition, in these regions, soils had the highest CEC, which could be due to the high amount of SOC and clay content (Fig. 6A and D). In addition, (Fig. 6E), soil pH in most of the areas was higher than seven, which indicates an alkaline characteristic linked to carbonate content. The values of the *K*-factor in the Middle East ranged from 0.14 and 0.32 t h MJ^{-1} mm^{-1} (Fig. 7). Highly sensitive areas to erosion are mainly observed in the central, northern, and northwestern parts of the Middle East. Moreover, slightly erodible areas were exposed in the southwest, south, and southeast (Fig. 7). The *K*-factor map (Fig. 7) has the same spatial distribution as the maps of coarse fragments (Fig. 6F), clay (Fig. 6A), and silt (Fig. 6C), which indicates the correlation between these two parameters and the *K*-factor.

4 Conclusion

Soil erodibility (K-factor) is one of the major factors influencing erosion in arid and semi-arid regions. Our results showed that the *K*-factor of the Middle East varied from 0.14 and 0.32 t h MJ^{-1} mm^{-1}. Among the soil particles, silt had the highest positive correlation with *K*-factor, while clay showed a weak relationship. Sand content had a strong negative correlation with *the K*-factor. In addition, organic carbon had an important effect on the *K factor*. Investigation of the RUSLE, USLE and some developed PTFs in the Middle East showed that these PTFs have a better performance than the RUSLE and USLE *K*-factor models for predicting the *K*-factor because they significantly considered carbonate as an input parameter in the developed models. However, the USLE *K*-factor did not consider lime and calcium carbonate contents. The results also revealed that these PTFs could be used to predict soil erodibility in the Middle East at a reasonable cost and are not too time-consuming. The spatial distribution map of the *K*-factor in the Middle East showed that the highest erodible areas were found in the central, northern, and northwestern parts of the area. The map of the *K*-factor had a similar distribution to the maps of the clay and silt contents.

Acknowledgment

We thank Ms. Elizabeth Hamzi-Schmidt at TUM English Writing Center for her advice, feedback, and support. The first author wishes to express his sincere thanks to Ms. Asieh Rasti for her excellent support and help in conducting the research.

References

1. Sentis IP. A soil water balance model for monitoring soil erosion processes and effects on steep lands in the tropics. *Soil Technol.* 1997;11(1):17–30. https://doi.org/10.1016/S0933-3630(96)00112-2.
2. Smith P, House JI, Bustamante M, et al. Global change pressures on soils from land use and management. *Glob Change Biol.* 2016;22(3):1008–1028. https://doi.org/10.1111/gcb.13068.
3. Ostovari Y, Ghorbani-Dashtaki S, Kumar L, Shabani F. Soil erodibility and its prediction in semi-arid regions. *Arch Agron Soil Sci.* 2019;65 (12):1688–1703. https://doi.org/10.1080/03650340.2019.1575509.
4. Ozsoy G, Aksoy E, Dirim MS, Tumsavas Z. Determination of soil erosion risk in the Mustafakemalpasa River Basin, Turkey, using the revised universal soil loss equation, geographic information system, and remote sensing. *Environ Manag.* 2012;50(4):679–694. https://doi.org/10.1007/s00267-012-9904-8.
5. FAO. *Agroecology to reverse soil degradation and achieve food security.* Rome: Food and Agriculture Organization of the United Nations; 2015:650.
6. FAO. *The multi-faced role of soil in the near East and North Africa region – policy brief.* Rome, Italy: FAO; 2019:36.
7. Hussein MH, Kariem TH, Othman AK. Predicting soil erodibility in northern Iraq using natural runoff plot data. *Soil Tillage Rese.* 2007; 94(1):220–228. https://doi.org/10.1016/j.still.2006.07.012.
8. Ostovari Y, Ghorbani-Dashtaki S, Bahrami HA, et al. Towards prediction of soil erodibility, SOM and CaCO3 using laboratory Vis-NIR spectra: a case study in a semi-arid region of Iran. *Geoderma.* 2018;314:102–112. https://doi.org/10.1016/j.geoderma.2017.11.014.
9. Ostovari Y, Ghorbani-Dashtaki S, Bahrami HA, Naderi M, Dematte JA, Kerry R. Modification of the USLE K factor for soil erodibility assessment on calcareous soils in Iran. *Geomorphology.* 2016;273:385–395. https://doi.org/10.1016/j.geomorph.2016.08.003.
10. Panagos P, Meusburger K, Ballabio C, Borrelli P, Alewell C. Soil erodibility in Europe: a high-resolution dataset based on LUCAS. *Sci Total Environ.* 2014;479:189–200. https://doi.org/10.1016/j.scitotenv.2014.02.010.
11. Wischmeier WH. A rainfall erosion index for a universal soil-loss equation. *Soil Sci Soc Am J.* 1959;23(3):246–249. https://doi.org/10.2136/sssaj1959.03615995002300030027x.
12. Wischmeier WH, Smith DD. *Predicting rainfall erosion losses: a guide to conservation planning.* Department of Agriculture, Science and Education Administration; 1978.
13. Shabani F, Kumar L, Esmaeili A. Improvement to the prediction of the USLE K factor. *Geomorphology.* 2014;204:229–234. https://doi.org/10.1016/j.geomorph.2013.08.008.
14. Wischmeier WH, Johnson CB, Cross BV. Soil erodibility nomograph for farmland and construction sites. *J Soil Water Conserv.* 1971;26:189–193.
15. Vaezi AR, Sadeghi SH, Bahrami HA, Mahdian MH. Modeling the USLE K-factor for calcareous soils in northwestern Iran. *Geomorphology.* 2008;97 (3–4):414–423. https://doi.org/10.1016/j.geomorph.2007.08.017.
16. Römkens M, Roth C, Nelson D. Erodibility of selected subsoils in relation to physical and chemical properties. *Soil Sci Soc Am J.* 1977;41:954–960.
17. Römkens MJ, Young RA, Poesen JW, McCool DK, El-Swaify SA, Bradford JM. oil erodibility factor (K). Predicting soil erosion by water: a guide to conservation. In: Renard KG, Foster GR, Weesies GA, McCool DK, Yoder DC, eds. *Planning with the revised universal soil loss equation (RUSLE).* 703. US Department of Agriculture, Agriculture Handbook; 1997:65–99.
18. Hassan K, Agha H. Effects of calcium carbonate on the erodibility of some calcareouse soils by water erosion. *Mesopotamia J Agric.* 2012; 40(4):11–19. https://doi.org/10.33899/magrj.2012.60183.
19. Vaezi AR, Bahrami HA, Sadeghi SH, Mahdian MH. Spatial variability of soil erodibility factor (K) of the USLE in North West of Iran. *J Agric Sci Technol.* 2010;12(2):241–252.
20. Bahrawi JA, Elhag M, Aldhebiani AY, Galal HK, Hegazy AK, Alghailani E. Soil erosion estimation using remote sensing techniques in Wadi Yalamlam Basin, Saudi Arabia. *Adv Mater Sci Eng.* 2016;9585962:1–8.
21. Qaryouti SL, Guertin DP, Taany RA. GIS modeling of water erosion in Jordan using "RUSLE". *Assiut Univ Bull Environ Res.* 2014;17(1):57–75.
22. Abdullah M, Feagin R, Musawi L. The use of spatial empirical models to estimate soil erosion in arid ecosystems. *Environ Monit Assess.* 2017;189:78. https://doi.org/10.1007/s10661-017-5784-y.
23. Ostovari Y, Ghorbani S, Bahrami H, Naderi M, Dematte JAM. Soil loss prediction by an integrated system using RUSLE, GIS and remote sensing in semi-arid region. *Geoderma Region.* 2017;11:28–36.
24. Ebrahimi M, Nejadsoleymani H, Sadeghi A, Mohammad Reza Mansouri-Daneshvar MR. Assessment of the soil loss-prone zones using the USLE model in northeastern Iran. *Paddy and Water Environm.* 2020;19:71–86. https://doi.org/10.1007/s10333-020-00820-9.

25. Dutal H, Reis M. Determining the effects of land use on soil erodibility in the Mediterranean highland regions of Turkey: a case study of the Korsulu stream watershed. *Environ Monit Assess.* 2020;192(3):1–5. https://doi.org/10.1007/s10661-020-8155-z.

26. Al-abadi AM, Ghalib HB, Al-qurnawi WS. Estimation of soil erosion in northern Kirkuk governorate, Iraq using RUSLE, remote sensing, and GIS. *Carpathian J Earth Environ Sci.* 2016;11(1):153–166.

27. Keya DR. Integration of GIS with USLE in assessing soil loss from Alibag catchment, Iraqi Kurdistan region. *Polytech J.* 2018;8(1):12–16.

28. Santos FL, Reis JL, Martins OC, Castanheira NL, Serralheiro RP. Comparative assessment of infiltration, runoff and erosion of sprinkler irrigated soils. *Biosyst Eng.* 2003;86(3):355–364. https://doi.org/10.1016/S1537-5110(03)00135-1.

29. Ostovari Y, Moosavi AA, Mozaffari H, Pourghasemi HR. RUSLE model coupled with RS-GIS for soil erosion evaluation compared with T value in Southwest Iran. *Arab J Geosci.* 2021;14(2):1–5. https://doi.org/10.1007/s12517-020-06405-4.

30. Ostovari Y, Moosavi AA, Pourghasemi HR. Soil loss tolerance in calcareous soils of a semiaridregion: evaluation, prediction, and influential parameters. *Land Degrad Dev.* 2020;31(15):2156–2167. https://doi.org/10.1002/ldr.3597.

31. Tayebi M, Tayebi M, Sameni A. Soil erosion risk assessment using GIS and CORINE model: a case study from western Shiraz. *Iran Arc Argon Soil Sci.* 2017;63(8):1163–1175. https://doi.org/10.1080/03650340.2016.1265106.

32. Fick SE, Hijmans RJ. WorldClim 2: new 1-km spatial resolution climate surfaces for global land areas. *Int J Climatol.* 2017;37(12):4302–4315.

33. JAXA EORC. *ALOS Global Digital Surface Model "ALOS World 3D-30m (AW3D30)*; 2016. https://www.eorc.jaxa.jp/ALOS/en/aw3d30/ index. htm. Accessed 4 March 2020.

34. Safanelli JL, Chabrillat S, Ben-Dor E, Demattê JAM. Multispectral models from bare soil composites for mapping topsoil properties over Europe. *Remote Sens.* 2020;12:1369. https://doi.org/10.3390/rs12091369.

35. Friedl M, Sulla-Menashe D. *MCD12Q1 MODIS/Terra+Aqua land cover type yearly L3 global 500m SIN Grid V006 NASA EOSDIS land processes DAAC*; 2019.

36. Sousa LM, Poggio L, Batjes NH, et al. SoilGrids 2.0: producing quality-assessed soil information for the globe. In: *Soil discussions*; 2020:1–37. https://doi.org/10.5194/soil-2020-65.

37. Hartmann J, Moosdorf N. The new global lithological map database GLiM: a representation of rock properties at the Earth surface. *Geochem Geophys Geosyst.* 2012;13(37).

38. Batjes NH, Ribeiro E, VanOostrum A. Standardised soil profile data to support global mapping and modelling (WoSIS snapshot 2019). *Earth Syst Sci Data.* 2020;12:299–320. https://doi.org/10.17027/isric-wdcsoils.20190901.

39. USDA. Agricultural Research Service USDA-ARS, "CREAMS, a field scale model for chemicals, runoff, and erosion from agricultural management systems". In: *Conservation research report.* USDA; 1980:26.

40. Gorelick N, Hancher M, Dixon M, Ilyushchenko S, Thau D, Moore R. Google Earth Engine: planetary-scale geospatial analysis for everyone. *Remote Sens Environ.* 2017;202:18–27. https://doi.org/10.1016/j.rse.2017.06.031.

41. Saygın SD, Basaran M, Ozcan AU, et al. Land degradation assessment by geo-spatially modeling different soil erodibility equations in a semi-arid catchment. *Environ Monit Assess.* 2011;180(1–4):201–215. https://doi.org/10.1007/s10661-010-1782-z.

42. Addis HK, Klik A. Predicting the spatial distribution of soil erodibility factor using USLE nomograph in an agricultural watershed. *Ethiopia Int Soil Water Cons Res.* 2015;3:282–290. https://doi.org/10.1016/j.iswcr.2015.11.002.

43. Bonilla CA, Johnson O. Soil erodibility mapping and its correlation with soil properties in Central Chile. *Geoderma.* 2012;189:116–123. https://doi.org/10.1016/j.geoderma.2012.05.005.

44. Veihe A. The spatial variability of erodibility and its relation to soil types: a study from northern Ghana. *Geoderma.* 2002;106(1–2):101–120. https://doi.org/10.1016/S0016-7061(01)00120-3.

45. Duiker SW, Flanagan DC, Lal R. Erodibility and infiltration characteristics of five major soils of Southwest Spain. *Catena.* 2001;45(2):103–121. https://doi.org/10.1016/S0341-8162(01)00145-X.

46. Di Stefano C, Ferro V. Linking clay enrichment and sediment delivery processes. *Biosys Eng.* 2002;81(4):465–479. https://doi.org/10.1006/bioe.2001.0034.

47. Ampontuah EO, Robinson JS, Nortcliff S. Assessment of soil particle redistribution on two contrasting cultivated hillslopes. *Geoderma.* 2006;132 (3–4):324–343. https://doi.org/10.1016/j.geoderma.2005.05.014.

48. Romero C, Stroosnijder L, Baigorria G. Interril and rill erodibility in the northern Andean highlands. *Catena.* 2007;70(2):105–113. https://doi.org/10.1016/j.catena.2006.07.005.

49. Dimoyiannis DG, Tsadilas CD, Valmis S. Factors affecting aggregate instability of Greek agricultural soils. *Commun Soil Sci Plan.* 1998;29 (9–10):1239–1251. https://doi.org/10.1080/00103629809370023.

50. Zhang K, Li S, Peng W, Yu B. Erodibility of agricultural soils on the loess plateau of China. *Soil Tillage Res.* 2004;76(2):157–165. https://doi.org/10.1016/j.still.2003.09.007.

51. Rodríguez AR, Arbelo CD, Guerra JA, Mora JL, Notario JS, Armas CM. Organic carbon stocks and soil erodibility in Canary Islands Andosols. *Catena.* 2006;66(3):228–235. https://doi.org/10.1016/j.catena.2006.02.001.

52. Charman PEV, Murphy BW. Soils (their properties and management). In: *Land and water conservation.* 2nd ed. New South Wales: Oxford; 2000:206–212.

53. Hoyos N. Spatial modeling of soil erosion potential in a tropical watershed of the Colombian Andes. *Catena.* 2005;63(1):85–108. https://doi.org/10.1016/j.catena.2005.05.012.

54. Jin K, Cornelis W, Gabriels D, et al. Residue cover and rainfall intensity effects on runoff soil organic carbon losses. *Catena.* 2009;78(1):81–86. https://doi.org/10.1016/j.catena.2009.03.001.

55. Tejada M, Gonzalez JL. The relationships between erodibility and erosion in a soil treated with two organic amendments. *Soil Tillage Res.* 2006;91 (1–2):186–198. https://doi.org/10.1016/j.still.2005.12.003.

56. Mirzaee S, Ghorbani-Dashtaki S, Mohammadi J, Asadi H, Asadzadeh F. Spatial variability of soil organic matter using remote sensing data. *Catena.* 2016;145:118–127. https://doi.org/10.1016/j.catena.2016.05.023.

57. Mirzaee S, Zolfaghari AA, Gorji M, Dyck M, Ghorbani-Dashtaki S. Evaluation of infiltration models with different numbers of fitting parameters in different soil texture classes. *Arch Argon Soil Sci.* 2014;60(5):681–693. https://doi.org/10.1080/03650340.2013.823477.

58. Tashayo B, Honarbakhsh A, Akbari M, Ostovari Y. Digital mapping of Philip model parameters for prediction of water infiltration at the watershed scale in a semi-arid region of Iran. *Geoderma Region.* 2020;22. https://doi.org/10.1016/j.geodrs.2020.e00301, e00301.

59. Evrendliek F, Celik I, Kilic S. Changes in soil organic carbon and other physical soil properties along adjacent Mediterranean forests, grassland and cropland ecosystems. *J Arid Environ.* 2004;59:743–752.

60. Barthès B, Roose E. Aggregate stability as an indicator of soil susceptibility to runoff and erosion; validation at several levels. *Catena.* 2002;47:133–149.

61. Gupta OP. *Water in relation to soils and plants: with special reference to agriculture.* Delhi: Agrobios; 2002. 164 p.

62. Castro C, Logan TJ. Liming effects on the stability and erodibility of some Brazilian Oxisols. *Soil Sci Soc Am J.* 1991;55(5):1407–1413. https://doi.org/10.2136/sssaj1991.03615995005500050034x.

63. Orts JW, Sojka RE, Glenn GM. Biopolymer additives to reduce erosion-induced soil losses during irrigation. *Ind Crops Prod.* 2000;11:19–26.

64. Pérez-Rodríguez R, Marques MJ, Bienes R. Spatial variability of the soil erodibility parameters and their relation with the soil map at subgroup level. *Sci Total Environ.* 2007;378(1–2):166–173. https://doi.org/10.1016/j.scitotenv.2007.01.044.

33

Non-carcinogenic health risk assessment of fluoride in groundwater of the River Yamuna flood plain, Delhi, India

Shakir Ali[a], Shashank Shekhar[a], and Trupti Chandrasekhar[b]

[a]Department of Geology, University of Delhi, Delhi, India [b]Department of Earth Sciences, IIT Bombay, Mumbai, India

1 Introduction

Rivers provide water to humans for sustenance and are a vital component of livelihood. In India, rivers are an essential source of aquifer recharge and water supply. The Yamuna River is one of the largest tributaries of the Ganga and traverses around 1376 km in north India before it meets the Ganga River.[1] Various scientific communities working on the Yamuna water quality reported that the water is highly polluted in the plains,[2–9] and the existing sewage treatment plants (STPs) in Delhi have sub-optimal treatment capacity in comparison to the sewage load.[1] In addition, antibiotics have also been reported in the effluents of STPs.[10–12]

Delhi contributes approximately 79% of the pollution load to the Yamuna, leaving behind Uttar Pradesh (16%) and Haryana (5%) states of India.[2,9] However, the monsoon flood of the river recharges floodplain aquifers on an annual basis and can sustain its exploitation for desired uses.[13–15] Furthermore, the bank filtrate of the river can also provide safe drinking water.[16] The study area is entirely residential, except for a small area under agricultural activity in the northern part of Delhi.

The high concentration of fluoride (F) in groundwater in parts of India and its consumption by humans can have serious short (dental fluorosis) and long (skeletal fluorosis) consequences.[17–26]

Thus, this study was conducted to evaluate the groundwater quality with respect to F^- in the Yamuna flood plains (YFP) from a geochemical perspective. Furthermore, in this study, the health risk assessment of humans due to the consumption of F-contaminated groundwater was evaluated. In general, it was observed that the groundwater is safe with respect to F^- in YFP and has the potential to serve as a source of sustainable water supply in the future, provided that other organic and inorganic pollutants are within safe limits. The hazard quotient (HQ_{ORAL}) in the area shows that children consuming groundwater in Kherakalan and Narela localities are likely to have adverse health effects (Fig. 1). This study is the first integrated attempt to gain insight into F contamination and the major ion chemistry of the groundwater of YFP from the perspective of local sediment geochemistry. It also incorporates health hazard assessment in the context of the prevailing F concentration in the groundwater of the study area.

2 Study area

The aquifer system of Delhi is a part of the highly stressed regional aquifer system of NW India, where the solution to the groundwater crisis is an hour.[27,28] In the case of Delhi, groundwater quality is a major concern in addition to groundwater overexploitation.[25,29,30] The YFP aquifers are the most potential aquifers of Delhi.[13,29] An insight into the geological and hydrogeological aspects of the study area is as follows.

FIG. 1   Geological map showing groundwater and sediments sampling locations on Yamuna flood plains.

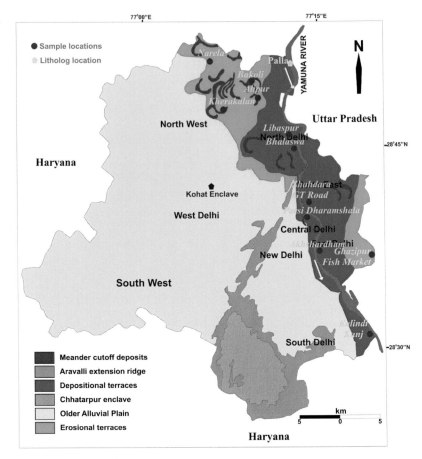

2.1 Geology

The Yamuna flood plain (YFP) was deposited on both sides of the Yamuna River in the region. The YFP is broadly divided into the Older Yamuna flood plain and the active Yamuna flood plain (Fig. 1). Gray color medium sand, silt, and clay define the lithology of the area.[31] The occurrence of coarse sand mixed with kankars (calcareous nodules) at a shallow level with yellowish-brown sand at a deeper level was also observed by author SA in older Yamuna flood plains.

2.2 Hydrogeology

The aquifer along the river is predominantly composed of fine-to-medium unconsolidated sand. The average discharge of tube wells of newer alluviums ranges from 150 to 300 m³/h, with an average transmissivity of 730–2100 m²/d.[29] The aquifer is unconfined in nature and hence receives in situ rainfall/flood recharge by direct infiltration.[32,33]

During peak floods, groundwater velocity in the sands of flood plains can reach up to 2.12 m/d.[14] The depth to water level (DTWL) map suggests a shallow level of groundwater in the study area, with the DTWL varying in the range of 10 m below ground level (mbgl) (Fig. 2). The inhabitants of Delhi have their drinking wells on the plain and drink water without any treatment.

3 Material and methods

3.1 Groundwater and sediment sampling

In this study, a total of eleven groundwater samples were collected from the YFP. The groundwater samples were acidified by nitric acid during the field to maintain a water pH below two for the analysis of major cations. Unacidified

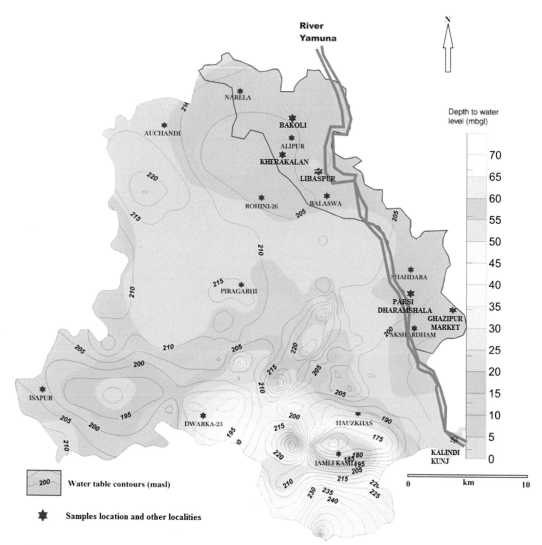

FIG. 2 Water table and depth to water level map of Delhi for non-monsoon (May) season.[34] The *redline* in the figure shows the area under the Yamuna flood plain. Sampling stations and other important locations are shown in the figure.

samples were collected for analysis of major anions. Therefore, samples were collected in duplicates in poly-lab bottles. The bottles were thoroughly rinsed three times with water for analysis. The sediment along the river was collected at a depth of 0.3 m from the river bed in a transparent zip-lock bag from the Palla locality (Fig. 1) and opened only during the bulk sediment analyses by XRD.

3.2 Lab analysis

Eleven groundwater samples were collected along the Yamuna stretch during the pre-monsoon (dry) season in 2016 and analyzed for major ions such as Na^+, K^+, Ca^{2+}, Mg^{2+}, Cl^-, SO_4^{2-}, HCO_3^-, and CO_3^{2-} at IIT Bombay, India. F^- Concentration was measured at the University of Delhi using a fluoride meter. Electrical conductivity (EC) and pH were measured during the field using an EC and a pH meter (Hanna). The geographical location was mapped using GPS and an Android phone app (Samsung 7562). For confirmation, the EC was again measured in the lab at IIT Bombay. The major cations, such as Na^+, K^+, Ca^{2+}, and Mg^{2+}, were analyzed using ICP-AES (Perkin-Elmer, France) at IIT Bombay. The SO_4^{2-} concentration was measured using a spectrophotometer (Shimadzu UV-visible spectrophotometer 160), alkalinity by titration, and chloride (Cl^-) using an Expandable Ion Analyzer 940A with a combination of the Orion ion plus 9817 BN available at IIT Bombay. For XRD, bulk sediments of the Yamuna flood plain were mounted on a sample holder using the back-loading technique and scanned from 5° to 70° (2θ) with a step size of 0.01° and a scan speed of

38 s/step, using Cu-Kα radiation from an Empyrean X-ray diffractometer (PANalytical) equipped with a Pixel 3D detector.

3.3 Human health risk assessment

The non-carcinogenic human health risk assessment due to the consumption of F-enriched groundwater expressed as hazard quotient (HQ$_{\text{ORAL}}$), was calculated using Eq. (1).

$$HQ_{\text{ORAL}} = EDI_{\text{ORAL}}/RfD_{\text{ORAL}} \tag{1}$$

HQ$_{\text{ORAL}}$ was calculated by estimating the daily intake due to the ingestion of F enriched groundwater (EDI$_{\text{ORAL}}$) and RfD$_{\text{ORAL}}$ (oral reference dose). EDI$_{\text{ORAL}}$ (Eq. 2) was estimated using the concentration of F(c), ingestion rate (IR), exposure frequency (EF), exposure duration (ED), body weight (BW), and average time (AT) for children and adults.

$$EDI_{\text{ORAL}} = (C \times IR \times EF \times ED)/(BW \times AT) \tag{2}$$

The RfD$_{\text{ORAL}}$ values of the Delhi region were obtained from Ali et al.[19] An estimated HQ$_{\text{ORAL}}$ value greater than 1 was considered unsafe.[19,35]

The method used in this study is summarized as a flow chart (Fig. 3).

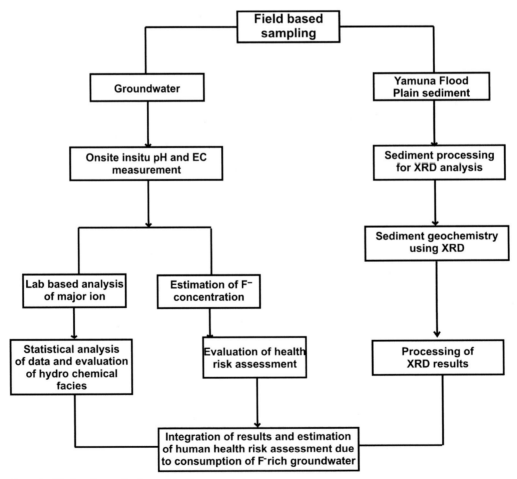

FIG. 3   A flow chart elucidating the methods used in the present study.

4 Results and discussion

4.1 Major ions chemistry and hydro-chemical facies

The Na^+ and Cl^- ions in the groundwater of the study area were comparatively higher than those of the other ions and varied from 97 to 667 mg/L (average: 382 mg/L) and 150 to 1350 mg/L (average: 750 mg/L; Fig. 4, Table 1).

Except for groundwater from the Bakoli locality (Fig.1), the concentration of SO_4^{2-} ions are comparatively lower than the groundwater of the Older Alluvium Plains (OAP) and ranges from 3.9 to 904 mg/L.[26,36] Similarly, Ca^{2+} (35–202 mg/L) and Mg^{2+} (23.7–137 mg/L) ion and electrical conductivity values of the groundwater in the Yamuna flood plains are comparatively lower than the OAP.[36] However, the HCO_3^- ion concentration was comparatively higher than that of the OAP groundwater.[26] Very high K^+ (153.2 mg/L) and HCO_3^- concentrations were observed in the Bhalaswa sampling locality near the landfill site (Fig. 1; Table 1). Thus, the major ions in the landfill locality clearly show their effect on the quality of groundwater.

The hydro-chemical facies of the Yamuna floodplain groundwater suggests the dominance of the Na-Cl facies (Fig. 5). In addition, groundwater also showed the Na-Ca-Cl-HCO3 water type. The variation of facies in the groundwater in the area is obvious owing to different water-sediment interactions.

4.2 Statistical analysis

Statistical correlation is a powerful tool and is best known for evaluating the relationship between various parameters (Table 2). The correlation suggests a positive correlation between F and pH and a negative correlation with Ca^{2+} and Mg^{2+}. This indicates that the high pH of the groundwater is responsible for triggering F from sediment to the aqueous phase in the study area. A similar relationship was also observed in earlier studies.[17]

4.3 Bivariate plot

After statistical correlation, bivariate plots were created for a deeper understanding of the interrelationship between the major ion and F (Fig. 6). The bivariate plot of F with pH showed a positive correlation, while F with HCO_3^- and Ca^{2+} showed a negative correlation.

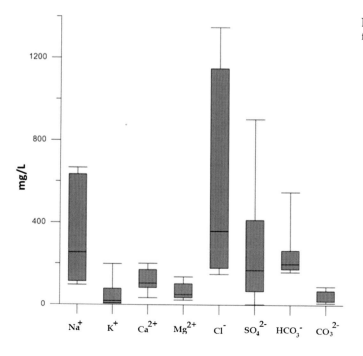

FIG. 4   Box and whisker plot of major ion of groundwater of Yamuna flood plains.

TABLE 1 Major ions chemistry and F⁻ level in groundwater along with the aquifers of River Yamuna stretch, Delhi.[36]

S. No	Locality	Latitude	Longitude	Source	Depth[a]	EC	pH	Na⁺	K⁺	Ca²⁺	Mg²⁺	Cl⁻	SO₄²⁻	HCO₃⁻	CO₃²⁻	F⁻	Water type
					m	μS/cm							mg/L				
1	Akshardham	28.616	77.273	SS	75	968	7.97	114	6.7	105.8	36	180	54	200	20	0.24	Na-Ca-Mg-Cl-HCO₃
2	Alipur	28.797	77.134	HP	7.5	2233	7.79	236	199	131	61.6	360	144	255	50	0.65	Na-Ca-Cl-HCO₃
3	Bakoli	28.815	77.146	HP	NA	4880	7.71	654	11.3	202	135.9	1080	904	175	10	0.18	Na-Ca-Mg-Cl-SO₄
4	Bhalaswa	28.741	77.166	HP	4.5	4440	7.94	634	153.2	171.8	137	1150	109	550	90	0.37	Na-Mg-Cl-HCO₃
5	Ghazipur Fish Market	28.628	77.327	SS	90	2422	7.98	343	7.6	93	63	1350	176	160	20	0.44	Na-Mg-Cl
6	Kalindi Kunj	28.549	77.303	SS	24	917	7.88	102	10.2	84	36	150	3.9	250	10	0.53	Na-Ca-Mg-Cl-HCO₃
7	Kherakalan	28.772	77.115	HP	15	1720	8.15	254	74	67	44.7	220	170	310	80	1.2	Na-Ca-Cl-HCO₃-SO₄
8	Libaspur	28.75	77.143	TW	24	4660	7.69	559	20	180	103.7	1215	447	185	20	0.6	Na-Mg-Cl
9	Narela	28.842	77.088	SS	42	3100	8.25	667	5.5	35	23.7	430	415	180	20	1.57	Na-Cl-SO₄
10	Parsi Dharamshala	28.638	77.24	SS	60	1647	8.18	119	36	99	51	230	172	165	30	0.7	Na-Ca-Mg-Cl-SO₄
11	Shahdara GT Road	28.673	77.283	HP	15	1085	7.96	97	79	114	43	150	68	265	70	0.26	Na-Ca-Mg-Cl-HCO₃

[a] Depth information was noted during the field by nearby residents, which may not be accurate. NA, not available; SS, submersible; HP, hand pump; TW, tube-wells.

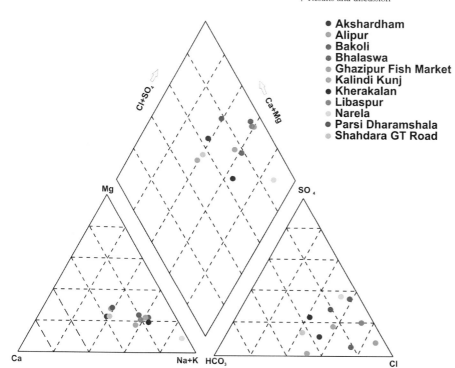

FIG. 5 Hydro-chemical facies of groundwater of Yamuna flood plains for the non-monsoon season (2016).

TABLE 2 Statistical correlation of major ion and other parameters of non-monsoon season groundwater.

	EC	pH	Na$^+$	K$^+$	Ca^{2+}	Mg^{2+}	Cl$^-$	SO$_4{}^{2-}$	HCO$_3{}^-$	CO$_3{}^{2-}$	F
EC	1										
pH	−0.46	1									
Na$^+$	**0.92**	−0.19	1								
K$^+$	0.10	−0.16	0.01	1							
Ca^{2+}	**0.72**	**−0.82**	0.41	0.24	1						
Mg^{2+}	**0.85**	**−0.63**	**0.63**	0.27	**0.92**	1					
Cl$^-$	**0.80**	−0.47	**0.69**	−0.03	**0.62**	**0.76**	1				
SO$_4{}^{2-}$	**0.73**	−0.35	**0.70**	−0.29	0.49	**0.53**	0.46	1			
HCO$_3{}^-$	0.21	0.00	0.21	**0.66**	0.23	0.42	0.12	−0.34	1		
CO$_3{}^{2-}$	0.09	0.22	0.08	**0.75**	0.04	0.23	−0.01	−0.35	**0.84**	1	
F	−0.15	**0.69**	0.13	−0.02	**−0.72**	**−0.55**	−0.38	−0.05	−0.09	0.18	1

4.4 Control of bulk sediments chemistry on major ions

The XRD patterns of the bulk sediments of the YFP sediments reveal the dominance of quartz and biotite minerals with albite and fluorite as an accessory (Fig. 7). The high amounts of quartz and biotite in the sediments were due to the presence of sandy soils.

From sediment chemistry, it was observed that the Na$^+$ ions in the groundwater may be due to the chemical weathering of albite (NaAlSi$_3$O$_8$), whereas biotite (KMg$_3$AlSi$_3$O$_{10}$(OH)$_2$ weathering may contribute to Mg^{2+} and K$^+$ ions in the water.[26] The irrigation practices in the area are confined only to the Palla locality (Fig. 1), while the entire study area is residential. Except for Bakoli, it is assumed that the anthropogenic pollution of SO$_4{}^{2-}$ ions are not sufficient to push the SO$_4{}^{2-}$ concentration beyond the safe limits (Table 1).

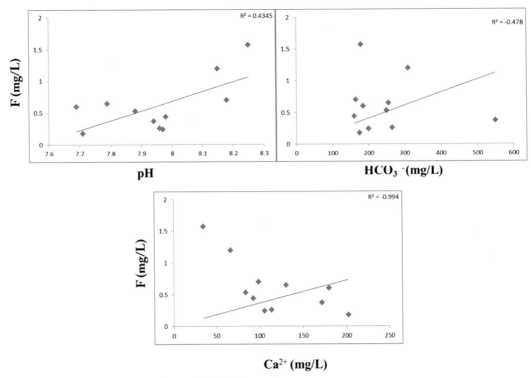

FIG. 6 Bivariate plots between F with pH, HCO_3^- and Ca^{2+} with zero intercept.

4.5 Fluoride level and human health risk assessment

The F level in the Narela drinking well was observed to be 1.57 mg/L, while in the rest of the samples, the F concentration was below 1.5 mg/L (permissible limit). This shows that in the majority of drinking wells, the F level is within the safe limit. This is further substantiated by the low to moderate EC values of groundwater, which suggest shorter water-sediment interactions. The significant positive correlation between pH and F indicates that a high pH triggers F in groundwater.

In this study, F exposure was observed in children and adults in the two age groups. The HQ_{ORAL} of children and adults are given in Table 3 and shown as box and whisker diagrams in Fig. 8.

The estimated HQ_{ORAL} shows that children drinking groundwater in Kherakalan and Narela (Fig. 1) are likely to have adverse health effects (HQ_{ORAL} more than 1[35]), while children and adults in other localities are safe (Table 3 and Fig. 8). The HQ_{ORAL} values for children are higher than those for adults because of their lower weight. This has also been observed in many other studies conducted on the non-carcinogenic health risk assessment of F.[37–41] This indicates that the problem is more serious for the health of children than for adults.

Fig. 9 explains the possible causes for the low level of F in YFP.

The Yamuna River, like all other major rivers, flows through a valley, which under normal circumstances is a natural sink for surface and groundwater flow. Flood plain sediments are dominated by sand with higher values of hydraulic conductivity. This facilitates regular flushing of the groundwater system and annual recharge from the monsoon floods, thereby resulting in low F and EC concentrations in the groundwater of the study area. Such a hydrogeological setting makes the YFP a safe option for water supply during stress periods.

5 Conclusions

The F^- level in the groundwater from the study area ranges from 0.18 to 1.57 mg/L. Statistical correlation and bivariate plots revealed that the alkaline water in the study area was responsible for triggering F^- release in the groundwater. However, because of continuous flushing and recharge through the permeable formations, the sediment-groundwater interaction is for a limited duration, leading to a lower F^- concentration in the groundwater. The bulk sediments of the Yamuna flood plain suggest the dominance of quartz and biotite minerals with albite and fluorite as accessory minerals. The existing sediment chemistry is significant for controlling the hydro-chemical facies in the area.

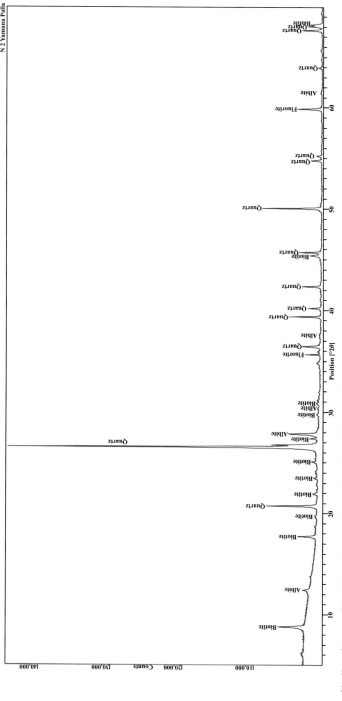

FIG. 7 XRD of bulk sediment of the River Yamuna flood plain from 0.3 m depth.

TABLE 3 HQ$_{ORAL}$ for children and adults in the study area for all sampling locations.

S. No.	Locality	HQ$_{CHILDREN}$	HQ$_{ADULTS}$
1	Akshardham	0.24	0.13
2	Alipur	0.64	0.34
3	Bakoli	0.18	0.09
4	Bhalaswa	0.37	0.19
5	Ghazipur Fish Market	0.44	0.23
6	Kalindi Kunj	0.52	0.28
7	Kherakalan	**1.19**	0.63
8	Libaspur	0.59	0.31
9	Narela	**1.55**	0.82
10	Parsi Dharamshala	0.69	0.37
11	Shahdara GT Road	0.26	0.14

FIG. 8 Hazard quotient (HQ$_{ORAL}$) via drinking for children and adults.

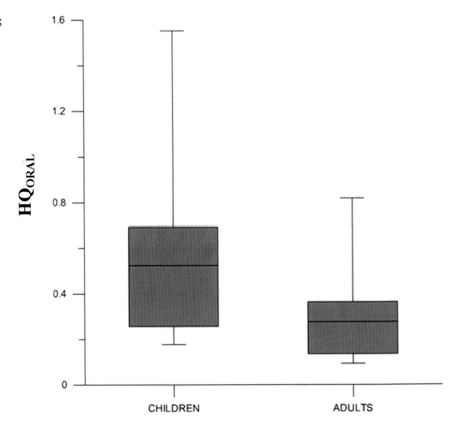

The HQ$_{ORAL}$ values suggest that the children in the Kherakalan and Narela localities are at adverse health risks due to the intake of groundwater (HQ$_{ORAL}$ > 1). In other localities, no adverse health risks were observed (HQ$_{ORAL}$ < 1). In the context of F-related health hazards, this study suggests that except for two localities, the use of groundwater is safe for human consumption. Thus, it is advised to undertake scientific planning for the exploitation of the flood plain groundwater resources and formulate strategies for the adoption of a bank filtration approach to augment the drinking water needs of Delhi.

FIG. 9 A schematic figure explaining the possible causes for a low level of $\bar{F}$ in the YFP.

Rain

Yamuna flood
plain

River
Yamuna

Regional groundwater
flow direction

Local groundwater
flow direction

Low level
of fluoride

High flushing
of groundwater

Acknowledgments

It is duly acknowledged that the present work is a part of the Ph.D. of SA. This research was based on the datasets of Ph.D. and M.Phil. of SA. This work was supported by the R&D project (2015-16) of the University of Delhi granted to SS and is duly acknowledged. SA acknowledges the non-NET fellowship provided by the University of Delhi. It is to acknowledge that an abstract entitled "Major ions chemistry and non-carcinogenic health risk assessment of fluoride in groundwater of the alluvial plains of River Yamuna, Delhi, India" using partially the dataset of the article was accepted for the 2nd International Conference on Water Resources in Arid Areas, Muscat, Oman" but could not be presented due to disruptions caused by the COVID-19 pandemic.

References

1. Upadhyay A, Rai RK. Brief overview of the Yamuna river basin and issues. In: *Water management and public participation. Springer briefs in earth sciences.* Dordrecht: Springer; 2013.
2. CPCB. *Water quality status of Yamuna River*; 2006. Available at http://www.yamunariverproject.org/assets/cpcb_2006-water-quality-status.pdf. Accessed 9 October 2019.
3. Katyal D, Qader A, Ismail AH, Sarma K. Water quality assessment of Yamuna River in Delhi region using index mapping. *Interdisc Environ Rev.* 2012;13(2–3):170–186.
4. Misra AK. A river about to die: Yamuna. *J Water Resour Protect.* 2010;2(5):489.
5. Sharma D, Kansal A. Water quality analysis of River Yamuna using water quality index in the national capital territory, India (2000–2009). *Appl Water Sci.* 2011;1(3–4):147–157.
6. Singh RB, Chandna V. Spatial analysis of Yamuna River water quality in pre-and post-monsoon periods. *IAHS-AISH Publ.* 2011;348:8–13.
7. Srivastava SK, Ramanathan AL. Geochemical assessment of groundwater quality in vicinity of Bhalswa landfill, Delhi, India, using graphical and multivariate statistical methods. *Environ Geol.* 2008;53(7):1509–1528.
8. Zafar M, Alappat BJ. Landfill surface runoff and its effect on water quality on river Yamuna. *J Environ Sci Health A.* 2004;39(2):375–384.
9. Upadhyay R, Dasgupta N, Hasan A, Upadhyay SK. Managing water quality of River Yamuna in NCR Delhi. *Phys Chem Earth Parts A/B/C.* 2011;36(9–11):372–378.
10. Mutiyar PK, Mittal AK. Occurrences and fate of selected human antibiotics in influents and effluents of sewage treatment plant and effluent-receiving river Yamuna in Delhi (India). *Environ Monit Assess.* 2014;186(1):541–557.
11. Sarafraz M, Ali S, Sadani M, et al. A global systematic, review-meta analysis and ecological risk assessment of ciprofloxacin in river water. *Int J Environ Anal Chem.* 2020;1–16.
12. Velpandian T, Halder N, Nath M, et al. Un-segregated waste disposal: an alarming threat of antimicrobials in surface and ground water sources in Delhi. *Environ Sci Pollut Res.* 2018;25(29):29518–29528.
13. Shekhar S, Prasad RK. The groundwater in the Yamuna flood plain of Delhi (India) and the management options. *Hydrogeol J.* 2009;17(7):1557.
14. Soni V, Shekhar S, Singh D. Environmental flow for the Yamuna river in Delhi as an example of monsoon rivers in India. *Curr Sci.* 2014;106:558–564.

15. Soni V, Shekhar S, Rao SV, Kumar S, Singh D. A new solution for city water: quality drinking water from the river floodplains. *Curr Sci.* 2018;114 (3):452–461.

16. Lorenzen G, Sprenger C, Taute T, Pekdeger A, Mittal A, Massmann G. Assessment of the potential for bank filtration in a water-stressed megacity (Delhi, India). *Environ Earth Sci.* 2010;61(7):1419–1434.

17. Ali S, Thakur SK, Sarkar A, Shekhar S. Worldwide contamination of water by fluoride. *Environ Chem Lett.* 2016;14:291–315. https://doi.org/10.1007/s10311-016-0563-5.

18. Ali S, Shekhar S, Bhattacharya P, Verma V, Chandresekhar T, Chandrashekhar AK. Elevated fluoride in groundwater of Siwani Block, Western Haryana, India: a potential concern for sustainable water supplies for drinking and irrigation. *Groundw Sustain Dev.* 2018;7:410–420. https://doi.org/10.1016/j.gsd.2018.05.008.

19. Ali S, Fakhri Y, Golbini M, et al. Concentration of fluoride in groundwater of India: a systematic review, meta-analysis and risk assessment. *Groundwater Sustain Dev.* 2019;9:100224. https://doi.org/10.1016/j.gsd.2019.100224.

20. Kashyap CA, Ghosh A, Singh S, et al. Distribution, genesis and geochemical modeling of fluoride in the water of tribal area of Bijapur district, Chhattisgarh, central India. *Groundwater Sustain Dev.* 2020;11:100403. https://doi.org/10.1016/j.gsd.2020.100403.

21. Kumar S, Sarkar A, Ali S, Shekhar S. Groundwater system of National Capital Region Delhi, India. In: *Groundwater of South Asia*. Singapore: Springer; 2018:131–152.

22. Saha D, Shekhar S, Ali S, Elango L, Vittala S. Recent scientific perspectives on the Indian hydrogeology. *Proc Indian Natn Sci Acad.* 2020;86 (1):459–478. https://doi.org/10.16943/ptinsa/2020/49790.

23. Saha D, Shekhar S, Ali S, Vittala SS, Raju NJ. Recent hydrogeological research in India. *Proc Indian Nat Sci Acad.* 2016;82(3):787–803. https://doi.org/10.16943/ptinsa/2016/48485.

24. Podgorski JE, Labhasetwar P, Saha D, Berg M. Prediction modeling and mapping of groundwater fluoride contamination throughout India. *Environ Sci Technol.* 2018;52(17):9889–9898.

25. Sarkar A, Ali S, Kumar S, Shekhar S, Rao SVN. Groundwater environment in Delhi, India. In: *Groundwater environment in Asian cities: concepts, methods and case studies*. 1st ed. Elsevier Inc; 2016:77–108. https://doi.org/10.1016/B978-0-12-803166-7.00005-2.

26. Ali S, Shekhar S, Chandrasekhar T, et al. Influence of the water–sediment interaction on the major ions chemistry and fluoride pollution in groundwater of the Older Alluvial Plains of Delhi, India. *J Earth Syst Sci.* 2021;130:98. https://doi.org/10.1007/s12040-021-01585-3.

27. Shekhar S, Kumar S, Densmore AL, et al. Modelling water levels of northwestern India in response to improved irrigation use efficiency. *Sci Rep.* 2020;10(1):1–15.

28. van Dijk WM, Densmore AL, Jackson CR, et al. Spatial variation of groundwater response to multiple drivers in a depleting alluvial aquifer system, northwestern India. *Prog Phys Geogr Earth Environ.* 2020;44(1):94–119.

29. Chatterjee R, Gupta BK, Mohiddin SK, Singh PN, Shekhar S, Purohit R. Dynamic groundwater resources of National Capital Territory, Delhi: assessment, development and management options. *Environ Earth Sci.* 2009;59:669–686.

30. Shekhar S. An approximate projection of availability of the fresh groundwater resources in the south west district of NCT Delhi, India: a case study. *Hydrgeol J.* 2006;14(7):1330–1338.

31. Thussu JL. *Geology of Haryana and Delhi*. Bangalore: Geological Society of India; 2006.

32. Kumar S, Sarkar A, Thakur SK, Shekhar S. Hydrogeological characterization of aquifer in palla flood plain of Delhi using integrated approach. *J Geol Soc India.* 2017;90(4):459–466.

33. Shekhar S. An approach to interpretation of step drawdown tests. *Hydrgeol J.* 2006;14(6):1018–1027.

34. CGWB. *Data set taken from CGWB*; 2014. Available at www.cgwb.gov.in.

35. Yousefi M, Ghoochani M, Mahvi AH. Health risk assessment to fluoride in drinking water of rural residents living in the Poldasht city, Northwest of Iran. *Ecotoxicol Environ Saf.* 2018;148:426–430.

36. Ali S. *Fluoride contamination and major ions chemistry of groundwater in NCT Delhi, India*. Unpublished M. Phil. thesis, University of Delhi; 2017.

37. Adimalla N, Li P, Qian H. Evaluation of groundwater contamination for fluoride and nitrate in semi-arid region of Nirmal Province, South India. a special emphasis on human health risk assessment (HHRA). *Human and Ecol Risk Assess.* 2018;25:1107–1124.

38. Fallahzadeh RA, Miri M, Taghavi M, et al. Spatial variation and probabilistic risk assessment of exposure to fluoride in drinking water. *Food Chem Toxicol.* 2018;113:314–321.

39. Karunanidhi D, Aravinthasamy P, Roy PD, et al. Evaluation of non-carcinogenic risks due to fluoride and nitrate contaminations in a groundwater of an urban part (Coimbatore region) of South India. *Environ Monit Assess.* 2020;192(2):1–16.

40. Qasemi M, Afsharnia M, Zarei A, Farhang M, Allahdadi M. Non-carcinogenic risk assessment to human health due to intake of fluoride in the groundwater in rural areas of Gonabad and Bajestan, Iran: a case study. *Human Ecol Risk Assess.* 2018;25:1–12.

41. Yadav KK, Kumar V, Gupta N, Kumar S, Rezania S, Singh N. Human health risk assessment: study of a population exposed to fluoride through groundwater of Agra city, India. *Regul Toxicol Pharmacol.* 2019;106:68–80.

CHAPTER

34

Digital soil mapping of organic carbon at two depths in loess hilly region of Northern Iran

Sedigheh Maleki[a,b], Farhad Khormali[a], Songchao Chen[c],
Hamid Reza Pourghasemi[d], and Mohsen Hosseinalizadeh[e]

[a]Department of Soil Science, Gorgan University of Agricultural Sciences and Natural Resources, Gorgan, Iran [b]Department of Soil Science, Faculty of Agriculture, Ferdowsi University of Mashhad, Mashhad, Iran [c]INRAE, Unité InfoSol, Orléans, France [d]Department of Natural Resources and Environmental Engineering, College of Agriculture, Shiraz University, Shiraz, Iran [e]Department of Arid Zone Management, Gorgan University of Agricultural Sciences and Natural Resources, Gorgan, Iran

1 Introduction

Soil is a vital key and manageable factor in agricultural production,[1,2] controlling the geochemical cycles of water and organisms,[3] and biodiversity security.[4] Among different soil properties, soil organic carbon (SOC) has an important role in soil health, soil quality and productivity,[5] growth vegetation,[6,7] prevention of soil erosion, biological characteristics, and soil quality especially in agricultural lands.[8–10] Indeed, soil fertility is recognized as a function of SOC and subsequent microbial activity.[1] Therefore, monitoring the spatial variability of SOC, especially in arid and semiarid regions, is critically important for providing information on soil and water conservation, carbon storage, soil fertility, and climate change.[11]

Considering these issues, the demand to produce SOC maps has been recently increased at fine and national scales.[5,7,9,11–19] Meanwhile, because of the high spatial variability of SOC at different scales (large and fine scales), very dense sampling is required to achieve precise estimates.[10,20] Therefore, many studies have been conducted on digital soil mapping (DSM) approaches.[6,10,13,18,20–27]

Regarding "scorpan" model,[21] to predict the spatial distribution of soil properties in every region, it is important to determinate the relationship between environmental covariates and soil properties[13,19,27,28]; then, a significant number of linear and nonlinear techniques were applied to predict soil properties. Numerous studies used different algorithms and models to predict the variety of soil characteristics at a specified region through DSM framework such as Cubist (Cu),[12,18,23] K-nearest neighbors (KNN),[23,29] regression tree (RT),[18,30] multiple linear regression (MLR),[18,31,32] random forest (RF),[18,25,26,32] and support vector machine/regression (SVM/SVR).[17,20,25,33] Several recent studies have applied different prediction algorithms for different soil property mapping and reported RF as a good performance algorithm.[18,28,32,34] As a result, it is feasible to apply topographic attributes, remote sensing, soil legacy data, and DSM methods[28] to describe the spatial variability of soil properties in different regions.

The Iranian Loess Plateau (ILP) is a unique hilly landscape in the semiarid region, where nearly 80% of the land is devoted to hills and valleys.[35] Owing to the physical and chemical properties of loess materials, these eolian sediments are fertile and highly prone to piping and gully erosion.[36] In addition, this region is known as an important rangeland/cropland in Golestan Province and has been affected by intense human activities/disturbances to produce wheat, especially in valley landscapes. Therefore, knowledge of SOC spatial variation is important for soil health in this region and is linked to ecological issues. However, most studies on DSM methods to predict and map SOC was commonly restricted to the upper 30 cm of soils,[34] due to the state of erosion and thinning of the soil surface layer in the ILP, investigation of SOC in the subsurface soil layer is needed. Hence, the main aims of this study were (i) to predict the spatial variability of SOC in two soil layers (0–20 and 20–40 cm), (ii) to explore the controlling environmental covariates affecting the spatial distribution of SOC in two soil layers, and (ii) to evaluate the potential of RF in digital mapping of SOC over two soil depths.

2 Materials and methods

2.1 Description of the study area

The study area is a part of the ILP, situated in the Golestan Province (Fig. 1), with a land area of 5390 ha (55°13 26″–55°09 36″ E and 37°36 37″–37°41 41″ N) (Fig. 1B), with a mean annual rainfall of 350 mm and mean annual temperature of 17°C.[35] The Dry Xeric soil moisture regime and Thermic soil temperature regime are dominant in the region[37] and cultivated lands and rangeland are the major land uses. While the parent material varies from north to south, its major component is marl, marl dominated by shale and loess, limestone, clay deposits, reddish-brown lower Pleistocene loess (LPL), loess deposit, and reworked loess.[35]

2.2 Environmental covariates

Topography is the most important parameter in the "*scorpan*" model in arid and semiarid areas.[38,39] The digital elevation model (DEM) is therefore believed to be useful for deriving the main topographic attributes to predict soil properties,[17,18] mainly in regions with high topographic variations. To obtain topographic attributes, a high-resolution "ALOS PALSAR DEM" was downloaded from "https://asf.alaska.edu/." The DEM's spatial resolution was 12.5 m and was resampled to a 10 m spatial resolution. The terrain attributes extracted from the DEM listed in Table 1 were prepared using the SAGA GIS 2.2 software. Also, "Sentinel 2 imagery" data (top of the atmosphere reflectance) with 10 m spatial resolution, downloaded from https://scihub.copernicus.eu/, and were used for extraction remote sensing indices (Table 1) using ENVI 4.4 and ArcGIS 10.2 software. It should be noted that several soil property maps (i.e., electrical conductivity (EC), bulk density (BD), sand, silt, and clay percentages) were included in the SOC modeling. The mentioned soil property maps were prepared using geostatistical methods[40] in ArcGIS 10.2 software.

2.3 Sampling design and soil analysis

A total of 252 soil samples were selected according to conditioned Latin hypercube sampling (cLHS[48]) based on a MATLAB script[49] from two soil layers in 126 soil sampling sites (126 sites × 2 layers = 252 samples), as illustrated in Fig. 1B. It should be noted that topographic attributes and remote sensing indices mentioned in Table 1 were used to

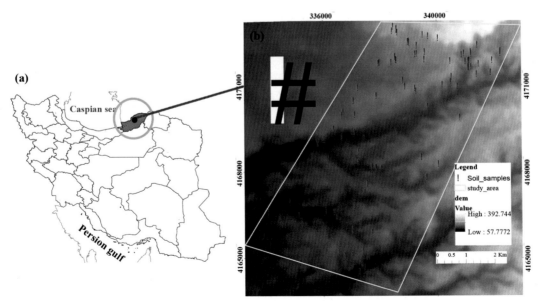

FIG. 1   Location of the study area. (A) The black dot shows the study area in Golestan Province, Iran; and (B) the digital elevation model (ALOS PALSAR DEM downloaded through www.asf.alaska.edu) and the location of sampling points in the study area.

TABLE 1   Environmental covariates utilized as predictors for SOC mapping in the study area.

Covariates	Covariate name (abbreviation)	Reference
Topographic attributes	Elevation (ELV)	[41,42]
	Slope (Slope)	
	Aspect (ASP)	
	Profile curvature (Curv)	
	Plan curvature (Plancurv)	
	Convergence (Converg)	
	Topographic roughness index (TRI)	
	Vertical distance to channel network (VDTCN)	
	Stream power index (LS)	
	Topographic wetness index (TWI)	
	Valley depth (valley)	
	Multi-resolution valley bottom flatness index (Mrv)	
	Multi-resolution of ridge top flatness index (Mrr)	
	Topographic position index (TPI)	[43]
Remote sensing attributes	Soil adjusted vegetation index (SAVI)	[44]
	Normalized difference vegetation index (NDVI)	[45]
	Ratio Vegetation Index (RVI)	[23]
	Brightness Index (BI)	[46]
	Carbonate index (CaI)	[47]
	Coloration Index (CI)	[46]
	Hue Index (HI)	[46]
	Grain size index (GSI)	[47]
	Saturation Index (SI)	[46]
Soil data	Clay	–
	Sand	–
	Silt	–
	Electrical conductivity (EC)	–
	Bulk density (BD)	–

run the cLHS. After air-drying and sieving (<2 mm) soil samples, the Walkley and Black method[50] was used to measure SOC.

2.4 Random forest (RF) model and assessment of model performance

RF is a nonparametric model and an ensemble of randomized decision trees[51] that aggregates them to improve prediction accuracy.[26,52] Approximately 2/3 of the training data were used for bootstrap sampling, which formed each decision tree.[53] Additionally, the "Out-of-Bag (OOB)" error (Eq. 1) is estimated randomly by changing the values of the covariates, which is considered as variable importance. The variable importance changes between 0 and 100% while it is below 15%, the environmental covariate is recognized as unimportant and omitted from the model.[26] Determining the importance of variables is known as one of the most important tasks in the prediction and modeling of soil properties,[54] and a larger variable importance value indicates a higher contribution to the model for a covariate.[55]

$$\text{OOB error} = \frac{1}{N}\sum_{i=1}^{N} I[Y_{OOB}(X_i) \neq Y_i] \tag{1}$$

where $I[Y_{OOB}(X_i) \neq Y_i]$ is known as an indicator function equal to 0 when the predicted and actual classes are the same and equal to 1 otherwise.

The *mtry* (the number of environmental covariates used in each random subset) and *ntree* (the number of trees used in the forest) are two user-defined parameters[11,52] that are important in RF. "Both parameters were optimized by iterating over *mtry* values ranging from 1 to the total number of covariates, while *ntree* values ranged from 100 to 10,000 in increments of 100[56]".

To assess the prediction capability of the RF model, 10-fold cross-validation[23,52] was utilized. Reliable performance and unbiased performance on smaller datasets are the advantages of this procedure. Two indices, "root mean square error (RMSE)" and "coefficient of determination (R^2)," were selected to assess the accuracy of the model's performance[28] as follows:

$$RMSE = \sqrt{\frac{1}{n}\sum_{i=1}^{n}(Z_i - Z^*_i)^2} \tag{2}$$

$$R^2 = \left[\frac{\sum_{i=1}^{n}(Z_i - Z_{avg})(Z_i^* - Z^*_{avg})}{\sum_{i=1}^{n}(Z_i - Z_{avg})^2(Z_i^* - Z^*_{avg})^2}\right]^2 \tag{3}$$

where Z_i and Z_i^* are the observed and predicted values, Z_{avg} and Z_{avg}^* are the mean of the observed and predicted values, respectively, and n is the number of observations.

SOC modeling and assessment of covariate importance were implemented using the "randomForest" package[57] and "caret" package[58] in R.[59]

3 Results and discussion

3.1 Descriptive statistics of SOC

Summary statistics of SOC at the two soil depths are presented in Table 2. SOC ranged between 0.08 and 2.40% and 0.01%–1.60% at 0–20 cm and 20–40 cm, respectively. The mean value of SOC value decreased with increasing depth. According to Carter and Gregorich,[60] the coefficient of variation (CV) can be classified as <15%, 15%–35%, 35%–75%, and 75%–150% as low, medium, high, and very high, respectively. Therefore, SOC showed high variability at both depths, with a CV value of 67% (Table 2). The high variability of SOC indicated their higher sensitivity to changes in environmental factors such as topography, vegetation cover, and the effect of landscape position on erosion and sedimentation.[17] Similarly, Taghizadeh-Mehrjardi et al.[23] and Zeraatpisheh et al.[18] reported high CV values for SOC in Iran.

3.2 Model performance and evaluation

The Assessment of the RF model using 10-fold cross-validation is presented in Table 3. In modeling of SOC, RMSE and R^2 values differed between 0.36 and 0.38 for first depth and 0.30 and 0.16 for second depth. This finding supports earlier research by Tajik et al.[17] who indicated decreasing R^2 values (accuracy) with increasing soil depth for predicted SOC at two soil depths. Similarly, Taghizadeh-Mehrjardi et al.[23] reported high accuracy of SOC prediction for the topsoil (0–30-cm) in different models, while the lowest accuracy of prediction was for 60–100-cm. An R^2 value <0.5, which

TABLE 2 Descriptive statistics of soil organic carbon (SOC) in different depths.

Depth (cm)	Mean (%)	Minimum (%)	Maximum (%)	CV (%)	Skewness	SD
0–20	0.68	0.08	2.40	67.0	1.87	0.46
20–40	0.49	0.01	1.60	67.0	1.15	0.33

SD, standard deviation; *CV*, coefficient of variation.

TABLE 3 Summary of RF model for prediction of SOC in calibration and validation data in the study area.

Soil property	Depth (cm)	mtry	Calibration		Validation	
			RMSE	R^2	RMSE	R^2
SOC	0–20	29	0.15	0.93	0.36	0.38
	20–40	15	0.13	0.94	0.3	0.16

may be related to the high variability of SOC (67% of CV value) in the study area and complex topography. Taghizadeh-Mehrjardi et al.[30] stated the R^2 of 0.5 or less and over 0.7 as the more common and the less common ones. Overall, these findings are in line with many studies that suggested the RF model had a good performance in SOC prediction with the lowest RMSE.[7,11,13,17,18,34,61,62]

3.3 Environmental covariates importance

The relative importance of the different covariates and their significance produced by RF is shown in Fig. 2. For both soil depths, slope aspect, silt, clay, elevation, and remote sensing indices, including carbonate (CI), grain size (GSI), coloration (CI), and saturation (SI) were the most important covariates affecting SOC prediction. It should be noted that aspect and EC had the highest influence on the model prediction of SOC at 0–20 and 20–40 cm, respectively. This result is not surprising and strongly suggests that SOC accumulation in ILP is also mainly restricted directly or indirectly by slope aspect in the soil surface layer, which is highly consistent with the expert knowledge in the field of soil sciences in this region, as illustrated in previous studies.[35,63–65]

The elevation is another topographic attribute that controls the spatial variation of SOC, which has been widely reported in previous studies.[13,15,16,18,35,62,66] The higher effect of elevation on SOC content could be related mainly to its effect on vegetation, slope, water, and nutrient transport along the landscape, which explains most of the SOC variation in sediment transfer and erosion.

The importance of silt, clay and GSI at both soil depths (Fig. 2) indicates the importance of soil texture in the preservation of organic carbon[67,68] because of the large surface area of the lower particle size (i.e., clay and silt) provides more surface for maintaining SOC.[69] Schillaci et al.[68] and Pahlavan-Rad et al.[11] reported that soil texture is one of the

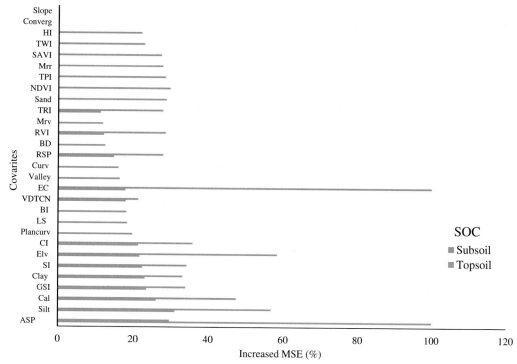

FIG. 2 Covariate importance of random forest (RF) model for topsoil (0–20 cm) and subsoil (20–40 cm). The variable importance <10% is not shown here. Symbols for covariates are given in Table 1.

most important variables through SOC prediction in Italy and Iran, respectively. In addition, the amount of SOC is a function of the entry of plant residuals and its losses from the soil; thus, EC can change its amount in the soil.[70] In other words, it seems to produce less biomass under saline conditions (high EC) and consequently, less return to soil leads to the loss of SOC. Therefore, SOC content decreased with increasing EC and was considered an important covariate for SOC prediction at the second depth.

We should mention that remote sensing indices had an important role in the prediction of SOC content especially in the studied area (dryland). In this regard, the results of the current study agree with those of Minasny et al.,[22] Schillaci et al.,[68] Wang et al.,[62] Bhunia et al.,[5] Pouladi et al.,[7] and Tajik et al.[17]

3.4 Spatial prediction

The spatial prediction maps of the SOC contents are presented in Fig. 3 for 0–20 and 20–40 cm. The highest SOC values were predicted in the southern part of the study area, where loess deposits are higher, and lower in the west, northwest, and northeast parts of the study area where outcrop rocks of limestone and marl with a high elevation and eroded soils are located. This could explain the formation of shallow Entisols on a steep south hillslope with scarce vegetation cover[63,65] that experience high erosion and subsequently lower SOC. However, deeper soils (Inceptisols) have been observed in cultivated lands on valley landforms and north-facing slopes with higher SOC (Fig. 4). In addition, some parts of the northern and central areas have a high content of sand (50%) with low SOC (0.1%). This result confirmed the effect of slope aspect, elevation, soil texture, and color index on the variability of SOC, which is consistent with covariate importance. Evidently, the spatial distribution of SOC is the same in both the surface and subsurface layers of soils.

Additionally, the predicted maps differed somewhat from the observed SOC, whereas the maximum and minimum SOC predictions were 1.82% and 0.21% at the first depth (Fig. 3A), while the maximum and minimum SOC observations were 2.40 and 0.08%, respectively (Table 2). At the second depth, the minimum and maximum varied from 0.22 to 1.04% on the predicted map of SOC (Fig. 3B), while SOC observations were 0.01% to 1.60% (Table 2). Nevertheless, based on the predicted maps, there was a satisfactory level of compatibility between the predicted and observed values of SOC (Fig. 3). In agricultural lands (valley landform), some parts are seen as spots with maximum values of SOC, possibly due to the addition of manure and leaving crop residues on the agricultural areas. Lacoste et al.[71] found a higher mean value of predicted SOC in cultivated lands than in forested areas. Regarding most of the areas predicted as low to moderate content of SOC (based on the mean value of observed SOC), it can be concluded that the RF model was capable of providing reasonable predictions of SOC at both depths.

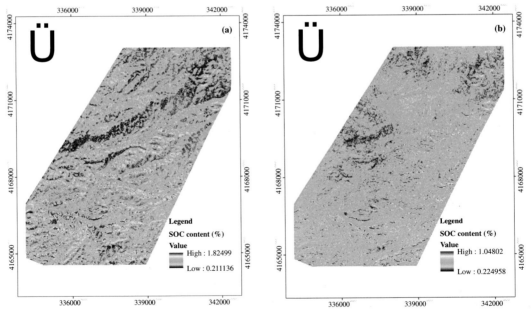

FIG. 3 Spatial distribution of SOC contents using the RF model at (A) 0–20 cm and (B) 20–40 cm.

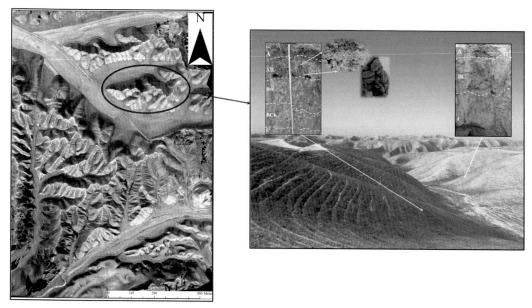

FIG. 4   The orthoimagery of unmanned aerial vehicle (UAV) of the study area[65] on the left side of the image and showing the effect of slope aspect on soil development as formed Haploxerepts on north-facing slope with vegetation cover and Xerorthents on south-facing *(right side of the image)*.

4 Conclusions

The spatial variability of SOC was investigated in a semiarid region of loess-derived soils in Iran using the DSM framework. The results showed that the covariates used, including soil layer maps, topographic attributes, and remote sensing indices, could accurately capture the surface and subsurface variation of SOC. The highly complex relief landscape in this study area and the preparation of a reliable environmental covariate with the utilization of expert person knowledge provide suitable tools for the prediction and recognition of the relationship between SOC and a set of independent environmental variables. The results of the relative importance of the RF model and comparison with expert person knowledge and pedagogical information confirmed the high performance of this algorithm. It is important to match the expert's personal knowledge with the modeling results of relative importance because of the increasing trend in using machine learning (ML) approaches for DSM. Expert knowledge may be lost among many covariates applied to construct predictions and ML methods "blindly replace expert person knowledge." Moreover, more research and high-resolution data are needed to improve the quality of the covariates used as input data in DSM in regions with complex topography in the future.

References

1. Keesstra SD, Bouma J, Wallinga J, et al. The significance of soils and soil science towards realization of the United Nations sustainable development goals. *Soil.* 2016;2(2):111–128.
2. Rodrigo-Comino J, Lopez-Vicente M, Kumar V, et al. Soil science challenges in a new era: a transdisciplinary overview of relevant topics. *Air Soil Water Res.* 2020;13:1–17.
3. Brevik EC, Cerdà A, Mataix-Solera J, et al. The interdisciplinary nature of soil. *Soil.* 2015;1(1):117–129.
4. Bouma J, McBratney A. Framing soils as an actor when dealing with wicked environmental problems. *Geoderma.* 2013;200–201:130–139.
5. Bhunia GS, Shit PK, Pourghasemi HR. Soil organic carbon mapping using remote sensing techniques and multivariate regression model. *Geocarto Int.* 2019;34(2):215–226.
6. Pham TG, Nguyen HT, Kappas M. Assessment of soil quality indicators under different agricultural land uses and topographic aspects in Central Vietnam. *Int Soil Water Conserv Res.* 2018;6(4):280–288.
7. Pouladi N, Bjørn Møller A, Tabatabai S, Grevea MH. Mapping soil organic matter contents at field level with cubist, random forest and kriging. *Geoderma.* 2019;342:85–92.
8. Castro FC, Araújoa JF, dos Santos AM. Susceptibility to soil salinization in the quilombola community of Cupira - Santa Maria da Boa Vista - Pernambuco – Brazil. *Catena.* 2019;179:175–183.
9. Chen S, Arrouays D, Angers DA, et al. National estimation of soil organic carbon storage potential for arable soils: a data-driven approach coupled with carbon-landscape zones. *Sci Total Environ.* 2019;666:355–367.
10. Xu Y, Wang X, Bai J, Wang D, Wang W, Guan Y. Estimating the spatial distribution of soil total nitrogen and available potassium in coastal wetland soils in the Yellow River Delta by incorporating multi-source data. *Ecol Indic.* 2020;111:106002.

11. Pahlavan-Rad MR, Dahmardeh K, Brungard C. Predicting soil organic carbon concentrations in a low relief landscape, eastern Iran. *Geoderma Reg*. 2018;, e00195.

12. Akpa SI, Odeh IO, Bishop TF, Hartemink AE, Amapu IY. Total soil organic carbon and carbon sequestration potential in Nigeria. *Geoderma*. 2016;271:202–215.

13. Lamichhane S, Kumar L, Wilson B. Digital soil mapping algorithms and covariates for soil organic carbon mapping and their implications: a review. *Geoderma*. 2019;352:395–413.

14. Liang Z, Chen S, Yang Y, Zhou Y, Shi Z. High-resolution three-dimensional mapping of soil organic carbon in China: effects of SoilGrids products on national modeling. *Sci Total Environ*. 2019;685:480–489.

15. Long J, Liu Y, Xing S, et al. Optimal interpolation methods for farmland soil organic matter in various landforms of a complex topography. *Ecol Indic*. 2020;110:105926.

16. Ramifehiarivo N, Brossard M, Grinand C, et al. Mapping soil organic carbon on a national scale: towards an improved and updated map of Madagascar. *Geoderma Reg*. 2017;9:29–38.

17. Tajik S, Shamsollah Ayoubi S, Zeraatpisheh M. Digital mapping of soil organic carbon using ensemble learning model in Mollisols of Hyrcanian forests, northern Iran. *Geoderma Reg*. 2020;20, e00256.

18. Zeraatpisheh M, Ayoubi S, Jafari A, Tajik S, Finke P. Digital mapping of soil properties using multiple machine learning in a semi- arid region, Central Iran. *Geoderma*. 2019;338:445–452.

19. Zhang S, Lu X, Zhang Y, Nie G, Li Y. Estimation of soil organic matter, Total nitrogen and Total carbon in sustainable coastal wetlands. *Sustainability*. 2019;11(3):667.

20. Forkuor G, Hounkpatin OK, Welp G, Thiel M. High resolution mapping of soil properties using remote sensing variables in South-Western Burkina Faso: a comparison of machine learning and multiple linear regression models. *PLoS One*. 2017;12, e0170478.

21. McBratney AB, Mendonc Santos ML, Minasny B. On digital soil mapping. *Geoderma*. 2003;117:3–52.

22. Minasny B, McBratney AB, Malone BP, Wheeler I. Digital mapping of soil carbon. In: *Advances in Agronomy*. Elsevier; 2013:1–47.

23. Taghizadeh-Mehrjardi R, Nabiollahi K, Kerry R. Digital mapping of soil organic carbon at multiple depths using different data mining techniques in Baneh region, Iran. *Geoderma*. 2016;266:98–110.

24. Byrne JM, Yang M. Spatial variability of soil magnetic susceptibility, organic carbon and total nitrogen from farmland in northern China. *Catena*. 2016;145:92–98.

25. Keskin H, Grunwald S, Harris WG. Digital mapping of soil carbon fractions with machine learning. *Geoderma*. 2019;339:40–58.

26. Pahlavan-Rad MR, Akbarimoghaddam A. Spatial variability of soil texture fractions and pH in a flood plain (case study from eastern Iran). *Catena*. 2018;160:275–281.

27. Ye L, Tan W, Fang L, Ji L. Spatial analysis of soil aggregate stability in a small catchment of the Loess Plateau, China: II. Spatial prediction. *Soil Tillage Res*. 2019;192:1–11.

28. Malone B, Minasny B, McBratney AB. *Using R for Digital Soil Mapping*. Springer; 2017. https://doi.org/10.1007/978-3-319-44327-0.

29. Mansuy N, Thiffault E, Paré D, et al. Digital mapping of soil properties in Canadian managed forests at 250 m of resolution using the k-nearest neighbor method. *Geoderma*. 2014;235:59–73.

30. Taghizadeh-Mehrjardi R, Minasny B, Sarmadian F, Malone B. Digital mapping of soil salinity in Ardakan region, Central Iran. *Geoderma*. 2014;213:15–28.

31. Angelini ME, Heuvelink GBM, Kempen B. Multivariate mapping of soil with structural equation modelling. *Eur J Soil Sci*. 2017;68(5):575–591.

32. Chagas CS, de Carvalho JW, Bhering SB, Calderano FB. Spatial prediction of soil surface texture in a semiarid region using random forest and multiple linear regressions. *Catena*. 2016;139:232–240.

33. Kovačević M, Bajat B, Gajić B. Soil type classification and estimation of soil properties using support vector machines. *Geoderma*. 2010;154 (3–4):340–347.

34. Khaledian Y, Miller BA. Selecting appropriate machine learning methods for digital soil mapping. *App Math Model*. 2020;81:401–418.

35. Maleki S, Khormali F, Bagheri Bodaghabadi M, Mohammadi J, Hoffmeister D, Kehl M. Role of geomorphic surface on the above-ground biomass and soil organic carbon storage in a semi-arid region of Iranian loess plateau. *Quat Int*. 2020;552:111–121.

36. Hosseinalizadeh M, Kariminejad N, Chen W, et al. Gully headcut susceptibility modeling using functional trees, naïve Bayes tree, and random forest models. *Geoderma*. 2019;342:1–11.

37. Soil Survey Staff. *Keys to Soil Taxonomy*. 12th ed. U.S. Department of Agriculture, Natural Resources Conservation Service; 2014:372.

38. Jafari A, Ayoubi S, Khademi H, Finke PA, Toomanian N. Selection of a taxonomic level for soil mapping using diversity and map purity indices: a case study from an Iranian arid region. *Geomorphology*. 2013;201:86–97.

39. Zeraatpisheh M, Ayoubi S, Jafari A, Finke P. Comparing the efficiency of digital and conventional soil mapping to predict soil types in a semi-arid region in Iran. *Geomorphology*. 2017;285:186–204.

40. Godwin RJ, Miller PCH. A review of the technologies for mapping within field variability. *Biosyst Eng*. 2003;84:393–407.

41. Wilson JP, Gallant JC. *Terrain Analysis*. New York: Wiley & Sons; 2000.

42. Gallant JC, Dowling TI. A multi resolution index of valley bottom flatness for mapping depositional areas. *Water Resour Res*. 2003;39:1347–1360.

43. Weiss AD. Topographic position and landforms analysis. In: *Proceedings of the ESRI User Conference, 9–13 July, 2001; San Diego, CA, USA*; 2001.

44. Huete AR. A soil adjusted vegetation index (SAVI). *Remote Sens Environ*. 1988;25:295–309.

45. Rouse JW, Hass RH, Schell JA, Deering DW. Monitoring vegetation systems in the great plaints with ERTS. In: *Proceedings 3rd Earth Resource Technology Satellite (ERTS) Symposium*. vol. 1; 1974:48–62.

46. Ray SS, Singh JP, Das G, Panigrahy S. Use of high resolution remote sensing data for generating site-specific soil management plan. *Int Arch Photogramm Remote Sens Spatial Inf Syst*. 2004;35(7):127–131.

47. Loiseau T, Chen S, Mulder VL, et al. Satellite data integration for soil clay content modelling at a national scale. *Int J Appl Earth Obs Geoinf*. 2019;82:101905.

48. Minasny B, McBratney AB. A conditioned Latin hypercube method for sampling in the presence of ancillary information. *Comput Geosci*. 2006;32:1378–1388.

49. MathWorks. *Matlab*. Natick, MA: The Math Works. Inc; 2009.

50. Nelson DW, Sommers LE. Total carbon, organic carbon, and organic matter. In: Page AL, ed. *Methods of Soil Analysis. Agron. Monger.* vol. 9. Madison, WI: ASA and SSSA; 1982:539–577.

51. Breiman L. Random forests. *Mach Learn.* 2001;45(1):5–32.

52. Hengl T, Heuvelink GB, Kempen B, et al. Mapping soil properties of Africa at 250 m resolution: Random forests significantly improve current predictions. *PLoS One.* 2015;10(6), e0125814.

53. Breiman L, Cutler A. *Random Forests Homepage*; 2004. Retrieved April 23rd.

54. Behrens T, Zhu AX, Schmidt K, Scholten T. Multi-scale digital terrain analysis and feature selection for digital soil mapping. *Geoderma.* 2010;155:175–185.

55. Chen S, Mulder VL, Martin MP, et al. Probability mapping of soil thickness by random survival forest at a national scale. *Geoderma.* 2019;344:184–194.

56. Zhi J, Zhang G, Yang F, et al. Predicting mattic epipedons in the northeastern Qinghai-Tibetan plateau using random forest. *Geoderma Reg.* 2017;10:1.

57. Liaw A, Wiener M. Classification and regression by random forest. *R News.* 2002;2(3):18–22.

58. Kuhn M. *Caret: Classification and Regression*; 2020. Training. R package version 6.0-86.

59. R Development Core Team. *R: A Language and Environment for Statistical Computing.* Vienna, Austria: R Foundation for Statistical Computing; 2017. Retrieved from http://www.R-project.org.

60. Carter MR, Gregorich EG. *Soil Sampling and Methods of Analysis.* 2nd ed. CRC Press; 2007:60.

61. Gomes LC, Faria RM, De Souza E, et al. Modelling and mapping soil organic carbon stocks in Brazil. *Geoderma.* 2019;340:337–350.

62. Wang B, Waters C, Orgill S, et al. High resolution mapping of soil organic carbon stocks using remote sensing variables in the semiarid rangelands of eastern Australia. *Sci Total Environ.* 2018;630:367–378.

63. Kehl M, Khormali F. *Excursion Book of International Symposium on Loess, Soils & Climate Change in Southern Eurasia. 15–19 October.* Gorgan, Iran: Gorgan University of Agricultural Sciences and Natural Resources; 2014.

64. Kramm T, Hoffmeister D, Curdt C, Maleki S, Khormali F, Kehl M. Accuracy assessment of landform classification approaches on different spatial scales for the Iranian loess plateau. *ISPRS Int J Geo-Inf.* 2017;6(366):1–22.

65. Maleki S, Khormali F, Mohammadi J, Bogaert P, Bodaghabadi MB. Effect of the accuracy of topographic data on improving digital soil mapping predictions with limited soil data: an application to the Iranian loess plateau. *Catena.* 2020;195:104810.

66. Fissore C, Dalzell BJ, Berhe AA, Voegtle M, Evans M, Wu A. Influence of topography on soil organic carbon dynamics in a Southern California grassland. *Catena.* 2017;149:140–149.

67. Angers DA, Arrouays D, Saby NPA, Walter C. Estimating and mapping the carbon saturation deficit of French agricultural topsoils. *Soil Use Manage.* 2011;27:448–452.

68. Schillaci C, Acutis M, Lombardo L, et al. Spatio-temporal topsoil organic carbon mapping of a semi-arid Mediterranean region: the role of land use, soil texture, topographic indices and the influence of remote sensing data to modeling. *Sci Total Environ.* 2017;601–602:821–832.

69. Konen ME, Burras CL, Sandor JA. Organic carbon, texture and quantitative color measurement relationships for cultivated soils in North Central Iowa. *Soil Sci Soc Am J.* 2003;67:1823–1830.

70. Cai ZC, Qin SW. Dynamics of crop yields and soil organic carbon in a long-term fertilization experiment in the Huang-Huai-Hai Plain of China. *Geoderma.* 2006;136:708–715.

71. Lacoste M, Minasny B, McBratney A, Michot D, Viaud V, Walter C. High resolution 3D mapping of soil organic carbon in a heterogeneous agricultural landscape. *Geoderma.* 2014;213:296–311.

35

Hydrochemistry and geogenic pollution assessment of groundwater in Akşehir (Konya/Turkey) using GIS

Erhan Şener[a], Simge Varol[b], and Şehnaz Şener[b]

[a]Remote Sensing Center, Suleyman Demirel University, Isparta, Turkey [b]Department of Geological Engineering, Suleyman Demirel University, Isparta, Turkey

1 Introduction

In the 21st century, all over the world is increasing the demand for high-quality and safe water sources for human needs (drinking, domestic, etc.), agricultural activities, and industrial applications in rural, semi-urban and urban areas.[1] Groundwater is widely preferred as a freshwater resource in the world.[2] It is more difficult for pollutants to affect groundwater quality compared to other water sources such as streams, lakes, and ponds. For this reason, groundwater is used as a source of drinking water worldwide.[3] Establishing an adequate understanding of the current quality status of available water supplies and the several factors that influence their chemistry is fundamental for making wise plans and decisions for water resource development, protection, management, and sustainability.

In groundwater management studies, it is necessary to know the water quality, availability, and potential risks to human health. Therefore, there is a need for research on the hydrogeochemical processes of groundwater.[4,5] Generally, geogenic effects and anthropogenic human effects directly affect the chemical properties and quality of groundwater. Processes such as dissolution, decomposition, and ion exchange of rocks determine the presence of ions in water. In addition, point and diffuse pollutants that exist in humans control the water quality.[6–14] The chemical characteristics of the groundwater are determined depending on the chemical composition of the rocks and/or anthropogenic inputs in the aquifer environment and the duration of contact with these factors. As the contact time increases, ion exchange increases.[15–18] Chemical reactions that occur as a result of geogenic effects can be determined from the groundwater discharged. This situation sometimes causes geogenic pollution in groundwater, in addition to anthropogenic pollution. As with anthropogenic pollution, the use of groundwater is limited because of geogenic pollution. For this reason, it is important to know the hydrochemical properties of groundwater and the processes that occur during rock-water interactions.

Many researchers have investigated the chemistry of groundwater in different regions and the hydrochemical processes that are effective in the form of these chemical properties. Yadav et al.[43] investigated the relationship between the spread of fluorosis diseases and groundwater in central India. This study focused on the hydrochemical properties of groundwater and F⁻ source. The authors stated that the high amount of F ions in the groundwater resulted from the geogenic origin of rhyolite mineralization. Yang et al.[13] studied the chemical properties of Ordos Basin (China) groundwaters and their usability as irrigation water. In this study, it was revealed that water hydrochemistry directly affects water quality and develops with rock-water interaction. Groundwater hydrochemistry in the Dagu River Basin (China) was investigated by Yin et al.[19] In this study, it is stated that geogenic and anthropogenic factors directly affect groundwater chemistry, especially in the long term. In addition, the importance of groundwater chemistry research for groundwater resource management is emphasized. Knowing the hydrochemical properties of groundwater, which is used as a drinking water source, is essential for proper water management. Groundwater cannot be potable when the ion content, which increases with the rock-water interaction, exceeds the drinking water limit values.[20,21]

Accordingly, the hydrochemical properties of groundwater determine its quality characteristics. About this, groundwater resources in Akşehir and its surroundings were selected as the study area. The groundwater of Akşehir is used for domestic usage, drinking, and irrigation, and its surroundings have different aquifer environments and therefore different geochemical properties. Together with aquifer environments, anthropogenic pollutants affect groundwater chemistry. The main anthropogenic pollution is caused by residential areas and industrial and agricultural activities, while geological units of different ages constitute the natural sources of geogenic pollution in the region. Therefore, determining the hydro-chemical properties of groundwater with different usage areas with their quality and the processes that determine them is also important in terms of determining the effects on human health. However, no research has been conducted on the chemical properties and quality of these waters. The study includes the hydro-chemical evaluations required to prevent the pollution of groundwater due to natural and anthropogenic origins in the region and the health and environmental problems that may arise due to the use of this water. In addition, a GIS environment was used to visualize the quantity and regional distribution of dominant ions in groundwater and to provide a better understanding.

2 Materials and methods

2.1 The study area

The study area lies between North Latitudes 38°17′49″–38°30′04″ and East Longitudes 31°15′13″–31°42′07″ in the Central Anatolia region of Turkey (Fig. 1). Akşehir Lake is located north of the Akşehir District. The area of the district

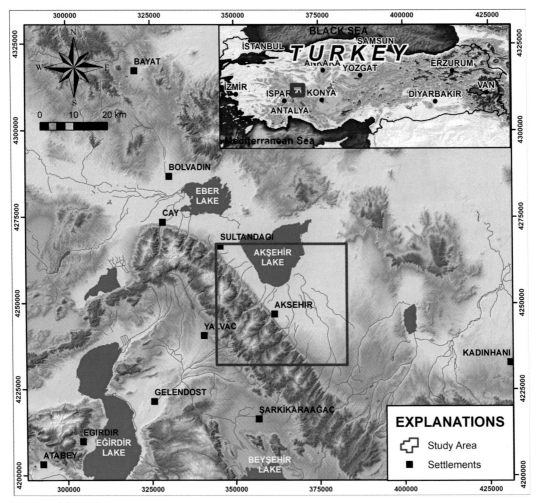

FIG. 1   Location map of the study area.

is 955 km^2, and it is 1050 m above sea level. Wet and dry agriculture was carried out in Akşehir. Vegetable and fruit growing has been developed and cereal products, beets, grains, legumes, industrial plants, and animal feeds also have an important place. While the Continental climate is dominant in the region, Central Anatolia and Inner West Anatolian climates are also effective. The annual precipitation amount was recorded as 546 mm, and the total annual potential evaporation amount was 690.5 mm.[22] The Sultan Mountains to the south and southwest of the study area and the Akşehir Lake to the north form the morphological structure of the region. Akşehir Lake has a maximum surface area of 6000–7000 ha and a depth of 1 m. The lake is 953.50 m above sea level. All surface waters (Karabulut Stream, Adıyan Stream, Nadir Stream, Akşehir Stream, and Saray Stream) in the study area were discharged to Akşehir Lake. Akşehir Lake is used for irrigation of agricultural areas in the region. Agricultural activities are very common in the region and the total irrigation area is 115.8 km^2.In addition, the main water source in the study area for domestic, industrial, and agricultural uses is groundwater.

2.2 Geology and hydrogeology

Several geological units of different ages were mapped in the study area, and a geological map is shown in Fig. 2. Autochthonous lithological units are observed in the region to the north of the Isparta angle which is the major tectonic structure in Turkey. The age, tectonostratigraphy, and lithological properties of all lithological units are detailed in Table 1. To determine the hydrogeological properties of the study area, each lithological unit was evaluated according to aquifer capacity, and four groups were defined: impermeable (aquifuge), semipermeable (aquitard), permeable

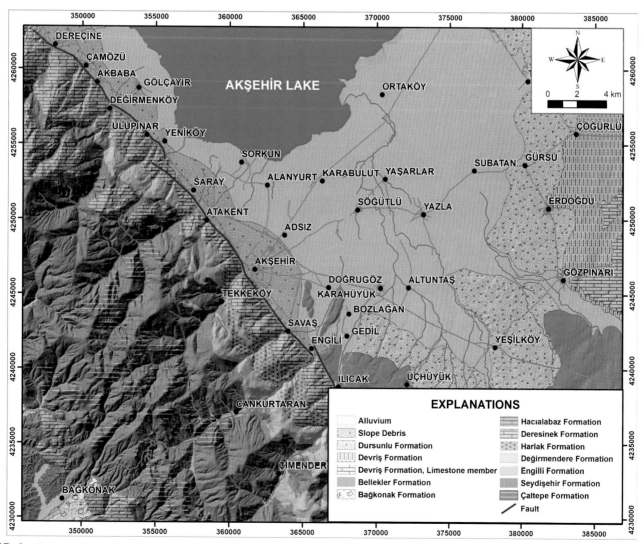

FIG. 2   Geological map of the study area.

TABLE 1   Hydrogeological properties of lithological units (Umut 2009a; Umut 2009b).

Formation	Age	Tectonostratigraphy	Lithology	Hydrogeological properties
Alluvium (Qal)	Quaternary	Autochthonous	Clay, silt, sand, gravel and block size material	Permeable (Granular Aquifer)
Slope debris (Qym)	Quaternary	Autochthonous	Attached to the loose clay, silt, sand, gravel and block size material	Permeable (Granular Aquifer)
Dursunlu Formation (plQd)	Plio.-Quaternary	Autochthonous	Clay, marl, sandstone, gravel and lignite	Semipermeable (Aquitard-1)
Devriş Formation (Tmpld)	Up. Miocene-Pliocene	Autochthonous	Clay, marl, sandstone, limestone, tuff	Semipermeable (Aquitard-1)
Devriş Formation, Limestone member (Tmpldk)	Up. Miocene-Pliocene	Autochthonous	Limestone	Permeable (Karstic Aquifer)
Bellekler Formation (Tmbe)	Up. Miocene	Autochthonous	Conglomerate, sandstone, mudstone	Semipermeable (Aquitard-1)
Bağkonak Formation (Tmb)	Miocene	Autochthonous	Conglomerate, sandstone, claystone	Semipermeable (Aquitard-1)
Hacıalabaz Formation (JKh)	Jurassic-Cretaceous	Autochthonous	Limestone, dolomite	Permeable (Karstic Aquifer)
Deresinek Formation (Pd)	Permian	Autochthonous	Limestone, dolomite, recrystallized limestone, quartzite	Semipermeable (Aquitard-1)
Harlak Formation (Cha)	Carboniferous	Autochthonous	Slate, quartzite, dolomite, limestone	Impermeable (Aquifuge)
Değirmendere Formation (DCd)	Devonian-Carboniferous	Autochthonous	Slate, schist, dolomite, quartzite, metasandstone, metamilestone, limestone	Impermeable (Aquifuge)
Engilli Formation (De)	Upper Devonian	Autochthonous	Quartzite, schist, dolomite, limestone	Impermeable (Aquifuge)
Seydişehir Formation (ɛOs)	Up. Cambrian-L. Ordovician	Autochthonous	Quartzite, schist, metashale, metasandstone, metamilestone, dolomite, limestone	Impermeable (Aquifuge)
Çaltepe Formation (ɛç)	M. Cambrian	Autochthonous	Dolomite, crystallized limestone	Permeable (Karstic Aquifer)

granular aquifer, and permeable karstic aquifer. The hydrogeological properties determined for each formation are presented in Table 1 and Fig. 3. The most important aquifer unit in terms of groundwater is the granular aquifer consisting of alluvium and slope debris. The area of the alluvium is approximately $720.3 \, km^2$. Groundwater wells were drilled on the alluvial unit in the region, and groundwater was mainly taken from the alluvium. The thickness of the sediments, which store a significant part of the groundwater potential, is more than 300 m in the plain. However, the carbonated rocks mapped in the study area were classified as karstic aquifers. The melting and broken cracks in the carbonated rocks contain a significant amount of water. Harlak, Değirmendere, Engilli, and Seydişehir formations were classified as impermeable units owing to their metamorphic content such as quartzite, schist, metashale, meta-sandstone, etc. The Deresinek, Bağkonak, Bellekler, Devriş, and Dursunlu formations are classified as semipermeable aquifers. These units may contain water in the sandstone and limestone layers.

2.3 Data collection and analysis methods

A flowchart of the methodology is shown in Fig. 4. 31 groundwater samples were collected in October 2018. Two separate water samples were taken from each water sampling point to be used in anion and cation analysis. A few

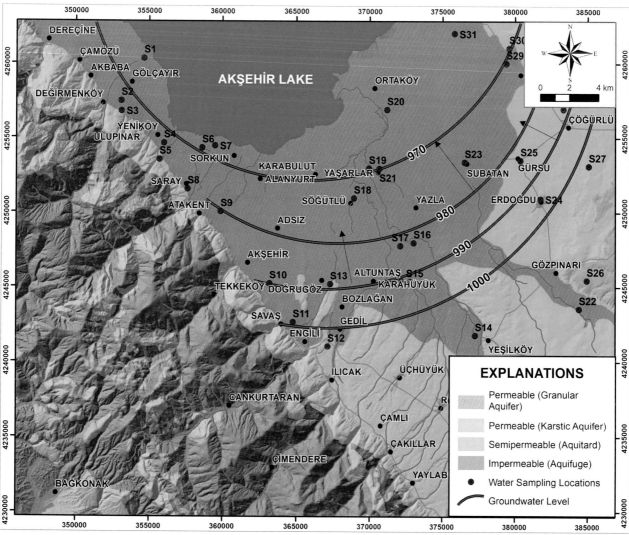

FIG. 3 Hydrogeological and groundwater level maps of the study area.

drops of 0.5% nitric acid (HNO_3) were added to one of the bottled water samples for cation analysis, and its acidity was reduced. The water sample in the other bottle was not acidified for anion analysis. pH, Eh, temperature (T; °C), electrical conductivity (EC; µS/cm), and total dissolved solids (TDS; mg/L) measurements were performed using a YSI Professional Plus handheld multi-parameter instrument. The samples were sent to the Bureau Veritas Mineral Laboratory (Canada-ISO 9002 Accredited Co.) for cation analysis. In addition, the samples were delivered to the Geothermal Energy Laboratory of Geology Engineering Department, SDÜ (Isparta, Turkey) for sulfate, chloride, carbonate, and bicarbonate analyses. Here, analyses were performed using titrimetric (for CO_3, HCO_3, and Cl) and barium sulfate turbidity (for SO_4) methods. The inductively coupled plasma mass spectrometry (ICP-MS) was used for all cation analyses. AquaChem software was used to analyze the chemical data. The calculated charge-balance error result was less than 5% within the limits of acceptability. The spatial distribution maps of the main ions were prepared using a geographical information system. In this context, ArcGIS 10.7 software was used. Piper and Gibbs diagrams were used for the hydrogeochemical evaluation. The chemical structure and water types of the waters in the region were defined using a Piper diagram prepared using the chemical analysis results of the waters.[23] Gibbs[24] Diagrams were used to determine the function and/or mechanism of the water-rock interaction in the aquifer environment.

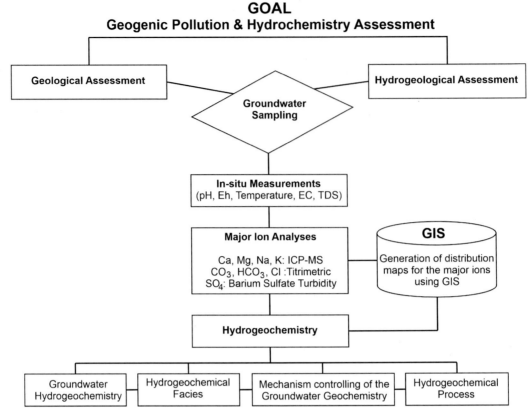

FIG. 4   Flowchart of the methodology.

3 Results and discussion

3.1 Groundwater hydrogeochemistry

The major hydro-chemical parameters such as temperature (°C), pH, EC, Eh, TDS, Ca, Mg, Na, K, HCO_3, SO_4, Cl, and CO_3 are listed as statistically summarized and presented in Table 2. The water temperature (°C) ranged from 12.40–16.00°C to °C. The pH values in the water samples were measured to be between 7.15 and 8.10 in water samples. This indicates that water has the features of alkaline water. In the study area, the Eh values of groundwater samples ranged from 260.00 to 350.00. While EC values of groundwater samples range from 290.00 (μS/cm) to 2900.00 (μS/cm), TDS values were found to be between 150.00 and 1780.00 (mg/L). As the temperature and concentration of ions increase in water, the electrical conductivity also increases. There is generally a linear relationship between the total amount of ions dissolved in water and electrical conductivity.[25] At the same time, the EC value refers to the change in solids dissolved in water.[26]

Ca, Na, Mg, K, HCO_3, SO_4, and Cl are the main ions forming water chemistry, and they take place in water chemistry under the influence of natural and/or pollutant factors.[14] The chemical data of the major ions were analyzed using AquaChem 3.7. According to the analysis results, the Ca ion contents in the water samples ranged from 40.00 mg/L to 175.50 mg/L. The Mg ion concentration of the water samples ranged from 5.31 mg/L to 82.42 mg/L. Na ion contents in the water samples range from 2.00 mg/L to 285.90 mg/L, and the K ion concentration of water samples range from 0.78 mg/L to 18.51 mg/L. Accordingly, Ca, Na, and Mg ions were determined as the dominant cations in the study area. According to the distribution map given in Fig. 5, Ca ions increase in regions where carbonate rocks are located. Carbonate rocks such as limestones and dolomite cause an increase in Ca and Mg ions in groundwater. High Na ion concentration may be more related to cation exchange and dissolution of silicate minerals in the study area.[27] In addition, a high Na^+ concentration is an indicator of anthropogenic pollution associated with agricultural activities.

The HCO_3 ion concentration of the water samples range from 105.60 mg/L to 333.00 mg/L. The SO_4 ion concentration of the water samples ranged from 3.98 mg/L to 482.60 mg/L. The Cl ion concentration of the water samples ranged from 8.80 mg/L to 519.80 mg/L. The CO_3 ion concentration in the water samples in the study area was 2 mg/L. Accordingly, the dominant anions in the study area were HCO_3, SO_4, and Cl. The alkalinity of water increased in the

TABLE 2   Statistical summary of the physical and chemical parameters of the groundwater.

Parameters	Unit	N	Min.	Max.	Mean	Std. Deviation
Temp	(°C)	31	12.40	16.00	14.21	.87
pH		31	7.15	8.10	7.72	.22
EC	(µS/cm)	31	290.00	2900.00	811.19	582.10
Eh	(mV)	31	260.00	350.00	287.96	22.08
TDS	(mg/L)	31	150.00	1780.00	488.09	361.10
Ca	(mg/L)	31	40.00	175.50	79.54	32.39
Mg	(mg/L)	31	5.31	82.42	22.40	18.00
Na	(mg/L)	31	2.00	285.90	56.55	77.32
K	(mg/L)	31	.78	18.51	5.27	4.91
HCO_3	(mg/L)	31	105.60	333.00	189.09	47.25
SO_4	(mg/L)	31	3.98	482.60	99.56	109.71
Cl	(mg/L)	31	8.80	519.80	88.78	113.79
CO_3	(mg/L)	31	2	2	2	0
Valid N (listwise)		31				

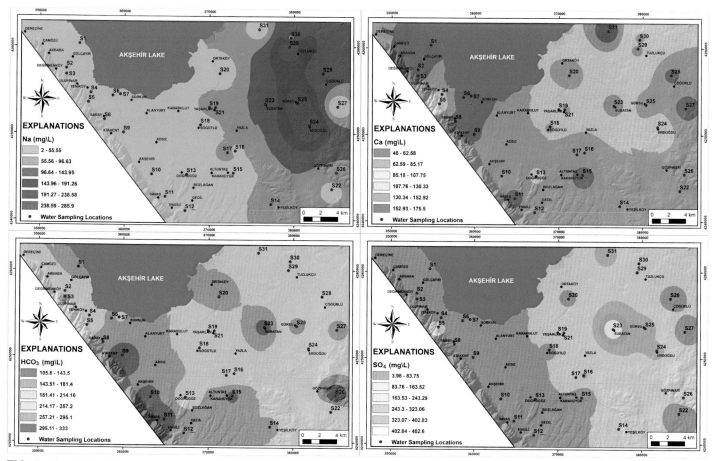

FIG. 5   Spatial distribution map of the major ions.

presence of HCO₃. In general, carbonate dissolution and/or carbonic acid and silicate reactions cause the presence of HCO₃ in groundwater.[28–30] Accordingly, a high HCO₃ ion concentration in the water samples was considered as the interaction of groundwater with carbonated rocks (Fig. 5). High sulfate ion concentrations in water samples have also been associated with mineral dissolution, especially agricultural activities in the study area (Fig. 5). Water with high chloride (> 250 mg/L) tastes salty and this situation causes serious problems in drinking water usage.[31] 16.12% of the samples (S23, S24, S25, S29, and S31) have more than 250 mg/L Cl content. These high Cl contents are mainly related to human activities (domestic wastewater, septic tanks), especially agricultural activities (fertilizers and pesticides), as well as natural sources such as precipitation of chloride-bearing minerals by precipitation.[6,32] In this case, the Cl concentrations in groundwater can have both natural and anthropogenic origins.

3.2 Hydrogeochemical facies

Hydrogeochemical facies are important in hydrogeochemical investigations and are used to classify groundwater and identify the origins of groundwater. Hydrogeochemical facies determined by the dominant major ions in the groundwater samples were used to present similar and different types of water. Piper diagram was used to determined hydrogeochemical facies of the groundwater samples.[23] The Piper diagrams prepared using the analysis results (meq/L units) of the major anions and cations are presented in Fig. 6.

Accordingly, Ca-Mg-HCO₃ and Ca-HCO₃ water types were determined as the major water types in the region. The Ca-Mg-HCO₃ and Ca-HCO₃ facies represented 19.35% and 12.90% of the total water samples, respectively. Apart from these, there are many different water types, such as Na-SO₄, Ca-Mg-SO₄, Na-Ca-HCO₃ in the study area (Table 3). The observation of different water types may be related to the interaction time of groundwater with anthropogenic pollutants and/or lithological units and their degree of exposure.

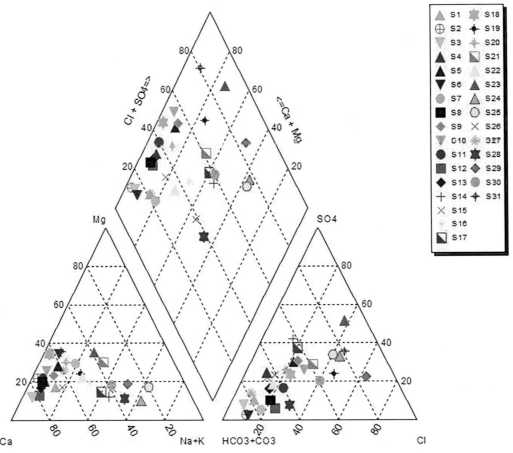

FIG. 6   Piper diagram.

TABLE 3 The major ion sequences of water samples.

| No. | Sample Type | (Oct 2018) | |
		Cation Sequence	Anion Sequence
S1	Well water	$Ca>Na>Mg>K$	$HCO_3>SO_4>Cl>CO_3$
S2	Well water	$Ca>Mg>Na>K$	$HCO_3>Cl>SO_4>CO_3$
S3	Well water	$Ca>Mg>Na>K$	$HCO_3>SO_4>Cl>CO_3$
S4	Well water	$Ca>Mg>Na>K$	$HCO_3>SO_4>Cl>CO_3$
S5	Well water	$Ca>Mg>Na>K$	$HCO_3>SO_4>Cl>CO_3$
S6	Well water	$Ca>Mg>Na>K$	$HCO_3>Cl>SO_4>CO_3$
S7	Well water	$Ca>Na>Mg>K$	$HCO_3>Cl>SO_4>CO_3$
S8	Well water	$Ca>Mg>Na>K$	$HCO_3>Cl>SO_4>CO_3$
S9	Well water	$Ca>Mg>Na>K$	$HCO_3>SO_4>Cl>CO_3$
S10	Well water	$Ca>Mg>K>Na$	$HCO_3>SO_4>Cl>CO_3$
S11	Well water	$Ca>Mg>Na>K$	$HCO_3>Cl>SO_4>CO_3$
S12	Well water	$Ca>Na>Mg>K$	$HCO_3>Cl>SO_4>CO_3$
S13	Well water	$Ca>Mg>Na>K$	$HCO_3>SO_4>Cl>CO_3$
S14	Well water	$Na>Ca>Mg>K$	$HCO_3>SO_4>Cl>CO_3$
S15	Well water	$Ca>Na>Mg>K$	$HCO_3>SO_4>Cl>CO_3$
S16	Well water	$Ca>Na>Mg>K$	$HCO_3>SO_4>Cl>CO_3$
S17	Well water	$Ca>Na>Mg>K$	$HCO_3>SO_4>Cl>CO_3$
S18	Well water	$Ca>Mg>Na>K$	$HCO_3>SO_4>Cl>CO_3$
S19	Well water	$Ca>Na>Mg>K$	$HCO_3>Cl>SO_4>CO_3$
S20	Well water	$Ca>Mg>Na>K$	$HCO_3>SO_4>Cl>CO_3$
S21	Well water	$Ca>Na>Mg>K$	$HCO_3>SO_4>Cl>CO_3$
S22	Well water	$Ca>Na>Mg>K$	$HCO_3>SO_4>Cl>CO_3$
S23	Well water	$Ca>Na>Mg>K$	$SO_4>Cl>HCO_3>CO_3$
S24	Well water	$Na>Ca>Mg>K$	$SO_4>Cl>HCO_3>CO_3$
S25	Well water	$Na>Ca>Mg>K$	$SO_4>HCO_3>Cl>CO_3$
S26	Well water	$Na>Ca>Mg>K$	$HCO_3>SO_4>Cl>CO_3$
S27	Well water	$Ca>Mg>Na>K$	$HCO_3>SO_4>Cl>CO_3$
S28	Well water	$Na>Ca>Mg>K$	$HCO_3>Cl>SO_4>CO_3$
S29	Well water	$Na>Ca>Mg>K$	$Cl>SO_4>HCO_3>CO_3$
S30	Well water	$Na>Ca>Mg>K$	$HCO_3>Cl>SO_4>CO_3$
S31	Well water	$Ca>Mg>Na>K$	$SO_4>Cl>HCO_3>CO_3$

3.3 Mechanism controlling of the groundwater geochemistry

Ion concentrations of water samples were placed on Gibbs diagrams to determine the mechanisms that affect the formation of chemical properties of water resources in the study area. In these diagrams, TDS and $(Na+K)/(Na+K+Ca)$ ion concentrations and TDS and $Cl/(Cl+HCO_3)$ concentrations were compared.[24] In the widely used Gibbs diagram, different regions such as "precipitation dominant," "rock dominant" and "evaporation dominant" are defined. Gibbs diagrams prepared to determine the main mechanism controlling groundwater chemistry in the region are presented in Fig. 7.

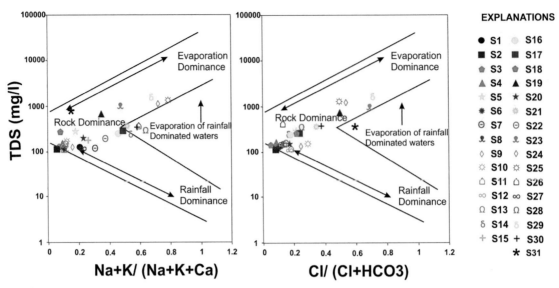

FIG. 7   Gibbs diagram.

Most of the groundwater samples fall into the "rock-dominated" zone. In these water samples, water chemistry develops mainly due to rock-water interactions and is related to hydrogeochemical processes. Ion reactions, dissolution, and weathering processes were dominant in these samples. On the other hand, in the water samples falling into the "evaporation dominant" region, the main process in the development of water chemistry is evaporation. High amounts of Na and TDS are observed in the waters in this region because of the precipitation of Ca and bicarbonate ions.[33] In addition, samples that fall outside the rock dominance and evaporation areas may indicate that anthropogenic activities cause changes in groundwater chemistry.[34]

3.4 Hydrogeochemical process

Sodium, calcium, magnesium, bicarbonate, and sulfate have been determined as the dominant ions in groundwater. The enrichment of these ions is caused by different hydrogeochemical processes in the aquifer environment. Various graphs were prepared to explain these processes and are presented in Fig. 8A–H. In the study area, limestone and silicate were the major sources of mineralization. A plot of the $(Ca^{2+}+Mg^{2+})$ versus $(HCO_3^-+SO_4^{2-})$ diagram shows the dominant weathering type in the groundwater. The points falling along the equiline (1:1) suggest that weathering of carbonates and the points below the 1:1 line are indicators of silicate weathering.[35] According to Fig. 8A, weathering of carbonates and silicate weathering contributed to the dominant ions.

The Ca^{2+}/Mg^{2+} ratio of groundwater is the determination of calcite and dolomite dissolution. If the molar Ca^{2+}/Mg^{2+} ratio was higher than 1, the calcite effect was greater. If the ratio is greater than 2, Ca and Mg are present owing to the decomposition of silicate minerals. The Ca^{2+}/Mg^{2+} ratio was found to be greater than 2 in 22 groundwater samples, indicating silicate minerals (Fig. 8B). According to the $Ca^{2+}+Mg^{2+}$ versus $HCO_3^-+CO_3^-$ diagram, the water samples fall above the equilibrium (1:1) (Fig. 8C). This suggests that calcium and magnesium are derived from non-carbonate sources. As shown in Fig. 8D and E, silicate weathering occurred in the aquifer. However, in the plot of $Ca^{2+}+Mg^{2+}$ versus TC (total cation), sample points are located between the 1:1 line and the 1:0.5 line, in the plot of (Na^++K^+) versus TC (total cation), the points fall below the equilibrium line 1:0.5. At the same time, with the increase in alkalis, an increase in Cl^- and SO_4^{2-} ions is observed (Fig. 8F), which supports the ion enrichment that occurs as a result of the dissolution of soil salts.[36] In the Na^+ versus Cl^- diagram, some of the samples are plotted below and above the 1:1 line, and the molar ratio of Na^+/Cl^- ranged from 0.1 to 3.1 (Fig. 8G and H).

The groundwater samples from S1, S7, S14, S15, S16, S17, S21, S22, S24, S25, S26, S27, S28, and S30 boreholes were located in the Plio-Quaternary and Miocene-Pliocene sediments, mainly clay, marl, sandstone, and silt. The Na^+/Cl^- ratio was determined to be greater than 1 in these samples, indicating silicate weathering and/or ion exchange. Salt dissolution in the soil is more dominant in groundwater samples from other regions.[35,37,38] The chloro-alkaline indices (CAI) I and II are used to determine the ion exchange mechanism. These indices were calculated using the meq/L values of ions and equations given below (Eqs. 1, 2).

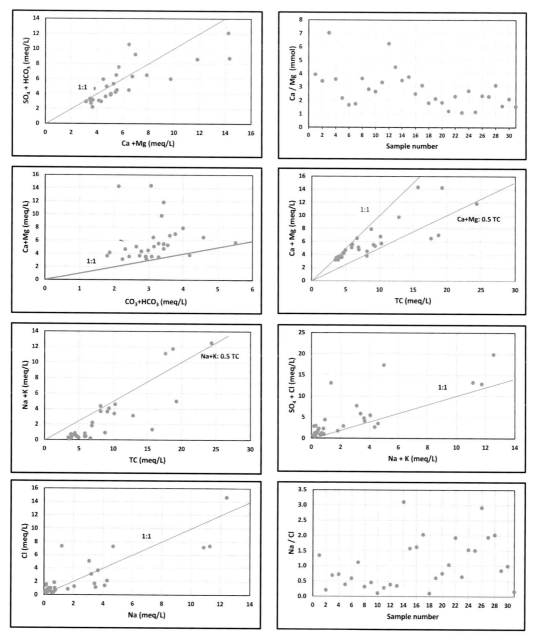

FIG. 8 Relationship and-or distribution graphs of major ions in groundwater.

$$CAI-1 = [Cl-(Na+K)]/Cl \qquad (1)$$

$$CAI-2 = [Cl-(Na+K)]/(SO_4 + HCO_3 + NO_3) \qquad (2)$$

Groundwater samples with negative CAI-I and CAI-II values had high Na^+/Cl^- ratios. The CAI-I and CAI-II values of groundwater samples taken from areas where carbonate rocks are commonly located are positive. This result shows that in the hydrogeochemical development of groundwater, both reverse ion exchange and normal ion exchange are observed depending on the geogenic formation and lithological characteristics in the region.

Similar results were obtained from hydrogeochemical studies of groundwater in the central region of Shandong Province, North China. In the study carried out by Liu et al.,[4] carbonate/silicate weathering and ion exchange were shown to be the dominant factors affecting the chemistry of groundwater in this region. In addition, human effects affect water chemistry. According to the hydro-chemical investigation of groundwater in the Chithar River basin (Southern India) by Subramani et al.,[38] Ca, Mg, and HCO3 are the dominant ions in the water. It is also stated that major hydro-chemical processes are carbonate mineral dissolution, silicate weathering, and ion exchange. A hydrogeochemical investigation of groundwater with Ca-Mg-HCO$_3$ water type was conducted by Tay.[39] In this study, calcite, dolomite, silicate dissolution, ion exchange reactions were identified as the dominant hydrogeochemical processes affecting the groundwater quality of the Savelugu-Nanton district. Hydro-chemical investigations of groundwater in different parts of Turkey demonstrate that weathering, dissolution, and ion exchange reactions are the main processes determining the water chemistry.[12,37,40–42] In hydrochemical studies, interpreting the relationships between ions on graphics provides very useful results. However, if the water chemistry is evaluated together with the analysis results of the rocks in the aquifer environment and the mineral saturation rates, more accurate results will be obtained. Although water chemistry has been revealed in the present study, the water quality and usability characteristics of groundwater have not been evaluated. In addition, the effects of these waters on human health should be investigated in detail.

Today, water consumption in the drinking, irrigation, industry, and energy sectors is constantly increasing in line with population growth. For this reason, future population projections should be made by determining the population in the study area, and water needs should be determined. In addition to increasing water consumption, climate change and pollution factors create difficulties in providing water to meet these needs. To protect existing water resources, local people should be made aware of water use and prevention of pollution through training/awareness studies. In addition, the expected climate change in the region and its effects on water resources should be revealed. In addition, the management of water resources should be given maximum importance within the scope of climate change adaptation studies. To realize sustainable groundwater management in the region, the proposed studies should be carried out, and alternative water resources should be investigated considering the obtained results. For the sustainable usage of groundwater, it is necessary to anticipate the reactions of aquifers to natural and/or human-induced effects and to take the necessary precautions. When water is drawn from a well in an aquifer environment in the study area, it is necessary to monitor whether the water levels in other wells are affected. In addition, it is important to understand the local mix processes of pollutants that influence groundwater. Performing online monitoring will allow rapid interventions in terms of water management.

It is not sufficient to consider only the aquifer mass in the assessment of both the quantity and quality of groundwater. In water management, the terrestrial ecosystem is extremely important, along with the surface waters with which the aquifer interacts. In terms of pollution, pollution in the terrestrial environment will inevitably mix into groundwater with the effect of leaching over time and rainfall. Similarly, the decrease in surface water decreases groundwater recharge over time for an aquifer environment fed from surface waters. Therefore, to achieve good groundwater status in terms of both quantity and quality, the aquatic and terrestrial ecosystems in the region must be in good condition. For this, it is necessary to carry out the necessary controls and take precautions by local authorities.

4 Conclusion

The hydro-chemical properties of groundwater and also and potential geogenic pollution were investigated in the Akşehir district and surrounding areas using GIS. The main usage areas of groundwater in the region are drinking water, using, and irrigation water. 31 water samples were collected from well waters in October 2018 and their chemical measurements and analyzes were carried out. In this study, the physicochemical and major ion concentrations of the water samples were evaluated using a Piper diagram to determine the hydrogeochemical facies of the samples taken from the wells. According to the results obtained from the piper diagram, Ca-Mg-HCO$_3$ and Ca-HCO$_3$ were determined to be the major water types in the study area. This shows that water-rock interaction and anthropogenic inputs are powerful factors in the formation of the chemical composition of water samples. According to Gibbs diagrams, the main processes that are effective in the formation of water chemistry are the decomposition of minerals due to rock–water interactions and evaporation. In addition, the samples fell inside the rock dominance and evaporation areas. The chemical structure of groundwater has developed under the influence of multiple hydrochemical processes such as calcite dissolution, silicate decomposition, and ion exchange. As a result, the use of groundwater in the study area is not suitable because of geogenic ion pollution, such as Na, Cl, and SO$_4$. Using these water sources as drinking water adversely affects human health. All these should be taken into consideration by the authorities, and alternative sources of drinking water should be investigated.

References

1. Egbueri JC, Ezugwu CK, Unigwe CO, Onwuka OS, Onyemesili OC, Mgbenu CN. Multidimensional analysis of the contamination status, corrosivity and hydrogeochemistry of groundwater from parts of the Anambra Basin, Nigeria. *Anal Lett.* 2020;54:1–31.
2. Redwan M, Moneim AAA, Amra MA. Effect of water–rock interaction processes on the hydrogeochemistry of groundwater west of Sohag area, Egypt. *Arab J Geosci.* 2016;9(2):111.
3. Li F, Zhu J, Deng X, Zhao Y, Li S. Assessment and uncertainty analysis of groundwater risk. *Environ Res.* 2018;160:140–151.
4. Liu J, Peng Y, Li C, Gao Z, Chen S. An investigation into the hydrochemistry, quality and risk to human health of groundwater in the central region of Shandong Province, North China. *J Clean Prod.* 2021;282:125416.
5. Rao NS, Chaudhary M. Hydrogeochemical processes regulating the spatial distribution of groundwater contamination, using pollution index of groundwater (PIG) and hierarchical cluster analysis (HCA): a case study. *Groundw Sustain Dev.* 2019;9:100238.
6. Adimalla N. Groundwater quality for drinking and irrigation purposes and potential health risks assessment: a case study from semi-arid region of South India. *Expos Health.* 2019;11(2):109–123.
7. González-Valoys AC, Vargas-Lombardo M, Higueras P, García-Navarro FJ, García-Ordiales E, Jiménez-Ballesta R. Hydrochemistry of groundwater from Tocumen sector, Panamá city: an assessment of its possible usage during emergency events. *Environ Earth Sci.* 2021;80(5):1–11.
8. Ijioma UD. Delineating the impact of urbanization on the hydrochemistry and quality of groundwater wells in Aba, Nigeria. *J Contam Hydrol.* 2021;240:103792.
9. Kazi TG, Arain MB, Jamali MK, et al. Assessment of water quality of polluted lake using multivariate statistical techniques: a case study. *Ecotoxicol Environ Saf.* 2009;72:301–309.
10. Li P, Tian R, Xue C, Wu J. Progress, opportunities and key fields for groundwater quality research under the impacts of human activities in China with a special focus on western China. *Environ Sci Pollut Res.* 2017;24(15):13224–13234.
11. Şener Ş, Şener E, Nas B, Karagüzel R. Combining AHP with GIS for landfill site selection: a case study in the Lake Beyşehir catchment area (Konya, Turkey). *Waste Manag.* 2010;30(11):2037–2046.
12. Varol S, Davraz A. Assessment of geochemistry and hydrogeochemical processes in groundwater of the Tefenni plain (Burdur/Turkey). *Environ Earth Sci.* 2014;71(11):4657–4673.
13. Yang Q, Li Z, Xie C, Liang J, Ma H. Risk assessment of groundwater hydrochemistry for irrigation suitability in Ordos Basin, China. *Nat Hazards.* 2020;101(2):309–325.
14. Zhang Q, Kang S, Wang F, Li C, Xu Y. Major ion geochemistry of Nam Co Lake and its sources, Tibetan Plateau. *Aquat Geochem.* 2008;14(4):321–336.
15. Busico G, Cuoco E, Kazakis N, et al. Multivariate statistical analysis to characterize/discriminate between anthropogenic and geogenic trace elements occurrence in the Campania Plain, Southern Italy. *Environ Pollut.* 2018;234:260–269.
16. Nadiri AA, Moghaddam AA, Tsai FT, Fijani E. Hydrogeochemical analysis for Tasuj plain aquifer, Iran. *J Earth Syst Sci.* 2013;122(4):1091–1105.
17. Sargın AH. *Groundwaters.* Ankara: General Directorate of State Hydraulic Works (SHW) Geotechnical Services and Groundwater Department; 2010:200.
18. Yetiş AD. *Determination of Ceylanpınar Plain Groundwater Quality and Pollution Potential.* Adana, Turkey: Çukurova University, Institute of Science, Department of Environmental Sciences; 2013. PhD thesis. Ç. Ü. Supported by Research Projects Unit. Project No: MMF2012D5.
19. Yin Z, Lin Q, Xu S. Using hydrochemical signatures to characterize the long-period evolution of groundwater information in the Dagu River Basin, China. *Front Environ Sci Eng.* 2021;15(5):1–13.
20. He X, Wu J, He S. Hydrochemical characteristics and quality evaluation of groundwater in terms of health risks in Luohe aquifer in Wuqi County of the Chinese Loess Plateau, Northwest China. *Hum Ecol Risk Assess Int J.* 2019;25(1–2):32–51.
21. Wagh VM, Mukate SV, Panaskar DB, Muley AA, Sahu UL. Study of groundwater hydrochemistry and drinking suitability through water quality index (WQI) modelling in Kadava river basin, India. *SN Appl Sci.* 2019;1(10):1–16.
22. Abolished Ministry of Forestry and Water Affairs. *Akarçay Basin Hydrogeological Study Report*; 2013. Ankara, Turkey.
23. Piper AM. A graphic procedure in the chemical interpretation of water analysis. *Am Geophys Union Trans.* 1944;25:914–923.
24. Gibbs RJ. Mechanisms controlling worlds water chemistry. *Science.* 1970;170:1088–1090.
25. Şahinci A. *Geochemistry of Natural Waters.* Izmir: The Reform Printing Office; 1991.
26. Şener Ş, Karakuş M. Investigating water quality and arsenic contamination in drinking water resources in the Tavşanlı District (Kütahya, Western Turkey). *Environ Earth Sci.* 2017;76(21):750.
27. He S, Wu J. Hydrogeochemical characteristics, groundwater quality, and health risks from hexavalent chromium and nitrate in groundwater of Huanhe formation in Wuqi County, Northwest China. *Expos Health.* 2019;11(2):125–137.
28. Jeong CH. Effect of land use and urbanization on hydrochemistry and contamination of groundwater from Taejon area, Korea. *J Hydrol.* 2001;253:194–210.
29. Ranjan RK, Ramanathan AL, Parthasarthy P, Kumar A. Hydrochemical characteristics of groundwater in the plains of Phalgu river in Gaya, Bihar, India. *Arab J Geosci.* 2013;6:3257–3267.
30. Varol S, Köse İ. Effect on human health of the arsenic pollution and hydrogeochemistry of the Yazır Lake wetland (Çavdır-Burdur/Turkey). *Environ Sci Pollut Res.* 2018;25(16):16217–16235.
31. WHO (World Health Organization). *Guidelines for Drinking-Water Quality.* 4th ed. Geneva: WHO; 2011.
32. Kumar A, Abhay T, Sing K. Hydrogeochemical investigation and groundwater quality assessment of Pratapgarh district, Uttar Pradesh. *J Geol Soc India.* 2014;83:329–343.
33. Laxmankumar D, Satyanarayana E, Dhakate R, Saxena PR. Hydrogeochemical characteristics with respect to fluoride contamination in groundwater of Maheshwarm Mandal, RR district, Telangana state, India. *Groundw Sustain Dev.* 2019;8:474–483.
34. Annapoorna H, Janardhana MR. Assessment of groundwater quality for drinking purpose in rural areas surrounding a defunct copper mine. *Aquat Procedia.* 2015;4:685–692.
35. Kumar M, Kumari K, Singh UK, Ramanathan AL. Hydrogeochemical processes in the groundwater environment of Muktsar, Punjab:conventional graphical and multivariate statistical approach. *Environ Geol.* 2009;57:873–884.

36. Sarin MM, Krishnaswamy S, Dilli K, Somayajulu BLK, Moore WS. Major ion chemistry of the Ganga-Brahmaputra river system: weathering processes and fluxes to the Bay of Bengal. *Geochim Cosmochim Acta*. 1989;53:997–1009.

37. Bozdağ A. Assessment of the hydrogeochemical characteristics of groundwater in two aquifer systems in Çumra Plain, Central Anatolia. *Environ Earth Sci*. 2016;75(8):1–15.

38. Subramani T, Rajmohan N, Elango L. Groundwater geochemistry and identification of hydrogeochemical processes in a hard rock region, Southern India. *Environ Monit Assess*. 2010;162:123–137.

39. Tay CK. Hydrochemistry of groundwater in the Savelugu–Nanton District, Northern Ghana. *Environ Earth Sci*. 2012;67:2077–2087.

40. Davraz A, Varol S, Şener E, et al. Assessment of water quality and hydrogeochemical processes of Salda alkaline lake (Burdur, Turkey). *Environ Monit Assess*. 2019;191(11):1–18.

41. Kuşcu M, Şener Ş, Tuncay EB. Recharge sources and hydro geochemical evaluations of Na2SO4 deposits in the Acıgöl Lake (Denizli, Turkey). *J Afr Earth Sci*. 2017;134:265–275.

42. Şener Ş, Şener E, Davraz A, Varol S. Hydrogeological and hydrochemical investigation in the Burdur Saline Lake Basin, Southwest Turkey. *Geochemistry*. 2020;80(4):125592.

43. Yadav A, Nanda A, Sahu BL, et al. Groundwater hydrochemistry of Rajnandgaon district, Chhattisgarh, Central India. *Groundw Sustain Dev*. 2020;11:100352.

44. Umut M. *1:100,000 Scale Geological Map of Turkey, L-26 Layout*. Ankara: The General Directorate of Mineral Research and Exploration; 2009.

45. Umut M. *1:100,000 Scale Geological Map of Turkey, L-27 Layout*. Ankara: The General Directorate of Mineral Research and Exploration; 2009.

36

Comparison of the frequency ratio, index of entropy, and artificial neural networks methods for landslide susceptibility mapping: A case study in Pınarbaşı/Kastamonu (North of Turkey)

Enes Taşoğlu and Sohaib K.M. Abujayyab

Department of Geography, Karabuk University, Karabuk, Turkey

1 Introduction

Landslides are the motion of soil, rock, and debris that occur on a downward slope due to gravity. Landslide occurrence is caused by various drivers, such as geological, geomorphological, topographical, and seismic features of the area. These properties are separated into two categories: trigger and conditioning. The first category, which is the condition of the surface, involves slope degree, slope aspect, elevation, land use, soil types, stream density, etc. The second, which is associated with the triggering of a mass movement, contains rainfall, earthquakes, and anthropogenic changes in the area.[1–3]

Landslides are among the most damaging natural threats in several parts of the world, and their damage affects forests, agricultural areas, mines, roads, buildings, and infrastructure.[4] To reduce these damages, susceptibility maps are used in the risk-based land-use planning of the area.[5,6] Landslide susceptibility mapping (LSM) is described as the spatial likelihood of landslide occurrence.[7–9] The predicted landslide susceptibility map shows a classified area into zones according to different levels of susceptibility classes using the prediction method.[10,11] Many quantitative approaches have been used to detect landslide susceptibility. These approaches can be categorized into four main categories: landslide inventory-based heuristic, probabilistic, statistical, and deterministic methods.[12–14]

The deterministic techniques used in LSM are based on the laws of physical conservation of mass,[12] and the considerations utilized in these methods can be concluded during fieldwork or in the laboratory.[15] Generally, deterministic methods are locally and do not consider the spatial input parameters.[16] In addition, and deterministic models require knowledge of the mechanical properties of the surface and the degree of soil saturation. Thus, it is not applicable for LSM in medium-and large-scale areas.[14,17]

Heuristic methods are based on expert knowledge and field observations. The existing landslides in the area and the factors affecting them are well known.[9,13] Heuristic techniques also require comprehensive and long-term information about the area; therefore, they are difficult to apply.[1]

Statistical and probabilistic methods such as frequency ratio,[18–20] logistic regression,[21,22] linear regression,[23] tree-based,[24,25] fuzzy sets,[26,27] artificial neural networks,[28–31] weights of evidence,[32,33] and entropy-based methods,[10,34,35] are frequently used by researchers in LSM. These methodologies provide susceptibility maps based on the knowledge that can be measured quantitatively.[35–37] In the literature, several studies were employed only one technique for LSM, while there are some studies that compared these techniques.[6,14,16,19,25,35,38–46]

The main aim of this analysis was to compare the index of entropy (IoE), frequency ratio (FR), and artificial neural network (ANN) models in the LSM, in terms of their accuracy for predicting susceptibility maps. In line with the previous studies mentioned above, there is a deficiency in the existing literature on the comparison of the ANN and IoE models in LSM studies. This deficiency is considered to be caused by the relatively new use of the IoE method in LSM studies. To the best of our knowledge, none of the earlier literature has compared these three methods. In this way, it is thought that this study will provide a different perspective on methodology selection in LSM.

2 Materials and methods

2.1 Study area

Pınarbaşı is the district of Kastamonu Province in the Black Sea region of Turkey. The study area is located north of Turkey and approximately 13 km from the Black Sea (Fig. 1). The study area is geographically located at 41.6661° in the north and 33.0350° in the east. The total administrative area of Pınarbaşı is 545 km², with a population of 5841 people in 2019. The altitude ranges varied between 326 and 1475 m. One of the most important geomorphological units is the Küre Mountains, located east of the study area. In addition to the Küre Mountains National Park and Valla Canyon, the study area has many tourist sites such as canyons, caves, waterfalls, and historical places.[47–49] warm and lukewarm weather conditions in summer and rainfall are observed during the year (Cfa, Köppen classification); therefore, summer droughts do not occur in the study area.[50,51] Devrekani Creek is the most important fluvial factor for an area with an existing geomorphological landscape. Landslides generally occur on the slopes of the valleys and in the high mountainous areas of the study area, which have been eroded by deep valleys by the Devrekani Creek and its tributaries.

2.2 Geospatial database

To develop an LSM, it is necessary to know about landslides and their conditioning factors.[12,19] To obtain these explanatory factors, a geospatial database involving spatial data representing the area is required.[20] including topographic elevation, topographic wetness index (TWI), aspect, slope, curvature, profile curvature, roads, lithology, soil types, streams, and plan curvature, were the principal landslide occurrence factors considered for the LSM of the study

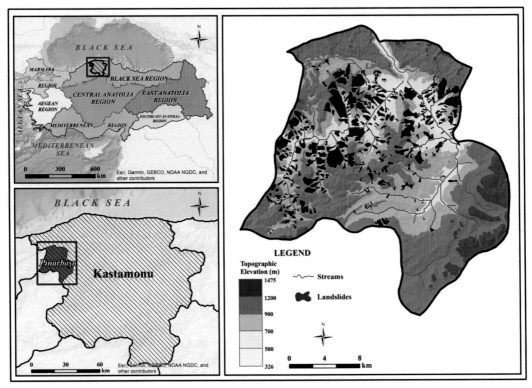

FIG. 1   Location of the study area.

TABLE 1 Selected landslide conditioning parameters, their sources, descriptions, and the studies that used them.

Factors	Sources	Explanations	Used by
Landslide inventory	MTA	involve landslide areas	–
Topographic elevation	ASTER	30 m DEM	52–54
Aspect	ASTER	extract from DEM	55–57
Slope	ASTER	extract from DEM	58–60
Curvature	ASTER	extract from DEM	61–63
Plan curvature	ASTER	extract from DEM	33,42,43
Profile curvature	ASTER	extract from DEM	53,64,65
TWI	ASTER	calculated using DEM	66–68
Distance to streams	ASTER	Euclidean distance from streams	69–71
Distance to roads	OSM	Euclidean distance from roads	20,71,72
Lithology	MTA	Distribution of rocks	10,67,73
Soil types	Agriculture and Forest Ministry	Distribution of soil types	74,75

DEM is the digital elevation model data.
MTA is the General Directorate of Mineral Research and Exploration (Maden Tetkik Arama Genel Müdürlüğü).
OSM is the Open Street Map.
ASTER is the Advanced Spaceborne Thermal Emission and Reflection Radiometer.

area (Table 1). The table shows a list of employed conditioning factors and an example from their application in previous studies. The entire layer has a 30 m spatial resolution of 30 × 30 m based on the WGS 84 UTM Z36 N coordinate system. The study area represented 373,185 raster cells in a geographic information system (GIS) framework. In addition, considering the distribution of landslides in the study area, all parameters were divided into classes (Table 2).

2.2.1 Inventory map of landslide

The inventory map contains landslides that happened previously and was used to create the LSM. Considering that "the past and present are keys to the future" or uniformitarianism concepts, the former landslide data are the basic data for LSM prediction.[12,52] In addition, the inventory contributes to understanding the characterization of landslide formation in the area and prediction of future landslides.[19] The landslide inventory map used in this study was collected from the General Directorate of Mineral Research and Exploration (MTA). Then, the polygons representing the landslide areas were geometrically corrected using Google Earth images. Consequently, 293 active landslide areas were identified within the study area. The total landslide area was 70 km^2 (Fig. 2). Based on investigations from satellite images, debris slides and light slides are the main types of landslides.

2.2.2 Topographic elevation

Topographic altitude is described as the altitude of a surface area from the mean sea level. An elevation map was obtained from the digital elevation model (DEM).[53,54] The elevation map indicates the level at which landslides are clustered. For this reason, the use of topographic elevation factors is frequently preferred by researchers in LSM.[55–57] The topographic elevation values varied between 326 and 1475 m in the study area (Fig. 3A), and 47% of the landslide areas were observed at altitudes between 900 and 1200 m. In this study, topographic elevation values were divided into five subclasses (−356 to 500, 500–700, 700–900, 900–1200, and 1200–1475) considering the distribution of landslides (Table 2).

2.2.3 Aspect

The aspect values of the study area were produced using DEM. Aspect is an important criterion for LSM because it is associated with exposure to sunlight, wind, and rainfall.[54,58] The aspect parameters were divided into ten subclasses (aspect map) according to the compass angle: flat, northwest, west, southwest, south, southeast, east, northeast, and north. Most landslides occurred in the northeast, east, and southeast (46%) of the study area (see Fig. 3B; Table 2).

TABLE 2 Landslide conditioning factors and their FR and IoE values.

Data layers	Class	Class pixels	Landslide pixels	FR	IoE
Elevation (m)	356–500	14,527	3026	1.612	0.087
	500–700	48,174	12,989	2.086	
	700–900	110,404	21,593	1.513	
	900–1200	175,946	10,626	0.467	
	1200–1475	24,134	0	0.000	
Total				**5.678**	
Aspect	Flat	23	1	0.336	0.095
	North	25,541	3201	0.970	
	Northeast	46,589	7095	1.178	
	East	43,844	8590	1.516	
	Southeast	43,236	6838	1.224	
	South	48,134	5009	0.805	
	Southwest	45,068	4964	0.852	
	West	46,925	6082	1.003	
	Northwest	49,113	4346	0.685	
	North	24,712	2108	0.660	
Total				**9.229**	
Slope (degree)	0–2	5465	131	0.185	0.091
	2–6	37,362	1598	0.331	
	6–12	102,254	11,238	0.850	
	12–20	143,571	23,802	1.283	
	20–30	70,041	10,657	1.177	
	30–71	14,492	808	0.431	
Total				**4.258**	
Curvature	Concave (<−0.05)	179,111	24,355	1.052	0.096
	Flat (−0.05–0.05)	16,533	2065	0.966	
	Convex (>0.05)	177,541	21,814	0.951	
Total				**2.969**	
Plan curvature	Concave (<−0.05)	156,887	22,203	1.095	0.096
	Flat (−0.05–0.05)	53,015	6747	0.985	
	Convex (>0.05)	163,283	19,284	0.914	
Total				**2.993**	
Profile curvature	Concave (<−0.05)	165,735	21,195	0.989	0.096
	Flat (−0.05–0.05)	36,356	4415	0.940	
	Convex (>0.05)	171,094	22,624	1.023	
Total				**2.952**	
TWI	2–5	42,462	5463	0.995	0.096
	5–7	224,976	26,457	0.910	
	7–9	76,716	11,236	1.133	

TABLE 2 Landslide conditioning factors and their FR and IoE values—cont'd

Data layers	Class	Class pixels	Landslide pixels	FR	IoE
	9–11	22,284	4073	1.414	
	11–18	6747	1005	1.152	
Total				**5.605**	
Distance to streams (m)	0–300	68,538	8467	0.956	0.096
	300–600	64,328	10,274	1.236	
	600–900	60,076	9794	1.261	
	900–1200	53,634	7930	1.144	
	1200–1500	42,734	6264	1.134	
	1500–5175	83,875	5505	0.508	
Total				**6.239**	
Distance to roads (m)	0–100	105,009	18,197	1.341	0.092
	100–200	70,054	11,994	1.325	
	200–300	41,973	7174	1.322	
	300–400	30,991	4730	1.181	
	400–500	20,317	2561	0.975	
	500–1000	46,309	3365	0.562	
	1000–4655	58,532	213	0.028	
Total				**6.734**	
Lithology	Limestone	103,300	126	0.009	0.073
	Sandstone-Mudstone	244,903	48,019	1.514	
	Alluvium	7843	106	0.104	
	Sandstone	1805	0	0.000	
	Riodasite	1739	71	0.315	
	Pebble	9015	0	0.000	
	Pebble-Sandstone-Mudstone	240	0	0.000	
	Andesite-Dasite-Latite	93	12	0.996	
	Olistostrome	4311	18	0.032	
Total				**2.971**	
Soil types	non-soil	1142	405	2.738	0.081
	Aluvial	9694	146	0.116	
	Gray-Brown Podsolic	265,120	47,123	1.372	
	Koluvial	2392	211	0.681	
	Brown Forest	94,636	423	0.035	
Total				**4.942**	

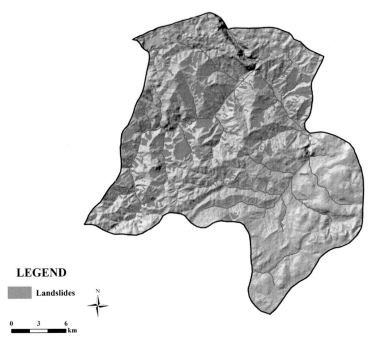

LEGEND

Landslides

N

0   3   6
km

FIG. 2   Landslide inventory map of the study area.

2.2.4 Slope

The slope map is the main parameter for the LSM, as it is directly related to the mass movement. Therefore, slopes are frequently used in LSM studies.[59,60] The slope values varied between 0 and 71 degree in the study area (Fig. 3C). In this study, the slope values were divided into six subclasses: −0 to 2, 2–6, 6–12, 12–20, 20–30, and 30–71. Landslide pixels were generally (49%) in the 12–20 degree slope class (Table 2).

2.2.5 Curvature, plan curvature, and profile curvature

The curvature values, morphologically flat (−0.05–0.05), convex (>0.05), or concave (<−0.05), are represented. Plan curvature is the slope curving with a flat plane, that is, the curvature of the contour lines. The hillsides, which have a planar plane curvature, are more susceptible to landslide occurrence, especially in clay soil.[61] Profile curvature is defined as curving in the downward direction of the slope and the slope change rate.[54,61] According to Jebur et al.,[60] landslides are more likely to occur because water can remain longer in flat areas. However, according to Lee et al.,[59] the more strongly negative (concave) or strongly positive (convex) curvature value, the higher the probability of landslide occurrence. According to Ohlmacher,[61] the more concave or convex curvature, the possibility of landslide occurrence decreases. In this study, the distribution of the landslide pixels was roughly equal in curvature, plan curvature, and profile curvature in the study area (Fig. 3; Table 2).

2.2.6 Topographic wetness index (TWI)

The topographic wetness index (TWI) explains runoff formation's impact on the location and size of the moisturized zones. The TWI is the percentage among the slope and specific catchment area.[38,62,63] Additionally, TWI indicates the tendency for water to accumulate anywhere in the area. Although the infiltration of water into the soil relies on material properties such as permeability and porosity, TWI provides a topographical explanation of infiltrating water control.[10] TWI has been calculated using the following equation (Eq. 1).

$$TWI = \ln\left(\frac{A_s}{\beta}\right),\qquad(1)$$

where

A_s = area of a particular catchment (m/m^2), and
β = angle of the slope using degrees units.

In this study, the TWI results were separated into five groups: as −2 to 5, 5–7, 7–9, 9–11, and 11–18 (Fig. 4G; Table 2).

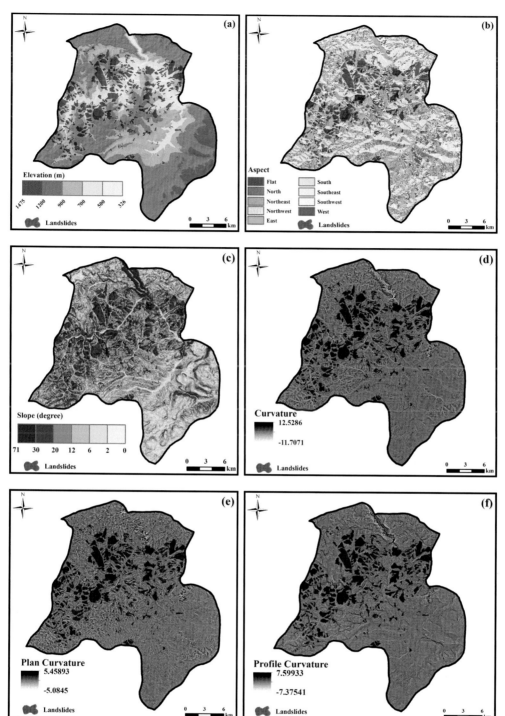

FIG. 3   Geospatial database. (A) topographic elevation, (B) aspect, (C) slope, (D) curvature, (E) plan curvature, (F) profile curvature.

2.2.7 Distance to streams

Distance to streams is described as the closeness of the drainage lines and streams. Distance to streams is another important factor for LSM. closeness from streams increases the water saturation of the mass on the hillsides and changes the balance of the mass. Therefore, as the balance of the mass is disturbed, susceptibility to landslides increases.[18,64] The distance to the stream layer is determined utilizing the Euclidean distance function in the GIS. Euclidean distance map was then divided into six groups: −0 to 300, 300–600, 600–900, 900–1200, 1200–1500, and >1500 (Fig. 4H; Table 2).

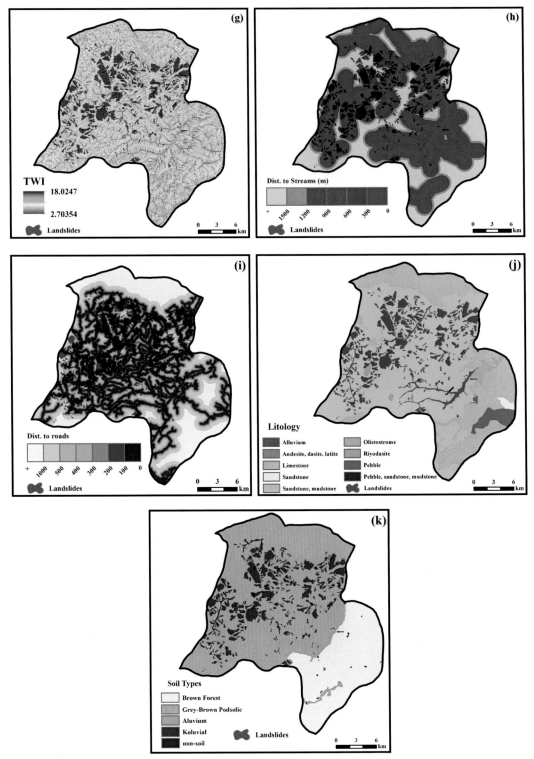

FIG. 4 Geospatial database. (G) TWI, (H) distance to streams, (I) distance to roads, (J) lithology, (K) soil types.

2.2.8 Distance to roads

The roads can cause a barrier or corridor for water flow depending on the extension line. In addition, they can disrupt the load balance of the mass on the slope.[18,32] Roads are used in LSM because man-made factors trigger the occurrence of landslides by changing the original appearance of the topography of the area. In this study, the road data are provided from the Open Street Map (OSM), which then produces the distance to the road layer using the Euclidean distance function. After that, it was divided into seven classes: −0 to 100, 100–200, 200–300, 300–400, 400–500, 500–1000, and >1000 (Fig. 4I; Table 2).

2.2.9 Lithology

There is a strong relationship between lithology and landslide occurrence areas. The lithology indicates the properties of the ground, such as the hardness of the rock and clay content. Therefore, the lithology layer is a crucial factor in explaining the LSM of the study area.[9,40,65] The distribution of lithological units in the study area was obtained from the MTA. Nine different lithological units exist in the Pınarbaşı area: limestone, sandstone-mudstone, alluvial, sandstone, riodasite, pebble, pebble-sandstone-mudstone, andesite-dasite-latite, and olistostrome. The sandstone and mudstone units tended to be landslides in the study area (Fig. 4J; Table 2).

2.2.10 Soil types

The soil type layer includes the ground structural features such as the depth, clay ratio, water holding capacity, porosity, and permeability for each soil type. Hence, the soil type factor is frequently used in LSM.[66–68] Soil type data were obtained from the Agriculture and Forest Ministry of Turkey. According to the data, there are four main soil types in Pınarbaşı: alluvial, gray-brown podsolic, colluvial, and brown forest soils. Additionally, the data also included non-soil areas, such as settlements and bare rocks (Fig. 4K; Table 2).

2.3 Landslide susceptibility mapping (LSM)

2.3.1 Frequency ratio (FR) model

Landslide occurrence occurred under different conditions in each area. According to common thought, the occurrence factors of previous landslides that took place in the area provide an understanding of possible future landslide conditions.[64,69,70] In line with the aforementioned approach, the frequency ratio (FR) is utilized to illuminate the relationships between landslides and their occurrence factors. The FR was calculated using the landslide area percentage to the non-landslide area[64,69]. FR was calculated using the following equations (Eqs. 2, 3).[20,71]

$$FR_{ij} = \frac{LP_{ij}}{TP_{ij}}. \tag{2}$$

where

FR_{ij} = frequency ratio value of group i of factor j,
LP_{ij} = landslide areas pixel number of group i of factor j, and
TP_{ij} = total pixel number of group i of factor j.

$$LSM = \sum_{j=1}^{n} FR_{ij} \tag{3}$$

The FR_{ij} values represent the weights of landslide occurrence factors, and these values are summed to produce the LSM.

FR values greater than 1 indicate a strong association between the relevant criteria and landslide occurrence; on the contrary, FR values less than 1 indicate a weak relationship.[20]

2.3.2 Index of entropy (IoE) method

Entropy is described as the changeability, indecision, unbalance, indefiniteness of any approach and measuring the tendency of prediction improvement.[72] In addition, entropy is the quantitative determination of the relationship between the causes and results of any related topic.[20] The entropy value of the system is correlated with the degree of complexity. The correlation called the Boltzmann principle represents the thermodynamic conditions of a method.[10] In which LSM the value of entropy indicates the potential of each factor to cause landslide occurrence. The greater the entropy rate of the specified criteria, the greater is the impact of the criteria on landslide occurrence. On the other hand, a lower entropy rate represent a lower impact of the factor on landslide occurrence.[73] as shown in Eqs. (4)–(8) are used to calculate the index of entropy (IoE) of each factor [20]:

$$E_{ij} = \frac{FR}{\sum_{i=1}^{M_j} FR}. \tag{4}$$

where FR is the frequency ratio and E_{ij} is the likelihood intensity for each category:

$$H_j = -\sum_{i=1}^{M_j} E_{ij} \log_2 E_{ij}, j = 1, \ldots, n \tag{5}$$

$$H_{jmax} = \log_2 M_j, M_{j-number\ of\ classes} \tag{6}$$

$$I_j = \left(H_{jmax} - \frac{H_j}{H_{jmax}} \right), I = (0, 1), j = 1, \ldots \tag{7}$$

$$V_j = I_j FR \tag{8}$$

where

H_j and H_{jmax} are the values of entropy,
I_j is information coefficient,
M_j is the number of classes in each factor, and
V_j is the weight value obtained for a specified factor. The range of values was between 0 and 1. A value close to 1 indicates higher uncertainty and instability.

2.3.3 Artificial neural networks (ANN) model

Artificial neural networks (ANNs) have an exceptional capability to derive meaning from complicated and even missing datasets. With this capability, the ANN can recognize and classify complex patterns.[6] The structure of an ANN, which is a data-driven technique, comprises three basic stages: inputs, hidden neurons, and output.[30] The learning process carried out through training and test datasets was implemented with the neurons connected to each other.[31] Learning occurs when the weight values held in the connections between neurons reach optimal values. The landslide susceptibility of an area can be addressed as a classification problem. Accordingly, the outputs of the ANN model can be regarded as landslide susceptibility classes in the area.[28]

In this study, 1000 sample points were generated in the GIS framework for the training process of the ANN. The sample points were randomly distributed in the landslide and non-landslide areas. The samples were then extracted from the values of eleven landslide conditioning criteria (aspect, elevation, curvature, plan curvature, lithology, soil types, profile curvature, TWI, water streams, slope, and roads), which were later utilized as inputs for the ANN model. Sample points were randomly separated into training (70%), testing (15%), and validation (15%), so that the training and validation processes could be carried out. Additionally, the ANN was designed using 35 hidden neurons and by using the scaled conjugate gradient (SCG) algorithm to predict the landslide susceptibility classes of the Pınarbaşı district (Fig. 5).

Statistical validation, such as the LSM, is a critical issue in predictive modeling. Unless statistically validated the results of any prediction model, the prediction model was deprived of scientific assessment.[37,74] In this study, the LSM models were assessed using receiver operating characteristic (ROC) analysis. ROC expresses the true-positive values of past landslides as a percentage of the percentage of false-positive values in cumulatively reducing order.[37] Then and the area under the ROC curve (AUC) value was obtained. AUC accuracy indicated the success of the predictive model. AUC values varying between 0.5 and 1 indicate that the closer to 1, the better the model.[10,46,75]

3 Results and discussion

3.1 Results of the FR model

The spatial relationship between the landslide inventory and landslide conditioning factors was assessed using the FR method. Statistical analyses were performed using Excel. Then, Eq. (9) was obtained and implemented in the GIS framework.

INPUTS

HIDDEN
NEURONS

OUTPUT

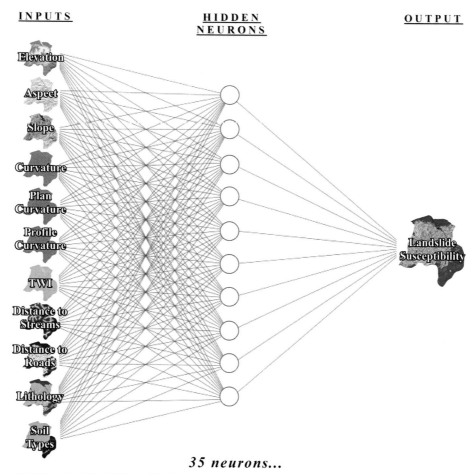

35 neurons...

FIG. 5   Architecture of ANN method for LSM used in the study area.

$$FR = (elevation_{FR}*5.678) + (aspect_{FR}*9.229) + (slope_{FR}*4.258) + (curvature_{FR}*2.969) + (plancurvature_{FR}*2.993)$$
$$+ (profilecurvature_{FR}*2.952) + (TWI_{FR}*5.605) + (distostreams_{FR}*6.239) + (distoroads*6.734) + (lithology_{FR}*2.971)$$
$$+ (soil_{FR}*4.942)$$

(9)

The results of the FR analysis were mapped to produce LSM and categorized into five distinct landslide susceptibility classes using the Natural Breaks (Jenks) algorithm (Fig. 6). According to the LSM of the FR method, the study area was graded into five categories; (1) very low susceptibility, (2) low susceptibility, (3) moderate susceptibility, (4) high susceptibility, (5) very high susceptibility, and their share of the total area was 18%, 13%, 9%, 27%, and 33%, respectively (Table 3).

3.2 Results of the IoE model

The relevant equations were applied to produce the LSM of the study area using the IoE method. The IoE values obtained from the results represent the weights of each landslide conditioning factor, and the FR values represent the weights of the subclasses of the factors. The mapping process was carried out in the GIS framework by utilizing all the weights obtained as the last step of LSM production. For this purpose, Eq. (10) was performed using the raster calculator.

$$IoE = (elevation_{FR}*0.087) + (aspect_{FR}*0.095) + (slope_{FR}*0.091) + (curvature_{FR}*0.96) + (plancurvature_{FR}*0.096)$$
$$+ (profilecurvature_{FR}*0.096) + (TWI_{FR}*0.096) + (distostreams_{FR}*0.096) + (distoroads*0.092) + (lithology_{FR}*0.073)$$
$$+ (soil_{FR}*0.081)$$

(10)

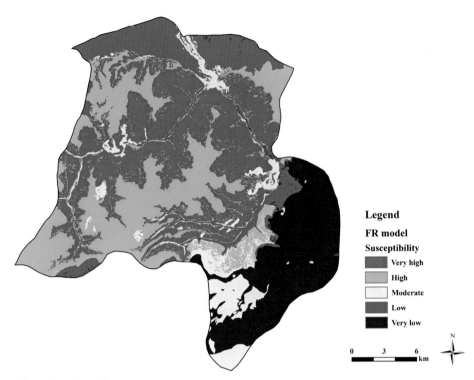

FIG. 6 LSM obtained from the FR model.

The LSM acquired from the analysis was categorized into five different landslide susceptibility classes (very low, low, moderate, high, and very high) using the Natural Breaks (Jenks) algorithm (Fig. 7). The shares of the classes representing the susceptibility zones in the total area are 13% for very low, 18% for low, 14% for moderate, 26% for high, and 28% for very high (Table 4).

3.3 Results of the ANN model

After creating the ANN model, all the raster cells were transformed into points. The study area had 373,185 points. All landslide conditioning factors were extracted from the points within the GIS framework. The attribute table of the points was presented to the ANN model, and the landslide susceptibility prediction process was completed. In the last step, the predicted value of each point was converted to a raster format, and the LSM of the ANN model was produced (Fig. 8).

In accordance with the LSM of the ANN model, the study area revealed five landslide susceptibility classes based on the natural breaks (Jenks) algorithm. The first category, representing very low susceptibility, shares 27% of the total area, following the categories of low, moderate, high, and very high susceptibility with shares of 12%, 11%, 22%, and 27%, respectively (Table 5).

TABLE 3 Percentage of the susceptibility classes of the FR model.

Susceptibility classes	Area (km^2)	Percentage (%)
Very low	95	18
Low	72	13
Moderate	47	9
High	150	27
Very high	181	33
Total	**545**	

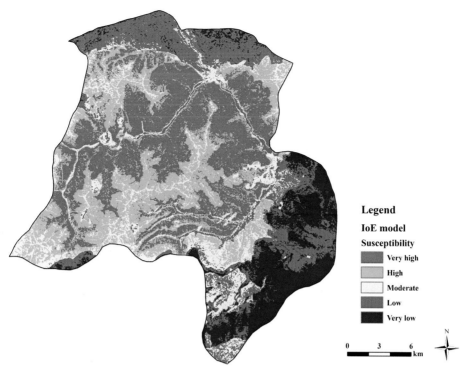

FIG. 7 LSM of the IoE model.

3.4 Validation and comparison of the models

According to the calculations of the AUC for the FR, IoE, and ANN methods, the success rates were 0.873, 0.869, and 0.962, correspondingly. Outcomes revealed that the ANN method had the maximum performance accuracy. In addition, all the models applied have satisfactorily high accuracy for predicting landslide susceptibility in the Pınarbaşı district (Fig. 9).

Several methods have been used in LSM, and in some previous studies, one or two of the three methods used in this research have been compared. In this study, parameters such as slope, elevation, aspect, plan curvature, water streams, profile curvature, TWI, roads, curvature, lithology, and soil types were used, and the ANN model had the highest prediction rate.

According to previous studies, the predictive performance of the FR model is quite similar to that of the IoE model. For example, Hong et al. used 15 factors (slope aspect, slope angle, plan curvature, elevation, roads, faults, land use, rivers, STI, normalized difference vegetation index, topographic wetness index, SPI, lithology, and rainfall) to analyze the susceptibility of landslide in Chongren area (China) using FR and IoE methods.[69] The models' accuracy results were FR (0.801) and IoE (0.817). Jaafari et al. used factors as slope angle, elevation, forest canopy, SPI, faults, slope aspect, topographic wetness index, lithology, STI, streams, plan curvature, normalized difference vegetation index, timber volume, and plant community.[38] According to their study, the prediction rates of FR and IoE models have been shown 0.726, 0.755, respectively. Nohani et al. reported prediction rates of 0.83 and 0.82 for FR and IoE models,

TABLE 4 Percentage of the landslide susceptibility classes of the IoE model.

Susceptibility classes	Area (km^2)	Percentage (%)
Very low	72	13
Low	99	18
Moderate	78	14
High	142	26
Very high	153	28
Total	**545**	

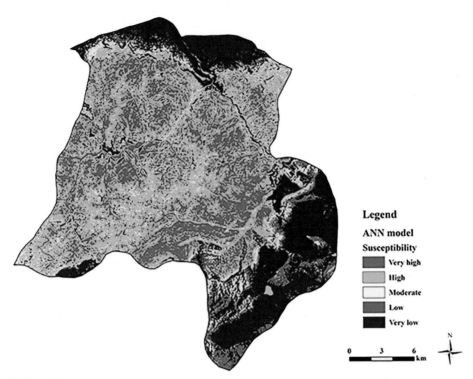

FIG. 8   LSM of the ANN model.

TABLE 5   Percentage of the landslide susceptibility classes of the ANN model.

Susceptibility classes	Area (km²)	Percentage (%)
Very low	147	27
Low	65	12
Moderate	62	11
High	122	22
Very high	149	27
Total	**545**	

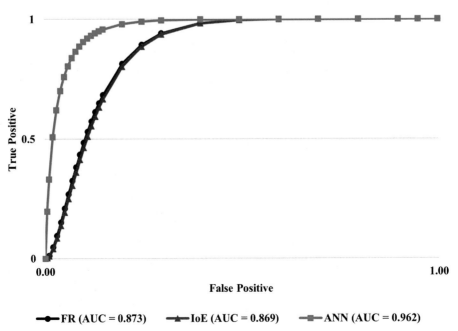

FIG. 9   AUC for the LSM produced by the three models.

respectively, and used factors such as curvature, river, lithology, faults, normalized difference vegetation index, elevation, slope, road, aspect, and land use.[20] On the other hand, Qui et al. compared FR and ANN methods and have been reached 0.89, 0.88 AUC values for the models, respectively.[76] Işık reported AUC outcomes of 0.826 and 0.852 AUC values for the FR and ANN models, correspondingly.[6] Pradhan and Lee used factors such as distance to lineament, soil, slope, curvature, aspect, distance to drainage, land use, and geology, NDVI, precipitation, and reported AUC outcomes of 0.864 for FR method and 0.835 for ANN method.[44] Park et al. reported AUC outcomes of 0.794 for the FR method and 0.806 for the ANN method.[16]

According to previous studies, the ANN model generally exhibits the best performance accuracy. In this study, the accuracy of the ANN model achieved the best accuracy of 0.962. Similar to the literature, the FR and IoE models achieved almost the same performance accuracy (0.873 and 0.869). However, ANN modeling requires advanced skills, such as coding, and is difficult to understand. In addition, ANNs require massive computation capacity compared with FR and IoE models, and it is very difficult to apply it in GIS software as well as required, for example, MATLAB software. In this case, the FR and IoE models are preferred and are easy to apply. The FR and IoE models require less computation and can be used directly in the GIS environment. In other words, the FR and IoE models are considered qualitative methods, which can be affected by the bias of human interference and their limited knowledge. In contrast, the ANN model is quantitative and can be more trusted. The use of an ANN quantitative model is suitable for landslide susceptibility mapping, especially in remote, isolated, and unexplored areas for which there is no information about physical, human, or spatial characteristics.

The scope of the present study was limited to a comparison of the three methods. Hence, more methods such as weight of evidence, logistic regression, support vector machine, random forest, fuzzy logic, and hybrid models can be functional for future studies to improve the capacity of comparisons.

4 Conclusions

The objective of this article was to compare the FR, IoE, and ANN methods according to their predictive performance in LSM. Eleven landslide conditioning considerations, for instance, elevation, curvature, distance to roads, profile curvature, TWI, water streams, slope, lithology, aspect, plan curvature, and soil types were selected and analyzed using the relevant methods. The obtained results were then mapped in the GIS framework. After all the analysis and mapping processes, the accuracy of the results was assessed using the ROC-AUC technique. The AUC outcomes of FR, IoE, and ANN models were 0.873, 0.869, and 0.962, correspondingly. Regarding the AUC outcomes, all models showed reasonable results. Additionally, the AUC values showed that the ANN method had the maximum precision. The AUC value of the FR model was identical to that of the IoE model. Therefore, the ANN model can be used in LSM research to obtain a high-accuracy LSM.

References

1. Dai FC, Lee CF, Li J, Xu ZW. Assessment of landslide susceptibility on the natural terrain of Lantau Island, Hong Kong. *Environ Geol.* 2001; 40(3):381–391.
2. Çevik E, Topal T. GIS-based landslide susceptibility mapping for a problematic segment of the natural gas pipeline, Hendek (Turkey). *Environ Geol.* 2003;44:949–962. https://doi.org/10.1007/s00254-003-0838-6.
3. Pourghasemi HR, Yansari ZT, Panagos P, Pradhan B. Analysis and evaluation of landslide susceptibility: a review on articles published during 2005–2016 (periods of 2005–2012 and 2013–2016). *Arab J Geosci.* 2018;11(193):4–12. https://doi.org/10.1007/s12517-018-3531-5.
4. Reichenbach P, Busca C, Mondini AC, Rossi M. The influence of land use change on landslide susceptibility zonation: the briga catchment test site (Messina, Italy). *Environ Manag.* 2014;54:1372–1384. https://doi.org/10.1007/s00267-014-0357-0.
5. Fell R, Corominas J, Bonnard C, Cascini L, Leroi E, Savage WZ. Guidelines for landslide susceptibility, hazard and risk zoning for land use planning. *Eng Geol.* 2008;102(3–4):85–98. https://doi.org/10.1016/j.enggeo.2008.03.022.
6. Yilmaz I. Landslide susceptibility mapping using frequency ratio, logistic regression, artificial neural networks and their comparison: a case study from Kat landslides (Tokat-Turkey). *Comput Geosci.* 2009;35(6):1125–1138. https://doi.org/10.1016/j.cageo.2008.08.007.
7. Guzzetti F, Reichenbach P, Cardinali M, Galli M, Ardizzone F. Probabilistic landslide hazard assessment at the basin scale. *Geomorphology.* 2005;72:272–299. https://doi.org/10.1016/j.geomorph.2005.06.002.
8. Constantin M, Bednarik M, Jurchescu MC, Vlaicu M. Landslide susceptibility assessment using the bivariate statistical analysis and the index of entropy in the Sibiciu Basin (Romania). *Environ Earth Sci.* 2011;63(2):397–406. https://doi.org/10.1007/s12665-010-0724-y.
9. Tseng CM, Lin CW, Hsieh WD. Landslide susceptibility analysis by means of event-based multi-temporal landslide inventories. *Nat Hazards Earth Syst Sci Discuss.* 2015;3:1137–1173. https://doi.org/10.5194/nhessd-3-1137-2015.
10. Pourghasemi HR, Mohammady M, Pradhan B. Landslide susceptibility mapping using index of entropy and conditional probability models in GIS: Safarood Basin, Iran. *Catena.* 2012;97:71–84. https://doi.org/10.1016/j.catena.2012.05.005.

11. Gaprindashvili G, Van Westen CJ. Generation of a national landslide hazard and risk map for the country of Georgia. *Nat Hazards*. 2016;80 (1):69–101. https://doi.org/10.1007/s11069-015-1958-5.

12. Guzzetti F, Carrara A, Cardinali M, Reichenbach P. Landslide hazard evaluation: a review of current techniques and their application in a multi-scale study, Central Italy. *Geomorphology*. 1999;31:181–216.

13. Chacon J, Irigaray C, Fernandez T, Hamdouni EE. Engineering geology maps: landslides and geographical information systems. *Bull Eng Geol Environ*. 2006;65:341–411. https://doi.org/10.1007/s10064-006-0064-z.

14. Akgun A. A comparison of landslide susceptibility maps produced by logistic regression, multi-criteria decision, and likelihood ratio methods: a case study at İzmir, Turkey. *Landslides*. 2012;9(July 2011):93–106. https://doi.org/10.1007/s10346-011-0283-7.

15. Terlien MTJ, Van Asch TWJ, Van Westen CJ. Deterministic modelling in GIS-based landslide hazard assessment. In: Carrara A, Guzzetti F, eds. *Geographical Information Systems in Assessing Natural Hazards*. Dordrecht: Springer; 1995:57–77. https://doi.org/10.1007/978-94-015-8404-3_4.

16. Park S, Choi C, Kim B, Kim J. Landslide susceptibility mapping using frequency ratio, analytic hierarchy process, logistic regression, and artificial neural network methods at the Inje area, Korea. *Environ Earth Sci*. 2013;68(5):1443–1464. https://doi.org/10.1007/s12665-012-1842-5.

17. Armaş I, Vartolomei F, Stroia F, Braşoveanu L. Landslide susceptibility deterministic approach using geographic information systems: application to Breaza town, Romania. *Nat Hazards*. 2014;70(2):995–1017. https://doi.org/10.1007/s11069-013-0857-x.

18. Pradhan B, Lee S. Delineation of landslide hazard areas on Penang Island, Malaysia, by using frequency ratio, logistic regression, and artificial neural network models. *Environ Earth Sci*. 2010;60(5):1037–1054. https://doi.org/10.1007/s12665-009-0245-8.

19. Ramesh V, Anbazhagan S. Landslide susceptibility mapping along Kolli hills Ghat road section (India) using frequency ratio, relative effect and fuzzy logic models. *Environ Earth Sci*. 2014;73(12):8009–8021. https://doi.org/10.1007/s12665-014-3954-6.

20. Nohani E, Moharrami M, Sharafi S, et al. Landslide susceptibility mapping using different GIS-based bivariate models. *Water (Switzerland)*. 2019;11(7). https://doi.org/10.3390/w11071402.

21. Lee S. Application of logistic regression model and its validation for landslide susceptibility mapping using GIS and remote sensing data. *Int J Remote Sens*. 2005;26(7):1477–1491. https://doi.org/10.1080/01431160412331331012.

22. Bai SB, Wang J, Lü GN, Zhou PG, Hou SS, Xu SN. GIS-based logistic regression for landslide susceptibility mapping of the Zhongxian segment in the three gorges area, China. *Geomorphology*. 2010;115(1–2):23–31. https://doi.org/10.1016/j.geomorph.2009.09.025.

23. Onagh M, Kumra VK, Rai PK. Landslide susceptibility mapping in a part of Uttarkashi District (India) by multiple linear regression method. *Int J Geol Earth Environ Sci*. 2012;2(2):102–120.

24. Nefeslioglu HA, Sezer E, Gökçeoğlu C, Bozkır AS, Duman TY. Assessment of landslide susceptibility by decision trees in the metropolitan area of Istanbul, Turkey. *Math Probl Eng*. 2010;2010. https://doi.org/10.1155/2010/901095.

25. Park S, Kim J. Landslide susceptibility mapping based on random forest and boosted regression tree models, and a comparison of their performance. *Appl Sci*. 2019;9(5):1–19. https://doi.org/10.3390/app9050942.

26. Kayastha P, Bijukchhen SM, Dhital MR, De Smedt F. GIS based landslide susceptibility mapping using a fuzzy logic approach: a case study from Ghurmi-Dhad Khola area, Eastern Nepal. *J Geol Soc India*. 2013;82(3):249–261. https://doi.org/10.1007/s12594-013-0147-y.

27. Aydın A, Eker R. Fuzzy rule-based landslide susceptibility mapping in Yığılca Forest District (Norhtwest of Turkey). *J Fac For Istanbul Univ*. 2016;66(2):559–571. https://doi.org/10.17099/jffiu.48480.

28. Ermini L, Catani F, Casagli N. Artificial neural networks applied to landslide susceptibility assessment. *Geomorphology*. 2005;66:327–343. https://doi.org/10.1016/j.geomorph.2004.09.025.

29. Bhardwaj A, Venkatachalam G. *Landslide Hazard Evaluation Using Artificial Neural Networks and GIS*. 2. Springer International Publishing; 2014. https://doi.org/10.1007/978-3-319-05050-8.

30. Can A, Dağdelenler G, Ercanoğlu M, Sönmez H. Landslide susceptibility mapping at Ovacık-Karabu using different artificial neural network models: comparison of training algorithms. *Bull Eng Geol Environ*. 2019;78:89–102. https://doi.org/10.1007/s10064-017-1034-3.

31. Sameen MI, Pradhan B, Lee S. Application of convolutional neural networks featuring Bayesian optimization for landslide susceptibility assessment. *Catena*. 2020;186(September):104249. https://doi.org/10.1016/j.catena.2019.104249.

32. Dahal RK, Hasegawa S, Nonomura A, Yamanaka M, Masuda T, Nishino K. GIS-based weights-of-evidence modelling of rainfall induced landslides in small catchments for landslide susceptibility mapping. *Environ Geol*. 2008;54(2):311–324. https://doi.org/10.1007/s00254-007-0818-3.

33. Pradhan B, Oh HJ, Buchroithner M. Weights-of-evidence model applied to landslide susceptibility mapping in a tropical hilly area. *Geomat Nat Hazards Risk*. 2010;1(3):199–223. https://doi.org/10.1080/19475705.2010.498151.

34. Sharma LP, Patel N, Ghose MK, Debnath P. Development and application of Shannon's entropy integrated information value model for landslide susceptibility assessment and zonation in Sikkim Himalayas in India. *Nat Hazards*. 2015;75(2):1555–1576. https://doi.org/10.1007/s11069-014-1378-y.

35. Tsangaratos P, Ilia I, Hong H, Chen W, Xu C. Applying information theory and GIS-based quantitative methods to produce landslide susceptibility maps in Nancheng County, China. *Landslides*. 2017;14(3):1091–1111. https://doi.org/10.1007/s10346-016-0769-4.

36. Reichenbach P, Rossi M, Malamud BD, Mihir M, Guzzetti F. A review of statistically-based landslide susceptibility models. *Earth Science Rev*. 2018;180(March):60–91. https://doi.org/10.1016/j.earscirev.2018.03.001.

37. Nsengiyumva JB, Luo G, Amanambu AC, et al. Comparing probabilistic and statistical methods in landslide susceptibility modeling in Rwanda/ Centre-Eastern Africa. *Sci Total Environ*. 2019;659(818):1457–1472. https://doi.org/10.1016/j.scitotenv.2018.12.248.

38. Jaafari A, Najafi A, Pourghasemi HR, Rezaeian J, Sattarian A. GIS-based frequency ratio and index of entropy models for landslide susceptibility assessment in the Caspian forest, northern Iran. *Int J Environ Sci Technol*. 2014;11(4):909–926. https://doi.org/10.1007/s13762-013-0464-0.

39. Wang Q, Li W, Xing M, et al. Landslide susceptibility mapping at Gongliu county, China using artificial neural network and weight of evidence models. *Geosci J*. 2016;20. https://doi.org/10.1007/s12303-016-0003-3.

40. Pourghasemi RH, Rahmati O. Prediction of the landslide susceptibility: which algorithm, which precision? *Catena*. 2017. https://doi.org/10.1016/j.catena.2017.11.022.

41. Gayen A, Reza H, Saha S, Keesstra S, Bai S. Science of the Total environment gully erosion susceptibility assessment and management of hazard-prone areas in India using different machine learning algorithms. *Sci Total Environ*. 2019;668:124–138. https://doi.org/10.1016/j.scitotenv.2019.02.436.

42. Hosseinpoor Milaghardan A, Ali Abbaspour R, Khalesian M. Evaluation of the effects of uncertainty on the predictions of landslide occurrences using the Shannon entropy theory and Dempster–Shafer theory. *Nat Hazards*. 2020;100(1):49–67. https://doi.org/10.1007/s11069-019-03798-8.

43. Brenning A. Spatial prediction models for landslide hazards: review, comparison and evaluation. *Nat Hazards Earth Syst Sci*. 2005;5(6):853–862. https://doi.org/10.5194/nhess-5-853-2005.

44. Pradhan B, Lee S. Landslide susceptibility assessment and factor effect analysis: backpropagation artificial neural networks and their comparison with frequency ratio and bivariate logistic regression modelling. *Environ Model Software*. 2010;25(6):747–759. https://doi.org/10.1016/j.envsoft.2009.10.016.

45. Erener A, Duzgun S. Landslide susceptibility assessment: what are the effects of mapping unit and mapping method? *Environ Earth Sci*. 2011;66:859–877. https://doi.org/10.1007/s12665-011-1297-0.

46. Devkota KC, Regmi AD, Pourghasemi HR, et al. Landslide susceptibility mapping using certainty factor, index of entropy and logistic regression models in GIS and their comparison at Mugling-Narayanghat road section in Nepal Himalaya. *Nat Hazards*. 2013;65(1):135–165. https://doi.org/10.1007/s11069-012-0347-6.

47. Öztürk S. Kastamonu-Bartın Küre Dağları Milli Parkı'nın Rekreasyonel Kaynak Değerlerinin İrdelenmesi. *SDÜ Orman Fakültesi Derg*. 2005;A(2):138–148.

48. İbret BÜ, Cansız E. Kanyon turizmi ve ekoturizm açısından değerlendirilmesi gereken bir yöre: Küre Ersizlerdere-Karacehennem Kanyonu. *Marmara Coğrafya Derg*. 2016;34:107–117.

49. Tanrisever C, İbret BÜ, Aydınözü D, Cansiz E. Geomorphologic features and tourism potential of the Valla Canyon. *Karadeniz Araştırmaları*. 2016;50:191–202.

50. Coşkun M, Akbaş V. Karadeniz kıyısından iç kesime: Kastamonu çevresinin iklim parametreleri. *J Soc Sci*. 2017;11:46–86. https://doi.org/10.16990/sobider.3486.

51. Öztürk M, Çetinkaya G, Aydın S. Köppen-Geiger İklim Sınıflandırmasına Göre Türkiye'nin İklim Tipleri. *Istanbul Univ J Geogr*. 2017;35:17–27. https://doi.org/10.26650/JGEOG330955.

52. Guzzetti F, Mondini AC, Cardinali M, Fiorucci F, Santangelo M, Chang KT. Landslide inventory maps: new tools for an old problem. *Earth Sci Rev*. 2012;112(1–2):42–66. https://doi.org/10.1016/j.earscirev.2012.02.001.

53. Çellek S. Morphological parameters causing landslides: a case study of elevation. *Bull Miner Res Explor*. 2020;162:197–224. https://doi.org/10.19076/mta.19369.

54. Huang F, Yao C, Liu W, Li Y, Liu X. Landslide susceptibility assessment in the Nantian area of China: a comparison of frequency ratio model and support vector machine. *Geomat Nat Hazards Risk*. 2018;9(1):919–938. https://doi.org/10.1080/19475705.2018.1482963.

55. Marjanović M, Kovačević M, Bajat B, Voženílek V. Landslide susceptibility assessment using SVM machine learning algorithm. *Eng Geol*. 2011;123(3):225–234. https://doi.org/10.1016/j.enggeo.2011.09.006.

56. Acharya S, Pathak D. Landslide hazard assessment between Besi Sahar and Tal area in Marsyangdi River basin, West Nepal. *Int J Adv Remote Sens GIS*. 2017;5(1):29–38.

57. Sahin EK, Colkesen I, Kavzoglu T. A comparative assessment of canonical correlation forest, random forest, rotation forest and logistic regression methods for landslide susceptibility mapping. *Geocarto Int*. 2020;35(4):341–363. https://doi.org/10.1080/10106049.2018.1516248.

58. Yalcin A, Reis S, Aydinoglu AC, Yomralioglu T. A GIS-based comparative study of frequency ratio, analytical hierarchy process, bivariate statistics and logistics regression methods for landslide susceptibility mapping in Trabzon, NE Turkey. *Catena*. 2011;85(3):274–287. https://doi.org/10.1016/j.catena.2011.01.014.

59. Lee S, Choi J, Min K. Probabilistic landslide hazard mapping using GIS and remote sensing data at Boun, Korea. *Int J Remote Sens*. 2004;25(11):2037–2052. https://doi.org/10.1080/01431160310001618734.

60. Jebur MN, Pradhan B, Tehrany MS. Optimization of landslide conditioning factors using very high-resolution airborne laser scanning (LiDAR) data at catchment scale. *Remote Sens Environ*. 2014;152:150–165. https://doi.org/10.1016/j.rse.2014.05.013.

61. Ohlmacher GC. Plan curvature and landslide probability in regions dominated by earth flows and earth slides. *Eng Geol*. 2007;91(2–4):117–134. https://doi.org/10.1016/j.enggeo.2007.01.005.

62. Moore ID, Grayson RB, Ladson AR. Digital terrain modelling: a review of hydrological, geomorphological, and biological applications. *Hydrol Process*. 1991;5(1):3–30. https://doi.org/10.1002/hyp.3360050103.

63. Wang Q, Li W, Chen W, Bai H. GIS-based assessment of landslide susceptibility using certainty factor and index of entropy models for the Qianyang county of Baoji city, China. *J Earth Syst Sci*. 2015;124(7):1399–1415. https://doi.org/10.1007/s12040-015-0624-3.

64. Arca D, Keskin Citiroglu H, Tasoglu IK. A comparison of GIS-based landslide susceptibility assessment of the Satuk village (Yenice, NW Turkey) by frequency ratio and multi-criteria decision methods. *Environ Earth Sci*. 2019;78(81):4–13. https://doi.org/10.1007/s12665-019-8094-6.

65. Henriques C, Zêzere JL, Marques F. The role of the lithological setting on the landslide pattern and distribution. *Eng Geol*. 2015;189:17–31. https://doi.org/10.1016/j.enggeo.2015.01.025.

66. Sharma LP, Patel N, Debnath P, Ghose MK. Assessing landslide vulnerability from soil characteristics – a GIS-based analysis. *Arab J Geosci*. 2012;5(4):789–796. https://doi.org/10.1007/s12517-010-0272-5.

67. Fonseca LDM, Lani JL, Filho EIF, dos Santos GR, Ferreira WPM, Santos AMRT. Variabilidade dos atributos físicos do solo em áreas suscetíveis ao deslizamento de terra. *Acta Sci Agron*. 2017;39(1):109–118. https://doi.org/10.4025/actasciagron.v39i1.30561.

68. Schilirò L, Poueme Djueyep G, Esposito C, Scarascia MG. The role of initial soil conditions in shallow landslide triggering: insights from physically based approaches. *Geofluids*. 2019;2019. https://doi.org/10.1155/2019/2453786.

69. Hong H, Chen W, Xu C, Youssef AM, Pradhan B, Tien BD. Rainfall-induced landslide susceptibility assessment at the Chongren area (China) using frequency ratio, certainty factor, and index of entropy. *Geocarto Int*. 2016;32(2):139–154. https://doi.org/10.1080/10106049.2015.1130086.

70. Ramos-Bernal RN, Vázquez-Jiménez R, Tizapa SS, Matus RA. Characterization of susceptible landslide zones by an accumulated index. In: Ray R, Lazzari M, eds. *Landslides*. Rijeka: IntechOpen; 2020. https://doi.org/10.5772/intechopen.89828.

71. Youssef AM, Pradhan B, Jebur MN, El-Harbi HM. Landslide susceptibility mapping using ensemble bivariate and multivariate statistical models in Fayfa area, Saudi Arabia. *Environ Earth Sci*. 2015;73:3745–3761. https://doi.org/10.1007/s12665-014-3661-3.

72. Roodposhti MS, Aryal J, Shahabi H, Safarrad T. Fuzzy Shannon entropy: a hybrid GIS-based landslide susceptibility mapping method. *Entropy*. 2016;18(10). https://doi.org/10.3390/e18100343.

73. Zhao H, Yao L, Mei G, Liu T, Ning Y. A fuzzy comprehensive evaluation method based on AHP and entropy for a landslide susceptibility map. *Entropy*. 2017;19(8):1–16. https://doi.org/10.3390/e19080396.

74. Dou J, Yamagishi H, Pourghasemi HR, et al. An integrated artificial neural network model for the landslide susceptibility assessment of Osado Island, Japan. *Nat Hazards*. 2015;78(3):1749–1776. https://doi.org/10.1007/s11069-015-1799-2.

75. Hong H, Shahabi H, Shirzadi A, et al. *Landslide Susceptibility Assessment at the Wuning Area, China: A Comparison between Multi-Criteria Decision Making, Bivariate Statistical and Machine Learning Methods*. Vol. 96. The Netherlands: Springer; 2019. https://doi.org/10.1007/s11069-018-3536-0.

76. Qiu H, Cui P, Regmi AD, Hu S, Hao J. Loess slide susceptibility assessment using frequency ratio model and artificial neural network. *Q J Eng Geol Hydrogeol*. 2019;52(1):38–45. https://doi.org/10.1144/qjegh2017-056.

CHAPTER

37

Remote sensing technology for postdisaster building damage assessment

Mohammad Kakooei[a], Arsalan Ghorbanian[b], Yasser Baleghi[a], Meisam Amani[c], and Andrea Nascetti[d]

[a]Department of Electrical and Computer Engineering, Babol Noshirvani University of Technology, Babol, Iran
[b]Department of Photogrammetry and Remote Sensing, Faculty of Geodesy and Geomatics Engineering, K. N. Toosi University of Technology, Tehran, Iran [c]Wood Environment & Infrastructure Solutions, Ottawa, ON, Canada
[d]Geoinformatics Division, KTH Royal Institute of Technology, Stockholm, Sweden

1 Introduction

Extreme adverse phenomena resulting from the natural process of the Earth, which causes devastating destruction to the surrounding environment, are called natural disasters.[1] Natural disasters include geophysical (e.g., tsunami and earthquake), meteorological (e.g., hurricane and tornado), hydrological (e.g., floods and landslides), and climatological (e.g., drought and wildlife) events.[2] Every year, natural disasters affect different geographical extents and cause physical (e.g., fatalities, injuries, and property damages) and nonphysical (e.g., mental) disturbances.[3]

Several studies have reported a dramatic increase in the number of global natural disaster events in recent decades.[4–6] Consequently, the total number of deaths/injuries, economic losses, and severe environmental disturbances due to these events have been increased.[7, 8] Therefore, it is necessary to acquire a profound understanding of their behavior and, consequently, develop robust methods to monitor their spatial and temporal impacts. Moreover, obtaining reliable information about the corresponding damages is required for governments, policy-makers, and insurance organizations to provide efficient disaster management and financial supports.[9]

Postdisaster damage assessment requires the acquisition, consolidated investigations, and dissemination of information to allow effective emergency management.[10] Ground surveys are a common approach for collecting postdisaster information and monitoring damaged areas. Although this method provides highly accurate information, it is time-consuming and resource-intensive.[11] An alternative solution is to employ remote sensing (RS) technology, including satellite, airborne, and unmanned aerial vehicle (UAV) imagery along with image processing and machine learning (ML) techniques to extract the desired information.[12] Moreover, the existence of many operational and publicly available RS systems enabled the near real-time and cost-effective data acquisition for postdisaster studies.[13] For instance, both active and passive RS systems have been effectively employed for study earthquakes, landslides, floods, hurricanes, and tsunamis.[14–20]

Generally, two common active RS data sources—synthetic aperture radar (SAR)[21] and light detection and ranging (LiDAR)[22]—were employed for postdisaster studies. For instance,[23] employed a multitemporal Advanced SAR (ASAR) sensor onboard Environmental Satellite (ENVISAT) data to tackle the rapid building damage detection challenge of an earthquake in Bam, Iran. Furthermore,[24] took advantage of the spotlight submeter data acquisition mode of the TerraSAR sensor to extract individual building damage information in old Beichuan County, China. In another study, a sensitivity analysis was conducted to reveal the capability of building damage detection based on dual-polarization features provided by the European Sentinel-1 data.[25] The results showed that the inter-channel coherency data provided comparable information to the traditional single-polarization method with lower false alarm rates.

Another source of RS data is optical satellite/aerial imagery, which is captured vertically. In contrast to active RS data, the interpretation of optical images is much easier and, consequently, is more attractive. Building damage detection methods using optical data vary from visual interpretation,[26] image enhancement,[27] pixel- and object-based processing,[28] and postclassification comparison techniques.[29, 30] For instance,[31] proposed a novel segment-by-segment comparison algorithm for building damage detection using high-resolution optical satellite imagery. Furthermore,[32] integrated WorldView-1 and QuickBird-2 satellite images to map earthquake damage in Haiti. Moreover,[33] produced an earthquake building damaged map using GeoEye-1 data over Varzaghan, Iran. Additionally,[34] compared histograms of pre- and post-disaster Landsat 7 and Landsat 8 data to investigate damage and recovery maps. To this end, they used the random forest (RF) algorithm to classify pre- and post-disaster images into five classes: forest/trees, built-up, cropland, waterbody, and others. Finally, they investigated the temporal variation of the histogram bins to analyze the damage and recovery maps.

Similar to vertical images, oblique images acquired by RS systems were also employed for building damage identification and assessment. Despite the beneficial capability of vertical imagery for building damage mapping, it is limited in its ability to collect only roof information. Therefore, oblique aerial and UAV imagery were also employed to quantify building facade damage. In particular, oblique airborne images are identified as the most suitable data because they facilitate roof and lateral building damage detection.[35] For instance,[36] evaluated the capability of the Bag of Visual Words (BOVW) algorithm for building damage identification using four different oblique image datasets. In another study, Kakooei and Baleghi[37, 38] identified building irregularities for the postdisaster assessment using very high resolution (VHR) oblique imagery. In the first step, the shadow information and the gray level co-occurrence matrix (GLCM) features were extracted from preevent vertical images and then fed to a supervised classifier to extract building maps. Subsequently, the preevent images, building maps, and color-equalized postevent data were integrated and applied to spectral and geo-spectral RF classifiers. Finally, the probability classification maps were combined in a decision-level fusion approach to provide a highly accurate building irregularity detection map with a significantly low false-alarm rate.

Most damage assessment researchers rely on their specific imagery device and the study area, which makes it difficult to provide a fair comparison between different methods. To organize a comprehensive investigation and a reliable comparison, this study considers different damage assessment aspects, including different disasters, different RS devices, various imagery angles, and pixel- and object-based analysis in supervised and unsupervised methods.

Considering the information provided above, the main objective of this study is to follow the trend of optical RS technology in damage assessment. Three regions in the USA that were affected by tornado and hurricane disasters were selected as study areas and are described in the following subsection. We explained the theoretical aspects of the study in the Methods section, which is mainly based on Refs. 28, 34, 37, 38. Finally, the results are discussed visually and investigated analytically in the Results and Discussion section.

2 Study area and data

The study area includes three different regions affected by Joplin MO Tornado (2011) located between longitudes of $94°25'$ to $94°34'$E and latitudes of $37°03'$ to $37°05'$N, Hurricane Harvey (2017) located between longitudes of $97°01'30''$ to $97°01'50''$W and latitudes of $28°02'$ to $28°03'$N, and Hurricane Michael (2018) located between longitudes of $85°24'90''$ to $85°25'50''$W and latitudes of $29°56'40''$ to $29°57'20''$N. Fig. 1 shows an overview of the study areas and their approximate geographical locations. High-resolution images were provided by the National Agricultural Imagery Program (NAIP).

Joplin MO tornado (2011) was a catastrophic disaster that was classified as EF5 according to the Enhanced Fujita (EF) scale. It damaged about 30.2% of building structures in Joplin, and 162 fatalities were occurred.[39] Reports show that 1048 individual structures were damaged during the Joplin MO Tornado.[40] Hurricane Harvey (2017) and Hurricane Michael (2018) were classified as category 4 and 5 storms, respectively. They both affected large areas and caused extensive damage[41, 42] on USA coastal areas.

2.1 Data collection and processing

In this study, we utilized different RS datasets and techniques to illustrate various image-processing aspects. A list of employed data and their information are provided in Table 1. We used pre- and post-disaster optical satellite images, including Surface Reflectance (SR) Landsat 5 in Joplin MO Tornado, and Sentinel 2 in Hurricane Harvey

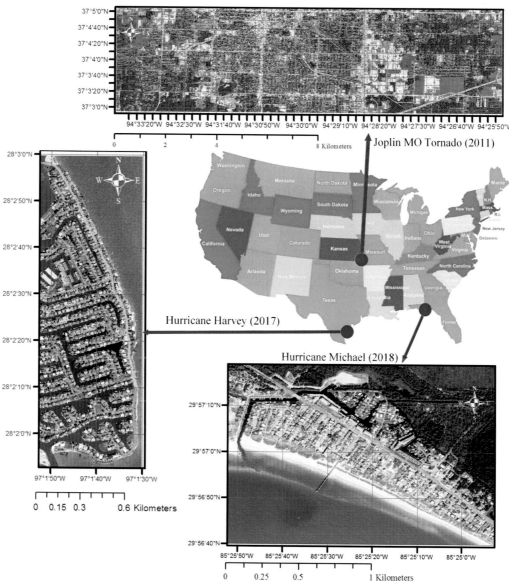

FIG. 1 Study areas that were affected by natural disasters.

TABLE 1 List of remote sensing datasets used in this study.

Disaster	Event date	Remote sensing system	Spatial resolution	Preevent date	Postevent date
Joplin MO Tornado	May 21–26, 2011	Landsat 5	30 m	April 29, 2011	July 2, 2011
		NAIP	1 m	June 4, 2010	June 25, 2012
		NOAA	35 cm	–	May 24, 2011
Hurricane Harvey	Aug 17-Sep 2, 2017	Sentinel 2	10 m	July 21, 2017	Aug 30, 2017
		NAIP	1 m	Sep 30, 2016	–
		NOAA	50 cm	–	Aug 28, 2017
Hurricane Michael	Oct 7–16, 2018	Sentinel 2	10 m	Aug 23, 2018	Oct 12, 2018
		NAIP	1 m	Oct 24, 2017	–
		NOAA	30 cm	–	Oct 11, 2018

The first and the second columns indicate the disasters' names and data, respectively. The third column shows the utilized remote sensing system for each event, and the fourth column shows its spatial resolution. The two last columns show the pre- and post-event dates.

and Michael. Landsat 5 and Sentinel 2 are multispectral satellites that provide images at 30 and 10 m spatial resolutions, respectively. Furthermore, 1 m resolution NAIP images were used, which were captured through the aerial imagery program during the growing season and updated every 2 years.[43] Additionally, we employed the VHR National Oceanic and Atmospheric Administration (NOAA)[44] postdisaster images captured during and after natural disasters for rapid damage assessment and analysis. Aerial NOAA images were captured at either the vertical or oblique angles.

In this study, the proposed methods were implemented using Google Earth Engine (GEE) cloud computing platform.[45] GEE provides data collections and high processing power to analyze geospatial data[46] at the global scale.[47] Landsat and Sentinel datasets and NAIP images were freely accessible through GEE. NOAA data were also downloaded from the NGS website and manually uploaded to the GEE servers.

3 Methodology

We followed two different scenarios to investigate the development of the algorithm following RS technology improvements. In this section, several contrary subjects such as pixel-based and object-based analysis, coarse-resolution and fine-resolution images, spectral and spatial features, and vertical and oblique imagery angles are discussed. These methods are based on spectral distance and classification, the details of which are provided in the following subsections.

3.1 Spectral distance

In the first scenario, a spectral distance-based change detection method was used. This method is relatively simple and easy to interpret.[28] Therefore, the earth moving distance (EMD)[48] is employed as the spectral distance between the pre- and post-disaster images. EMD is based on the minimum cost of transforming one distribution into another. Furthermore, we used simple noniterative clustering (SNIC)[49] superpixel segmentation algorithm to compare pixel-based and object-based analysis. Fig. 2 shows the flowchart of the spectral distance method, in which the importance of SNIC segmentation is discussed according to the resolution of the input RS data.

3.2 Postdisaster classification

Postclassification analysis provides more meaningful class-based information for the study area. Therefore, in the second scenario, Fig. 3, we combined the methods proposed by Refs. 34, 37, 38 to produce pre- and post-disaster landcover maps using the RF supervised classification method. Finally, the land covers were compared according to their histograms.

The supervised RF algorithm requires training samples. In this study, the samples were considered for five classes: tree, built-up, nontree vegetation, waterbody, and others. The Others class mainly refers to impervious surfaces and barren in the predisaster image. It also includes damaged and debris-covered areas in the postdisaster image.

In this study, both spectral and spatial features are utilized. The spectral features are the spectral bands and the spatial features are derived from the GLCM. The RF classifier was trained using the predisaster samples to create a predisaster landcover map. On the other hand, a combination of pre- and post-disaster samples is used to train the RF classifier[37, 38] and create a postdisaster landcover map. It is worth noting that the postdisaster image requires a color-matching preprocessing step before classification.

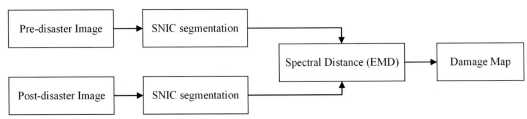

FIG. 2   Creating damage map according to the spectral distance.

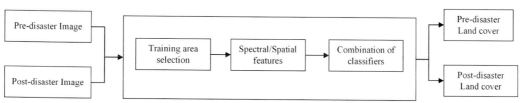

FIG. 3 Damage map according to the supervised classification.

4 Results and discussion

In this section, the trend of damage analysis, including pixel- to object-based analysis, regional to building-based analysis, and vertical to oblique angle analysis are investigated. To this end, the Joplin MO Tornado, Hurricane Harvey, and Hurricane Michael case studies were employed.

4.1 Pixel-based versus object-based analysis

The availability of high-resolution imagery has encouraged researchers to conduct building damage assessments using object-based analysis instead of traditional pixel-based methods. In this subsection, we compare the results of building damage assessment using pixel- and object-based image analysis applied to the Landsat 5 and NAIP images with 30 m and 1 m spatial resolutions, respectively. Fig. 4 shows the pixel- and object-based spectral distance analysis, in which the object-based investigation is based on the SNIC segmentation algorithm.

Although the object-based analysis in coarse-resolution images provides valuable information for large-scale applications, such as land cover classification[50] and cropland classification[51] over large areas, it does not improve the results of building damage assessment. This is because each pixel value is related to the mixture of the buildings and their surrounding areas in the 30 m resolution of the Landsat 5 image. Therefore, the object-based analysis did not enhance the postdisaster damage map.

The importance of object-based analysis in building-level change detection in high spatial resolution imagery, such as NAIP data, has been proven.[52] Object-based analysis decreases the amount of noise, which is common in VHR images; thus, an object cannot be classified into multiple classes.[53, 54] Additionally, other unwanted pixels, such as shadows that exist in VHR images, can be efficiently refined by object-based image analysis.[37, 38] For example, the zoomed scenes in Fig. 5 illustrate these concepts by comparing pixel- and object-based analyses. As is clear, although SNIC degrades damage assessment maps using Landsat 5, it improves NAIP-based spectral distance maps.

It should be noted that in the Joplin MO Tornado event, most buildings were destroyed completely in the tornado path. Therefore, in the next subsection, we analyzed damages during the Hurricane Harvey event, in which there were many moderately damaged buildings in the study area.

4.2 Regional versus building-based analysis

In this section, the 10 m spatial resolution Sentinel 2 data from Hurricane Harvey was used for postdisaster damage analysis. Moreover, NOAA provided postevent aerial images in 30 cm resolution. Sentinel 2 pre- and post-disaster images (Fig. 6A and B) were used to produce the spectral distance in Fig. 6C. According to visual interpretation, the spectral distance cannot highlight the damaged areas where buildings are moderately damaged. Thus, it was concluded that (1) coarse-resolution images can show heavily damaged areas (Joplin MO, 2011), but they cannot show moderately damaged buildings (Harvey, 2017); (2) although 10 m spatial resolution Sentinel 2 images contain more detailed data than 30 m spatial resolution Landsat 5 images, they still do not provide a damage map at the building level.

As discussed, moving from coarse- to fine-resolution imagery, the importance of object-based analysis and spatial features. NAIP predisaster and NOAA postdisaster images are shown in Fig. 6D and E, respectively. To illustrate the zoomed scenes, six areas were selected and are shown by red rectangles in these figures. We classified the NOAA image into five classes using pixel-based spectral-only features, as shown in Fig. 6F. Then, the SNIC segmentation algorithm was applied to provide object-based spectral classification in Fig. 6G to verify the importance of object-based analysis in VHR images, which were previously investigated in the spectral distance analysis for the Joplin MO Tornado event. In Fig. 6H, we extended the investigation to visualize the role of the spatial *imcorr1* feature generated from GLCM.

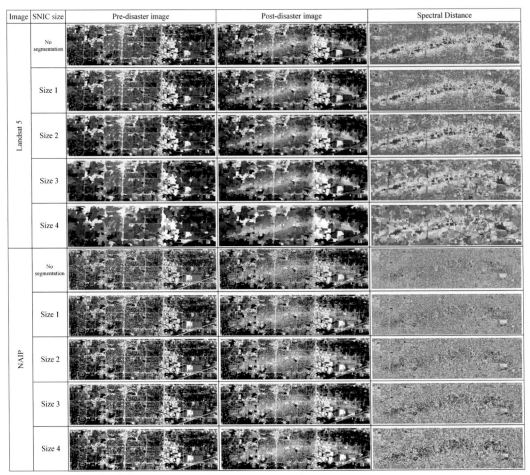

FIG. 4 The Joplin MO Tornado event. Landsat 5 and NAIP pre- and post-disaster images and their spectral distance images. The structure of images in the Landsat 5 and NAIP are the same in which the first row shows pixel-based spectral distance images, and the other rows show object-based analysis using different segment sizes generated from the SNIC algorithm. Moreover, the pre- and post-disaster images are illustrated in the first and the second columns. The third column shows the spectral distance image in which green color to red color shows the undamaged to highly-damaged areas.

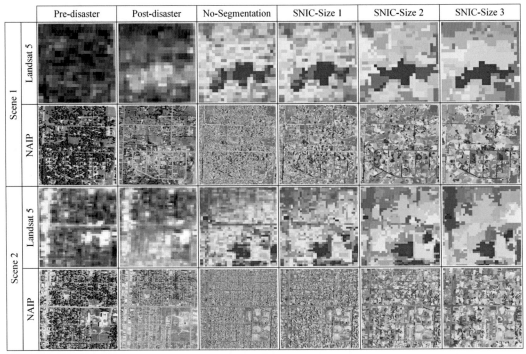

FIG. 5 Zoomed scenes of the Joplin MO Tornado event. Landsat 5 and NAIP pre- and post-disaster images and their spectral distance images are shown. The structure of images in scenes 1 and 2 are the same, in which the first and the second columns show the pre- and post-disaster images of Landsat 5 and NAIP. Pixel-based spectral distance images are shown in the third column. Moreover, the last three columns show the object-based spectral distance with three different SNIC sizes.

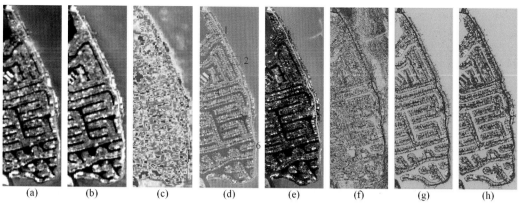

FIG. 6 Hurricane Harvey damage analysis. (A) Predisaster Sentinel 2 image. (B) Postdisaster Sentinel 2 image. (C) Sentinel 2 spectral distance. (D) Predisaster NAIP image. (E) Postdisaster NOAA image. (F) Postdisaster NOAA pixel-based spectral classification. (G) Postdisaster NOAA object-based spectral classification. (H) Postdisaster NOAA object-based spectral-spatial classification.

Comparing spectral-only and spectral-spatial classified maps can reveal the degree of improvement. The results of Hurricane Harvey's damage analysis for the six zoomed scenes (see Fig. 6D) are shown in Fig. 7. These regions include pre- and post-disaster VHR images from the NAIP and NOAA.

The results obtained from Sentinel 2 imagery can reveal destroyed buildings at a regional level; however, it cannot show moderately damaged buildings. The Spectral distance shows relatively high values around the destroyed buildings in scenes 1, 4, 5, and 6. On the other hand, moderately damaged buildings were not detected in scenes 1, 2, and 3. This verifies that the coarse-resolution images can neither provide the damage map at the building level nor detect moderately damaged buildings.

Using NOAA VHR images, a building damage map was provided at the building level. We evaluated the spectral-only and spectral-spatial classification of postdisaster NOAA images to investigate the role of neighborhood information in VHR image analysis. Moderately damaged buildings in all scenes were detected and the damaged parts were classified into the Others class.

Spatial features consider texture information in VHR images and can increase building-based damage assessment analysis. We labeled 150 points in each class to analytically investigate the effect of the GLCM *imcorr1* feature. Tables 2 and 3 show the confusion matrices of the spectral-only and spectral-spatial classifications, respectively. This shows that the overall accuracy is increased by approximately 3.3% when this texture feature is employed. Thus, the Kappa coefficient also increased from 0.78 to 0.82.

4.3 Vertical versus oblique angle images

Although deploying object-based spectral-spatial features in analyzing the vertical VHR postdisaster images provides a building-level damage map, the results might be overestimated or underestimated because the corresponding images only provide rooftop information. However, oblique imagery can provide both rooftop and facade information in postdisaster analysis and, thus, provide a more accurate building damage assessment.

Comparing pre- and post-disaster landcover maps leads to damage map creation. Here, we used two strategies in which coarse-resolution Sentinel 2 and fine-resolution NAIP and NOAA images are analyzed to provide a histogram-based landcover analysis. Fig. 8 shows the pre- and post-disaster Sentinel 2 images and their landcover maps. Moreover, VHR NAIP and NOAA images and their landcover maps are provided for comparison.

Several zoomed images are illustrated in Fig. 9 to show that the object-based spectral-spatial RF classification has very high accuracy in classifying nondamaged and destroyed buildings into the Built-up and Others classes, respectively.

The fusion of different RS imagery can improve the accuracy of the damage map. In this study, we fused VHR pre-disaster vertical NAIP and postdisaster oblique NOAA images and assessed the building damage. Zoomed images show that nondamaged building facades play a role in postdisaster landcover classification. Therefore, it is visually illustrated that the information of both rooftops and facades was considered in the oblique landcover map.

Pre- and post-disaster landcover maps were analyzed through histogram interpretation in Fig. 10. The comparison between pre- and post-disaster Sentinel 2 histograms shows that there is a decrease in the Tree and Built-up classes, and an increase in the Others class. It is expected that broken trees, damaged buildings, and debris areas are classified

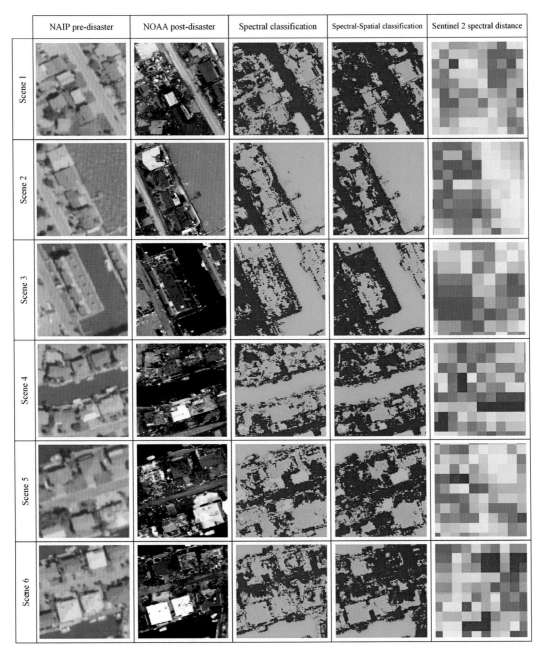

FIG. 7 The six zoomed images of the Hurricane Harvey event to compare the results of spectral and spectral-spatial classification. The pre- and post-disaster images from NAIP and NOAA are shown in the first and the second column, respectively. The postdisaster classified image using spectral-only and spectral-spatial features are shown in the third and fourth columns. The last column shows the zoomed spectral distance image that is based on Sentinel 2 imagery.

TABLE 2 Confusion matrix of the object-based spectral-only classification.

	Tree	**Built up**	**Grass**	**Water**	**Other**
Tree	128	0	38	7	2
Built up	1	137	3	0	22
Nontree vegetation	12	0	90	1	0
Water	2	0	15	137	0
Other	7	13	4	5	126
Producer accuracy (%)	85.33	91.33	60	91.33	84
User accuracy (%)	73.14	84.05	87.38	88.96	81.29

Overall accuracy: 82.4% Kappa coefficient = 0.78.

TABLE 3   Confusion matrix of the object-based spectral-spatial classification.

	Tree	Built up	Grass	Water	Other
Tree	142	0	52	2	0
Built up	0	141	1	0	12
Nontree vegetation	4	0	81	3	4
Water	1	0	4	145	0
Other	3	9	12	0	134
Producer accuracy (%)	94.67	94	54	96.67	89.33
User accuracy (%)	72.45	91.56	88.04	96.67	84.81

Overall accuracy: 85.73% Kappa coefficient = 0.82.

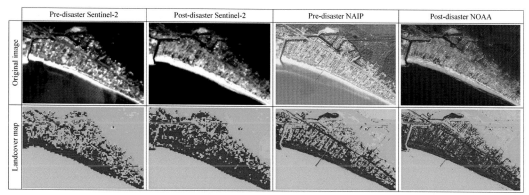

FIG. 8   Pre- and post-disaster landcover maps for the Hurricane Michael case study. The first row of this figure shows the original pre- and post-disaster images, and the second row shows landcover maps.

FIG. 9   Zoomed images of postdisaster NOAA data and landcover maps in the Hurricane Michael event. The first row shows VHR zoomed images from NOAA. The second row shows the landcover maps.

as belonging to the Others class. Furthermore, the percentage of built-up areas decreased from 17.6% to 11%, which means that 37.5% of building areas are destroyed or damaged in the hurricane. The comparison between the landcover histograms of VHR images in predisaster NAIP and postdisaster NOAA shows a more accurate result, in which the percentages of Built-up and Others classes decrease and increase, respectively. Building debris was classified as the Others class in the postdisaster image. Furthermore, decreasing the built-up percent from 8.8 to 5.3% shows that 39.8% of building areas are damaged.

5   Discussion

Damage assessment techniques are different in both the study area and the utilized data, and thus it is hard to provide a fair comparison. However, our research provides a comprehensive study on different aspects of damage assessment that are included in the columns of Table 4, which also contains a summary of previous damage assessment studies.

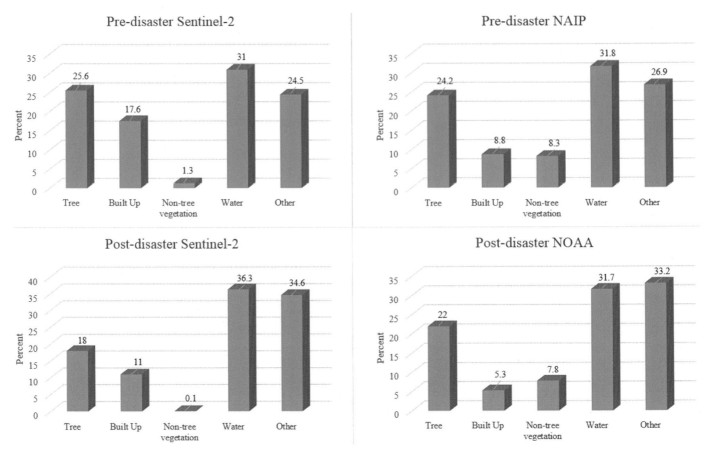

FIG. 10 Histograms of pre- and post-disaster land covers. Two histograms on the left are related to the pre- and post-landcover maps of Sentinel 2. Two histograms on the right are related to the pre- and post-landcover maps of VHR images, NAIP and NOAA.

TABLE 4   Comparison table to summarize damage assessment studies.

	Study area	Data type	Pixel- or object-based	Features	Damage level	Imagery angle
[32]	Port-au-Prince, Haiti, 2011	WorldView-1 Quickbird-2	Pixel-based	Spectral Spatial	Building-level	Vertical
[33]	Varzaghan, Iran, 2011	GeoEye-1	Pixel-based	Spectral Spatial	Building-level	Vertical
[31]	Port-au-Prince, Haiti, 2011 Bam, Iran, 2004	WorldView-2 QuickBird	Object-based	Spectral Spatial	Building-level	Vertical
[34]	Tacloban, Philippines, 2013	Landsat 7 Landsat 8	Pixel-based	Spectral	Regional	Vertical
[37, 38]	Fort Morgan, US, 2017 Key Allegro, US, 2017	Aerial NAIP Aerial NOAA	Object-based	Spectral Spatial	Building-level	Oblique
This study	Joplin, US, 2011 Key Allegro, US, 2017 Mexico beach, US, 2018	Landsat-5 Sentinel-2 Aerial NAIP Aerial NOAA	Both	Spectral Spatial	Regional Building-level	Vertical Oblique

Our study confirms the positive role of the object-based analysis in VHR postdisaster image processing, which goes along with,[31, 37, 38] studies. On the other hand, although[32, 33] did not deploy object-based analysis to process VHR postdisaster data, they extracted and utilized several spatial features. However, their methods could be improved by including object-based image analysis.

By now, mid-resolution images have become the only free global RS data, and thus, pixel-based analysis is also in demand. In this study, we showed pixel-based analysis outperforms object-based analysis in mid-resolution images.

This also goes along with[34] study in which pixel-based spectral classification was applied to mid-resolution Landsat series images.

Finally, we showed, the VHR oblique imagery provides the most accurate postdisaster building-level damage map by using spectral-spatial features in an object-based classification. However, we showed the mid-resolution image analysis provides a proper damage map at the regional level. To this end, mid- and high-resolution damage maps were compared at the regional level through a histogram analysis.

6 Conclusion

RS technology provides more detailed information owing to the availability of high-resolution images. Therefore, damage assessment is transferring from region-level to building-level maps, which requires the development of advanced feature extraction and machine learning algorithms. In this study, we defined different scenarios to study three disasters, including the Joplin MO Tornado, Hurricane Harvey, and Hurricane Michael, using Landsat 5 (30 m), Sentinel 2 (10 m), NAIP (1 m), and NOAA (submeter) images.

Landsat and Sentinel imagery were too coarse to show standalone buildings. Therefore, the value of each pixel is a mixture of the spectral responses of the buildings and their surrounding environment. Thus, object-based spectral distance does not enhance damage assessment maps using coarse-resolution imagery. However, object-based analysis and the relation between neighborhood pixels can improve damage maps in fine-resolution images. We created damage maps for the Joplin MO Tornado event using coarse resolution Landsat 5 and fine-resolution NAIP data and compared pixel-based and object-based analyses. Therefore, the comparison verifies the growing role of object-based analysis in building damage assessment.

We compared coarse- and fine-resolution damage maps for the Hurricane Harvey event. This shows that moderately damaged buildings were not detected in coarse-resolution imagery, as they could not show standalone buildings. However, the VHR images reveal small changes in the building structure. Therefore, there is a growing trend in the use of VHR images in postdisaster damage analysis.

Coarse RS imagery, such as those captured by Landsat and Sentinel 2, cannot illustrate the texture of standalone buildings. However, texture information, such as GLCM features, plays an important role in the processing of fine-resolution images. To this end, we used texture information to analyzing the Hurricane Harvey event and compared it with the spectral-only investigation. Therefore, it is verified that feature extraction techniques consider more spectral-spatial information.

The number of RS systems has increased significantly. Therefore, the fusion of data acquired by different systems provides a great opportunity to support rapid saving and rescue missions. We analyzed the Hurricane Michael event using pre- and post-disaster Sentinel 2 data and compared it with vertical predisaster NAIP and oblique postdisaster NOAA data. Then, we compared the two methods and showed that the fusion of NAIP and NOAA provides reliable results.

In summary, there are currently many RS systems that provide images with high spatial resolution images. Therefore, it is expected to provide a building-level damage map instead of regional analysis. Thus, the importance of object-based analysis and spatial information is increasing. Moreover, there is a growing interest in the fusion of different types of imagery for accurate damage assessment.

References

1. Bayissa Y, Tadesse T, Demisse G, Shiferaw A. Evaluation of satellite-based rainfall estimates and application to monitor meteorological drought for the Upper Blue Nile Basin, Ethiopia. *Remote Sens (Basel)*. 2017;9(7):669.
2. Shen G, Hwang SN. Spatial-temporal snapshots of global natural disaster impacts revealed from EM-DAT for 1900–2015. *Geomat Nat Haz Risk*. 2019;10(1):912–934.
3. Lindell MK. Disaster studies. *Curr Sociol*. 2013;61(5–6):797–825.
4. Coronese M, Lamperti F, Keller K, Chiaromonte F, Roventini A. Evidence for sharp increase in the economic damages of extreme natural disasters. *Proc Natl Acad Sci U S A*. 2019;116(43):21450–21455.
5. Hoeppe P. Trends in weather related disasters-consequences for insurers and society. *Weather Clim Extrem*. 2016;11:70–79.
6. Nel JL, Le Maitre DC, Nel DC, et al. Natural hazards in a changing world: a case for ecosystem-based management. *PLoS One*. 2014;9(5), e95942.
7. Botzen WW, Deschenes O, Sanders M. The economic impacts of natural disasters: a review of models and empirical studies. *Rev Environ Econ Policy*. 2019;13(2):167–188.
8. Padli J, Habibullah MS, Baharom AH. The impact of human development on natural disaster fatalities and damage: panel data evidence. *Ecol Res*. 2018;31(1):1557–1573.

9. Shen G, Zhou L, Wu Y, Cai Z. A global expected risk analysis of fatalities, injuries, and damages by natural disasters. *Sustainability*. 2018;10 (7):2573.

10. Adib A, Jenab VH. Post disaster information management: issues related to mitigation activities in Iran. In: *Advances in Environmental Geotechnics*. Springer; 2010:879–882.

11. Kakooei M, Baleghi Y. Fusion of satellite, aircraft, and UAV data for automatic disaster damage assessment. *Int J Remote Sens*. 2017;38 (8–10):2511–2534.

12. Novellino A, Jordan C, Ager G, Bateson L, Fleming C, Confuorto P. Remote sensing for natural or man-made disasters and environmental changes. In: *Geological Disaster Monitoring Based on Sensor Networks*. Springer; 2019:23–31.

13. Williams JG, Rosser NJ, Kincey ME, et al. Satellite-based emergency mapping using optical imagery: experience and reflections from the 2015 Nepal earthquakes. *Nat Hazards Earth Syst Sci*. 2018;18(1):185–205.

14. Cao QD, Choe Y. Building damage annotation on post-hurricane satellite imagery based on convolutional neural networks. *Nat Hazards*. 2020;103(3):3357–3376.

15. Cui Y, Cheng D, Choi CE, Jin W, Lei Y, Kargel JS. The cost of rapid and haphazard urbanization: lessons learned from the Freetown landslide disaster. *Landslides*. 2019;16(6):1167–1176.

16. Jiménez-Jiménez SI, Ojeda-Bustamante W, Ontiveros-Capurata RE, Marcial-Pablo MJ. Rapid urban flood damage assessment using high resolution remote sensing data and an object-based approach. *Geomat Nat Haz Risk*. 2020;11(1):906–927.

17. Koshimura S, Moya L, Mas E, Bai Y. Tsunami damage detection with remote sensing: a review. *Geosciences*. 2020;10(5):177.

18. Rahman MS, Di L. A systematic review on case studies of remote-sensing-based flood crop loss assessment. *Agriculture*. 2020;10(4):131.

19. Song D, Tan X, Wang B, Zhang L, Shan X, Cui J. Integration of super-pixel segmentation and deep-learning methods for evaluating earthquake-damaged buildings using single-phase remote sensing imagery. *Int J Remote Sens*. 2020;41(3):1040–1066.

20. Valentijn T, Margutti J, van den Homberg M, Laaksonen J. Multi-hazard and spatial transferability of a cnn for automated building damage assessment. *Remote Sens (Basel)*. 2020;12(17):2839.

21. Endo Y, Adriano B, Mas E, Koshimura S. New insights into multiclass damage classification of tsunami-induced building damage from SAR images. *Remote Sens (Basel)*. 2018;10(12):2059.

22. Zhou Z, Gong J, Hu X. Community-scale multi-level post-hurricane damage assessment of residential buildings using multi-temporal airborne LiDAR data. *Autom Constr*. 2019;98:30–45.

23. Gamba P, Dell'Acqua F, Trianni G. Rapid damage detection in the Bam area using multitemporal SAR and exploiting ancillary data. *IEEE Trans Geosci Remote Sens*. 2007;45(6):1582–1589.

24. Gong L, Wang C, Wu F, Zhang J, Zhang H, Li Q. Earthquake-induced building damage detection with post-event sub-meter VHR TerraSAR-X staring spotlight imagery. *Remote Sens (Basel)*. 2016;8(11):887.

25. Ferrentino E, Nunziata F, Migliaccio M, Vicari A. A sensitivity analysis of dual-polarization features to damage due to the 2016 Central-Italy earthquake. *Int J Remote Sens*. 2018;39(20):6846–6863.

26. Yamazaki F, Yano Y, Matsuoka M. Visual damage interpretation of buildings in bam city using quickbird images following the 2003 bam, Iran, earthquake. *Earthq Spectra*. 2005;21(1_suppl):329–336.

27. Miura H, Modorikawa S, Chen SH. Texture characteristics of high-resolution satellite images in damaged areas of the 2010 Haiti earthquake. In: *Paper Presented at Proceedings of the 9th International Workshop on Remote Sensing for Disaster Response, Stanford, CA, USA*; 2011.

28. Hussain M, Chen D, Cheng A, Wei H, Stanley D. Change detection from remotely sensed images: from pixel-based to object-based approaches. *ISPRS J Photogramm Remote Sens*. 2013;80:91–106.

29. Dong L, Shan J. A comprehensive review of earthquake-induced building damage detection with remote sensing techniques. *ISPRS J Photogramm Remote Sens*. 2013;84:85–99.

30. Gusella L, Adams BJ, Bitelli G, Huyck CK, Mognol A. Object-oriented image understanding and post earthquake damage assessment for the 2003 Bam, Iran, earthquake. *Earthq Spectra*. 2005;21(1_suppl):225–238.

31. Khodaverdizahraee N, Rastiveis H, Jouybari A. Segment-by-segment comparison technique for earthquake-induced building damage map generation using satellite imagery. *Int J Disaster Risk Reduct*. 2020;46:101505.

32. Cooner AJ, Shao Y, Campbell JB. Detection of urban damage using remote sensing and machine learning algorithms: revisiting the 2010 Haiti earthquake. *Remote Sens (Basel)*. 2016;8(10):868.

33. Ranjbar HR, Ardalan AA, Dehghani H, Saradjian MR. Using high-resolution satellite imagery to provide a relief priority map after earthquake. *Nat Hazards*. 2018;90(3):1087–1113.

34. Ghaffarian S, Rezaie Farhadabad A, Kerle N. Post-disaster recovery monitoring with Google Earth Engine. *Appl Sci*. 2020;10(13):4574.

35. Fernandez Galarreta J, Kerle N, Gerke M. UAV-based urban structural damage assessment using object-based image analysis and semantic reasoning. *Nat Hazards Earth Syst Sci*. 2015;15(6):1087–1101.

36. Vetrivel A, Gerke M, Kerle N, Vosselman G. Identification of structurally damaged areas in airborne oblique images using a visual-bag-of-words approach. *Remote Sens (Basel)*. 2016;8(3):231.

37. Kakooei M, Baleghi Y. A two-level fusion for building irregularity detection in post-disaster VHR oblique images. *Earth Sci Inf*. 2020;13 (2):459–477.

38. Kakooei M, Baleghi Y. Shadow detection in very high resolution RGB images using a special thresholding on a new spectral-spatial index. *J Appl Remote Sens*. 2020;14(1), 016503.

39. Prevatt DO, van de Lindt JW, Back EW, et al. Making the case for improved structural design: tornado outbreaks of 2011. *Leadersh Manag Eng*. 2012;12(4):254–270.

40. Roueche DB, Prevatt DO. Residential damage patterns following the 2011 Tuscaloosa, AL and Joplin, MO tornadoes. *J Disaster Res*. 2013;8 (6):1061–1067.

41. Wurman J, Kosiba K. The role of small-scale vortices in enhancing surface winds and damage in Hurricane Harvey (2017). *Mon Weather Rev*. 2018;146(3):713–722.

42. Kennedy A, Copp A, Florence M, et al. Hurricane Michael in the area of Mexico Beach, Florida. *J Waterw Port Coast Ocean Eng*. 2020;146(5), 05020004.

43. Office U-F-AAPF. National Agriculture Imagery Program (NAIP), vol. 2019. United States Department of Agriculture: Farm Service Agency, n.d. https://www.fsa.usda.gov/programs-and-services/aerial-photography/imagery-programs/naip-imagery/index.

44. Program AS. NOAA. vol. 2019. Aeronautical Survey Program: Emergency Reponse, n.d. https://storms.ngs.noaa.gov.

45. Gorelick N, Hancher M, Dixon M, Ilyushchenko S, Thau D, Moore R. Google Earth Engine: planetary-scale geospatial analysis for everyone. *Remote Sens Environ.* 2017;202:18–27.

46. Amani M, Ghorbanian A, Ahmadi SA, et al. Google Earth Engine cloud computing platform for remote sensing big data applications: a comprehensive review. *IEEE J Sel Top Appl Earth Obs Remote Sens.* 2020;13:5326–5350.

47. Kakooei M, Nascetti A, Ban Y. Sentinel-1 global coverage foreshortening mask extraction: an open source implementation based on Google Earth Engine. In: *Paper presented at: IGARSS 2018–2018 IEEE International Geoscience and Remote Sensing Symposium*; 2018.

48. Rubner Y, Tomasi C, Guibas LJ. The earth mover's distance as a metric for image retrieval. *Int J Comput Vis.* 2000;40(2):99–121.

49. Achanta R, Süsstrunk S. Superpixels and polygons using simple non-iterative clustering. In: *2017 IEEE Conference on Paper Presented at: Computer Vision and Pattern Recognition (CVPR)*; 2017.

50. Ghorbanian A, Kakooei M, Amani M, Mahdavi S, Mohammadzadeh A, Hasanlou M. Improved land cover map of Iran using sentinel imagery within Google Earth Engine and a novel automatic workflow for land cover classification using migrated training samples. *ISPRS J Photogramm Remote Sens.* 2020;167:276–288.

51. Amani M, Kakooei M, Moghimi A, et al. Application of Google Earth Engine cloud computing platform, sentinel imagery, and neural networks for crop mapping in Canada. *Remote Sens (Basel).* 2020;12(21):3561.

52. Kakooei M, Baleghi Y. Spectral unmixing of time series data to provide initial object seeds for change detection on Google Earth Engine. In: *Paper Presented at: 2019 27th Iranian Conference on Electrical Engineering (ICEE)*; 2019.

53. Amani M, Salehi B, Mahdavi S, Granger JE, Brisco B, Hanson A. Wetland classification using multi-source and multi-temporal optical remote sensing data in Newfoundland and Labrador, Canada. *Can J Remote Sens.* 2017;43(4):360–373.

54. Amani M, Mahdavi S, Berard O. Supervised wetland classification using high spatial resolution optical, SAR, and LiDAR imagery. *J Appl Remote Sens.* 2020;14(2), 024502.

38

Doing more with less: A comparative assessment between morphometric indices and machine learning models for automated gully pattern extraction (A case study: Dashtiari region, Sistan and Baluchestan Province)

Aiding Kornejady[a], Abbas Goli Jirandeh[a], Hadi Alizadeh[a], Alireza Sarvarinezhad[b], Abdollah Bameri[b], Luigi Lombardo[c], Christian Conoscenti[d], Amir Alizadeh[a], Mahdi Karimi[a], Mahmood Samadi[a], and Esmaeil Silakhori[a]

[a]Spatial Sciences Innovators Consulting Engineering Company, Tehran, Iran [b]Natural Resources and Watershed Management Organization of Sistan and Baluchestan Province, Zahedan, Iran [c]Faculty of Geo-Information Science and Earth Observation (ITC), University of Twente, Enschede, Netherlands [d]Department of Earth and Marine Sciences (DISTEM), University of Palermo, Palermo, Italy

1 Introduction

Erosion is a global concern that occurs in different forms because of the complex, interlaced connection of different causative factors. Once a raindrop touches the soil surface, water erosion commences, forming a wide range of geomorphological facies.[1] As a common form of water erosion, gullies occur mainly under the interaction of soil chemistry, land use, climate, slope, and the governing hydrological system.[2–4] Large amounts of soils are washed away along the unstable channels characterized as gullies, mostly formed on sparsely vegetated and unprotected lands. Hence, extracting the gully pattern as a substantial part of spatial modeling and, thereby, risk analysis is of prime importance.

Humans have become pattern seekers for their survival and their societal development. In parallel, pattern recognition techniques have undoubtedly obviated that need. To date, different algorithms have been developed to differentiate well the natural patterns, examples of which are supervised and unsupervised machine/deep learning algorithms (ML and DL hereafter) and geographic object-based image analysis (GOBIA).[5–9] However, these techniques operate on big data to attain an acceptable level of success. Particularly for gully delineation, OBIA techniques are a bottom-up trial-and-error optimization technique that entails several object features/variables such as mean, standard deviation, length/width, and the gray level co-occurrence matrix (GLCM) (i.e., frequency of different gray-level pixel combinations in a remotely sensed image).[8] Additionally, GOBIA techniques require a tedious process for learning and prediction/classification tasks. On the other hand, visual interpretation and extraction are cumbersome procedures diluted by boundless human errors and may not be an efficient choice in case of emergency due to the time-consuming procedures involved.

Morphometric indices with singular and multilateral connotations (e.g., topographic, topo-hydrologic, edaphic, and botanic) can signify erosional processes subjected to different runoff mechanisms, especially in areas where lack of data is a major concern. Most morphometric indices are DEM derivatives and follow practical, yet straightforward functions. Many studies have focused on the application of morphometric indices in the context of spatial modeling of different natural hazards, such as landslide susceptibility assessment,[10–12] flood hazard analysis,[13] and gully erosion susceptibility mapping.[14–16] However, the literature review attests that the single use of these indices (i.e., not fed to other standalone models) for automated classification and pattern extraction of natural features has not yet been addressed. Hence, the main idea behind this work is to build a simple yet practical conceptual model based on which different single and combined morphometric indices are used to extract gully boundaries. Compared to the advance and complicated models being used in literature, our proposed framework adopts a back-to-basics routine.

Based on these premises, this study sets out to fill this study gap and test the potential use of morphometric indices for gully pattern extraction, using as a test site the Dashtiari region (Iran). We also took it a step further to compare the results with a powerful family of machine learning models and classification trees and discuss their similarities and differences.

2 Study area

The Dashtiari region is located in Sistan and Baluchestan Province, southeast of Iran (Fig. 1). It lies between the latitudes of 25°41′17″–25°43′10″ N and the longitudes of 60°56′36″–60°59′38″ E. The selected parcel extends for 597.7 ha, 14.2% of which (about 85 ha) is affected by gullies. Elevation ranges between −0.1 and 18.7 m a.s.l. Maximum and minimum precipitations are 125 and 150 mm. Maximum and minimum temperatures fluctuate between 21°C and 35°C. As with many gully-prone areas, massive gullies have deeply incised the entire region of Dashtiari. Silty soil has made the area highly sensitive to erosional processes such that a small amount of rainfall can substantially change the landscape due to discernible longitudinal and lateral extension of gullies. Successively widened and deepened, gullies pose a direct threat to the infrastructures, and casualties are anticipated once they reach the villages. For instance, one of the main gully branches have moved toward the Kajoo stream, the primary water source of residents' drinking and agricultural demands. Once gully branches and streams meet, the gullies' intertwined network will drain out the water and make it out of the locals' reach.

3 Materials and methods

3.1 Data compilation

A DJI Phantom 4 Pro (P4P) V2.0, was used to capture an area of approximately 6 km^2 (Fig. 2), from which 3137 UAV images were acquired and processed using Agisoft Metashape v1.6 and Inpho UASMaster v7.1 image processor tool. Five hours of flight mission with a GSD of 3.88 cm per pixel was accomplished on April 21–22, 2019, based on which an orthophoto and a digital surface model (DSM) with a 15 cm × 15 cm pixel resolution were produced for a Dashtiari parcel, as presented in Fig. 1. Furthermore, Fig. 3 presents two excerpts of the gully branches.

A specialized team of cartographers and photogrammetrists manually drew the gully affected boundaries using 3Dstereo anaglyph glasses, which took approximately 6 months to be completed.

3.2 Morphometric indices

Morphometric indices are valuable DEM derivatives that express indirect connotations of earth processes, such as flood generation mechanisms and erosional features. Hence, they can be useful in areas characterized by a lack of data. In this study, seven morphometric indices, including valley depth (VD), topographic position index (TPI), positive openness (PO), red relief image map (RRIM), elevation, slope degree, and coupled PO-DEM (i.e., multiplication), were purposively selected so that gully patterns could be differentiated and extracted. Table 1 lists the implications of each index. The implementation process was carried out in SAGA-GIS,[17] which offers many useful indices with a straight-forward execution process.[18] Additionally, the pixel resolution of the DSM layer was resized to 1 m to expedite the production process of the morphometric indices.

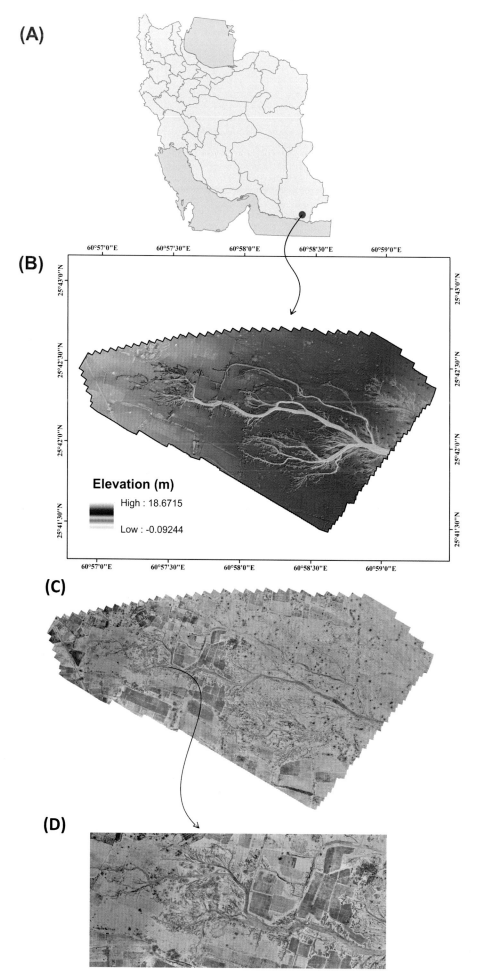

FIG. 1 Location of the study area in Iran (A), UAV-derived digital surface model (B), and the orthophoto of the region (C) with an excerpt site (D).

FIG. 2   Phantom 4 Pro V2.0—DJI.

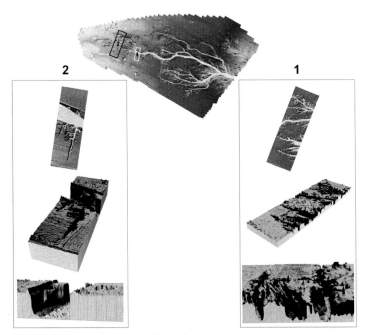

FIG. 3   Cross-sectional excerpts of the gullies sprawled across the Dashtiari region.

TABLE 1   Functional role of the applied morphometric indices for gully pattern extraction.

Factor	Function
Valley depth (VD)	Delineation of ridges from valleys
Topographic position index (TPI)	Delineation of upper, middle, and lower slopes
Positive openness (PO)	Surface concavities from a given zenith
Red relief image map (RRIM)	Effective visualization of subtle topographic features without shading (positive openness-negative openness/2)
Elevation	Differentiating highlands from lowlands
Slope degree	Sudden topographical deflections such as gully edges
Multiplication of PO and DEM (PO-DEM)	A data manipulation technique to incorporate surface concavity and elevation

3.3 Classification tree

In a nutshell, a classification tree (CT) is a way to unifying and classifying the most homogeneous sets from the training samples. To attain this goal, CT adopts different techniques, including splitting, labeling, and pruning. For classification purposes, the process starts with a random binary splitting guess and moves forward to consider different partitions. This is performed through an iterative recursive partitioning process. Each time, the CT algorithm selects the splitting threshold that leads to the largest increase in homogeneity. The Gini index measures such inequality with a fluctuating value between 0 and 1. A Gini value of 1 indicates that each record at the node belongs to a different category, while a Gini value of 0 indicates that all the records at a node belong to the same category.[19] The pruning process removes the leaves and branches that are not used to better predict real cases (unknown classes). The leaves at the root comprise larger records and information and are trained on known classes. Moving forward, the idiosyncrasy of the training records becomes more peculiar to those records and may offer no use for the prediction task. Hence, the CT starts removing the small branches that fail to generalize using the validation dataset. The latter helps the model circumvent the overfitting issue (i.e., being highly accustomed to the training samples such that the model has poor generalization power). More mathematical details can be found in Breiman et al.[19] In this work, the produced morphometric indices and the rasterized training and validation presence:absence samples were used as inputs to the CT model in ModEco software.[20]

3.4 Model training and validation

The Classification of morphometric indices and the employed machine learning model (i.e., CT) both operate in the model training and validation phases. In this regard, machine learning models are specialized in training on partial data (i.e., supervised learning), based on which the model can extract the emerging pattern for the remaining areas in an unsupervised manner (i.e., cast off from the model training phase). Antithetically, the classification of a standalone morphometric index entails finding an optimal threshold, which generally involves trial and error. Hence, we developed a new automated geospatial tool in ArcGIS called the Best Threshold Selector (BTS) to expedite this process. The BTS operates on a pixel-based cross-validation learning scheme in which the raw (unclassified) morphometric index together with the ground truth (presence-absence gully pixels) in the training zone are used as inputs (Fig. 4A). Accordingly, the BTS initiates a recursive threshold-exploring process (Fig. 4B). This process starts with a random guess (50% as a threshold cutoff), whose classification success would be compared to 25% threshold addition (i.e., 75%) and subtraction (i.e., 25%). This process continues until the BTS reaches an unchanged status in the success value (i.e., a plateau). As a rule, BTS consistently follows a direction that leads to a higher success rate.

The success rate corresponds to the so-called optimization goal, represented by three success metrics: true skill statistics (TSS), Area under the receiver operating characteristic curve (AUC), and Cohen's kappa. The tool automatically calculates the optimal threshold value by selecting the desired metric and, accordingly, classifies the morphometric index. For model training and validation, the region was arbitrarily partitioned into two representative areas. In particular, the easternmost part of the region, which is affected by two parallel gully branches, was selected for model training, and the remaining area was kept apart to validate the model results (Fig. 5). Because of the abundance of presence and absence pixels involved in the training and validation stages, obliged to the pixel-based cross-validation technique embedded in BTS, the conventional sample balance issues (i.e., training:validation partitions) are obviated.

3.5 Performance assessment

The performance of the adopted morphometric indices and machine learning model in gully pattern extraction was based on four different metrics: precision, TSS, Cohen's kappa, and Matthews correlation coefficient (MCC). Using four main elements of the confusion matrix (i.e., TP, true positive; TN, true negative; FP, false positive; and FN, false negative), these metrics can assess the performance of models from different aspects.[21-23]

Precision, also termed positive predictive value, is the proportion of correct predictions of presence locations (Eq. 1). As its name implies, precision disregards errors emanating from incorrect predictions of absence locations.

$$\text{Precision} = \frac{TP}{TP + FP} \tag{1}$$

The TSS, referred to as Pierce's skill score, uses more arguments of the confusion matrix, representing the model's ability to distinguish presence from absence (Eq. 2).

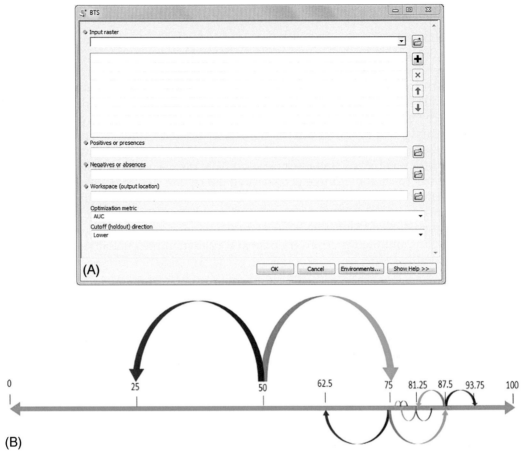

FIG. 4 The graphical user interface of the BTS tool in ArcGIS (A) and the embedded recursive threshold-exploring technique (B—*larger arrows* represent preliminary search steps. *Green arrows*, as opposed to *red ones*, lead to higher success rates).

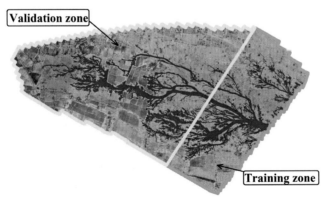

FIG. 5 The selected zones for model training and validation.

$$TSS = \frac{TP}{TP + FN} - \frac{FP}{FP + TN} = \text{Sensitivity} + \text{Specificity} - 1 \tag{2}$$

Cohen's kappa enables the user to compare the performance of the model for the outcome of random success (Eq. 3).

$$Kappa = \frac{(TP + TN) - [(TP + FN)(TP + FP) + (FN + TN)(FP + TN)]/T}{T - [\{(TP + FN)(TP + FP) + (FN + TN)(FP + TN)\}/T]} \tag{3}$$

The MCC is a useful correlation coefficient that compares binary classification success, incorporating true and false positive and negative elements (Eq. 4).

$$MCC = \frac{(TP \times TN) - (FP \times FN)}{\sqrt{(TP + FP)(TP + FN)(TN + FP)(TN + FN)}} \tag{4}$$

The calculation of the metrics mentioned above was carried out in a modified version of an ArcGIS tool called performance metric tool (PMT). The new version of the PMT is capable of using rasterized samples of presence and absence instead of considering representative point samples. Compared to the older version, PMT-Modified would better reflect the models' true success by creating an all-inclusive matrix of samples. Additionally, PMT-Modified is capable of presenting the elements of the confusion matrix (i.e., TP, TN, FP, and FN) in the form of a raster map to spatially pinpoint the strength of the models (i.e., TP and TN) and weaknesses (i.e., FP and FN).

4 Results and discussion

Fig. 6 presents the classifications derived from morphometric indices (using the BTS tool) and classification trees. A visual comparison between the classification results indicates that all the maps performed well in the training zone,

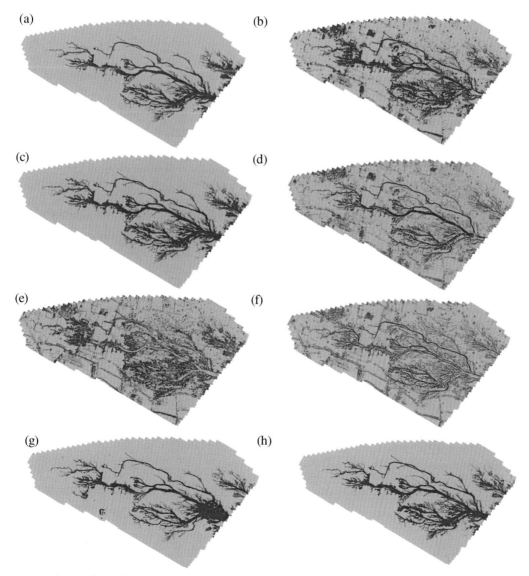

FIG. 6 Gully presence-absence classes derived from morphometric indices (A: DEM, B: positive openness, C: positive openness × DEM, D: RRIM, E: slope, F: TPI, G: valley depth) and classification tree model (H).

while asymmetries became more transparent in the validation zone. From the eight classification maps presented in Fig. 6, it is evident that elevation (DEM), coupled PO-DEM, valley depth, and classification tree performed far better than the other indices. The reason may be the functions presented by each index (Table 1).

Elevation as the main proxy for various morphometric indices may seem highly beneficial for gully extraction, as shown by the classification presented in Fig. 6. However, as previously mentioned, two gully branches with different geometries (i.e., a shallow, less-digitated gully at the upper part and a deep, large gully at the lower part) are discernible in the easternmost sector of the study area, which makes it difficult to assign a single threshold value to each morphometric index. In other words, a single elevation point may not effectively differentiate the gully affected pattern from nongully areas because two gullies with different geometrical features co-exist in the same site. Similarly, valley depth operates on elevation difference and vertical distance to the channel network (referred to as ridge-level interpolation), which explains its small deficiencies. Positive openness represents surface concavities, which not only characterize gullies but also almost anything that exhibits a form of concavity, including a small pit. Hence, conceptualizing the functions each index presented and taking a glance at the gullies together with some trial-and-error factor combination sessions led us to the coupled PO-DEM. The elevation difference helps PO better distinguish gully patterns and exclude small concavities. Testing two arithmetic operations (i.e., addition and multiplication) and investigating the pixel histogram showed that the multiplied PO-DEM exhibits a better breaking point for classification than the added PO-DEM (Fig. 7). In essence, multiplication poses a more restrictive function on the combined indices, while addition creates a relatively smooth map that may not serve best for pattern extraction. On the other hand, the CT also leads to an outstanding classification. Fig. 8 depicts the graphical success and error rates posed by PO, DEM, PO-DEM,

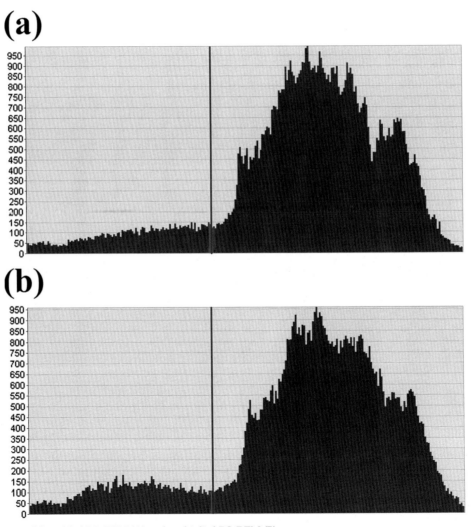

FIG. 7   Histogram created for added PO-DEM (A) and multiplied PO-DEM (B).

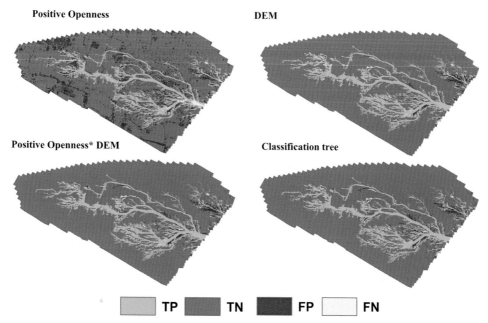

FIG. 8 Calculated true positive (TP), true negative (TN), false positive (FP), and false-negative (FN) values for some of the employed morphometric indices (PO, DEM, and PO × DEM) and the machine learning model (CT) using PMT-modified.

and CT, making it evident that the combination of DEM and PO results in a better gully pattern extraction by reducing the errors encountered in the validation zone.

More dissimilarities between the CT and the well-performing morphometric indices can be derived from the confusion matrix and its derivative performance metrics (Tables 2 and 3, Figs. 9 and 10). The elements of the confusion matrix and the averaged value of the performance metrics are unanimously in line with our previous visual check, based on which PO-DEM, CT, elevation, and valley depth perform better than other indices. More significantly, the PO-DEM that is derived from a simple factor multiplication outperforms a powerful machine learning model, CT.

Although the outstanding performance of PO-DEM is indebted to the automated threshold detection made possible by the BTS tool, the CT follows more complicated recursive training sessions, yet a simple factor combination supersedes its result. In general, a timewise comparison of manually and automatically extracted gully patterns justifies the fact that the latter presents far more promising results. The precision of the extracted gullies using PO-DEM and CT, which may take several minutes to some hours, compared to those manually drawn in 6 months, is beyond expectation. Lastly, this work does not intend to diminish the outstanding results provided by CT. Instead, it encourages the critical role of conceptual models, the merits of simple morphometric indices, and their notable standalone performances, rather than being inputted into complicated black-box models.

TABLE 2 Calculated confusion matrix elements and performance metrics for different employed morphometric indices and machine learning models in the training stage (*bold values* represent the highest performance).

Factor/ Models	Optimal threshold value detected by BTS	Value	Training				Precision	TSS	Cohen's Kappa	MCC	Average
			TP	TN	FP	FN					
PO	75.00	1.45	333,923	1,434,641	215,398	42,068	0.608	0.758	0.6431	0.6625	0.6678
PO_DEM	30.86	9.66	355,124	1,531,949	118,090	20,867	0.751	0.873	0.7937	0.8021	**0.8048**
TPI	41.76	−0.17	259,001	1,499,992	150,047	116,990	0.633	0.598	0.5783	0.5791	0.5971
Valley	37.50	3.95	313,301	1,523,913	126,126	62,690	0.713	0.757	0.7105	0.7139	0.7236
Slope	9.12	7.15	236,656	1,285,795	364,244	139,335	0.394	0.409	0.332	0.3478	0.3706
Elevation	34.90	6.46	332,128	1,552,398	97,641	43,863	0.773	0.824	0.781	0.7838	0.7905
RRIM	43.65	−0.03	272,017	1,526,467	123,572	103,974	0.688	0.649	0.6358	0.6361	0.6520
CT	Self-detected		357,841	1,527,886	122,153	18,150	0.746	0.878	0.793	0.8025	0.8047

TABLE 3 Calculated confusion matrix elements and performance metrics for different employed morphometric indices and machine learning models in the validation stage (*bold values* represent the highest performance).

Factor/ Models	Optimal threshold value detected by BTS	Value	Training				Precision	TSS	Cohen's Kappa	MCC	Average
			TP	TN	FP	FN					
PO	75.00	1.45	466,292	2,992,790	466,958	8492	0.5	0.847	0.598	0.6487	0.6484
PO_DEM	30.86	9.66	414,638	3,443,812	15,936	60,146	0.963	0.869	0.9051	0.9064	**0.9108**
TPI	41.76	−0.17	357,538	3,058,175	401,573	117,246	0.471	0.637	0.5062	0.5258	0.5350
Valley	37.50	3.95	361,922	3,430,942	28,806	112,862	0.926	0.754	0.8163	0.8212	0.8295
Slope	9.12	7.15	340,708	2,816,383	643,365	134,076	0.346	0.532	0.3635	0.3999	0.4103
Elevation	34.90	6.46	344,254	3,455,822	3926	130,530	0.989	0.724	0.818	0.8303	0.8402
RRIM	43.65	−0.03	367,886	3,109,528	350,220	106,898	0.512	0.674	0.5517	0.5681	0.5764
CT	Self-detected		411,425	3,426,648	33,100	63,359	0.926	0.857	0.8812	0.8818	0.8864

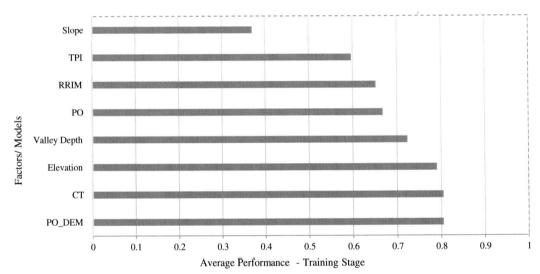

FIG. 9 Average performance of the employed morphometric indices and machine learning models in the training stage.

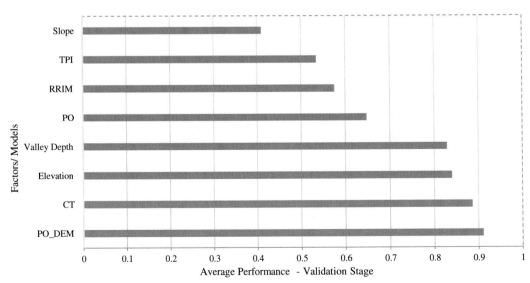

FIG. 10 Average performance of the employed morphometric indices and machine learning models in the validation stage.

5 Comparison, limitations, and future works

Literature review shows that a range of methods has been applied to extract gullies across different areas. Sheshukov et al. tested several compound topographic index models in two paired catchments.[14] Their results attest to the high applicability of topographic indices in extracting the gully patterns, which is in line with our results. Shruthi et al. investigated the application of object-based image analysis techniques for gully delineation.[5] They found that the differences between the gully pattern extracted by OBI methods and the ones manually extracted respectively account for 0.03% and 1.77% for two representative areas. Moreover, they found that gully-related edges are well detected by OBIA methods and can considerably reduce the time spent on manual extraction, which is in complete accordance with our findings. Rahmati et al. also tested the reliability of OBI techniques in gully extraction, which resulted in an overall accuracy value of 92.4%.[8] Phinzi et al. studied the accuracy of machine learning models for gully extraction and found that the random forest and support vector machine models can result in outstanding performance values of 98.7% and 98.01%,[9] which, in accordance with our results, justifies the use of machine learning models in unsupervised feature classification.

In order to give a more reliable factor combination scheme in the Dashtiari region, we would need to apply different techniques (especially OBIA) on the same dataset, which is considered one of the limitations of this work. Additionally, the proposed combined indices (i.e., PO-DEM) should be tested in different areas and gully types. Also, various data configurations and spatial resolution may alter the defined classification thresholds. These points should be addressed and tested in future studies and compared with the findings of this work.

6 Conclusion

This work encourages back-to-basics conceptual modeling using available and straightforward data. Morphometric indices, as simple they may look, can reflect critical hydrological and erosional processes. According to our goal (i.e., gully pattern extraction), the application of morphometric indices was found to be highly beneficial and successful in single-handedly extracting the gully pattern in a short time. Moreover, we presented how knowledge of a phenomenon can help form new and informative factors that would consequently lead to a highly representative model of nature.

The comparative assessment of the coupled positive openness and elevation with a powerful machine learning model justified that a simple factor combination can outperform a complicated machine learning model in terms of both goodness-of-fit and generalization capacity (i.e., prediction). Moreover, the time spent producing PO-DEM or implementing a machine learning model is beyond comparison, considering the 6 months spent on the manual extraction of gullies. Finally, relying on the outstanding precision of morphometric indices in differentiating gully affected from nongully areas, incorporating their results with further manual modifications as a semiautomated procedure may yield the best possible results.

Acknowledgment

We would like to thank Sadaf Amani and Minoo Haghani Shirazi for their meticulous efforts in the manual delineation of gully patterns and data curation.

References

1. Morgan RPC. *Soil Erosion and Conservation.* John Wiley & Sons; 2009.
2. Piest RF, Bradford JM, Spomer RG. Mechanisms of erosion and sediment movement from gullies. In: *Present and Prospective Technology for Predicting Sediment Yields and Sources. Proceeding of Sediment Yield Workshop.* Oxford, Mississippi: USDA sedimentation Lab; 1972:162–176.
3. Chaplot V. Impact of terrain attributes, parent material and soil types on gully erosion. *Geomorphology.* 2013;186:1–11.
4. Day SS, Gran KB, Paola C. Impacts of changing hydrology on permanent gully growth: experimental results. *Hydrol Earth Syst Sci.* 2018;22 (6):3261–3273.
5. Shruthi RB, Kerle N, Jetten V. Object-based gully feature extraction using high spatial resolution imagery. *Geomorphology.* 2011;134(3–4):260–268.
6. d'Oleire-Oltmanns S, Eisank C, Drăgut L, Blaschke T. An object-based workflow to extract landforms at multiple scales from two distinct data types. *IEEE Geosci Remote Sens Lett.* 2013;10(4):947–951.
7. d'Oleire-Oltmanns S, Marzolff I, Tiede D, Blaschke T. Detection of gully-affected areas by applying object-based image analysis (OBIA) in the region of Taroudannt, Morocco. *Remote Sens (Basel).* 2014;6(9):8287–8309.
8. Rahmati O, Tahmasebipour N, Haghizadeh A, Pourghasemi HR, Feizizadeh B. Evaluating the influence of geo-environmental factors on gully erosion in a semi-arid region of Iran: an integrated framework. *Sci Total Environ.* 2017;579:913–927.

9. Phinzi K, Abriha D, Bertalan L, Holb I, Szabó S. Machine learning for gully feature extraction based on a pan-sharpened multispectral image: multiclass vs. binary approach. *ISPRS Int J Geo Inf*. 2020;9(4):252.

10. Kornejady A, Ownegh M, Bahremand A. Landslide susceptibility assessment using maximum entropy model with two different data sampling methods. *Catena*. 2017;152:144–162.

11. Kornejady A, Ownegh M, Rahmati O, Bahremand A. Landslide susceptibility assessment using three bivariate models considering the new topo-hydrological factor: HAND. *Geocarto Int*. 2018;33(11):1155–1185.

12. Lombardo L, Opitz T, Huser R. Numerical recipes for landslide spatial prediction using R-INLA: a step-by-step tutorial. In: *Spatial Modeling in GIS and R for Earth and Environmental Sciences*. Elsevier; 2019:55–83.

13. Choubin B, Moradi E, Golshan M, Adamowski J, Sajedi-Hosseini F, Mosavi A. An ensemble prediction of flood susceptibility using multivariate discriminant analysis, classification and regression trees, and support vector machines. *Sci Total Environ*. 2019;651:2087–2096.

14. Sheshukov AY, Sekaluvu L, Hutchinson SL. Accuracy of topographic index models at identifying ephemeral gully trajectories on agricultural fields. *Geomorphology*. 2018;306:224–234.

15. Garosi Y, Sheklabadı M, Conoscenti C, Pourghasemi HR, Van Oost K. Assessing the performance of GIS-based machine learning models with different accuracy measures for determining susceptibility to gully erosion. *Sci Total Environ*. 2019;664:1117–1132.

16. Conoscenti C, Rotigliano E. Predicting gully occurrence at watershed scale: comparing topographic indices and multivariate statistical models. *Geomorphology*. 2020;359:107123.

17. Conrad O, Bechtel B, Bock M, et al. System for automated geoscientific analyses (SAGA) v. 2.1. 4. *Geosci Model Dev*. 2015;8(7):1991–2007.

18. Olaya V, Conrad O. Geomorphometry in SAGA. *Dev Soil Sci*. 2009;33:293–308.

19. Breiman L, Friedman J, Stone CJ, Olshen RA. *Classification and Regression Trees*. CRC press; 1984.

20. Guo Q, Liu Y. ModEco: an integrated software package for ecological niche modeling. *Ecography*. 2010;33(4):637–642.

21. Benjamini Y, Hochberg Y. Controlling the false discovery rate: a practical and powerful approach to multiple testing. *J R Stat Soc Ser B Methodol*. 1995;57(1):289–300.

22. Powers DM. *Evaluation: From Precision, Recall and F-Measure to ROC, Informedness, Markedness and Correlation*; 2011. http://hdl.handle.net/2328/27165.

23. Rahmati O, Kornejady A, Samadi M, et al. PMT: new analytical framework for automated evaluation of geo-environmental modelling approaches. *Sci Total Environ*. 2019;664:296–311.

39

Identification of land subsidence prone areas and their mapping using machine learning algorithms

Zeynab Najafi, Hamid Reza Pourghasemi, Gholamabbas Ghanbarian, and Seyed Rashid Fallah Shamsi

Department of Natural Resources and Environmental Engineering, College of Agriculture, Shiraz University, Shiraz, Iran

1 Introduction

Human habitats have always faced the challenges of "natural and man-made disasters," and the extent of vulnerability resulting from these disasters varies widely. Subsidence is defined as an unexpected falling or even descend sinking surface of the ground's, sometimes accompanied by a slight horizontal displacement vector and can be created under the effect of overuse of aquifers, activities to mining, underground corridors and tunnels, karst event, bursting of water pipes in urban areas, and even faults.[1] Also, excavation of unauthorized wells, illegal construction, development of settlements without planning and measuring the geological capacity of the area, land-use changes, organic soil drainage, loading of engineering structures, and pumping oil and water from the ground are other factors on LS occurrence.[2]

Undoubtedly, such incidents can cause a great deal of financial damage and even loss of life, which makes studying in this field more important. Urban expansion in various sectors of industry, commerce, and transportation has negative environmental impacts. These effects lead to a subsidence phenomenon that can affect the urban development and planning process; so, it can be said that urban developments can contribute to land subsidence that is a bilateral connection.[3] Perrin et al.[4] studied LS in karst lands using the "WoE (weights-of-evidence)" model. This study proved the susceptibility of karst areas to LS events. Pourghasemi and Mohseni Sarvi[5] indicated that a "random forest (RF)" ML model has high accuracy (AUC value of 93.9%) for preparing an LS map in Kerman Province, Iran. In addition, variables importance "variables importance" based on "mean decrease accuracy-MDA" showed "groundwater level changes" groundwater level change data and elevation were the most significant factor.

Oh et al.[6] used Bayesian functional and ensemble models at Hwajeon in Taebaek. The results showed that the highest accuracy was related to the ensemble model, where its "AUC" value was 91.44%.

Thus, according to previous research, a spatial modeling approach using statistical and ML algorithms with a Geographic Information system (GIS) can be very effective in predicting the occurrence of land subsidence and its zonation.

Therefore, the foremost characteristics of this research are (1) considering the spatial relationship between LS and effective factors using the BSI model, (2) LS modeling using SVM, BRT, and RF ML models in the Gharebolagh region, and (3) prioritizing effective factors for LS occurrence using the Bagged CART method.

2 Description of the study area

The Gharebolagh Region (Fig. 1), with an area of 448 km², is situated southeast of Fars Province, Iran (53° 55′–54° 10′ E and 28° 5′–29° 5′ N). The climate of this area is semiarid. The average height of the plain is 1463 m. The average

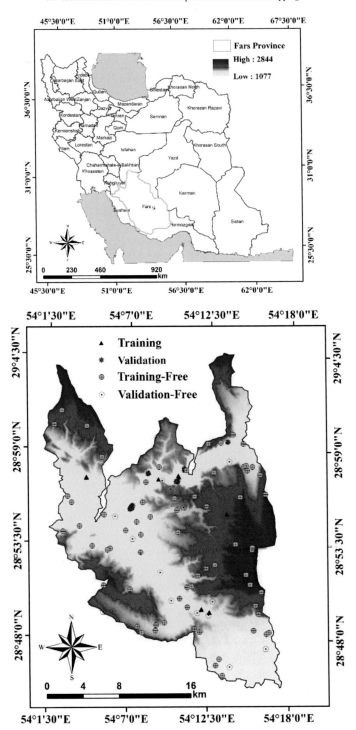

FIG. 1   Location of the Gharebolagh Region in Fars Province, Iran.

rainfall is 295.2 mm per year. The mean temperature in the plain is 17.6°C, whereas the formations in the area are more karstic with good permeability.[7]

3 Materials and methods

The flowchart of the present study consists of several steps and is presented in Fig. 2 as follows:

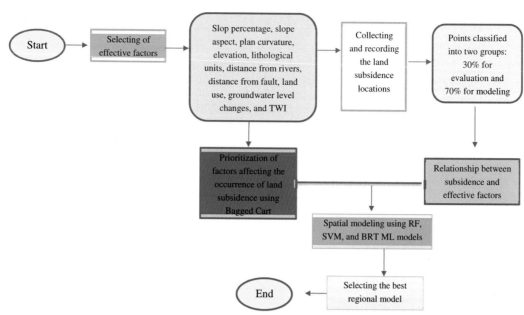

FIG. 2 Flowchart of this study.

3.1 LS "inventory mapping"

Given the enormous financial damages that the subsidence may cause in different areas, it was attempted to record the subsidence locations So, the LS locations in Gharebolagh Region were diagnosed using extensive field visits, Google Earth images, and also data collected from" Regional Water Company of Fars". For this purpose, 96 LS locations were first recorded with the help of the GPS and the "LS inventory map" was prepared in ArcGIS 10.3. Next, 70% of these subsidence points were randomly chosen for modeling and the remaining 30% were chosen for evaluation.

3.2 Factors affecting the occurrence of LS

Considering the previous studies[4] and with attention to objectives of this research, ten effective factors on LS are consist of, slope percentage, slope aspect, plan curvature elevation, lithological units, "distance from river," groundwater level changes, "distance from faults," land use, and "TWI (topographic wetness index)" were selected as their details are given in following: After preparation of "DEM (Digital Elevation Model)," four morphometric factors namely, slope percentage, slope aspect, plan curvature, and TWI were prepared in ArcGIS and SAGA GIS software (http://www.saga-gis.org/en/index.html) in a resolution of 12.5 m * 12.5 m. The land use map was prepared at a 1:100,000-scale by "Forest, Range, Watershed Management Organization of Fars Province" and rechecked using Google Earth images. The Gharebolagh Region's Geological map on a scale of 1:100,000 prepared by the Unk >"Geological Survey of Iran" and "lithological units" (Table 1) and" distance from faults" mapped based on it. Distance from rivers map was created using "Euclidean Distance Tool" using the topographic map of the studied region. Map of groundwater level changes was prepared using 15 piezometric wells data during the years 1998–2016 base on the "IDW (Inverse Distance Weight) interpolation" model in ArcGIS 10.3.

In the following section, we examine the classification methods for various factor maps. The slope percentage map is classified into four classes: 0–2, 2–5, 5–12, and >12 classes.[8,9] The plan curvature, which denotes the convex, flat, or concave terrain, is divided into 3 floors less than −0.01 (convex), −0.01 to +0.01 (smooth) and larger than +0.01 (concave).[10] The map of elevation classes was divided into four classes based on natural breaks: 1077–1512, 1512–1807, 1807–2190, and 2190–2844 m.[11]

Map of groundwater level changes was also divided into 4 classes of 31.49–47.89, 47.89–60.13, 60.13–71.74, and 71.74–97.88 m based on natural breaks. The map of distance from the river also provided by Euclidean Distance Function in ArcGIS software contains four classes of less than 50–50-100 50–150 and greater than 150. The map of distance from the fault is also divided into four classes using the natural break method: 0–561.42, 561.42–1160.17, 1160.17–1834.14, 1834.14–2669.25, and 2669.25–4525.68 (Table 2).

TABLE 1 Lithological units of Gharebolagh Region.

Class	Unit name	Lithology	Formation
A	Qft2	Low level pediment fan and valley terrace deposits	–
B	Eja	Gray and brown weathered, massive dolomite, low weathered thin to medium -bedded dolomite, and massive, feature forming, buff dolomitic limestone	Jahrum
C	PeEsa	Pale red marl, marlstone, limestone, gypsum, and dolomite	Sachun
D	EK	Well bedded green tuff and tuffaceous shale	
E	MuPlaj	Brown to gray, calcareous, feature-forming sandstone and low weathering, gypsum- veined, red marl, and siltstone	Karaj
F	Ktb	Massive, shelly, cliff-forming partly anhydrite limestone	Tarbur
G	Ksv	Gray, thick-bedded to massive limestone with thin marl intercalations in the upper part	Sarvak
H	OMr	Red, gray, and green silty marls interbedded with subordinate silty limestone and minor sandstone ribs	Razak

TABLE 2 The values of SIM.

Factors	Classes	Pixels number	Land-subsidence number	SIM value
Aspect	Flat	1,065,209	21	0.41
	North	920,597	12	−0.01
	East	898,436	1	−2.47
	South	1,171,344	20	0.26
	west	1,122,057	14	−0.05
Plan curvature (100/m)	<−0.01	1,400,248	14	−0.27
	−0.01 to 0.01	2,203,117	41	0.35
	>0.01	1,574,278	13	0.46
Slope percent	0–2	1,220,023	29	0.59
	2–5	317,511	10	0.87
	5–12	567,021	25	1.21
	>12	3,073,088	4	−2.31
Altitude (m)	1077–1512	2,330,782	67	−5.12
	1512–1807	1,560,341	0	−2.52
	1807–2190	853,057	1	1.16
	2190–2844	433,463	0	1.67
TWI	0–5.7	2,546,453	20	−0.51
	5.7–9.41	1,649,274	21	−0.03
	9.41–21.43	981,916	27	0.74
Distance from rivers (m)	0–50	1,195,317	27	0.54
	50–100	997,477	16	0.20
	100–150	827,317	9	−0.19
	>150	2,157,532	16	−0.57

TABLE 2   The values of SIM—cont'd

Factors	Classes	Pixels number	Land-subsidence number	SIM value
Distance from faults (m)	0–561.42	1,515,667	8	−0.91
	561.42–1160.17	1,466,644	31	0.48
	1160.17–1834.14	1,081,170	23	0.48
	1834.14–2669.35	727,012	3	−1.16
	2669.35–4525.68	387,150	3	−0.54
Land use type	Agriculture	1,441,803	38	0.70
	Urban	51,773	17	3.22
	Range	3,429,286	13	−1.24
	Lank forest	245,128	0	0.00
	Rock land	9653	0	0.00
Piezometric data (m)	31.49–47.89	1,834,914	8	−1.10
	47.89–60.13	1,103,534	28	0.66
	60.13–74.71	1,769,842	27	0.15
	74.71–97.88	469,353	5	−0.21
Lithological units	Qft2	2,053,101	39	0.37
	Eja	146,596	25	2.56
	PeEsa	6252	0	0.00
	EK	24,894	0	0.00
	MuPlaj	501,518	0	0.00
	Ktb	25,423	4	2.48
	Ksv	5891	0	0.00
	OMr	2,413,968	0	0.00

3.3 Prioritizing effective factors using "bagged CART"

CART is one of the most popular "machine learning" methods obtained by developing R software.[12] This model is also less commonly used in LS modeling. In this model, data splitting occurs using repeated splitting of nodes, and this continues as long as similar or limited values are less than the user-defined observations.[13] The model works evenly and the different scales of the model remain unchanged.[14,15] The pruning process is used to avoid problems that are too appropriate or insignificant from a binary decision tree, we can point to the result of this model that the predicted space is divided into regions by homogeneous response factor values.[16]

The Bagged CART method, which reduces the variance associated with predictions, makes predictions more accurate and can be used in classification and regression methods. The performance of this method is that, by averaging the simple results, it increases the accuracy of the prediction, while it is more complex than the predictions that are made normally, so the prediction models are better using unstable classification and regression.[17]

3.4 Statistical index model

van Westen[18] proposed a statistical index method for preparing landslide susceptibility maps. Different weights for each class of each factor were obtained as a natural logarithm of subsidence density in that class divided by the subsidence density in the entire map.[19]

3.5 Land subsidence spatial modeling

3.5.1 Random forest

RF is a nonparametric classification method in the late 19th century.[20] was obtained from a set of decision trees. This algorithm replaces and changes the effective and related factors with the goal continuously; it can generate many decision trees. Finally, all decision trees are combined for prediction. In this tree structure, the point of division into two embranchments is called a node. The first node is called the root and the last node is known as leave, and by dividing the space of the explanatory variables into two smaller subspaces, the decision tree grows with the help of one node that is divided into two internal nodes and. The random forest algorithm propagates the tree without using all available data; only 66% of the data are used, whereas the remaining data (33%) are used to evaluate the practiced tree. Mentioned procedure has been applied several times and finally, the average of the values obtained as the predicted value of the algorithm. To determines the significance of factors affecting the subsidence two indices are used namely "mean decrease accuracy and mean Gini accuracy." To prioritize the effective factors, especially when environmental factors interact with each other, the mean decrease accuracy is better to use.[21,22]

After obtaining the weights of each factor from the RF algorithm, they were transferred from R to the ArcGIS software. Finally, land subsidence maps were prepared for the study area and the map was divided into four classes (low, medium, high, and very high).

3.5.2 Boosted regression tree

The Ensemble of statistical and machine learning techniques can lead to a BRT model that helps to improve the performance of a single model using a combination of two algorithms.[23] These algorithms are: (1) classification and regression tree (CART) and (2) boosting. To increase the accuracy of the model and its working boosting, which is based on the construction, combination, and meaning of a countless number of better models because it is more precise than creating a model. BRT dominates the biggest shortcoming of a single decision tree by a relatively poor practice.[24] In this model, only the first tree will be constructed of total educational data and the remaining data can be created following trees that are on the tree before itself. Trees will not be created using all data, and only a small amount of data is used. The basic idea of this method is to combine a set of poor forecasting models (high forecast error) to achieve a strong prediction (low prediction error). In each step, each data set is categorized and used to prepare the next tree.[23,25]

3.5.3 Support vector machine

In Vapnic statistical learning theory, SVM and its root are classification and regression tools.[26] Unlike ANN, which tries to explain the complexity of functions from the input space, the SVM takes nonlinear data into a higher dimensional space.[27,28] It then uses simple linear functions to create linear delimiter boundaries in the new space. The most significant feature of the support vector machine is that its regression formulation is based on structural risk minimization rather than empirical risk minimization so that it performs better than other common algorithms, such as neural networks, whose structure is based on experimental risk minimization.[27–29]

3.6 Validation of models

Estimating the precision of LS "susceptibility" maps is not possible using the training dataset. Therefore, to solve this problem, two-thirds of LS points are used for modeling and the remaining one-third is applied to assess the accuracy of the models. Therefore, after preparing three MLM (machine learning models), to validate these built models, 30% of LS points (25 points) and the "ROC" curve were used too. This method can quantitatively calculate the accuracy of the models. The ideal model has an AUC value of close to 1, whereas an "AUC" value of close to 0.5 is known as the worst model.[11]

4 Results and discussion

4.1 Prioritizing the effective factors on LS event

Prioritizing and determining the importance of variables on LS occurrence was performed using the bagged CART ML method (Fig. 3). The results in Fig. 3 show that "groundwater level changes" are the most important and consequently the most effective factor for the occurrence of LS in the study region. Also, the variables of elevation (DEM),

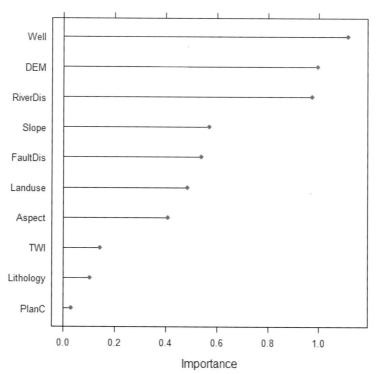

FIG. 3 Prioritizing and determining the importance of variables using Bagged Cart ML model.

distance from rivers (RiverDis), slope percentage (Slope), distance from faults (FaultDis), land use (Landuse), slope aspect (Aspect), TWI, lithological units (Lithology), and plan curvature (PlanC) ranked from second to tenth in terms of importance, respectively (Fig. 3). Therefore, it can be concluded that the factors of "groundwater level changes (Well)" and "plan curvature" have the highest and lowest impact on the occurrence of LS in the Gharebolgh area, respectively. The results obtained for the most important variables are in line with those of Abdollahi et al.[11] and Ghorbanzadeh et al.[8,9]

4.2 Outcomes of the relationship between the LS events and effective factors

The relationship between classes of affective factors (10 variables) and the occurrence of LS identified in the study region was calculated using the BSI (Table 2). The Results show that most LS events occurred in the flat slope direction, with a weight value of 0.41. In addition, most LS events were observed in flat curvatures (−0.01 to 0.01) with a value of SI = 0.35. This result is in line with Mohammady et al.[30] On the other hand, with regard to the slope factor, the third class of the slope percentage (5%–12%), the highest LS event was observed (1.21). In general, on lower slopes, subsidence is more likely to occur.[31] This relationship for altitude factors can be observed at high altitudes. At altitudes of 2190–2844 m, the weight of the SI is the highest.

In terms of the TWI, the highest weight of LS occurrences was given in the third class of the TWI factor. This means that the higher the moisture index, the more "land subsidence" occurred in the study area (SI = 0.74).

The Results (Table 2) indicate that in the shortest distance of rivers, the highest occurrence of "land subsidence" was observed. In contrast, the greater the distance from the rivers, the lower is the weight of the SI. Ozdemir[32] and Abdollahi et al.[11] reported this result regarding distance from rivers. The results of the land use layer and its relationship with "land subsidence" showed that the highest weight (SI = 3.22) was observed in urban lands. This can be considered as excessive groundwater abstraction in urban areas because the drop in groundwater level is the most important reason for subsidence.[11]

However, the results of groundwater level changes indicated that the most LS occurred in the class of 47.89–60.13 m (SI = 0.66). In addition, most of the LS events in the study area were observed in Eja, Ktb, and Qft$_2$ "lithological units" most subsidence often occurs in urban and residential areas.

4.3 Susceptibility classes of the used ML models

The LS susceptibility maps produced by the three ML models (BRT, SVM, and RF) are shown in Fig. 4. Each final LS map was divided into four classes (Low, Moderate, High, and Very high) using the "Natural Break" method.[33,34] In addition, the percentage of "land subsidence susceptibility" area in each of the classes is given in Table 3. According to the RF algorithm, 62.52% of the area affected by LS was classified as "very low," followed by "moderate" (23.69%), "high" (9.20%), and "very high" (4.56%). Similarly, as shown in the SVM method, the highest LS susceptibility is related to "low" (36.78%), "moderate" (35.47%), "high" (17.77%), and "very high" (9.96%). Finally, the BRT method predicted the highest percentage of subsidence susceptibility as the low class (52.16%), followed by moderate (26.69%), very high (12.90%), and high (8.23%) classes, respectively (Table 3).

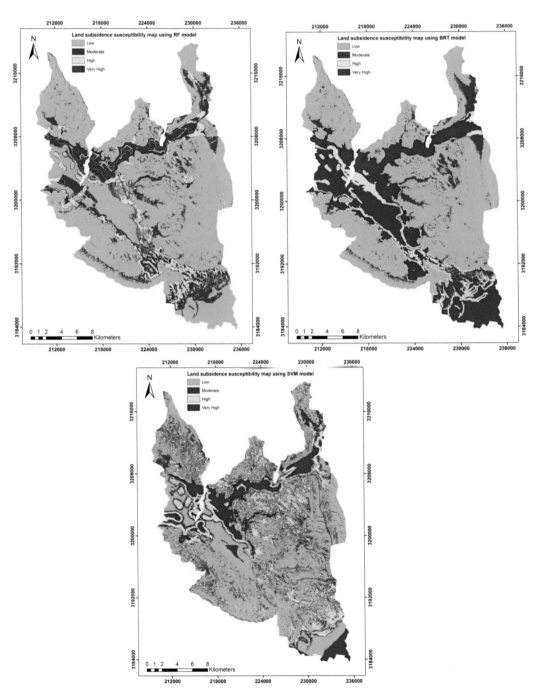

FIG. 4 "Land subsidence susceptibility" maps created using three ML models.

TABLE 3   Percentage of "land subsidence susceptibility" area.

% Area of BRT model	% Area of SVM model	% Area of RF model	No. of pixels in BRT model	No. of pixels in SVM model	No. of pixels in RF model	classes	Groups
52.16	36.78	62.53	2,700,780	1,904,534	3,237,550	Low	1
26.70	35.48	23.69	1,382,184	1,836,833	1,226,778	Moderate	2
8.24	17.78	9.21	426,577	920,546	476,735	High	3
12.90	9.96	4.57	668,102	515,730	236,580	Very high	4

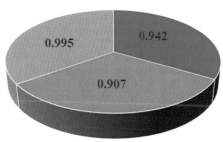

FIG. 5   Evaluation of three ML models used in the present study.

4.4 Evaluation of the accuracy of ML algorithms

The results of evaluating the ML algorithms used in this study showed that the highest accuracy was observed in the RF, BRT, and SVM models with AUC values of 0.99, 0.942, and 0.907, respectively (Fig. 5). Thus, the RF model is the best algorithm, according to the main goal of this study.

Generally, "land subsidence" causes significant human and financial losses in communities every year. Most LS events resulted in a drop in the groundwater levels. Changes in the surface and volume of aquifers can be one of the basic agents for "land subsidence" events.[35] In arid and semiarid regions, groundwater is the main source of water supply; Therefore, it is reasonable that groundwater level changes are an important factor in the occurrence of "land subsidence" in these areas. This has been proven in previous studies.[30,35] As a big suggestion, the modeling process is an important and valuable approach for management and decision making on LS susceptibility areas, especially in high and very high regions. In general, there are tools for expanding the knowledge of environmental hazards. Machine learning methods are currently a sustainable tool in environmental hazard studies.[36] One of the advantages of ML algorithms is their high ability to spatially predict environmental hazards[37] such as "land subsidence." On the other hand, it can easily process big data.

5 Conclusion

In this study, three ML models (BRT, SVM, and RF) were used to investigate and map the LS hazard in the Qarabolagh Region, Fars Province, Iran. In addition, a BSI method was applied to evaluate the relationship between the classes of each effective factor and LS locations. The results of this study showed that the RF, BRT, and SVM models had excellent accuracy in terms of area under the curve" AUC values of 0.99, 0.942, and 0.907, respectively. The Results indicated that "groundwater level changes" were chosen as the most important and effective variable in the occurrence of land subsidence in the studied region. Finally, "land subsidence susceptibility" maps developed using ML models and planners and policy-makers can take advantages of these useful tools for planning and managing sustainable development, environmental and natural hazards especially "land subsidence."

References

1. Khorsandi Aghai A. Survey of land subsidence—case study: the land subsidence formation in artificial recharge ponds at South Hamadan Power Plant, northwest of Iran. Indian Academy of Sciences. *J Earth Syst Sci.* 2015;124(1):261–268.
2. Putra DPE, Setianto A, Keokhampui K, Fukuoka H. Land subsidence risk assessment case study: Rongkop, Gunung Kidul, Yogyakarta-Indonesia. In: *The 4th AUN/SEED-Net Regional Conference on Geo- Disaster Mitigation in ASEAN "Past Tragedies are the Lessons for Future Mitigation".* Phuket, Thailand: The Royal Paradise Hotel & Spa; 2011.
3. Park I, Choi J, Jin Lee M, Lee S. Application of anadaptive neuro fuzzy inference system to ground subsidence hazard mapping. *Comput Geosci.* 2012;48:228–238.
4. Perrin J, Cartannaz C, Noury G, Vanoudheusden E. A multicriteria approach to karst subsidence hazard mapping supported by weights-of-evidence analysis. *Eng Geol.* 2015;197:296–305.
5. Pourghasemi HR, Mohseni Sarvi M. Land-subsidence spatial modeling using the random forest data-mining technique. In: *Spatial Modeling in GIS and R for Earth and Environmental Sciences Book.* Elsevier; 2019:147–159.
6. Oh H, Syifa M, Lee C, Lee S. Land subsidence susceptibility mapping using bayesian, functional, and meta-ensemble machine learning models. *Appl Sci.* 2019;9:1248. https://doi.org/10.3390/app9061248.
7. Yamani M, Najaf E, Abedini MH. The relationship between ground subsidence and groundwater drop in Ghare- Bolgh Plain (Fars Province). *Geography.* 2001;3(8–9):9–27 [In Persian].
8. Ghorbanzadeh O, Blaskche T, Feizizadeh B. An interval matrix method used to optimize the decision matrix in AHP technique for land subsidence susceptibility mapping. *Environ Earth Sci.* 2018;77:584.
9. Ghorbanzadeh O, Gholamnia K, Blaskche T, Jagannath A. A new GIS-based data mining technique using an adaptive neurofuzzy inference system (ANFIS) and k-fold cross validation approach for land subsidence susceptibility mapping. *Nat Hazards.* 2018;94:497–517.
10. Rahmati O, Pourghasemi HR. Identifcation of critical food prone areas in data-scarce and ungauged regions: a comparison of three data mining models. *Water Resour Manage.* 2017;31:1473–1487.
11. Abdollahi S, Pourghasemi HR, Ghanbarian GA, Safaeian R. Prioritization of effective factors in the occurrence susceptibility mapping using an SVM model and their different Kernel functions. *Bull Eng Geol Environ.* 2018. https://doi.org/10.1007/s1006 4-018-1403-6.
12. Breiman L, Friedman JH, Olshen RA, Stone CJ. *Classification and Regression Trees.* New York: Routledge; 2017. https://doi.org/10.1201/9781315139470.
13. Naghibi SA, Pourghasemi HR, Dixon B. GIS-based groundwater potential mapping using boosted regression tree, classification and regression tree, and random forest machine learning models in Iran. *Environ Monit Assess.* 2016;188:44. https://doi.org/10.1007/s10661-015-5049-6.
14. Gayen A, Pourghasemi HR. Spatial modeling of gully erosion: a new ensemble of CART and GLM data-mining algorithms. In: *Spatial Modeling in GIS and R for Earth and Environmental Sciences.* Elsevier; 2019:653–669. https://doi.org/10.1016/B978-0-12-815226-3.00030-2.
15. Gayen A, Pourghasemi HR. Spatial modeling of gully erosion: a new ensemble of CART and GLM data-mining algorithms. In: *Spatial Modeling in GIS and R for Earth and Environmental Sciences.* Elsevier; 2019:653–669. https://doi.org/10.1016/B978-0-12-815226-3.00030-2.
16. Therneau T, Atkinson B, Ripley B. *Recursive Partitioning And Regression Trees*; 2019. https://cran.r-project.org/package=rpart. Accessed 7 December 2019.
17. Breiman L. Heuristics of instability and stabilization in model selection. *Ann Stat.* 1996;24(6):2350–2383. https://doi.org/10.1214/aos/1032181158.
18. Westen C. Statistical landslide hazard analysis. *ILWIS.* 1997;2:73–84.
19. Rautela P, Lakhera RC. Landslide risk analysis between Giri and Tons rivers in Himachal Himalaya (India). *Int J Appl Earth Obs Geoinf.* 2000;2(3):153–160.
20. Dietterich T. An experimental comparison of three methods for constructing ensembles of decision trees: bagging, boosting and randomization. *Mach Learn.* 2000;40:139–157.
21. Nistor C, Church M. Suspended sediment transport regime in a debris-flow gully on Vancouver Island, British Columbia. *Hydrol Process.* 2005;19:861–885.
22. Diaz-Uriarte R, Alvares D, Andres S. Gene selection and classification of microarray data using random forest. *BMC Bioinf.* 2006;7:3.
23. Elith J, Leathwick JR, Hastie T. A working guide to boosted regression trees. *J Anim Ecol.* 2008;77(4):802–813.
24. Aertsena W, Kinta V, Orshovena J, Özkanb K, Muysa B. Comparison and ranking of different modelling techniques for prediction of site index in Mediterranean mountain forests. *Ecol Model.* 2010;221:1119–1130.
25. Main AR, Michel NL, Headley JV, Peru KM, Morrissey CA. Ecological and landscape drivers of neonicotinoid insecticide detections and concentrations in Canada's Prairie Wetlands. *Environ Sci Technol.* 2015;49:8367–8376.
26. Vapnik VN. *The Nature of Statistical Learning Theory.* Springer Verlag; 1995.
27. Li Y, Fang T. Wavelet and support vector machines for short-term electrical load forecasting. In: *Proceedings of the Third International Conference on Wavelet Analysis and its Applications.* vol. 1; 2003:399–404.
28. Taheri K, Gutiérrez F, Mohseni H, Raeisi E, Taheri M. Sinkhole susceptibility mapping using the analytical hierarchy process (AHP) and magnitude–frequency relationships: a case study in Hamadan province, Iran. *Geomorphology.* 2015;234:64–79.
29. Verikas A, Gelzinis A, Bacauskiene M. Mining data with random forests: a survey and results of new tests. *Pattern Recognit.* 2011;44:330–349.
30. Mohammady M, Pourghasemi HR, Amiri M. Land subsidence susceptibility assessment using random forest machine learning algorithm. *Environ Earth Sci.* 2019;78:503. https://doi.org/10.1007/s12665-019-8518-3.
31. Chen W, Pourghasemi HR, Kornejady A, Zhang N. Landslide spatial modelling: introducing new ensembles of ANN, MaxEnt, and SVM machine learning techniques. *Geoderma.* 2017;305:314–327.
32. Ozdemir A. Investigation of sinkholes spatial distribution using the weights of evidence method and GIS in the vicinity of Karapinar (Konya, Turkey). *Geomorphology.* 2016;245:40–50.
33. Falaschi F, Giacomelli F, Federici PR, et al. Logistic regression versus artificial neural networks: landslide susceptibility evaluation in a sample area of the Serchio River valley, Italy. *Nat Hazards.* 2009;50:551–569.
34. Kavzoglu T, Sahin EK, Colkesen I. Landslide susceptibility mapping using GIS-based multi-criteria decision analysis, support vector machines, and logistic regression. *Landslides.* 2014;11:425–439.

35. Mohammady M, Pourghasemi HR, Amiri M, Tiefenbacher J. Spatial modeling of susceptibility to subsidence using machine learning techniques. *Stochastic Environ Res Risk Assess*. 2021. https://doi.org/10.1007/s00477-020-01967-x.

36. Nasiri F, Huang G. A fuzzy decision aid model for environmental performance assessment in waste recycling. *Environ Model Software*. 2008;23 (6):677–689. https://doi.org/10.1016/j.envsoft.2007.04.009.

37. Stamatopoulos C, Petridis P, Parcharidis I, Foumelis M. A method predicting pumping-induced ground settlement using back-analysis and its application in the Karla region of Greece. *Nat Hazards*. 2018;92(3):1733–1762. https://doi.org/10.1007/s11069-018-3276-1.

40

Monitoring of spatiotemporal changes of soil salinity and alkalinity in eastern and central parts of Iran

Sedigheh Maleki[a], Hassan Fathizad[b], Alireza Karimi[a], Ruhollah Taghizadeh-Mehrjardi[c,d,e], and Hamid Reza Pourghasemi[f]

[a]Department of Soil Science, Faculty of Agriculture, Ferdowsi University of Mashhad, Mashhad, Iran [b]Department of Arid and Desert Regions Management, College of Natural Resources and Desert, Yazd University, Yazd, Iran [c]Department of Geosciences, Soil Science and Geomorphology, University of Tübingen, Tübingen, Germany [d]CRC 1070 Resource Cultures, University of Tübingen, Tübingen, Germany [e]Faculty of Agriculture and Natural Resources, Ardakan University, Ardakan, Iran [f]Department of Natural Resources and Environmental Engineering, College of Agriculture, Shiraz University, Shiraz, Iran

1 Introduction

Soil is a vital key and manageable factor in producing food and biomass, controlling the geochemical cycles of water and organisms.[1,2] In addition, the soil is an important part of the protection of the environment and the production of bio-energy.[3] Land-use change and intensive and long-term cultivation have been shown to affect soil properties on a large scale and likely change the physical, chemical, and biological properties of the soil.[4] Therefore, soil and its properties are continuously exposed to various degradation processes and unsustainable management practices.[5,6] According to Foley et al.,[7] more than 50% of agricultural soils are moderately or highly exposed to degradation processes. Soil degradation includes the reduction of soil fertility, soil erosion, loss of soil organic carbon (SOC), alkalization, and salinization,[7–10] resulting in a decline in crop yields of cultivated lands.[6]

Soil salinity and alkalinity are among the major soil limitations for agriculture and the main causes of land degradation in arid and semiarid regions[5,11] such as Iran.[12–14] The area of primary salinity in the world's soils is approximately 955 million hectares and the area of saline land resulting from human activities (secondary salinization) is approximately 77 million hectares, 58% of which is in irrigated agriculture.[5] In Iran, the area of saline-affected soils is estimated at 15.5 to 18 million hectares, which negatively affects agriculture, soil fertility, and water resources.[15] Based on the results of a literature review,[16] soil salinity is mainly related to natural and human-induced factors in Iran.

Dry climate (low rainfall and high evaporation) and topography are the main natural factors of soil salinity. In arid regions of Iran, salts are transported to large depressions, especially in playas and, due to higher evaporation and transpiration, the salts accumulate in the soils. According to Krinsley,[17] 60 playas have been identified in arid regions of Iran, which are the result of the influence of these two factors. Furthermore, Alavipanah and Ddaper[18] explained that in some parts of Iran, the concentration of soluble salts at the soil surfaces usually results in strong salt enrichment in the playa margin. Geologic formations, such as salt domes[19] and evaporates,[13] cause natural soil salinity in some areas. Given the above conditions, the presence of high soil salinity in the arid environments of Iran is not surprising. In recent years, secondary salinity has increased due to human activities, including a lack of suitable drainage infrastructure and irrigation with saline waters with inadequate management.

Appropriate management of saline areas requires accurate and quantitative information to be updated. The use of laboratory methods to estimate soil salinity is generally time-consuming and costly, especially in extent areas.[10] In addition, because of the high spatial variation of soil salinity at different spatial scales, it is better to use soil salinity maps and remote sensing (RS) data to determine and monitor saline soils. Therefore, the first step for monitoring and management of soil salinity is to identify saline areas and then prepare salinity maps.

A review of articles published in international journals on soil salinity in Iran in the last 20 years (2000–20) showed that the studies were divided into two categories: (i) studies that have detected and generated soil salinity maps in different parts of the country using statistical analysis, RS, and digital soil mapping "DSM"[20] methods and (ii) few studies have examined the temporal and spatial variation of soil salinity in different years (Table 1).

As emphasized in Table 1, articles published in international journals in 2020 have increased in relation to soil salinity mapping, indicating the need for larger-scale maps in different parts of Iran at the local to regional levels. In recent decades, satellite images and RS techniques have been applied to study soil salinity due to the high ability of these images to detect changes all over the world.[13,15,27,29,38–43] Nevertheless, salinity is a dynamic phenomenon whose dimensions change over time and space. Therefore, it is necessary to monitor the conditions of saline soils over the long-term annually, and by weighing and lightening the factors involved in salinization and desalination, to predict future changes in the salinity of agricultural soils by identifying the effective factors. In this way, an appropriate understanding of spatio-temporal soil salinity and alkalinity could be a valuable guide for better environmental management and solve various problems concerning agricultural management.

In this chapter, we select two regions located in arid regions to (i) show the temporal and spatial soil salinity and alkalinity changes in the eastern and central parts of Iran and (ii) indicate the role of human activities in increasing the soil salinity, alkalinity, and expansion of saline areas.

2 Materials and methods

In this study, two sites in eastern (Bajestan, site 1) and central (Yazd-Ardakan plain, site 2) of Iran were selected. At site 1, the temporal variation of soil salinity and alkalinity was investigated in different land uses on different geomorphic surfaces under an arid climate from 2004 to 2018 (continuous cultivation for 14 years). The study on site 2 builds on the investigation of the temporal soil alkalinity trends in the Yazd-Ardakan Plain, using the RS and DSM framework to predict soil alkalinity maps for past, present, and future years using soil sample data, land use, and groundwater change from 1986 to 2016.

2.1 Study areas

2.1.1 Site 1

Site 1 is situated in the Bajestan region, southwest of Khorasan Razavi between 57°57′56″ and 58°00′40″ E, and 34° 17′91″–34°33′79″ N (Fig. 1). As an arid region, it experiences mild winters with hot, dry summers. The mean annual temperature and rainfall are 17.3°C and 193 mm, respectively.[44] This region has an Aridic moisture regime and Thermic temperature regime.[45] Aridisols and Entisols are the dominant soil orders in this area. In addition, the most commonly planted species are pomegranate (*Punica granatum* L.), pistachio (*Pistacia vera* L.), barley (*Hordeum vulgare* L.), and saffron (*Crocus sativus* L.), which are located on different landforms such as pediments, alluvial fans, playa clay flats, and dune fields. Overall, the area's geologic material through the mountain to flatlands consists of lower Cretaceous undifferentiated rocks, partly massive and bedded limestone, and recent and old alluvial deposits. The elevation ranges from 786 to 2283 m.a.s.l. (m above sea level) in the study area.

2.1.2 Site 2

Site 2 is located in the central plateau of Iran, Yazd province, between 53°08′36′ and 54° 85′32′ E and 31° 21′59′ and 32° 61′02′ N (Fig. 1). The minimum and maximum elevations were 997 and 2684 m.a.s.l., respectively. The annual mean rainfall and evaporation are 118 mm and 2200–3200 mm, respectively.[33] The vegetation in the study area appeared as shrubs. The major soil types were classified as Entisols and Aridisols. The crops in the study area mainly include wheat (*Triticum aestivum* L.), alfalfa (*Medicago sativa* L.), pomegranate (*P. granatum* L.), pistachio (*P. vera* L.), and barley (*H. vulgare* L.). The geological units consist of dense limestone, salty sediment, red gypsiferous marls, shale, dolomite, sandstone rocks, and conglomerate, which have a wide range of ages from the Precambrian to the Holocene. "Alluvial

TABLE 1 Soil salinity and alkalinity relevant case studies in different regions of Iran.

References	Region	Spatial extent (km²)	Methodology	Climate	Temperature (°C)	Rainfall (mm)	Elevation (m.a.s.l.)	Geological units
Taghizadeh-Mehrjardi et al.[10]	Isfahan	100,000	DSM	Warm-dry	–	113	1590	Cretaceous limestone that overlays Mesozoic shale and sandstone
Abedi et al.[19]	Darab plain in Fars	964	DSM	Arid	22	255	1000–1830	Quaternary sediments and the Jahrom formation
Akbari et al.[21]	Golestan	5000	Modeling (IMDPA), geostatistical technique	Semiarid	18	502	32–3088	Shale deposits and Quaternary alluvial sediments
Bahmaei et al.[22]	Ahvaz	Pilot design	Statistical analysis	Semiarid	–	213	15	–
Delavar et al.[23]	Urmia Lake	560	RS	Semiarid	13.7	267.5	1290–1320	Salty lakes
Farahmand and Sadeghi[24]	Urmia Lake	560	RS	Semiarid	15	300	1267	Salty lakes
Fathizad et al.[12]	Yazd-Ardakan plain	4829	RS DSM	Warm-dry	19.3	118	1279	Red gypsiferous marls and brown to gray limestone
Gorji et al.[13]	Urmia Lake	18	RS	Semiarid	15	360	1267	Salty lakes
Habibi et al.[25]	Sharifabad plain	3290	RS ML	Warm-dry	–	150	–	–
Merrikhpour and Rahimzadegan[15]	Iran	1.65×10^6	RS	Mostly arid to semiarid	–	–	–	–
Mirzavand et al.[26]	Femenin-Ghahavand plain	3172	RS, geostatistical technique	Semiarid	–	250	–	Oligocene limestone, Qom formation, and Miocene Marl, with an interlayer of halite and gypsum
Seifi et al.[27]	Eastern coast of Urmia Lake	–	RS	Semiarid	18	288.9	1292–1443	Salty lakes
Eishoeei et al.[28]	Dizaj-e Dowl area (southwest of Urmia Lake)	–	Modeling (SaltMod) with the Geo-Hydrological data	Semiarid	15	360	Flat	Salty lakes
Taghadosi et al.[29]	Kuh Se'īd	–	RS ML	Arid	–	Little rainfall	–	–
Hamzehpour and Bogaert[30]	West Urmia Lake	50	Geostatistical technique	Semiarid	15	264.7	1270–1300	Playa and alluvial deposits
Hamzehpour and Rahmati[31]	Bonab Plain	40	Geostatistical technique	Semiarid	15	264.7	1270–1300	Playa and alluvial deposits
Taghizadeh-Mehrjardi et al.[32]	Chah-Afzal	800	RS DSM	Arid	18.3	80	953–1900	Red gypsiferous marls and brown to gray limestone

Continued

TABLE 1 Soil salinity and alkalinity relevant case studies in different regions of Iran—cont'd

References	Region	Spatial extent (km²)	Methodology	Climate	Temperature (°C)	Rainfall (mm)	Elevation (m.a.s.l.)	Geological units
Taghizadeh-Mehrjardi et al.[33]	Ardakan	720	RS DSM	Warm-dry	18.5	75	944–1944	Red gypsiferous marls and brown to gray limestone
Matinfar et al.[34]	Ardakan	720	RS	Warm-dry	18.5	75	944–1944	Red gypsiferous marls and brown to gray limestone
Pakparvar et al.[35]	Darab plain in Fars	1060.70	RS and statistical analysis	Arid	22	212–287	1015–1150	Quaternary sediments and the Jahrom formation
Masoudi and Asrari[36]	Mond basin	47,835	Statistical model	Arid and semiarid	19.4	150–700	190	–
Masoudi et al.[37]	Payab basin	–	Preparing risk map at GIS	Arid and semiarid	19.4	150–700	190	–

DSM, digital soil mapping; RS, remote sensing; IMDPA, Iranian model for desertification potential assessment; ML, machine learning; m.a.s.l., meters above sea level.

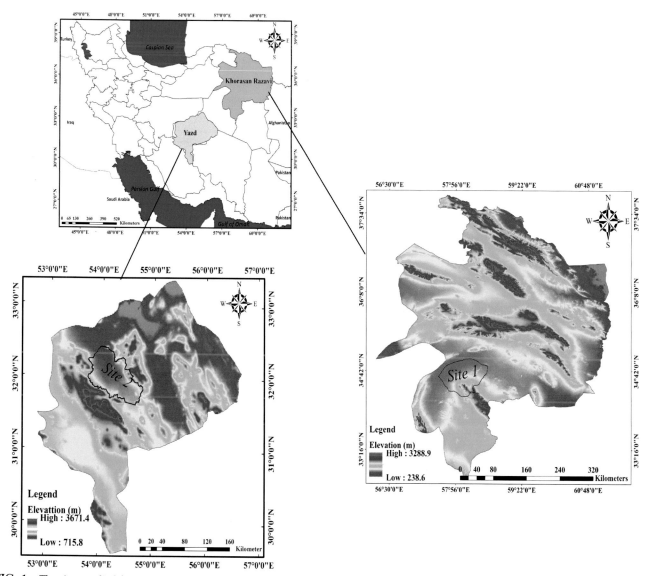

FIG. 1   The sites studied for investigation of spatio-temporal soil salinity and alkalinity in eastern and central Iran.

fans," "coalescing alluvial fans (bajadas)," "salt plain," and "gypsiferous hills" are the main landforms from east to west direction of the study area.[33]

2.2 Methodology

2.2.1 Soil sampling and laboratory analysis

2.2.1.1 Site 1

For site 1, 160 soil samples (0–30 cm depth) were collected from different geomorphic surfaces (Fig. 2) using stratified random sampling in both 2004 and 2018. It should be noted that the soil samples were taken at the same location at two time periods with land covers including *pomegranate, pistachio, saffron,* and *barley.*

2.2.1.2 Site 2

For site 2, 201 soil samples (0–30 cm depth) were selected according to conditioned Latin hypercube sampling (cLHS[46]) based on a MATLAB script[47] in 2016.

Soil samples at both sites were air-dried and passed through a 2 mm sieve. The electrical conductivity (EC) was determined in the extracted saturated soil paste using an EC meter.[48] Soluble cations, i.e., Ca^{2+}, Mg^{2+}, and Na^{+}, were

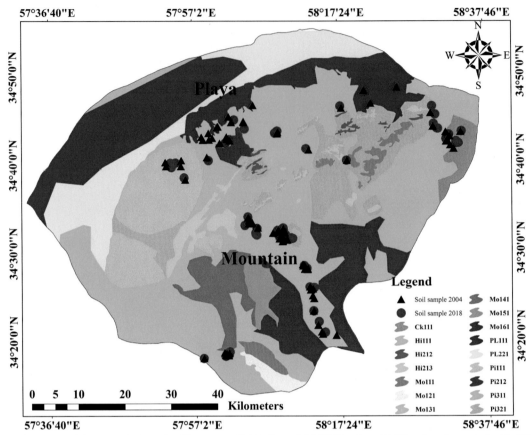

FIG. 2 The landform map of the site 1. Codes in landform map: Ck111: Dune field, Hi: Developed hill land partly eroded with different lithology, and Mo: Mountain with rock outcrop and different lithology, PL111: Clay flats of playa, PL221: Salt flats of playa, Pi111: Alluvial fan, Pi212: Dissected alluvial fan, Pi311: Pediment with recent alluvial deposits, Pi321: Pediment with old alluvial deposits.

extracted with 1-MM neutral ammonium acetate (for more detailed information, see Ref. 49). Afterward, "sodium adsorption ratio (SAR)" was calculated by Eq. (1)[50]:

$$SAR = \frac{Na^+}{\sqrt{\frac{Ca^{2+} + Mg^{2+}}{2}}} \tag{1}$$

2.2.2 Investigation of spatio-temporal soil salinity

At site 1, statistical analysis was used to analyze the temporal variation in soil salinity and alkalinity in the different geomorphic surfaces under continuous irrigated cultivation from 2004 to 2018. To analyze the significance of geomorphic surface and time on each of the studied soil properties, "one-way analysis of variance (ANOVA) was used with the Post hoc test (Duncan's test, with a critical significance level of 0.05)." In addition, Pearson's correlation coefficients were calculated to evaluate the relationships among different soil properties. The SRTM 1 arc-second digital elevation model (DEM) for site 1 was downloaded from the website (http://earthexplorer.usgs.gov) and then projected to the UTM coordinate system zone 40N, using ArcGIS 10.2. Then, the topographic wetness index (TWI) was extracted from DEM in SAGA GIS 2.2, to illustrate the spatial trend of soil salinity changes in the studied region, which will be discussed in Section 3.2.

In site 2, to predict the soil alkalinity maps between 1986 and 2030, the DSM method was applied based on *"scorpan"* model.[20]

$$S_c = f(s, c, o, r, p, a, n) + e \tag{2}$$

where S_c is a soil property or class and where the different inputs including "soil information, climate, organisms, relief, parent materials, time and spatial position," respectively and e is an error term. In this study at site 2, a random forest (RF) model was used to establish a relationship between soil SAR data and a series of environmental covariates.[51–55] It should be noted that the environmental covariates calculated for four-time intervals (1986–2016) included geological and land use maps, groundwater quality parameters (i.e., Na^+, K^+, Ca^{2+}, Mg^{2+}, HCO_3^-, Cl^-, SO_4^2 , EC, TDS, SAR, and pH), vegetation and soil salinity indices, DEM derivatives, distance from mines, distance from the road, distance from the river, distance from the residential points, rainfall, and population.

First, the soil alkalinity map was predicted using the RF model in 2016. Then, soil alkalinity maps for 1986, 1999, and 2010 were predicted based on the relationship obtained from salinity modeling in 2016 using the RF model. Furthermore, to forecast soil alkalinity maps in 2030, the trained RF model was applied to the simulated land use map and groundwater quality maps. It should be noted that land-use change and groundwater quality maps in 2030 were simulated by cellular automata (CA)-Markov and cellular automata models. It should be noted that time series are applied as linear models[12,56] to prepare groundwater quality data and predict for 2030.

2.2.3 Random forest model

RF is a nonparametric model and an ensemble of randomized decision trees that aggregate them to improve prediction accuracy.[57] Approximately 2/3 of the training data were used for bootstrap sampling, which formed each decision tree.[12,57] Additionally, the "Out-of-Bag (OOB)" error (Eq. 3) is estimated randomly by changing the values of the covariates, which is considered as variable importance. The variable importance changes between 0 and 100% while it is below 15%, the environmental covariate is recognized as unimportant and omitted from the model.[52]

$$OOB\,error = \frac{1}{N}\sum_{i=1}^{N} I[Y_{OOB}(X_i) \neq Y_i] \tag{3}$$

where $I[Y_{OOB}(X_i) \neq Y_i]$ is recognized as an indicator function equal to 0 when the predicted and actual classes are the same and equal to 1 otherwise.

The *mtry* (the number of environmental covariates used in each random subset) and *ntree* (the number of trees used in the forest) are two user-defined parameters that are important in RF. "Both parameters were optimized by iterating over *mtry* values ranging from 1 to the total number of covariates, while *ntree* values ranged from 100 to 10,000 in increments of 100.[58]"

To assess the prediction performance of the RF model, 10-fold cross-validation[32,59] was applied. Reliable performance and unbiased performance on smaller datasets are the advantages of this procedure. Two indices, root mean square error (RMSE) and coefficient of determination (R^2), were selected to evaluate the accuracy of the model's performance and to choose the best models[60] as follows:

$$RMSE = \sqrt{\frac{1}{n}\sum_{i=1}^{n} \left(Z_i - Z_i^*\right)^2} \tag{4}$$

$$R^2 = \left[\frac{\sum_{i=1}^{n} \left(Z_i - Z_{avg}\right)\left(Z_i^* - Z_{avg}^*\right)}{\sum_{i=1}^{n} \left(Z_i - Z_{avg}\right)^2 \left(Z_i^* - Z_{avg}^*\right)^2}\right]^2 \tag{5}$$

where Z_i and Z_i^* are the observed and predicted values, Z_{avg} and Z_{avg}^* are the mean of the observed and predicted values, respectively, and n is the number of observations.

Modeling of soil alkalinity and assessment of variable importance was implemented using the "randomForest" package[61] in R version 3.4.3[62] and RStudio version 1.1.383.[63]

3 Results and discussion

3.1 Temporal changes of soil salinity and alkalinity

The results clearly showed that increasing salinity levels and decreasing soil quality at sites 1 and 2 across 14 and 30 years, respectively. In detail, the results of site 1 indicated an increasing trend in average values of EC, SAR, and soluble cations of Ca^{2+}, Mg^{2+}, and Na^+, respectively, 16.13%, 38.72%, 36.81%, 44.76%, and 60.80%, over 14 years. Comparison between different classes of EC and SAR trends from 2004 to 2018 showed an increase in all of EC classes, except for 4–8 dS m^{-1} (Fig. 3A and B). No sample was located in the extreme limitations of >32 dS m^{-1} classes in 2004, but it

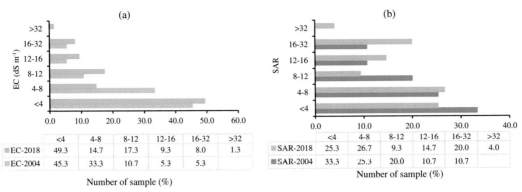

FIG. 3 Graph of percentage sample with different classes of (A) EC and (B) SAR in 2004 and 2018 at site 1. Grades or classes of electrical conductivity (EC) and sodium adsorption ratio (SAR) are defined as: <4: very low; 4–8: low; 8–12: moderate; 12–16: high; 16–32: very high; >32: extremely high.

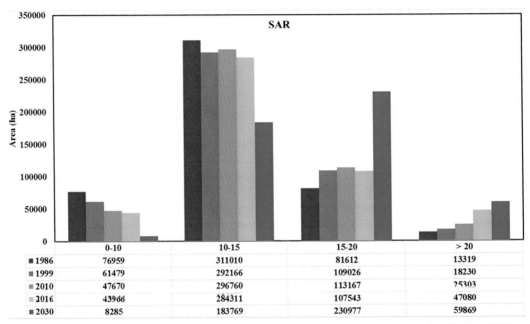

FIG. 4 Area of different sodium adsorption ratio (SAR) classes and the trend of their changes in 1986, 1999, 2010, 2016, and 2030 at site 2. Grades or classes of SAR defined as: 0–10: no restrictions; 10–15: low restrictions; 15–20: high restrictions; >20: very high restrictions.

increased to 1.3% in 2018. In addition, the results indicated that the vast extent of the moderate (8–12 dS m^{-1}), high (12–16 dS m^{-1}), and very high (16–32 dS m^{-1}) salinity classes also increased by approximately 7%, 4%, and 3%, respectively, over 14 years (Fig. 3A). In line with these findings, Fathizad et al.[12] reported that the area of EC class <32 has increased to 68,255 ha (14.13%) over 1986–2030, while classes <4 were eliminated by 2030.

As illustrated in Fig. 3B, the very low (<4) and moderate (8–12) SAR classes indicate decreasing trends of −8% and −11%, respectively at site1. However, there was an increasing trend in classes of SAR > 12 (38.7% of the samples). A decreasing trend in the area of no restrictions and low restrictions classes was also observed by 68,674 and 127,241 ha (14.22% and 26.34%), respectively, at site 2 (Fig. 4). In contrast, the area of the relatively high and high restriction classes increased by 149,365 and 46,550 ha (30.93% and 9.63%), respectively, from 1986 to 2030 (Fig. 4). This indicates an increase in soil alkalinity and progress toward poor soil quality in the future.

3.2 Spatial and temporal changes of soil salinity and alkalinity

As shown in Table 2, there was a significant increase in the EC, Ca^{2+}, Mg^{2+}, and Na$^+$, and SAR of all landforms (except T$_1$L$_3$) over 14 years, approximately two-fold at Site 1. For example, the mean value of EC, Na$^+$, and SAR were 4.58 dS m^{-1}, 12.37 meq L^{-1}, and 5.38 in 2004, respectively in T$_1$L$_6$, and increased 2, 4.6, and 3 times in 2018 (T$_2$L$_6$).

TABLE 2 Comparison of the mean values of the studied soil properties in different geomorphic surfaces through two times.

Treatment	EC (dS m^{-1})	Ca^{2+} (meq L^{-1})	Mg^{2+}	Na^{+}	SAR
T$_1$L$_1$	5.45dc	18.66cb	9.03cba	26.58dc	7.49c
T$_1$L$_3$	14.42a	29.85ba	21.42a	110.97a	22.20a
T$_1$L$_4$	8.03dcba	17.62cb	9.32cba	19.85dc	8.10c
T$_1$L$_5$	5.97dc	8.34c	3.48c	24.67dc	9.70cb
T$_1$L$_6$	4.58d	13.78cb	5.82cb	12.37d	5.38c
T$_2$L$_1$	6.34dcb	25.10cba	13.12cba	42.60dcb	10.27cba
T$_2$L$_3$	12.23cba	30.26ba	15.60cba	74.26cba	15.53cba
T$_2$L$_4$	12.42cba	39.13a	16.06ba	91.10ba	20.72ba
T$_2$L$_5$	13.63ba	30.60ba	17.56ba	70.38cba	14.27cba
T$_2$L$_6$	8.55dcba	13.06cb	15.30cba	56.33dcba	16.24cba

T = Time (T$_1$ = 2004; T$_2$ = 2018), L = Geomorphic surfaces (L$_1$ = Pediment with recent alluvial deposits (Pi311); L$_3$ = Dune field (Ck111); L$_4$ = clay flats of playa (PL111); L$_5$ = alluvial fan (Pi111); L$_6$ = Pediment with old alluvial deposits (Pi321).

This suggests that large-scale salinization changes had occurred at site 1 during the 14 years. The highest value of Ca^{2+} was also observed in T$_2$L$_4$, with high amounts of EC, Na$^+$, and SAR. Mg^{2+} showed no considerable differences between the different treatments.

The EC and SAR classes emphasized the increasing trends in the soil salinity and alkalinity in the northern area, which is located in most *pistachio* orchards adjacent to the salt crust zone of Bajestan playa (Fig. 2) at site 1. On the other hand, the mentioned units are associated with parts of the playa landform, which had larger values of TWI (25) located on the playa landform and the lowest values of TWI in the mountain (Fig. 5). This is similar to the results reported by Jafari et al.[64] and Taghizadeh-Mehrjardi et al.,[33] who showed a significant correlation between soil salinity, low elevation, and TWI. In addition, Alavipanah and Ddaper,[18] Matinfar et al.,[34] Taghizadeh-Mehrjardi et al.,[33] and Delavar et al.[23] claimed that the cultivated land adjacent to the playa has a high potential for the accumulation of salty sediments in an arid region of Iran.

At site 2, as illustrated in Fig. 6, high variability of soil alkalinity, especially from the eastern to western parts of the study area, was observed. Specifically, the highest amount of alkalinity was located in the center, southeast, and northwest of the region, and the lowest was observed in the west and southwest, where mountainous areas with higher elevation and slope, coarser texture, and fewer salts are located. This finding indicates that the trend of alkalinity changes in the region from east to west shows an increasing trend, consistent with the trend of changes in the most important environmental parameters identified in the study area.

Furthermore, the correlation analysis of RF models between observed and predicted values of SAR (Fig. 7) confirmed the high performance of model prediction maps of 2016, which is consistent with the results of several studies.[52,53,65] Given the values of RMSE and R^2 (Fig. 7), it can be concluded that the RF model predicts SAR well for 2016. Fathizad et al.[12] also reported the high performance of the RF model with an R^2 value of 0.73, for prediction of the soil EC map in the Yazd-Ardakan plain. After the accuracy assessment of the RF model, SAR maps were predicted for 2016 (Fig. 6). The relationship between soil modeling data and environmental data for 2016 was used to prepare the SAR maps for 1986, 1999, and 2010 (Fig. 6). After preparing these maps, soil SAR is graded and the soil maps of SAR are forecast for 2030 based on the Markov model. The results of the accuracy assessment of predicted maps for 2030 indicated the usefulness of the Markov model and its ability to simulate soil groundwater parameters and land-use changes.[12]

3.3 The most important parameters related to soil salinity and alkalinity

According to the results of sites 1 and 2, soil salinity and alkalinity were related to natural and human-induced factors in these regions over the 14 and 30 years. The primary salinity is mainly related to natural factors, including geology or salt-rich parent materials, dry climate (arid region of the studied area), the lowest elevation and highest TWI

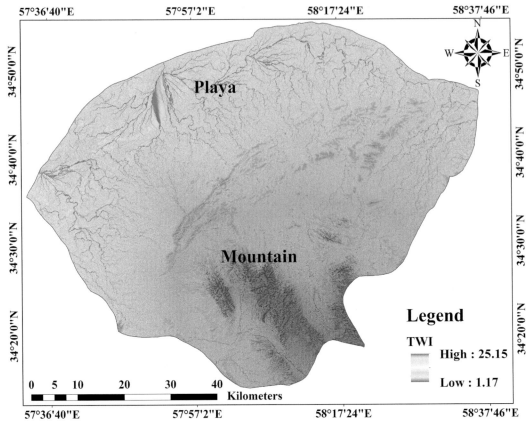

FIG. 5   Topographic wetness index (TWI) of site 1.

at site 1, and landform situation (i.e., playa or playa margins) in these areas. As previously explained in Section 2.1, the parent materials are rich in salts and gypsum layers in some parts of the study areas. Nonetheless, it is a natural factor that can cause an increase in groundwater salinity and surface sediments, ultimately adding to the intensity of soil salinity and desertification.

The results of the relative importance covariates of the RF model show the key role of groundwater parameters, for example, anions of water, Na^+, and SAR (Fig. 8) for predicting SAR maps at site 2. The results of the Pearson correlation analysis also confirmed the strong association between EC of soil EC and SAR of irrigation water ($r^2 = 0.76$, $P = .01$) at site 1. In addition, there were no large differences between CK111 (dune field) and PL111 (clay flat landform) for EC and SAR (Fig. 2) at site 1. Among the studied geomorphic surfaces (alluvial fan, dune field, pediment, and clay flats of playa), the CK111 landform had significantly higher EC (13.48 Â dSm^{-1}), Ca^{2+} (30.02 Â $meqL^{-1}$), Mg^{2+} (18.92 Â $meqL^{-1}$), Na^+ (95.24 Â $meqL^{-1}$) and SAR (19.34). The soil texture was sandy loam or loamy sand in this landform. The extensive irrigation of agricultural lands with water in high salinity is mainly due to an increase in soil salinity. Furthermore, rainfall is low in this area, and it is not sufficient to wash the salt further down the soil profile. As the results of the chemical analysis of five underwater resources in this unit indicate the mean value of 9.22 dSm^{-1} and 30.69 for EC and SAR, respectively. These results indicate the important role of groundwater quality in soil salinity and alkalinity in the studied area. The high temperature and evaporation of the areas cause the groundwater to rise to the surface and evaporate, leaving solutes on the soil surface. Therefore, frequent use of this water can intensify soil salt aggregation in agricultural lands.

Other reasons for high alkalinity in these areas can be improper human activities, including urban planning and road construction, on the SAR of soil (Fig. 8). Additionally, there are many mines and factories in the studied site (about 280); thus, their highest presence in the center of the region and on the road from Yazd to Ardakan can be another reason at site 2. The most important mining effects are related to industrial pollutants and ceramic residuals of tile factories that enter the soil and increase soil alkalinity. The main reason for the increase in soil salinity is improper disposal of waste, which reduces soil quality and subsequently increases soil alkalinity.

However, the role of human-induced factors, such as agricultural lands, has been more pronounced in recent years. For example, the EC of lands under *pistachio* cultivation is 32 dSm^{-1} in Meybod and Ardakan (site 2^{12}), with mean and maximum values of 11.23 and 37.20 dSm^{-1} in Bajestan due to inadequate irrigation water. In the future, soil salinity

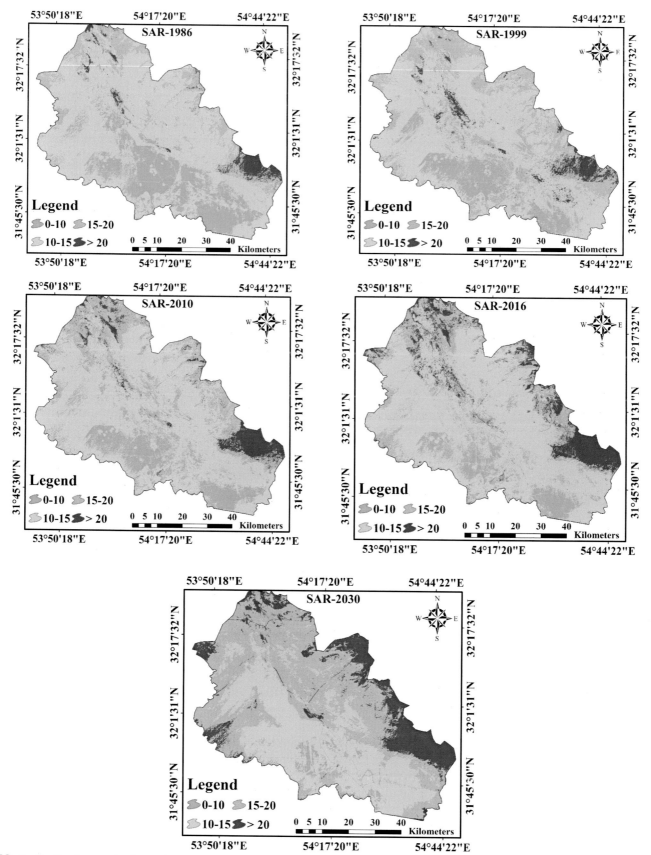

FIG. 6 Predicted maps of sodium adsorption ratio (SAR) at site 2 through 1986–2030. Grades or classes of SAR are defined as: 0–10: no restrictions; 10–15: low restrictions; 15–20: high restrictions; >20: very high restrictions. Note: The grade is assigned according to the sensitivity and response of crops to alkalinity.

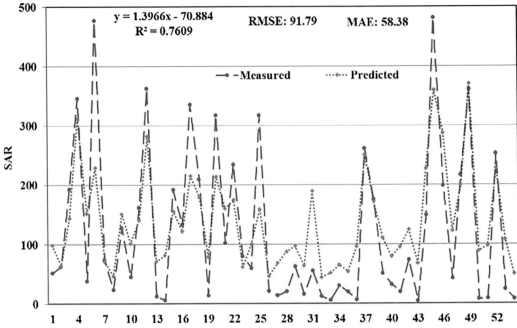

FIG. 7 Correlation between measured and predicted values of sodium adsorption ratio (SAR) using random forest (RF) model for 2016 at site 2.

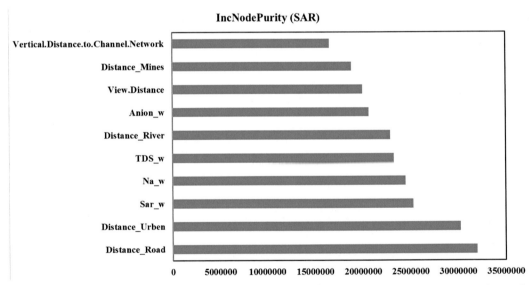

FIG. 8 Covariates importance of the random forest (RF) for sodium adsorption ratio (SAR) at the site 2. SAR-w: SAR of groundwater; Anion-w: anion (HCO_3^-, Cl^-, NO_3^-, SO_4^{2-}) of water groundwater; Na-w: sodium of groundwater; TDS_w: total dissolved solids of groundwater.

will limit the yield of this product and the advancement of soil degradation. However, in these agricultural lands, some parts are seen as spots with less salinity, possibly due to *pistachio* cultivation and irrigation time during sampling. Moreover, a decreasing trend was observed in cultivated lands of saffron by 47.20%[44] at site 1, which resulted in decreasing fertile lands and the salinity of water and soil resources.

After salinization and alkalization of agricultural lands as a result of irrigation with water of poor quality water, these lands are abandoned (pasture and desert); thus, cultivation spreads to the surrounding lands. High surface salinity decreases the soil osmotic potential; in other words, the lack of sufficient moisture in the soil reduces available water to absorb plant roots, while more than half of the area is covered, it is a poor plant; as a result, the wind easily comes into contact with the soil surface and causes soil erosion.

4 Conclusion

The present study indicated that the soil salinity and alkalinity changes increased during long-term cultivation in the eastern and central regions of Iran, which is consistent with the trends of changes in the most important parameters identified. The results indicated that if the intensity of soil degradation and changes in soil chemical properties (e.g., EC, Na^+, SAR) will increase in the regions are greater, its harmful effects on agricultural lands and production crops. These changes resulted from natural factors, such as high evaporation, low precipitation, and topographic features. The other reason for the high salinity and alkalinity in these areas is anthropogenic factors such as poor irrigation water and irrigation with saline water in the agricultural land, especially during *pistachio* orchards. There are a large number of mines and factories at site 2, which results in increased soil alkalinity in the center of the region. This finding showed that the zonation map of soil salinity and alkalinity at 1 year using satellite data and field data could not provide an accurate decision making for agricultural planning and management in the future. In addition, the protocols need to assess agricultural practices in arid regions and playa margins to be more sustainable areas and prevent soil degradation.

Salinity is a dynamic phenomenon in which dimensions change over time and space. Therefore, it is necessary to examine the conditions of saline soils during the long-term annually, and by weighing and lightening the factors involved in salinization and desalination, to ensure the process of changes in the salinity of agricultural soils and forecast trends of future changes. Since Iran is a country with different climates and the intensity of soil salinity varies in different regions, it is necessary to investigate spatio-temporal soil salinity to identify saline and non-saline soils. In future studies, it is recommended that the combination of terrestrial, RS, and DSM methods could be used as a powerful tool for monitoring the spatio-temporal soil salinity in sustainable agriculture.

References

1. Brevik EC, Cerdà A, Mataix-Solera J, et al. The interdisciplinary nature of soil. *Soil.* 2015;1(1):117–129.
2. Bogunovic I, Trevisani S, Seput M, Juzbasic D, Durdevic B. Short-range and regional spatial variability of soil chemical properties in an agroecosystem in eastern Croatia. *Catena.* 2001;154:50–62.
3. Chen S, Lin B, Li Y, Zhou S. Spatial and temporal changes of soil properties and soil fertility evaluation in a large grain-production area of subtropical plain, China. *Geoderma.* 2020;357:113937.
4. Chen L, Qi X, Zhang X, Li Q, Zhang Y. Effect of agricultural land use changes on soil nutrient use efficiency in an agricultural area, Beijing, China. *Chin Geogr Sci.* 2011;21(4):392.
5. Jabbar MT, Zhou J. Assessment of soil salinity risk on the agricultural area in Basrah Province, Iraq: using remote sensing and GIS techniques. *J Earth Sci.* 2012;23(6):881–891.
6. Shankar V, Evelin H. Strategies for reclamation of saline soils. In: *Microorganisms in Saline Environments: Strategies and Functions.* Cham: Springer; 2019.
7. Foley JA, DeFries R, Asner GP, et al. Global consequences of land use. *Science.* 2005;309(5734):570–574.
8. Prosdocimi M, Cerdà A, Tarolli P. Soil water erosion on Mediterranean vineyards: a review. *Catena.* 2016;141:1–21.
9. Nehrani SH, Askari MS, Saadat S, Delavar MA, Taheri M, Holden NM. Quantification of soil quality under semi-arid agriculture in the northwest of Iran. *Ecol Indic.* 2020;108:105770.
10. Taghizadeh-Mehrjardi R, Schmidt K, Toomanian N, et al. Improving the spatial prediction of soil salinity in arid regions using wavelet transformation and support vector regression models. *Geoderma.* 2021;383:114793.
11. Akça E, Aydin M, Kapur S, et al. Long-term monitoring of soil salinity in a semi-arid environment of Turkey. *Catena.* 2020;193:104614.
12. Fathizad H, Ardakani MA, Sodaiezadeh H, Kerry R, Taghizadeh-Mehrjardi R. Investigation of the spatial and temporal variation of soil salinity using random forests in the central desert of Iran. *Geoderma.* 2020;365:114233.
13. Gorji T, Yildirim A, Hamzehpour N, Tanik A, Sertel E. Soil salinity analysis of Urmia Lake Basin using Landsat-8 OLI and Sentinel-2A based spectral indices and electrical conductivity measurements. *Ecol Indic.* 2020;112:106173.
14. Qadir M, Qureshi AS, Cheraghi SA. Extent and characterisation of salt-affected soils in Iran and strategies for their amelioration and management. *Land Degrad Dev.* 2008;19(2):214–227.
15. Merrikhpour MH, Rahimzadegan M. A synergistic use of AMSR2 and MODIS images to detect saline soils (Study Area: Iran). *Compt Rendus Geosci.* 2020;352(2):127–138.
16. Moameni A, Stein A. Modeling spatio-temporal changes in soil salinity and waterlogging in the Marvdasht Plain, Iran. In: *17th World Congress of Soil Science*; 2002.
17. Krinsley DB. *A Geomorphological and Paleoclimatological Study of the Playas of Iran.* Washington, DC: Geological Survey U.S. Department of Interior; 1970.
18. Alavipanah K, Ddaper M. Caracterization of some soil salintty parameters in the playa margin (case study: yazd province). *Iran Agric Res.* 2001;20(2):189–200.
19. Abedi F, Amirian-Chakan A, Faraji M, et al. Salt dome related soil salinity in southern Iran: prediction and mapping with averaging machine learning models. *Land Degrad Dev.* 2020;26.
20. McBratney AB, Santos MM, Minasny B. On digital soil mapping. *Geoderma.* 2003;117(1–2):3–52.

21. Akbari M, Shalamzari MJ, Memarian H, Gholami A. Monitoring desertification processes using ecological indicators and providing management programs in arid regions of Iran. *Ecol Indic.* 2020;111:106011.

22. Bahmaei A, Albaji M, Naseri AA, Varjavand P. Effect of irrigation type and interval on soil salinity in clay soils in Ahvaz, Iran. *Arab J Geosci.* 2020;13(21):1.

23. Delavar MA, Naderi A, Ghorbani Y, Mehrpouyan A, Bakhshi A. Soil salinity mapping by remote sensing south of Urmia Lake, Iran. *Geoderma Reg.* 2020;22, e00317.

24. Farahmand N, Sadeghi V. Estimating soil salinity in the dried lake bed of Urmia lake using optical sentinel-2 images and nonlinear regression models. *J Ind Soc Remote Sens.* 2020;1–3.

25. Habibi V, Ahmadi H, Jafari M, Moeini A. Quantitative assessment of soil salinity using remote sensing data based on the artificial neural network, case study: Sharif Abad Plain, Central Iran. *Model Earth Syst Environ.* 2020;1.

26. Mirzavand M, Sadeghi S, Bagheri R. Groundwater and soil salinization and geochemical evolution of Femenin-Ghahavand plain, Iran. *Environ Sci Pollut Res.* 2020;27(34):43056–43066.

27. Seifi M, Ahmadi A, Neyshabouri MR, Taghizadeh-Mehrjardi R, Bahrami HA. Remote and Vis-NIR spectra sensing potential for soil salinization estimation in the eastern coast of Urmia hyper saline Lake, Iran. *Remote Sens Appl Soc Environ.* 2020;20:100398.

28. Eishoeei E, Nazarnejad H, Miryaghoubzadeh M. Temporal soil salinity modeling using SaltMod model in the west side of Urmia hyper saline Lake, Iran. *Catena.* 2019;176:306–314.

29. Taghadosi MM, Hasanlou M, Eftekhari K. Retrieval of soil salinity from sentinel-2 multispectral imagery. *Eur J Remote Sens.* 2019;52(1):138–154.

30. Hamzehpour N, Bogaert P. Improved spatiotemporal monitoring of soil salinity using filtered kriging with measurement errors: an application to the West Urmia Lake, Iran. *Geoderma.* 2017;295:22–33.

31. Hamzehpour N, Rahmati M. Investigation of soil salinity to distinguish boundary line between saline and agricultural lands in Bonab Plain, southeast Urmia Lake, Iran. *J Appl Sci Environ Manag.* 2016;20(4):1037–1042.

32. Taghizadeh-Mehrjardi R, Ayoubi S, Namazi Z, Malone BP, Zolfaghari AA, Sadrabadi FR. Prediction of soil surface salinity in arid region of central Iran using auxiliary variables and genetic programming. *Arid Land Res Manag.* 2016;30(1):49–64.

33. Taghizadeh-Mehrjardi R, Minasny B, Sarmadian F, Malone BP. Digital mapping of soil salinity in Ardakan region, central Iran. *Geoderma.* 2014;213:15–28.

34. Matinfar HR, Panah SK, Zand F, Khodaei K. Detection of soil salinity changes and mapping land cover types based upon remotely sensed data. *Arab J Geosci.* 2013;6(3):913–919.

35. Pakparvar M, Gabriels D, Arabi K, Edraki M, Raes D, Cornelis W. Incorporating legacy soil data to minimize errors in salinity change detection: a case study of Darab Plain, Iran. *Int J Remote Sens.* 2012;33:6215–6238. https://doi.org/10.1080/01431161.2012.676688.

36. Masoudi M, Asrari E. A prediction model for soil salinity using its indicators: a case study in Southern Iran. *Nat Environ Pollut Technol.* 2009;8(1):13–19.

37. Masoudi M, Patwardhan AM, Gore SD. A new methodology for producing of risk maps of soil salinity, case study: Payab Basin, Iran. *J Appl Sci Environ Manag.* 2006;10(3):9–13.

38. Abuelgasim A, Ammad R. Mapping soil salinity in arid and semi-arid regions using Landsat 8 OLI satellite data. *Remote Sens Appl Soc Environ.* 2019;13:415–425.

39. Afrasinei GM, Melis MT, Buttau C, Bradd JM, Arras C, Ghiglieri G. Assessment of remote sensing-based classification methods for change detection of salt-affected areas (Biskra area, Algeria). *J Appl Remote Sens.* 2017;11(1), 016025.

40. Davis E, Wang C, Dow K. Comparing Sentinel-2 MSI and Landsat 8 OLI in soil salinity detection: a case study of agricultural lands in coastal North Carolina. *Int J Remote Sens.* 2019;40(16):6134–6153.

41. Ding J, Yu D. Monitoring and evaluating spatial variability of soil salinity in dry and wet seasons in the Werigan-Kuqa Oasis, China, using remote sensing and electromagnetic induction instruments. *Geoderma.* 2014;235:316–322.

42. Ding J, Yu D. Monitoring and evaluating spatial variability of soil salinity in dry and wet seasons in the Werigan-Kuqa Oasis, China, using remote sensing and electromagnetic induction instruments. *Geoderma.* 2014;235:316–322.

43. Morgan RS, El-Hady MA, Rahim IS. Soil salinity mapping utilizing sentinel-2 and neural networks. *Indian J Agric Res.* 2018;52(5):524–529.

44. Feizi H, Maleki S, Poozeshi R. Impact of vegetation cover on soil carbon storage and CO_2 fixation in long-term land uses in Bajestan, Khorasan Razavi. *Appl Soil Res.* 2020. In press.

45. Banaei MH. *The Map of Resources and Land Capability of Iran Soils.* Karaj, Iran: Soil and Water Research Institute; 2000.

46. Minasny B, McBratney AB. A conditioned Latin hypercube method for sampling in the presence of ancillary information. *Comput Geosci.* 2006;32(9):1378–1388.

47. MathWorks. *Matlab.* Natick, MA: The Math Works. Inc; 2009.

48. Rhoades JD. Salinity: electrical conductivity and total dissolved solids. In: *Methods of Soil Analysis: Part 3 Chemical Methods.* vol. 5. Madison: American Society of Agronomy Inc & Soil Science; 1996:417–435.

49. Alarima CI, Annan-Afful E, Obalum SE, et al. Comparative assessment of temporal changes in soil degradation under four contrasting land-use options along a tropical toposequence. *Land Degrad Dev.* 2020;31(4):439–450.

50. Vasu D, Singh SK, Ray SK, et al. Soil quality index (SQI) as a tool to evaluate crop productivity in semi-arid Deccan plateau, India. *Geoderma.* 2016;282:70–79.

51. Maleki S, Khormali F, Mohammadi J, Bogaert P, Bodaghabadi MB. Effect of the accuracy of topographic data on improving digital soil mapping predictions with limited soil data: an application to the Iranian loess plateau. *Catena.* 2020;195:104810.

52. Pahlavan-Rad MR, Akbarimoghaddam A. Spatial variability of soil texture fractions and pH in a flood plain (case study from eastern Iran). *Catena.* 2018;160:275–281.

53. Teng H, Rossel RA, Shi Z, Behrens T. Updating a national soil classification with spectroscopic predictions and digital soil mapping. *Catena.* 2018;164:125–134.

54. Wang S, Jin X, Adhikari K, et al. Mapping total soil nitrogen from a site in northeastern China. *Catena.* 2018;166:134–146.

55. Vågen TG, Winowiecki LA, Tondoh JE, Desta LT, Gumbricht T. Mapping of soil properties and land degradation risk in Africa using MODIS reflectance. *Geoderma.* 2016;263:216–225.

56. Chiaudani A, Di Curzio D, Palmucci W, Pasculli A, Polemio M, Rusi S. Statistical and fractal approaches on long time-series to surface-water/groundwater relationship assessment: a central Italy alluvial plain case study. *Water.* 2017;9(11):850.

57. Breiman L. Random forests. *Mach Learn.* 2001;45:5–32.

58. Zhi J, Zhang G, Yang F, et al. Predicting mattic epipedons in the northeastern Qinghai-Tibetan Plateau using random forest. *Geoderma Reg.* 2017;10:1.

59. Hengl T, Heuvelink GB, Kempen B, et al. Mapping soil properties of Africa at 250 m resolution: random forests significantly improve current predictions. *PLoS One.* 2015;10(6), e0125814.

60. Malone B. *Use R for Digital Soil Mapping.* Sydney: Soil Security Laboratory, The University of Sydney; 2013. Available at: www.clw.csiro.Au/aclep/documents/DSM_R_manual_2013.pdf (Accessed 1 November 2013).

61. Liaw A, Wiener M. Classification and regression by random forest. *R News.* 2002;2(3):18–22.

62. R Development Core Team. *R: A Language and Environment for Statistical Computing.* Vienna, Austria: R Foundation for Statistical Computing; 2017. Retrieved from: http://www.R-project.org.

63. RStudio. *RStudio: Integrated Development Environment for R*; 2017. Boston, MA http://www.r-studio.com.

64. Jafari A, Finke PA, Vande Wauw J, Ayoubi S, Khademi H. Spatial prediction of USDA-great soil groups in the arid Zarand region, Iran: comparing logistic regression approaches to predict diagnostic horizons and soil types. *Eur J Soil Sci.* 2012;63(2):284–298.

65. Camera C, Zomeni Z, Noller JS, Zissimos AM, Christoforou IC, Bruggeman A. A high resolution map of soil types and physical properties for Cyprus: a digital soil mapping optimization. *Geoderma.* 2017;285:35–49.

41

Kernel-based granulometry of textural pattern measures on satellite imageries for fine-grain sparse woodlands mapping

Seyed Rashid Fallah Shamsi[a], *Sara Zakeri-Anaraki*[b], *and Masoud Masoudi*[a]

[a]Department of Natural Resources and Environmental Engineering, College of Agriculture, Shiraz University, Shiraz, Iran
[b]Desert Region Management, College of Agriculture, Shiraz University, Shiraz, Iran

1 Introduction

Among the vast variety of earth surface features, sparse woodlands are the most complicated for automatic segmentation and thematic mapping of remotely sensed images. In semi-arid regions, sparse woodlands are technical of a particular natural composition of highly frequent living objects (i.e., individual trees) and non-living ones (i.e., bare soil openings) in a scene. Therefore, sparse woodland segmentation is very different from detecting either trees or bare soil openings individually. In conventional multispectral classification and segmentation methods, attention is mostly paid to the pixel-based spectral response of each theme,[1] which presents its own spectral signature. Therefore, the classifiers yield unsatisfactory results, especially when they face thematic objects presenting weak spectral separability[2] and interfering spectral response from the background (i.e., bare soil), widely occur in sparse woodland segmentation.

According to the biometry textbook of natural resource engineering, partitioning (i.e., segmenting) of sparse woodlands, based on density classes of tree populations, is an important information layer for managing non-commercial thin forests. Here, woodland segmentation is a bi-conditional decision, including a detection operation on one hand and a spatial pattern recognition process on the other hand.[3] Detecting either an individual tree or bare soil openings is technically achievable using high-spectral-resolution satellite images. It generally follows a processing procedure to increase the spectral separability of target objects through pixel-based image processing and pixel-wise classification methods. These data sets are generally criticized because of their low and moderate spatial resolutions, which cause mixing phenomena of the spectral response from the target objects. Spatial pattern recognition is achieved by using contextual processing of image content of data sets offering higher spatial resolution.[4] It generally follows a processing procedure to calculate the textural attributes of the target objects, such as frequency, size, shape, and pattern, through kernel-based neighborhood processing and object-based segmentation methods. This capability is generally accompanied by the low spectral resolution of the limited spectral bands. It only yields excessive sub-segments of the target objects because of too many details of the surface features.

Image texture analysis is to explain the local variation of image spectral brightness (the digital number-DN),[5,6] using measures representing characteristics of the variation in a neighborhood.[7] Multi-dimensionality of texture concept results from variety of representations.[8] It is described by its probabilistic, functional, or structural dependencies among gray levels in random, pairwise, or multiple clusters.[9] There is a wide range of texture descriptors to quantify the textural information content of an image,[7] employed in statistical, geometrical, filter-kernel based, and model-based techniques. A descriptor is generally used to measure smoothness, coarseness, and regularities. The probabilistic descriptors such as gray level co-occurrence obtain the image texture measures by calculating spatial dependency of gray-level tones, while structural ones describe the texture as *"primitives"*.[9]

Comparing the capabilities of these techniques is of interest for differentiating contextual content.[10] The Kernel-based method adjuncts spatial and spectral information[11] using an adaptive neighborhood system. The neighbors are a set of connected pixels with an identical DN. A "Window" is defined as a square either sub-image or a certain rectangular pixel array (i.e., a matrix) on an image. The kernel is also a moving window to calculate the weighted mean of the matrix using a set of pre-defined coefficients[12] used in image texture analysis.[13] Larger kernel sizes are recommended for finer spatial resolution than coarser kernels.[14] To achieve higher accuracy, generally, non-continuous uneven target objects require larger kernel sizes than spectrally even target objects.[15]

Choosing proper kernel window sizes depends on the resolution of input satellite images and the size of the target objects, as discussed in mathematical morphology and the granulometry concept.[11] The concept lays back on the neighboring system definition for image pixels and the structuring element.[16] To determine the shape and size of objects in an image, it is necessary to use a range of structuring elements. In addition, the structuring elements present different object definitions according to a predetermined scale. Granulometry searches for objects based on the size distribution of openings according to the circular structuring element on the pixel neighborhood in a grid format.

Recent studies have reported the usefulness of adding contextual information as ancillary data.[14,17,18] Many previous studies indicated that textural measures had significant effects on the accuracy of classification outputs compared to original multi-spectral data.[1,19–23] Developing contextual spatial-spectral classifiers is also recommended for better classification results in remote sensing.[11] Researchers have also focused on texture indices plus kernel size, for example, using the homogeneity measure of texture indices for a kernel size of 7×7 pixels.[24] More research has indicated that the final accuracy of classification outputs may vary depending on the size of the kernel window[25,26] and the spatial resolution of satellite images.[21,27] Therefore, it has been recommended to choose kernel windows of different sizes according to the size of the target objects to be extracted.[24,28] The first- and second-order statistical texture features[29] are also used to detect tree species on high-resolution satellite images.[30]

The current research hypothesizes that kernel-based textural granulometry on multi-sensor satellite images will increase the accuracy of fine-grain sparse woodland segmentation using a pixel-based classification method. The main objective is to improve the accuracy of supervised semantic segmentation results by applying textural analysis to Quick-Bird, RapidEye, and ASTER satellite images. In this research, three main contributions have been offered: (a) using kernel-based textural granulometry for processing satellite images on the pattern measures, (b) developing multi-object ROIs as the training areas for supervised segmentation and (c) using a supervised maximum likelihood (ML) rule of probabilistic pixel-wise classification to delineate density classes of individual tree populations in sparse woodlands.

2 Materials and methods

2.1 Study area

The study area is a part of the Barm biosphere reserve, Iran, located at 51° 49′ 59″ to 51° 51′ 05″ N and 29° 33′ 41″ to 29° 35′ 36.02″ E, covering approximately 617.7 ha, as shown in Fig. 1. It is a semi-arid region, 270–350 mm precipitations including winter snowfall, in the southern part of Iran. The plain is the main habitat of the Persian oak (*Querqus brantti* L.) sparse woodlands. Thin non-commercial protected forests are the dominant scene in the study area, including substitutional seasonal farming under forest crown cover and annual grazing. The woodlands were facing severe drought and core-borer beetle outbreaks in recent years. Therefore, regular mapping of forest density is of interest to natural resource managers in Fars province. According to the woodland natural structure and the size of individual trees on the satellite images, the trees were presented as a fine-grain evenly distributed dot object throughout the scene.

2.2 Research data

In this research, three data sets of satellite images were used, including QuickBird, RapidEye, and ASTER, acquired in May 2011, June 2011, and September 2006, respectively. QuickBird images are multi-spectral image data covering a spectral range from 450 to 900 nm. As very high-resolution satellite imagery, its ground-level pixel size is approximately 2.4×2.8 m.[31] The RapidEye data set has five spectral bands ranging from 400 to 850 nm. Its spatial resolution was 5×5 m.[32] The Advanced Space borne Thermal Emission and Reflection Radiometer (ASTER) is also a multi-spectral imager, covering a wide electromagnetic domain including 14 bands including visible RGB, near-infrared (NIR), short-wave infrared (SWIR), and thermal infrared (TIR). Its spatial resolution varies from 15 m to 15 m for RGB color and NIR, 30×30 m for SWIR and 90×90 m for TIR.[33]

In this study, two major domains of VNIR and SWIR were used based on their sensitivity to the green plant cover spectral response. All standard image pre-processing, including radiometric and atmospheric corrections, was implemented on the imageries beforehand at the standard level.

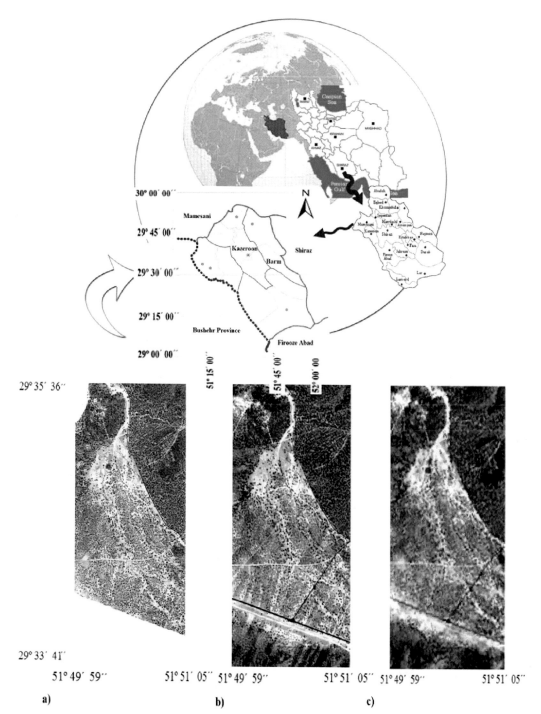

FIG. 1 The study area on multi-resolution satellite imageries, Iran. (A) QuickBird image (2.4 × 2.4); (B) RapidEye image (5 × 5); (C) ASTER image (15 × 15).

2.3 Research method

2.3.1 *Image pre-processing*

To achieve accurate geo-registration results, geo-ortho-rectification was performed for the image data sets. The overall quality of the results was also checked by overlaying the vector layer of Iranian 1:25,000 topographic maps of the study area. According to the technical limits of textural analysis to work with 11-bit data, a simple linear stretching with saturation (1%) was applied to the data to bring them in the range of 8-bit gray levels.[34]

2.3.2 Fieldworks and the target object definition

According to the forest density classes, defined by the headquarters of natural resources and watershed management for partitioning sparse woodlands, intensive fieldwork, and previous studies,[35] three forest density classes were considered as the target objects over the study area, including dense, semi-dense and open forests. Manufactured objects, such as roads, buildings, orchards, and bare soil, were also considered as additional classes, facilitating the detection process of the target objects to reduce spectral confusion among them.

2.3.3 Image processing

Spectral pixel-based processing: Various spectral image processing was applied to the image data set to better delineate the target objects, including contrast enhancement on gray-level images, true/false color compositing, and multiband image arithmetic. The results of the spectral processing were used to delineate the super-pixels[36] of the target object through a fast rough estimation of object boundaries by skilled operator visual interpretations.

Textural kernel-based processing: Seven pattern measures were considered in this research, including: (a) relative richness (RR),[37] which is a simple ratio of the number of pixels representing the target objects in a kernel to the number of objects considered over the entire image. (b) Diversity (H)[37] works based on the proportion of the number of pixels, presenting the target objects in a pre-defined kernel window multiplied by the natural logarithm of the proportion. (c) The Dominance index (DI) is a product of the diversity measures, calculated by subtracting the maximum diversity through the entire image by the diversity (H) in a certain kernel window, (d) the Fragmentation index (FI),[38] which is calculated by dividing the number of target objects in a kernel by the size of the kernel. (e) Number of different classes (NDC), represents the number of target objects, (f) Class versus Neighbors (CVN) represents the number of target objects different from the central cell, and (g) Binary Comparison Matrix (BCM) represents the number of target object pairs in a kernel.[39]

In this research, three kernel sizes were defined for calculating pattern measures,[40] including 3×3, 5×5, and 7×7 pixels of a rectangular window. The kernel windows were designed based on odd dimensions to account for having a central pixel containing calculation results during textural analysis and Granulometry.

In this step, first, a fixed-size circle-shaped region of interest (ROI) was overlaid on the target objects through a visual interpretation process, as shown in Fig. 2. Then, the pattern measures of the predefined kernel sizes were extracted for the samples. The extracted measures were entered into a statistical famous t-test to investigate whether the kernel sizes resulted in significantly different measures for the target objects.

According to the statistical analysis of the kernel size effects on the pattern measures, the granule size of the target objects was statistically calculated using the mean values of the measures on the multi-resolution satellite images. The mean values were calculated for the sampled super-pixels of dense, semi-dense, and open forests as the target objects in this study.

2.3.4 Unsupervised segmentation

To develop precise ROIs as training areas for supervised semantic segmentation, a quick unsupervised segmentation procedure was followed on the results of the textural pattern measures using K-means and ISO-data rules.[41] It provided the super-pixels on the pattern measures, facilitating defining ROIs based on their coincidence by the physical partitions of the sparse woodlands on the earth's surface.

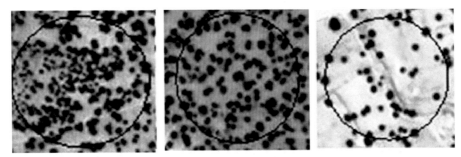

FIG. 2   Sample ROIs for the dense, semi-dense and open forest on QuickBird satellite imagery.

2.3.5 ROI development

According to the intensive fieldwork, the results of the unsupervised segmentation, and previous studies, the ROIs have defined on the images and the pattern measures as the training data sets. The same set of ROI was considered for all image data sets. The ROIs were selected in a circular shape because of their maximum compactness and the highest similarity to the shape of the image granules. In the training process, ROIs representing multiple objects were used instead of pure samples, which are conventionally used in supervised classifications. Approximately 15,000, 4500, and 400 pixels were sampled as the training data sets, covered by the fixed circle shape ROIs over the target objects on QuickBird, RapidEye, and ASTER images, respectively.

2.3.6 Separability analysis of ROIs

The ROI separability was investigated using the Jeffrey-Matusita distance (JMD)[42] on the spectral and textural measures. The data sets providing the highest separability index were entered into the supervised semantic segmentation process.

2.3.7 Supervised semantic segmentation

Then, a supervised semantic segmentation was performed using the ROIs, spectral images, textural results, and maximum likelihood (ML) rule of probabilistic pixel-wise classification,[43] which classifies the image pixels to form the super-pixels of the target objects.

2.3.8 Accuracy assessment

A confusion matrix was used for the accuracy assessment.[44] The ground truth samples, collected through a random sampling method during the fieldwork, were used as an independent validation data set for the accuracy assessment of the segmentation results. The overall accuracy and kappa coefficients were of the accuracy heuristics used in this study.

2.3.9 Z-test of accuracy heuristics

Finally, to investigate whether the differences in the heuristics are statistically significant, a Z-test of the accuracy heuristics was employed.[45] Fig. 3 briefly shows the flowchart of the steps followed in this research.

3 Results and discussion

As shown in Fig. 4C, the ASTER images presented good separability for the target objects, particularly for the dense, semi-dense and open forests. This capability goes back to the higher spectral resolution of the ASTER data in comparison to the higher spatial resolution of QuickBird and RapidEye images (Table 1).

Fig. 5 indicates the graphical representations of the ROI separability, using the pattern measures for the visible range spectral domain, as an instance. As shown in Fig. 5, the BCM and NDC offer higher separability among the pattern measures.

The mean values of the pattern measures for QuickBird, RapidEye, and ASTER data sets are shown in Tables 2, 3, and 4, respectively. Table 5 lists the results for the SWIR spectral domain of the ASTER images.

The results of Granulometry for SWIR spectral domains of the ASTER data set have been indicated in Fig. 4. According to the kernel structure for calculating BCM and NDC, both are based on the local co-occurrence of the measures in a kernel window of neighboring pixels than using the global parameters of an entire image, such as RR. Therefore, sparse woodlands as a physical structure composed of tree-openings co-occurrence were successfully defined by BCM and NDC, in this research. The results indicate that moving from very high-resolution QuickBird data toward moderate resolution ASTER in parallel to use larger kernels of 7×7 pixels have increased separability of ROIs on the pattern measures of the images, as shown in Fig. 6.

Based on the JM distance, as shown in Table 6, BCM, NDC, CVN, and RR are pattern measures that successfully discriminate the target objects by applying kernel sizes of 5×5 and 7×7 on QuickBird images. The more the JM distance is closer to two, the more separable the target objects.

According to Fig. 6, RapidEye images have not presented a considerable capability of separating the ROIs of the target objects based on the pattern measures. This is generally because of its high sensitivity to the crown cover spectral response of green trees in the composition of sparse woodland density. As shown in Fig. 7, the RapidEye dataset clearly discriminated open forest for both BCM and NDC using all kernel windows, but for the other target objects.

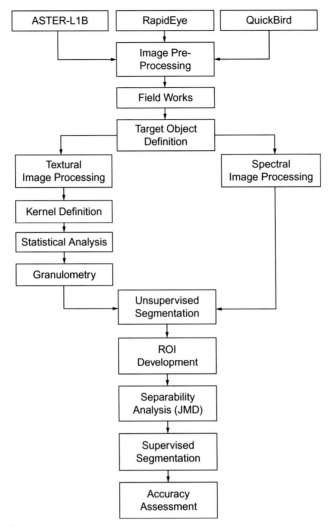

FIG. 3   The flowchart of the research steps.

As shown in Fig. 7, the results indicate that NDC also increases the separability of ROIs for ASTER images, particularly for the visible spectral domain. In addition to considering the co-occurrence of the tree-opening to define the target objects, NDC is concentrated on the number of different classes in a sub-image window. Therefore, the measure increased for QuickBird images, offering more details and features over a certain kernel window of neighboring pixels. However, the difference decreased because of the convolution effects of the larger kernel and lower resolution on the ASTER images. Consequently, the NDC decreased after applying a 7×7 kernel window to the ASTER images. Table 7 indicates the maximum ROI separability of the target objects for BCM and NDC by applying kernel sizes of 5×5 and 7×7 on RapidEye images.

It seems that using lower-resolution images in addition to using larger kernels has resulted in a local convolution operation on sub-image kernel windows. The local convolution offers an identical pattern measure for tree-opening co-occurrence in larger parts of the segments over sparse woodlands. By reducing the frequency of the pattern measures in a larger locality of sub-images, ROIs' separability was increased for the forest density classes using 7 by 7 kernel window of BCM pattern measures on the ASTER imageries, as shown in Fig. 7. In addition, Fig. 7 also indicates that the increasing separability belongs to the textural analysis of the pattern measures on the NIR and SWIR spectral domains of the ASTER data set, offering better discrimination of green crown cover versus bare soil background throughout the sparse woodlands.

Table 8 indicates the results of separability analysis for ROIs on the pattern measure of ASTER images. As shown in Table 8, BCM is the pattern measure that offers the highest separability for all kernel sizes. NDC is the second rank of the measures, just for kernel sizes of 5×5 and 7×7. While RR is the third choice only for the kernel size of 7×7, which includes enough pixels for calculating RR on the ASTER sub-image.

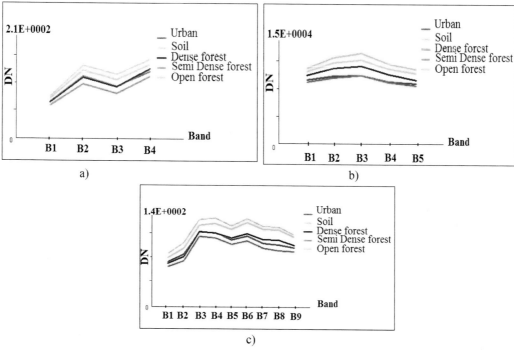

FIG. 4 Spectral separability of the ROIs on the original bands of the sensors. (A) QuickBird image; (B) RapidEye image; (C) ASTER image.

Fig. 8 shows the results of semantic segmentation of sparse woodland density using the ML rule of pixel-wise classification on the textural granulometry results of the textural pattern measures. Table 9 indicates the overall accuracy and kappa coefficient, respectively, for segmentation results on spectral pixel-wise image classification using the ML rule.

Table 10 indicates the accuracy heuristics for supervised segmentation using spectral-textural pixel-wise classification using the ML rule over the pattern measures derived from applying the kernels to the image data sets.

This indicates that the inclusion of the pattern measures in the segmentation process has increased the overall accuracy of the output results from 10% to 40% for QuickBird images. It is while it has enhanced kappa coefficients from a minimum of 30% to 50%. For QuickBird imageries, the best result belongs to spectral-textural segmentation of NDC products by kernel size of 7 by 7, offering 0.68 and 0.6 for overall accuracy and kappa coefficient, respectively. By inclusion of the pattern measures into segmentation processes, the rise of the heuristics is from a minimum of

TABLE 1 The main target objects, examples from the satellite images.

Data set / ROI	Urban (U)	Bare soil (BS)	Dense forest (D)	Semi-dense forest (SD)	Open forest (O)
QuickBird					
RapidEye					
ASTER					

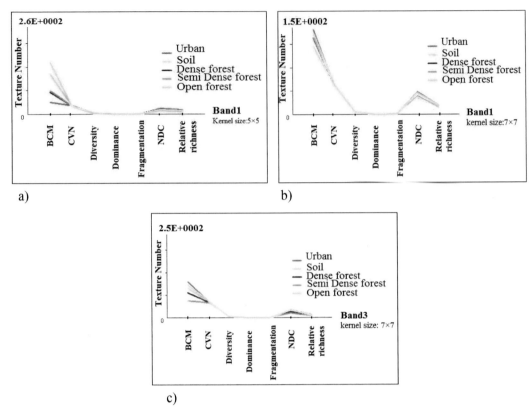

a)   b)   c)

FIG. 5   Spectral separability of the ROIs on the textural measures of the images. (A) QuickBird image (Band1, Kernel size 5×5); (B) RapidEye image (Band1, Kernel size 7×7); (C) ASTER image (Band3, Kernel size 7×7).

TABLE 2   Mean values of the pattern measures for the target objects on QuickBird data (kernel size: 7×7).

QickBird	B_1			B_2			B_3			B_4		
P.M./T.O.	D	S	O	D	S	O	D	S	O	D	S	O
BCM	94.21	88.95	107.79	110.2	105.2	104.06	111.5	107.4	105.89	96.81	86.11	101.7
CVN	43.85	43.46	2.262	45.85	45.6	2.739	45.9	45.65	2.729	44.92	44.35	43.46
NDC	19.11	17.22	14.788	27.41	25.61	21.256	27.86	26.63	21.227	21.73	18.61	18.1
RR	13.84	12.48	10.716	12.075	11.28	9.364	15.22	14.55	11.59	10.6	9.07	8.83
FI	0.377	0.338	0.287	0.55	0.512	0.422	0.55	0.53	0.421	0.43	0.36	0.35
H	2.652	2.562	2.262	3.092	3.018	2.739	3.109	3.054	2.729	2.82	2.66	2.57
DI	0.262	0.258	0.313	0.187	0.19	0.227	0.185	0.201	0.23	0.22	0.23	0.23

TABLE 3   Mean values of the pattern measures for the target objects on RapidEye data (kernel size: 7×7).

RapidEye	B_1			B_2			B_3			B_4			B_5		
P.M./T.O.	D	S	O	D	S	O	D	S	O	D	S	O	D	S	O
BCM	121.2	122	128	124.6	128	131	123.9	130.8	129.1	126.5	131.3	131.5	120	122	123.4
CVN	46.68	46.7	46.9	46.82	46.98	47.03	46.79	47.12	46.92	46.89	47.09	47.06	46.5	46.7	46.68
NDC	28.95	29.4	33.6	30.95	31.96	34.7	30.7	33.37	37.84	31.81	33.45	35	28.9	29.4	31.52
RR	11.35	11.5	13.1	12.13	12.53	13.61	12.03	13.08	13.27	12.47	13.11	13.72	11.3	11.5	12.36
FI	0.58	0.59	0.67	0.62	0.64	0.7	0.61	0.67	0.68	0.64	0.67	0.7	0.58	0.59	0.63
H	3.22	3.24	3.38	3.29	3.34	3.42	3.28	3.39	3.39	3.32	3.4	3.43	3.21	3.24	3.3
DI	0.12	0.12	0.11	0.12	0.11	0.1	0.12	0.1	0.11	0.11	0.1	0.1	0.13	0.12	0.13

TABLE 4 Mean values of the pattern measures for the target objects on ASTER data; visible & NIR (kernel size: 7×7).

ASTER	B_1			B_2			B_3		
P.M./T.O.	D	S	O	D	S	O	D	S	O
BCM	60.355	41.24	86.37	89.166	73.57	102.8	69.628	47.16	91.99
CVN	43.12	42.56	45.19	44.44	44.07	45.82	43.58	42.44	45.42
NDC	14.695	12.16	18.51	19.457	16.35	23.15	16.107	12.7	19.91
RR	5.762	4.77	7.26	7.63	6.41	9.078	6.316	4.98	7.81
FI	0.297	0.24	0.364	0.397	0.33	0.461	0.327	0.25	0.394
H	2.445	2.26	2.73	2.743	2.57	2.96	2.542	2.281	2.807
DI	0.219	0.217	0.173	0.198	0.2	0.172	0.216	0.245	0.174

TABLE 5 Mean values of the pattern measures for the target objects on ASTER data; SWIR (kernel size: 7×7).

ASTER	B_4			B_5			B_6			B_7			B_8			B_9		
P.M./T.O.	D	S	O	D	S	O	D	S	O	D	S	O	D	S	O	D	S	O
BCM	113	119	104	135	142	109	110	130	87.9	134	148	86.4	119	160	86.7	124	143	122
CVN	38.8	34.46	40.6	37.7	34.6	40.5	39	35.3	40.9	38.6	35.8	41.6	38.8	36.3	41.4	37.7	35.5	40.7
NDC	8.78	5.74	9.59	8.64	5.75	9.53	9.4	6.48	10.3	8.4	5.9	10.2	8.7	6.49	10.2	8.5	5.86	9.3
RR	3.44	2.25	3.76	3.39	2.25	3.73	3.7	2.54	4.03	3.2	2.31	4.01	3.4	2.54	4	3.3	2.3	3.64
FI	0.16	0.103	0.17	0.16	0.1	0.17	0.18	0.11	0.19	0.16	0.1	0.19	0.16	0.12	0.19	0.16	0.1	0.17
H	1.93	1.49	2.06	1.89	1.51	2.06	2.01	1.6	2.14	1.91	1.57	2.15	1.94	1.64	2.14	1.88	1.54	2.05
DI	0.2	0.2	0.17	0.22	0.2	0.17	0.18	0.21	0.17	0.18	0.19	0.16	0.19	0.19	0.16	0.21	0.2	0.15

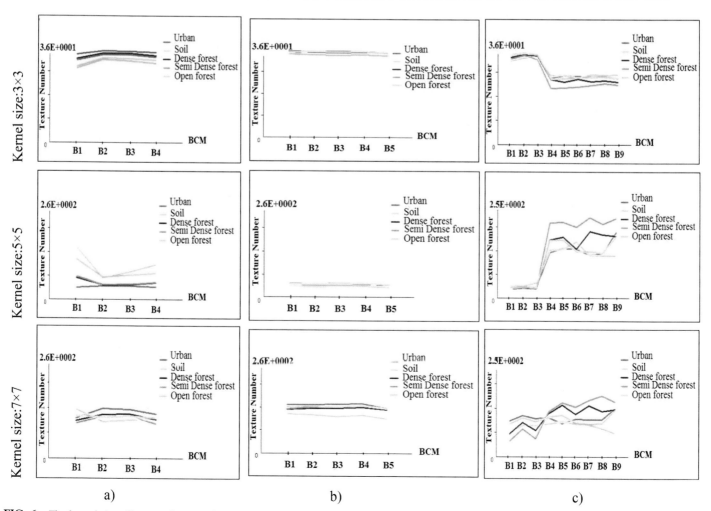

FIG. 6 The kernel size effects on the granule measures of the target objects using BCM. (A) QuickBird; (B) RapidEye; (C) ASTER.

TABLE 6   Results of the Jeffrey-Matusita distance analysis measures: (a) 3 × 3, (b) 5 × 5 and (c) 7 × 7 (QuickBird image), urban (U), bare soil (BS), semi-dense forest (SD), dense forest (D), open forest (O).

	The pattern measures																				
	BCM			CVN			NDC			RR			FI			H			DI		
	a	b	c	a	b	c	a	b	c	a	b	c	a	b	c	a	b	c	a	b	c
Class pairs	Jeffrey-Matusita distance																				
U~S	2	2	2	1.6	2	2	2	2	2	1.5	2	2	0.1	0.2	0.4	0.3	0.4	0.7	0.5	0.5	0.5
D~U	2	2	2	0.4	1.8	2	1.3	2	2	0.3	2	2	0.1	0.1	0.3	0.1	0.1	0.2	0.2	0.2	0.3
SD~U	2	2	2	0.6	2	2	1.8	2	2	0.5	2	2	0.1	0.2	0.4	0.2	0.1	0.2	0.3	0.3	0.4
O~U	2	2	2	1.5	2	2	2	2	2	1.4	2	2	0.2	0.3	0.4	0.3	0.4	0.6	0.5	0.5	0.6
D~BS	2	2	2	1.2	2	2	2	2	2	1.3	2	2	0.1	0.1	0.1	0.1	0.3	0.5	0.2	0.2	0.3
BS~SD	2	2	2	1	2	2	2	2	2	1	2	2	0	0.1	0.2	0	0.2	0.4	0.1	0.1	0.2
O~BS	2	2	2	0.2	1.9	2	1.5	2	2	0.3	2	2	0.1	0.1	0.1	0	0.1	0.3	0	0	0.1
D~SD	2	2	2	0.1	0.7	1.8	0.5	2	2	0	1	2	0	0	0.1	0	0	0.1	0	0	0
O~D	2	2	2	0.9	2	2	2	2	2	0.8	2	2	0	0	0.1	0	0.1	0.2	0.1	0.1	0.1
SD~O	2	2	2	0.6	2	2	1.9	2	2	0.5	2	2	0	0.1	0.3	0	0.1	0.2	0	0.1	0.1
Mean Val.	2	2	2	0.8	1.8	2	1.7	2	2	0.8	1	2	0.1	0.1	0.2	0.1	0.2	0.3	0.2	0.2	0.2

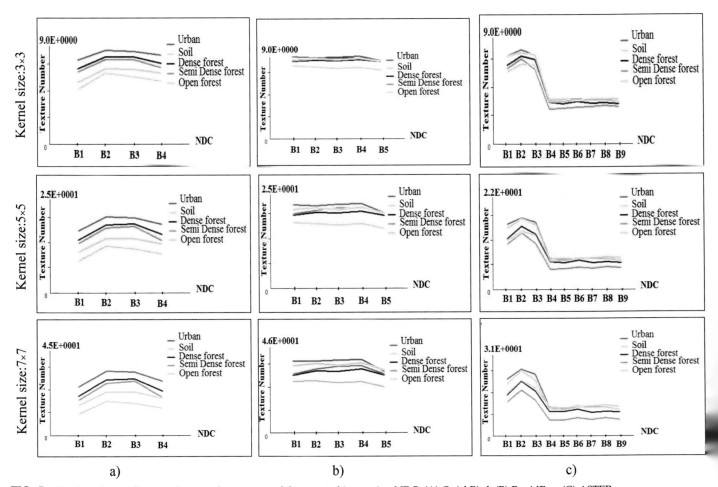

FIG. 7   The kernel size effects on the granule measures of the target objects using NDC. (A) QuickBird; (B) RapidEye; (C) ASTER.

TABLE 7 Results of the Jeffrey-Matusita distance analysis measures: (a) 3 × 3, (b) 5 × 5 and (c) 7 × 7 (RapidEye image), urban (U), bare soil (BS), semi-dense forest (SD), dense forest (D), open forest (O).

	The pattern measures																				
	BCM			CVN			NDC			RR			FI			H			DI		
	a	b	c	a	b	c	a	b	c	a	b	c	a	b	c	a	b	c	a	b	c
Class pairs	Jeffrey-Matusita distance																				
U~S	2	2	2	0.6	1.6	2	1.8	2	2	0.4	2	2	0.4	0.4	0.4	0.5	0.5	0.6	0.5	0.5	0.7
D~U	0.6	2	2	0.1	0.3	0.7	0.3	2	2	0.1	1.1	2	0.1	0.1	0.1	0.1	0.1	0.1	0.2	0.2	0.2
SD~U	0.2	2	2	0.1	0.1	0.2	0.1	2	2	0	0.5	2	0	0	0.1	0.1	0.5	0	0.1	0.1	0.1
O~U	0.6	2	2	0.2	0.3	0.4	0.2	2	2	0.1	0.6	1.9	0.1	0.2	0.2	0.2	0.2	0.2	0.2	0.3	0.3
D~BS	2	2	2	0.3	1.1	2	1.4	2	2	0.1	1.9	2	0.1	0.2	0.2	0.2	0.3	0.3	0.2	0.3	0.3
BS~SD	2	2	2	0.4	1.4	2	1.6	2	2	0.3	2	2	0.3	0.4	0.5	0.4	0	0.7	0.3	0.5	0.7
O~BS	2	2	2	0.3	1.3	2	1.6	2	2	0.2	2	2	0.1	0.1	0.1	0.2	0.2	0.3	0.2	0.2	0.2
D~SD	0	2	2	0	0.1	0.2	0.1	2	2	0	0.3	1.7	0	0.1	0.2	0	0.3	0.2	0	0.1	0.2
O~D	0.1	2	2	0	0	0.2	0.1	2	2	0	0.6	2	0	0.4	0	0	0	0	0	0.1	0.1
SD~O	0.3	2	2	0.1	0.1	0.2	0.1	2	2	0	0.4	2	0.1	0.2	0.2	0.1	0.2	0.2	0.1	0.2	0.3
Mean Val.	1	2	2	0.2	0.6	1	0.7	2	2	0.1	1.1	1.9	0.1	0.2	0.2	0.2	0.2	0.3	0.2	3	0.3

20 to 40 for RapidEye images. The best result for RapidEye images also belongs to the spectral-textural segmentation of the NDC product with a kernel size of 7 × 7, offering 0.70 and 0.61 for overall accuracy and kappa coefficient, respectively. However, the FI and RR provided comparable results. The rise in the heuristics is from 20% to 50% for ASTER images, offered by the inclusion of the textural pattern measures in the processes. The heuristics of the segmentation outputs are generally over 95% when using pattern measures. The best result of the ASTER images was derived from the spectral-textural segmentation of the H product by a kernel size of 7 × 7, offering 0.98 and 0.98 overall accuracies

TABLE 8 Results of the Jeffrey-Matusita distance analysis measures (a) 3 × 3, (b) 5 × 5 and (c) 7 × 7 (Aster image), urban (U), bare soil (BS), semi-dense forest (SD), dense forest (D), open forest (O).

	The pattern measures																				
	BCM			CVN			NDC			RR			FI			H			DI		
	a	b	c	a	b	c	a	b	c	a	b	c	a	b	c	a	b	c	a	b	c
Class pairs	Jeffrey-Matusita distance																				
U~S	2	2	2	0.4	1.8	2	1.3	2	2	0.3	1.3	2	0.2	0.6	0.8	0.3	0.8	1	0.2	0.4	0.9
D~U	2	2	2	0.4	2	2	0.7	2	2	0.3	1.3	2	0.3	0.5	0.6	0.3	0.7	1	0.2	0.5	1.3
SD~U	2	2	2	1	2	2	1.2	2	2	0.2	1.7	2	0.2	0.5	0.6	0.3	0.6	1.1	0.3	0.5	1.4
O~U	2	2	2	0.1	1.5	2	0.2	1.6	2	0.1	0.4	1	0.1	0.3	0.4	0.1	0.4	0.5	0.1	0.2	0.5
D~BS	2	2	2	0.6	2	2	0.8	2	2	0.4	0.7	1.4	0.3	0.5	0.7	0.4	0.6	1	0.3	0.4	0.9
BS~SD	2	2	2	1.1	2	2	0.6	2	2	0.3	1	2	0.3	0.7	0.8	0.5	0.8	1.2	0.5	0.5	1
O~BS	2	2	2	0.5	1.9	2	1.2	2	2	0.3	1.2	1.8	0.2	0.5	0.6	0.3	0.6	1	0.3	0.3	0.6
D~SD	2	2	2	0.7	2	2	0.6	2	2	0.2	0.9	2	0.2	0.3	0.3	0.2	0.4	0.4	0.2	0.4	0.4
O~D	2	2	2	0.3	2	2	0.5	2	2	0.1	0.9	2	0.1	0.2	0.4	0.1	0.2	0.6	0.1	0.3	1
SD~O	2	2	2	0.9	2	2	1	2	2	0.1	1.6	2	0.1	0.4	0.4	0.2	0.3	0.7	0.3	0.4	1
Mean Val.	2	2	2	0.6	1.9	2	0.8	2	2	0.2	1.1	1.8	0.2	0.5	0.5	0.3	0.5	0.9	0.2	0.4	0.9

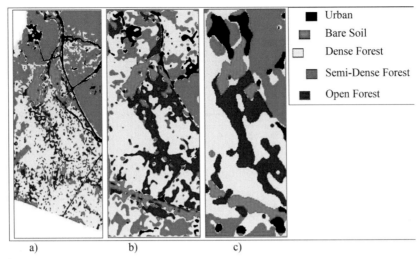

FIG. 8 Spectral-texture pixel-wise classification of forest density on multi-resolution images. (A) QuickBird; (B) RapidEye; (C) ASTER.

TABLE 9 Result of classification on QuickBird, RapidEye & ASTER Images Using only spectral information in pixel-wise classification.

Data Set		The pattern measures ~ Kernel sizes																				
		BCM			CVN			NDC			RR			FI			H			DI		
		a	b	c	a	b	c	a	b	c	a	b	c	a	b	c	a	b	c	a	b	c
QuickBird	OA	21	19	24	20	21	23	24	**35**	35	24	35	35	24	35	35	23	32	43	19	23	28
	KC	7	9	11	6	8	9	11	21	48	11	21	48	11	21	48	10	18	29	6	9	13
RapidEye	OA	24	30	39	21	22	26	25	41	**51**	25	41	51	25	41	51	25	37	48	21	31	36
	KC	11	16	26	7	9	12	12	27	38	12	27	38	12	27	38	11	23	35	9	17	22
ASTER	OA	37	35	36	38	35	34	45	61	73	45	61	**73**	45	61	**73**	42	60	72	34	37	47
	KC	23	22	45	22	20	26	32	51	65	32	51	65	32	51	65	28	50	65	19	23	34

TABLE 10 Result of spectral classification in QuickBird, RapidEye & ASTER images inclusion of the textural measures in the classification process.

Data Set		The pattern measures ~ Kernel sizes																				
		BCM			CVN			NDC			RR			FI			H			DI		
		a	b	c	a	b	c	a	b	c	a	b	c	a	b	c	a	b	c	a	b	c
QuickBird	OA	54	28	45	57	57	57	62	65	**68**	62	65	68	62	65	68	61	65	68	58	59	61
	KC	43	18	34	46	46	46	52	55	59	52	55	59	52	55	59	51	55	60	47	49	51
RapidEye	OA	59	55	64	58	59	60	61	65	70	61	65	**70**	61	65	**70**	61	64	69	59	62	64
	KC	48	43	54	48	48	50	51	56	61	51	56	61	51	56	61	51	55	61	49	52	54
ASTER	OA	90	88	96	94	96	93	95	96	97	95	96	97	95	96	97	95	97	**98**	93	94	97
	KC	90	85	95	92	95	91	93	94	96	93	94	96	93	94	96	93	96	98	92	93	96

and kappa coefficient, respectively. It is also the best result of the current research among the output products of semantic segmentation derived from the three data sets.

4 Conclusions

In this study, textural pattern measures and granulometry were used to increase the accuracy of semantic segmentation of sparse woodlands on satellite images. Kernel-based textural granulometry was used to calculate pattern measures. The pattern measures, considering the local co-occurrence of accompanying target objects, are more successful in high-and moderate-resolution images. The pattern measure represents an identical vector value for each kernel in a small neighborhood. The convolutional effect of the larger kernels on the lower spatial resolution images leads to better performance of those measures, describing an individual target object without paying attention to the occurrence of the accompanying objects. The statistical *t*-test indicated that the pattern measures were significantly different for the target objects for a fixed circle-shaped ROI over segments of sparse woodlands. The size of the target objects, their spatial frequency, and co-occurrence patterns on the earth's surface are still challenging issues in choosing appropriate satellite imagery for object-based image segmentation and pixel-based spectral classification. Achieving suitable results using the technique used in this research depends solely on the pre-knowledge of the users from spectral and textural attributes of the target objects, such as spatial co-occurrence, frequency, and coincidence of defined segments to physical partitions of the target objects on the Earth's surface. The method presented in this study is easy to accomplish and retrieve. It is very helpful for using image contextual information for the semantic segmentation of complex objects, either presenting spatial patterns or pattern recognition studies. There are many opportunities for putting the method into the test of pattern measures on a wide range of satellite images and target objects in remote sensing.

References

1. Zhang Y. Optimization of building detection in satellite images by combining multispectral classification and texture filtering. *J Photogramm Remote Sens*. 1999;54:50–60.
2. Lee J-H, Philpot WD. Spectral texture pattern matching: a classifier for digital imagery. *IEEE Trans Geosci Remote Sens*. 1991;29(4):545–554.
3. Wulder MA, Franklin SE. *Understanding Forest Disturbance and Spatial Pattern: Remote Sensing and GIS Approaches*. London: Taylor and Francis; 2007.
4. Hammouche K, Diaf M, Postaire JGA. Clustering method based on multi-dimensional texture analysis. *J Pattern Recogn*. 2006;39:1265–1277.
5. Wechsler H. Texture analysis – a survey. *J Signal Process*. 1980;2(3):271–282.
6. Berberoglu S, Curran PJ, Lloyd CD, Atkinson PM. Texture classification of Mediterranean land cover. *J Appl Earth Observ Geo-inform*. 2007;9:322–334.
7. Srinivasan GN, Shobha G. Statistical texture analysis. In: *Proceedings of World Academy of Science, Engineering and Technology*. 36; 2008:1264–1269.
8. Chen CH, Pau LF, Wang PSP. *Handbook of Pattern Recognition and Computer Vision*. 2nd ed. World Scientific Publishing; 1993.
9. Haralik RM. Statistical and structural approaches to texture. *Proc IEEE*. 1979;67(5):786–812.
10. Singh M, Singh S. Texture algorithms: performance variability across data sets. *Cybern Syst*. 2003;34:1–17.
11. Fauvel M, Chanussot V, Benediktsson V. A spatial–spectral kernel-based approach for the classification of remote-sensing images. *J Pattern Recogn*. 2012;45:381–392.
12. Franklin SE. *Remote Sensing for Sustainable Forest Management*. New York: Lewis Boca Raton; 2001.
13. Franklin SE, Wulder MA, Lavigne MB. Automated derivation of geographic window sizes for use in remote sensing digital image texture analysis. *J Comput Geosci*. 1996;22(6):665–673.
14. Chen D, Stow DA, Gong P. Examining the effect of spatial resolution and texture window size on classification accuracy: an urban environment case. *J Remote Sens*. 2004;25(11):2177–2192.
15. Fallah Shamsi SR. *Using Sampling Method in Accuracy Assessment of LU/LC Map, derived from Satellite Imageries* [M.Sc. Dissertation]. Iran: University of Tehran; 2004.
16. Matheron R. *Random Sets and Integral Geometry*. NY: John Wiley and Sons; 1975.
17. Gebreslasie MT, Ahmed FB, van Aardt JAN. Extracting structural attributes from IKONOS imagery for Eucalyptus plantation forests in KwaZulu-Natal, South Africa, using image texture analysis and artificial neural networks. *J Remote Sens*. 2011;32(22):7677–7701.
18. He DC, Wang L. Texture unit, texture spectrum, and texture analysis. *IEEE Trans Geo-Sci Remote Sens*. 1990;28(4):509–512.
19. Marceau D, Howarth PJ, Dubois JM, Gratton D. Evaluation of the Grey-level co-occurrence matrix method for land-cover classification using SPOT imagery. *IEEE Trans Geosci Remote Sens*. 1990;28(4):513–519.
20. Franklin SE, Maudie AJ, Lavlgne MB. Using spatial co-occurrence texture to increase forest structure and species composition classification accuracy. *J Photogramm Eng Remote Sens*. 2001;67(7):849–855.
21. Narayanan RM, Desetty MK, Reichenbach SE. Effect of spatial resolution on information content characterization in remote sensing imagery based on classification accuracy. *J Remote Sens*. 2002;23(3):537–553.
22. Zhang Q, Wang J, Gong P, Shi P. Study of urban spatial patterns from SPOT panchromatic imagery using textural analysis. *Int J Remote Sens*. 2003;24(21):4037–4060.
23. Coburn CA, Roberts CB. A multiscale texture analysis procedure for improved forest stands classification. *J Remote Sens*. 2004;25(20):4287–4308.
24. Pussant A, Hirsch J, Weber C. The utility of texture analysis to improve per-pixel classification for high to very high spatial resolution imagery. *J Remote Sens*. 2005;26(4):733–745.
25. Hodgson ME. What size window for image classification? A cognitive perspective. *J Photogramm Eng Remote Sens*. 1998;64(8):797–807.
26. Kwast J, Voorde T, Canters F, Uljee I, Van Looy S, Engelen G. Inferring urban land use using the optimized spatial reclassification kernel. *J Environ Model Softw*. 2011;26:1279–1288.
27. Li W, Ouyang Z, Zhou W, Chen Q. Effects of spatial resolution of remotely sensed data on estimating urban impervious surfaces. *J Environ Sci*. 2011;23(8):1375–1383.
28. Wu J. Effects of changing scale on landscape pattern analysis: scaling relations. *J Landsc Ecol*. 2004;19:125–138.
29. Tsaneva MG, Krezhova DD, Yanev TK. Development and testing of a statistical texture model for land cover classification of the Black Sea region with MODIS imagery. *Adv Space Res*. 2010;46:872–878.
30. Kim C, Hong S. The characterization of a forest covers through shape and texture parameters from QuickBird imagery. *IEEE Trans*. 2008;6:295–692.
31. QuickBird Imagery Products. *Product Guide Revision 4.7.1*. Colorado: Digital Globe Inc.; 2006.
32. Sandau R. Status and trends of small satellite missions for Earth observation. *Acta Astronaut*. 2010;66:1–12.

33. Abrams M, Hook S. *ASTER User Guide*. NY: NASA; 2003.
34. Pesaresi M, Gerhardinger A, Haag F. Rapid damage assessment of built-up structures using VHR satellite data in tsunami affected area. *Int J Remote Sens*. 2007;28(13):13–30.
35. Fallah Shamsi SR, Erfanifard SY, Negahban M, et al. Mapping spatial pattern of defoliation in Persian oak forest, using remote sensing techniques. In: *Proc. 2012, 1st National Symposium on Zagross Forests (Challenges, Threatens and Opportunities), Iran*; 2012.
36. Xiao Feng R, Malik J. Learning a classification model for segmentation. In: *Proc. 2003; ICCV '03, Nice-France*. 1; 2003:10–17.
37. Turner MG. Landscape ecology: the effect of pattern on the process. *Annu Rev Ecol Syst*. 1989;20:171–179.
38. Monmonier MS. Measures of pattern complexity for choropleth maps. *Am Cartogr*. 1974;1(2):159–169.
39. Murphy DL. Estimating neighborhood variability with a binary comparison matrix. *Photogramm Eng Remote Sens*. 1985;51(6):667–674.
40. Pratt WK. *Digital Image Processing*. 2nd ed. NY: John Wiley and Sons; 1991.
41. Koonsanit K, Jaruskulchai C, Eiumnoh A. Determination of the initialization number of clusters in K-means clustering application using co-occurrence statistics techniques for multispectral satellite imagery. *Int J Inf Electron Eng*. 2012;2:785–789.
42. Sá Junior JM, Backes AR, Rossatto DR, Kolb RM, Bruno OM. Measuring and analyzing color and texture information in anatomical leaf cross sections: an approach using computer vision to aid plant species identification. *Botany*. 2011;89:467–479.
43. Miles N, Wernick G, Morris M. Maximum-likelihood image classification: digital and optical shape representation and pattern recognition. In: *Proc. 1988; SPIE 0938*; 1988.
44. Congalton RG. A review of assessing the accuracy of classification of remotely sensed data. *Remote Sens Environ*. 1991;37:35–46.
45. Congalton RG, Green K. A practical look at the sources of confusion in error matrix generation. *Photogramm Eng Remote Sens*. 1993;59:641–644.

42

Badland erosion mapping and effective factors on its occurrence using random forest model

Majid Mohammady[a], Hamid Reza Pourghasemi[b], and Saleh Yousefi[c]

[a]Faculty of Natural Resources, Semnan University, Semnan, Iran [b]Department of Natural Resources and Environmental Engineering, College of Agriculture, Shiraz University, Shiraz, Iran [c]Soil Conservation and Watershed Management Research Department, Chaharmahal and Bakhtiari Agricultural and Natural Resources Research and Education Center, AREEO, Shahrekord, Iran

1 Introduction

Soil degradation and erosion are global problems associated with climate change and global warming.[1] Fluvial erosion and badlands are the results of many hydrogeomorphic processes. Generally, soil erosion is a two-step process consisting of the detachment of particles from the soil mass and transport by erosive factors such as runoff. When there is not enough energy to transport sediment particles, sedimentation occurs.[2,3] In addition, because of human activities in the last few decades, erosion rates and badland development have increased at an alarming rate in many parts of the world.[4] Erosion may decrease soil organic matter, soil fertility, and rooting depth. An increase in pH and EC (electrical conductivity) values also induce land degradation.[5] Linear erosion processes, including rills, gullies, and badlands, are important indicators of severe erosion, regardless of geographical setting.[6] Badland areas are characterized by rills that are associated with severe erosion rates and very high sediment yields[7]; therefore, they are not suitable for agriculture. Badland areas are characterized by no vegetation cover, steep slopes, or severe erosion processes.[8] The origin of badlands is very complex because many variables are interacting together, such as the high clay content of the soil, deforestation, and other land-use changes.[9] Badlands is a very important phenomenon that has been studied in many parts of the world, including Canada,[10] South Africa,[11] the USA,[12] Ethiopia,[13] Europe,[14–16] and India.[17,18] Therefore, the aim of this research is BLSM using the random forest (RF) method. The susceptibility of a region to a phenomenon is explained by the likelihood of a certain phenomenon occurring in the region.[19] The susceptibility mapping of badlands is an important subject for soil erosion and land management, especially for planning control activities.

Stochastic models are useful tools for a "susceptibility" mapping at the watershed and regional scales for water erosion and badlands. Such algorithms are commonly used to assess and model the response of a binary outcome (presence/absence) concerning a set of conditioning factors. The random forest machine learning" (RFML) technique is one of these models whose accuracy has been confirmed in many past studies.

2 Study area

The Firozkuh Watershed is located in the Tehran Province and covers an area of approximately 1450 sq. km. The Firozkuh Watershed lies within the eastern longitudes of 52°19′ and 53° 07′ and the northern latitudes of 35°40′ and 35° 57′. The "annual mean rainfall" ranges from 400 to 800 mm/year. The dominant climate in this watershed is semi-arid and known as the Turani Climate, although sometimes it is cold and Mediterranean.[20] The altitude ranges from 1712 m a.s.l. at the outlet to 3941 m a.s.l. at the highest point. The location of the Firozkuh watershed and the identified badlands in this region are indicated in Fig. 1.

FIG. 1 Location of the Firozkuh
Watershed and identified badlands in
this area.

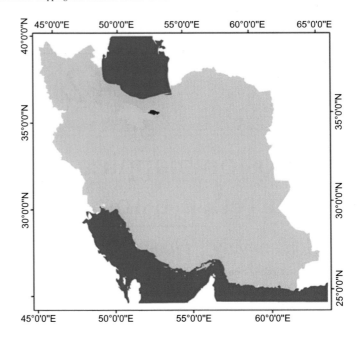

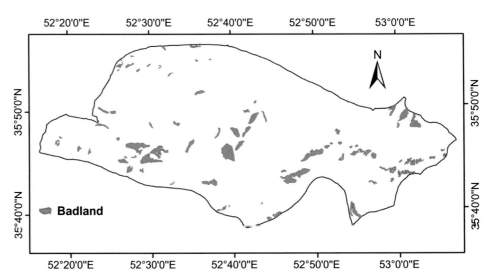

3 Materials and methods

3.1 Data preparation

The first step in "badland susceptibility assessment-BDSA" is data collection on the history of badlands in the study area. The Locations of the badlands were extracted using Landsat 7 satellite images and field surveys (Fig. 1). The badlands inventory (BLI) was randomly split into two groups: the modeling set (70% of the badlands) and the validation set (30%). The modeling set was applied for susceptibility mapping, and the validation set was used to validate the final map.

A "digital elevation model (DEM)" was created using "contour lines and elevation points" from maps produced by the "Iranian National Cartographic Center (INCC)."[21] Elevation, aspect, slope, "topographic wetness index-TWI", and "plan curvature" were extracted from the "DEM" in ArcGIS 10.3 and SAGA-GIS. A geology map was prepared at a scale of 1:100,000.[22] To prepare a land-use map, a synthetic classification method[23] was applied to Landsat images from 2019. Rivers and drainage network maps were extracted from the topography map and buffer zones were set along rivers. A Drainage density map was created using a drainage network and area calculation as km/km². Meteorological organization data from Tehran Province were used to prepare the rainfall map. To obtain soil characteristics, the

TABLE 1 Data specifications.

Maps	Specifications
badlands inventory map	115 polygons
DEM (Elevation)	Range: 1712–3941 m
Aspect	Flat, North, South, West, East, Northeast, Northwest, Southeast, Southwest
Slope	Range: 0–69.9 degree
Topographic wetness index (TWI)	Range: 2–25
Plan curvature (PlanC)	Range: −8 to 11 100/m
Geology	Eleven lithology classes
Land use (LU)	Land use types: forest, range, agriculture, residential and bare soil
Distance from river (RiverDistance)	Range: 0–6600 m
Drainage density (DrainageDensity)	Range: 0.1–2.5 km/km^2
Rainfall	Range: 400–800 mm
Silt percent (Silt)	Range: 27–52
Clay percent (Clay)	Range: 14–33
Soil hydrology groups (SoilHydrology)	Three classes (B, C, and D)
Soil pH (pH)	Range: 6–8.1
Soil depth (SoilDepth)	Range: 57–242
Permeability	Five classes including; "very low, low, moderate, high, and very high permeability"
Bulk density (BD)	Range: 0.949–1.589

samples were collected and sent to the laboratory for analysis. The silt percentage, clay percentage, pH, and bulk density were computed using the Inverse Distance Weight (IDW)" method in a GIS environment. In addition, soil depth, permeability, and soil hydrologic group maps were prepared in the study region. Table 1 lists the data specifications used for BLSA.

3.2 Random forest

RFMLT derived from classification and regression trees, first developed by Breiman,[24] is a very useful classifier. The RF algorithm is a widely used technique in different fields of geo-environmental phenomena evaluation. The RF algorithm is a collection of decision trees, and each decision tree is made from a set of input records where the subset used is sampled randomly.[24] A random forest can handle missing data in the training and validation steps. In addition, because of its group plan, this technique can predict when some of the input values are missing. In addition, the prioritization of variables can be obtained through a modeling process.[25,26] Random forest applies random binary trees that use a set of observations through boot-strapping techniques. The RF algorithm estimates the priority of a factor by looking at how much the prediction error goes up when "OOB (out-of-bag)" data for that factor is transferred while other factors are left unchanged.[27] This ability can be applied to study the relative priorities of various explanatory factors.[26] Two types of errors are calculated in this process: "mean decrease in Gini and mean decrease in accuracy". These two measures can be applied to prioritize the variables.[28] After prioritizing factors, the extracted weights in R 3. 2 software were converted to ArcGIS 10.3, and finally, the BLSM was prepared and classified into four susceptibility classes.

3.3 Validation of the BLSM

Validation and accuracy assessment is an important step in modeling and "susceptibility mapping" to test the effectiveness and scientific significance of the models.[29] For accuracy assessment of the BLSM prepared by the random forest algorithm, the "ROC (receiver operating characteristic) curve" was used. In this curve, the "AUC (the area under the curve)" value ranges from 0.5 to 1.0.[30] In this accuracy assessment, the specificity and sensitivity were measured on the x-axis and y-axis, respectively. Higher AUC values indicate a higher accuracy of the model.[31]

4 Results and discussion

To prioritize the conditioning factors of badland susceptibility-BLS, "mean decrease Gini and mean decrease accuracy errors" were calculated using a random forest algorithm (Fig. 2). Mean decrease accuracy and mean decrease Gini errors in random forest algorithm have been widely used in many fields showing good performance for variable selection.[32–34] The Results indicated that drainage density (DrainageDensity), elevation, rainfall, and distance from rivers (RiverDistance) have the greatest impact on badland occurrence in the Firozkuh watershed.

Some previous researchers have pointed out the importance of these factors in erosion and badland occurrence. For example, Vergari[35] applied a method to assess soil erosion and badland hazards in Italy by combining the susceptibility of soil erosion and erosion rate prediction. The results showed that the most important factors of water erosion processes were elevation and drainage density. Drainage density is one of the most important conditioning factors in areas affected by badland, as high drainage density shows areas where runoff increases. The elevation and slope are related to the physiographic conditions that may affect hydrological processes and soil moisture. Because of temperature changes and rainfall with an increase in elevation, this factor has a direct effect on land use types as conditioning and controlling factor of badlands.[26] In many previous studies, rainfall rate and its distribution are considered as a factor to model the "erosion susceptibility" assessment.[36] In fact, badland formation and dynamics are the results of climatic factors, such as the Mediterranean rainfall regime.[37] Adequate moisture and soil saturation are very important for initiating and intensifying erosion. Rivers are one of the main sources of moisture; therefore, distance from rivers is considered an important condition factor. Deshmukh et al.[38] and Bianchini et al.[19] also mentioned the importance of this factor. The results also showed that soil properties had a direct effect on erosion. For example, Sinha and Joshi[39] explained that coarse texture and very fine texture (clay) can reduce erosion. The results of their research showed that fine sand and silt are more erodible.[39]

As mentioned above, a random forest was used to map the BLS in the Firozkuh watershed. The RF model was applied because of its high efficiency in using input layers with different natures and its ability to prioritize conditioning factors.[40] Fig. 3 shows the BLSM of the Firozkuh watershed created by the RF model.

As previously mentioned for the accuracy assessment of the BLS created by the RF technique, the "ROC curve" was used. The AUC indicates the accuracy of the mentioned model. Using the "ROC" curve, the AUC value was calculated as 0.97, indicating the excellent precision of the RF algorithm (Fig. 4 and Table 2).

FIG. 2 Variables prioritization calculated by the RF model.

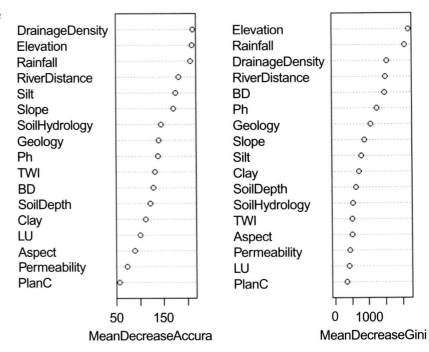

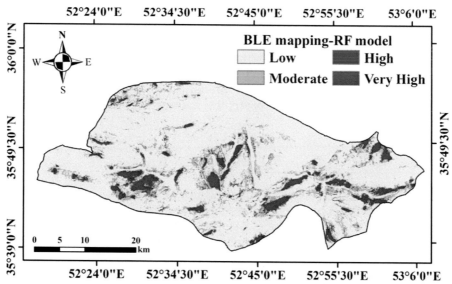

FIG. 3 "Badland erosion susceptibility map-BLESM" of Firozkuh Watershed.

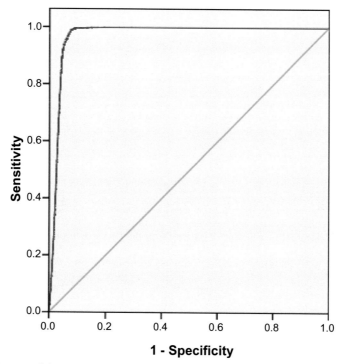

FIG. 4 ROC curve for random forest model.

TABLE 2 Statistical indices for the random forest.

Model	Area	Standard error	Asymptotic significant	Asymptotic 95% confidence interval	
				Lower bound	Upper bound
Random forest	0.970	0.001	0.000	0.968	0.972

Previous studies have also reported the ability of the RFMLT to study various geo-environmental phenomena.[41–45] The complexity of processes originating badland landscapes and their variability in space and time suggests that further studies are needed to establish new models able to explain the evolution of this very astonishing type of landform. The information supplied and the BLSM are very important for planners and decision-makers involved in managing watersheds affected by this erosion and can improve decisions for limiting and controlling erosion. Investigation of erosion "susceptibility" and classification into several susceptibility zones is a useful tool for understanding vulnerability to erosion and badland for the improvement of suitable management programs and control strategies.

5 Conclusions

"Badland" is a very important and problematic geo-hydrological phenomenon in the Firozkuh watershed. In this research, the RFMLT was used for BLSA and prioritizing the condition factors of "badland" occurrence. Seventeen conditioning factors were selected and input as spatial layers in a GIS environment. The weights of each factor were calculated using R software, and then the BLSM was created using ArcGIS software. The results of variable importance explained that drainage density, elevation, rainfall, and distance from rivers have the greatest impact on badland occurrence in the Firozkuh watershed. Also, random forest showed high efficiency to map "badland susceptibility" in the study area. The results of this research and "badland susceptibility map" are very useful for decision-makers involved in managing watersheds affected by badlands and can improve decisions for limiting and control erosion.

References

1. Pruski FF, Nearing MA. Runoff and soil loss responses to changes in precipitation: a computer simulation study. *J Soil Water Conserv.* 2002;57 (1):7–16.
2. Ghosh S, Bhattacharya K. Multivariate erosion risk assessment of lateritic badlands of Birbhum (West Bengal, India): a case study. *J Earth Syst Sci.* 2012;121(6):1441–1454.
3. Morgan RPC. *Soil Erosion and Conservation.* Malden: Blackwell Publishing; 2005:67–157.
4. Joshi VU. Soil loss estimation by field measurements in the Badlands along Pravara River (Western India). *J Geol Soc India.* 2014;83:613–624.
5. Abdallah SM, Massoud EE. Land degradation risk assessment in Al-Sawda terraces. *Kingdom Saudi Arabia Arab J Geosci.* 2018;11:599. https://doi.org/10.1007/s12517-018-3956-x.
6. Guney Y. The geomorphosite potential of the Badlands around Küpyar, Manisa. *Turkey Geoheritage.* 2020;12:21. https://doi.org/10.1007/s12371-020-00433-y.
7. Nadal-Romero E, Martínez-Murillo JF, Vanmaercke M, Poesen J. Scale-dependency of sediment yield from badland areas in Mediterranean environments. *Prog Phys Geogr.* 2011;35:297–332.
8. Howard AD. Badland morphology and evolution: interpretation using a simulation model. *Earth Surf Process Landforms.* 1997;22:211–227.
9. Caraballo-Arias NA, Ferro V. Assessing, measuring and modelling Erosion in Calanchi areas: a review. *J Agric Eng.* 2016;47(4):181–190.
10. Bryan RB, Campbell IA, Yair A. Postglacial geomorphic development of the Dinosaur Provincial Park badlands. *Alberta Can J Earth Sci.* 1987;24:135–146.
11. Boardman J, Parsons A, Holland R, Holmes PJ, Washington R. Development of badlands and gullies in the Sneeuberg, Great Karoo, South Africa. *CATENA.* 2003;50:165–184.
12. Howard AD. Badlands and gullying. In: Parsons AJ, Abrahams AD, eds. *Geomorphology of Desert Environments.* The Netherlands: Springer; 2009:265–299.
13. Feoli E, Vuerich LG, Woldu Z. Processes of environmental degradation and opportunities for rehabilitation in Adwa. *Northern Ethiopia Landsc Ecol.* 2002;17:315–325.
14. Gallart F, Marignani M, Perez-Gallego N, Santi E, Maccherini S. Thirty years of studies on badlands, from physical to vegetational approaches. A succinct review. *Catena.* 2013;106:4–11.
15. Phillips C. The badlands of Italy: a vanishing landscape? *Appl Geogr.* 1998;18:243–257.
16. Torri D, Calzolari C, Rodolfi G. Badlands in changing environments: an introduction. *Catena.* 2000;40:119–125.
17. Haigh MJ. Ravine erosion and reclamation in India. *Geoforum.* 1984;15:543–561.
18. Pani P, Carling P. Land degradation and spatial vulnerabilities: a study of inter-village differences in Chambal Valley. *India Asian Geogr.* 2013;30:65–79.
19. Bianchini S, Soldato MD, Solari L, Nolesini T, Pratesi F, Moretti S. Badland susceptibility assessment in Volterra municipality (Tuscany, Italy) by means of GIS and statistical analysis. *Environ Earth Sci.* 2016;75:889. https://doi.org/10.1007/s12665-016-5586-5.
20. Mohammadzadeh MJ, Emam JZ, Safari M, Mousavi M, Ghanbarzadeh B, Philips GO. Physicochemical and emulsifying properties of Barijeh (Ferula gumosa) gum. *Iran J Chem Chem Eng Int English Ed.* 2007;26(3):81–88.
21. Iranians National Cartographic Center (INCC). https://www.ncc.gov.ir.
22. Geology Survey of Iran (GSI); 1997. http://www.gsi.ir/Main/Lang_en/index.html.
23. Mohammady M, Morady HR, Zeinivand H, Temme AJAM. A comparison of supervised, unsupervised and synthetic land use classification methods in the North of Iran. *Int J Environ Sci Technol.* 2015;12(5):1515–1526.
24. Breiman L. Random forests. *Mach Learn.* 2001;45(1):5–32.

25. Ball RL. Comparison of random forest, artificial neural network, and multi-linear regression: a water temperature prediction case. In: *Seventh Conference on Artificial Intelligence and its Applications to the Environmental Sciences*; 2009:1–6.
26. Mohammady M, Pourghasemi HR, Amiri M. Land subsidence susceptibility assessment using random forest machine learning algorithm. *Environ Earth Sci*. 2019;75:503.
27. Catani F, Lagomarsino D, Segoni S, Tofani V. Landslide susceptibility estimation by random forests technique: sensitivity and scaling issues. *Nat Hazards Earth Syst Sci*. 2013;13:2815–2831.
28. Calle ML, Urrea V. Letter to the editor: stability of random forest importance measures. *Brief Bioinform*. 2010;12(1):86–89.
29. Chung CJ, Fabbri AG. Validation of spatial prediction models for landslide hazard mapping. *Nat Hazards*. 2003;30:451–472.
30. Nandi A, Shakoor A. A GIS-based landslide susceptibility evaluation using bivariate and multivariate statistical analyses. *Eng Geol*. 2010;110:11–20.
31. George CM, Anu VV. Predicting piping erosion susceptibility by statistical and artificial intelligence approaches – a review. *Int J Res Eng Technol*. 2018;5(12):243–249.
32. Cutler DR, Edwards TC, Beard KH, et al. Random forests for classification in ecology. *Ecology*. 2007;88(11):2783–2792.
33. Lawrence RL, Wood SD, Sheley RL. Mapping invasive plants using hyperspectral imagery and Breiman Cutler classifications (random forest). *Remote Sens Environ*. 2006;100:356–362.
34. Stumpf A, Kerle N. Object-oriented mapping of landslides using random forests. *Remote Sens Environ*. 2011;115:2564–2577.
35. Vergari F. Assessing soil erosion hazard in a key badland area of Central Italy. *Nat Hazards*. 2015;79:71–95.
36. Vijith H, Dodge-Wan D. Modelling terrain erosion susceptibility of logged and regenerated forested region in northern Borneo through the analytical hierarchy process (AHP) and GIS techniques. *Geoenviron Disast*. 2019;6(8):1–18.
37. Moretti S, Rodolfi G. A typical "calanchi" landscape on the Eastern Apennine margin (Atri, Central Italy): geomorphological features and evolution. *Catena*. 2000;40:217–228.
38. Deshmukh DS, Chaube UC, Tignath S, Pingale SM. Geomorphological analysis and distribution of badland around the confluence of Narmada and Sher River. *India Eur Water*. 2011;35:15–26.
39. Sinha D, Joshi VU. Application of universal soil loss equation (USLE) to recently reclaimed Badlands along the Adula and Mahalungi Rivers, Pravara Basin. *Maharashtra J Geol Soc India*. 2012;80:341–350.
40. Hastie TJ, Tibshirani RJ, Friedman JJH. *The Elements of Statistical Learning*. New York: Springer; 2009.
41. Chen W, Xie X, Wang J, et al. A comparative study of logistic model tree, random forest, and classification and regression tree models for spatial prediction of landslide susceptibility. *Catena*. 2017;151:147–160.
42. Golkarian A, Naghibi SA, Kalantar B, Pradhan. Groundwater potential mapping using C5.0, random forest, and multivariate adaptive regression spline models in GIS. *Environ Monit Assess*. 2018;190:149. https://doi.org/10.1007/s10661-018-6507-8.
43. Oliveira S, Oehler F, San-Miguel-Ayanz J, Camia A, Pereira JMC. Modeling spatial patterns of fire occurrence in Mediterranean Europe using multiple regression and random Forest. *For Ecol Manage*. 2012;275:117–129.
44. Vorpahl P, Elsenbeer H, Märker M, Schröder B. How can statistical models help to determine driving factors of landslides? *Ecol Model*. 2012;239:27–39.
45. Wang Q, Nguyen TT, Huang JZ, Nguyen TT. An efficient random forests algorithm for high dimensional data classification. *Adv Data Anal Classif*. 2018;12:953–972.

43

Application of machine learning algorithms in hydrology

Hamidreza Mosaffa[a], Mojtaba Sadeghi[b], Iman Mallakpour[b],
Mojtaba Naghdyzadegan Jahromi[d], and Hamid Reza Pourghasemi[c]

[a]Department of Water Engineering, Shiraz University, Shiraz, Iran [b]Department of Civil and Environmental Engineering, University of California, Irvine, CA, United States [c]Department of Natural Resources and Environmental Engineering, College of Agriculture, Shiraz University, Shiraz, Iran [d]Department of Water Engineering, School of Agriculture, Shiraz University, Shiraz, Iran

1 Introduction

Hydrology is the science of Earth's water movement and properties within each step of the hydrological cycle. The hydrological model is considered as a simplified mathematical description of the hydrological cycle[1] and provides practical applications for understanding, predicting, and managing the occurrence, movement, and distribution of water on Earth.[2–4] Since the advent of hydrological models, many changes have taken place in modeling because of the use of limitless computing resources, effective data gathering, and remote sensing techniques, the rate of these changes has been growing.[5] Therefore, several hydrologic models that were not feasible in the past are possible, including two-or three-dimensional modeling, large-scale modeling, water quality modeling, and optimization techniques in water resource management activities. Despite rapid progress in hydrologic modeling, drawbacks remain. According to Beven,[6] hydrology requires macroscale theories for different catchment scales to address the problem that is affected by the heterogeneities and nonlinearities of hydrological processes. This issue is still a challenge in hydrology.[7] In addition, there are limitations to utilizing the physical-based hydrological model, mainly because of the challenges in fully understanding the complex, nonlinear, and interconnected hydrology.[8,9] Moreover, because of deficiencies in the accuracy, complexity, and computational cost, the use of a simple data-driven model for hydrological process modeling also has problems.[10]

The problems and limitations of physical and simple data-driven models promote other advanced models, such as machine learning (ML). ML is considered as part of artificial intelligence, and its algorithms find a pattern and build a model through empirical data to make a prediction. Because of the effective computation power of ML algorithms,[11] they are used in a vast variety of research and applications such as hydrology. Precipitation estimation and forecasting,[12,13] flood forecasting,[14] groundwater modeling,[15] and water quality monitoring[16] are some of the ML applications in hydrology. The Science Direct and Springer databases were used to investigate the number of studies in the subfield of hydrology with machine learning and hydrology keywords for the period up to 2020 (Fig. 1). A comparison of the prediction results of traditional hydrological models and ML showed that ML made a better prediction and could address hydrological model problems such as scale issues.[17] Moreover, ML methods improve modeling when their training covers a variety of data; otherwise, they need to be carefully used.[10]

ML algorithms are divided into three categories: (1) supervised learning, where there are input and output variables. This algorithm is used to estimate the mapping function between the input and output based on sample data, known as training data. The training process stops when the desired degree of accuracy of the training data is reached by the model. The aim is to use this function to predict the new input data. Supervised learning algorithms can be grouped into classification and regression which some of their examples are as follows: Linear Regression (LR), Support Vector Machine

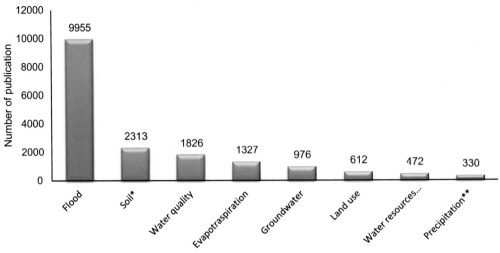

FIG. 1   Hydrology subfields research papers up to 2020.

(SVM), Random Forest (RF), Logistic Regression (LR), and Gradient Boosting (GB). (2) Unsupervised learning involves only the input data. The goal of the unsupervised learning algorithm is to find the hidden pattern of data based on their structures and similar features. Clustering and dimensionality reduction techniques are two groups of unsupervised learning algorithms. (3) The reinforcement learning algorithm allows an agent to learn by trial and error in a dynamic environment to receive the goal. This algorithm improves based on reward and punishment feedback, whereas data are training.[18] Deep learning (DL) is a part of ML-based on artificial neural networks that use multiple layers to learn. Its learning can be supervised, semi-supervised, or unsupervised. Different architectures of DL, including recurrent neural networks (RNNs), long short-term memory (LSTM), and convolutional neural networks (CNNs) have been applied in the field of hydrology.

This book chapter presents the applications of ML and DL in the subfields of hydrology. Sections 2–5 provide an overview of the flood, precipitation estimation, water quality, and groundwater, respectively. Finally, the conclusions are summarized in Section 6.

2   Flood

Floods are a widespread natural hazard that has a devastating socioeconomic toll worldwide. Flood prediction modeling and monitoring are powerful tools for mitigating flood damage. The ongoing development of ML methods over the past decades, along with the availability of long-term hydrological parameters such as precipitation and streamflow, has shown a high level of ability to predict floods.[10,19,20] Indeed, comparing the performance of physical and traditional statistical models with ML prediction models to flood prediction shows that the ML models have greater accuracy.[21] In this section, an overview of some common ML algorithms, including LSTM, RF, SVM, and CNN, for flood prediction and monitoring are presented.

A considerable amount of literature has been published on the use of ML algorithms to predict flood and streamflow magnitudes at different temporal scales, including hourly,[22–24] daily,[14,25,26] and monthly.[27–29] For instance, an LSTM model was chosen to predict the daily streamflow over 531 basins in the United States. They compared their results with the output of the Sacramento Soil Moisture Accounting (SAC-SMA) physical model and found that LSTM improved prediction.[19] This is consistent with what has been the findings of another study.[20] Furthermore, to improve the performance of the ML modeling procedure, some researchers have used a hybrid method that combines ML algorithms.[10,30] Ni et al.[31] used a coupled LSTM with CNN and wavelet-LSTM to forecast monthly streamflow. Their results demonstrated that both coupled models improved the accuracy of a single LSTM. In addition, the hybridization of the ML method with the physical model can improve streamflow prediction performance. Yang et al.[32] used the hybrid physical-ML (LSTM) method to predict daily discharge. They used the daily discharge output of the physical model as an input to the LSTM model. The results are better when a single LSTM is used.

In the literature, there are examples of applying ML and DL to detect and forecast flood water levels. Because the data being used is an image, most of these studies applied the CNN method. Moy de Vitry et al.[33] applied a deep convolutional neural network (DCNN) model to identify floods from surveillance camera systems in urban areas. Their results showed that flood water was well captured, with an 85% correlation. Wang et al.[34] forecasted the water level for a lead time of 1 to 6 h using a dilated causal convolutional neural network (DCCNN). They compared their model with multilayer perceptron (MLP) and SVM models, and their results showed the priority of DCCNN.

Flood inundation and susceptibility maps are necessary for mitigating, planning, and managing flood incidents.[35] ML methods with higher performance than physical and numerical methods have been used to map flood susceptibility.[10] El-Haddad et al.[36] used four different ML methods, including the functional data analysis (FDA), boosted regression tree (BRT), multivariate discriminant analysis (MDA), and general linear model (GLM) to produce flood-prone zones in Egypt, where the FDA shows better performance than other methods. Moghaddam et al.[37] produced a flood susceptibility map for the Tashan watershed in southwestern Iran using six different advanced ML methods. The classification and regression tree (CART) has the best skill among RF, generalized additive model (GAM), multivariate adaptive regression splines (MARS), and maximum entropy (ME). Vafakhah et al.[38] and Pourghasemi et al.[39] applied three (FR, adaptive network-based fuzzy inference system (ANFIS, and RF) and two (BRT and GLM) algorithms to generate flood susceptibility maps, respectively. Rahmati and Pourghasemi[40] used a bivariate statistical analysis model, the evidential belief function (EBF), and two ML algorithms, RF and BRT, to produce flood maps. Their results showed that EBFs exhibited better performance. In addition, other methods such as ANN and ANIFIS have been used in different studies to map flood-prone zones.[41,42]

A few studies have used optimization algorithms to improve the performance of ML techniques. Termeh et al.[42] used an ensemble of different optimization methods, such as genetic algorithm, particle swarm optimization (PSO), and ant colony optimization with ANFIS. In this study, an ensemble of ANFIS-PSO was the best model. Yuan et al.[43] forecasted monthly runoff using LSTM and the ant lion model (ALO), which is an optimization algorithm. Their results demonstrated that LSTM-ALO performed better than LSTM.

3 Precipitation estimation (case study: PERSIANN family developments)

Precipitation plays an important role as the most fundamental input of the water cycle in hydro-meteorological applications and climate science.[44] Precise and reliable precipitation estimates are crucial for hydrological applications, including flash flood monitoring and drought analysis.[45–47] During the last four decades, rapid advancements in remote sensing technologies, as well as machine learning techniques, have brought unprecedented opportunities for homogenous precipitation estimation at high spatiotemporal resolutions worldwide.[13,48–50] In 1997, one of the first attempts to investigate the application of machine learning in the field of precipitation estimation was published by Hsu et al.[51] They developed the Precipitation Estimation from Remotely Sensed Information using Artificial Neural Networks (PERSIANN), which offers rainfall estimates with 0.25 degree spatial and hourly temporal resolution since March 2000 for the latitude band 60°N–60°S. The PERSIANN architecture utilizes an artificial neural network (ANN) model based on a multilayer neural feed-forward network. This algorithm estimates the precipitation in two steps: First, an automatic clustering process, known as a self-organizing feature map (SOFM), is employed to transfer the IR information into the hidden layer. Second, discrete SOFM clusters in the hidden layer were used to estimate the rainfall rate. The parameters in both the input-hidden and hidden-output transformations were optimized by incorporating passive microwave (PMW) rainfall.[51]

The PERSIANN-Cloud Classification System (PERSIANN-CCS) algorithm was introduced by Hong et al.[52] to improve the lag time and spatiotemporal resolution of the PERSIANN algorithm. This machine-learning algorithm extracts predefined features of IR information under certain values. In the first step, the IR cloud information is separated into different patches by considering the temperature threshold. In the second step, useful cloud features are extracted and categorized using the SOFM method. Finally, the relationship between rain rate and cloud temperature was determined through historical matching. PERSIANN-CCS provides precipitation with 0.04 degree spatial and hourly temporal resolution with a lag time of approximately one hour.

During the last decade, the development of deep neural network (DNN) algorithms has provided exciting opportunities to further improve traditional precipitation estimation models. Among the different machine learning models, DNN models are most recognized for their flexibility and capacity to extract important features from a large amount of image data.[53] Both PERSIANN and PERSIANN-CCS extract useful features from inputs utilizing manually defined features, including geometry, texture, and coldness. Manual feature extraction limits the capability of rainfall retrieval models because of the complexity and nonlinear behavior of the precipitation phenomenon. One of the main advantages of DNNs is that they can extract useful features in the input automatically.

Tao et al.[54] conducted the first study to investigate the application of DNNs for the estimation of precipitation. Tao et al. developed the PERSIANN-stacked denoising autoencoder (PERSIANN-SDAE) algorithm, which utilizes IR and WV information in a fully connected DNN model, to estimate the precipitation. The SDAE algorithm, introduced by

Vincent et al.,[55] is an unsupervised pretraining algorithm for extracting important features from input data. PERSIANN-SDAE uses a fully connected neural network that utilizes greedy layer-wise pretraining based on stacked denoising autoencoders using IR and WV information. Their results suggested that PERSIANN-SDAE, as a DNN model, can show higher performance in capturing the spatial pattern as well as the amount of rainfall compared to the PERSIANN-CCS dataset. Although PERSIANN-SDAE improves the PERSIANN-CCS estimates by automatically extracting features from inputs, this model cannot efficiently leverage neighborhood information for precipitation estimation. In other words, because of the large number of parameters and inefficient structure of PERSIANN-SDAE, this model tends to estimate the rain rate of each pixel by using information from the corresponding pixel of the input datasets (pixel-to-pixel relationship instead of using neighborhood information). Nevertheless, in addition to the pixel-to-pixel relation of inputs and amount of precipitation, local spatial variations of inputs (IR information) offer useful information for accurate precipitation retrieval. This drawback in the PERSIANN-SDAE algorithm was improved by Sadeghi et al.,[56] who utilized a CNN algorithm for precipitation estimation.

CNN architectures are recognized as one of the most popular and efficient types of DNN frameworks. One of the important features of CNNs is weight sharing, which is achieved by sharing the same kernels (learned parameters) across the images. Weight sharing reduces the number of parameters and enables CNNs to extract local features efficiently. Sadeghi et al.[56] investigated the application of CNN for improving near-real-time precipitation by proposing a model referred to as PERSIANN-CNN. PERSIANN-CNN provides near-real-time precipitation with hourly temporal and 0.04 degree spatial resolutions using IR and WV information. Their study highlights that PERSIANN-CNN offers more accurate precipitation estimates than PERSIANN-SDAE at various spatial and temporal resolutions. Sadeghi et al.[57] further improved PERSIANN-CNN by including geographical information to IR information and using a U-Net-based convolutional neural network. Recent studies investigating the application of DNNs for precipitation estimation suggest that applying a DNN architecture to geographical and IR information can provide an opportunity to improve near-real-time precipitation estimates.

4 Water quality

The physical, biological, chemical and microbiological properties of the water are known as water quality. Monitoring water quality helps to manage water resources and make decisions for the future. Several ML methods have been applied to monitor and predict the quality of water.[16,58] Wang et al.[59] used a support vector regression (SVR) model to predict the water quality index (WQI). In addition, the PSO algorithm helps SVR to optimize parameter values. Bui et al.[60] used 12 hybrid ML models and four single models, including random tree (RT), RF, reduced error pruning tree (REPT), and M5P to evaluate the water quality of the river. The results show that hybrid bagging (BA)-RT has the best performance among the single and hybrid models.

Few studies have been published on the application of ML to estimate chlorophyll a concentrations. Yu et al.[61] applied a combination of wavelet mean fusion (WMF), LSTM, and wavelet domain threshold denoising (WDTD) to predict chlorophyll-a concentration. The results show that the model has good predictive output. Liang et al.[62] used LSTM to forecast the concentration of chlorophyll at 1–31 days of lead time. Lee and Lee[63] applied three deep learning models (MLP, RNN, and LSTM) to predict chlorophyll-a in four major rivers in South Korea. Two standalone LSTM and CNN models, and their hybrid, were applied by Barzegar et al.[64] to predict chlorophyll-a and dissolved oxygen (DO), and similar to other research, the hybrid model outperformed the single models.

5 Groundwater

Identifying the groundwater level (GWL) is necessary for sustainable management and planning, especially for irrigation scheduling and water supply. In recent years, ML has helped to address groundwater complexity modeling.[9] Rahman et al.[65] coupled two ML methods, including extreme GB and RF with wavelet transforms, to forecast GWL forecasts for one to 1–3 months ahead of southern Japan. Panahi et al.[15] used SVR and CNN methods to map groundwater potential. Their results showed that CNN performs better than SVR. Al-Fugara et al.[66] used five different ML methods, RF, BRT, SVM, MDA, and MARS to map groundwater potential over the Wadi az-Zarqa watershed in Jordan. Their evaluation showed that MDA had the highest performance among the five ML methods. Prasad et al.[67] developed an LM model including RF, BRT, and an ensemble of RF and BRT to predict groundwater potential maps on the west coast of India. Contrary to the findings of previous ML models, they showed that RF has a better performance hybrid model.

Groundwater quality modeling using ML methods has grown rapidly in recent years. Bedi et al.[68] applied three ML models, including ANN, SVM, and XGB, to predict contamination levels over 12 states in the USA. Mo et al.[69] applied a deep autoregressive neural network to the identification of a groundwater contaminant source and achieved better performance than physical models. To predict the groundwater nitrate (NO_3) concentrations, Rahmati et al.[70] developed a model by using RF, SVM, and KNN. Their results demonstrated that the RF model performed better than the others. In addition, Sajedi-Hosseini et al.[71] used BRT, MDA, and SVM and their ensemble methods to evaluate groundwater nitrate concentrations. In addition, other authors, such as Rokhshad et al.[72] and Knoll et al.[73] used these methods to predict nitrate in groundwater nitrate.

6 ML issue and recommendation

ML will rapidly be an important application technique in hydrology, while there are challenges in this way, including (1) the hydrology process has different spatial and temporal scales, and it is difficult to integrate these processes; (2) although ML requires high quality and sufficient data to train, uncertainty, and missing data in hydrology cause problems of overfitting and underfitting.[74]

Developing ML in hydrology has some limitations; therefore, further work needs to be performed to overcome these limitations. Hydrological modeling, especially predictions of processes such as floods, requires real-time and reliable data. However, hydrological data are generated in dispersed organizations, and even in some cases, the recording time is not the same.[75] Therefore, in the future, it will be necessary to try to cooperate with different organizations to develop an up-to-date database. Moreover, the future of hydrology is in the use of ML and physical models, and for this, tools need to be developed for hydrologists to easily use ML techniques.[17] However, at present, no significant investment has been made in the use of ML in hydrology.

7 Conclusion

This chapter presents an overview application of ML and DL in the subfield of hydrology, including flood, precipitation estimation, water quality, and groundwater. In each subfield, some recently published papers that use ML have been presented. In summary, this chapter highlights the importance of applying ML to different parts of hydrology. Our review shows the high performance of ML compared to physical and simple data-driven models and addresses some common hydrological model problems, including complexity and scale issues. For instance, there is a challenge in short-term flood forecasting when physical and simple data-driven models are used, while ML can predict floods on an hourly time scale with a high level of accuracy. According to our review, using an optimization algorithm and hybridization of two or more ML methods can improve the quality of ML prediction, while few studies have shown that a single ML method performs better than the hybrid method. Therefore, the application and improvement of ML methods depend highly on their correct use. Finally, the application of ML in this area is still ongoing, and future research is needed to focus on applying state-of-the-art ML methods.

References

1. Sorooshian S, Hsu K-L, Coppola E, Tomassetti B, Verdecchia M, Visconti G, eds. *Hydrological Modelling and the Water Cycle: Coupling the Atmospheric and Hydrological Models.* Springer; 2010.
2. Devia GK, Ganasri BP, Dwarakish GS. A review on hydrological models. *Aquat Procedia.* 2015;4:1001–1007.
3. Moradkhani H, Sorooshian S. General review of rainfall-runoff modeling: model calibration, data assimilation, and uncertainty analysis. In: *Water Science and Technology Library.* Berlin, Heidelberg: Springer; 2008:1–24.
4. Raghavendra NS, Deka PC. Support vector machine applications in the field of hydrology: a review. *Appl Soft Comput.* 2014;19:372–386.
5. Singh VP. Hydrologic modeling: progress and future directions. *Geosci Lett.* 2018;5(1):15.
6. Beven K. *Towards a New Paradigm in Hydrology;* 1987.
7. Blöschl G, Bierkens MF, Chambel A, et al. Twenty-three unsolved problems in hydrology (UPH) – a community perspective. *Hydrol Sci J.* 2019;64 (10):1141–1158.
8. Kim B, Sanders BF, Famiglietti JS, Guinot V. Urban flood modeling with porous shallow-water equations: a case study of model errors in the presence of anisotropic porosity. *J Hydrol (Amst).* 2015;523:680–692.
9. Sahoo S, Russo TA, Elliott J, Foster I. Machine learning algorithms for modeling groundwater level changes in agricultural regions of the US. *Water Resour Res.* 2017;53(5):3878–3895.
10. Mosavi A, Ozturk P, Chau K-W. Flood prediction using machine learning models: literature review. *Water (Basel).* 2018;10(11):1536.

11. Mewes B, Oppel H, Marx V, Hartmann A. Information-based machine learning for tracer signature prediction in karstic environments. *Water Resour Res.* 2020;56(2). https://doi.org/10.1029/2018WR024558.

12. Asanjan AA, Yang T, Hsu K, Sorooshian S, Lin J, Peng Q. Short-term precipitation forecast based on the PERSIANN system and LSTM recurrent neural networks. *J Geophys Res Atmos.* 2018;123(22):12–543.

13. Sadeghi M, Asanjan AA, Faridzad M, et al. PERSIANN-CNN: precipitation estimation from remotely sensed information using artificial neural networks–convolutional neural networks. *J Hydrometeorol.* 2019;20(12):2273–2289.

14. Zhu S, Luo X, Yuan X, Xu Z. An improved long short-term memory network for streamflow forecasting in the upper Yangtze River. *Stoch Environ Res Risk Assess.* 2020;34(9):1313–1329.

15. Panahi M, Sadhasivam N, Pourghasemi HR, Rezaie F, Lee S. Spatial prediction of groundwater potential mapping based on convolutional neural network (CNN) and support vector regression (SVR). *J Hydrol (Amst).* 2020;588(125033):125033.

16. Zou Q, Xiong Q, Li Q, Yi H, Yu Y, Wu C. A water quality prediction method based on the multi-time scale bidirectional long short-term memory network. *Environ Sci Pollut Res.* 2020;27(2):1–12.

17. Nearing GS, Kratzert F, Sampson AK, et al. What role does hydrological science play in the age of machine learning? *Water Resour Res.* 2021;57(3), e2020WR028091.

18. Bishop CM. *Pattern Recognition and Machine Learning*. New York: Springer; 2006.

19. Kratzert F, Klotz D, Herrnegger M, Sampson AK, Hochreiter S, Nearing GS. Toward improved predictions in ungauged basins: exploiting the power of machine learning. *Water Resour Res.* 2019;55(12):11344–11354.

20. Kratzert F, Klotz D, Shalev G, Klambauer G, Hochreiter S, Nearing G. Towards learning universal, regional, and local hydrological behaviors via machine learning applied to large-sample datasets. *Hydrol Earth Syst Sci.* 2019;23(12):5089–5110.

21. Abbot J, Marohasy J. Input selection and optimisation for monthly rainfall forecasting in Queensland, Australia, using artificial neural networks. *Atmos Res.* 2014;138:166–178.

22. Kourgialas NN, Dokou Z, Karatzas GP. Statistical analysis and ANN modeling for predicting hydrological extremes under climate change scenarios: the example of a small Mediterranean agro-watershed. *J Environ Manage.* 2015;154:86–101.

23. Lohani AK, Goel NK, Bhatia KKS. Improving real time flood forecasting using fuzzy inference system. *J Hydrol (Amst).* 2014;509:25–41.

24. Yu P-S, Yang T-C, Chen S-Y, Kuo C-M, Tseng H-W. Comparison of random forests and support vector machine for real-time radar-derived rainfall forecasting. *J Hydrol (Amst).* 2017;552:92–104.

25. Aichouri I, Hani A, Bougherira N, Djabri L, Chaffai H, Lallahem S. River flow model using artificial neural networks. *Energy Procedia.* 2015;74:1007–1014.

26. Kratzert F, Klotz D, Brenner C, Schulz K, Herrnegger M. Rainfall–runoff modelling using long short-term memory (LSTM) networks. *Hydrol Earth Syst Sci.* 2018;22(11):6005–6022.

27. Lin J-Y, Cheng C-T, Chau K-W. Using support vector machines for long-term discharge prediction. *Hydrol Sci J.* 2006;51(4):599–612.

28. Rezaeian-Zadeh M, Tabari H, Abghari H. Prediction of monthly discharge volume by different artificial neural network algorithms in semi-arid regions. *Arab J Geosci.* 2013;6(7):2529–2537.

29. Sridharam S, Sahoo A, Samantaray S, Ghose DK. *Assessment of Flow Discharge in a River Basin through CFBPNN, LRNN and CANFIS*. Springer; 2021:765–773.

30. Al-Juboori AM. A hybrid model to predict monthly streamflow using neighboring rivers annual flows. *Water Resour Manage.* 2021;35(2):729–743.

31. Ni L, Wang D, Singh VP, et al. Streamflow and rainfall forecasting by two long short-term memory-based models. *J Hydrol (Amst).* 2020;583 (124296):124296.

32. Yang T, Sun F, Gentine P, et al. Evaluation and machine learning improvement of global hydrological model-based flood simulations. *Environ Res Lett.* 2019;14(11):114027.

33. Vitry M, Kramer S, Wegner JD, Leitão JP. Scalable flood level trend monitoring with surveillance cameras using a deep convolutional neural network. *Hydrol Earth Syst Sci.* 2019;23(11):4621–4634.

34. Wang J-H, Lin G-F, Chang M-J, Huang I-H, Chen Y-R. Real-time water-level forecasting using dilated causal convolutional neural networks. *Water Resour Manage.* 2019;33(11):3759–3780.

35. Yousefi S, Pourghasemi HR, Emami SN, et al. Assessing the susceptibility of schools to flood events in Iran. *Sci Rep.* 2020;10(1):1–15.

36. El-Haddad BA, Youssef AM, Pourghasemi HR, Pradhan B, El-Shater A-H, El-Khashab MH. Flood susceptibility prediction using four machine learning techniques and comparison of their performance at Wadi Qena Basin, Egypt. *Nat Hazards (Dordr).* 2021;105(1):83–114.

37. Moghaddam DD, Pourghasemi HR, Rahmati O. Assessment of the contribution of geo-environmental factors to flood inundation in a semi-arid region of SW Iran: comparison of different advanced modeling approaches. In: *Natural Hazards GIS-Based Spatial Modeling Using Data Mining Techniques. Advances in Natural and Technological Hazards Research*. Springer; 2019:59–78.

38. Vafakhah M, Mohammad Hasani Loor S, Pourghasemi HR, Katebikord A. Correction to: Comparing performance of random forest and adaptive neuro-fuzzy inference system data mining models for flood susceptibility mapping. *Arab J Geosci.* 2020;13(14). https://doi.org/10.1007/s12517-020-05637-8.

39. Pourghasemi HR, Amiri M, Edalat M, et al. Assessment of urban infrastructures exposed to flood using flood susceptibility map and Google earth engine. *IEEE J Sel Top Appl Earth Observ Remote Sens.* 2020;14:1923–1937.

40. Rahmati O, Pourghasemi HR. Identification of critical flood prone areas in data-scarce and ungauged regions: a comparison of three data mining models. *Water Resour Manage.* 2017;31(5):1473–1487.

41. Falah F, Rahmati O, Rostami M, Ahmadisharaf E, Daliakopoulos IN, Pourghasemi HR. *Artificial Neural Networks for Flood Susceptibility Mapping in Data-Scarce Urban Areas*. Elsevier; 2019:323–336.

42. Termeh SVR, Kornejady A, Pourghasemi HR, Keesstra S. Flood susceptibility mapping using novel ensembles of adaptive neuro fuzzy inference system and metaheuristic algorithms. *Sci Total Environ.* 2018;615:438–451.

43. Yuan X, Chen C, Lei X, Yuan Y, Adnan RM. Monthly runoff forecasting based on LSTM–ALO model. *Stoch Environ Res Risk Assess.* 2018;32 (8):2199–2212.

44. Flato G, Jochem Marotzke BA, Braconnot P, et al. Climate change 2013: the physical science basis. In: *Contribution of Working Group I to the Fifth Assessment Report of the Intergovernmental Panel on Climate Change*. Cambridge University Press; 2013.

45. Beck HE, van Dijk AIJM, Levizzani V, et al. MSWEP: 3-hourly 0.25° global gridded precipitation (1979–2015) by merging gauge, satellite, and reanalysis data. *Hydrol Earth Syst Sci.* 2017;21(1):589–615.

46. Mosaffa H, Sadeghi M, Hayatbini N, et al. Spatiotemporal variations of precipitation over Iran using the high-resolution and nearly four decades satellite-based PERSIANN-CDR dataset. *Remote Sens (Basel).* 2020;12(10):1584.

47. Yilmaz KK, Hogue TS, Hsu K-L, Sorooshian S, Gupta HV, Wagener T. Intercomparison of rain gauge, radar, and satellite-based precipitation estimates with emphasis on hydrologic forecasting. *J Hydrometeorol.* 2005;6(4):497–517.

48. He Y, Zhang Y, Kuligowski R, Cifelli R, Kitzmiller D. Incorporating satellite precipitation estimates into a radar-gauge multi-sensor precipitation estimation algorithm. *Remote Sens (Basel).* 2018;10(2):106.

49. Kidd C, Levizzani V. Status of satellite precipitation retrievals. *Hydrol Earth Syst Sci.* 2011;15(4):1109–1116.

50. Sorooshian S, AghaKouchak A, Arkin P, et al. Advanced concepts on remote sensing of precipitation at multiple scales. *Bull Am Meteorol Soc.* 2011;92(10):1353–1357.

51. Hsu K-L, Gao X, Sorooshian S, Gupta HV. Precipitation estimation from remotely sensed information using artificial neural networks. *J Appl Meteorol.* 1997;36(9):1176–1190.

52. Hong Y, Gochis D, Cheng J-T, Hsu K-L, Sorooshian S. Evaluation of PERSIANN-CCS rainfall measurement using the NAME event rain gauge network. *J Hydrometeorol.* 2007;8(3):469–482.

53. Rumelhart DE, Hinton GE, Williams RJ. Learning representations by back-propagating errors. *Nature.* 1986;323(6088):533–536.

54. Tao Y, Hsu K, Ihler A, Gao X, Sorooshian S. A two-stage deep neural network framework for precipitation estimation from bispectral satellite information. *J Hydrometeorol.* 2018;19(2):393–408.

55. Vincent P, Larochelle H, Lajoie I, Bengio Y, Manzagol PA, Bottou L. Stacked denoising autoencoders: learning useful representations in a deep network with a local denoising criterion. *J Mach Learn Res.* 2010;11(12):3371–3408.

56. Sadeghi M, Akbari Asanjan A, Faridzad M, et al. Evaluation of PERSIANN-CDR constructed using GPCP V2.2 and V2.3 and A comparison with TRMM 3B42 V7 and CPC unified gauge-based analysis in global scale. *Remote Sens (Basel).* 2019;11(23):2755.

57. Sadeghi M, Nguyen P, Hsu K, Sorooshian S. Improving near real-time precipitation estimation using a U-Net convolutional neural network and geographical information. *Environ Model Software.* 2020;134(104856):104856.

58. Wang P, Yao J, Wang G, et al. Exploring the application of artificial intelligence technology for identification of water pollution characteristics and tracing the source of water quality pollutants. *Sci Total Environ.* 2019;693(133440):133440.

59. Wang X, Zhang F, Ding J. Evaluation of water quality based on a machine learning algorithm and water quality index for the Ebinur Lake watershed, China. *Sci Rep.* 2017;7(1):1–18.

60. Bui DT, Khosravi K, Tiefenbacher J, Nguyen H, Kazakis N. Improving prediction of water quality indices using novel hybrid machine-learning algorithms. *Sci Total Environ.* 2020;721(137612):137612.

61. Yu Z, Yang K, Luo Y, Shang C. Spatial-temporal process simulation and prediction of chlorophyll-a concentration in Dianchi Lake based on wavelet analysis and long-short term memory network. *J Hydrol (Amst).* 2020;582(124488):124488.

62. Liang Z, Zou R, Chen X, Ren T, Su H, Liu Y. Simulate the forecast capacity of a complicated water quality model using the long short-term memory approach. *J Hydrol (Amst).* 2020;581(124432):124432.

63. Lee S, Lee D. Improved prediction of harmful algal blooms in four major South Korea's rivers using deep learning models. *Int J Environ Res Public Health.* 2018;15(7):1322.

64. Barzegar R, Aalami MT, Adamowski J. Short-term water quality variable prediction using a hybrid CNN–LSTM deep learning model. *Stoch Environ Res Risk Assess.* 2020;34(2):415–433.

65. Rahman ATMS, Hosono T, Quilty JM, Das J, Basak A. Multiscale groundwater level forecasting: coupling new machine learning approaches with wavelet transforms. *Adv Water Resour.* 2020;141(103595):103595.

66. Al-Fugara A, Pourghasemi HR, Al-Shabeeb AR, et al. A comparison of machine learning models for the mapping of groundwater spring potential. *Environ Earth Sci.* 2020;79(10):206. https://doi.org/10.1007/s12665-020-08944-1.

67. Prasad P, Loveson VJ, Kotha M, Yadav R. Application of machine learning techniques in groundwater potential mapping along the west coast of India. *GIsci Remote Sens.* 2020;57(6):735–752.

68. Bedi S, Samal A, Ray C, Snow D. Comparative evaluation of machine learning models for groundwater quality assessment. *Environ Monit Assess.* 2020;192(12):776.

69. Mo S, Zabaras N, Shi X, Wu J. Deep autoregressive neural networks for high-dimensional inverse problems in groundwater contaminant source identification. *Water Resour Res.* 2019;55(5):3856–3881.

70. Rahmati O, Choubin B, Fathabadi A, et al. Predicting uncertainty of machine learning models for modelling nitrate pollution of groundwater using quantile regression and UNEEC methods. *Sci Total Environ.* 2019;688:855–866.

71. Sajedi-Hosseini F, Malekian A, Choubin B, et al. A novel machine learning-based approach for the risk assessment of nitrate groundwater contamination. *Sci Total Environ.* 2018;644:954–962.

72. Rokhshad AM, Khashei Siuki A, Yaghoobzadeh M. Evaluation of a machine-based learning method to estimate the rate of nitrate penetration and groundwater contamination. *Arab J Geosci.* 2021;14(1). https://doi.org/10.1007/s12517-020-06257-y.

73. Knoll L, Breuer L, Bach M. Large scale prediction of groundwater nitrate concentrations from spatial data using machine learning. *Sci Total Environ.* 2019;668:1317–1327.

74. Zaidi SMA, Chandola V, Allen MR, et al. Machine learning for energy-water nexus: challenges and opportunities. *Big Earth Data.* 2018;2(3):228–267.

75. Sit M, Demiray BZ, Xiang Z, Ewing GJ, Sermet Y, Demir I. A comprehensive review of deep learning applications in hydrology and water resources. *Water Sci Technol.* 2020;82(12):2635–2670.

44

Digital soil mapping of soil bulk density in loess derived-soils with complex topography

Narges Kariminejad[a], Mohsen Hosseinalizadeh[a], and Hamid Reza Pourghasemi[b]

[a]Department of Arid Zone Management, Gorgan University of Agricultural Sciences and Natural Resources, Gorgan, Iran
[b]Department of Natural Resources and Environmental Engineering, College of Agriculture, Shiraz University, Shiraz, Iran

1 Introduction

Sustainable management of natural resources requires reliable information about the spatial distribution of soil chemical and physical properties related to eco-systemic functions and geomorphological processes. It is of fundamental importance to model the relationship between environmental variables and soil properties because of the effect of environmental attributes effect on soil fertility and soil formation/loss. The physical and chemical soil properties and their spatial distribution at different scales with a high sampling density are needed to obtain precise estimates, mainly because of their high spatio-temporal variability.[1] Due to topographic complexity and the lack of suitable access paths, soil sampling in the Iranian Loess Plateau (ILP) is quite difficult.[2] Therefore, a large number of field observations and laboratory analyses of soil samples should be conducted to obtain an accurate data set.[3,4]

To predict and evaluate the spatial distribution of different soil properties in various regions, it is vital to predicting the interaction between soil properties (BD with aggregate stability and organic matter) and environmental covariates.[5,6] BD is a key indicator of soil compaction and health. It affects rooting depth/restrictions, water content/infiltration, soil porosity, plant nutrient availability, soil microorganism activity, and biodiversity, which influence key soil processes and productivity. Several studies have argued that BD is one of the essential factors that affect the spatial distribution of soil properties (aggregate stability and organic matter), soil development, and vegetation communities[2,7] and gully head development.[8] It is also expected that the use of topographic factors and digital elevation models (DEMs) will help predict soil properties[9] in regions with complex topographic variations. Moreover, the data-driven by UAVs and derivative indices can provide efficient information on plant communities with various spatiotemporal resolutions.[10,11] Recently, digital soil mapping (DSM), as a soil spatial estimation method, has been commonly applied by soil scientists, underscoring available derivative data to map soil attributes.[5,6,12–16] Furthermore, it is feasible to use soil properties, UAV images, and DSM methods to qualify the spatial variability of soil chemical and physical properties in the studied area.

Although the use of machine learning (ML) techniques for DSM has been growing in recent years, few studies have been conducted to identify areas capable of low and high values of soil properties at the regional scale. Numerous reports applied different models and methods to predict the variety of soil properties at a given location through DSM framework including regression tree (RT),[6,17] K-nearest neighbors (KNN),[18–20] multiple linear regression (MLR),[6,18,21] random forest (RF),[6,15,18,22] and support vector regression (SVR).[14,15,18,23] Therefore, less effort has been devoted to the land and environmental variables related to soil properties in DSM techniques. In addition, DSM methods were frequently surrounded to the upper 30 cm of soils to map soil properties.[24]

The loess deposits in Golestan Province provide valuable information on soil properties and are extremely prone to piping and gully erosion.[10,25] Furthermore, the identification of soil properties is vital for the health of this study area and is linked to eco-geomorphological issues. Some physical soil properties, including BD, are important for evaluating soil quality and ameliorating soil management systems.[3,26] They have a fundamental impact on the growth of vegetation and soil fertility.[4] This means that BD has a principal role in soil quality and fertility, particularly in agricultural lands and it can be changed by management practices. It had a direct or indirect effect on the environmental covariates, which is exceedingly consistent with the knowledge of soil scientists who are familiar with the study area. Moreover, identification of the precise spatial distribution of BD can guide the prediction of soil degradation. Furthermore, preparing spatial soil prediction models of BD by applying UAV and DEM derivatives has a noteworthy role in sustainable land management. The methodology applied here is consigned to be innovative concerning the three aspects: (1) to map the distribution of BD in topsoil layer (0–20 cm) through developing two ML techniques (RF and SVR), (2) to compare the results of two applied techniques in evaluating BD, and (3) to explore the first controlling variables affecting the BD distribution in 0–20 cm of the soil. Overall, this proposed research at such a very high spatial resolution driven from UAV imagery could hopefully be applicable to produce soil property probability maps in other study regions.

2 Materials and methods

2.1 Description of the study area

The study area is a complex landscape and hilly region, where most of the landscapes are covered by valleys and hills.[2] These eolian sediments are highly fertile, fragile, and sensitive to different types of erosion due to their specific soil properties and terrain attributes.[10] The study region covers an area of 2730 ha located in northeast Iran, Golestan Province (55°36 and 55°40 E longitude 37°37 and 37°40 N latitude (Figs. 1 and 2). The mean annual rainfall and temperature were 365 mm and 18°C, respectively.[8] The temperature regimes and soil moisture are thermic and dry xeric, respectively.[27] In addition, the area of interest is known as an important agricultural land use that has exposed intensive agricultural practices, particularly in flat regions and valley floors for wheat and sunflower production.

2.2 Affecting variable and soil sampling scheme

In this study, affecting covariates, including UAV images[8] and topographic attributes were applied in ArcGIS and SAGA-GIS software. We applied a DEM with a spatial resolution of 1×1 m.

A soil sampling scheme of 0–20 cm was carried out and a total of 273 soil samples were collected from the topsoil layer.[25] BD was analyzed using the clod method.[28]

2.3 Modeling approaches

Two non-linear ML algorithms, RF and SVR, were utilized for the digital mapping of BD.

RF is a fast and accurate technique as an ensemble of regression trees and classifications that are aggregated to produce the terminal prediction.[29] Each node of the decision tree is created by applying bootstrap sampling from the training data, which is approximately 2/3 of them. The number of considered trees and effective factors in each random subset used in the forest are determined as "ntree" and "mtry," respectively. Also, modeling of soil BD was performed applying the "randomForest" package[30] in RStudio v1.1.383 software (RStudio, 2009–2017) and R 4.0.3 software. The variable importance and all modeling processes were implemented using the "randomForest" package[30] in RStudio v1.1.383 software (RStudio, 2009–2017) and R 4.0.3. In this part, variable importance was also calculated to select variables that were more effective than the others in the prediction of soil properties.

SVR is applied as one of the best controlling learning algorithms for regression, which continues to the point that the lowest level error can arise in the grouping dataset.[14,15,23] The algorithm is based on mathematical learning theory and applies structural risk minimization to provide an optimal overall response.[31] It contains various types of regression functions for the calculation of errors and the extension of new information that requires the smallest value of model tuning. In this study, the SVR algorithm was considered in R 4.0.3 statistical software using "e1071," "caret," and "raster" packages.[32–34] where d is a polynomial degree, r is a measurable factor of the kernel functions, and γ is the kernel width.

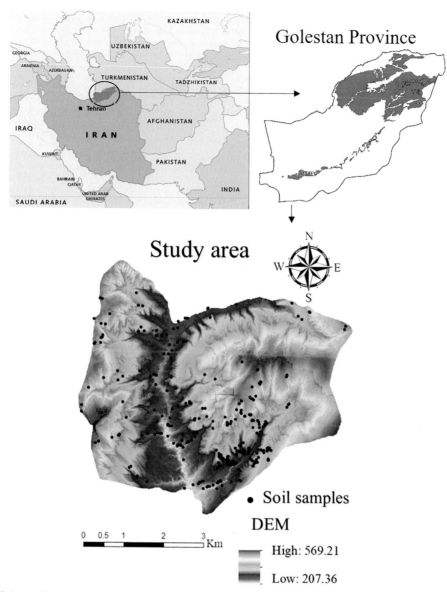

FIG. 1   (A) Maps of Golestan Province in Iran (B) loess areas (red polygon) in Golestan Province (C) the study area.

2.4 Validation soil properties maps

The main type of error index that is commonly applied in DSM was used here. It was the root mean square error or RMSE. The RMSE shows the accuracy of the algorithm predictions.[20] The coefficient of determination (R^2) and mean square error (MSE) were also used to evaluate the ability to provide accurate forecasts and model performance.[35]

3 Results and discussion

3.1 Controlling of variable importance analysis

According to the soil samples gathered from the field, the mean value of BD was 1.52, the maximum was 1.99, and the minimum was $1.01\,g/cm^3$. The average values of BD in the rangeland and agricultural land are 1.51 and $1.54\,g/cm^3$. The variable importance of the different contributing factors and their significant effects on the distribution

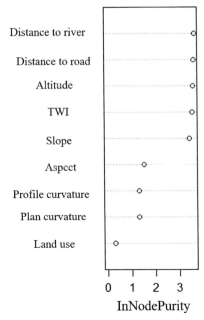

FIG. 2   The relative importance of environmental factors for BD using RF model.

of BD in the study area provided by RF, which is shown in Fig. 2. For the topsoil depth, distance to a river, distance to road, and altitude were the three most vital variables affecting BD probabilities according to their relative importance. This result confirmed that this is not surprising due to the construction activities and suggests that BD accumulation in the study area, which is one portion of ILP, is highly restricted by the hydrological processes in conjunction with human activists. This result is in complete accordance with the knowledge of soil experts who work in this region.[2,8,36] Based on the distribution of BD values in this area, the BD was higher near roads, rivers, and the rangelands (Fig. 3). This result could be better indicated by the fact that flat hillslopes with less or no vegetation cover were commonly expected in silty loam soils[7] that result from different types of soil erosion, especially gully head erosion at the research site.

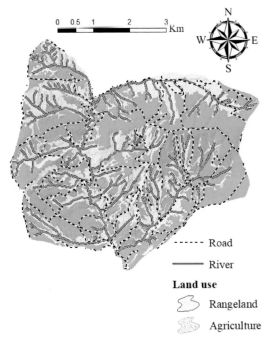

FIG. 3   The schematic map of three affecting factors of the study area.

These results come from differences in both plant (e.g., root system type, species) and soil characteristics (e.g., soil BD, soil texture). As far as we know, the higher vegetation cover and the percentage of soil organic matter, and the more developed soil structure can exceedingly decrease erosion rates by affecting on BD of soils. This sustainable condition is more visible in areas far from rivers and roads. Thus, we can explain that hydrological attributes especially distance to rivers had more essential in topsoils regarding BD values. In addition, the significant interaction between vegetation indices and soil properties was proved by soil scientists and reflected the strong role of land cover in the alteration of soil attributes, particularly BD, in this study. In addition to vegetation cover, some researchers have also reported the impact of topographic factors on the variability of soil properties.[25,37,38]

3.2 Probability maps of soil BD

In this study, we considered different driving forces that impact the variability of BD in the study area. The results of the RF and SVR models showed clear ranges of probability values for the variations in soil BD. Fig. 4 showed the probability maps of BD as one of the main factors of soil properties at the topsoil layer (0–20 cm) which obtained from the RF and SVR models. Based on Fig. 5, the four probability classes for the two maps were defined as "low, medium-high, and very high" Fig. 5 shows the area percentages of soil BD for each of the four classes. The probability maps of BD showed that the high class had the largest area in both applied models. The results of probability zoning based on the two models indicated that the low class had the lowest area in the studied region.

One of the important parts of this study was the investigation of the potential impacts of BD on the erosion-decreasing potential from the pedologic point of view. As stated by Vannoppen et al.,[39] soil bulk density (abbreviated as BD in this chapter), also known as dry bulk density, was essential to be considered, because an enhancement of soil BD reduced soil erosion rates and soil erodibility. This can be undoubtedly described by an enhancement of soil cohesion.[40] Erosion rates commonly reduce with enhanced soil cohesion. However, there is no way to say that the environmental changes are under the influence of a special soil property or one can lonely decrease or increase the rate of soil erosion, as we know that changes in landforms are the results of many factors which are affecting together.[8] For instance, soil texture and dry soil bulk density played a notable role, when they were considered together. It is because they were both related to soil cohesion.[39]

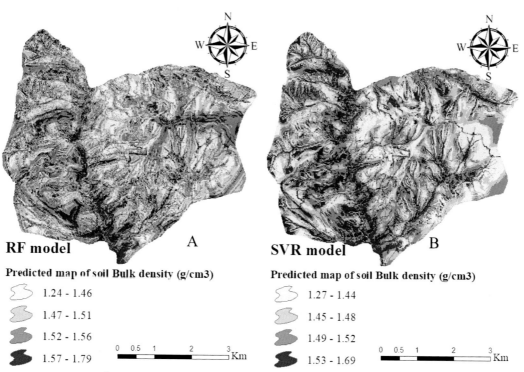

FIG. 4   Predicted maps of BD (g/cm^3) across the study area using RF (A) and SVR (B) models.

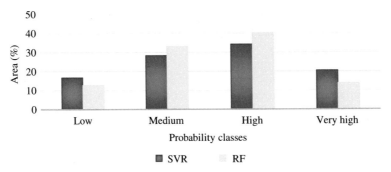

FIG. 5 Spatial distribution of BD using the RF and SVR models.

TABLE 1 Validation criteria for BD prediction.

	Models	MSE	RMSE	R^2
Training	RF	0.0063	0.08	0.9473
	SVR	0.0257	0.16	0.2113
Testing	RF	0.0306	0.17	−0.0011
	SVR	0.0306	0.17	−0.0029

3.3 Validation of the probability maps of soil BD

The results of the training dataset showed that the probability map of BD in the topsoil layer had the lowest RMSE and MSE, and the highest R^2 values for the RF model, as shown in Table 1. However, both models had acceptable RMSE and R^2 values, as well as the best capability to predict soil BD mapping in the topsoil layer. Furthermore, the results showed that the two methods considered for the surface variation of soil BD were very good. The applicability of using DSM methods for predicted soil properties at different soil depths was supported earlier in the research done by Zhang et al.[26] Consequently, by the accurate prediction of soil properties using DSM models, sustainable land planning and decision-making will be accessible to combat desertification.

4 Conclusions

The spatial prediction of BD was modeled in some parts of the ILP in Golestan Province through RF and SVR digital soil mapping algorithms. The results of the variable importance of the RF algorithm showed the highest importance of distance to the river as one of the commonly used hydrologic factors. The RF and SVR algorithms carefully predicted soil BD and showed a non-linear relationship between the probability of soil BD and its affective factors. The model's evaluation indicated the good performance of RMSE, MSE, and R^2. Furthermore, the methodology used represented accurate results for the surface soil layer. However, more research is needed to analyze other types of soil properties as input datasets in DSM in areas with complex topography.

References

1. Bellamy PH, Loveland PJ, Bradley RI, Lark RM, Kirk GJD. Carbon losses from all soils across England and Wales 1978-2003. *Nature*. 2011;437:245–248.
2. Maleki S, Khormali F, Bagheri Bodaghabadi M, Mohammadi J, Hoffmeister D, Kehl M. Role of geomorphic surface on the above-ground biomass and soil organic carbon storage in a semi-arid region of Iranian loess plateau. *Quat Int*. 2018.
3. Khaledian Y, Brevik EC, Pereira P, Cerdà A, Fattah MA, Tazikeh H. Modeling soil cation exchange capacity in multiple countries. *Catena*. 2017;158:194–200.
4. Xu Y, Wang X, Bai J, Wang D, Wang W, Guan Y. Estimating the spatial distribution of soil total nitrogen and available potassium in coastal wetland soils in the Yellow River Delta by incorporating multi-source data. *Ecol Indic*. 2020;111:106002.
5. Ye L, Tan W, Fang L, Ji L. Spatial analysis of soil aggregate stability in a small catchment of the loess plateau, China: II. Spatial prediction. *Soil Tillage Res*. 2019;192:1–11.

6. Zeraatpisheh M, Ayoubi S, Jafari A, Tajik S, Finke P. Digital mapping of soil properties using multiple machine learning in a semiarid region, Central Iran. *Geoderma*. 2018;338:445–452.
7. Kehl M, Khormali F. *Excursion Book of International Symposium on Loess, Soils & Climate Change in Southern Eurasia. 15–19 October*; 2014.
8. Kariminejad N, Hosseinalizadeh M, Pourghasemi HR, Bernatek-Jakiel A, Campetella G, Ownegh M. Evaluation of factors affecting gully head-cut location using summary statistics and the maximum entropy model: Golestan Province, NE Iran. *Sci Total Environ*. 2019;677:281–298.
9. Bagheri Bodaghabadi M, Salehi MH, Martinez-Casasnovas JA, Mohammadi J, Toomanian N, Esfandiarpoor BI. Using canonical correspondence analysis (CCA) to identify the most important DEM attributes for digital soil mapping applications. *Catena*. 2011;86:66–74.
10. Hosseinalizadeh M, Kariminejad N, Chen W, et al. Gully headcut susceptibility modeling using functional trees, naïve Bayes tree, and random forest models. *Geoderma*. 2019;342:1–11.
11. Rasaei Z, Bogaert P. Spatial filtering and Bayesian data fusion for mapping soil properties: a case study combining legacy and remotely sensed data in Iran. *Geoderma*. 2019;344:50–62.
12. Aitkenhead M, Coull M. Mapping soil profile depth, bulk density and carbon stock in Scotland using remote sensing and spatial covariates. *Eur J Soil Sci*. 2020;71(4):553–567.
13. Fan NQ, Zhu A, Qin CZ, Liang P. Digital soil mapping over large areas with invalid environmental covariate data. *ISPRS Int J Geo Inf*. 2020; 9(2):102–112.
14. Forkuor G, Hounkpatin OK, Welp G, Thiel M. High resolution mapping of soil properties using remote sensing variables in South-Western Burkina Faso: a comparison of machine learning and multiple linear regression models. *PLoS One*. 2017;12, e0170478.
15. Keskin H, Grunwald S, Harris WG. Digital mapping of soil carbon fractions with machine learning. *Geoderma*. 2019;339:40–58.
16. Mosleh Z, Salehi MH, Jafari A, Borujeni IE, Mehnatkesh A. The effectiveness of digital soil mapping to predict soil properties over low-relief areas. *Environ Monit Assess*. 2016;188:1–13.
17. Taghizadeh-Mehrjardi R, Minasny B, Sarmadian F, Malone B. Digital mapping of soil salinity in Ardakan region, Central Iran. *Geoderma*. 2014;213:15–28.
18. Khaledian Y, Bradley AM. Selecting appropriate machine learning methods for digital soil mapping. *App Math Model*. 2020;81:401–418.
19. Mansuy N, Thiffault E, Paré D, et al. Digital mapping of soil properties in Canadian managed forests at 250 m of resolution using the k-nearest neighbor method. *Geoderma*. 2014;235:59–73.
20. Taghizadeh-Mehrjardi R, Nabiollahi K, Kerry R. Digital mapping of soil organic carbon at multiple depths using different data mining techniques in Baneh region, Iran. *Geoderma*. 2016;266:98–110.
21. Angelini ME, Heuvelink GBM, Kempen B. Multivariate mapping of soil with structural equation modelling. *Eur J Soil Sci*. 2017;68(5):575–591.
22. Nabiollahi K, Taghizadeh-Mehrjardi R, Shahabi A, et al. Assessing agricultural salt-affected land using digital soil mapping and hybridized random forests. *Geoderma*. 2021;385, 114858.
23. Kovačević M, Bajat B, Gajić B. Soil type classification and estimation of soil properties using support vector machines. *Geoderma*. 2010;154 (3–4):340–347.
24. Khaledian Y, Miller BA. Selecting appropriate machine learning methods for digital soil mapping. In: *Applied Mathematical Modelling*; 2019.
25. Kariminejad N, Rossi M, Hosseinalizadeh M, Pourghasemi HR, Santosh M. Gully head modelling in Iranian loess plateau under different scenarios. *Catena*. 2020;194104769–194104778.
26. Zhang X, Zhang F, Wang D, et al. Effects of vegetation, terrain and soil layer depth on eight soil chemical properties and soil fertility based on hybrid methods at urban forest scale in a typical loess hilly region of China. *PLoS ONE*. 2018;13(10), e0205661.
27. Soil Survey Staff. *Soil survey laboratory methods manual*. Report No. 42, USDA, NRCS, NCSS; 1996.
28. Blake GR, Hartge KH. Bulk density. In: *Methods of Soil Analysis: Part 1 Physical and Mineralogical Methods*. vol. 5; 1986:363–375.
29. Breiman L, Cutler A. In: *RFtools—for predicting and understanding data. Technical Report*. Berkeley: Berkeley University; 2004.
30. Liaw A, Wiener M. Classification and regression by randomForest. *R News*. 2002;2(3):18–22.
31. Vapnik V, Guyon I, Hastie T. Support vector machines. *Mach Learn*. 1995;20(3):273–297.
32. Hijmans RJ. *Raster: Geographic Data Analysis and Modeling R Package Version 2.6-7*; 2017.
33. Kuhn M. Caret: classification and regression training. In: *Astrophysics Source Code Library*; 2015:1500–1505.
34. Meyer D, Dimitriadou E, Hornik K, et al. Misc functions of the Department of Statistics (e1071), TU Wien. *R Package Version*. 2014;1(3):e1071.
35. Shahabi M, Jafarzadeh AA, Neyshabouri MR, Ghorbani MA, Valizadeh Kamran K. Spatial modeling of soil salinity using multiple linear regression, ordinary kriging and artificial neural network methods. *Archiv Agron Soil Sci*. 2017;63(2):151–160.
36. Hosseinalizadeh M, Kariminejad N, Campetella G, Jalalifard A, Alinejad M. Spatial point pattern analysis of piping erosion in loess-derived soils in Golestan Province, Iran. *Geoderma*. 2018;328:20–29.
37. Kramm T, Hoffmeister D, Curdt C, Maleki S, Khormali F, Kehl M. Accuracy assessment of landform classification approaches on different spatial scales for the Iranian loess plateau. *ISPRS Int J Geo-Inf*. 2017;6(366):1–22.
38. Maleki S, Khormali F, Mohammadi J, Bogaert P, Bodaghabadi MB. Effect of the accuracy of topographic data on improving digital soil mapping predictions with limited soil data: an application to the Iranian loess plateau. *Catena*. 2020;195:104810.
39. Vannoppen W, De Baets S, Keeble J, Dong Y, Poesen J. How do root and soil characteristics affect the erosion-reducing potential of plant species? *Ecol Eng*. 2017;109:186–195.
40. Geng R, Zhang GH, Li ZW, Wang H. Spatial variation in soil resistance to flowing water erosion along a regional transect in the loess plateau. *Earth Surf Process Landf*. 2015;40(15):2049–2058.

CHAPTER

45

Landslide susceptibility mapping along the Thimphu-Phuentsholing highway using machine learning

Manju Sara Dahal[a], Asheer Chhetri[a], Hemant Ghalley[a], Sangey Pasang[a,b], and Moujhuri Patra[a]

[a]Civil Engineering Department, College of Science and Technology, Royal University of Bhutan, Rinchending, Bhutan
[b]Department of Geography, Masaryk University, Brno, Czech Republic

1 Introduction

A natural disaster is the result of unforeseen shifts in environmental conditions (such as earthquakes, tsunamis, and floods) that cause considerable financial, environmental, or human loss.[1] Landslides are described as the motion of a mass of boulder, debris, or earth down a steep hill. In Bhutan, landslides caused by road cutting are relatively frequent. Throughout the May-September monsoon months, landslides pop up in multiple locations blocking the roads. Bhutan's lifeline, the road from Phuentsholing to Thimphu, is frequently disrupted by debris from landslides. Whenever this road is blocked, Thimphu is disconnected from India, Bhutan's primary trading partner[2] which brings us to the need for assessment of landslide susceptibility in this region. Landslide susceptibility maps (LSMs) highlight areas where it is most probable that landslides might occur.[3] To create a reliable LSM, high-quality data are required and appropriate steps for analysis and modeling should be carried out.

Machine learning is a branch of artificial intelligence that focuses on the conceptualization and implementation of algorithms that enable computers to learn and adapt their behavior based on empirical data where the models are created.[4] Statistical methods are designed to interpret the correlation between the variables but machine learning is designed to make accurate predictions.[5] There are many different machine learning techniques used for susceptibility modeling like ANN, linear regression, nearest neighbor, decision trees, random forest (RF), logistic regression (LR), and many more. The two methods of machine learning which were used for this study are RF and LR. Both the models have shown accuracy greater than 0.85 on the AUC curve.

The major aim of this study was to analyze and compare the performance of two methods namely, LR and RF for landslide susceptibility analysis along the Phuentsholing-Thimphu highway. Although both the methods come under supervised machine learning, RF is a classification analysis and LR is a probabilistic analysis. The outcomes of the model were compared using the receiver operating characteristic (ROC) curve and area under curve the ROC curve (AUC).

2 Study area

The study area consists of a 2 km buffer along the AH48 highway, a part of the Great Asian Highway, inside Bhutan which covers a region of 47,287 ha. The AH48 highway begins in Changrabandha, India, and ends in Thimphu, Bhutan with a total length of 276 km, but for this study, only the portion which lies inside of Bhutan have been taken into

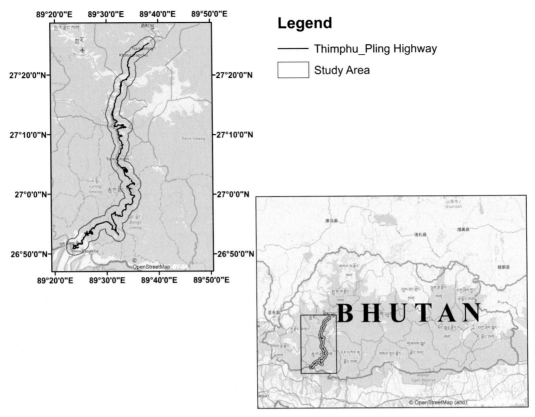

FIG. 1   Study area.

account, which is of a length of 165 km. The Phuentsholing-Thimphu highway connects Phuentsholing Dzongkhag (26°51′5.9″N, 89°23°18.13″E) and Thimphu Dzongkhag (27°36°0.00″N, 89°34′59.99″E) (Fig. 1).

3 Methodology

After a detailed literature review and discussion, the methodology has been prepared which acts guideline for the study to complete and achieve the set goal. It helps to estimate the duration of the study and to distribute the duration for each work (Fig. 2).

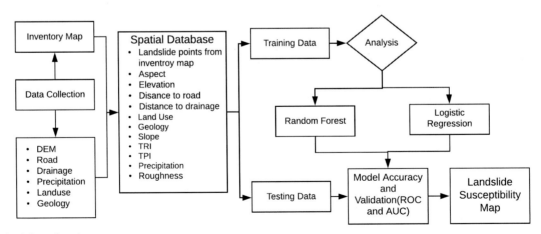

FIG. 2   Methodology flowchart.

3.1 Data collection

Remote sensing data were obtained from the National Land Commission and using Google Earth satellite data. ASTER DEM with a resolution of 30 m was obtained from the National Land Commission, which is a three-dimensional illustration of the surface of the terrain. A DEM can either be in the form of a raster or a vector-based triangular irregular network (TIN). Aspect, slope, TPI, TRI, roughness, and elevation data were obtained by analyzing the DEM. The land cover map was obtained from the Ministry of Agriculture and Forest. Road and drainage maps were obtained by the National Land Commission. Precipitation data for the 3 years 2016, 2017, and 2018 were obtained from the National Centre of Hydrology and Meteorology, Bhutan.

3.2 Inventory map

An inventory map of landslides provides the location and details of landslides.[6] Inventory maps are important for explaining the relationship between the distributions of landslides and the variables of conditioning. The inventory map used for this study has been produced through the combination of satellite imageries as well as field surveys. Although a typical landslide inventory map would show the location and outlines of landslides, for the sake of this model creation, only points with locations were noted on the map.

The inventory map shown below contains 217 points of landslide data, that is the location of a point of interest as well as whether or not a landslide has occurred there.

3.2.1 Parameters

Landslides are caused by a combination of different parameters and sometimes one can be more dominating than others. The selection of these parameters is vital to work for landslide assessments.[7] However, there are no universal guidelines or set rules to select the parameter.

Aspect

Aspect describes the orientation of the ground surface that is measured clockwise in degrees from 0° to 360°, where 0° is north-facing, 90° is east-facing, 180° is south-facing and 270° is west-facing. Aspect affects the rainfall direction and exposure to solar radiation causes the moisture and vegetation distribution uneven.[8] Even though the relationship between the aspect and mass motion has often been examined, there could have been no specific decision regarding the aspect-landslide relationship.[9]

Elevation

The elevation is the height from sea level. It is among the major elements in the landslide occurrence because elevation directly impacts the load-carrying capacity on the slope and therefore amplifies the landslides whenever the sliding plain seems to have a dip (orientation) toward such open excavation. Elevation can trigger various circumstances of the climate, including various slopes, vegetation, and precipitation. The weathering profile also relies upon the elevation of region.[10] As the height also affects the loading on a slope which enhances the landslide, elevation can be considered as an important parameter (Fig. 3).

Distance from the roads

The distance from the roads largely affects the likelihood of landslides since many landslides are reported near road.[7] The stability of the slope near the road can become unstable by road construction and vehicle motions. In soil and rock, irregularity is rendered by removing the slope hills for building highways in the slopes more than 10 degrees. The road-related effects were noted to disappear from a distance of 150 m from road.[11] The road built on slopes causes loss of toe support.[12]

Distance from the drainage

The distance from the drainage is a variable that can adversely influence the instability of the slope in the hillside motions.[13] Close proximity to drainage influences erosive processes caused by the impact on the slope of the terrain, causing instability and increased water depth when the flow is plugged.[14] The distance to the drainage can affect the landslide occurrence as it affects the saturation of underground material.

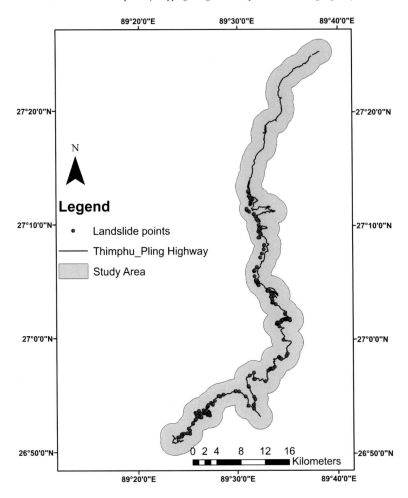

FIG. 3 Inventory map.

Land cover

Landcover describes the physical state of the land surface such as cropland, forest, or mountains.[15] The landcover map provides information on the prevalent activities at a certain location, thus it can be considered as an important variable in the stability of slope.[12] The forest-covered land controls the constant flow of water and water infiltrates periodically while cultivated land impacts the stability of the slope due to the saturation of the covered soil.[16]

Geology

Geology represents the type of rocks and minerals in the region and can be considered as an important parameter that may influence the slope stability as the study area lies in the Himalayas which is found to be fragile. Consequent landslides, which seem to be widespread earth slides and earth flows, usually have taken place in rock and ground formed on bedrock, and almost every rock unit plotted in the region consisted of at least one recent landslide.

Slope

The slope angle being the outline of any portion of the earth's face and a horizontal datum[17] emphasizes that slope is a key feature that triggers landslides. The slope plays a crucial role in steep mountainous terrain as it regulates the mass movement processes. Slope instability increases with an increase in slope.[18]

Surface roughness

Surface roughness shows the degree of irregularity of the surface which reflects landform characteristics, degree of erosivity, and distribution of crenulations. Roughness maps were obtained by calculating the topographical variation in the local area surrounding each grid cell in DEM.[19] Although fine-resolution LiDAR DEMs are preferred for producing a roughness map, the current map obtained was deemed sufficient for this study. Hence roughness data can be utilized by machine learning algorithms to build an accurate model.

Terrain ruggedness index

The Terrain Ruggedness Index (TRI) expresses the amount of elevation difference between adjacent cells of a DEM. TRI taken via DEM enables the characterization of the terrain as smooth or rough landforms and the local variation of gradients or curvatures of the surface. Originally, TRI processing was proposed, to measure topographic heterogeneity by measuring elevation or slope variation in a particular neighborhood. Their model simply calculated TRI values for every DEM grid cell using a "DOCELL" Data Only Cells order in ArcGIS, which determines the amount of elevation fluctuation between a grid cell and its eight neighboring grid cells.[20] Also, each of the eight differential values of elevations is squared to convert them to all positive and averages the squares obtained. The topographic index of ruggedness would calculate from the average of the square root.

Topographic position index

The topographic position index (TPI) correlates the difference between every cell's height in a DEM to the mean height of a given neighborhood around that cell. The TPI is measured as the variation between the cell height and the average neighboring cell elevation.[21] Classifying current topographic landforms, that is valley, slope, and ridge, specific threshold values are needed to be described.[20] Positive TPI values indicate places at a greater height than the mean of their surroundings, just like it is described by neighborhood (ridges). Negative TPI values depict locations below their surroundings (valleys). TPI values close to zero have either been flat areas (in which the slope is close to zero), or constant slope areas.

Precipitation

Precipitation is one of the prominent factors that cause landslides in this region. As rainfall infiltrates the soil, it saturates the soil which changes the pressure within the slope, contributing to instability in the slope. If the rainfall occurs at a higher intensity for a limited duration; the landslide is mainly shallow soil slides with higher severity of debris flows. Rainfall occurring for a longer time at a lower intensity, leads to larger debris avalanches and slumps. Landslides usually occur during monsoons when the rainfall is maximum, therefore a precipitation map was obtained by taking an average of a maximum monthly rainfall of 3 years of four stations that; Phuentsholing, Chapcha, Chukha, and Semotokha. By plotting these four stations in the GIS platform and using IDW (Inverse distance weighting) interpolation of these points we can effectively estimate the amount of rainfall throughout the entire map (Figs. 4 and 5).

3.2.2 Analysis

Input data

For landslide assessment and analysis, the data were collected in the form of landslide area occurred or in the form of the points. In this case, the data were collected in the form of points and it required both landslide and nonlandslide points to train the model and subsequently test it, which is easier to analyze when the data is in point form. A total of 300 points were collected of which 217 points were landslide points and 83 were nonlandslide points. After the parameter data were collected, they were analyzed using the GIS software. With the help of the *point sampling tool* plugin in QGIS, an attribute table containing all the parameter data for each point was obtained.

Training and testing data

Arthur Samuel, who coined the term machine learning in 1959, hypothesizes that machine learning is a computer science subfield that provides computers with the power to gain knowledge without explicit programming. For the computer to gain knowledge without being explicitly programmed, the model needs to be trained by providing data that is inclusive of the correct outcome. Once the model is trained, we then observe how accurately it can predict the correct outcome by comparing its prediction with the corresponding outcome from the testing data. Generally speaking, the more the training data, the better the model performs in prediction although that may not always be the case. In this study, we have 300 points, whereby the data are divided into portions of 70% and 30% for training and testing for both models.

Logistic regression

LR is a technique used in machine learning classification algorithms based on Statistics. The LR model establishes the linkage between independent factors (parameters causing landslides) and dependent factors (landslides). The dependent variable is always a binary variable, i.e., It usually has only two options, yes/no, true/false or 0/1. LR models the likelihood of landslides or other independent values. LR has been widely used for LSM because it has been proven to be the most reliable approach. The predicted value, i.e., the likelihood of landslide happening is mostly in the range of 0 and 1.

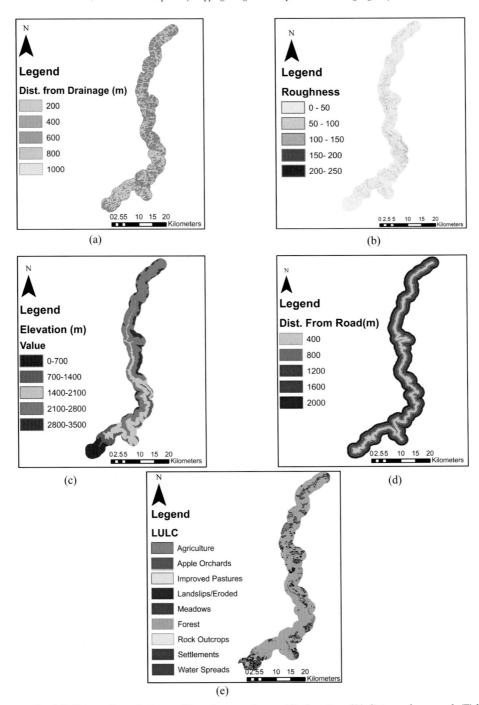

FIG. 4 Parameter maps for (A) distance from drainage; (B) surface roughness; (C) elevation; (D) distance form road; (E) landcover.

LR is a derived linear regression method, but it does not require independent and dependent data to be linear. It can handle both linear and nonlinear data because the prediction is based on a logistic function. The logistic function also called the sigmoid function, expresses the logistic distribution. It is derived from the linear regression; and the logistic function is given by;

$$P = \frac{1}{(1 + e^{-z})}$$
$$z = b_0 + b_1 x_1 + b_2 x_2 + \cdots b_n x_n \tag{1}$$

where P = probability of landslide occurrence. b_0 = intercept of the model. $x_1, x_2, \ldots, x_n$ = landslide causative factor (parameters). $b_1, b_2, \ldots b_n$ = weights of each parameter.

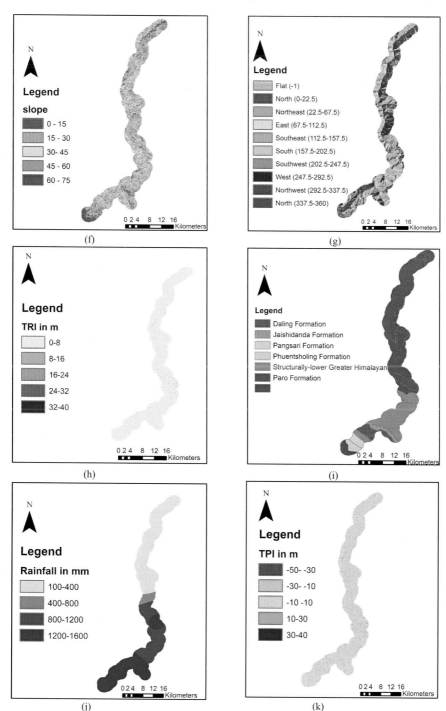

FIG. 5   Parameter maps of (F) slope; (G) aspect; (h) TRI; (i) geology; (g) rainfall; (k) TPI.

3.2.3 *Multicollinearity and variance inflation factor*

Multicollinearity exists when independent variables in a regression model are connected. The regression coefficient obtained for any variable indicates a change in the dependent variable for each unit change in that independent variable while keeping all independent variables constant. However, when the independent variable is correlated, changes made in one independent variable change the other independent variable. The greater the connection, the harder it is to modify one variable without modifying the other. This correlation is a problem because the independent variables are no longer independent which causes a problem in fitting the model and the coefficient obtained may not be accurate.

This problem can be solved in various ways. In this case, a simple test was performed. The test is the variance inflation factor (VIF), which defines the connection between the independent variables and determines the strength of the connection. VIF is calculated using the formula:

$$VIF = \frac{1}{(1 - R^2)} \tag{2}$$

R^2 is the coefficient of determination which is calculated by regressing one independent variable with other independent variables.

The VIF was calculated with ease in statistical software. The VIF values start from 1, which shows that there is no correlation between the independent variable and the others. VIF has no upper limit, the higher the VIF value, the stronger the correlation. The acceptable range of VIF was 1–5. A VIF greater than 5 represents the correlation that can cause an error in model fitting, so corrective measures are required. Corrective measures include removing the independent variable which has high VIF which decreases the VIF of other independent variables.

3.2.4 Random forest

RF is also one of the supervised machine learning algorithms based on ensemble learning. Ensemble learning is a type of learning in which the same algorithm is used multiple times or different algorithms are combined to create an influential forecast model. The RF comprises a large number of individual decision trees that function as an ensemble (collaboratively). Every tree in the RF gives out the prediction made, and the model's predictions are made on the majority of the predictions of trees. For example, in a case where the statement is true or false, it will depend on the majority of the predictions made by the number of decision trees. A significant proportion of comparatively uncorrelated models (trees) can yield a prognostication of the ensemble which is far more precise than any single prediction. RF works well with both numerical and categorical data. A RF can be used for classification and regression. In this study, we extensively used the RF classifier model (Figs. 6 and 7).

3.2.5 Classification and regression tree (CART)

In a random forest, the final decision is based on majority voting. When the input data runs into a single decision tree, it formulates certain rules and conditions to make forecasts. There are two key decision-tree algorithms utilized in RF to describe those rules that assess the quality of a split, and iteratively separate the dataset into areas as follows:

● Classification and regression tree (CART)
● Iterative Dichotomiser (ID₃)

In our case, we used CART which is a simple and powerful approach to make the prediction. CART does not develop any equation similar to those in the LR. In this case, the data are divided into various subsets with many if-else questions to make predictions. Thus, the algorithm can capture the nonlinearity in the dataset. When taking more than one attribute to make the prediction, it is important to make the most relevant or most important attribute to be placed as the root node and further traversing down by splitting the nodes. As we move down in a decision tree, the level of uncertainty should decrease. Such tasks are performed by separating actions such as information gain or the Gini index. For CART, the Gini index was used as a splitting measure.

The Gini index or Gini impurity calculates the likelihood of incorrect classification of a particular variable when selected at random. When all the elements belong to the same group, they are called Pure. Gini index varies from

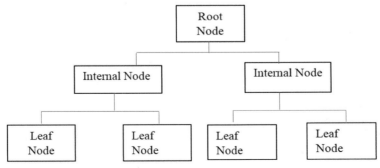

FIG. 6 A single decision tree.

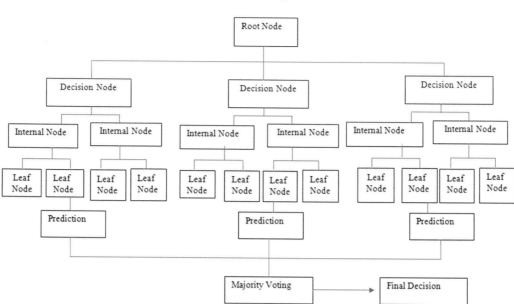

FIG. 7 Random forest (representation).

0 to 1.0 showing that all the elements belong to a certain group and 1 that one element is randomly distributed among the various groups. Usually, for the root node, the Gini index should be the least.

$$\text{Gini index} = 1 - \sum_{i=1}^{n} (Pi)^2 \qquad (3)$$

where Pi is the probability of an object being classified into a particular class.

3.2.6 Bootstrap aggregation (Bagging)

Bootstrap aggregation or bagging refers to the creation of several subsets of data from the training sample chosen randomly with replacement. Bagging reduces the variance and increases the prediction accuracy of statistical learning.[22] The model with higher variance might perform well with training data but cannot predict well with testing data[23] which in turn reduces the predictive capacity of the model. This step is useful especially when there are less data.

3.2.7 Feature importance

Feature importance refers to the technique of ranking the independent variable based on how important/useful they are at predicting a dependent variable. There are two different methods to rank the feature that is mean decrease impurity (MDI) and mean decrease accuracy (MDA).[24] In MDI, the decrease in weighted impurity in the individual tree is calculated. Using the Gini index or information gain the features are ranked by averaging the total decrease in node impurity over all trees in the forest. MDA predicts the importance of features based on errors it makes while testing them through out-of-bag datasets. The out-of-bag data sets are those data that are not considered in bootstrap sampling. The process is run through all decision trees and the average error is taken for each feature. The features with the least error are ranked as the most important.

The respective weights obtained from the LR which shows the relationship between the parameters and landslide are shown below. When some of the parameters increase, the risk of landslides decreases, and these parameters have negative weights. As for the RF, analyses were performed as mentioned in the methodology and the Feature importance was obtained.

3.2.8 Model building in regression logistic

The logistic model (or logit model) is a statistical model typically used for binary dependent variable applications. The logistic model is one where the log-odds of the likelihood/probability of an event are a linear combination variable. The two possible outcomes are often labeled 0 and 1 which signify pass/fail. LR in Python was built by fitting the training data into LR using the library *sklean.linear model* and we obtained the coefficient of the parameters and intercept.

3.2.9 Variance inflation factor

The VIF is a measure of multicollinearity among independent variables. In Python, it is calculated rapidly using codes of VIF codes from the Stats model which is a Python library. The VIF is checked and the parameter having a VIF greater than 5 is removed (Tables 1 and 2).

3.2.10 Coefficient and y-intercept parameters after VIF check

The weights of LR are different from those linear regressions as the weights are affected linearly by the outcome. The negative coefficient indicates that the factor is inversely proportional to the dependent variable. The y-intercept for most of the time is neglected but if there is no constant present in the equation or some residual, it would mean that independent and dependent variables are becoming 0 simultaneously. The y-intercept should not be attributed if it is negative as it has no meaning.[25] In our cases, parameters such as precipitation and geology have a higher value of coefficient from which it can be concluded that these parameters have a higher impact on the dependent variable, and the parameters such as distance from the road, distance from road and slope have a negative impact which implies that the when the value of this parameter is inversely proportional to the probability of landslides (Table 3).

3.2.11 Model building in random forest

Random forest is a collection of a large number of decision trees, which makes a final prediction based on the decision made by the majority of trees. Once the data are preprocessed and divided into training and testing the data are ready to be passed for build the model. One of the methods to build a RF in Python is by using the Python library *sklearn.ensemble.* A RF is built differently for different sets of data. To make the best prediction in a RF, it is the hyperparameters that have to be tuned.

TABLE 1 Initial parameters with VIF values.

	Features	VIF
0	TRI	3.075
1	TPI	1.012
2	Slope	3.841
3	Roughness	2.237
4	Dist. to drainage	2.294
5	Dist. to road	3.148
6.	Land cover	11.072
7	Geology	2.819
8	Elevation	30.612
9	Rainfall	4.508
10	Aspect	6.933

TABLE 2 Parameters after VIF check.

	Features	VIF
0	TRI	3.075
1	TPI	1.012
2	Slope	3.841
3	Roughness	2.237
4	Dist. to drainage	2.294
5	Dist. to road	3.148
6	Geology	2.819
7	Rainfall	4.508

TABLE 3 Coefficients and y-intercept.

Sl. no.	Parameter	Weights
1	TRI	0.449
2	TPI	−0.002
3	Dist. to the road	−0.684
4	Geology	1.55
5	Slope	−0.048
6	Roughness	−0.096
7	Dist. to the drainage	−0.14
8	Precipitation	1.329
	Constant	1.298

3.2.12 Tuning hyperparameters

A set of hyperparameters was chosen to obtain the best performance of the RF model. *Sklearn* has a default value of hyperparameters values. To obtain the best set, the model must be tested with different values and criteria of the hyperparameters, and the best possible combination of hyperparameters is used for making predictions. The hyperparameters were as follows:

1. The criterion is a function for measuring the quality of a division. Accepted criteria are "gini" for Gini impurity and "entropy" for obtaining information gain.
2. n_estimators are the numbers of decision trees generated. It is recommended to have a larger number of decision trees as better performance and an increase in the accuracy of the model can be obtained. Occasionally, having a large number of decision trees can lead to overfitting as there will be a decrease in randomness and a slower training process. This may decrease the performance of the model.
3. max_depth is the maximum extent up to which the tree spreads or the maximum depth of the tree. To capture more information, it is preferable to have a deeper tree. When "None" is used the nodes are expanded until all leaves are pure or until all leaves contain less than min_samples_split samples.
4. min_sample_split is the least number of samples required to split an internal node. This can differ from at least one sample at each node for all of the samples. When the number of min_sample_split increase in each tree in the forest becomes more controlled as it has to consider more samples at each node.
5. min_samples_leaf is the least number of samples required to be at a leaf node. A split is rejected if, after the split, one of the leaves would hold less than min_samples_leaf samples.
6. class_weight is the weights aligned with the class. It is beneficial to set the class_weight to balance as a class is automatically weighted inversely relative to how often they appear in the data. By default, the class weights were set to 1.
7. Bootstrap refers to creating multiple different models using a single dataset that produces high variance and powerful classifiers. Some RF models do not use the bootstrap sampling, in which case the bootstrap is set to False

3.2.13 Feature importance

Feature importance ranks the importance of the parameter in making a decision. Feature importance is calculated using inbuilt function importance from the *sklearn*. Precipitation was ranked highest followed by geology and elevation (Fig. 8 and Table 4).

3.3 Validation and comparison of the model

Performance measurement is an essential task in machine learning. A ROC curve is a technique for visualizing and selecting classifiers with respect to their performance.[26] True positive (TP) is an observation that is positive and is predicted to be positive and false positive (FP) is that observation is negative but is predicted positive. ROC curve plot of true positive rate versus false-positive rate. True positive rate is also called sensitivity and the false positive rate is 1-Specificity. Specificity is the extent to which a negative is classified as negative. The model will have better performance if the ROC curve is closer to the Left corner of the plot, i.e., the FP rate is lower and the TP rate is higher (Table 5).

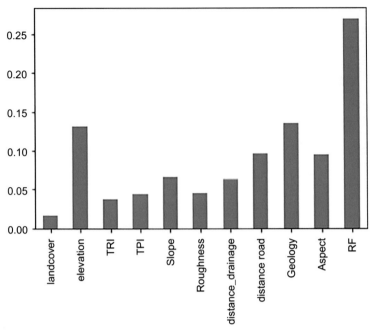

FIG. 8 Graphical representation of feature importance.

TABLE 4 Hyperparameters.

Sl. no.	Hyperparameter	Final value
1	Criterion	Gini
2	n_estimators	50
3	max_depth	None
4	min_sample_split	2
5	min_samples_leaf	1
6	class_weight	Balanced
7	Bootstrap	True

TABLE 5 Feature importance.

Sl. no.	Parameter	Feature Importance
1	Land cover	0.017
2	Elevation	0.131
3	TRI	0.037
4	TPI	0.044
5	Dist. to the road	0.096
6	Geology	0.134
7	Aspect	0.094
8	Slope	0.066
9	Roughness	0.046
10	Dist. to the drainage	0.064
11	Precipitation	0.27

$$TruePositiveRate = \frac{Positive\ Correctly\ Classified\ Observation}{Total\ Positives}$$

$$FalsePositiveRate = \frac{Negatives\ Incorrectly\ classified\ Observation}{Total\ Negatives}$$

After training the model, the predicted result for the testing dataset was obtained. ROC is obtained from the plot of the true positive rate and false-positive rate for different threshold settings. The threshold ranges from 0 to 1. In this study, the ROC was plotted using the Jupyter Notebook function. The area under the curve (AUC) was the area under the ROC curve. AUC determines the accuracy of the model. For the perfect model, AUC is 1, and AUC less than 0.5 is worthless.[27] The AUC values of the LR model are 0.86 and 0.91 for a RF which indicates both models have higher accuracy and the susceptible map will be more reliable.

4 Result and discussion

The coefficients and intercept of the LR and feature importance of RF are given in Tables 4 and 6 respectively. The LSM is created using the Raster Calculator in ArcGIS. The following formula was used to calculate the landslide susceptibility (Figs. 9 and 10).
For logistic regression:

$$\mathbf{P} = \text{Distance from the road} \times -0.684) \pm (\text{Geology} \times 1.55) \pm (\text{Slope} \times -0.048) \pm (\text{Roughness} \times -0.096)$$
$$\pm (\text{Distance from drainage} \times -0.14) \pm (\text{Precipitation} \times 1.329) \pm (\text{TRI} \times 0.449) \pm (\text{TPI} \times -0.002) \pm 1.298$$

TABLE 6 Percentage of landslide points on each susceptibility zones.

	LR (%)	RF (%)
Very low	11.52	14.29
Low	18.89	5.07
Moderate	28.57	15.21
High	32.72	23.96
Very high	8.29	41.47

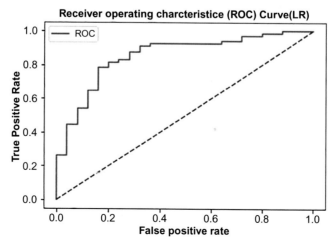

FIG. 9 ROC curve for logistic regression.

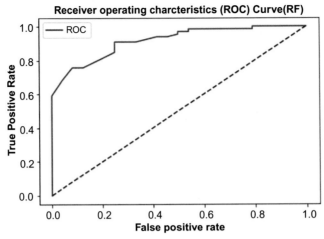

FIG. 10   ROC curve for random forest.

For random forest:

$$\mathbf{P} = (\text{Geology} \times 0.134) \pm (\text{Aspect} \times 0.094) \pm (\text{Slope} \times 0.066) \pm (\text{Roughness} \times 0.046) \pm (\text{Dist.from drainage} \times 0.064)$$
$$\pm (\text{Dist.from the road} \times 0.096) \pm (\text{Precipitation} \times 0.27) \pm (\text{Land Cover} \times 0.017) \pm (\text{Elevation} \times 0.131)$$
$$\pm (\text{TRI} \times 0.037) \pm (\text{TPI} \times 0.044)$$

where P is the probability of occurrence of landslide.

The LSM that was generated based on the data obtained from each algorithm was divided into five levels of susceptibility ranging from very low to very high. Both algorithms were allocated a high weighting for geology and precipitation, which has greatly influenced the maps below. Most landslides occur during the monsoon in this region when the precipitation is maximum and continuous. Our study area mostly includes Chukha dzongkhag which has heavily fractured and weathered rocks of phyllites, slates, and schists which contains a high amount of clay.[28] Clay tends to be soft in water and liquefies.[29] There is the presence of shear zones including weathered and folded phyllite rocks turning into clay.[30] The study area has about 65% of geology as Paro Formation and about 58% of landslide points from inventory map found at Structurally lower Greater Himalayan which mostly composes of quartzite, muscovite, and schist. Muscovite and schist have low physical strength.[31] However, landslide susceptibility cannot be limited to these factors. As this highway is important and occupies most of the time, this highway needs to undergo frequent changes and widening which could be a cause of the landslide.

Comparatively, the model obtained from the RF algorithm can be said to be more reliable as it obtained a higher AUC than the LR model. While the RF model had an AUC of 0.91, LR had an AUC of only 0.86. The following charts show a more detailed account of the various levels of susceptibility as compared to the entire study area. The LSM obtained from LR showed more than 50% of the total area in the low and very low susceptibility zones and 27% of the area in the high and very high susceptibility zones. The LSM obtained from RF shows that 44% of the area is in the low and very low susceptibility zones and 38% in the high and very high susceptibility zones.

By overlaying the inventory map over the susceptibility maps obtained, it can be observed that a large number of landslides are present in the highest susceptibility zone and decreases subsequently in the following zones with a spike in the very low zone on the map produced by the RF model. However, on the LR susceptibility map, a noticeably low percentage of landslides are present in the highest susceptible zone after which a regular decrease in landslides in the following zones was observed. Hence, it can be concluded that RF is indeed the better algorithm as per the findings of this study notwithstanding the irregularity in the lowest susceptible zone (Figs. 11 and 12).

Comparatively, the map obtained from the RF algorithm can be said to be more reliable as its model obtained a higher AUC than the LR model. The following charts show a more detailed account of the various levels of susceptibility as compared to the entire study area. The LSM obtained from the LR showed 50% of the area in the low susceptibility zone and 27% of the area in the highly susceptible zone. The LSM obtained from RF shows that 44% of the area is under a low susceptible zone and 38% in a high susceptible zone. Most of the area under high risk can be observed in the route between Phuentsholing and Kamji, which can be supported by the numerous landslides that can be seen on site (Fig. 13).

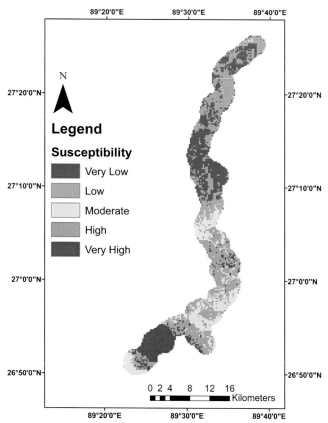

FIG. 11 Landslide susceptibility map from random forest.

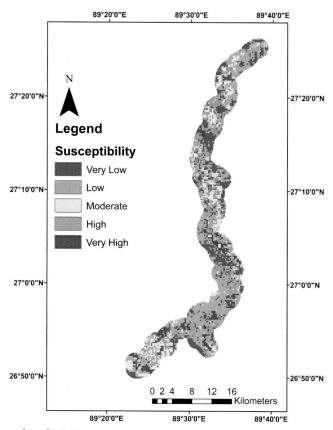

FIG. 12 Landslide susceptibility map from logistic regression.

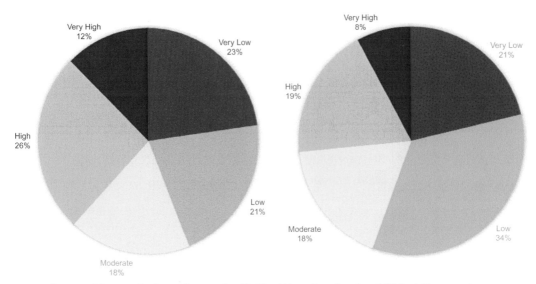

FIG. 13 Percentage of susceptible zones in the study area classified by (A) random forest and (B) logistic regression.

The zones that have been recognized as low-risk are well suited for any developmental project. High-risk zones should be avoided as much as possible, as these areas mostly consist of unstable slopes that may be susceptible to failure. In cases where construction of high-risk zones is unavoidable, appropriate measures can be taken to prevent or mitigate landslides after conducting proper site investigations and geological studies.

5 Conclusion

In Bhutan, the Phuentsholing to Thimphu highway is the lifeline of the country's trade and commerce which is prone to blockage by landmass and rocks, usually during the monsoon season. Landslides act as a massive threat, rising to their peak during the summer season; bringing mass destruction of private and cooperative properties, and posing a threat to the lives of the travelers. Based on the consequences of these hazards, this study aimed on creating LSMs along the Phuentsholing to Thimphu highway using the two algorithms of machine learning which were selected based on the ability of the algorithm to adapt to diverse data as well as its capability to handle categorical data well.

Throughout the study, it was found that geology and precipitation played a significant role in the occurrence, followed by elevation. As such, locations with poor geological formations, especially on higher altitudes are prone to landslides during storms, which is a frequent triggering factor for most nonseismically included landslides. The zones in the susceptibility map that were identified with a very high susceptibility to landslides should be thoroughly investigated in the field, and proper mitigation measures should be implemented to prevent any possible loss of life and property. The Thimphu-Phuentsholing highway is vital to the country's economy, and any delays would cause massive inconveniences and hamper local businesses.

GIS-based techniques have been implemented extensively globally for landslide susceptibility. During the research, machine learning-based algorithms, LR, and RF were implemented for LSM in the study area. The dataset of the landslide explanatory parameters was trained and tested to construct models for each of the algorithms; 70% for training and 30% for testing. To validate the results of models, prediction accuracy rates were determined using the ROC curve and the area under the ROC curve (AUC). The ROC and AUC curves were obtained using 30% of the test dataset. The validation results indicated that the LSM produced by the RF algorithm provided a better accuracy compared to LR with an AUC value of 0.91 and 0.86, respectively. While overlaying the landslide inventory map with the LSM generated from the models each algorithm, a higher range of similarity was observed with the map generated from the RF map; where a large number of the landslides were identified in the southern region of the study area and were considered to be high landslide risk zone. Hence, it can be concluded that the RF algorithm performs better than the LR algorithm but the prediction accuracy rate for both models is greater than 80%. Therefore, the RF algorithm could be recommended for use in land use planning and slope management in the upcoming years.

References

1. Kavzoglu T, Sahin EK, Colkesen I. Landslide susceptibility mapping using GIS-based multi-criteria decision analysis, support vector machines, and logistic regression. *Landslides.* 2014;11(3):425–439. https://doi.org/10.1007/s10346-013-0391-7.
2. Kuenza K, Dorji Y, Wangda D. Landslide in Bhutan. *Risk Manag South Asia.* 2012;126.
3. Oh H-J, Lee S, Hong S-M. *Landslide Susceptibility Assessment Using Frequency Ratio Technique With Iterative Random Sampling*; 2017. https://doi.org/10.1155/2017/3730913.
4. Musawi. Chapter One : introduction to machinelearning 1. In: *Mach Learn UTQ, CSM, CSD*; 2018. February.
5. Stewart M. *The Actual Difference Between Statistics and Machine Learning | by Matthew Stewart.* Towards Data Science; 2019. https://towardsdatascience.com/the-actual-difference-between-statistics-and-machine-learning-64b49f07ea3. PhD Researcher (Accessed 23 July 2020).
6. Quantin C, Allemand P, Delacourt C. Morphology and geometry of valles marineris landslides. *Planet Space Sci.* 2004;52(11):1011–1022. https://doi.org/10.1016/j.pss.2004.07.016.
7. Ayalew L, Yamagishi H. The application of GIS-based logistic regression for landslide susceptibility mapping in the Kakuda-Yahiko Mountains, Central Japan. *Geomorphology.* 2005;65(1–2):15–31. https://doi.org/10.1016/j.geomorph.2004.06.010.
8. Sun X, Chen J, Bao Y, Han X, Zhan J, Peng W. Landslide susceptibility mapping using logistic regression analysis along the jinsha river and its tributaries close to derong and Deqin county, Southwestern China. *ISPRS Int J Geo-Inf.* 2018;7(11):438. https://doi.org/10.3390/ijgi7110438.
9. Ercanoglu M, Gokceoglu C. Use of fuzzy relations to produce landslide susceptibility map of a landslide prone area (West Black Sea Region, Turkey). *Eng Geol.* 2004;75(3–4):229–250. https://doi.org/10.1016/j.enggeo.2004.06.001.
10. Nguyen VV, Pham BT, Vu BT, et al. Hybrid machine learning approaches for landslide susceptibility modeling. *Forests.* 2019;10(2):1–27. https://doi.org/10.3390/f10020157.
11. Reichenbach P, Busca C, Mondini AC, Rossi M. The influence of land use change on landslide susceptibility zonation: the briga catchment test site (Messina, Italy). *Environ Manag.* 2014;54(6):1372–1384. https://doi.org/10.1007/s00267-014-0357-0.
12. Devkota KC, Regmi AD, Pourghasemi HR, et al. Landslide susceptibility mapping using certainty factor, index of entropy and logistic regression models in GIS and their comparison at Mugling-Narayanghat road section in Nepal Himalaya. *Nat Hazards.* 2013;65(1):135–165. https://doi.org/10.1007/s11069-012-0347-6.
13. Ortiz JAV, Martínez-Graña AM. A neural network model applied to landslide susceptibility analysis (Capitanejo, Colombia). *Geomat Nat Hazards Risk.* 2018;9(1):1106–1128. https://doi.org/10.1080/19475705.2018.1513083.
14. Gökceoglu C, Aksoy H. Landslide susceptibility mapping of the slopes in the residual soils of the Mengen region (Turkey) by deterministic stability analyses and image processing techniques. *Eng Geol.* 1996;44(1–4):147–161. https://doi.org/10.1016/s0013-7952(97)81260-4.
15. Parveen S, Basheer J, Praveen B. A literature review on land use land cover changes. *Int J Adv Res.* 2018;6(7):1–6. https://doi.org/10.21474/ijar01/7327.
16. Igor F, Denis G, Daria S, Fedor B. Landslide processes as a risk factor for Russian cultural heritage objects. In: *Proc 2018 IPL Symp Landslides, Kyoto, Japan*; 2018:36–40.
17. Gómez H, Kavzoglu T. Assessment of shallow landslide susceptibility using artificial neural networks in Jabonosa River Basin, Venezuela. *Eng Geol.* 2005;78(1–2):11–27. https://doi.org/10.1016/j.enggeo.2004.10.004.
18. Chawla A, Chawla S, Pasupuleti S, Rao ACS, Sarkar K, Dwivedi R. Landslide susceptibility mapping in Darjeeling Himalayas, India. *Adv Civ Eng.* 2018;2018. https://doi.org/10.1155/2018/6416492.
19. Lindsay J, Newman D. *Hyper-Scale Analysis of Surface Roughness*; 2018:1–4. https://doi.org/10.7287/peerj.preprints.27110.
20. Saleem N, Enamul Huq M, Twumasi NYD, Javed A, Sajjad A. Parameters derived from and/or used with digital elevation models (DEMs) for landslide susceptibility mapping and landslide risk assessment: a review. *ISPRS Int J Geo-Inf.* 2019;8(12). https://doi.org/10.3390/ijgi8120545.
21. Guisan A, Weiss SB, Weiss AD, Ecology SP, Weiss D. GLM versus CCA spatial modeling of plant species distribution GLM versus CCA spatial modeling of plant species distribution. *Plant Ecol.* 2011;143(1):107–122. https://doi.org/10.1023/A:1009841519580.
22. James G, Witten D, Hastie T, Tibshirani R. In: James G, ed. *An Introduction to Statistical Learning—With Applications in R.* Springer; 2013. https://www.springer.com/gp/book/9781461471370%0Ahttp://www.springer.com/us/book/9781461471370.
23. Herrera Herrera M. *Landslide Detection Using Random Forest Classifier [Master Thesis].* Delft University of Technology; 2016.
24. Biau G, Scornet E. A random forest guided tour. *Test.* 2016;25(2):197–227. https://doi.org/10.1007/s11749-016-0481-7.
25. Frost J. *How to Interpret the Constant (Y Intercept) in Regression Analysis—Statistics By Jim*; 2020. https://statisticsbyjim.com/regression/interpret-constant-y-intercept-regression/. Accessed 10 July 2020.
26. Fawcett T. An introduction to ROC analysis. *Pattern Recognit Lett.* 2006;27(8):861–874. https://doi.org/10.1016/j.patrec.2005.10.010.
27. Xu J. *ROC (Receiver Operating Characteristic) Curve Analysis*; 2017. November.
28. Gariano SL, Sarkar R, Dikshit A, et al. Automatic calculation of rainfall thresholds for landslide occurrence in Chukha Dzongkhag, Bhutan. *Bull Eng Geol Environ.* 2019;78(6):4325–4332. https://doi.org/10.1007/s10064-018-1415-2.
29. Bin JA. *Improving the Stability of Clay for Construction.* Phys.org; 2014. https://phys.org/news/2014-07-stability-clay.html. [Accessed 23 July 2020].
30. Dikshit A, Sarkar R, Pradhan B, Acharya S, Dorji K. Estimating rainfall thresholds for landslide occurrence in the Bhutan Himalayas. *Water (Switzerland).* 2019;11(8):1–12. https://doi.org/10.3390/w11081616.
31. King HM. *Schist: Metamorphic Rock—Pictures, Definition & More*; 2005. Geology.com. https://geology.com/rocks/schist.shtml. [Accessed 23 July 2020].

46

Drought assessment using the standardized precipitation index (SPI) in GIS environment in Greece

Demetrios E. Tsesmelis[a,b], *Constantina G. Vasilakou*[b], *Kleomenis Kalogeropoulos*[c], *Nikolaos Stathopoulos*[d], *Stavros G. Alexandris*[b], *Efthimios Zervas*[a], *Panagiotis D. Oikonomou*[e,f], *and Christos A. Karavitis*[b]

[a]Laboratory of Technology and Policy of Energy and Environment, School of Science and Technology, Hellenic Open University, Patra, Greece [b]Department of Natural Resources Development and Agricultural Engineering, Agricultural University of Athens, Athens, Greece [c]Department of Geography, Harokopio University of Athens, Athens, Greece [d]Institute for Space Applications and Remote Sensing, National Observatory of Athens, BEYOND Centre of EO Research & Satellite Remote Sensing, Athens, Greece [e]Vermont EPSCoR, University of Vermont, Burlington, VT, United States [f]Gund Institute for Environment, University of Vermont, Burlington, VT, United States

1 Introduction

Natural resource management has focused more on the operation, observation, mitigation, and adaptation of ecological and environmental problems, rather than on their theoretical design.[1–3] Although osmosis with contingency planning is desirable, the management of natural resources is mainly based on the consideration of the relationship between humanity, culture, and natural processes, heading to the science application to solve any problems that arise each time. In this context, nature can present risky alterations in the variables and features of human systems. Such undesirable alterations and/or hazards, such as earthquakes, droughts, and floods, the so-called natural hazards, can present intractable difficulties and complications to human systems.[4–9]

Today, natural resource degradation generates pressure in the environment, including qualitative and quantitative impacts on water resources, overexploitation, desertification, soil erosion, deforestation, and environmental degradation. This degradation is of increasing societal concern.[10–12] In addition, human activities may pressure these delicate ecological systems and further load the status of natural resources.[8, 13]

In this context, it is obvious that droughts are among the extreme natural hazards that can affect urban and industrial water supply and irrigation, and in general human life.[14, 15] Droughts usually score from a mixture of environmental principles that can increase due to human intervention. The initial reason for any drought event is a lack of precipitation values and, especially, the tempospatial intensity and distribution of this shortage about the currently available water resources and water demand. This scarcity may lead to water shortages necessary for the operation of ecosystems and/or anthropogenic interventions.[16] Drought definition can rise universally in high and low rainfall areas for any season.[17, 18] No drought definition can be explained to all drought aspects, making it difficult to define the starting and endpoints accurately.[19, 20] Thus, the definition of drought remains a complicated state, which means that it is not unambiguous.[18, 19, 21–24]

The current trend among politicians, administrators, and policymakers, and commonly between citizens, is to regard drought as an impermanent, random, and remote risk that involves only emergency mobilization.[4, 25]

However, the available knowledge arising from the scientific observations and explorations of recent periods indicates that drought phenomena are unavoidable, as these events appear to be inevitable and perpetual facts of the global or local climate.[17, 19, 25–28]

History has shown that today's drought usually becomes tomorrow's water resource crisis, and these issues are going to play a fundamental role in the next years worldwide.[19, 29–32] In Europe, there have been a plethora of drought incidents during the last 50 years, with little change in the variation of spatial extent, severity, frequency, and duration, as well as impacts.[20] Specifically, the drought in Greece in 1989–93, in France and Spain, in 2005 and 2003, again in Greece in 2007–08, but also in northern Europe in the summer of 2018, are examples that confirm the phenomenon. Therefore, there is a need to take measures and strategies to mitigate the consequences not only for the Mediterranean environment but also for the whole of Europe.[16, 18–20, 33–40]

Drought impacts and their costs should be considered from the initial phases of water resource management efforts.[19, 41–45] Otherwise, the high economic cost of drought may increase.[46]

Therefore, any action to mitigate droughts should begin by learning the magnitude of the phenomenon.[18, 47–49] Practical explanations allow the determination of the stages and the degree of severity of drought,[50, 51] which are categorized into four different key lines: meteorological, hydrological, agricultural, and socioeconomic drought.[8, 9, 52, 53] Drought is a provisional condition (months/years) compared to aridity (enduring climate state).[54, 55]

The continuous development of studies related to drought indicators improves the methods and tools used, but also provides specific criteria for the implementation of policies (for drought management development, critical area recognition, comparability, threshold characterizations, monitoring improvement), and precision planning and mobilization of resources and moderation approaches.[56, 57] This study presents the Spatiotemporal variability of drought events using the Standardized Precipitation Index (SPI) with time steps of 6 and 12 months in Greece. In addition, the transformation from points (gauges) to spatial distribution used ordinary kriging.

2 Materials and methods

2.1 Datasets and application area

Greece is in Southern/Southeast Europe (Fig. 1) and is peninsular, whose mainland is covered by mountains (the peninsula is dominated by the Pindus Mountain range). It extends to the sea at the Peloponnese peninsula and has almost 3000 islands archipelago. The topography is mainly mountainous, with almost 30 peaks over 2000 m. Additionally, about 30% of the total area of the country is plains. Greece has the 11th longest coastline in the world, with almost 14,000 km (8700 miles) due to its extremely jagged coastline and many islands. Seventy percent of Greece contains hills and mountains. Therefore, it is among the most mountainous countries in Europe. Western Greece has a large number of lakes and wetlands.

The climatic conditions are particular in the South-East Mediterranean, with most of the precipitation events occurring during the winter period, and the summers are hot and very dry. There is a high range of average yearly rainfall from 350 to approximately 2000 mm/year (the highest values in the northwest part). The average interannual precipitation was approximately 760 mm/year.[58, 59] Due to its topography, Greece has many microclimates, which are reflected in the above precipitation values.

Greece is highly dependent on annual rainfall and local patterns. Since large cities and rural areas, systems rely on reservoir capacity and their operation on an annual basis.[4, 18, 60, 61] In the island environment, aquifers also have a robust yearly recovery cycle. In Greece, due to the seasonality of rainfall and intense relief, it is practically impossible to achieve measurements in mountainous areas, which constitute more than 70% of the country. Such restrictions generate an almost superlative situation for the implementation of SPI in a work to recognize drought characteristics in Greece.[16]

For the needs of this work, drought in Greece from 1981 to 2010 was portrayed using the SPI. The drought years were spatially visualized using geostatistical methods using the commercial software ArcGIS 10.8. Precipitation data were collected from 33 meteorological stations at monthly time steps in collaboration with the National Meteorological Service of Greece (HNMS) and the National Observatory of Athens, covering different periods from 1858 to 2010 (Figs. 1 and 2). The stations cover almost all of the different topographies in Greece. The following figures present the mean annual precipitation for the period 1981–2010.

2.2 The methodology

The SPI was described in 1993 as an index that might be used as a useful tool for drought analysis and monitoring procedures.[62] Since then, many studies have used SPI as a tool for identifying drought.[15, 16, 63–66] Fig. 3 presents a general flowchart of this methodology.

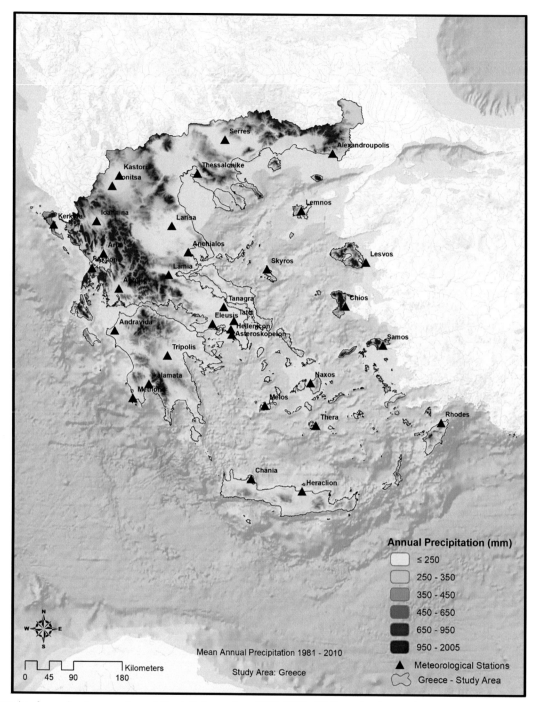

FIG. 1 Greece (study area)—Average Annual Precipitation values (1981–2010), (Hellenic Geodetic Reference System 1987—HGRS87).

Based on Guttman,[63] the first step in calculating the SPI is to compute the probability density function for the precipitation data over different time scales. The index can select various time scales on monthly steps, such as 3, 6, 12, and 24, which represent random monthly steps for deficits of precipitation with index application premises. Each of these datasets was tailored to the gamma probability density function to determine the relationship between the probability of precipitation.

Monthly rainfall time series were investigated as input for the tool calculation algorithm programmed in Fortran 95.[67] A calibration cycle was applied from 1981 to 2010, and the SPI results were computed.

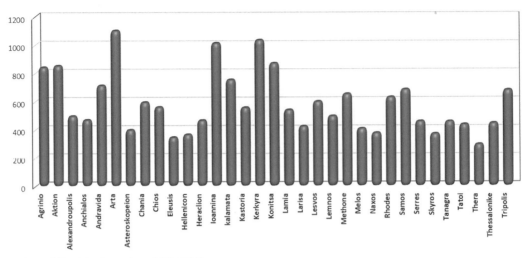

FIG. 2   Average Annual Precipitation values (1981–2010).

The methodology was implemented in a GIS environment. Therefore, all the spatial analysis and spatial visualization of the SPI values were performed with the Geostatistical Analyst ArcGIS 10.8 (ESRI, Redlands, California, USA). The kriging method, ordinary kriging (drought maps), and co-kriging (precipitation map) methods were used.

In this context, the spatial analysis of the data provides tools to define relationships, make predictions for the future, and finally to understand the phenomenon of drought.

Summarizing the current effort focuses on drought events on a tempospatial scale for Greece, and the current approach is distributed into three separate phases:

- Based on the precipitation data gathered from 33 meteorological stations and for an application period from 1981 to 2010, the output results of SPI6 and SPI12 were calculated on a monthly scale.
- The results of indices for the reference period were used as input data for the recognition of drought events.
- The third phase is a comparative analysis between the spatial analysis and SPI results from meteorological stations. Finally, drought events are shown in the following maps (using the GIS environment).

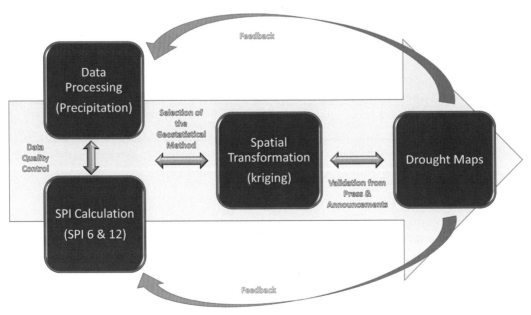

FIG. 3   Flowchart of the used methodology.

3 Results

The problem of drought events will not cease to exist even if human beings become more friendly to the environment but depend on multiple factors. Adequate forecasting and management of droughts are necessary to mitigate them and address their consequences. The policies that will be implemented should be following scientific principles and not based on urgent management and time pressures. Policymakers and stakeholders must give suggestions, advice, and orders to vulnerable areas depending on the situation.

In the present study, a geostatistical method was applied to capture the SPI results with time steps of six and 12 months. This time choice was made because precipitation occurs in Greece during the period October–March and is represented by SPI 6. In the case of heavier precipitation, such as those observed in Corfu, Ioannina, and Arta, SPI 12 was also chosen. For the examined period, the years with the most important drought events were 1989–90, 1992–93, 2000, and 2007–08. Additionally, this method was applied for the years 2009 and 2010, initially chosen as reference years to show the differences with the other years. However, droughts have been observed at various locations in the Hellenic territory.

The drought event of 1989–90, according to SPI6 (Fig. 4), started from the previous year, and specifically in April 1989, and its intensity peaked in June–August 1990. This phenomenon was considered to have ended in December 1990 due to low precipitation from August to December.

However, with SPI12 (Fig. 5), the phenomenon started in November 1989 and ended in January 1991. There is an expected delay at the end of the phenomenon because the index with a 12-month step takes into account more precipitation values.

Based on SPI6 (Fig. 6), the event of 1992–93 started in February 1992, culminating in January 1993 and ending in December of the same year. Although this phenomenon was not as severe as that of the years 1989–90, it severely affected all of Greece, specifically from January to March 1993.

SPI12 (Fig. 7) portrays the drought conditions of 1992–93. The drought began in December 1992 and ended in January 1994. Throughout the period 1992–93, according to this index, there were no large fluctuations; on the contrary, there was a stable situation until October 1993.

The drought of 2000, based on SPI6 (Fig. 8), appeared in January in extreme form in Samos, the Dodecanese, the eastern part of Crete, and in Pindos. In June of the same year, this phenomenon was eliminated from Rhodes, Karpathos, Santorini, and Pindos. However, it affected the rest of Greece, until October. Drought appears more intensely in Central Greece, the Dodecanese, Crete, Ioannina, Arta, and milder in Thrace.

In 2007–08, other severe dry conditions were recorded in Greece. It was more intense than in 2000 and 1992–93, but less than that in 1990. SPI6 (Fig. 8) showed that this drought event started in December of the previous year (2006) (Fig. 9).

The situation is similar for SPI12 (Fig. 10) of 2007–08, which affected almost all of Greece. April 2007 was the starting month of January 2008. For this lack of water, both indicators (SPI6 and SPI12) yielded similar results.

4 Discussion

The current study examines the most important drought events in Greece over the last 30 years. As mentioned in the previous chapters, the years with the most severe drought events in Greece were 1989–90, 1992–93, 2000, and 2007–08. The most severe event, both in duration and intensity, was in 1989–90, as was also recorded in the pertinent literature.[4, 16, 18, 19] Initially, it created serious problems in the agricultural sector through irrigation limitations and the production of hydroelectric energy at the beginning of the year, specifically in March, when crop irrigation began. After just 1 month, Karavitis[4] reported that only 10% of the crops would survive, which would create a serious economic problem for farmers and national resources. The 1989–90 drought cost to the Greek economy was 10^9 USD in 1990 prices.[18] According to the results of SPI6, the worst month for Greece was July. However, this is superficial, as the average precipitation in July is usually extremely low (approximately 10 mm); thus, an 80% decrease may lead to extreme drought, whereas it is not significant for the annual average of 700 mm falling predominantly during the winter.[7, 16, 20] Attica, the region of Athens, had very low available water quantities for supplying westerly reservoirs that created tremendous problems in the water supply. To mitigate this problem, plans were announced for water hauling by ships from the Acheloos River to Athens.[4, 18] This was clearly a means to pacify the public, as the peak of the problem for Athens surfaced in October 1990, when in Athens, water supplies were only enough for 56 days, and there were not many alternatives left.[18, 52] Nevertheless, the November rains solved drought problems.[4]

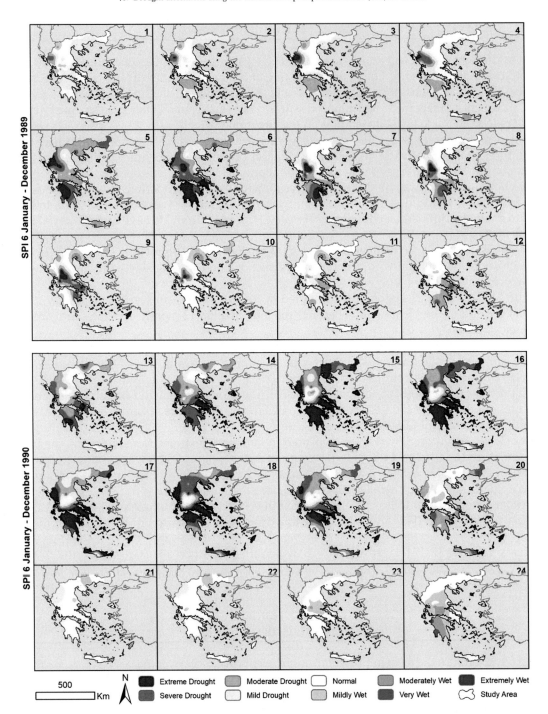

FIG. 4 SPI 6, period: 1989–90.

In the following years, there were three more drought events: that of 2000, the more intense, and that of 1992–93 and 2007–08. These three droughts cannot be compared with those of 1989–90, but they also created significant problems, mainly in the agricultural sector. The droughts of 1992–93 mainly affected Western and Northern Greece, Thrace, the islands of the North-Eastern Aegean, and less the Dodecanese. Crisis measures were taken: drilling water from wells in the Boieticos Kefissos plain and their incorporation into the Athens system, as well as the announcement of the creation of a dam on the river Euenos.[19] In the islands, the problem became more intense during the summer months due to tourism. In 2000, the phenomenon of water scarcity was more intense in the region of Central Greece and on the islands of the North East Aegean. In August 2000, this phenomenon intensified throughout Greece; it weakened considerably

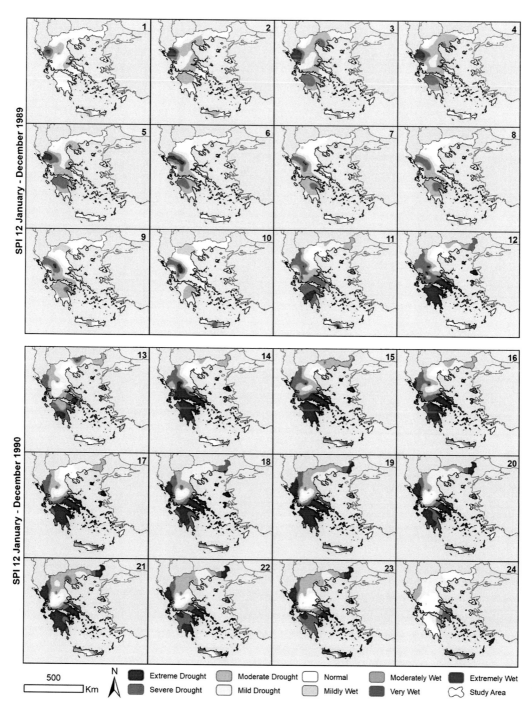

FIG. 5 SPI 12, period: 1989–90.

in October of the same year, and normal conditions returned at the end of the year. The phenomenon of 2000 was the most severe after 1989–90; it created significant problems in the crops of Northern Greece, but no significant problems were encountered in the water supply of large urban centers.[20]

Significant problems were created by the water shortage of 2007–08, where the phenomenon intensified in April and May and created problems in agriculture. This led politicians to take urgent measures to avoid problems similar to those in 1989–90. Some small water storage projects were carried out on the islands, and studies on desalination plants were announced together with even small land improvement projects, mainly in the region of Thessaly, where

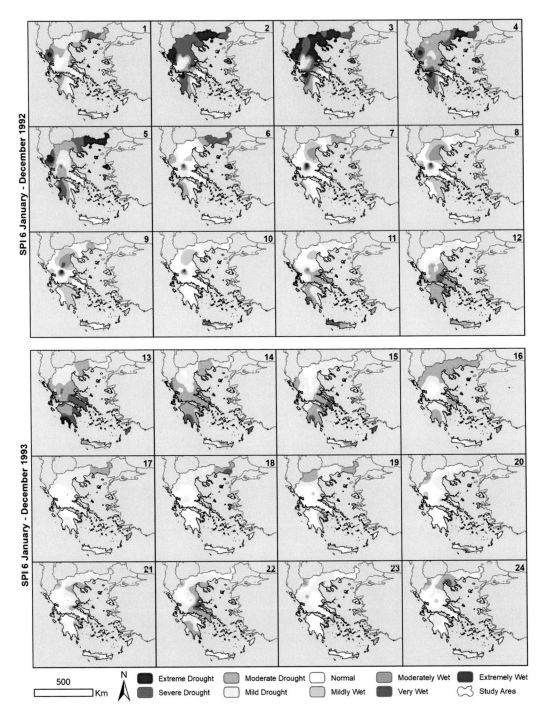

FIG. 6 SPI 6, period: 1992–93.

extensive groundwater withdrawals have led to land subsidence in many areas. This was due to intensive irrigated cultivation, which makes it vulnerable to drought. In the summer months the islands become more intense in the islands because of the tourism industry and, thus, the high water demand. This phenomenon weakened significantly after the end of summer (Fig. 11).

In 2009, the same procedure was performed to show the differences between a wet year and a year of extreme drought. The difference was obvious, where almost all Greece was wet, in contrast to the years of drought. The results for 2009 showed that the areas of Crete and Heleia region in Peloponnesus are affected by water shortages, and press

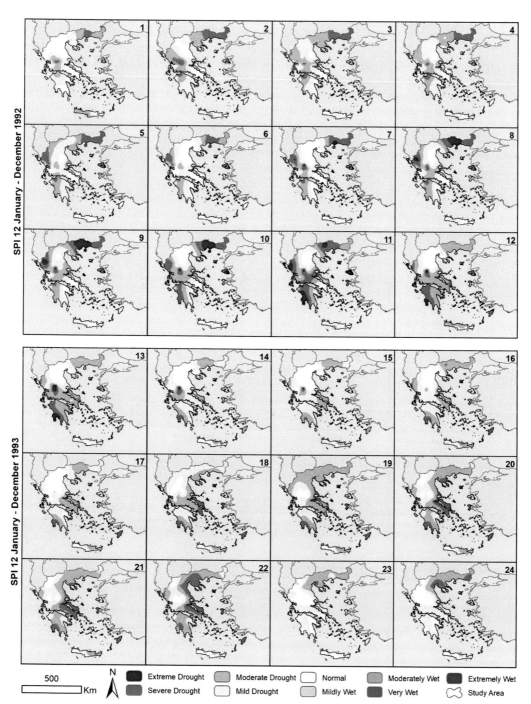

FIG. 7 SPI 12, period: 1992–93.

reports in May 2010 reported that available water supplies in the Crete region were running low. Ithaca Island, which is geographically opposite to Astakos, announced that during the summer months, when water demand would be high due to tourism, water supply interruptions would take place. Such an option is not considered a drought mitigation response, but rather as incompetent management because water supply system operational seizures are one of the prime reasons for water quality and system deterioration. The formula and conditions in both areas confirmed that the results of the indicators were correct and proved that one could rely on them.

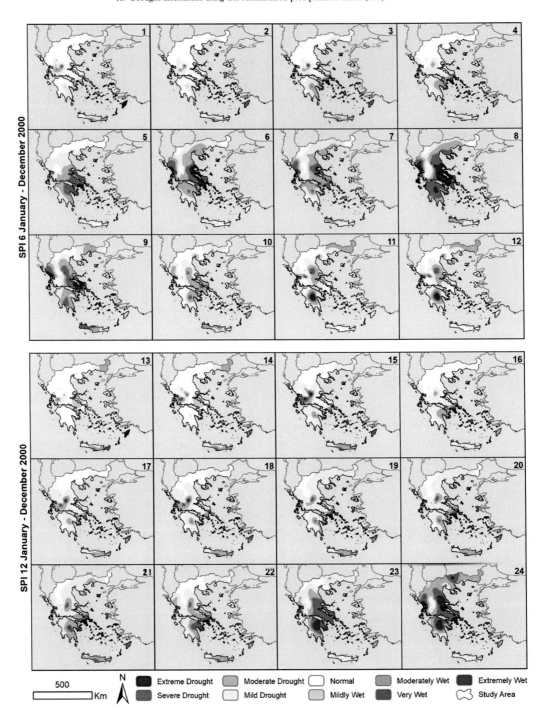

FIG. 8   SPI 6 and SPI 12, period: 2000.

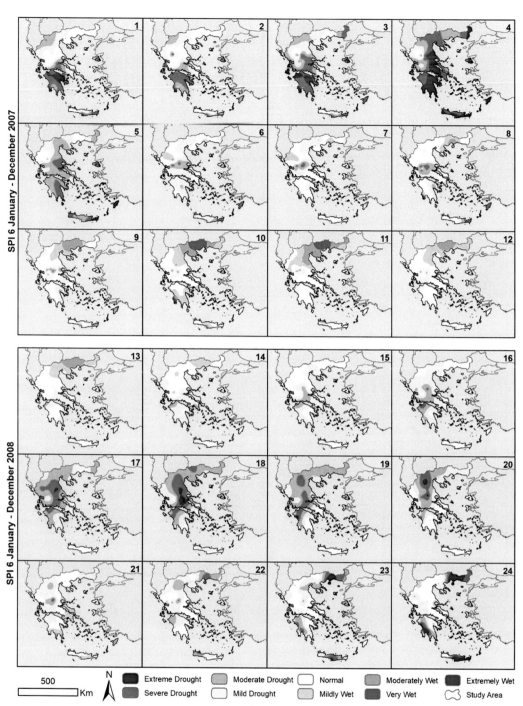

FIG. 9 SPI 6, period: 2007–08.

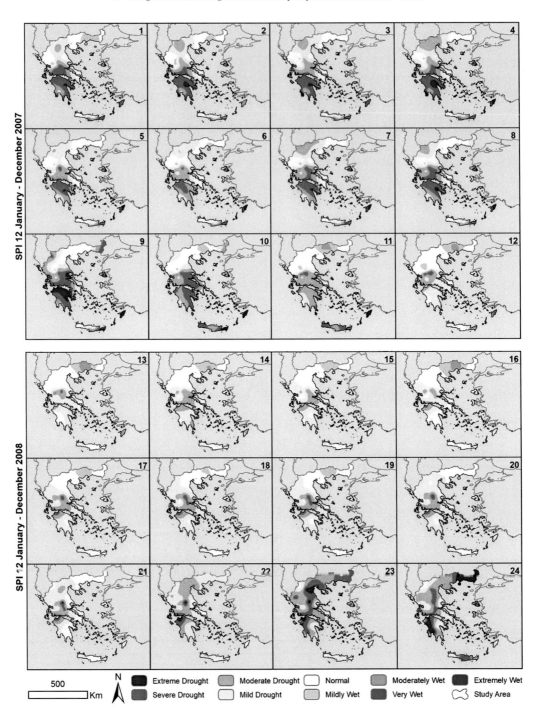

FIG. 10   SPI 12, period: 2007–08.

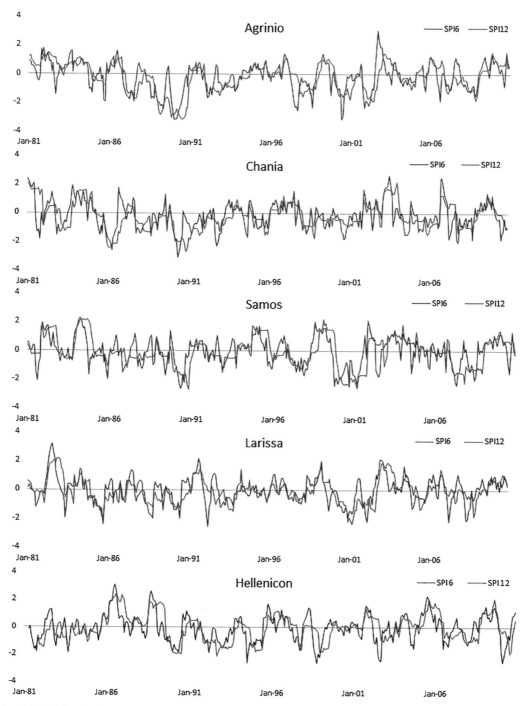

FIG. 11   SPI 6 and SPI 12 for five locations, period: January 1981—December 2008.

5 Conclusions

The use of drought indicators is an important part of integrated water resource management in a region or nation-wide. Even if the uncertainty of the event occurrence is stochastic, a combination of indicators can lead to a more satisfactory description of decision-making policies. For greater safety and better results, it is necessary to use more than one indicator. In addition, the quality of the meteorological data used is very important for obtaining satisfactory results. In this study, the number of meteorological stations used was quite high, but the greater the number of stations and the larger the area covered, the better the results.

To protect Greece and other countries from droughts and water shortages, a master plan with precise goals and objectives concerning the management of water resources must be established and ready to operate. This initial plan was submitted to the Central Water Agency of Greece in 2008.[7, 68] However, it has never been implemented and needs to be updated. Unfortunately, this is also the case in several countries; anxiety and panic appear during droughts, while apathy comes as soon as rain returns.

References

1. Gray R, Walters D, Bebbington J, Thompson I. The greening of Enterprise: an exploration of the (NON) role of environmental accounting and environmental accountants in organizational change. *Crit Perspect Account.* 1995;6:211–239.
2. Berkes F, Folke C, Colding J. *Linking Social and Ecological Systems: Management Practices and Social Mechanisms for Building Resilience.* Cambridge, UK: Cambridge University Press; 2000.
3. Barrow C. *Environmental Management for Sustainable Development.* UK: Routledge, Abingdon; 2006.
4. Karavitis C. Drought and urban water supplies: the case of metropolitan Athens. *Water Policy.* 1998;1:505–524.
5. Kalabokidis KD, Karavitis C, Vasilakos C. Automated fire and flood danger assessment system. In: *Proceedings of the Proceedings of the International Workshop on Forest Fires in the Wildland-Urban Interface and Rural Areas in Europe*; 2004:143–153.
6. Kalabokidis K, Kallos G, Karavitis C, et al. Automated fire and flood hazard protection system. In: *Proceedings of the Proceedings of the 5th International Workshop on Remote Sensing and GIS Applications to Forest Fire Management: Fire Effects Assessment.* Spain: Universidad de Zaragoza; 2005:167–172.
7. Karavitis C, Tsesmelis DE, Skondras NA, et al. Linking drought characteristics to impacts on a spatial and temporal scale. *Water Policy.* 2014;16:1172–1197.
8. Tsesmelis DE, Karavitis CA, Oikonomou PD, Alexandris S, Kosmas C. Assessment of the vulnerability to drought and desertification characteristics using the standardized drought vulnerability index (SDVI) and the environmentally sensitive areas index (ESAI). *Resources.* 2019;8(6).
9. Tsesmelis DE, Oikonomou PD, Vasilakou CG, et al. Assessing structural uncertainty caused by different weighting methods on the standardized drought vulnerability index (SDVI). *Stoch Environ Res Risk Assess.* 2019;33:515–533.
10. Horrigan L, Lawrence RS, Walker P. How sustainable agriculture can address the environmental and human health harms of industrial agriculture. *Environ Health Perspect.* 2002;110:445–456.
11. Laraus J. The problems of sustainable water use in the Mediterranean and research requirements for agriculture. *Ann Appl Biol.* 2004;144:259–272.
12. Lal R. Restoring soil quality to mitigate soil degradation. *Sustainability.* 2015;7:5875–5895.
13. Amanat A. Environment and culture: an introduction. *Iran Stud.* 2016;49:925–941.
14. Wilhite DA. *Chapter 1 Drought as a Natural Hazard: Concepts and Definitions.* Drought Mitig. Cent. Fac. Publ; 2000.
15. Sönmez F, Kömüscü K, Erkan A, Turgu E. An analysis of spatial and temporal dimension of drought vulnerability in Turkey using the standardized precipitation index. *Nat Hazards.* 2005;35:243–264.
16. Karavitis C, Alexandris S, Tsesmelis DE, Athanasopoulos G. Application of the standardized precipitation index (SPI) in Greece. *Water.* 2011;3:787–805.
17. Bordi I, Fraedrich K, Petitta M, Sutera A. Large scale assessment of drought variability based on NCEP/NCAR and ERA- 40 re- analyses. *Water Resour Manag.* 2006;20:889–915.
18. Karavitis C. *Drought Management Strategies for Urban Water Supplies: The Case of Metropolitan Athens.* Ph.D. Dissertation, Fort Collins, Co., USA: Department of Civil Engineering, Colorado State University; 1992.
19. Karavitis C. Decision support Systems for Drought Management Strategies in metropolitan Athens. *Water Int.* 1999;1(24).10–21.
20. Oikonomou PD, Karavitis CA, Tsesmelis DE. Drought characteristics assessment in Europe over the past 50 years. *Water Resour Manag.* 2020;34:4757–4772.
21. Cancelliere A, Salas JD. Drought length properties for periodic-stochastic hydrologic data. *Water Resour Res.* 2004;40, wo2503.
22. Mishra AK, Desai VR. Drought forecasting using stochastic models. *Stoch Environ Resour Risk Assess.* 2005;19:326–339.
23. Grubić A. The astronomic theory of climatic changes of Milutin Milankovich. *Episodes.* 2006;3(29):197–203.
24. Andreu J, Rossi G, Vagliasindi F, Vela A, eds. *Drought Management and Planning for Water Resources.* N.Y: Taylor & Francis Group; 2006.
25. Grigg NS, Vlachos EC. *Drought Water Management. International School for Water Resources.* Fort Collins, Colorado, USA: Department of Civil Engineering, Colorado State University; 1990:48.
26. Yevjevich V, Cunha L, Vlachos E. *Coping with Droughts.* Water Resources Publications; 1983.
27. Mishra AK, Singh VP. A review of drought concepts. *J Hydrol.* 2010;391:202–216.
28. Oikonomou PD, Tsesmelis E, Waskom RM, Grigg NS, Karavitis CA. Enhancing the standardized drought vulnerability index by integrating spatiotemporal information from satellite and in situ data. *J Hydrol.* 2019;569:265–277.
29. Adger WN. Vulnerability. *Glob Environ Chang.* 2006;16:268–281.
30. Horton G, Hanna L, Kelly B. Drought, drying and climate change: emerging health issues for ageing Australians in rural areas. *Australas J Ageing.* 2010;29:2–7.
31. Wilhite DA, Sivakumar MVK, Pulwarty R. Managing drought risk in a changing climate: the role of national drought policy. *Weather Clim Extrem.* 2014;3:4–13.
32. Kalogeropoulos K, Stathopoulos N, Psarogiannis A, et al. An integrated GIS-hydro modeling methodology for surface runoff exploitation via small-scale reservoirs. *Water.* 2020;12(11):3182.
33. Karavitis C, Chortaria C, Alexandri S, Vasilakou CG, Tsesmelis DE. Development of the standardised precipitation index for Greece. *Urban Water J.* 2012;9:401–417.
34. Pedro-Monzonís M, Ferrer J, Solera A, Estrela T, Paredes-Arquiola J. Key issues for determining the exploitable water resources in a Mediterranean river basin. *Sci Total Environ.* 2015;503–504:319–328.

35. Pedro-Monzonís M, Solera A, Ferrer J, Estrela T, Paredes-Arquiola J. A review of water scarcity and drought indexes in water resources planning and management. *J Hydrol.* 2015;527:482–493.

36. Ciais P, Reichstein M, Viovy N, et al. Europe-wide reduction in primary productivity caused by the heat and drought in 2003. *Nature.* 2005;437:529–533.

37. Tsesmelis DE. *Application of the Standardized Precipitation Index (SPI) in Greece for the Integrated Management of Droughts.* M.Sc, Department of Natural Resources Management & Agricultural Engineering, Agricultural University of Athens; 2010.

38. Tsesmelis DE. *Development, Implementation and Evaluation of Drought and Desertification Risk Indicators for the Integrated Management of Water Resources.* Ph.D, Athens, Greece: Dissertation, Department of Natural Resources Management & Agricultural Engineering, Agricultural University of Athens; 2017.

39. EC. *Drought in Europe*; 2018. https://ec.europa.eu/commission/news/drought-europe-2018-aug-30_en.

40. Di Liberto T. A hot, dry summer has led to drought in Europe in 2018. In: *NOAA Clim*; 2018. https://www.climate.gov/news-features/event-tracker/hot-dry-summer-has-led-drought-europe-2018.

41. Loucks DP. Sustainable water resources management. *Water Int.* 2000;25:3–10.

42. Wilhite DA, Haye MJ, Knutson C, Smith KH. Planning for drought: moving from crisis to risk management. *J Am Water Resour Assoc.* 2007;697–710.

43. Kampragou E, Apostolaki S, Manoli E, Froebrich J, Assimacopoulos D. Towards the harmonization of water-related policies for managing drought risks across the EU. *Environ Sci Policy.* 2011;14:815–824.

44. Skondras NA, Karavitis CA, Gkotsis II, Scott PJB, Kaly UL, Alexandris SG. Application and assessment of the environmental vulnerability index in Greece. *Ecol Indic.* 2011;11:1699–1706.

45. Skondras N. *Decision Making in Water Resources Management: Development of a Composite Indicator for the Assessment of the Social-Environmental Systems in Terms Resilience and Vulnerability to Water Scarcity and Water Stress.* Ph.D. Dissertation, Athens, Greece: Department of Natural Resources Management and Agricultural Engineering, Agricultural University of Athens; 2015.

46. EC. *Addressing the Challenge of Water Scarcity and Droughts in the European Union: Impact Assessment.* Brussels: European Commission; 2007.

47. Salas JD, Fu C, Cancelliere A, et al. Characterizing the severity and risk of drought in the Poudre River, Colorado. *J Water Resour Plan Manag.* 2005;131:383–393.

48. Priscoli JD. Keynote address: clothing the IWRM emperor by using collaborative modeling for decision support. *J Am Water Resour Assoc.* 2013;49:609–613.

49. Cancelliere A, Mauro GD, Bonaccorso B, Rossi G. Drought forecasting using the standardized precipitation index. *Water Resour Manag.* 2007;21:801–819.

50. Vlachos E. Prologue: water peace and conflict management. *Water Int.* 1990;15:185–188.

51. Vlachos E, Braga B. The challenge of urban water management. In: *Proceedings of the Frontiers in Urban Water Management: Deadlock or Hope.* London: IWA Publishing; 2001:1–36.

52. Grigg NS. Water resources management. In: *Water Encyclopedia.* John Wiley & Sons, Inc; 1996.

53. Swathandran S, Aslam MAM. Assessing the role of SWIR band in detecting agricultural crop stress: a case study of Raichur district, Karnataka, India. *Environ Monit Assess.* 2019;191:442.

54. Vlachos C. *Drought Management Interfaces.* 15. Las Vegas, Nevada, USA: ASCE; 1982.

55. AghaKouchak A, Feldman D, Hoerling M, Huxman T, Lund J. Water and climate: recognize anthropogenic drought. *Nat News.* 2015;524:409.

56. USACE. *Managing Water for Drought. National Study of Water Management During Drought.* September; 1994. IWR Report 94-NDS-8.

57. Bonaccoroso B, Bordi I, Cancelliere A, Rossi G, Sutera A. Spatial variability of drought: an analysis of the Spi in Sicily. *Water Resour Manag.* 2003;17:273–296.

58. Karavitis C. Scientific Co-ordinator, *Technical Support to the Central Water Agency of Greece for the Development of a Drought Master Plan for Greece and an Immediate Drought Mitigation Plan, Technical Report (in Greek) on Contract No. 10889/11/07 /2007 with Agricultural University of Athens (AUA).* Athens, Greece: Department of Natural Resources and Agricultural Engineering, Water Resources Management Sector, Ministry of Planning, Public Works and the Environment; 2008.

59. Stathopoulos N, Skrimizeas P, Kalogeropoulos K, Louka P, Tragaki A. Statistical Analysis and Spatial correlation of rainfall in Greece for a 20-year time period. In: *14th International Conference on Meteorology, Climatology and Atmospheric Physics, Proceedings of the EasyChair Preprints.* EasyChair; 2019.

60. Kalogeropoulos K, Chalkias C, Pissias E, Karalis S. Application of the SWAT model for the investigation of reservoirs creation. In: Lambrakis N, Stournaras G, Katsanou K, eds. *Advances in the Research of Aquatic Environment.* Berlin: Springer; 2011. Environmental Earth Sciences; vol. II.

61. Pissias E, Psarogiannis A, Kalogeropoulos K. Water savings-a necessity in a changing environment. The case of small reservoirs. In: *WIN4life Conference, Tinos Isl., Greece*; 2013.

62. McKee TB, Doesken NJ, Kleist J. The relationship of drought frequency and duration to time scales. In: *Proceedings of the 8th Conference on Applied Climatology.* California, USA: Anaheim; 1993. 17–22 January.

63. Guttman NB. Accepting the standardized precipitation index: a calculation algorithm. *J Am Water Resour Assoc.* 1999;2(35):311–322.

64. Lana X, Serra C, Burgueño A. Patterns of monthly rainfall shortage and excess in terms of the standardized precipitation index for Catalonia (NE Spain). *Int J Climatol.* 2001;21:1669–1691.

65. Tsakiris G, Vangelis H. Establishing a drought index incorporating evapotranspiration. *Eur Water.* 2005;9–10:1–9.

66. Livada I, Assimakopoulos VD. Spatial and temporal analysis of drought in Greece using the standardized precipitation index (SPI). *Theor Appl Climatol.* 2007;89(3–4):143–153.

67. DMCSEE Project. *Drought Management Centre for South Eastern Europe.* European Commission Funded Project, EU; 2009.

68. Chortaria C. *The Development of the SPI Drought Index for Greece Using Statistical Methods and GIS.* Thesis (M.Sc.), in Greek: Agricultural University of Athens; 2008.

CHAPTER

47

COVID-19: An analysis on official reports in Iran and the world along with some comparisons to other hazards

Soheila Pouyan[a], Mojgan Bordbar[b], Mojdeh Mohammadi Khoshoui[c], Soroor Rahmanian[d], Zakariya Farajzadeh[e], Bahram Heidari[f], Sedigheh Babaei[a], and Hamid Reza Pourghasemi[a]

[a]Department of Natural Resources and Environmental Engineering, College of Agriculture, Shiraz University, Shiraz, Iran [b]Department of GIS/RS, Faculty of Natural Resources and Environment, Science and Research Branch, Islamic Azad University, Tehran, Iran [c]Watershed Management Engineering, Faculty of Natural Resources and Desert Studies, Yazd University, Yazd, Iran [d]Quantitative Plant Ecology and Biodiversity Research Lab, Department of Biology, Faculty of Science, Ferdowsi University of Mashhad, Mashhad, Iran [e]Department of Agricultural Economics, College of Agriculture, Shiraz University, Shiraz, Iran [f]Department of Plant Production and Genetics, School of Agriculture, Shiraz University, Shiraz, Iran

1 Introduction

The COVID-19 epidemic has faced the world with a health and economic crisis that has not been seen since the Spanish flu epidemic in 1918. The spread of the coronavirus has been unprecedented for the past 100 years, and even crises such as cholera and SARS outbreaks have not been as widespread and destructive as COVID-19. Previous infectious diseases have generally persisted in several countries or are controlled with the help of drugs, while COVID-19 is a different story. This virus was first identified on December 31, 2019, in Wuhan, China, and the World Health Organization called it COVID-19.[1] The COVID-19 spread was rapidly started in China and moved to the rest of the world and affected more than 218 countries until January 31, 2021. From the beginning of infection until January 31, 2021, the number of registered cases infected by this virus has reached more than 83,055,202 cases, of which 1,811,358 have died (https://www.worldometers.info/coronavirus). The effects of the COVID-19, as a major socio-economic crisis referred to by the United Nations, have had a negative impact on the world and even developed countries, so that its continued spread has faced the World Health Organization with serious problems.[2, 3] The transmission rate for COVID-19 is unknown; however, there is strong evidence of efficient human-to-human transmission. This virus is claiming more lives by affecting health systems, shaking the global economic foundations, and creating lasting geopolitical changes. Globally, strict efforts are being made to restrain what has become a profoundly destructive epidemic.[4] Following the outbreak of this disease, countries, including Iran, have increased their planning and monitoring to quickly identify new possible cases of the disease and work to break the transmission chain. Iran is also one of the countries with the highest incidence statistics, with COVID-19.[5] China has made significant progress in coronavirus control when other countries faced significant challenges with the spread of the pandemic. Centralized patient management has played an important role in the Chinese epidemic, thereby effectively reducing the transmission chain.[6] It is noteworthy that in the United States and Europe, crisis management is based on the principles of immediate isolation of symptomatic or suspected individuals, prevention of gatherings, especially in indoor areas, observance of social distance, cancelation of unnecessary travel, closure of schools, and educational centers, as well as providing care for the elderly and children done.[6] Until January 31, 2021, the highest mortality rates were reported in the United States (441,324), Brazil (224,504), Mexico (158,536), India (154,392), the United Kingdom (106,367), Italy (88,516), France (76,201), Russia (72,029), Spain (58,319), and Iran (57,959), respectively. On January 31, 2021, Iran ranked 15th in terms of the total number of cases (1218, 752) and 9th in terms of the number of deaths (57,959). At this time, China ranked 81st (87,027) and 45th (4634) in terms of the number of cases and deaths, respectively (https://www.worldometers.info/coronavirus). Currently, the COVID-19 pandemic is one of the most important health issues in the world and Iran.[3] Given the severity of the incidence of this disease, several researchers around the world have focused on the behavior of the virus and the effects of its far-reaching dimensions on human life. Therefore, reviewing the studies published in the Scopus profile has shown the allocation of 2.5% of all articles in the world to COVID-19 (https://www.scopus.com).

Studies conducted on COVID-19 have mainly focused on the clinical aspects and immunopathogenesis[7–14] of the virus, cities, and urbanism in the future,[15–17] economic effects,[18–23] and social consequences of this virus.[24–27] Considering the high number of cases and mortality caused by COVID-19 in Iran as well as the issues remaining unknown related to this virus, it is necessary to develop and conduct applied research in this field. This study, from a statistical-analytical point of view, it was tried to examine the COVID-19 virus epidemic about the importance of the outcomes of the treatment of this disease and the risks and costs it imposes to the people and the health field of the world and Iran. Moreover, the rate of cases and mortality due to the COVID-19 virus is evaluated as the most important cause of mortality in Iran and the world in different continents.

2 Methodology

2.1 Statistical analysis of COVID-19

The present work is an applied and descriptive-analytical study. The statistical population of the study includes all the infected and deaths due to COVID-19 in Iran and the world. For statistical analysis of the prevalence of the COVID-19 virus, the official data reported for the world and Iran were used. This dataset was obtained from https://www.worldometers.info/coronavirus and https://github.com/owid/covid-19-data/tree/master/public/data, and the Ministry of Health and Medical Education. First, the high amount of monthly data for corona-caused infections and deaths in Iran were analyzed based on the months of Shamsi Hijri. Then, the data for the infected and dead cases of the world were examined for each continent. Finally, a comparison was made between the death rates due to COVID-19 and the 34 major causes of death in the world. The study data consisting of long-term data of the major deaths from 1990 to 2017 were obtained on an annual average from https://ourworldindata.org.statistical analysis of available data is the most basic principle for planning and formulating control strategies, policy-making, and management.

2.2 Models for death cases trend

The behavior of the variable death cases was captured using a fourth-degree polynomial specification as follows:

$$\text{Total death}\,(t) = \alpha_0 + \alpha_1 t + \alpha_2 t^2 + \alpha_3 t^3 + \alpha_4 t^4 \tag{1}$$

where Total death (t) represents the total number of deaths on day t and t denotes the days starting from 19th of February and 22 of January 22, 2020, for Iran and the world as a whole, respectively. Other specifications, including quadratic and third-degree polynomial specifications, were also examined and it was found that the fourth-degree form has more accurate predictions. The cubic form of the specification was used by Aik et al.[28] to examine the Salmonellosis incidence in Singapore and for COVID-19 death cases in Iran by Pourghasemi et al.[29]

We also used an ARMA model to compare the process generating the variables for Iran and the World. This model includes two processes: autoregressive (AR) and moving average (MA). An ARMA model of order (p,q) can be written as[30]

$$y(t) = \beta_0 + \sum_{i=1}^{p} \beta_i y_{t-i} + \sum_{j=1}^{q} \beta_j \varepsilon_{t-j} \tag{2}$$

where y is the dependent variable and ε is the white noise stochastic error term. In the applied model, y denotes the total number of deaths, and t is the number of days starting from the first day of death.

Regarding the significant fluctuations in the daily death cases, a volatility model is known as generalized autoregressive conditional heteroskedasticity (GARCH) or GARCH model was applied that in general form for order of q and p can be presented as:

$$\sigma^2(t) = \theta_0 + \sum_{j=1}^{q} \theta_j \sigma^2_{t-j} + \sum_{i=1}^{p} \theta_i \epsilon_{t-i} \qquad (3)$$

In the Eq. (3) $\sigma^2(t)$ is the one-period ahead forecast variance which is obtained from past information, θ_0 is a constant term and ϵ is residual obtained from the mean equation.

In building a time-series model, the data must be stationary.[31] In other words, ARMA model is based on the assumption that the data series is stationary. In brief, a time series process $x(t)$ is stationary if the mean and variance are constant and independent of time and the covariances given by $cov(x(t), x(t-s) = \gamma(s))$ is time-invariant, which is dependent only on the distance between the two time periods considered.[32, 33] Thus, if a time series has a time-varying mean or time-varying variance or both will be nonstationary. Using nonstationary time series for forecasting purposes has little practical value. If the applied time series data are not stationary, after differencing it'd time, we obtain a stationary time series. This series is called an integrated of order d. After differencing d times, the ARMA (p,q) model is called ARIMA(p,d,q).[32] In other words, an ARIMA(p,d,q) is an ARMA(p,q) that applies d-times differencing data. Benvenuto et al.[34] also applied an ARIMA model to predict the epidemiological trend of COVID-2019. In addition, Saba and Elsheikhb[35] used this model to forecast the outbreak of COVID-19 in Egypt.

3 Results

3.1 Analysis of official reports of coronavirus spread in Iran

The number of coronavirus infections and deaths from the onset of the pandemic (February 19, 2020) of the pandemic to January 19, 2021 (11 months) according to the Shamsi Hijri (between Bahman 30, 1398 and Dey 30, 1399) in the country indicates that the number of infected and dead people in Iran was 1,342,134 and 56,973, respectively. The maximum number of daily infected cases (1365 individuals) from February 19, 2020, to March 19, 2020, was related to March 14. During the same period (30 days), the maximum number of deaths on March 19, 2020, was 149 individuals (Fig. 1). In other words, the average daily number of deaths was 43 in average in this period of 30 days.

Analysis of COVID-19 outbreaks from March 23 to March 30, the number of infected cases increased, with the highest number of daily infected cases of 3186 individuals found on March 30 (Fig. 2). However, a declining trend was observed from March 30 to April 19. The highest daily death rate (158 cases) in this month (March 20 to April 19) was reported on April 4 (Fig. 2). In other words, the average number of deaths was 124. Therefore, compared to 30 days in 2020 (February 19 to March 19), the country encountered three peaks for the number of deaths.

Analysis of the COVID-19 epidemic from April 20 to May 20 showed that the number of infected individuals was 44,738. The highest daily number of infected individuals was related to May 20, with 2346 individuals (Fig. 3). In other words, the number of people infected with the coronavirus in this month is approximately 0.7 of the previous month. The number of deaths reported between April 20 and May 20 was 2065. The highest daily death rate in this month was reported on April 27, which was 96 individuals (Fig. 3). The number of deaths from March 20 to April 19 was 0.54 1.62, which was higher than those reported from February 19 to March 19, 2020.

Analysis of coronavirus from May 21 to June 20 indicated that the number of infected cases increased, reaching 75,646. The highest number of daily infections was identified on June 4, with 3574 individuals. The number of infected cases in June was 1.69 times more than that reported in May and 1.19 times more than that in April (Fig. 4). The total number of deaths in June was 2325. The highest daily death rate in this month was related to June 19, which was reported in 120 individuals (Fig. 4). From May 21 to June 20, an increasing trend in the number of deaths was found compared to April 20 to May 20, with 260 individuals. In other words, the number of deaths from May 21 to June 20 was 1.13 times more than that between April 20 and May 20, 0.61 times more than those from March 20 to April 19, and 1.82 times more than those from February 19 to March 19, 2020.

The number of coronavirus-infected cases from June 21, 2020, to July 21, 2020, in Iran was 76,243. The highest daily number of infected cases was found on July 8, with 2691 individuals (Fig. 5). It should be emphasized that the daily number of COVID-19 cases was over 2000 individuals every day in July. According to the analysis, the number of infected cases from June 21 to July 21 was 1.19 times more than that from March 20 to April 19, 1.70 times more than those from April 20 to May 20, and 1.01 times more than those from May 21 to June 20. In addition, the total number of

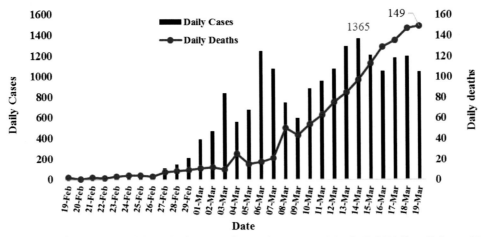

FIG. 1   The number of daily infected cases and daily deaths in Iran from February 19 to March 19, 2020 (from Bahman 30 to Esfand 29, 1398).

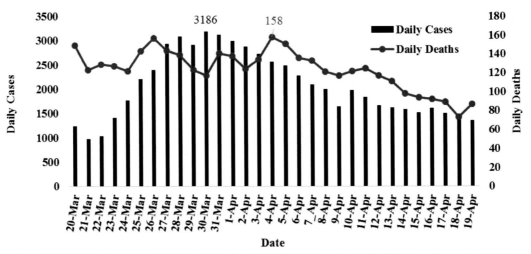

FIG. 2   The number of daily infected cases and daily deaths in Iran from March 20 to April 19, 2020 (from Farvardin 1 to Farvardin 31, 1399).

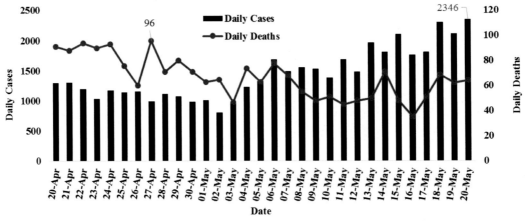

FIG. 3   The number of daily infected cases and daily deaths in Iran from April 20 to May 20, 2020 (from Ordibehesht 1, 1399 to Ordibehesht 31, 1399).

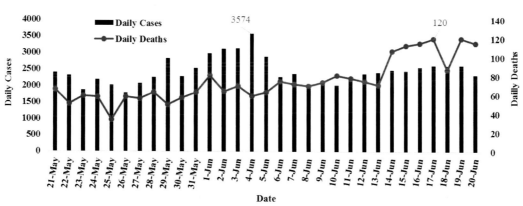

FIG. 4 The number of daily infected cases and daily deaths in Iran from May 21 to June 20, 2020 (from Khordad 1 to Khordad 31, 1399).

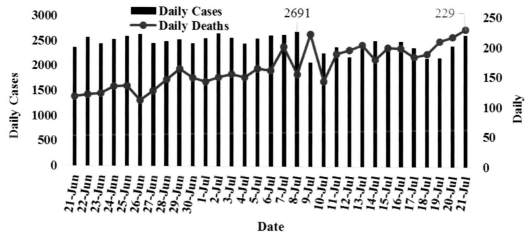

FIG. 5 The number of daily infected cases and daily deaths in Iran from June 21 to July 21, 2020 (from Tir 1 to Tir 31, 1399).

dead people in July in Iran was 5127, and the highest daily death rate in this month (June 21 to July 21) was related to July 21, which was reported to be 229 individuals (Fig. 5). The number of deaths from June 21 to July 21 was 2.21 times higher than those from May 21 to June 20, 2.48 times higher than those detected from April 20 to May 20, 1.34 times higher than those in the period from March 20 to April 19 and 4.02 times more than those in the period from February 19 to March 19, 2020.

Analysis of coronavirus spread between July 22 and August 21 illustrated that the total number of infected cases was 75,937 individuals and the highest daily number of infected cases was related to August 4, with 2751 individuals (Fig. 6). In August, the total number of dead people was 5742 and the highest daily death (235 individuals) in this month (July 22 to August 21) was found on July 28(Fig. 6). In Asia, of the 49 countries affected by the coronavirus

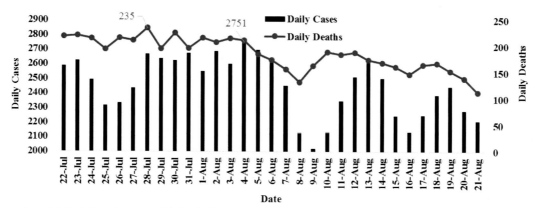

FIG. 6 The number of daily infected cases and daily deaths in Iran from July 22 to August 21, 2020 (from Mordad 1 to Mordad 31, 1399).

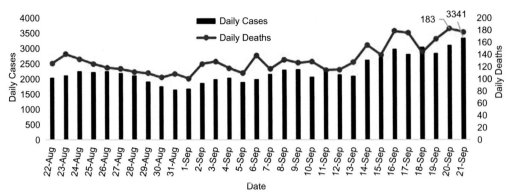

FIG. 7 The number of daily infected cases and daily deaths in Iran from August 22 to September 21, 2020 (from Shahrivar 1 to Shahrivar 31, 1399).

this month (July 22 to August 21), Iran has the second-highest place in terms of the number of infections and deaths, while it is ranked 10th in Asia in terms of population.

Analysis of coronavirus-infected cases from August 22 to September 21 illustrated that the total number of infected individuals was 70,717, and the highest daily number of infected individuals was related to September 21, with 3341 individuals (Fig. 7). Additionally, the total number of dead people was reported to be 4102 and the highest daily death rate in this month (from August 22 to September 21) was related to September 20, which was 183 individuals (Fig. 7).

Analysis of the COVID-19 outbreak between September 22 and October 21 illustrated that the total number of infected cases was 124,056 and 5616 on October 21, respectively (Fig. 8). In addition, the total number of dead people in this month was 6870 and the highest daily death rate in this month (from September 22 to October 21) was related to October 19, which was 337 individuals (Fig. 8).

The total number of infected people from October 22, 2020, to November 20, 2020, was 282,179. The highest daily number of infected individuals was related to November 16, with 13,421 individuals (Fig. 9). In addition, the total number of dead people in November in Iran was 12,550, and the highest daily death rate in this month (from October 22 to November 20) was related to November 16, which was reported to be 489 individuals (Fig. 9).

Analysis of the COVID-19 outbreak from November 21 to December 20 illustrated that the total number of infected individuals was 330,007, with the highest daily number of infected cases (14,051 individuals), which was related to November 27 (Fig. 10). In addition, the total number of dead people in this month was 9719 and the highest daily death rate in this month (November 21 to December 20) was related to November 24, which was 483 individuals (Fig. 10).

The number of COVID-19 cases from December 21, 2020, to January 19, 2020, in the country indicated that the total number of infected people was 183,750. The highest daily number of infected individuals was observed on January 15, with 6485 individuals (Fig. 11). In addition, the total number of dead people in January in Iran was 3348, and the highest daily death rate in this month (from December 21 to January 19) was related to December 21, which was reported to be 191 individuals (Fig. 11).

In this study, the number of infected patients and deaths from onset (February 19, 2020) to January 19, 2021 (11 months) in Iran was reported. The results of infected patients showed that the highest number of infected patients was related to the period of November 21 to December 20, followed by the period of October 22 to November 20, the period of December 21 to January 19, the period of September 22 to October 21, the period of June 21 to July 21, the

FIG. 8 The number of daily infected cases and daily deaths in Iran from September 22 to October 21, 2020 (from Mehr 1 to Mehr 30, 1399).

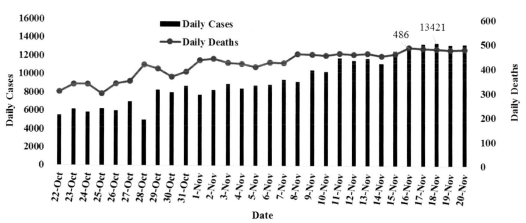

FIG. 9 The number of daily infected cases and daily deaths in Iran from October 22 to November 20, 2020 (from Aban 1 to Aban 30, 1399).

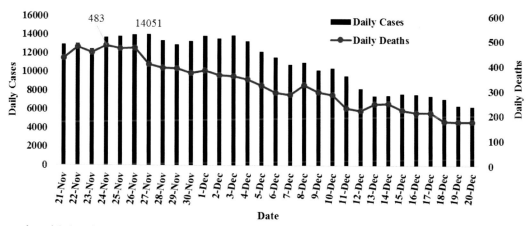

FIG. 10 The number of daily infected cases and daily deaths reported from November 19 to December 20, 2020, in Iran (from Azar 1 to Azar 30, 1399).

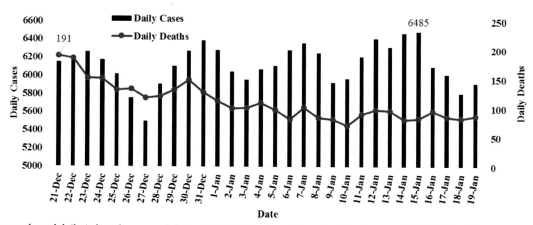

FIG. 11 The number of daily infected cases and daily deaths in Iran from December 21 to January 19, 2020 (from Dey 1 to Dey 30, 1399).

period of July 22 to August 21, the period of May 21 to June 20, the period of August 22 to September 21, the period of March 20 to April 19, the period of April 20 to May 20, and the period of February 19 to March 19, respectively (Fig. 12). Moreover, the result of deaths showed that the highest rate of deaths was related to the period of October 22 to November 20, followed by the period of November 21 to December 20, the period of September 22 to October 21, the period from July 22 to August 21, the period from June 21 to July 21, the period from August 22 to September 21, the period of March 20 to April 19, the period of December 21 to January 19, the period of May 21 to June 20, the period of April 20 to May 20, and the period of February 19 to March 19, respectively (Fig. 13).

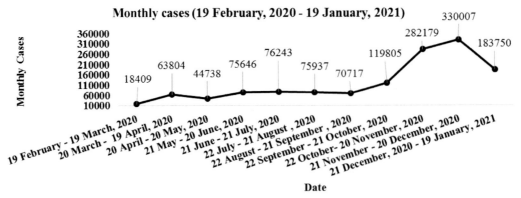

FIG. 12 The highest number of infected cases for 11 months from February 19, 2020, to January 19, 2021, in Iran (between Bahman 30, 1398 and Dey 30, 1399).

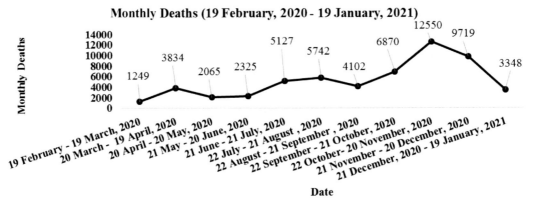

FIG. 13 The highest number of deaths in Iran for 11 months from February 19, 2020, to January 19, 2021 (between Bahman 30, 1398 and Dey 30, 1399).

The total number of coronavirus infections and deaths from January 2020 to January 2021 in the world indicates that the maximum number of infected and dead people were related to December 2020 and January 2021, which were 20,124,074 and 406,831, respectively. In addition, the minimum number of infections and deaths was related to January 2020 (Figs. 14 and 15). The majority of coronavirus cases were related to December 2020 (19.57%), followed by January 2021 (18.77%), and November 2020 (16.79%). In addition, the majority of coronavirus deaths were related to January 2021 (18.27%), followed by December 2020 (15.72%), and November 2020 (12.26%) (Fig. 16 and Table 1).

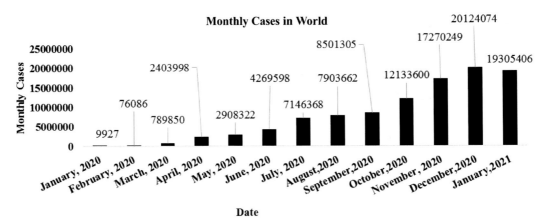

FIG. 14 Total monthly coronavirus infections in the world (from January 2020 to January 2021).

FIG. 15 Total monthly deaths in the world (from January 2020 to January 2021).

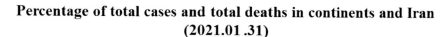

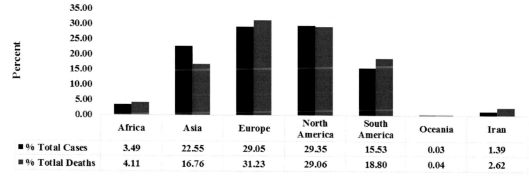

	Africa	Asia	Europe	North America	South America	Oceania	Iran
■ % Total Cases	3.49	22.55	29.05	29.35	15.53	0.03	1.39
■ % Total Deaths	4.11	16.76	31.23	29.06	18.80	0.04	2.62

FIG. 16 Percentage of total cases and total deaths in continents and Iran (2021.01.31).

TABLE 1 The percentage of coronavirus infections and deaths in the world (from January 2020 to January 2021).

Month	% Total cases	% Total deaths
January, 2020	0.01	0.01
February, 2020	0.07	0.12
March, 2020	0.77	1.86
April, 2020	2.34	8.67
May, 2020	2.83	6.18
June, 2020	4.15	5.99
July, 2020	6.95	7.51
August, 2020	7.69	7.88
September, 2020	8.27	7.36
October, 2020	11.80	8.15
November, 2020	16.79	12.26
December, 2020	19.57	15.72
January, 2021	18.77	18.27

3.2 Comparison of COVID-19 mortality rate and other hazards in Iran and the world

According to long-term data from 1990 to 2017, the most important causes of mortality in Iran and the world[36] were studied separately in different continents of the 34 major deaths worldwide, cardiovascular diseases and tumors are ranked first and second, respectively.

The death rate from coronavirus is the eighth leading cause of death worldwide (Fig. 17). in Europe, which showed 30% of coronavirus-caused deaths in the world, it ranked third after cardiovascular disease and tumors[36] (Fig. 18). Coronavirus mortality in Asia accounted for 18.6% of coronavirus-caused deaths worldwide, which ranked 15 among 34 fatalities in Asia (Fig. 19). In Oceania, corona-caused deaths account for approximately 0% of the world's corona-related deaths and it is ranked 32nd among the 34 leading causes of death in Oceania (Fig. 20). Coronavirus-caused deaths in Africa account for only 3.5% of coronavirus deaths and it is ranked 19th among the 34 leading causes of death in Africa (Fig. 21). South America accounts for 20%, North America accounts for 28% of the world's coronavirus-caused deaths, and COVID-19 related mortality is the third leading cause of death in South and North America after cardiovascular and tumors (Figs. 22 and 23). In Iran, coronavirus-related mortality accounts for 3% of the global mortality rate due to this virus, and this fatal factor is the second leading cause of death after cardiovascular and tumor diseases, road accidents, and neonatal disorders (Fig. 24).

3.3 Comparison of world and Iranian death cases trend

Fig. 25 and the ARIMA model results presented in Table 2 indicate that the Iranian death cases experience a relatively flatter trend compared to the world. As shown in Fig. 25, both cases are presented with a fourth-degree polynomial specification; however, those for the world case sound steeper. For both, the number of deaths is increasing exponentially, but for the World at a steeper rate. This fact has been examined more deeply and quantitatively in Table 2, where the ARIMA estimation results have been presented. In addition, the ARCH specification was applied to model volatility as well. Based on the polynomial model, the number of deaths increased over the selected horizon. The first derivative of the estimated model, which turns it into a third-degree polynomial equation, represents daily death cases. However, the current trend of death cases failed to show an acceptable turning point, while there is some

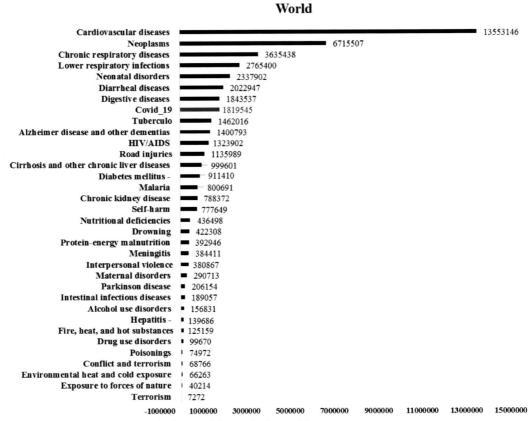

FIG. 17 The death rate from Coronavirus and other lethal factors in the world.

Europe

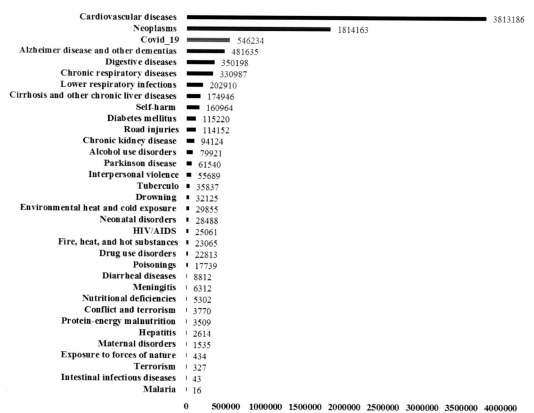

FIG. 18   The death rate from Coronavirus and other lethal factors in Europe.

Asia

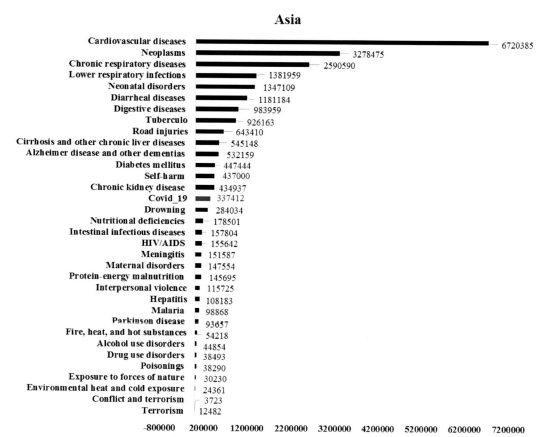

FIG. 19   The death rate from Coronavirus and other lethal factors in Asia.

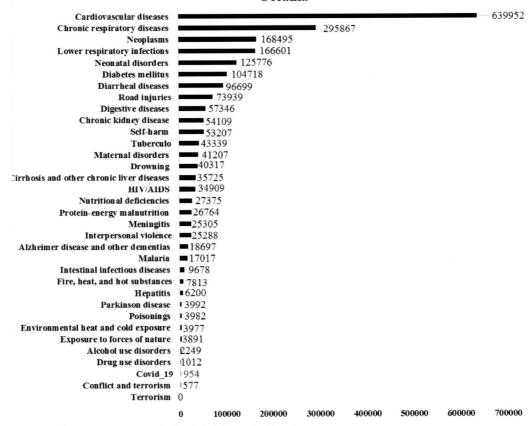

FIG. 20 The death rate from Coronavirus and other lethal factors in Oceania.

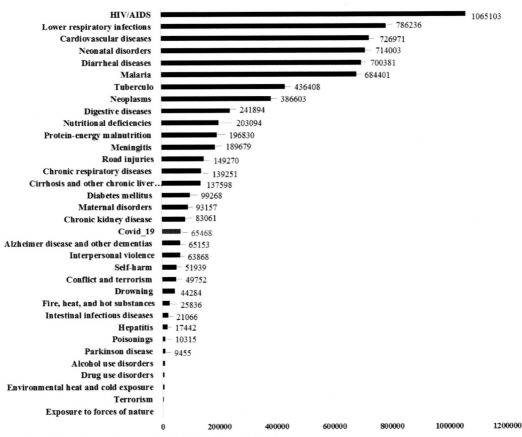

FIG. 21 The death rate from Coronavirus and other lethal factors in Africa.

North America

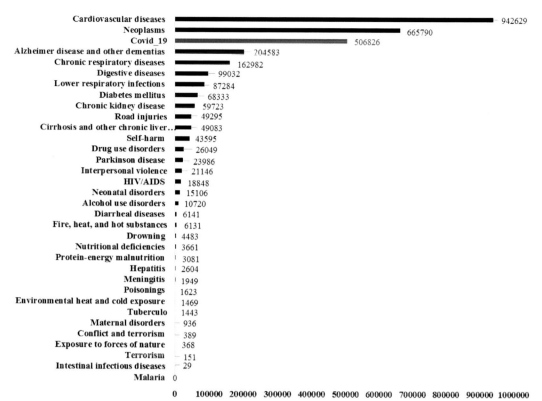

Cardiovascular diseases	942629
Neoplasms	665790
Covid_19	506826
Alzheimer disease and other dementias	204583
Chronic respiratory diseases	162982
Digestive diseases	99032
Lower respiratory infections	87284
Diabetes mellitus	68333
Chronic kidney disease	59723
Road injuries	49295
Cirrhosis and other chronic liver..	49083
Self-harm	43595
Drug use disorders	26049
Parkinson disease	23986
Interpersonal violence	21146
HIV/AIDS	18848
Neonatal disorders	15106
Alcohol use disorders	10720
Diarrheal diseases	6141
Fire, heat, and hot substances	6131
Drowning	4483
Nutritional deficiencies	3661
Protein-energy malnutrition	3081
Hepatitis	2604
Meningitis	1949
Poisonings	1623
Environmental heat and cold exposure	1469
Tuberculo	1443
Maternal disorders	936
Conflict and terrorism	389
Exposure to forces of nature	368
Terrorism	151
Intestinal infectious diseases	29
Malaria	0

FIG. 22 The death rate from Coronavirus and other lethal factors in North America.

South America

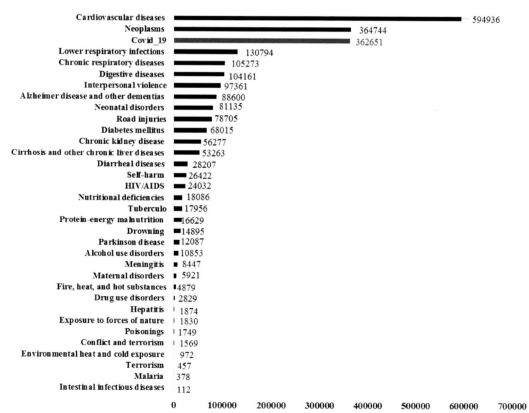

Cardiovascular diseases	594936
Neoplasms	364744
Covid_19	362651
Lower respiratory infections	130794
Chronic respiratory diseases	105273
Digestive diseases	104161
Interpersonal violence	97361
Alzheimer disease and other dementias	88600
Neonatal disorders	81135
Road injuries	78705
Diabetes mellitus	68015
Chronic kidney disease	56277
Cirrhosis and other chronic liver diseases	53263
Diarrheal diseases	28207
Self-harm	26422
HIV/AIDS	24032
Nutritional deficiencies	18086
Tuberculo	17956
Protein-energy malnutrition	16629
Drowning	14895
Parkinson disease	12087
Alcohol use disorders	10853
Meningitis	8447
Maternal disorders	5921
Fire, heat, and hot substances	4879
Drug use disorders	2829
Hepatitis	1874
Exposure to forces of nature	1830
Poisonings	1749
Conflict and terrorism	1569
Environmental heat and cold exposure	972
Terrorism	457
Malaria	378
Intestinal infectious diseases	112

FIG. 23 The death rate from Coronavirus and other lethal factors in South America.

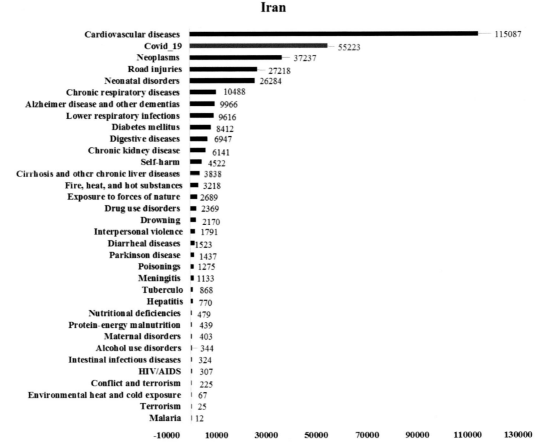

FIG. 24   The death rate from Coronavirus and other lethal factors in Iran.

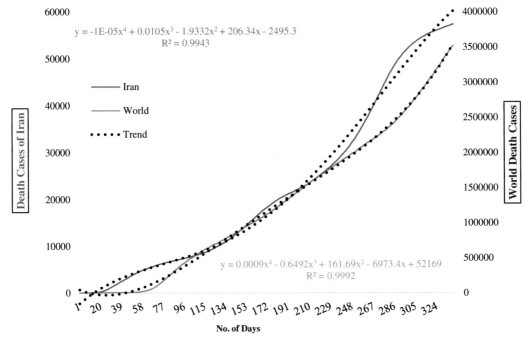

FIG. 25   Actual cases versus estimated cases in Iran and the World.

TABLE 2 The results of the ARMA model for COVID-19 death cases of World and Iran.

	Regressor	Coefficient	Standard error	t-statistics	Probability
World Mean Eq.	Constant	−5.436	443,247.3	−0.000	1.000
	AR(6)	0.256	0.025	10.318	0.000
	AR(7)	0.743	0.072	10.317	0.000
	MA(1)	−0.586	0.026	−22.466	0.000
	MA(8)	0.289	0.015	19.137	0.000
	Dummy	−592.949	353.304	−1.678	0.093
Variance Eq.					
	Constant	843.739	301.922	2.794	0.005
	Resid^2(−1)	0.558	0.098	5.646	0.000
	GARCH(−1)	0.672	0.042	15.757	0.000
	Dummy	116,803.7	64,816.40	1.802	0.071
	Adjusted R^2	0.540			
	Q(5)[a]	5.012			0.025
	Q(6)[a]	5.282			0.071
	Heteroskedasticity (ARCH)	0.002			0.960
Iran Mean Eq.	Constant	275.401	79,934.73	0.003	0.997
	AR(1)	0.999	0.000	325,463.5	0.000
	MA(1)	−0.395	0.060	−6.581	0.000
	MA(7)	0.153	0.058	2.619	0.008
	MA(8)	0.163	0.060	2.716	0.006
	Dummy	−105.274	52.225	−2.015	0.043
Variance Eq.					
	Constant	5254.662	1229.806	4.272	0.000
	Resid^2(−1)	0.206	0.083	2.471	0.013
	Dummy	−5051.708	1229.239	−4.109	0.000
	Adjusted R^2	0.976			
	Q(5)[a]	3.661			0.057
	Q(6)[a]	4.563			0.102
	Heteroskedasticity (ARCH)	0.646			0.421

[a] Q(p) is the significance level of the Ljung-Box statistics, in which the first p of the residual autocorrelations are jointly equal to zero.

evidence showing that a turning point in infection is expected. For instance, it has been reported for SARS incidence,[37] HAV,[38] ARI,[39] and A(H1N1)v.[40] It is worth noting that a turning point means that after passing the peak, it is expected to show a decreasing trend. This fact may also reveal the different nature of the COVID-19 virus compared to other viruses. Another possibility is that the measures taken up to now have not been enough and effective and other measures are needed. Regarding the leading coefficient of the polynomial model, we may expect a decreasing trend in Iran's daily cases in the future. It is worth noting that the comparison of the specified models is more appropriate for investigating the effectiveness of the measures taken by the corresponding health body rather than using it to predict future values.

The ARIMA time series models for the death variables of the whole world and Iran are presented in Table 2. These models may show the process of generating the variables in the time horizon. The data series stationarity was tested to evaluate whether the variable was stationary or not. The results of the unit root test for Iran showed that the death case series are not stationary at the level and the first difference showed stationarity; however, it was found that there was a break at the 9th month as well. Thus, a breakpoint unit root test was conducted. In addition, this breakpoint was included in the ARIMA model using dummy variables. The world's series was also difference-stationary, but the second-order differenced series was found to be stationary. This may also reveal the curvature of the steeper trend, as illustrated in Fig. 25.

As shown in Table 2, the world model is generated by an ARIMA $(2_{(6, 7)},2,2_{(1, 8)})$ process, while for the Iranian death trend, a simpler process was obtained, resulting in an ARIMA $(1,1,3_{(1, 7, 8)})$. Although the absolute values of the AR terms for both are similar, it is worth noting that the data for the world was applied after two different times, indicating a slower process of an increasing trend for Iran compared to those of the world.

In addition to the equation that is an ARIMA model, the variance of the series was examined using Generalized Autoregressive Conditional Heteroskedasticity (GARCH) models. For Iran's specification, GARCH (1,0) was estimated, while for the World's data, GARCH (1) was found to be appropriate. Given the significant coefficients for these variables, the death cases for the World tend to fluctuate more than those in Iran. This is a fact that is not easily captured in the trends shown in Fig. 25.

Another interesting result was a breakpoint that was found in the unit root test of data applied in Iran. It was observed that in the 9th month after the first death cases were reported, a steeper increasing trend was observed. The negative coefficient for the dummy variable shows lower death cases for the first 9 months in Iran. In addition, in the variance equation, lower volatility is observed for the first 9 months. In other words, after 9 months, an increasing wave of death cases is observed, while it has experienced higher volatility. This higher volatility may stem from the periodical efforts of the government to quarantine and impose more social restrictions. For the world's cases, a dampening trend was found in the last 37 days; however, the estimated coefficients were significant only at the 10% level.

Generally speaking, the diagnostic statistics indicate that the estimated models are acceptable because the Q-statistics indicate that the residuals are not significantly correlated. In addition, the ARCH effect was not significant, indicating insignificant volatility in the residuals of the estimated equations. In addition, except for dummy variables in the World's specification and the dummy variable in the mean Eq. of specifications presented for Iran, the coefficients are significant at 99%.

4 Discussion

The COVID-19 epidemic is a global challenge and threat in recent history and the need for timely, appropriate and cost-effective policies and measures to restrain and mitigate its deadly consequences is essential. Reviewing monthly statistics showed that from February 19, 2020, to January 19, 2021, Iran had been faced with three major peaks of COVID-19 infections. The first peak was 1 month after the onset of the disease between March 20 and April 19, 2020, the second peak was from May 21 to June 20, 2020, and the third peak was from September 21 to October 2020. The main problem in Iran is the large number of domestic trips that have contributed to the rapid spread of the virus and has faced the country with three peaks over a period of 11 months (February 19, 2020–January 19, 2021).[41-43] From April 20 to May 20, 2020, and December 21, 2020, to January 19, 2021, a declining trend was observed for the number of infected and deaths cases due to COVID-19. The reduction rate of the number of infected and deaths cases from April 20 to May 20, 2020, was 1.4 and 1.86, respectively, compared to the period of March 20 to April 19, 2020, and the same reduction rate in the period of December 21, 2020–19 January, 2021, were 1.80 and 2.90, respectively, compared to December. It is noteworthy that Iran has implemented two full quarantine periods, including the closure of shopping centers, educational centers, restaurants, hotels, and entertainment centers on February 21, 2020–April 13, 2020, and November 1, 2020–December 12, 2020. This has been the main reason for the decrease in the number of infected and deaths in the months after the quarantine period; especially in the second period of quarantine, the intercity and interprovincial restrictions of personal cars were also added to the previous restrictions and it has reduced domestic travel and concentrations of crowds in recreational centers. Therefore, there has been a further decline in the period from December 21, 2020, to January 19, 2021. Some demographers have reported that the rapid spread of COVID-19 in Italy was also due to high intergenerational contact, especially among young travelers who travel from Milan to their villages.[44, 45] In the Pourghasemi et al.[29] study, the comparison between the statistics of six continents from February 19 to June 14 showed that the highest mortality rate was related to Europe (42.21%), and the highest number of cases was observed in North America (31.12), which was consistent with the results of this study despite adding another period of 7 months. Studies on the COVID-19 virus in different countries have also shown that factors such as (1) age and family structure, (2) patterns of cohabitation, (3) individual characteristics such as socioeconomic status, and (4) ethnicity and mobility have great impacts on the incidence and spread of coronavirus.

Since the world has made significant progress in dealing with many of the deadly causes, high-income countries have viable strategies to cope with infectious diseases, whereas in low-income countries in South Africa, the majority of people die due to infection with the AIDS virus.[36] The use of contaminated water or the loss of children and mothers are among the factors commonly seen in African countries.[46] Coronavirus disease is one of the most dangerous and acute problems that the world has been unable to do much to treat. It should be noted that when faced with a new viral epidemic such as COVID-19 with complicated and unknown biology and transmission potential, it is highly important to estimate the impact of disease on public health in terms of disease severity and mortality. Under these circumstances, estimating the underlying cause of death using a standard approach throughout the country provides a powerful tool for rapidly achieving unbiased estimates of COVID-19 mortality and its impact on different age groups as well as different countries and regions.[47] As statistics show, high-income and populous countries show higher mortality rates for this virus.[48] As the data showed, the Corona-caused mortality rate in Europe, North America, and South America, as well as Iran, followed a similar trend. Contrary to popular belief, the death rate in African countries is significantly lower, but catastrophic death rates are seen in European and American countries. One of the causes of high mortality in European countries is the aging population[49] because most corona victims are elderly.[50-52] Other factors include the lack of a clear border and travel bans between EU countries, especially at the beginning of the pandemic, because there are no strict rules for travel between different countries in the EU and the disease is transmitted very quickly.

Africa accounts for 17% of the world's population, but it includes only 3.5% of the reported cases of COVID-19 worldwide. All deaths are important; therefore, we should not discount the seemingly small number, and certainly, the data collected in a wide range of countries is of variable quality.[48] Most African countries, even with less advanced health systems, still experience lower rates of COVID-19 mortality.[48] Although the highest mortality rate in Africa has been related to the HIV, Africa has been less vulnerable to the coronavirus compared to other countries in the world.[36]

In Lawal,[48] the average age and life expectancy were lower in most African countries than in other countries, and this has created a population pyramid consisting of a predominantly young population in Africa, while Western countries and other developed countries have a predominantly older population. This discrepancy may be explained by the higher birth rates in less developed African countries and more advanced health systems in developed countries, which allow their citizens to live longer. Additionally, Africa, with its less advanced health systems, continues to suffer from communicable and noncommunicable diseases that lead to reduced life expectancy. In this regard, mortality data showed that 65-year-old mortality is clearly higher in the West and other developed countries than in Africa. This is obviously because a much larger proportion of the population in the West and other industrialized countries tends to be over 65 years old in comparison with the population of Africa, and this rate is reflected in the COVID-19 mortality rate, as older people are most likely to be at risk of dying from COVID-19.

Moreover, the low mortality rate due to corona in the Pacific continent can be attributed to the isolation of this continent and the lack of relations with other parts of the world. Social distance, public health measures, and reduced international travel might have been effective in reducing the spread of the disease in Australian society.[53]

In the Yamamotoa and Bauer[54] study, fewer deaths in Asia that were less than those in Europe were compared. The results of the Yamamotoa and Bauer[54] study showed that (1) differences in social behaviors and culture of the people in the two regions; (2) the possible prevalence of dangerous viruses in Central Europe due to multiple viral infections, and the involvement of related viral immune factors; and (3) possible involvement of health factors, including cultural and behavioral differences between people in Central Europe and Eastern Asia, viral factors, and even anthropological issues such as human evolution were the main factors affecting the differences between the two continents. In the Sorci et al.[55] study, economic parameters played an equal role in the formation of COVID-19 mortality. With the increase in the number of severe cases during the epidemic, the healthcare system may encounter difficulties and may not be able to receive and treat all those in need of special care. Therefore, mortality may be due to health care systems that are not sufficient to deal with the large number of cases requiring simultaneous hospitalization in special care units. In this study, several proxies were used to describe each country's investment in the healthcare system and a negative relationship was found between the number of hospital beds and corona mortality rates. Contrary to this view, however, they found that the death rate from the coronavirus was highest in countries with high GDP per capita and high overall health expenditures as a share of GDP. This is a strange result, but it confirms the notion that the rich countries of Europe and North America have paid a lot of damage to this infection. In general, the relationship between investment in the healthcare system and coronavirus mortality appears to be more complex than expected. In Iran, COVID-19-related mortality is the second most common cause of death. One of the most important reasons for the difference in mortality rates in some countries, such as Iran, with other countries, is the lack of attention to the observance of health protocols.

Factors such as age, sex, underlying diseases, and broader social factors influence the understanding of the impact of coronavirus disease on mortality worldwide. Recall that the use of younger age structures in low-income countries may be partially compensated by health system weakness and a higher prevalence of HIV, tuberculosis, and other chronic diseases.[44] In fact, in countries with poorer health, we will see a change in the age distribution of COVID-19 related deaths in a younger population compared to rich and healthier countries with a higher rate of underlying diseases such as diabetes and hypertension.[56] Data from the United States and the United Kingdom also show that some ethnic minorities and communities with socio-economic disadvantages are at a high risk of mortality due to COVID-19 infection.[57–59] Undoubtedly, accelerating biological aging along with social and economic harm plays an important role in the incidence of COVID-19 world.[60] Naturally, the real world represents a more complex dynamic, but there is still a strong relationship between age and mortality due to COVID-19.

5 Conclusions

Analysis of the statistical reports of coronavirus infections and deaths in Iran showed that the spread of the virus in different months, from the start of the outbreak (February 19, 2020) to January 19, 2021, has an increasing trend, so that the highest number of coronavirus infections and deaths were reported in November 21 to December 20, 2020, and October 21 to November 20 and 2020, respectively. The highest number of coronavirus infections and deaths were detected in North America and Europe in December 2020 and January 2021, respectively. Among the 34 leading causes of death worldwide, coronavirus was ranked 8th, 19th, 15th, 3rd, 32nd, and 2nd in Africa, Asia, Europe, North America, South America, and Oceania, respectively.

A fourth-degree polynomial and an ARIMA model were applied to examine and compare the general trend of death cases in the world and Iran. While these models apply only one variable, that is, death cases, which are available daily, they are well-powered in depicting the general trend. These tools may be recommended for two reasons. First, they apply only one variable, that is, death cases, which are the most important and easily available. Second, they indicate the effectiveness of the measures taken by governments. The behavior of the estimated equations simply shows how the variable behavior would be if the current attempts to continue. In addition, the estimated models can be used to predict the number of cases in the following days; however, this contribution is less significant than the above-mentioned contributions.

Comparing the general trends of Iran and the world reveals that the trend of death is similar, while the Word's cases show significant volatility and a higher tendency to increase. The more significant volatility for the World may mean that some countries experience volatility in death cases in a direction different from the others, leading to more fluctuation in the death cases. Generally speaking, the significant volatility may mean that even though it is more than 1 year that several attempts have been made to cope with the virus outbreak, the attempts have not been properly and globally adopted, leading to trial-and-error measures and many different treatments. Another source of fluctuation is periodic widespread quarantine, which has been taken after an increasing outbreak. It is not expected to encounter a decreasing trend, indicating the urgent need to keep measures such as quarantine and even leaving room for other attempts with stronger restrictions. Although restricting attempts have been the main measure to cope with the outbreak, governments are always choosing a combination of social and economic restrictions and infection outbreaks because there is a tradeoff between them. The lower the social restrictions, the more the infection outbreak will be, and vice versa.

The Literature shows that the most important factors for coronavirus spread in Iran are the lack of social distance, travel, and high population density due to various events and holidays, which accelerated the spread of the disease. Although the months of complete quarantine have had a great impact on reducing the corona-caused infections and deaths, according to the results of this study, evaluating how the virus has spread and the impact of quarantine and public closure measures requires a comprehensive analysis beyond health indicators. Since the coronavirus is new and unknown, and studies and research have been unable to find a distinct treatment for the virus, analyses presented in the current study might be effective in strategic decision-making and provide solutions on how to deal with the virus epidemic and understand how it spreads.

References

1. Organization WH. *Coronavirus Disease 2019 (COVID-19): Situation Report*; 2020:51. https://www.worldometers.info/coronavirus.
2. World Health Organization (WHO). *Report of the WHO - China Joint Mission on Coronavirus Disease 2019 (COVID-19)*; 2020. Retrieved from https://www.who.int/docs/defaultsource/coronaviruse/who-china-joint-mission-oncovid-19-final-report.pdf.

3. Mollalo A, Vahedi B, Rivera K. GIS-based spatial modeling of COVID-19 incidence rate in the continental United States. *Sci Total Environ.* 2020;728:1–8.

4. World Economic Forum. *Strategic Intelligence: COVID-19;* 2020. *https://intelligence.weforum.org.*

5. World Health Organization. *WHO Report: Global Research Emergency Coronaviruses 2020.* WHO; 2020. [updated 2020 April 8; cited 2020 Mar 25]. Available at http://www.who.int/emergencies/diseases/novel-coronavirus-2019/global-research-on-novel-coronavirus-2019-ncov.

6. World Health Organization. *Novel Coronavirus (2019-nCoV): Strategic Preparedness and Response Plan.* WHO; 2019–2020. [updated 2020 Feb 3; cited 2020 Mar 4]. Available from: *https://www.who.int/docs/default-source/coronaviruse/srp04022020.pdf.*

7. Rithanya M, Brundha M. Molecular immune pathogenesis and diagnosis of COVID-19—a review. *Int J Cur Res Rev.* 2020;12(21):69.

8. Lai C-C, Liu YH, Wang C-Y, et al. Asymptomatic carrier state, acute respiratory disease, and pneumonia due to severe acute respiratory syndrome coronavirus 2 (SARS-CoV-2): facts and myths. *J Microbiol Immunol Infect.* 2020;53(3):404–412.

9. Sun J, He W, Wang L, et al. COVID-19: epidemiology, evolution, and cross-disciplinary perspectives. *Trends Mol Med.* 2020;26(5):483–495. https://doi.org/10.1016/j.molmed.2020.02.008.

10. Wan S, Yi Q, Fan S, et al. Characteristics of lymphocyte subsets and cytokines in peripheral blood of 123 hospitalized patients with 2019 novel coronavirus pneumonia (NCP*). MedRxiv.* 2020.

11. Huang C, Wang Y, Li X, et al. Clinical features of patients infected with 2019 novel coronavirus in Wuhan, China. *Lancet.* 2020;395 (10223):497–506.

12. Wang P, Lu J, Jin Y, Zhu M, Wang L, Chen S. Epidemiological characteristics of 1212 COVID-19 patients in Henan, China. *medRxiv.* 2020.

13. Filatov A, Sharma P, Hindi F, Espinosa PS. Neurological complications of coronavirus disease (COVID-19): encephalopathy. *Cureus.* 2020;12(3).

14. Ye M, Ren Y, Lv T. Encephalitis as a clinical manifestation of COVID-19. *Brain Behav Immun.* 2020;85:945–946. https://doi.org/10.1016/j.bbi.2020.04.017.

15. Batty M. *The Coronavirus Crisis: What Will the Post-Pandemic City Look like?* UK: London, England: SAGE Publications Sage; 2020.

16. Sharifi A, Khavarian-Garmsir AR. The COVID-19 pandemic: impacts on cities and major lessons for urban planning, design, and management. *Sci Total Environ.* 2020;142391.

17. Daneshpour ZA. *Out of the Coronavirus Crisis, A New Kind of Urban Planning Must be Born.* Accessed on 30; 2020.

18. Al-Ubaydli O. *Understanding How the Coronavirus Affects the Global Economy: A Guide for Non-Economists.* Munich Personal RePEc Archive (MPRA); 2020. https://mpra.ub.uni-muenchen.de/99642/.

19. Duffin E. *Impact of the Coronavirus Pandemic on the Global Economy-Statistics & Facts.* Statista Recuperado de; 2020. https://www.statista com/topics/6139/covid-19-impact-on-the-global-economy/#topFacts_wrapper.

20. Fernandes N. *Economic Effects of Coronavirus Outbreak (COVID-19) on the World Economy.* Available at SSRN 3557504; 2020.

21. Laing T. The economic impact of the coronavirus 2019 (Covid-2019): implications for the mining industry. *Extract Ind Soc.* 2020;7(2):580–582.

22. Canuto O. *The Impact of Coronavirus on the Global Economy;* 2020. https://www.africaportal.org/publications/impact-coronavirus-global-economy/.

23. Mele C, Russo-Spena T, Kaartemo V. The impact of coronavirus on business: developing service research agenda for a post-coronavirus world. *J Serv Theory Pract.* 2020.

24. Fuchs C. Everyday life and everyday communication in coronavirus capitalism. tripleC: communication, capitalism & critique open access. *J Glob Sustain Inf Soc.* 2020;18(1):375–399.

25. Ratten V. Coronavirus (covid-19) and social value co-creation. *Int J Sociol Soc Policy.* 2020.

26. Javadi SMH, Arian M, Qorbani-Vanajemi M. The need for psychosocial interventions to manage the coronavirus crisis. *Iran J Psychiatry Behav Sci.* 2020;14(1), e102546.

27. Proaño CR. On the macroeconomic and social impact of the coronavirus pandemic in Latin America and the developing world. *Intereconomics.* 2020;55:159–162.

28. Aik J, Heywood AE, Newall AT, Ng LC, Kirk MD, Turner R. Climate variability and salmonellosis in Singapore–a time series analysis. *Sci Total Environ.* 2018;639:1261–1267.

29. Pourghasemi HR, Pouyan S, Heidari B, et al. Spatial modeling, risk mapping, change detection, and outbreak trend analysis of coronavirus (COVID-19) in Iran (days between February 19 and June 14, 2020). *Int J Infect Dis.* 2020;98:90–108.

30. Enders WA. *Applied Econometric Time Series.* 2nd ed. New York (US): University of Alabama; 2004.

31. Ahmar AS, Del Val EB. SutteARIMA: short-term forecasting method, a case: Covid-19 and stock market in Spain. *Sci Total Environ.* 2020 Aug 10;729:138883.

32. Gujarati DN, Porter DC, Gunasekar S. *Basic Econometrics.* Tata McGraw-Hill Education; 2012.

33. Baltagi BH. *Econometrics.* 4th ed. Berlin: Springer; 2008.

34. Benvenuto D, Giovanetti M, Vassallo L, Angeletti S, Ciccozzi M. Application of the ARIMA model on the COVID-2019 epidemic dataset. *Data Brief.* 2020;29:105340.

35. Saba AI, Elsheikh AH. Forecasting the prevalence of COVID-19 outbreak in Egypt using nonlinear autoregressive artificial neural networks. *Process Saf Environ Protect.* 2020;141:1–8.

36. Roth GA, Abate D, Abate KH, et al. Global, regional, and national age-sex-specific mortality for 282 causes of death in 195 countries and territories, 1980–2017: a systematic analysis for the global burden of disease study 2017. *Lancet.* 2018;392(10159):1736–1788.

37. Wong G. Has SARS infected the property market? Evidence from Hong Kong. *J Urban Econ.* 2008;63(1):74–95.

38. Alberts CJ, Boyd A, Bruisten SM, et al. Hepatitis a incidence, seroprevalence, and vaccination decision among MSM in Amsterdam, the Netherlands. *Vaccine.* 2019;37(21):2819–2856.

39. Leonenko VN, Ivanov SV, Novoselova YK. A computational approach to investigate patterns of acute respiratory illness dynamics in the regions with distinct seasonal climate transitions. *Proc Comput Sci.* 2016;80:2402–2412.

40. Flasche S, Hens N, Boëlle PY, et al. Different transmission patterns in the early stages of the influenza A (H1N1) v pandemic: a comparative analysis of 12 European countries. *Epidemics.* 2011;3(2):125–133.

41. Takian A, Raoofi A, Kazempour-Ardebili S. COVID-19 battle during the toughest sanctions against Iran. *Lancet.* 2020;395(10229):1035. 28.

42. Ghelichkhani P, Esmaeili M. Prone position in management of COVID-19 patients; a commentary. *Arch Acad Emerg Med.* 2020;8(1).

43. Yousefifard M, Zali A, Ali KM, et al. Antiviral therapy in management of COVID-19: a systematic review on current evidence. *Arch Acad Emerg Med*. 2020;8(1).

44. Dowd JB, Andriano L, Brazel DM, et al. Demographic science aids in understanding the spread and fatality rates of COVID-19. *Proc Natl Acad Sci*. 2020;117(18):9696–9698.

45. Esteve A, Permanyer I, Boertien D. *La vulnerabilitat de les províncies espanyoles a la covid-19 segons estructura per edats i de co-residència: implicacions pel (des) confinament*; 2020.

46. Institute for Health Metrics and Evaluation, Global Burden of Disease Collaborative Network. *Global Burden of Disease Study 2016 (GBD 2016) Results*; 2016.

47. Vestergaard LS, Nielsen J, Richter L, et al. Excess all-cause mortality during the COVID-19 pandemic in Europe–preliminary pooled estimates from the EuroMOMO network, March to April 2020. *Eurosurveillance*. 2020;25(26):2001214.

48. Lawal Y. Africa's low COVID-19 mortality rate: a paradox? *Int J Infect Dis*. 2021 Jan 1;102:118–122.

49. Carone G, Costello D. Can Europe afford to grow old. *Finance Dev*. 2006;43(3):28–31.

50. Fu L, Wang B, Yuan T, et al. Clinical characteristics of coronavirus disease 2019 (COVID-19) in China: a systematic review and meta-analysis. *J Infect*. 2020;80(6).656–665.

51. Mueller AL, McNamara MS, Sinclair DA. Why does COVID-19 disproportionately affect older people? *Aging (Albany NY)*. 2020;12(10):9959.

52. West E, Moore K, Kupeli N, et al. Rapid review of decision-making for place of care and death in older people: lessons for COVID-19. *Age Ageing*. 2020;18.

53. COVID-19 National Incident Room Surveillance Team. COVID-19, Australia: epidemiology report 16 (reporting week to 23: 59 AEST 17 May 2020). *Commun Dis Intell*. 2018;44. 2020 May 22.

54. Yamamoto N, Bauer G. Apparent difference in fatalities between Central Europe and East Asia due to SARS-COV-2 and COVID-19: four hypotheses for possible explanation. *Med Hypotheses*. 2020;144:110160.

55. Sorci G, Faivre B, Morand S. *Why Does COVID-19 Case Fatality Rate Vary Among Countries?* medRxiv preprint; 2020. https://doi.org/10.1101/2020.04.17.20069393.

56. Burki T. COVID-19 in Latin America. *Lancet Infect Dis*. 2020;20(5):547–548.

57. Office of National Statistics. *Deaths Involving COVID-19 by Local Area and Socioeconomic Deprivation: Deaths Occurring Between 1 March and 17 April*; 2020. https://www.ons.gov.uk/peoplepopulationandcommunity/birthsdeathsandmarriages/deaths/bulletins/deathsinvolvingcovid19bylocalareasanddeprivation/. deathsoccurringbetween1marchand17april. Accessed 12 May 2020.

58. Office of National Statistics. *Coronavirus (COVID-19) Related Deaths by Ethnic Group, England and Wales: 2 March 2020 to 10 April 2020*; 2020. https://www.ons.gov.uk/Peoplepopulationandcommunity/birthsdeathsandmarriages/deaths/articles/coronavirusrelateddeathsbyethnicgroupenglandandwales/latest. Accessed 12 May 2020.

59. Kirby T. Evidence mounts on the disproportionate effect of COVID-19 on ethnic minorities. *Lancet Respir Med*. 2020;8(6):547–548.

60. Levine ME, Crimmins EM. Evidence of accelerated aging among African Americans and its implications for mortality. *Soc Sci Med*. 2014;118:27–32.

48

Multihazard risk analysis and governance across a provincial capital in northern Iran

Aiding Kornejady[a], Elham Kouchaki[b], Ali Boustan[c], Hamid Reza Pourghasemi[d], Majid Sadeghinia[e], and Anis Heidari[f]

[a]Spatial Sciences Innovators Consulting Engineering Company, Tehran, Iran [b]Water and Hydraulic Structure Department, Faculty of Civil Engineering, Islamic Azad University, Estahban Branch, Estahban, Iran [c]Civil Engineering Department, Islamic Azad University, Kerman Branch, Kerman, Iran [d]Department of Natural Resources and Environmental Engineering, College of Agriculture, Shiraz University, Shiraz, Iran [e]Department of Nature Engineering, Faculty of Agriculture and Natural Resources, Ardakan University, Ardakan, Iran [f]Department of Reclamation of Arid and Mountainous Regions, Faculty of Natural Resources, University of Tehran, Karaj, Iran

1 Introduction

Socioeconomic and environmental losses incurred from natural hazards are continuously increasing, mainly due to the accelerated rate of population growth and residing in highly susceptible regions.[1] Despite the current advanced technologies serving to facilitate the modeling procedure of natural phenomena, many politically and culturally critical cities and societies around the world have been partially damaged and, sometimes, extirpated due to natural hazards.[2-5] Northern Iran, among other highly prone regions of the world, annually experiences countless natural hazards; hence, it encounters severe monetary and human losses,[6-8] let alone the traumatic disorders of survivors.[9] Encompassing diverse climatic agents, inherently susceptible geological settings, active hydro-erosional processes, as well as poorly located cities in downstream areas that, in some cases, are highly prone to floods and seismic activities, are some of the factors that predispose the area to the occurrence of different natural disasters.

A part of such exorbitant exposure is because the current studies are substantially inadequate and technically do not satisfy the needs of modern societies. Cities expand as geo-topological and anthropogenic factors allow them. Encountering different natural hazards along with such expansion would require detailed studies to put a step forward from previous susceptibility assessments of solitary hazards. The latter can manifest in the form of multihazard assessment and, more precisely, multirisk analyses. Despite the merits of susceptibility assessment which have been extensively studied in recent years for different natural and human-made hazards in Iran such as landslides,[10-12] floods,[13-15] wildfire,[16-18] gully erosion,[19,20] and land subsidence,[21-23] risk analysis by integrating the elements at risk and their vulnerability provides more realistic insights regarding the imminent losses subjected to different hotspots.[24] In this regard, many studies have been conducted in which risk is solitarily analyzed for a particular natural hazard, such as a flood,[25,26] landslides,[27-29] and earthquakes.[30,31] Nonetheless, a solitary risk analysis by merely focusing on one hazard would distort the reality of nature, and the resultant unrealistic estimation would lead to an ineffective and ill-suited allocation of infrastructure and mitigation measures.[32,33] Few studies have analyzed multirisk issues,[34-37] while some studies have adopted a multisusceptibility assessment scheme.[33] In particular, northern Iran, as a region prone to multiple natural hazards, suffers from such a study gap. Hence, recognizing the main natural hazards and conducting a combined analysis of their quantitative/

qualitative risks subjected to the exposed elements are necessary. This task also expedites sustainable land use planning and emergency management of natural hazards that may occur solitarily or in tandem.[38] From the modeling viewpoint, a specific type of hazard dictates the type of model to be adopted by the end-user. For instance, when channelized floods and mass movements are under study, hydraulic models are more suitable because they capture the initiation and propagation dynamics. Sporadically occurring urban floods at different points due to intense rainfall can be modeled via point-pattern models, such as data mining or machine learning techniques.[13–15] Earthquakes should be modeled using algorithms specially formulated based on scenario-based seismic risk equations, although recent studies have been conducted on the application of the random forest model in predict shear stress and pile drivability.[39,40] In light of these premises and the main threats to the Sari district in northern Iran, urban floods and earthquakes as the main prevailing hazards were studied by using two sets of models. Random forest and maximum entropy as data mining models were comparatively used to produce flood inundation susceptibility across the city. The latter, together with the exposure of the elements and their vulnerability, was used in the risk equation to estimate flood risk. Earthquake risk was estimated directly by the RADIUS model, which has been designed explicitly for seismic risk for exposed infrastructures and residents in developing countries.[41] Combining the multirisk maps eventually gave us an outline to prioritize different districts of Sari, which was carried out using a technique for order of preference by similarity to ideal solution (TOPSIS) technique as multiattribute decision-making (MADM) method.[42] To the best of our knowledge, the proposed framework is unique in the study area and Iran because it integrates the multirisk concept and uses it as a multirisk platform to prioritize urban districts; hence, it can be a generalizable outline for risk management agencies in Iran's different provinces.

2 Study area

Sari, the provincial capital of Mazandaran, extends for an area of about $41.6\,km^2$ between the northern slopes of the Alborz Mountains and the southern coast of the Caspian Sea in northern Iran. It lies within 4,043,853–4,052,483 m N and 679,640–690,655 m E, UTM zone 39 N (Fig. 1). As the largest and most populous city of Mazandaran, Sari is home to 314,529 inhabitants. From an infrastructure standpoint, Sari encompasses a total of 52,000 buildings, four major transportation tunnels, eleven transportation bridges, five gasoline stations, one sewage pumping station, 6 km of water/sewage distribution lines, and approximately 338 km of roads (i.e., freeways and highways). Buildings and residential compounds accounted for most of the study area (i.e., 77%), followed by different land covers such as orchards (16%) and agriculture (7%), which are located within the city lots and the suburbs. A mixed humid subtropical and Mediterranean climate has prevailed in Sari, with an average annual precipitation of approximately 790 mm. A 15.7 km major fault runs tangentially above the north of Sari, while another 8 km major fault passes through the southern parts of the city. Both faults have been actively causing sizable seismic records, which, together with repetitive flood episodes, have put the infrastructures at high risk. In particular, many cultural heritage sites, ancient monuments, and religious sites have been damaged or destroyed mainly by historical earthquakes.

3 Methods and materials

The methodological process of this study is elaborated in Fig. 2. The Different steps of the workflow are explained in detail as follows.

3.1 Flood inundation susceptibility assessment

Flood inundation across cities is the result of the interaction between topo-climatic and geological factors, which manifests in the form of inundated water on impermeable surfaces.[43] the first spatial modeling stage and the promising results of data mining models in flood inundation susceptibility assessment in urban areas,[13,44] as well as the nature of their required data, inundated hotspots in Sari's districts from the excess of rainwater through canals and rivers acquired from the archived data of Mazandarran's Regional Water Company. Recorded events were transformed into centroid point features and used as inputs for the modeling process of the adopted data mining techniques. The second modeling stage considered the selection of flood inundation causative factors. Drawing extensively on the urban flood susceptibility literature in Iran,[13–15] and with a glance at the initiation mechanism

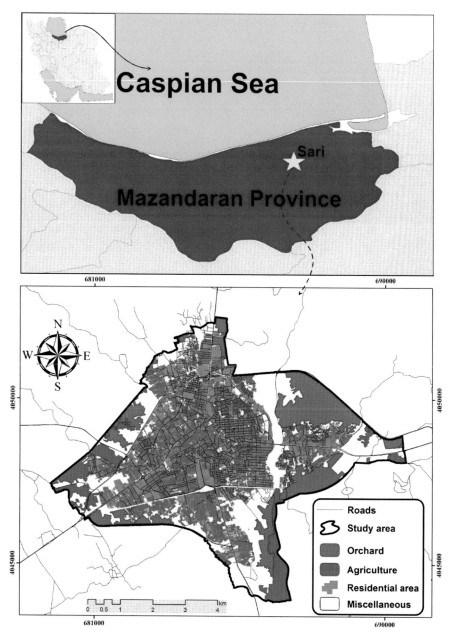

FIG. 1 Geographical location of the study area.

of flood inundation in the study area, the main causative factors of flood inundation were identified and further sieved through the process of multicollinearity analysis to select the optimal factors. Hence, six causative factors, including elevation (m), curve number (CN), distance to canals (m), distance to rivers (m), groundwater level (m), and slope percentage were initially selected (Fig. 3) and then subjected to the multicollinearity test, which was carried out using the variance inflation factor (VIF) (Table 1). VIF values higher than five indicate a strong correlation between two causative factors, which is identified as a critical multicollinearity issue and can lead to bias in the model results.[33,45,46]

The modeling scheme for flood inundation susceptibility was based on the application of the data mining technique. To this aim, random forest and maximum entropy models were used because of their promising capability in flood susceptibility assessment in previous studies.[47-50] Such a model application would also open up an interesting comparison between the performance of a presence-absence (i.e., random forest) and a presence-only model (i.e., maximum entropy). In summary, the random forest is a new version of tree models that accompanies two concepts of

658

48. Multihazard risk analysis

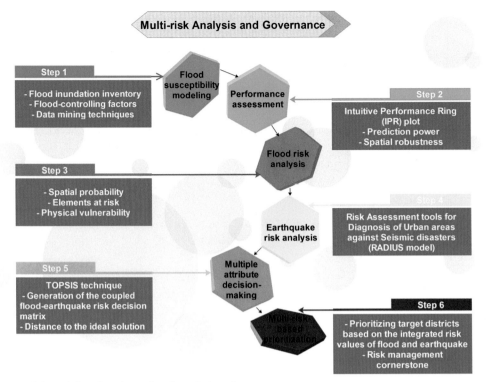

FIG. 2   Adopted methodological flowchart for multirisk analysis and governance.

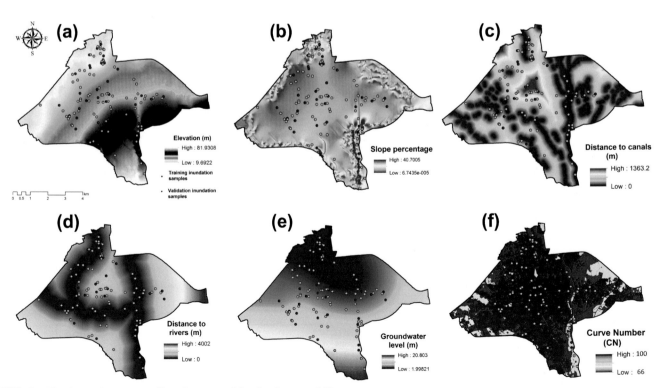

FIG. 3   Flood inundation controlling factors used for flood susceptibility assessment.

TABLE 1 VIF values representing the multicollinearity among flood inundation-controlling factors.

Factors	Sig.	Multicollinearity statistics	
		Tolerance	VIF
Slope angle	0.563	0.937	1.067
Groundwater level	0.037	0.362	2.762
Distance to rivers	0	0.704	1.420
Distance to canals	0.413	0.644	1.552
Elevation	0.020	0.401	2.492
Curve number	0	0.839	1.192

bagging and feature randomness to create multiple trees, each of which consists of a randomly selected training sample set (flood inundation presence and absence locations) and predictors (flood inundation causative factors). The final output of the model is an average of many previously created trees, which is a unique way of improving the model's performance. More details regarding the mathematical background of the random forest model can be found in Breiman,[51] Strobl et al.,[52] and Bachmair and Weiler.[53]

On the other hand, the maximum entropy (MaxEnt) is a generative presence-only model that avoids the inclusion of absence location ad; instead, it creates 10,000 pseudo-absence locations to better distinguish the presence from absences. As first coined by Shannon,[54,55] entropy signifies the stored information whose pattern can be distinguished. By maximizing entropy, more information is provided for the model, which facilitates pattern recognition.[56] Adopting the Gibbs equation underpinned by a generative-based entropy maximization scheme and through a random walk into the parameter space, MaxEnt can circumvent the need for prevalence (the ratio of true presence to true absence) estimation and accurately estimate the target distribution of the phenomenon of interest. More details are given by Phillips et al.[56,57] and Elith et al.[58]

3.2 Flood inundation risk estimation and mapping

Flood risk analysis was performed by adopting the risk estimation framework proposed by Varnes.[59] The risk equation consists of three constituents: hazard, exposure of elements at risk, and vulnerability.[24] Hazard analysis and the degree of its precision revolves around data availability, the required precision, the number of events under study, and more importantly, the scale of the study—more details would inevitably result in a more precise risk estimation of the phenomenon, which is termed as the fully quantitative risk estimation.[60,61] The latter can be attained when the phenomenon is studied at a site scale, which would require assessment of the spatial, temporal, and magnitude probability of the event. However, studying more events in more extensive areas (as in the present work) dictates to suffice only to spatial probability as translates into susceptibility assessment. Although precision-wise, it would only reach a semi-quantitative risk estimation of the phenomenon, it is efficient and suffices the multirisk-based prioritization need as the main objective of this work. To analyze the exposure of elements at flood risk, we listed the principal elements and monetized them to unify the quantitative unit of elements and the final risk map. Land uses/covers constituted the main human-made and natural elements at risk. The monetary values of buildings and roads as built-ups were acquired from local real state agencies and the Roads and Urban Development Office of Mazandaran Province, respectively. Pastures and forests were monetized based on their primary ecosystem services, including provisioning of food, wood, and medicinal resources, and services that contribute to climate and soil stability, control of agricultural pests, and purification of air and water (Table 2).

Moreover, the monetary values of orchards were acquired from local real estate agencies.

Physical vulnerability analysis reflects the degree to which physical elements can be affected by external excitements, such as natural hazards.[62,63] It is the most problematic part of risk estimation because it revolves around different variables that can be difficult to quantify.[64,65] For instance, the physical vulnerability of a building is a function of its materials and the angle of impact (i.e., external force).[24] Vulnerability values range between 0 and 1; hence, they can considerably change the final risk values. For instance, given constant values for exposure and hazard, a vulnerability value of 0.5, can decrease the final risk estimate by half. Although some field experiments have been proposed to quantify the physical vulnerability of the elements, this can be insurmountable given the extent and variety of the elements under study and the intrinsic subjectivity of its context. Above all, because the main objective of our work is to

Vulnerability and monetary values assigned to the land use/cover classes in the study area.

Land use/cover	Area (ha)	Vulnerability	Monetary value (USD per 1 m² pixel)
Critical centers	181.08	1	14,285.714
Residential	2425.87	0.95	1190.476
Road	607.16	0.9	9.762
Agriculture	268.23	0.85	47.619
Orchard	679.73	0.8	95.238

prioritize the districts based on multirisk values, vulnerability values should be identified in such a way that they reflect the inherent variation between different exposed elements. Such numerification was carried out based on a review of the most relevant literature, among which Saldivar-Sali and Einstein[66] and Guillard-Gonçalves et al.[63] are the most adaptable studies because they identify the vulnerability of diverse land uses/covers. However, most studies have categorized built-ups (i.e., buildings and roads) into one group in which they are assigned a single constant value of 1 (i.e., the highest vulnerability). Hence, a varied range of vulnerability values was applied to different residential units such as hospitals, cultural heritage sites, schools, and residential complexes to attain a more differentiated risk pattern across Sari's districts (Table 2).

3.3 Seismic risk estimation and mapping

Seismic risk analysis was carried out by the RADIUS model (Risk Assessment tools for Diagnosis of Urban areas against Seismic disasters).[41,67] To fulfill the needs of earthquake-prone areas in developing countries and delineating high-risk hotspots in urban structures, the United Nations launched the RADIUS with a simple, self-explanatory, and easy-to-use interface in Excel environment. RADIUS calculates the seismic risk parameters in a mesh delimiting the study area, which was designed to simplify the calculation process.[68] In each mesh cell, diverse information regarding the population, soil type, structure type, number of buildings, and lifelines are entered as inputs.[69] This tool enables end-users to simulate different earthquake scenarios in terms of the location of the epicenter, earthquake magnitude, depth, direction, type of attenuation function, and augmentations because of the local soil conditions.[70]

After mesh generation, an area ID is assigned to different districts; each consists of different mesh cells, which are embedded to seek a pattern across different districts of interest. Weights can also be assigned to different mesh cells about their importance (e.g., presence of vital lifelines or political/cultural/societal values) whose numerical range is between 0 (none) and 4 (very high). Different soil types should also be determined from the defined categories, including hard/soft rocks, average stiff soils, and landfill/reclaimed lands. Further, 10 construction types in terms of materials (from unfired rocks to reinforced masonry) and the use of buildings are provided, which should be defined for different area IDs in parentage. The Lifeline inventory is the next step, which consists of identifying the length of roads (either local or major roads), transportation bridges and tunnels, the number of major electrical and telecommunication transmission towers, major water/sewage pumping sites and distribution lines, and the number of gasoline stations for the entire area.

The final step is to identify the scenario earthquake information, which entails defining earthquake magnitude, depth, time of occurrence (day or night), type of attenuation equation, epicentral distance, and direction from the reference point. The final products of the model include the peak ground acceleration (PGA) and modified Mercalli intensity (MMI) magnitudes, damage to the lifelines and buildings, and deceased/injured individuals across the designed mesh, which are presented in both tabular and thematic map types.[71]

In this study, the entire city was dissected into 133 mesh cells within which the required data regarding the distribution of lifelines, infrastructure, and buildings were acquired from the Municipal Administration Office (MAO) and Roads and Urban Development Office (RUDO) of Mazandaran Province, which are presented in Tables 3–5. The Average stiff soil was selected as the common soil type (denoted as type 3) in the entire area based on the soil texture map archive obtained from Mazandaran's Office of Natural Resources and Watershed Management.

TABLE 3 Detailed information on the lifelines in the study area.

Lifeline	Total count	Unit	Definition
Road	338	km	Length of major roads such as freeways/ highways (in km)
Bridge	11	Count	Number of major transportation bridges (road and railway)
Tunnels	4	Count	Number of major transportation tunnels, for the concerned city or target region
Water-1	6	km	Length of major water & sewage trunk and distribution lines (km)
Water-2	1	Site	Number of water & sewage pumping stations
Gasoline	5	Count	Number of gasoline stations

TABLE 4 Percentage of different building classes across Sari's area IDs (see Table 5 for a description of the classes).

Area ID	RES1 (%)	RES2 (%)	RES3 (%)	RES4 (%)	EDU1 (%)	EDU2 (%)	MED1 (%)	MED2 (%)	COM (%)	IND (%)
1	21	0	39	40	0	0	0	0	0	0
2	21	0	39	40	0	0	0	0	0	0
3	21	0	39	40	0	0	0	0	0	0
4	21	0	39	40	0	0	0	0	0	0
5	21	0	34	34	5	0	0	0	6	0
6	21	0	34	11	17	0	0	17	0	0
7	15	0	34	6	28	0	0	12	5	0
8	15	0	19	6	17	0	0	18	25	0
9	15	0	0	6	22	0	0	29	28	0
10	15	0	34	18	6	0	0	12	15	0
11	21	0	34	35	0	0	0	6	4	0
12	21	0	34	22	6	0	0	6	11	0
13	21	0	45	30	0	0	0	0	4	0
14	21	0	39	40	0	0	0	0	0	0
15	21	0	39	40	0	0	0	0	0	0
16	21	0	39	40	0	0	0	0	0	0
17	21	0	39	40	0	0	0	0	0	0

To design earthquake scenarios recorded historical earthquake data recorded within a 100 km radius around Sari between 1990 and 2019 were acquired from Iran's International Institute of Earthquake Engineering and Seismology. From a total of 495 records, two historical records, together with two hypothetical earthquakes with more destructive powers, were considered as earthquake scenarios for seismic risk analysis. Table 6 provides detailed information on the designed earthquake scenarios.

3.4 Performance assessment

In this work, performance assessment only pertains to flood susceptibility modeling because the losses driven by floods, earthquakes, and generally most natural hazards are not well documented in Iran, and the archived data only

TABLE 5 Description of the building classes considered by the RADIUS model.

Building classes	Description
RES1	Informal construction—mainly slums, row housing, etc. made from unfired bricks, mud mortar, loosely tied walls, and roofs
RES2	URM-RC composite construction—substandard construction, not complying with the local code provisions. Height up to 3 stories. URM is Un-Reinforced Masonry and RC is a Reinforced Concrete building
RES3	URM-RC composite construction—old, deteriorated construction, not complying with the latest code provisions. Height 4–6 stories
RES4	Engineered RC construction—newly constructed multistoried buildings, for residential and commercial purposes
EDU1	School buildings, up to 2 stories
EDU2	School buildings, greater than 2 stories
MED1	Low to medium rise hospitals
MED2	High rise hospitals
COM	Shopping centers
IND	Industrial facilities, both low and high risk

TABLE 6 Description of the designed earthquake scenarios in the study area used in the RADIUS model.

Earthquake scenario		Magnitude (Richter)	Depth (km)	Time of occurrence (hrs.)	Epicentral distance (km)	Direction relative from a reference point (city center)
Optimistic scenario (historical data)	1	4.5	43	12:00	3	North
	2	4.5	15		5	Northwest
Worst-case scenario	3	7	5	24:00	4	Northwest
	4	7	5		3	Southeast

consider the most destructive hazards, casting out historical events with less critical damage to the vital centers of the city. The latter regards the results of both the flood and earthquake risk maps. Hence, flood susceptibility maps derived from the random forest and maximum entropy models were used in the performance assessment test. In this regard, different metrics have been suggested in the literature and can be categorized into two groups: probability cutoff-dependent (e.g., sensitivity, specificity, miss rate, accuracy, precision, true skill statistics, etc.) and -independent metrics (e.g., receiver operating characteristic curve and prediction/success rate curves).[72,73] Drawing on these methods, we suggested a probability cutoff-independent radar chart called the intuitive performance ring (IPR), which enables the end-users to select the superior model as well as to visually monitor models' under- and outperformance behaviors compared to their counterparts. The IPR applies to both the training and validation stages, which would, respectively, indicate the goodness-of-fit and prediction power of the models. The IPR plot is underpinned by the fact that each sample presence location (here flooded area) should have the highest susceptibility value, which is 1. Hence, predicted susceptibility values lower than 1 would diminish the performance of the model. Plotting IPR follows two main steps: (1) extracting and listing the susceptibility values at the flooded point features and plotting them in an Excel worksheet using the radar chart option (no sorting is required) and (2) transforming the values into a logarithmic scale. The latter, in fact, only smooths the susceptibility values to provide a more discernible pattern of models' performance trends. The proposed IPR plot is against other metrics that try to decisively select the superior model because even a superior model may exhibit turbulent and unstable rise and decline in its performance trend across the presence samples, which would affect the reliability of its results. In other words, end-users may lean toward a model with somewhat lower performance but a more stable pattern. Hence, the IPR can intuitively address the spatial robustness of susceptibility or potentiality models that have spatial connotations.

3.5 Multirisk governance technique

The adopted method for multirisk governance was a MADM technique named TOPSIS, which was first developed by Hwang and Yoon[42] and has been previously used for hazard-based target prioritization in the literature.[33,74-77] The TOPSIS method followed four main steps. The first step involves the creation of a matrix in which the rows (i.e., alternatives) and columns (i.e., criteria) signify Sari's districts and the average risk values emanated from flood and earthquake scenarios. The second step pertains to normalizing the matrix using the vector normalization method, following Eq. (1):

$$r_{ij} = \frac{x_{ij}}{\sqrt{\sum_{i=1}^{m} x_{ij}^2}}$$
(1)

where x_{ij} is the criterion value belonging to each alternative in the $(x_{ij})_{m \times n}$ matrix and r_{ij} is the new normalized value creating the new $(R_{ij})_{m \times n}$ matrix. Third, different weights can be assigned to different criteria. Since the objective of this work was to conglomerate the resultant impact of floods and earthquakes as well as to avoid prejudgments, we considered the same weights for both hazards, assuming that both are equally predominant. In the fourth step, the worst (i.e., the highest risk value) and the best conditions (i.e., the lowest risk value) of each criterion (e.g., flood risk) among different vicinities were identified. In the fifth step, the geometric distances between different vicinities from the best and worst conditions were calculated using Eqs. (2) and (3):

$$d_{iw} = \sqrt{\sum_{j=1}^{n} (t_{ij} - t_{wj})^2}$$
(2)

$$d_{ib} = \sqrt{\sum_{j=1}^{n} (t_{ij} - t_{bj})^2}$$
(3)

where d_{iw} and d_{ib} are the distances of the target vicinity (t_{ij}) from the best (t_{bj}) and the worst (t_{wj}), respectively. The sixth step regards the calculation of similarity to the worst condition, following Eq. (4), based on which vicinities can be prioritized.

$$S_{iw} = \frac{d_{iw}}{d_{iw} + d_{ib}}$$
(4)

where the S_{iw} value of 0 represents the closest similarity to the worst condition, and vice versa. Sorting S_{iw} values would give us the multirisk-based prioritization of Sari's vicinities.

4 Results and discussion

4.1 Flood susceptibility modeling

Table 1 provides in detail the multicollinearity values between the causative factors included in flood susceptibility modeling, based on which no critical correlation exists between the factors exists. This result permits the use of all the factors together in the modeling process. The obtained acceptable range of VIF values (i.e., lower than 5) can stem from the adopted purposive factor selection procedure in this study. In other words, despite the modeling procedure adopted by some scholars in which an agglomeration of all the available thematic layers, particularly the DEM derivatives, are involved, we strived to include only the factors that can directly influence the process of flood inundation and, more importantly, those that are spatially heterogeneous. Regarding the latter, it was decided to exclude the land use/cover map from the modeling process because most of the area is covered by residential land use (i.e., 77%), while a negligible proportion is shared between orchards and agriculture. Although some DEM derivatives may exist that their inclusion can indirectly influence flood inundation, there is a considerable chance of adding fuel to the multicol linearity issue. For instance, some DEM derivatives, despite providing different and source-irrespective information, are generated through simple equations that can wind up in a map with a somewhat identical pattern to its source and accordingly, a strong correlation and critical multicollinearity value.

The flood inundation susceptibility maps derived from the RF and MaxEnt models are shown in Fig. 4. The first glance at the susceptibility maps points to the main similarities and differences. The similarities regard the

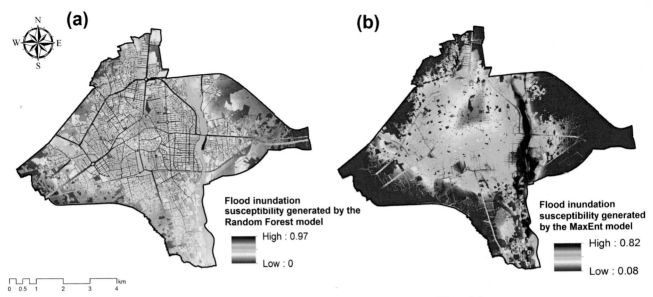

FIG. 4   Flood susceptibility maps derived from the random forest (A) and maximum entropy (B) models.

susceptibility patterns upon which both models are agreed, the circular pattern in the central parts of Sari, as well as the areas near the main river running through the city.

The differences, on the other hand, interestingly regard the overall looks of the maps in which the MaxEnt has followed a conservative modeling scheme in reflecting the flood susceptibility pattern because of the high concentration of both low and high susceptible pixels, which are represented by red and blue colors, respectively. Conversely, the RF model seems to apply more moderation, which can be translated into its high pattern recognition power through very low to very high susceptibility regions.

As previously mentioned, the performance of the susceptibility models regards both the training and validation stages. These two features should be interpreted concurrently as articulated in the following sentences: the average probability values of flood inundation throughout the historical episodes gave us a single value to identify the superior model, while the generated IPR plots indicated the models' behavior and their spatial robustness. As shown in Fig. 5, both the RF and MaxEnt models exhibit somewhat stable results, where negligible oscillation in their performance is evident; however, according to the performance classification proposed by Hosmer and Lemeshow,[78] a performance

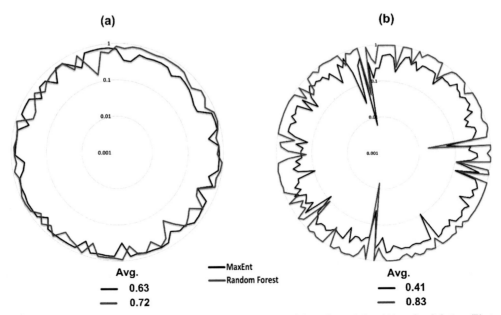

FIG. 5   The intuitive performance ring (IPR) plotted for flood susceptibility models in the training (A) and validation (B) stages.

range lower than 0.7, which pertains to MaxEnt (i.e., 0.63), is considered unsatisfactory and rather weak. On the other hand, the RF model with the same stability status gives higher and satisfactory performance (i.e., 0.72) compared to that of MaxEnt.

Further, the results of the validation stage, both visually and value-based, appear to better differentiate the performance of the adopted data mining models. Based on Fig. 5, it is evident that MaxEnt faces drastic ascents and descents, which, to a lesser degree, manifest in the results of the RF model. Nonetheless, the RF by giving beyond satisfactory performance value (i.e., 0.83) once again outperforms MaxEnt in terms of prediction power and generalization capacity. The higher prediction power also connotes the spatial transferability of the RF results.

Looking into the models' structures, it seems that the main reason for the underperformance of MaxEnt roots in its main modeling feature, which is being a presence-only model. Despite the previously mentioned merits of the presence-only modeling scheme, the MaxEnt is reportedly liable to overestimation or underestimation, which can result in either high false-positive (incorrectly predicting nonflooded areas as flooded) or false-negative rates (incorrectly predicting flooded areas as nonflooded).[73,79] To achieve higher performance, a model should be able to set a delicate balance between success and error rates. The pitfalls of not being equipped with a presence-absence algorithm that has possibly resulted in the underperformance of MaxEnt is that it has to randomly select 10,000 pseudo-absence locations across the study area. Although this is a generative mechanism to avoid handling the prevalence and enables MaxEnt to better distinguish the presence from absence, the pseudo-absence locations might have been mistakenly selected from areas that have high flood inundation susceptibility. However, the actual event has not yet occurred because of the absence of the triggering drivers and the overweighting of some predisposing factors (i.e., seemingly the ones that work in favor of noninundation such as steeper slopes and lower curve numbers) over others in the factor interaction pool.

In contrast, the absence locations for the RF model were selected based on expertise from the places where the flood could not possibly occur. Such an expert-based procedure resulted in an aerial filtering/constraining mechanism that could narrow down the domain to more reasonable absence locations. We avoided applying the same absence selection process for MaxEnt mainly because MaxEnt claims to be presence-only and functions on that mechanism.

Another structure-based explanation pertains to the tree-based algorithm of the RF, which is capable of perpetually minimizing the errors throughout the many generated trees via bootstrapping (also termed as bagging) and averaging processes. Such a mechanism also helps the model to avoid overfitting issues that often manifest in either overestimation or underestimation. Such deduction can also be inferred from Table 7, where the RF model categorizes 18.6% of the study area in the highly susceptible class to flood inundation, while MaxEnt presents the same class by almost a two-fold increase in spatial extent. The same applies to the very low susceptibility class, where MaxEnt insists on an areal exaggeration. This deduction supports previous inferences derived from a preliminary visual check of the susceptibility maps.

It is noteworthy that both models are judged under the same classification scheme, which was carried out by the natural break classifier because of the skewed susceptibility histogram derived from both models.[80]

Moreover, the judgment on the overestimation of MaxEnt is only made by considering the RF model as the benchmark and the reference model, which seems rational because of its superior performance in both the training and validation stages. Additionally, a realistic model tends to reduce the areal extent of highly susceptible areas, which aligns well with the natural patterns where hazardous areas at their highest intensity are spatially concentrated in a limited space. In contrast, the vast majority of the region may exhibit low and very low susceptibility to the occurrence of a phenomenon of interest. From the risk management perspective, models with areal exaggeration attributes, particularly in high susceptibility classes, are also considered to be less practical because the allocation of management

TABLE 7 The areal extent of the flood inundation susceptibility classes in Sari.

Susceptibility classes	Area (%)	
	RF	MaxEnt
Very low	21.88	38.58
Low	32.35	12.81
Moderate	27.15	18.60
High	5.31	19.52
Very high	13.31	10.50

practices in such large areas is cost-ineffective and technically infeasible. Hence, based on the overall assessment, the RF model was selected as the superior model for modeling flood inundation susceptibility across the Sari district. Further, the final product of the RF model as a connotation of the spatial probability of flood inundation occurrence was inputted into the risk estimation equation.

4.2 Flood risk estimation

In addition to the natural land covers located in the vicinity of the urbanized area of Sari (i.e., orchards and agriculture), the main elements at flood risk are infrastructural values (i.e., buildings and roads). The monetary values in Table 2 are calculated in units of $1 m^2$ pixel because they can better represent the level of details of buildings and roads, and we were indeed in favor of greater details than calculation space and simplicity. Hence, the susceptibility map of the superior model (i.e., the RF) in the previous stage was also resized from 10 to $1 m^2$ to be in line with the level of details associated with the exposure and vulnerability maps, although it would not add any more details to the susceptibility map. Monetary values of the elements are presented in USD based on Iran's foreign exchange rate reported by the central bank of Iran in February 2020 (i.e., 42,000 IRR/USD[a]). Among the monetized values reported in Table 2, critical centers, including hospitals, schools, banks, and cultural heritage sites, gained the highest monetary values, followed by residential buildings, orchards, agriculture, and roads, resulting in values ranging between 9 and 14,286 USD per 1 m pixel.

Regarding cultural heritage, the reported monetary values pertain only to the commercial land worth of these properties, excluding such inherent merits as their principal role in shaping the traditional, religious, and aspirational attributes. Although considering the inherent values of the critical sites would substantially increase their worth, monetizing such merits is problematic and rather infeasible, and doing so would have added to the uncertainty level of our study. Nonetheless, the monetary differences between the critical sites and residential buildings, as well as other elements at risk, fulfill our objective of differentiating the risk level in the urban fabric and its vicinities. This technique was also employed in assign vulnerability values, where the previously reported numbers were deliberately modified to attain a higher degree of differentiation and precision. Regarding the vulnerability assessment, the reviewed studies have placed residential areas and roads into one category as built-ups, which have the highest vulnerability value (i.e., 1) because they pose critical capital-intensive equipment and machinery, civil work, and economic activities.[66] Natural land covers gain lower vulnerability values, mainly due to the lower probability of people's presence. However, some modifications were executed following the same procedure adopted in the exposure analysis, where the critical sites were placed at the highest priority and had the highest vulnerability values. In particular, cultural heritage is among the most exposed elements at risk, mainly because the reallocation option as a counteraction to the occurring hazards would not be feasible; instead, they should be protected by maintenance and repair. The same applies to residential buildings, but with the difference that they have higher physical resistance than historical heritage and, in case of any destruction, insurance companies would cover most of the incurred damages to people's properties. Antithetically, damages to critical sites are somewhat irredeemable, particularly those incurred by cultural heritage sites. Moreover, in case of emergency, people can more easily evacuate their houses, while hospitalized patients and relatively inexperienced youths in local schools would decelerate the evacuation procedure.

Banks also hold sizable credits, investments, and other monetary reserves and assets that cannot be rectified and recovered. As for the agriculture and orchards as the main land-covers included in flood risk analysis, the former is more actively occupied by the residents and labor force, and crops have been more intensely destroyed by historical floods than orchards. Hence, a higher vulnerability value was assigned to agriculture, followed by orchards at the lower rank. It is noteworthy that such a subjective process of vulnerability numerification and assignment is only designed to disintegrate the risk between different elements and attain a more differentiated, precise, and practical flood risk map. Although this may undermine the final monetary risk values, it still fulfills our primary objective, a risk-based target-prioritizing scheme. The flood risk map, as the product of susceptibility, exposure, and vulnerability, is presented in Fig. 6, in which the highest risk values expectedly correspond to critical buildings because they possess higher exposure and vulnerability values. Altogether, the urban fabric shows a higher flood risk than its vicinities.

Fig. 7 portrays an interesting graphical analogy between susceptibility and risk, based on which areas with higher susceptibility necessarily do not impose higher risk. Hence, susceptibility and, more precisely, hazard in its broader sense, is not always in line with the risk entity.

---

[a] Source: https://www.cbi.ir/exrates/rates_en.aspx.

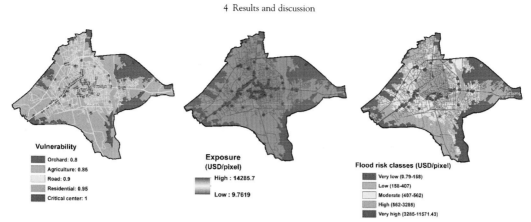

FIG. 6   Maps of physical vulnerability, exposure of the elements at risk, and flood monetary risk.

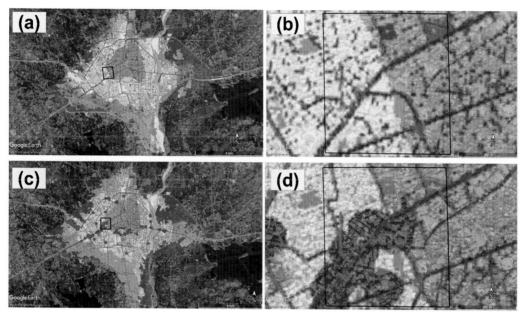

FIG. 7   Schematic comparison between the spatial patterns of flood susceptibility (A,B) and risk (C,D).

The reason rests in the fact that although some areas may exhibit high susceptibility to the occurrence of a phenomenon, other factors also play pivotal roles when risk is under consideration such as the absence of elements, presence of low-priced (i.e., monetarily) elements, or presence of high-priced elements yet with lower vulnerability. Based on the results of risk classification (Table 8), approximately 14% of Sari falls within high-and very high-risk classes, which places the vicinities within at the highest priority for any flood mitigation practice. However, as the aim was to adopt a holistic view in the form of multirisk-based prioritization, we continued to estimate the earthquake risk in addition to flood risk, as discussed below.

4.3 Earthquake risk estimation

As previously mentioned, the adopted earthquake risk estimation model (RADIUS) was specially designed for urban areas. Hence, we excluded the land covers (i.e., agriculture and orchards) from the equation and focused on the main urban fabric. Historical earthquake records were acquired for the scenario design from which two earthquakes were selected as the optimistic scenarios, accompanied by two severe earthquakes as the worst-case scenario (Fig. 8A, Table 6). Further, Sari was divided into 133 calculation mesh cells (Fig. 8B), for which the lifeline and infrastructural inputs were fed to the RADIUS model.

As shown in Fig. 9, there is a discernible direct pattern between the level of destruction and earthquake depth/intensity under the designed scenarios. For instance, in the case of the first and second scenarios (Fig. 9) where earthquake magnitude was assumed constant, earthquake depth plays a determinant role; that is, earthquake scenario 1,

TABLE 8   The areal extent of the flood inundation risk classes in Sari.

Risk classes	Area (ha)	Area (%)
Very low	1586.15	38.13
Low	1165.12	28.01
Moderate	827.66	19.90
High	476.99	11.47
Very high	104.14	2.50

compared to scenario 2, imposes less destruction because its energy faces more dissipation before reaching the surface. The same applies to the third and fourth earthquake scenarios (Fig. 9). Different earthquake epicentral distances and directions have expectedly resulted in different spatial patterns of earthquake intensity scales (i.e., PGA and MMI) as well as their emanated human and infrastructural losses.

The Calculated PGA values in Sari can be classified as light in the first scenario, moderate in the second scenario, and severe in the third and fourth scenarios according to the PGA classification proposed by the USGS (ShakeMap Scientific Background). The similarities in the third and fourth scenarios lie in their somewhat similar earthquake variables, as the fourth scenario accounts for less depth (i.e., 3 km) than that of the third scenario (i.e., 4 km).

The MMI scale addresses the severity of an earthquake in terms of its impact on people's lives and infrastructure. According to the MMI classification proposed by the USGS, the MMI values of Sari can be classified into three classes: light in the first scenario, moderate to strong in the second scenario, and severe to violent in the third and fourth scenarios. Based on the predictions made by the RADIUS model, the first and the second scenarios would affect the central and northern districts of Sari, where most of the incurred infrastructural and human losses are concentrated while no decrease has been caused. Central districts are mostly occupied by historical sites and old enclosed bazaars made of ancient lime mortars and low-strength materials liable to weathering and seismic shaking. Regarding the third and fourth scenarios, the spatial pattern of the damaged buildings is somewhat similar to those of the first and second scenarios, while a different distribution of injured and deceased people should be expected where the earthquake strikes the western, central, and eastern districts.

Fig. 10 plots an interesting relationship between the MMI scale as the best representative of an earthquake destructiveness and the destructed structures as well as the injured/perished lives, which are strongly correlated.

Based on this plot, it is evident that the increase in MMI can reciprocally increase the damaged buildings and the injuries by a power-law function, while the approximate death toll and MMI are rather linearly correlated. Hence, such an underlying regularity indicates the degree to which Sari's infrastructures and residents can be at risk of imminent earthquake events. Although strong cases may be rare, historical records show that somewhat similar events with magnitudes of 5 to 6 have previously occurred within the 100 km radius around Sari. Hence, instead of excluding the third and fourth scenarios, we incorporated all four earthquake risk maps together with the flood risk map

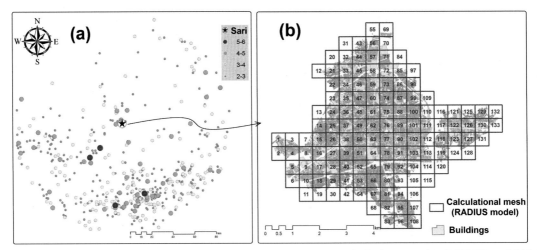

FIG. 8   Historical earthquake data within 100 km radius around Sari (A) and the generated calculation mesh in RADIUS model (B) for earthquake risk analysis.

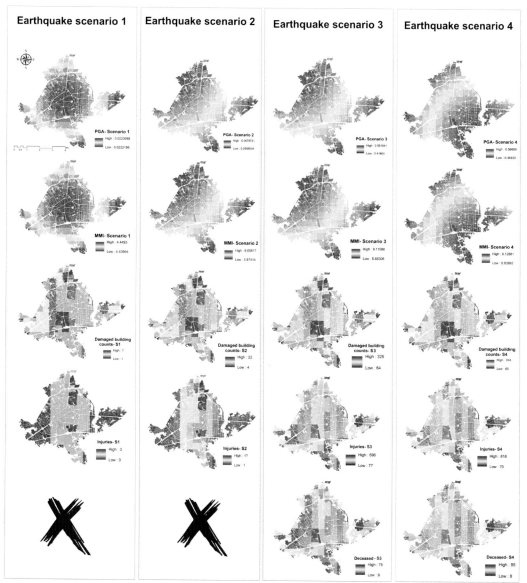

FIG. 9 Maps of PGA, MMI, and earthquake damages to buildings and residents of Sari under four different earthquake scenarios.

and made four broad bases for multirisk-based prioritization of Sari's districts: flood-earthquake scenario 1, flood-earthquake scenario 2, go forth.

4.4 Multirisk governance

The TOPSIS as a MADM method has long been used in different scientific branches where decision-making is tied to multiattributed conditions. After generating the decision matrix based on the average risk values of floods and earthquakes (summed damages to the buildings and residents) (Table 9), Sari's districts were prioritized under the coupled flood-earthquake scenarios (Fig. 11).

All four risk scenarios unanimously agreed upon the critical conditions of districts 4, 5, and 3, which represent the central and western parts of Sari. According to the *central place theory* expounded by Goodall,[81] most critical human settlements in terms of functionality, size, and number are distributed in the central parts, providing services to the surrounding areas. The latter rightly aligns with the urban texture of Sari, which is located in most administrative offices, banks, cultural heritage sites, schools, hospitals, and a large crowd of people sipped to major avenues and old bazaars. Hence, the high flood and earthquake risk values attributed to Sari's central part and its nearby districts put them in the highest priority for mitigation and reinforcement measures. It is also evident that the priority of district

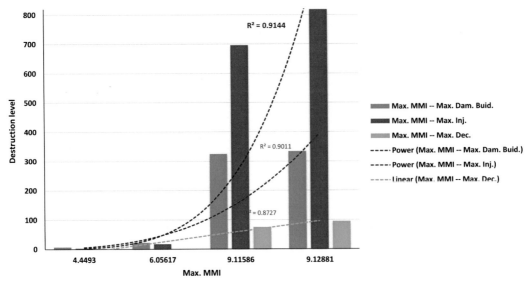

FIG. 10 Regression relationships between the maximum MMI and the maximum destructive levels incurred from four different earthquake scenarios.

2 profoundly alters owing to the changes in earthquake scenarios, given that the flood risk pattern remains intact throughout all four scenarios. District 2 is located in northern areas as the main target of the first, second, and third earthquake scenarios given their direction from the reference point (i.e., city center), out of which the second and third scenarios are designed to occur in less deep strata (i.e., 15 and 5 km, respectively). As opposed to these premises, the third scenario, as with the first and fourth scenarios, puts district 2 almost in the last priority.

TABLE 9   Generated decision matrix based on the average values of flood and earthquake risks in each district.

Districts	Average flood risk	Average earthquake risk (S1)	Average flood risk	Average earthquake risk (S2)
1	421.306	4.142	421.306	23.378
2	227.532	3.190	227.532	33.062
3	573.762	4.077	573.762	20.780
4	1001.243	4.886	1001.243	24.356
5	859.710	5.080	859.710	24.585
6	256.548	2.838	256.548	13.823
7	316.418	4.555	316.418	20.011
8	286.831	3.207	286.831	13.835
Districts	Average flood risk	Average earthquake risk (S3)	Average flood risk	Average earthquake risk (S4)
1	421.306	545.803	421.306	496.780
2	227.532	417.428	227.532	395.053
3	573.762	669.151	573.762	641.941
4	1001.243	775.225	1001.243	794.116
5	859.710	611.311	859.710	650.368
6	256.548	444.640	256.548	527.617
7	316.418	461.126	316.418	506.082
8	286.831	311.467	286.831	370.861

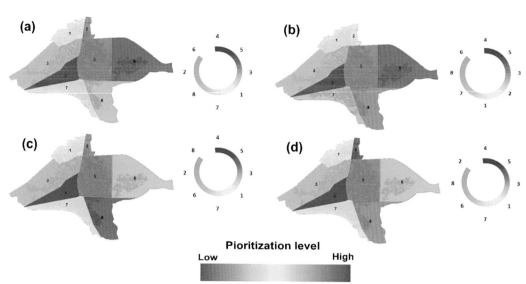

FIG. 11   Multirisk-based prioritization of Sari's districts (represented by numbers).

Further scrutinization revealed that although district 2 is among the main targets of the third earthquake scenario, it is less occupied by buildings and people, and hence possesses a lower number of injuries and deaths. This matter is pivotal, as risk management practices should not be merely focused on the most threatened districts, but the safe zones should be precisely identified. That is, as reinforcements and risk mitigation measures are designed and implemented in highly threatened districts, safe districts should be concurrently prepared to host the reallocated resources and residents in case of emergencies.

5   Conclusion

One-sided and superficial management leads to the distortion of nature's reality and unexpected damages incurred from augmented natural hazards. Hence, management paradigms need to be shifted toward undertaking a holistic view. In light of such holism, the threats emanating from all the prevailing natural hazards in a particular area should be considered. Hence, we propose a multirisk-based target prioritization that excels in previous susceptibility-based, multisusceptibility-based, and multihazard-based management schemes. The take-home achievements of this study are as follows: The risk and susceptibility clearly and expectedly showed different patterns. Hence, given the pivotal role of risk analysis before the implementation of pragmatic projects, risk-based management practices should supplement previous simplistic susceptibility-based endeavors. The Application of data mining models appeared to be compatible and successful for flood inundation susceptibility modeling, based on which random forest showed promising performance. The suggested IPR plot not only provides a single explicit value for model performance but also can successfully track the spatial oscillation/changes in their performance.

On the other hand, the RADIUS model with a self-explanatory and user-friendly interface can provide in detail invaluable insights into the level of infrastructural and human losses. Altogether, flood risk and the designed earthquake scenarios unanimously suggest a discernible pattern based on which the central-sociopolitical core and western parts of Sari are highly threatened, while northern, eastern, and the southernmost parts of Sari are identified as safe zones. The adopted holistic view suggests that the amendment of the highly threatened zones in the form of mitigation and reinforcement of infrastructure and capacity building at the individual and organizational levels should be concurrently implemented by safeguarding the less threatened zones to ensure their preparedness for reallocation of resources and residents, in case of imminent crises.

References

1. Singh AK. Bioengineering techniques of slope stabilization and landslide mitigation. *Disas Prev Manag Int J*. 2010;19(3):384–397.
2. Sanderson D. Cities, disasters and livelihoods. *Risk Manage*. 2000;2(4):49–58.
3. Taboroff J. Cultural heritage and natural disasters: incentives for risk management and mitigation. *Manag Disas Risk Emerg Econ*. 2000;2:71–79.
4. Pelling M. *The Vulnerability of Cities: Natural Disasters and Social Resilience*. Earthscan Press; 2003. 212 pp.

5. Gunderson L. Ecological and human community resilience in response to natural disasters. *Ecol Soc.* 2010;15(2):18.

6. Noji EK. Natural disasters. *Crit Care Clin.* 1991;7(2):271–292.

7. Bahrainy H. Natural disaster management in Iran during the 1990s—need for a new structure. *J Urban Plan Dev.* 2003;129(3):140–160.

8. Badri SA, Asgary A, Eftekhari AR, Levy J. Post-disaster resettlement, development and change: a case study of the 1990 Manjil earthquake in Iran. *Disasters.* 2006;30(4):451–468.

9. Sadeghi N, Ahmadi MH. Mental health preparedness for natural disasters in Iran. *Nat Hazards.* 2008;44(2):243–252.

10. Lombardo L, Opitz T, Huser R. Numerical recipes for landslide spatial prediction using R-INLA: a step-by-step tutorial. In: *Spatial Modeling in GIS and R for Earth and Environmental Sciences.* Elsevier; 2019:55–83.

11. Nhu VH, Hoang ND, Nguyen H, et al. Effectiveness assessment of Keras based deep learning with different robust optimization algorithms for shallow landslide susceptibility mapping at tropical area. *Catena.* 2020;188:104458.

12. Sameen MI, Pradhan B, Lee S. Application of convolutional neural networks featuring Bayesian optimization for landslide susceptibility assessment. *Catena.* 2020;186:104249.

13. Rahmati O, Darabi H, Haghighi AT, et al. Urban flood hazard modeling using self-organizing map neural network. *Water.* 2019;11(11):2370.

14. Tien Bui DT, Hoang ND, Martínez Álvarez F, et al. A novel deep learning neural network approach for predicting flash flood susceptibility: a case study at a high frequency tropical storm area. *Sci Total Environ.* 2020;701:134413.

15. Shahabi H, Shirzadi A, Ghaderi K, et al. Flood detection and susceptibility mapping using Sentinel-1 remote sensing data and a machine learning approach: hybrid intelligence of bagging ensemble based on K-nearest neighbor classifier. *Remote Sens (Basel).* 2020;12(2):266.

16. Ghorbanzadeh O, Valizadeh KK, Blaschke T, et al. Spatial prediction of wildfire susceptibility using field survey gps data and machine learning approaches. *Fire.* 2019;2(3):43.

17. Hong H, Jaafari A, Zenner EK. Predicting spatial patterns of wildfire susceptibility in the Huichang County, China: an integrated model to analysis of landscape indicators. *Ecol Indic.* 2019;101:878–891.

18. Tonini M, D'Andrea M, Biondi G, Degli Esposti S, Trucchia A, Fiorucci P. A machine learning-based approach for wildfire susceptibility mapping. The case study of the Liguria region in Italy. *Geosciences.* 2020;10(3):105.

19. Pourghasemi HR, Yousefi S, Kornejady A, Cerdà A. Performance assessment of individual and ensemble data-mining techniques for gully erosion modeling. *Sci Total Environ.* 2017;609:764–775.

20. Debanshi S, Pal S. Assessing gully erosion susceptibility in Mayurakshi river basin of eastern India. *Environ Dev Sustain.* 2020;22(2):883–914.

21. Abdollahi S, Pourghasemi HR, Ghanbarian GA, Safaeian R. Prioritization of effective factors in the occurrence of land subsidence and its susceptibility mapping using an SVM model and their different kernel functions. *Bull Eng Geol Environ.* 2019;78(6):4017–4034.

22. Pourghasemi HR, Saravi MM. Land-subsidence spatial modeling using the random forest data-mining technique. In: *Spatial Modeling in GIS and R for Earth and Environmental Sciences.* Elsevier; 2019:147–159.

23. Tsangaratos P, Ilia I, Loupasakis C. Land subsidence modelling using data mining techniques. The case study of Western Thessaly, Greece. In: *Natural Hazards GIS-Based Spatial Modeling Using Data Mining Techniques.* Cham: Springer; 2019:79–103.

24. Van Westen CJ, Van Asch TW, Soeters R. Landslide hazard and risk zonation—why is it still so difficult? *Bull Eng Geol Environ.* 2006;65 (2):167–184.

25. Albano R, Mancusi L, Sole A, Adamowski J. FloodRisk: a collaborative, free and open-source software for flood risk analysis. *Geomat Nat Haz Risk.* 2017;8(2):1812–1832.

26. Shivaprasad Sharma SV, Roy PS, Chakravarthi V, Srinivasa Rao G. Flood risk assessment using multi-criteria analysis: a case study from Kopili river basin, Assam, India. *Geomat Nat Haz Risk.* 2018;9:79–93.

27. Althuwaynee OF, Pradhan B. Semi-quantitative landslide risk assessment using GIS-based exposure analysis in Kuala Lumpur City. *Geomat Nat Haz Risk.* 2017;8(2):706–732.

28. Pellicani R, Argentiero I, Spilotro G. GIS-based predictive models for regional-scale landslide susceptibility assessment and risk mapping along road corridors. *Geomat Nat Haz Risk.* 2017;8(2):1012–1033.

29. Liu X, Miao C. Large-scale assessment of landslide hazard, vulnerability and risk in China. *Geomat Nat Haz Risk.* 2018;9(1):1037–1052.

30. Sinha N, Priyanka N, Joshi PK. Using spatial multi-criteria analysis and ranking tool (SMART) in earthquake risk assessment: a case study of Delhi region, India. *Geomat Nat Haz Risk.* 2016;7(2):680–701.

31. Wei B, Nie G, Su G, Sun L, Bai X, Qi W. Risk assessment of people trapped in earthquake based on km grid: a case study of the 2014 Ludian earthquake, China. *Geomat Nat Haz Risk.* 2017;8(2):1289–1305.

32. Komendantova N, Mrzyglocki R, Mignan A, et al. Multi-hazard and multi-risk decision-support tools as a part of participatory risk governance: feedback from civil protection stakeholders. *Int J Disas Risk Reduct.* 2014;8:50–67.

33. Sheikh V, Kornejady A, Ownegh M. Application of the coupled TOPSIS-Mahalanobis distance for multi-hazard-based management of the target districts of the Golestan Province, Iran. *Nat Hazards.* 2019;96(3):1335–1365.

34. Rafiq L, Blaschke T. Disaster risk and vulnerability in Pakistan at a district level. *Geomat Nat Haz Risk.* 2012;3(4):324–341.

35. Chen L, van Westen CJ, Hussin H, et al. Integrating expert opinion with modelling for quantitative multi-hazard risk assessment in the eastern Italian Alps. *Geomorphology.* 2016;273:150–167.

36. Liu X, Chen H. Integrated assessment of ecological risk for multi-hazards in Guangdong province in southeastern China. *Geomat Nat Haz Risk.* 2019;10(1):2069–2093.

37. Aksha SK, Resler LM, Juran L, Carstensen Jr LW. A geospatial analysis of multi-hazard risk in Dharan, Nepal. *Geomat Nat Haz Risk.* 2020;11 (1):88–111.

38. Marzocchi W, Garcia-Aristizabal A, Gasparini P, Mastellone ML, Di Ruocco A. Basic principles of multi-risk assessment: a case study in Italy. *Nat Hazards.* 2012;62(2):551–573.

39. Zhang W, Wu C, Zhong H, Li Y, Wang L. Prediction of undrained shear strength using extreme gradient boosting and random forest based on Bayesian optimization. *Geosci Front.* 2020. https://doi.org/10.1016/j.gsf.2020.03.007.

40. Zhang W, Wu C, Li Y, Wang L, Samui P. Assessment of pile drivability using random forest regression and multivariate adaptive regression splines. *Georisk Assess Manag Risk Eng Syst Geohazards.* 2019;15(1):27–40. https://doi.org/10.1080/17499518.2019.1674340.

41. Villacis CA, Cardona CN. *Guidelines for the Implementation of Earthquake Risk Management Projects.* Palo Alto, CA: Geo Hazards International; 1999.

42. Hwang CL, Yoon K. *Multiple Attribute Decision Making: Methods and Applications: A State-of-the-Art Survey. Lecture Notes in Economics and Mathematical Systems.* New York, NY: Springer-Verlag; 1981.

43. Støren EN, Dahl SO, Nesje A, Paasche Ø. Identifying the sedimentary imprint of high-frequency Holocene river floods in lake sediments: development and application of a new method. *Quat Sci Rev.* 2010;29(23–24):3021–3033.

44. Lee S, Lee S, Lee MJ, Jung HS. Spatial assessment of urban flood susceptibility using data mining and geographic information system (GIS) tools. *Sustainability.* 2018;10(3):648.

45. O'brien RM. A caution regarding rules of thumb for variance inflation factors. *Qual Quant.* 2007;41(5):673–690.

46. Colkesen I, Sahin EK, Kavzoglu T. Susceptibility mapping of shallow landslides using kernel-based Gaussian process, support vector machines and logistic regression. *J Afr Earth Sci.* 2016;118:53–64.

47. Wang Z, Lai C, Chen X, Yang B, Zhao S, Bai X. Flood hazard risk assessment model based on random forest. *J Hydrol.* 2015;527:1130–1141.

48. Lee S, Kim JC, Jung HS, Lee MJ, Lee S. Spatial prediction of flood susceptibility using random-forest and boosted-tree models in Seoul metropolitan city, Korea. *Geomat Nat Haz Risk.* 2017;8(2):1185–1203.

49. Siahkamari S, Haghizadeh A, Zeinivand H, Tahmasebipour N, Rahmati O. Spatial prediction of flood-susceptible areas using frequency ratio and maximum entropy models. *Geocarto Int.* 2018;33(9):927–941.

50. Chen W, Li Y, Xue W, et al. Modeling flood susceptibility using data-driven approaches of naïve bayes tree, alternating decision tree, and random forest methods. *Sci Total Environ.* 2020;701:134979.

51. Breiman L. Random forests. *Mach Learn.* 2001;45(1):5–32.

52. Strobl C, Boulesteix AL, Kneib T, Augustin T, Zeileis A. Conditional variable importance for random forests. *BMC Bioinf.* 2008;9(1):307.

53. Bachmair S, Weiler M. Hillslope characteristics as controls of subsurface flow variability. *Hydrol Earth Syst Sci.* 2012;16(10):3699.

54. Shannon CE. A mathematical theory of communication. *Bell Syst Tech J.* 1948;27(3):379–423.

55. Shannon CE. Prediction and entropy of printed English. *Bell Syst Tech J.* 1951;30(1):50–64.

56. Phillips SJ, Dudík M, Schapire RE. A maximum entropy approach to species distribution modeling. In: *Proceedings of the Twenty-First International Conference on Machine Learning*; 2004:83.

57. Phillips SJ, Anderson RP, Schapire RE. Maximum entropy modeling of species geographic distributions. *Ecol Model.* 2006;190(3–4):231–259.

58. Elith J, Phillips SJ, Hastie T, Dudík M, Chee YE, Yates CJ. A statistical explanation of MaxEnt for ecologists. *Divers Distrib.* 2011;17(1):43–57.

59. Varnes DJ. *Landslide Hazard Zonation: A Review of Principles and Practice.* Paris: United Nations International; 1984.

60. Carrara A, Cardinali M, Guzzetti F. Uncertainty in assessing landslide hazard and risk. *ITC J.* 1992;2:172–183.

61. Guzzetti F. Landslide hazard assessment and risk evaluation: limits and prospectives. In: *Proceedings of the 4th EGS Plinius Conference, Mallorca, Spain*; 2002:2–4.

62. Catani F, Casagli N, Ermini L, Righini G, Menduni G. Landslide hazard and risk mapping at catchment scale in the Arno River basin. *Landslides.* 2005;2(4):329–342.

63. Guillard-Gonçalves C, Zêzere J, da Silva Pereira S, Garcia RA. Assessment of physical vulnerability of buildings and analysis of landslide risk at the municipal scale: application to the Loures municipality, Portugal. *Nat Hazards Earth Syst Sci.* 2016;16(2):311–331.

64. Glade T. Vulnerability assessment in landslide risk analysis. *Erde.* 2003;134(2):123–146.

65. Uzielli M, Nadim F, Lacasse S, Kaynia AM. A conceptual framework for quantitative estimation of physical vulnerability to landslides. *Eng Geol.* 2008;102(3–4):251–256.

66. Saldivar-Sali A, Einstein HH. A landslide risk rating system for Baguio, Philippines. *Eng Geol.* 2007;91(2–4):85–99.

67. Villacis C. *RADIUS (Risk Assessment Tools for Diagnosis of Urban Areas against Seismic Disasters).* May 1999, San José, CR: UN. International Decade for Natural Disaster Reduction (IDNDR); 1999.

68. Deniz A, Korkmaz KA, Irfanoglu A. Probabilistic seismic hazard assessment for İzmir, Turkey. *Pure Appl Geophys.* 2010;167(12):1475–1484.

69. Barzinpour F, Esmaeili V. A multi-objective relief chain location distribution model for urban disaster management. *Int J Adv Manuf Technol.* 2014;70(5–8):1291–1302.

70. Boukri M, Farsi MN, Mebarki A, et al. Seismic vulnerability assessment at urban scale: case of Algerian buildings. *Int J Disas Risk Reduct.* 2018;31:555–575.

71. IDNDR, O. *RADIUS (Risk Assessment Tools for Diagnosis of Urban Areas Against Seismic Disasters) UN*; 2000. https://www.undrr.org/publication/radius-risk-assessment-tools-diagnosis-urban-areas-against-seismic-disasters.

72. Frattini P, Crosta G, Carrara A. Techniques for evaluating the performance of landslide susceptibility models. *Eng Geol.* 2010;111(1–4):62–72.

73. Rahmati O, Kornejady A, Samadi M, et al. PMT: new analytical framework for automated evaluation of geo-environmental modelling approaches. *Sci Total Environ.* 2019;664:296–311.

74. Levy JK, Hall J. Advances in flood risk management under uncertainty. *Stoch Environ Res Risk Assess.* 2005;19(6):375–377.

75. Klausner J. *Assessment of Physical Vulnerability of Buildings to an Earthquake Using Local TOPSIS and Global TOPSIS: A Case Study of the San Fernando Valley.* Doctoral Dissertation, Northridge: California State University; 2018.

76. Yang W, Xu K, Lian J, Ma C, Bin L. Integrated flood vulnerability assessment approach based on TOPSIS and Shannon entropy methods. *Ecol Indic.* 2018;89:269–280.

77. Alilou H, Rahmati O, Singh VP, et al. Evaluation of watershed health using fuzzy-ANP approach considering geo-environmental and topo-hydrological criteria. *J Environ Manage.* 2019;232:22–36.

78. Hosmer DW, Lemeshow S. *Applied Logistic Regression.* New York: John Wiley & Sons; 2000.

79. Tien Bui D, Hoang ND. A Bayesian framework based on a Gaussian mixture model and radial-basis-function fisher discriminant analysis (BayGmmKda V1. 1) for spatial prediction of floods. *Geosci Model Dev.* 2017;10(9):3391–3409.

80. Akgun A. A comparison of landslide susceptibility maps produced by logistic regression, multi-criteria decision, and likelihood ratio methods: a case study at İzmir, Turkey. *Landslides.* 2012;9(1):93–106.

81. Goodall B. *The Penguin Dictionary of Human Geography.* Puffin Books; 1987.

49

Distribution patterns in plants: Mapping and priorities for plant conservation

Ahmad Reza Mehrabian[a], Farzaneh Khajoei Nasab[a], and Hossein Mostafavi[b]

[a]Department of Plant Sciences and Biotechnology, Faculty of Life Sciences and Biotechnology, Shahid Beheshti University, Tehran, Iran [b]Department of Biodiversity and Ecosystems Management, Research Institute of Environmental Sciences, Shahid Beheshti University, Tehran, Iran

1 Introduction

The revolution of environmental information production has created wide-ranging challenges for its use and management.[1] In addition, the rising rate of permanent damage to natural ecosystems has strengthened the need for collecting biological diversity basic data (e.g., distribution patterns, diversity mapping, etc.) for conservation management.[2] Accordingly, extensive studies have been proposed to integrate and analyze these large amounts of data.[3] Spatial ecology (e.g., distribution patterns, ecological niche modeling, etc.) is progressively used to conserve ecology.

The ecological–biological dynamics of species occur over space and time. Accordingly, space affects biological species in several areas, such as resource harvesting, occupying space inside their ecological niche, dispersal, migration, and biological interactions. Space is related to conservation ecology in various ways. It is a necessary component in biodiversity mapping; it aids leadership in reducing the effects of ecological modification; it facilitates the priorities for conservation; and it is a fundamental component in modeling and ecological conservation.[4] A wide range of data on the distribution, diversity, and ecology of different species has resulted from field observations. A considerable amount of these data has been derived over the last few hundred years in uncountable herbaria as well as in the published literature.[5] The importance of the integration and coherence of species distribution data plays a key role in ecological analysis and reconstruction of the evolutionary history of life.[6] However, the data on the distribution patterns of numerous species are incomplete.[7] Distribution patterns showing the spatial arrangement of biological species depend mainly on the scale at which they are observed, from the arrangement of individuals inside a minor family and distribution patterns within a biological population to the full population of a species.[8] Indigenous (native), endemic, rare, introduced, and alien species are associated with distribution patterns. Determining the distribution patterns of biological taxa is considered to be the most important aim of ecological and biogeographical studies.[9] The distribution patterns of plant species mainly reflect a real ecological niche[10] and are also classified as a bioindicator of habitats.[11]

2 Biodiversity mapping

Biodiversity mapping is considered a vital component in biodiversity assessment, management, and conservation. These maps give a visual effect to biodiversity and are referred to as the mapping of species, populations or ecosystems, or communities as well as the degradation of biodiversity. Biodiversity maps can obviously communicate which biodiversity components are in critical states for protection and conservation. Mapping distribution as well as diversity patterns have been used to set conservation priorities on different scales.[2,12–14] Effective efforts to protect biodiversity in the biosphere have been made most specifically on distribution patterns of richness, endemism, and endangerment at the levels of both species and higher taxonomic categories. Several concepts (e.g., endemism, rarity, conservation status, priorities for conservation, protected areas, biodiversity hotspots, etc.) have been formed on the basis of distribution patterns. There are two approaches to mapping biodiversity: taxon-based and inventory-based mapping. Taxon-based mapping is shaped on the basis of extrapolation of ranges for individual taxa.[15,16] Range extrapolations use two species records: range-determining factors and abiotic and historical factors. Inventory-based approaches, however, use data associated with well-defined regions, ignoring the ranges of individual species. This method uses zoning maps of well-studied areas on smaller scales. In addition, historical aspects have less importance for diversity patterns with this technique. Nevertheless, inventory-based methods usually cover more particular factors than do taxon-based approaches and provide better conditions for calibrating range extrapolation.[17] Taxon-based phytodiversity mapping[18] has been used for plant taxa on regional, continental, and global scales.[12,13] Identifying the diversity centers of rare and endangered taxa,[19,20] endemism centers,[21] biodiversity hotspots,[22] and diversity zones of crop wild relatives[23] and medicinal plants constitutes major projects resulting from distribution patterns. In addition, these valuable data are a focal point in ecological modeling.[24,25] Today, several computer-based programs easily and rapidly calculate the mentioned mappings and perform ecological prominent analyses as discussed in the following sections of this chapter. Remote sensing (RS) is an effective tool for assessing the status and trends of ecosystems.[26] Geographic information systems (GISs) are valuable tools for the distribution mapping of phytodiversity, as they provide comprehensive analyses for conservation management.[27] There are two main forms of GIS data, i.e., raster and vector. With raster (or grid) data, a zone is separated into a collection of commonly formed cells. Each distinct cell is allocated a value for the variable being studied. However, vector files store mentioned data as polygons, lines, and points.[28] Because of the vastness of this scope, we have focused on mapping and modeling over time and space. Spatial or geographic distribution is the main component in conservation biology, and remote sensing is a valuable tool for providing data on this factor. RS data are most commonly used to document the environments of a specific zone at a specific time period in the form of a map. RS methods can monitor their changes over time.[29] Additionally, the creation as well as use of precise maps are vital tasks for biogeographers; they represent the environmental factors controlling the distribution patterns of plants, the anthropogenic impacts upon them, key ecological processes (e.g., succession, migration, speciation, etc.), and aid in developing management plans for protected areas.[30] In geoinformatics, the spatial, temporal, and thematic aspects are evaluated separately. Phenomena are categorized into lines, areas, points, and volumes according to their spatial features.[31] Currently, studies in ecology, biogeography, and conservation depend mainly on the analysis of grid maps of species richness. However, there is significant variety in the procedures as well as the production of phenomena maps[32–34] or predicted maps of distribution using the modeling method.[35,36] If sufficient distribution points of plant species occurrences are accessible, species richness (e.g., total species richness, endemic, surrogate, and endangered species richness, etc.) can be directly determined from these points by calculating the number of different species for cells on a grid.[37] However, species richness is usually highly influenced by volume as well as sampling method (i.e., the number of records for an area). This method is based on the area of occurrence. In addition, the exactness of the evaluation of species richness depends highly on grid cell size. Another method is covering expert-drawn range maps that are usually most used in field guides and are also known as range maps. Additionally, the data used to make such maps show variation. For example, some of them cover occurrence data, e.g., a polygon around known occurrences. Compared to point-to-grid richness maps, richness maps resulting from the range method are more powerful in terms of accuracy in local species richness because they are usually drawn on the area of occupancy.[36,38] Mapping distribution and species richness can also be accomplished by means of a circular area[37] or circular neighborhood.[39,40] ModestR is a user-friendly software for analyzing species distribution and taxonomic data.[41] It produces sample maps (occurrence records), range maps (areas), or a combination of both. ArcGIS has unique abilities in mapping, modeling, and analyzing ecological data. DIVA-GIS[39] includes free computer software for analyzing distribution mapping as well as geographic data, so it can produce maps of the world or of an extremely small area, such as the locations of plant species habitats. DMAP[42] is a mapping software designed mainly for creating distribution maps and coincidence maps, which are displayed in color on-screen with multiple zoom capabilities. LandScape Corridors simulates ecological corridors for different species with different ecological niches to determine the suitable lines for penetration into new habitats.[43] SAM is a program for surface pattern spatial analysis and is used worldwide in macroecology, biogeography, and conservation biology.[44] It is a valuable analytical tool for analyzing the geographical distribution of a species based on the presence/absence of data. In addition, SAM provides several modules for model selection and multimodel inference, logistic regression, and geographically weighted regression (GWR).[44]

Additionally, there are some comprehensive biodiversity datasets for plants such as GBIF[45] and USDA.[46] The USDA PLANTS database comprises standardized data about vascular plants, mosses, liverworts, hornworts, and lichens of the US and its territories. The GBIF-MAPA is a prominent web-based GIS tool for exploring biodiversity data globally, which was developed to effectively utilize large amounts of legacy biodiversity data served by GBIF. In addition, virtual herbaria including virtual herbariums of Vienna,[47] Kew,[48] and Australia[49] are the most prominent botanical databases that provide valuable data on plant taxa from diverse regions of the world.

3 Priority for conservation

In conservation-based approaches, not all taxa and habitats have similar significances; some are widespread, and others show restricted distribution. They also show differing levels of vulnerability to human activity.[50] Because of restricted resources, priorities seem to be necessary in polices of conservation management.[22,51] Rabinowitz[52] and Rabinowitz et al.[53] developed priorities on the basis of habitat specificity as well as local abundance to identify diverse kinds of rarities. However, other methods cover rule-based systems and scoring-and-ranking systems. The rule-based system follows the IUCN guidelines (e.g., Red List of species, Red List of ecosystems, key biodiversity areas, etc.). Scoring systems, however, are created on the basis of several scores for a wide range of criteria to calculate whole scores for every species[54] that categorize different biological taxa on the basis of their priorities for conservation.[55] Well-known ranking systems have been developed by the Natural Heritage Network and The Nature Conservancy.[56,57] Additionally, priorities have been identified by rarity threats, endemicity, and population decline.[58–60] However, others have categorized conservation priorities on the basis of rarity level,[61] distribution range, and habitat specialization.[60,62] In addition, economic criteria,[63] phylogenetic criteria,[64] habitat vulnerability,[65] and taxonomic uniqueness[66,67] are other factors in priorities for conservation.

4 Conservation status

There are five measurable criteria for classifying threatened levels: declining populations (past, present, and/or projected); geographic range size and fragmentation, decline, or fluctuations; small population size and fragmentation, decline, or fluctuations; extremely small population or highly restricted distribution; and quantitative analysis of extinction risk (e.g., population viability analysis). Accordingly, criteria B that cover geographic range and distribution patterns include B_1—extent of occurrence (EOO), and B_2—area of occupancy (AOO).[68]

B1	Extent of	$<100\,km^2$	$<5000\,km^2$	$<20,000\,km^2$	occurrence (EOO)
B2	Area of	$<10\,km^2$	$<500\,km^2$	$<2000\,km^2$	occupancy (AOO)

GeoCAT was established to use spatially referenced primary occurrence data to analyze two aspects of the geographic range of a taxon, which covers the area of occupancy (AOO) as well as the extent of occurrence (EOO). The mentioned measures are based on the 'B' criteria of the IUCN Red List system.[68] This online software can rapidly and easily analyze data to calculate both EOO and AOO[69] and was initially developed in the Avenue scripting language for ArcView 3.3 inside the Conservation Assessment Tools (CAT) extension (developed at the Royal Botanic Gardens, Kew).[70]

5 Endemism patterns

Endemism is a focal concept in ecological and biogeographical assessments.[71] Endemic species need more conservation attention because of their limited distributions and consequent vulnerability to endangerment. Thus, the mentioned taxa have been considered as surrogate species for identifying the priorities of conservation.[22] In areas with numerous endemic taxa, conservation biologists have used several procedures to guarantee the protection of all elements of biodiversity. Myers[58,59] surveyed endemic plant taxa globally and presented that protecting $746,400\,km^2$, a zone representing 0.5% of Earth's land surface in 18 sites worldwide, would conserve 50,000 taxa of endemic plants (20% of all known plant species). The distribution patterns of endemic species are considered fundamental data used for conservation plans, which include identifying biodiversity hotspots and shaping protected areas of diverse forms.[22,72] Mapping endemism centers has been used as an influential method for prioritization.[22,73–75] Moreover, endemism patterns show paleoecological events.[76] They can also be related to areas of evolutionary novelties.[77] Additionally, distribution patterns of endemics can be used to determine Alliance for Zero Extinction (AZE) sites.[50,78] Centers of endemism as key zones of biodiversity have attracted the attention of conservationists.[22] Up to now, two methodological approaches have been used to analyze and identify the phenomenon of endemism: (1) identifying areas or operational geographical units (OGUs) with several endemic species[71,79] and (2) phenetic and cladistic analyses to identify hierarchically structured, nonoverlapping AOEs.[80,81] To date, a variety of methods for identifying areas of endemism have been described: parsimony analysis of endemism (PAE)[80], cladistic analysis of distributions and endemism (CADE)[82], endemicity analysis (EA)[83,84], network analysis method (NAM)[85], analysis of biotic elements (BEs)[86], and analysis of geographical interpolation of endemism (GIE)[87] among others. From these, PAE and EA are the most commonly used methods for inferring areas of endemism.

The concept of **parsimony analysis of endemism (PAE)** is primarily based on vicariance, so extinction and dispersal play limited roles.[88] PAE is based on the cladistic analysis of presence–absence data matrices of species and supraspecific taxa. It analyzes raw distribution matrices based on species occurrences in geographical quadrants, and it shows testability when using large numbers of species distributions, which is highly significant to comparing assessments of diverse taxa in different areas.[80,81] A regularly used practical measure is the unique occurrence of at least two endemic taxa with congruent distribution in one AOE. However, another important criterion is that a maximum of endemic taxa of a study region is represented in the AOE.[64,89]

Endemicity analysis (EA) is another method used to infer areas of endemism. This method identifies areas of endemism using an endemicity index (EI) ranging from 0 to 1.[83,84] EA applies four criteria to the delimitation of areas of endemism. NDM/VNDM software is applied for EA.[84] NDM identifies areas of endemism by implementing an optimality criterion to assess the congruence between a species distribution and a given area using the endemicity index.[84]

Geographical interpolation of endemism (GIE) is shaped based on the assessment of the intersection between the distribution of species through a kernel interpolation of centroids of species distribution and areas of influence defined from the distance between the centroid and the farthest point of occurrence of each species.[87] GIE is a new method for the identification of areas of endemism and differs from other approaches such as PAE and EA in being independent of grid cells.[87]

6 Biodiversity hotspots

Biodiversity hotspots[58] are recognized as excellent centers of endemic species that face higher rates of habitat destruction.[19,22] Such areas must have a certain area. Accordingly, those areas with high values of both endemic species richness and habitat loss are engaged as hotspots.[34] Moreover, some criteria use a combination of distributions and phylogenetic factors.[90] Some approaches emphasize taxonomic differentiation or phylogenetic diversity (PD).[67] Other criteria merge geographic ranges and phylogenetic properties (e.g., phylogenetic endemism, etc.) as well as biogeographically weighted evolutionary distinctiveness (BED).[90] Recent criteria conservation priorities have shared the condition of having a limited resource.[34]

7 Important plant areas (IPAs) and alliance for zero extinction (AZE)

Protected areas are the most important focus in protecting biological diversity[91] and prioritizing conservation. Moreover, important plant areas (IPAs),[92] as a subset of key biodiversity areas (KBAs), "help set national priorities within the global context" and are used to establish a comprehensive system for protected areas.[50] These areas can fill the gaps in protected areas to improve conservation plans. Meanwhile, mapping distribution patterns provide an efficient and powerful tool for designing such areas.[73] To identify and explain KBAs, a wide range of datasets should be compiled, preferably in a geographic information system (GIS). These data cover distribution maps and priority zones as well as area records of the target species and background data layers (e.g., land use, geomorphology, habitats, protected areas, etc.).[50] An IPA covers "a natural or seminatural site exhibiting exceptional botanical richness and/or supporting an outstanding assemblage of rare, threatened, and/or endemic plant species and/or vegetation of high botanic value." In this concept, the term "plant" involves algae, fungi, lichens, liverworts, mosses, and wild vascular plants. An IPA, as a subsection of KBAs, is a site-based method for plant conservation at the national level.[92] Alliance for Zero Extinction (AZE) sites,[93] as a subset of KBAs, include "at least 95% of the known population of one or more Critically Endangered or Endangered species, and thus indicates where extinctions may be imminent."[50]

8 Land-cover monitoring

Land cover refers to the observed biotic and abiotic assemblage of Earth's surface and immediate subsurface.[94] Land-use/land-cover change monitoring is a key factor in understanding the dynamics of an ecosystem. Land-cover and land-use change (LCLUC) modifies distribution patterns of plant species and plant communities, thus affecting regional ecosystems.[95] The succession of plant communities is common. This phenomenon has occurred on a global scale over tens of millions of hectares. Monitoring land-use/land-cover (LULC) changes is considered a key action in conservation management. The monitoring and succession mapping of communities as well as biological species are the most important application aspects of geoinformatics techniques.[96] The growing population induces severe pressure on natural vegetation (e.g., rangelands, forests, wetlands, etc.). GIS and RS are powerful tools used to analyze land cover on different scales.[97] GIS shows high flexibility in assembling, storing, presenting, and analyzing environmental digital data for the detection of land-cover change.[98] Remote sensing imagery provides the most important data for GIS analysis. This satellite data imagery has been used to recognize the synoptic data of Earth's surface.[99] Vegetation is one of the most important land covers of Earth. With the exception of the Arctic, the Antarctic, and desert biomes that show scattered vegetation patterns, Earth is home to a prominent vegetation cover (e.g., formations, communities, etc.). Plant communities can be classified according to their properties based on diverse criteria following the properties of vegetation (e.g., physiognomy, structure, floristic, and phytosociology), external to vegetation (e.g., vegetation succession, habitat, or the environment, and geographical position of plant communities), and a combination of vegetation and the environment (independent analysis of vegetation, independent analysis of the environment, and combined analysis of vegetation and the environment).[100] Land covers change continuously. The change rate of vegetation cover can be either gradual or abrupt and dramatic,[101] resulting from long-term natural changes (e.g., climatological, geological, etc.) and short-term anthropogenic processes. Successful actions based on RS depend mainly on the selection of suitable data sources including spatial, temporal, spectral, and radiometric resolution.[102] Spatial resolution represents the scale of observation[103] and shows the size of the studied region in a pixel image. Temporal resolution represents the frequent visiting of a sensor in the same location on Earth's surface. Spectral resolution is related to spectral divergence at wavelength breaks that a sensor is capable of detecting.[104] NBR (normalized burn ratio), NDVI, (normalized difference vegetation index), and CVA (change vector analysis) are different methods for detecting land-cover changes. Multiindex integrated change (MIIC), also used to detect change, is an integration method using NBR, NDVI, CVA, and a relative of CVA.[29,105]

Remote sensors are the fundamental tools used for capturing data for mapping. Because different objects (e.g., vegetation) have their exclusive spectral properties (e.g., reflectance, structure, etc.), they can be distinguished by RS.[106] Remote sensing includes diverse wavelengths (visible to microwave) and different spatial resolutions ranging from submeter to kilometers. Spatial resolution has different definitions: (1) low (coarse) resolution, including pixels with a ground sampling distance (GSD) of 30 m or bigger; (2) medium resolution with a GSD in the domain expansion between 2.0 and 30 m; (3) high resolution (a GSD of 0.5–2.0 m); and (4) extremely high resolution that involves pixels shorter than 0.5 m GSD.[107] Accordingly, different GSDs are selected based on the aims of mapping, research budget, and climatic conditions as well as technical subjects.[108] Landsat TM and ETM+ (medium-to-coarse spatial resolution–multispectral data), SPOT (full range of medium spatial resolutions–multispectral data), MODIs (low spatial resolution–multispectral data), AVHER (1-km GSD–panchromatic data), IKONOS (high-resolution imagery–panchromatic and multispectral bands), QuickBird (high resolution–panchromatic and multispectral imagery), ASTER (medium spatial resolution–14 spectral bands), AVIRIS (airborne sensor collecting images with 224 spectral bands), and HYPERION (hyperspectral image with 220 bands) are the most well-known RS tools in the world.[109]

9 Recent studies on conservation priorities and distribution patterns of Iranian plant species

Iran has diverse flora, and many aspects of this flora have been studied by many researchers to date.[110] Several phytogeographical studies have been conducted on Iranian flora.[14,111–116] Species richness is one of the most widely used measures in evaluating biodiversity hotspots and conservation priorities in Iran. The species richness of plant taxa that have been mapped in Iran is as follows: endemic plants of the Khorassan-Kopet Dagh floristic province,[116] Solanaceae,[117] Convolvulaceae,[118] Iranian endemic trees and shrubs with the exception of Astragalus[14], monocots crop wild relatives (CWRs),[119] pteridophytes[120], Iranian endemic monocots[75], cudicots CWRs gene pool 1–2[121], parasitic plant species,[122] and aquatic plants[123]. Additionally, Alliance for Zero Extinction (AZE) of Iranian endemic trees and shrubs was identified by Mehrabian et al.[14]. Vegetation types in the Mond protected area were mapped based on unsupervised classification of the Spot XS bands [124] we have illustrated the maps of species richness of pteridophytes (Fig. 1), parasitic plant species (Fig. 2), endemic monocots (Fig. 3), and Boraginaceae (Fig. 4) in Iran.

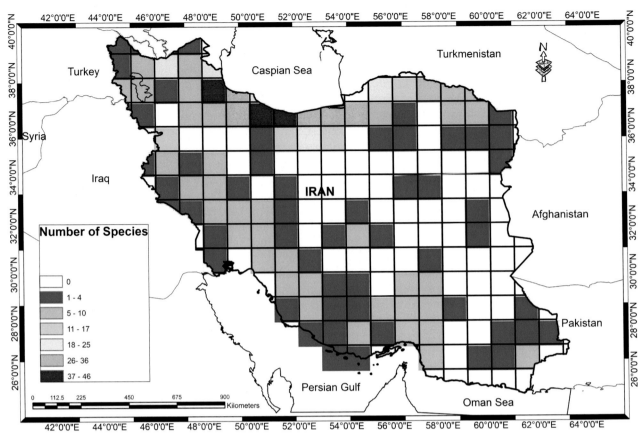

FIG. 1 Number of pteridophyte species per 1° × 1° grid cell.

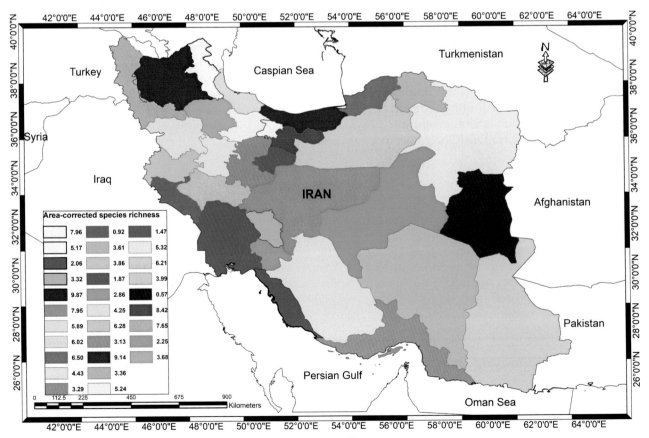

FIG. 2 Area-corrected species richness of parasitic plants in 31 provinces of Iran.

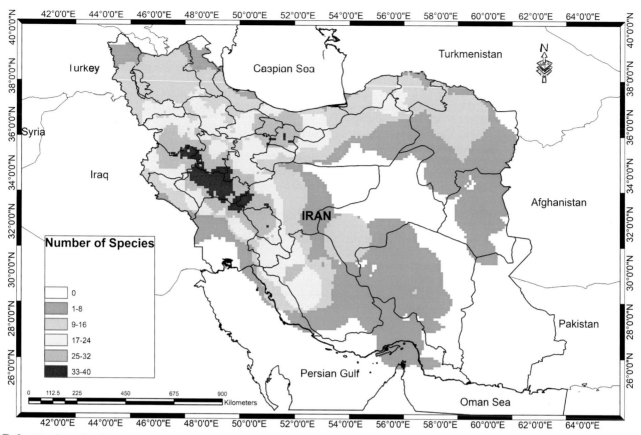

FIG. 3 Number of endemic monocots species per 10 × 10 km grid cell. A circular neighborhood with a radius of 25 km was used to assign observations to a grid cell.

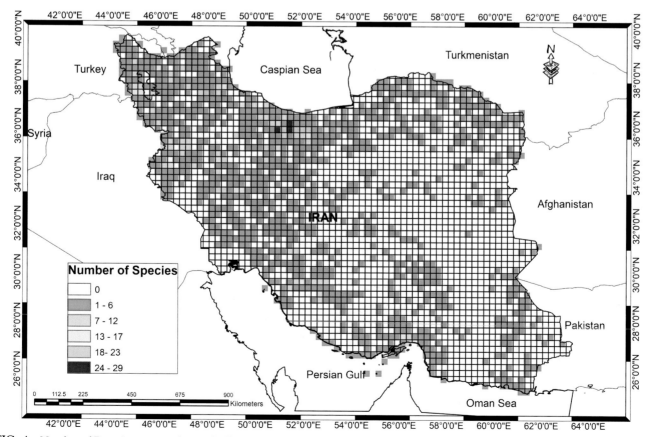

FIG. 4 Number of Boraginaceae species per 0.25° × 0.25° grid cell.

References

1. Arts K, van der Wal R, Adams WM. Digital technology and the conservation of nature. *Ambio*. 2015;44(S4):661–673. https://doi.org/10.1007/s13280-015-0705-1.
2. Mutke J, Barthlott W. 8 patterns of vascular plantdiversity at continental to global scales. *Biol Skr*. 2005;55:521–537.
3. Graham M, Kennedy J. Vesper: visualising species archives. *Eco Inform*. 2014;24:132–147. https://doi.org/10.1016/j.ecoinf.2014.08.004.
4. Fletcher R, Marie-Josée F. *Spatial Ecology and Conservation Modeling Applications with R*. Cham: Springer International Publishing; 2019.
5. Graham C, Ferrier S, Huettman F, Moritz C, Peterson A. New developments in museum-based informatics and applications in biodiversity analysis. *Trends Ecol Evol*. 2004;19(9):497–503. https://doi.org/10.1016/j.tree.2004.07.006.
6. Jetz W, McPherson JM, Guralnick RP. Integrating biodiversity distribution knowledge: toward a global map of life. *Trends Ecol Evol*. 2012;27(3):151–159. https://doi.org/10.1016/j.tree.2011.09.007.
7. Meyer C, Kreft H, Guralnick RP, Jetz W. Global priorities for an effective information basis of biodiversity distributions. *Nat Commun*. 2015;6:8221. https://doi.org/10.7287/peerj.preprints.856.
8. Turner WR. Interactions among spatial scales constrain species distributions in fragmented urban landscapes. *Ecol Soc*. 2006;11(2). https://doi.org/10.5751/es-01742-110206.
9. Vetaas OR, Grytnes J-A. Distribution of vascular plant species richness and endemic richness along the Himalayan elevation gradient in Nepal. *Glob Ecol Biogeogr*. 2002;11(4):291–301. https://doi.org/10.1046/j.1466-822x.2002.00297.x.
10. Jdos T, Kaminski J, Santanna MA, Santos DR. Tampão Santa Maria (TSM) como alternativa ao tampão SMP para medição da acidez potencial de solos ácidos. *Rev Bras Ciênc Solo*. 2012;36(2):427–435. https://doi.org/10.1590/s0100-06832012000200012.
11. Diekmann M. Species indicator values as an important tool in applied plant ecology—a review. *Basic Appl Ecol*. 2003;4(6):493–506. https://doi.org/10.1078/1439-1791-00185.
12. Barthlott W, Lauer W, Placke A. Global distribution of species diversity in vascular plants: towards a world map of phytodiversity. *Erdkunde*. 1996;50(1):317–328. https://doi.org/10.3112/erdkunde.1996.04.03.
13. Olmstead RG. Phylogeny and biogeography in Solanaceae, Verbenaceae and Bignoniaceae: a comparison of continental and intercontinental diversification patterns. *Bot J Linn Soc*. 2012;171(1):80–102. https://doi.org/10.1111/j.1095-8339.2012.01306.x.
14. Mehrabian AR, Sayadi S, Majidi Kuhbenani M, Hashemi Yeganeh V, Abdoljabari M. Priorities for conservation of endemic trees and shrubs of Iran: important plant areas (IPAs) and Alliance for zero extinction (AZE) in SW Asia. *J Asia Pac Biodivers*. 2020;13(2):295–305. https://doi.org/10.1016/j.japb.2019.09.010.
15. Jones PG, Gladkov A. *Floramap: A Computer Tool for Predicting the Distribution of Plants and Other Organisms in the Wild*. Cali, Colombia: Centro Internacional de Agricultura Tropical; 1999. CIAT CD-ROM Series, Version 1.
16. Guisan A, Zimmermann NE. Predictive habitat distribution models in ecology. *Ecol Model*. 2000;135(2–3):147–186. https://doi.org/10.1016/s0304-3800(00)00354-9.
17. Kier G, Barthlott W. Measuring and mapping endemism and species richness: a new methodological approach and its application on the flora of Africa. *Biodivers Conserv*. 2001;10(9):1513–1529. https://doi.org/10.1023/a:1011812528849.
18. Müller R, Nowicki C, Barthlott W, Ibisch PL. Iodiversity and endemism mapping as a tool for regional conservation planning—case study of the Pleurothallidinae (Orchidaceae) of the Andean rain forests in Bolivia. *Biodivers Conserv*. 2003;12(10):2005–2024. https://doi.org/10.1023/a:1024195412457.
19. Prendergast JR, Quinn RM, Lawton JH, Eversham BC, Gibbons DW. Rare species, the coincidence of diversity hotspots and conservation strategies. *Nature*. 1993;365(6444):335–337. https://doi.org/10.1038/365335a0.
20. Mittermeier RA, Gil PR, Hoffman M, et al. *Hotspots Revisited: Earth's Biologically Richest and Most Endangered Terrestrial Ecoregions. Hardcover*. CEMEX; 2005.
21. Linder HP. On areas of endemism, with an example from the African Restionaceae. *Syst Biol*. 2001;50(6):892–912. https://doi.org/10.1080/106351501753462867.
22. Myers N, Mittermeier RA, Mittermeier CG, da Fonseca GA, Kent J. Biodiversity hotspots for conservation priorities. *Nature*. 2000;403(6772):853–858. https://doi.org/10.1038/35002501.
23. Iriondo JM, De Hond L, Dulloo ME, et al. Genetic reserve management. In: *Plant Genetic Population Management*. Wallingford: CAB International; 2008:65–87.
24. Fuentes N, Pauchard A, Sánchez P, Esquivel J, Marticorena A. A new comprehensive database of alien plant species in Chile based on herbarium records. *Biol Invasions*. 2012;15(4):847–858. https://doi.org/10.1007/s10530-012-0334-6.
25. Droissart V, Hardy OJ, Sonké B, Dahdouh-Guebas F, Stévart T. Subsampling Herbarium collections to assess geographic diversity gradients: a case study with endemic Orchidaceae and Rubiaceae in Cameroon. *Biotropica*. 2011;44(1):44–52. https://doi.org/10.1111/j.1744-7429.2011.00777.x.
26. McPhearson PT, Wallace OC. *Remote Sensing Applications to Biodiversity Conservation*; 2008. Available at: http://ncep.amnh.org.
27. Zonneveld M, Thomas E, Galluzzi G. *Chapter 15/16: Mapping the Ecogeographic Distribution of Biodiversity and GIS Tools for Plant Germplasm Collectors*; 2011.
28. Guarino L., Jarvis A., Hijmans R.J., Maxted N.. Geographic information systems (GIS) and the conservation and use of plant genetic resources. In: Managing Plant Genetic Diversity. International Plant Genetic Resources Institute (IPGRI), Rome. 2002; pp. 387–404. Available online (Accessed 6 October 2011].
29. Horning N, Robinson JA, Sterling EJ, Turner W, Spector S. *Remote Sensing for Ecology and Conservation: A Handbook of Techniques*. Oxford: Oxford University Press; 2020.
30. Gilbertson DD, Kent M, Pyatt FB. *Practical Ecology for Geography and Biology, Survey, Mapping, and Data Analysis*. London: Hutchinson; 1985:320.
31. Slocum TA. *Thematic Cartography and Visualization*. Upper Saddle River, NJ: Prentice Hall; 2005.
32. Rahbek C, Graves GR. Multiscale assessment of patterns of avian species richness. *Proc Natl Acad Sci U S A*. 2001;98(8):4534–4539. https://doi.org/10.1073/pnas.071034898.
33. Burgess ND, Rahbek C, Larsen FW, Williams P, Balmford A. How much of the vertebrate diversity of sub-Saharan Africa is catered for by recent conservation proposals? *Biol Conserv*. 2002;107(3):327–339. https://doi.org/10.1016/s0006-3207(02)00071-x.

34. Orme CD, Davies RG, Burgess M, et al. Global hotspots of species richness are not congruent with endemism or threat. *Nature*. 2005;436 (7053):1016–1019. https://doi.org/10.1038/nature03850.

35. Ferrier S. In: Pigram PJ, Sundell RC, eds. *Biodiversity Data for Reserve Selection: Making Best Use of Incomplete Information*. *National Parks and Protected Areas: Selection, Delimitation and Management*. New England: Centre for Water Policy Research, University of New England; 1977:315–329.

36. Loiselle B, Howell C, Graham C, et al. Avoiding pitfalls of using species distribution models in conservation planning. *Conserv Biol*. 2003;17 (6):1591–1600. https://doi.org/10.1111/j.1523-1739.2003.00233.x.

37. Hijmans RJ, Spooner DM. Geographic distribution of wild potato species. *Am J Bot*. 2001;88(11):2101–2112. https://doi.org/10.2307/3558435.

38. Hurlbert AH, White EP. Disparity between range map- and survey-based analyses of species richness: patterns, processes and implications. *Ecol Lett*. 2005;8(3):319–327. https://doi.org/10.1111/j.1461-0248.2005.00726.x.

39. Hijmans RJ, Guarino L, Rojas E. *DIVAGIS a geographic information system for the analysis of biodiversity data*. *Manual*. Lima, Peru: International Potato Center; 2002.

40. Scheldeman X, van Zonneveld M. *Training Manual on Spatial Analysis of Plant Diversity and Distribution*. Rome, Italy: Biodiversity International; 2010.

41. García-Roselló E, Guisande C, González-Dacosta J, et al. ModestR: a software tool for managing and analyzing species distribution map databases. *Ecography*. 2013;36(11):1202–1207. https://doi.org/10.1111/j.1600-0587.2013.00374.x.

42. Morton A. *DMAP for Windows*. *Distribution Map Software 7.3*. Berkshire; 2009.

43. Ribeiro JW, Silveira dos Santos J, Dodonov P, Martello F, Brandão Niebuhr B, Ribeiro MC. LandScape corridors (lscorridors): a new software package for modelling ecological corridors based on landscape patterns and species requirements. *Methods Ecol Evol*. 2017;8(11):1425–1432. https://doi.org/10.1111/2041-210x.12750.

44. Rangel TF, Diniz-Filho JA, Bini LM. SAM: a comprehensive application for spatial analysis in macroecology. *Ecography*. 2010;33(1):46–50. https://doi.org/10.1111/j.1600-0587.2009.06299.x.

45. GBIF. http://www.gbif.org/. Accessed 15 February 2021.

46. PLANTS Database. *USDA PLANTS*. *Welcome to the PLANTS Database | USDA PLANTS*. https://plants.usda.gov/java/. Accessed 20 January 2021.

47. Herbarium WU. Institute of Botany, University Vienna. https://herbarium.univie.ac.at/. Accessed 1 January 2021.

48. Kew Herbarium Catalogue. http://apps.kew.org/herbcat/navigator.do. Accessed 14 February 2021.

49. Australian and New Zealand herbaria. *AVH*. https://avh.chah.org.au/. Accessed 5 April 2021.

50. Langhammer PF. *Identification and Gap Analysis of Key Biodiversity Areas: Targets for Comprehensive Protected Area Systems*; 2007. https://doi.org/10.2305/IUCN.CH.2006.PAG.15.en.

51. Brummitt N, Lughadha EN. Biodiversity: where's hot and where's not. *Conserv Biol*. 2003;17(5):1442–1448. https://doi.org/10.1046/j.1523-1739.2003. 02344.x.

52. Rabinowitz D. In: Synge H, ed. *Seven Forms of Rarity*. *The Biological Aspects of Rare Plant Conservation*. Chichester: John Wiley & Sons; 1981:205–217.

53. Rabinowitz D, Cairns S, Dillon T. Seven forms of rarity and their frequency in the flora of the British Isles. In: Soule ME, ed. *Conservation Biology, the Science of Scarcity and Diversity*. Sunderland, Mass: Sinauer; 1986:182–204.

54. Given DR, Norton DA. A multivariate approach to assessing threat and for priority setting in threatened species conservation. *Biol Conserv*. 1993;64(1):57–66. https://doi.org/10.1016/0006-3207(93)90383-c.

55. Ray JC, Hunter L, Zigouris J. *Setting Conservation and Research Priorities for Larger African Carnivores*. New York: Wildlife Conservation Society; 2005.

56. Master L. Assessing threats and setting priorities for conservation. *Conserv Biol*. 1991;5(4):559–563. https://doi.org/10.1111/j.1523-1739. 1991. tb00370.x.

57. Stein BA. Towards common goals: collections information in conservation databases. *ASC Newsl*. 1993;21(1):1–6.

58. Myers N. Threatened biotas: "hot spots" in tropical forests. *Environmentalist*. 1988;8(3):187–208. https://doi.org/10.1007/bf02240252.

59. Myers N. The biodiversity challenge: expanded hot-spots analysis. *Environmentalist*. 1990;10(4):243–256. https://doi.org/10.1007/bf02239720.

60. Sapir Y, Shmida A, Fragman O. Constructing red numbers for setting conservation priorities of endangered plant species: Israeli flora as a test case. *J Nat Conserv*. 2003;11(2):91–107. https://doi.org/10.1078/1617-1381-00041.

61. Williams PH, Gaston KJ, Humphries CJ. Mapping biodiversity value worldwide: combining higher-taxon richness from different groups. *Proc R Soc Lond [Biol]*. 1997;264(1378):141–148. https://doi.org/10.1098/rspb.1997.0021.

62. Sólymos P, Fehér Z. Conservation prioritization based on distribution of land snails in Hungary. *Conserv Biol*. 2005;19(4):1084–1094. https://doi.org/10.1111/j.1523-1739.2005.00193.x.

63. Bishop RC. Endangered species and uncertainty: the economics of a safe minimum standard. *Am J Agric Econ*. 1978;60(1):10–18. https://doi.org/10.2307/1240156.

64. Linder HP. Setting conservation priorities: the importance of endemism and phylogeny in the southern African orchid genus Herschelia. *Conserv Biol*. 1995;9(3):585–595. https://doi.org/10.1046/j.1523-1739.1995.09030585.x.

65. Tambutii M, Aldama A, Sánchez O, Medeellin R, Soberón J. La determinación del riesgo de extincion de especies silvestres en Mexico. *Gaceta Ecol*. 2001;61:11–21.

66. Vane-Wright RI, Humphries CJ, Williams PH. What to protect? —systematics and the agony of choice. *Biol Conserv*. 1991;55(3):235–254. https://doi.org/10.1016/0006-3207(91)90030-d.

67. Faith DP. Conservation evaluation and phylogenetic diversity. *Biol Conserv*. 1992;61(1):1–10. https://doi.org/10.1016/0006-3207(92)91201-3.

68. IUCN. *Red List Guidelines - ALCES*. https://www.alces.ca/references/download/267/Guidelines-for-Using-the-IUCN-Red-LIst-Categories-and-Criteria.pdf. Accessed 31 March 2021.

69. Willis F, Moat J, Paton A. Defining a role for herbarium data in red List assessments: a case study of Plectranthus from eastern and southern tropical Africa. *Biodivers Conserv*. 2003;12(7):1537–1552. https://doi.org/10.1023/a:1023679329093.

70. Bachman S, Moat J, Hill A, de la Torre J, Scott B. Supporting red list threat assessments with GeoCAT: geospatial conservation assessment tool. *ZooKeys*. 2011;150:117–126. https://doi.org/10.3897/zookeys.150.2109.

71. Crisp MD, Laffan S, Linder HP, Monro A. Endemism in the Australian flora. *J Biogeogr*. 2001;28(2):183–198. https://doi.org/10.1046/j.1365-2699.2001. 00524.x.

72. Anderson RP, Martinez-Meyer E. Modeling species' geographic distributions for preliminary conservation assessments: an implementation with the spiny pocket mice (Heteromys) of Ecuador. *Biol Conserv*. 2004;116(2):167–179. https://doi.org/10.1016/s0006-3207(03)00187-3.

73. Lombard AT, Cowling RM, Pressey RL, Rebelo AG. Effectiveness of land classes as surrogates for species in conservation planning for the cape floristic region. *Biol Conserv*. 2003;112(1–2):45–62. https://doi.org/10.1016/s0006-3207(02)00422-6.

74. Araujo MB, Pearson RG, Thuiller W, Erhard M. Validation of species-climate impact models under climate change. *Glob Change Biol*. 2005;11 (9):1504–1513. https://doi.org/10.1111/j.1365-2486.2005.01000.x.

75. Mehrabian AR, Amini RM. *The Map of Distribution Patterns and Geobotany of Iranian Endemic Monocotyledons*. Shahid Beheshti University; 2015:1.

76. Brown JH, Lomolino MV. *Biogeography*. 2nd ed. Sunderland, North East England: Sinauer Associates; 1998.

77. Bossuyt F, Meegaskumbura M, Beenaerts N, Gower DJ, Pethiyagoda R, Roela K. Local endemism within the Western Ghats-Sri Lanka biodiversity hotspot. *Science*. 2004;306(5695):479–481. https://doi.org/10.1126/science.1100167.

78. Ainsworth D. *Alliance for Zero Extinction and the Convention on Biological Diversity Join Forces, Convention on Biological Diversity*; 2011. https://www.cbd.int/doc/press/2010/pr-2010-06-11-zero-extinction-en.pdf. Accessed 14 January 2021.

79. Linder HP. Plant diversity and endemism in sub-Saharan tropical Africa. *J Biogeogr*. 2001;28(2):169–182. https://doi.org/10.1046/j.1365-2699.2001. 00527.x.

80. Morrone JJ. On the identification of areas of endemism. *Syst Biol*. 1994;43(3):438. https://doi.org/10.2307/2413679.

81. Morrone JJ, Escalante T. Parsimony analysis of endemicity (PAE) of Mexican terrestrial mammals at different area units: when size matters. *J Biogeogr*. 2002;29(8):1095–1104. https://doi.org/10.1046/j.1365-2699.2002. 00753.x.

82. Porzecanski AL, Cracraft J. Cladistic analysis of distributions and endemism (CADE): using raw distributions of birds to unravel the biogeography of the South American aridlands. *J Biogeogr*. 2005;32(2):261–275. https://doi.org/10.1111/j.1365-2699.2004. 01138.x.

83. Szumik CA, Cuezzo F, Goloboff PA, Chalup AE. An optimality criterion to determine areas of endemism. *Syst Biol*. 2002;51(5):806–816. https://doi.org/10.1080/10635150290102483.

84. Szumik CA, Goloboff PA. Areas of endemism: an improved optimality criterion. *Syst Biol*. 2004;53(6):968–977. https://doi.org/10.1080/10635150490888859.

85. Dos Santos DA, Fernández HR, Cuezzo MG, Domínguez E. Sympatry inference and network analysis in biogeography. *Syst Biol*. 2008;57 (3):432–448. https://doi.org/10.1080/10635150802172192.

86. Hausdorf B, Hennig C. Biotic element analysis in biogeography. *Syst Biol*. 2003;52(5):717–723. https://doi.org/10.1080/10635150390235584.

87. Oliveira U, Brescovit AD, Santos AJ. Delimiting areas of endemism through kernel interpolation. *PLoS One*. 2015;10(1). https://doi.org/10.1371/journal.pone.0116673.

88. Ron SR. Biogeographic area relationships of lowland Neotropical rainforest based on raw distributions of vertebrate groups. *Biol J Linn Soc*. 2000;71(3):379–402. https://doi.org/10.1111/j.1095-8312.2000.tb01265.x.

89. Morrone JJ, Crisci JV. Historical biogeography: introduction to methods. *Annu Rev Annu Rev Ecol Evol Syst*. 1995;26(1):373–401. https://doi.org/10.1146/annurev.es.26.110195.002105.

90. Cadotte MW, Jonathan DT. Rarest of the rare: advances in combining evolutionary distinctiveness and scarcity to inform conservation at biogeographical scales. *Divers Distrib*. 2010;16(3):376–385. https://doi.org/10.1111/j.1472-4642.2010. 00650.x.

91. Bruner AG. Effectiveness of parks in protecting tropical biodiversity. *Science*. 2001;291(5501):125–128. https://doi.org/10.1126/science.291.5501.125.

92. Anderson S, Kŭsik T, Radford E. *Important Plant Areas in Central and Eastern Europe*. Plantlife International; 2005. https://www.plantlife.org.uk/application/files/8214/8233/1761/IPAsinCEE-5mb.pdf. Accessed 31 March 2021.

93. Ricketts TH, Dinerstein E, Boucher T, et al. Pinpointing and preventing imminent extinctions. *Proc Natl Acad Sci U S A*. 2005;102(51):18497–18501. https://doi.org/10.1073/pnas.0509060102.

94. Meyer WB, Turner BL. Human population growth and global land-use/cover change. *Annu Rev Ecol Syst*. 1992;23(1):39–61. https://doi.org/10.1146/annurev.es.23.110192.000351.

95. Zhang F, H-te K, Johnson V. Assessment of land-cover/land-use change and landscape patterns in the two national nature reserves of Ebinur Lake watershed, Xinjiang, China. *Sustainability*. 2017;9(5):724. https://doi.org/10.3390/su9050724.

96. Szostak M. Automated land cover change detection and forest succession monitoring using LiDAR point clouds and GIS analyses. *Geosciences*. 2020;10(8):321. https://doi.org/10.3390/geosciences10080321.

97. Guerschman JP, Paruelo JM, Bella CD, Giallorenzi MC, Pacin F. Land cover classification in the argentine pampas using multi-temporal Landsat TM data. *Int J Remote Sens*. 2003;24(17):3381–3402. https://doi.org/10.1080/0143116021000021288.

98. Wu Q, H-qing L, R-song W, et al. Monitoring and predicting land use change in Beijing using remote sensing and GIS. *Landsc Urban Plan*. 2006;78(4):322–333. https://doi.org/10.1016/j.landurbplan.2005.10.002.

99. Ulbricht KA, Heckendorff WD. Satellite images for recognition of landscape and landuse changes. *ISPRS J Photogramm Remote Sens*. 1998;53 (4):235–243. https://doi.org/10.1016/s0924-2716(98)00006-9.

100. Giri CP. *Remote Sensing of Land Use and Land Cover: Principles and Applications*. Boca Ratón (Florida): CRC Press; 2012.

101. Verbesselt J, Hyndman R, Zeileis A, Culvenor D. Phenological change detection while accounting for abrupt and gradual trends in satellite image time series. *Remote Sens Environ*. 2010;114(12):2970–2980. https://doi.org/10.1016/j.rse.2010.08.003.

102. Lu D, Mausel P, Brondízio E, Moran E. Change detection techniques. *Int J Remote Sens*. 2004;25(12):2365–2401. https://doi.org/10.1080/0143116031000139863.

103. Woodcock CE, Strahler AH. The factor of scale in remote sensing. *Remote Sens Environ*. 1987;21(3):311–332. https://doi.org/10.1016/0034-4257(87)90015-0.

104. Lillesand T, Kiefer R. *Remote Sensing and Image Interpretation*. New York: John Wiley; 1994.

105. Fry JA, Xian G, Jin S, et al. Completion of the 2006 National Land Cover Database for the conterminous United States. *Photogramm Eng Remote Sens*. 2011;77(9):858–864.

106. Gallo KP, Daughtry CST, Bauer ME. Spectral estimation of absorbed photosynthetically active radiation in corn canopies. *Remote Sens Environ*. 1985;17(3):221–232. https://doi.org/10.1016/0034-4257(85)90096-3.

107. Navulur K. *Multispectral Image Analysis Using the Object-Oriented Paradigm*. Boca Raton, FL: CRC Press; 2007.

108. Soudani K, François C, le Maire G, Le Dantec V, Dufrêne E. Comparative analysis of IKONOS, SPOT, and ETM+ data for leaf area index estimation in temperate coniferous and deciduous forest stands. *Remote Sens Environ.* 2006;102(1–2):161–175. https://doi.org/10.1016/j.rse.2006.02.004.

109. Xie Y, Sha Z, Yu M. Remote sensing imagery in vegetation mapping: a review. *J Plant Ecol.* 2008;1(1):9–23. https://doi.org/10.1093/jpe/rtm005.

110. Zohary M. *Geobotanical Foundations of the Middle East.* Stuttgart: Fischer; 1973.

111. Hedge IC, Wendelbo P. *Patterns of Distribution and Endemism in Iran*; 1978.

112. Takhtadzhian AL, Crovello TJ, Cronquist A. *Floristic Regions of the World.* Dehra Dun: Bishen Singh, Mahendra Pal Singh; 2005.

113. White F, Leonard J. Phytogeographical links between Africa and Southwest Asia. In: *Flora et Vegetatio Mundi.* vol. 9; 1991:229–246.

114. Assadi M. Distribution patterns of the genus *Acantholimon*(Plumbaginaceae) in Iran. *Iran J Bot.* 2006;12:114–120.

115. Akhani H. Diversity, biogeography, and photosynthetic pathways of *Argusia* and *Heliotropium* (Boraginaceae) in south-West Asia with an analysis of phytogeographical units. *Bot J Linn Soc.* 2007;155(3):401–425. https://doi.org/10.1111/j.1095-8339.2007.00707.x.

116. Memariani F, Zarrinpour V, Akhani H. A review of plant diversity, vegetation, and phytogeography of the Khorassan-Kopet Dagh floristic province in the Irano-Turanian region (northeastern Iran–southern Turkmenistan). *Phytotaxa.* 2016;249(1):8. https://doi.org/10.11646/phytotaxa.249.1.4.

117. Sayadi S, Mehrabian AR. Diversity and distribution patterns of Solanaceae in Iran: implications for conservation and habitat management with emphasis on endemism and diversity in SW Asia. *Rostaniha.* 2016;17:136–160 [In Farsi].

118. Sayadi S, Mehrabian AR. Distribution patterns of Convolvulaceae in Iran: priorities for conservation. *Rostaniha.* 2017;18:181–197 [In Farsi].

119. Hosseini N, Mehrabian AR, Mostafavi H. The distribution patterns and priorities for conservation of monocots crop wild relatives (CWRs) of Iran. *J Wildl Biodiversity.* 2020. https://doi.org/10.22120/JWB.2020.136030.1186.

120. Mehrabian AR, Khajoei Nasab F, Fraser-Jenkins CR, Tajik F. Distribution patterns and priorities for conservation of Iranian pteridophytes. *Fern Gaz.* 2020;21(4):141–160.

121. Mehrabian AR, Sayadi S. *The Map of Distribution Patterns and Conservation Status of Iranian crop Wild Relatives (GP1).* Shahid Beheshti University; 2018:1.

122. Mehrabian AR, Khajoei NF. *The Map of Distribution Patterns, Diversity Centers, and Priorities for Conservation of Parasitic Plants in Iran.* Shahid Beheshti University; 2020:1.

123. Mehrabian AR, Khajoei NF. *The Map of Distribution Patterns, Diversity Centers, and Priorities for Conservation of Aquatic Plants in Iran.* Shahid Beheshti University; 2019:1.

50

Assessing agriculture land-use change using remote sensing data in the Gilan Province, Iran

Shilan Felegari[a], Alireza Sharifi[b], Kamran Moravej[a], and Ahmad Golchin[a]

[a]Department of Soil Science, Faculty of Agriculture, University of Zanjan, Zanjan, Iran
[b]Department of Surveying engineering, Faculty of civil engineering, Shahid Rajaee Teacher Training University, Tehran, Iran

1 Introduction

Land use consists of two words: land and use. "Land" refers to all the natural features, characteristics, and natural conditions of a place, such as climate, geology, topography, and hydrology. "Use" means using natural facilities, desirably and correctly. Land use and land cover describe the type of human exploitation of a piece of land for one or more purposes.[1] Land cover, under normal conditions, is the biophysical condition of Earth's surface and its subsurface crust, that is, land cover shows the biophysical condition of Earth's surface in terms of crops, mountains, or forests, but land use includes all the biophysical features of the land and how the land is used.[2] Land use is related to human activities and changes over time. Land-use change and land cover are widespread and rapid processes that, in many cases, negatively affect natural resources such as soil and water resources. Land-use change is caused by human activities and often affects the changes that human beings make. Land use can be changed in the following ways: (i) by converting one type of land use to another and (ii) by modifying a specific type of land use.[3]

Determining the effects of land use on soil is possible through the study and evaluation of the soil quality index. Land-use change causes changes in vegetation, hydrology, and many physical, chemical, and biological properties of soil, which, depending on environmental and climatic conditions, may have a positive and negative impact on soil quality.[4] According to the World Food Organization in 2012, the per-capita level of agricultural land in all parts of the world, except for some developed countries, declined between 1970 and 2009. This decrease, especially in third-world countries such as Iran, has been higher than the average of all parts of the world, whereas the level of agricultural land in Iran is lower than the global average. Part of this is due to population growth, and much of this is due to the change in agricultural land use to nonagricultural activities.[5] In the 2016 FAO report, the value added of the agricultural sector shows an increase in many parts of the world and in Asia, whereas in Iran, it has decreased from 1990 to 2014.[6] Knowledge of land cover, human activities, and how to use the land as basic information for various planning is of particular importance. Land-cover maps from satellite imagery play an important role in regional and national assessments. To better understand the landscape dynamics over time and for planning and management, it is important to examine the applied spatial patterns and land cover.[7]

Over the past few decades, remote sensing data have become a comprehensive data source for a variety of applications, including land-cover mapping, owing to periodic repetition, diversity and radiometry, integrated vision, and digital format suitable for computer processing.[8–11]

Currently, remote sensing technology, which has the highest speed and accuracy, is the best tool for monitoring environmental changes and extracting land use. Using multitime remote sensing data, land uses can be extracted at the lowest cost, and, then, the ratio of changes can be evaluated by comparing it in different periods. The use of satellite images and their digital processing with appropriate algorithms make it possible to identify and distinguish the details of phenomena that the human eye cannot distinguish while minimizing human error.[12]

Alkaradaghi et al.[13] assessed land-use change in Iraq's Sulaymaniyah Province using multipurpose satellite imagery. In this study, the development of settlements in the Sulaymaniyah Province from 2001 to 2017 using Landsat images was assessed. The results showed that the overall accuracy of classifying satellite images ranged from 78% to 90% from 2001 to 2017. Islam et al.[14] categorized land use and detected changes using multimedia images via satellite imagery for the Chanati protected area of Bangladesh. This study examined land-use changes in the Chanati Wildlife Sanctuary from 2005 to 2015 using Landsat images with TM and OLI/TIRS sensors. The results showed that the area of the degraded forest has increased from about 256 ha over 10 years and the annual rate of change has been 6.5%. Landsat satellite imagery from two different periods related to the TM sensor in 1990 and 2010 was obtained from land-cover and geological sites. The results of this study showed that during the last decade, vegetation and residential land increased by 51.3% and 55.3%, respectively. Nalina et al.[15] studied the dynamics of land-use change and land cover in an Indigenous region of India from 1990 to 2010 using remote sensing techniques and satellite imagery. Their results showed that the classification accuracy for the years 1990 and 2010 was estimated at 83% and 90%, respectively, and the kappa coefficients for these years were estimated to be 80% and 88%, respectively. Although much research has been conducted on land-use change and its impact factors, spatiotemporal analysis of land-use change by the principles of land management is an important research topic that has been less addressed so far. One of the important innovations of this research is the study of land-use change according to the characteristics and capacities of the study area as well as the use of satellite image processing during two time periods. This study aimed to determine the probability of land-cover change in gardens and agricultural land uses, residential areas, barren lands, and pastures between 1990 and 2015.

2 Methodology

2.1 Study area

The Gilan Province, with geographical coordinates of 36° 34′ to 38° 27′ north latitude and 48° 34′ to 50° 36′ east longitude and with an area of 13,952 square meters, constitutes 0.8% of the area of Iran. The average annual rainfall in the Gilan Province, excluding a small area in the south, is 1402 mm, which is much higher than the national average (255 mm). From the west to east of the province and from the north to south, the amount of rainfall decreases. The seasonal distribution of rainfall in the province shows that autumn receives the highest annual rainfall, with a total of 42%, and spring receives the lowest with a total annual rainfall of 15%.

The relative humidity of the province is between 40% and 100%, and its average temperature is 17.5°C (http://berenjamol.areeo.ac.ir/). The southernmost point of the Gilan Province differs from its northernmost part by almost two degrees of latitude. In meteorology, this value has little effect on temperature differences. However, the temperature conditions in this small range showed many differences. The main reason for this temperature difference is the role of the Alborz and Talesh ranges at different altitude levels. The Gilan Province in terms of topography has two parts that are almost equal and different from each other. Its flat part, with the name of the Gilan plain and also with the names of the Talesh and Astara plains, covers the northwest, east, and center of the province and generally includes the coastal and alluvial areas of the rivers leading to the sea. The rugged and mountainous part of Gilan also includes a series of mountains that stretch from the Astara River valley in the north of the province to the mountains in the east of the province. These roughness values are the result of orogenic movements during the late third geological period. Proper rainfall has long led to the emergence of good and extensive vegetation throughout the Gilan Province, and one of the natural features of this province throughout history has been the abundance of diverse forests and woodlands. However, the growth of urbanization and the indiscriminate cutting of forest trees in recent decades have led to the destruction of some forest areas in the province; however, about one-third of the Gilan Province is still covered with forest, which is estimated at 511,306 ha, and also has an important share in the agricultural production in the country and the production of rice; nearly 100% percent of tea and more than half of the country's silk cocoons are produced in this province.

2.2 Datasets

In this research, TM and ETM sensor images related to 1990 and 2015 from the Landsat satellite were used. The date of the images was related to the last day of September in 1990 and 2015. Provision of suitable satellite images of the region was one of the most challenging issues in the process of this research because, in the process of selecting the satellite images used, images should be selected such that the cloud coverage of each image is not more than 5%.

Owing to the specific climatic conditions of the region and the cloudiness of the region, there is a limit in the selection of images for most of the year. Landsat satellite imagery was geometrically and radiometrically corrected to reduce satellite imagery errors. To increase the accuracy of geometric correction, ground control points that were recorded from the area during the operation were used, and, by removing inappropriate points, it was attempted to make the amount of RMS less than 1 so that the RMS of the OLI reached 0.430. For atmospheric correction of ETM images, after correcting the OLI images using the image-to-image method and an error of 0.638, this correction was performed, which has acceptable accuracy. In this research, for processing images and monitoring changes, the combination of bands 5, 4, and 3 related to the ETM satellite and that of bands 7, 5, and 4 related to the OLI satellite was used in RGB, respectively.

2.3 Supervised method

In the supervised method for classifying pixels, the user must select several representative learning areas for each of the predefined classes. User experience is extremely useful in identifying and locating educational areas. The supervised method is preferred by most researchers because it usually provides a more accurate definition of classes than does the unsupervised method. Various methods of digital algorithms have been developed to detect land-use changes using remote sensing data. The most important methods for detecting changes are image differentiation, image division, principal component analysis, use of fuzzy logic, post-classification comparison, and time spectral classification. The algorithm used in this research is the maximum probability method, which is the most common method. The maximum probability method is a statistically supervised method for identifying the patterns.[7]

The probability of a pixel belonging to any of the predefined classes is calculated, and, then, these pixels are assigned to the class that has the highest probability. The existence of initial acquaintance with the region and field operations showed that the main uses in the region include lands with forest cover, agricultural lands, grasslands, sea areas, lands without vegetation, residential areas, and canebrake. Using the collected climatic samples, supervised classification was performed using the maximum probability method and the initial land-use map was extracted for each image.[16] To eliminate the noise pixels of the classified images, a 3×3 majority filter was used and the accuracy of the maps obtained from the maximum probability algorithm was estimated. The most common accuracy estimation parameters include overall accuracy, producer accuracy, user accuracy, and kappa coefficient, which are also used to calculate the kappa coefficient for pixels that are not properly classified. Hence, it is a good criterion for comparing the results of different classifications. The accuracy of the map classification is calculated using the kappa coefficient in equation[17]:

$$\text{Kappa} = \frac{P_0 - P_c}{1 - P_c} \times 100$$

After preparing the land-use map, the maps related to the two periods were stacked in the GIS environment and the change map was prepared using the postclassification comparison method.

3 Results

Using satellite imagery, the area was divided into seven user units (forest cover, agricultural lands, grasslands, marine areas, lands without vegetation, residential areas, and canebrake) and a user map was prepared separately for each date. For a better study of the region, which is the Gilan Province in northern Iran, this region was divided into two parts, west and east, and a land-use map was prepared for the two designated parts in 1990 and 2015 (Figs. 1 and 2).

One of the most widely used methods for assessing classification accuracy is the calculation of the error matrix, which has been used in many studies. In this study, land-use classification maps were evaluated with an overall accuracy of 80% and a kappa coefficient higher than 0.8 from the error matrix calculation. The kappa values and the overall accuracy above 80% indicate that there is generally a good agreement between the classification and the types of performance classes available on the ground, owing to the use of Google Earth software images in the preparation of training points, the high level of training points, and the careful selection of training points. Table 1 shows the accuracy of satellite image classification (Figs. 3–5).

In the next step, the maps in this study were compared over 15 periods. The area of land-use classes in the two time periods of 1990 and 2015 was determined as a percentage. According to the outputs of Chart 1, for the western part of the study area, the use of forest classes decreased from 70% in 1990 to 59% in 2015. Moreover, the use of grassland

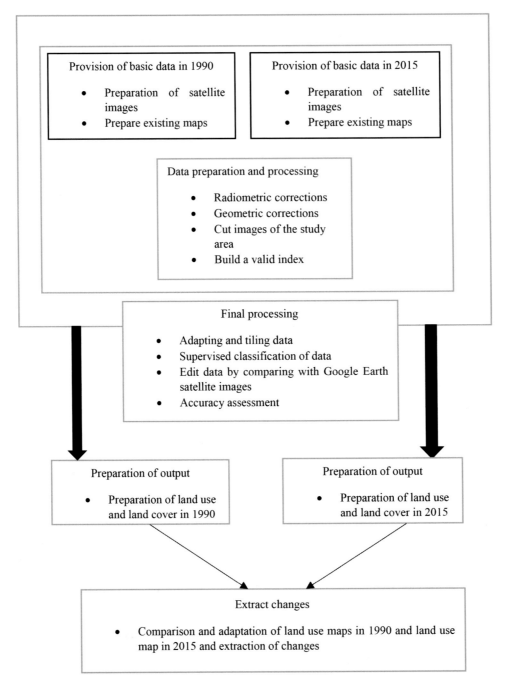

FIG. 1 Flowchart.

classes in the western region increased from 2% in 1990 to 15% in 2015. According to the outputs of Chart 2 for the eastern part of the study area, the use of forest classes did not change much, but the use of grassland decreased from 35% in 1990 to 31% in 2015.

The area of agricultural land in the western part of the study area did not change significantly for 15 years, but, in the eastern part of the study area, agricultural land decreased significantly by 17% in 2015 compared to 1990. The area of residential land has increased over time for both parts of the study area, but this increase was much larger in the eastern part of the study area than in the western part. In the western part, there was an increase of 66% and the eastern part faced a 90% increase in 2015 compared to 1990.

According to the results of this study, it can be said that in the western part of the study area, a significant decrease in forest land use occurred by 10% in 2015 compared to 1990, which indicates the destruction of natural ecosystems in this part of the study area, but, in the same part, grassland use was reported with a growth of 10% in 2015 compared to

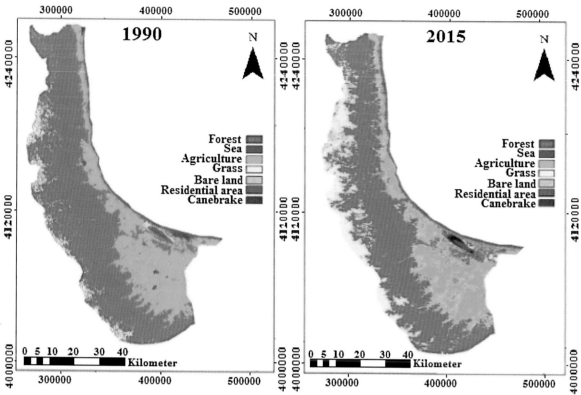

FIG. 2 Land-use map for 1990 and 2015 (western part).

TABLE 1 Results of the overall accuracy and Kappa coefficient for ETM + and OLI images.

Year	Image	Kappa coefficient	Accuracy coefficient
1990	TM	0.86	92.45
2015	ETM	0.88	92.3

1990. In the eastern part of the study area, most of the land has become a residential area, with a 10% increase reported in 2015 compared to 1990. In this part, in contrast to the west, grassland was reported to decrease by 10% compared to 1990.

4 Discussion

Today, sustainable management in the regions and achieving minimal environmental damage due to land-use change require predicting land-use changes, in the long run, to apply appropriate and sustainable management practices in the regions based on possible changes. The results of the overall accuracy and kappa coefficient obtained from the images to evaluate the number of changes in the area showed that satellite images are capable of detecting changes. In this study, high accuracy results were obtained because of the use of Google Earth software images in the preparation of educational points, the high level of educational points, and the accuracy in selecting educational points. This result is consistent with Johnson's theory that the acceptable accuracy of land-use classification using satellite imagery was 85%. During this time, land-cover and land-use patterns as a result of a change are fundamental and human factors can play the most important role in this process.

Factors affecting land-use change in the study area can be divided into economic, governance, demographic, and natural factors. Economic factors affecting land-use change in the region can be attributed to the increase in land prices after the land-use change and the lack of economic efficiency of agricultural activities due to agricultural imports.[18] In terms of governance factors, it can be said that insufficient government support for farmers, weakness of institutions

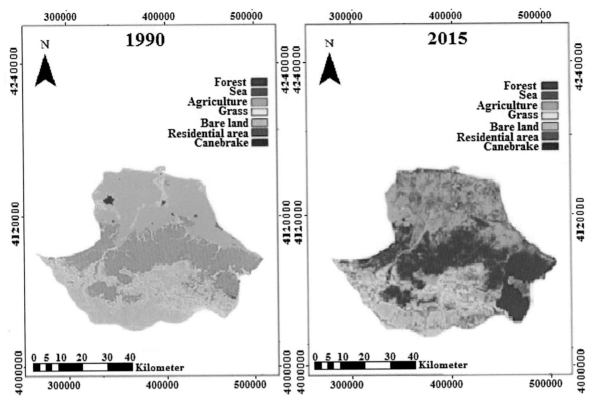

FIG. 3 Land-use map for 1990 and 2015 (eastern part).

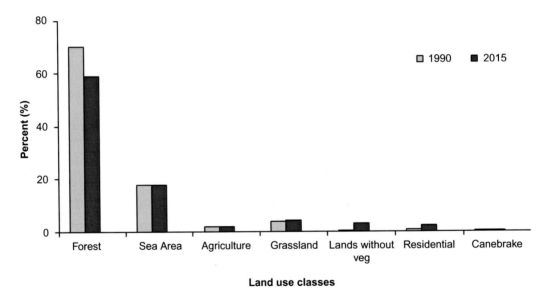

FIG. 4 Graph of land-use changes from 1990 to 2015 (western part).

and executive bodies in implementing and monitoring the law on land-use protection, and expansion of urban areas without considering the negative effects for this region are important factors.[19] In the study of the population factor, we concluded that the desire of urban dwellers to have villas in villages, the migration of villagers to cities, and the expansion of urban areas have all played an important and effective role in land change in this region. Climate change and drought that reduced land quality (due to water, soil pollution, and poor drainage) and increased environmental pollution are important natural factors that justify land-use change in this study.[11]

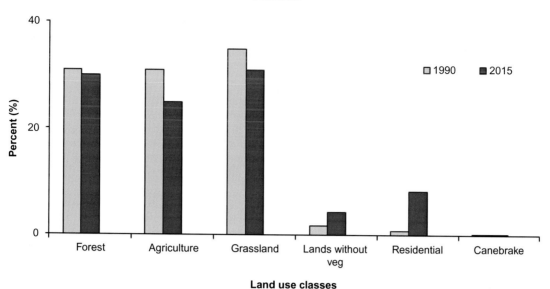

FIG. 5 Graph of land-use changes from 1990 to 2015 (eastern part).

In the study area, a significant part of the conversion of agricultural and forest lands is related to the replacement of residential land use with these lands, which is more in the eastern part of the study area. Therefore, it can be said that in the eastern part of the study area, more attention has been paid to urbanization, which is due to access to more facilities in this part of the city; the rapid growth of urbanization in the eastern part of the study area causes incoherence in spatial structure and polarization of one or more urban centers.

The reason for this can be attributed to the low price of agricultural land as well as the high demand for low-income migrant workers who migrate to this city to work; because of the lack of economic efficiency for agriculture, they try to sell these lands and make more profit.[12]

Karami,[20] in his studies on land-use change in the west of the country, obtained similar results; he announced that the level of forest land use in this part of the country has decreased in recent years. This significant reduction was caused by the marginalized people of the forest, by capturing the forest edge and turning it into illegal agricultural lands and abandoning these lands due to the loss of soil food reserves. Other reasons include population growth, which is a reason for the increase in residential land, and the influx of people into forest areas for tourist purposes, which increases residential land and deforestation.

The achievements of this research are consistent with the results of the following researchers:

Borri et al.[21] using IKONOS satellite images, studied land-cover changes in the Antalya National Park in Turkey. Based on the differences in spatial distribution and land-use patterns, these researchers studied land-use transformations in the area and concluded that the method of classification using satellite imagery produces better results than do traditional methods.

Zhou1 et al.[22] using satellite imagery, studied the green space of Baltimore and Maryland and obtained their area. In their work, they used high-resolution satellite images and digital aerial photographs, and, after preprocessing and processing in the image classification stage and analyzing the results based on the efficiency of the classification method, they used sensor images rather than the traditional method of monitoring changes to study the lands emphasized.

Davoodi et al.[23] in the study of land-use changes in Shahriar city, which was carried out with remote sensing data, pointed to the increase of urban lands and the decrease of agricultural lands.

Khani et al.[24] in the study of the construction of the Tehran-North freeway, pointed to the reduction of green space and biodiversity and the expansion of human activities in the study area.

The results of this study indicate the growth and development of urban physics and fluctuations in other land uses. This study also shows the usefulness of data and remote sensing techniques in assessing land-use change, and this technique can be used for urban planning and land management.

Today, sustainable management in the regions and achieving the minimum environmental damage caused by land-use change require predicting land-use changes, in the long run, to apply appropriate and sustainable management practices in the regions based on possible changes.

5 Conclusions

In this study, coverage changes in user classes in the region were utilized using satellite imagery from 1990 to 2015, and the maximum similarity algorithm of the monitored method was used. The amount of space changed from one user to another between 1990 and 2015.

In this regard, it was found that during the years 1990–2015, the land-use and residential areas increased sharply so that the forest area in the west of the study area decreased by 15%, but, in the east of the study area with 92% growth of the class, we were facing the use of a residential area. It was found that, from 1990 to 2015, the largest change in the land area in the west of the study area was related to the forest land-use class and, in the eastern part of the study area, the most significant area change was related to the land-use class of the residential area. Undoubtedly, without the use of new technologies in environmental studies, an accurate, fast, and economical estimation of these changes is impossible. In this regard, remote sensing plays an important role. In this research, satellite images were used and a map of land-use changes was extracted.

This study shows that remote sensing data have a high ability to extract a variety of land-use maps and evaluate land-use changes. In recent years, the inappropriate use of the existing lands, severe land-use changes, and the disappearance of forests, which in itself has become a problem in coordinated and sustainable development, have all affected the quality of life in the Gilan Province. The following measures have been proposed to prevent land-use changes and environmental degradation, especially in recent years.

- Strict enforcement of land-use protection laws.
- Satellite imagery with a higher spatial resolution is used to increase the accuracy of land-use extraction and thus to better monitor changes in the area.
- Comparing and adapting different existing uses with desirable uses and correcting the status of the undesirable uses.
- Conducting land management studies and necessary assessments to classify land resources and determine the country's land capabilities to determine the optimal land use.
- Prepare long-, medium-, and short-term plans to protect degraded resources.

References

1. Kogo BK, Kumar L, Koech R. Analysis of Spatio-temporal dynamics of land use and cover changes in Western Kenya. *Geocarto Int.* 2019;1–16.
2. Alemu B. The effect of land use land cover change on land degradation in the highlands of Ethiopia. *Environ Earth Sci.* 2015;5:1–12.
3. Kibet LP. Monitoring of land use/land-cover dynamics using remote sensing: a case of Tana River Basin, Kenya. *Geocarto Int.* 2019;1–20 [just-accepted].
4. Kidane D, Alemu B. The effect of upstream land-use practices on soil erosion and sedimentation in the upper Blue Nile Basin, Ethiopia. *Res J Agric Environ Manag.* 2015;4(2):55–68.
5. FAO. *Global Forest Land-Use Change 1990-2005.* Food and Agriculture or organization of the United Nations; 2012.
6. FAO. *State of the Worlds Forests. Forests and Agriculture: Land-Use Challenges and Opportunities;* 2016.
7. Tolessa T, Senbeta F, Kidane M. The impact of land use/land cover change on ecosystem services in the central highlands of Ethiopia. *Ecosyst Serv.* 2017;23:47–54.
8. Sharifi A. Using convolutional sparse representation and discrete wavelet decomposition. *J Electr Comput Eng Innov.* 2019;7(2):205–212.
9. Sharifi A. Flood mapping using relevance vector machine and SAR data: a case study from Aqqala, Iran. *J Indian Soc Remote Sens.* 2020;48(9):1289–1296.
10. Sharifi A. Evaluation of SAR sensor design parameters on remote sensing missions. *Aircr Eng Aerosp Technol.* 2020.
11. Sharifi A. Development of a method for flood detection based on sentinel-1 images and classifier algorithms. *Water Environ J.* 2020;35(3):924–929.
12. Singh D, Tsiang M, Rajaratnam B, Diffenbaugh NS. Observed changes in extreme wet and dry spell during the south Asian summer monsoon season. *Water Environ J.* 2014;4:456–461.
13. Alkaradaghi K, Ali S, Al-Ansari N, Laue J. Evaluation of land use & land cover change using multi-temporal landsat imagery: a case study Sulaimaniyah governorate, Iraq. *Geogr Inf Syst.* 2018;10(6):247–260.
14. Islam K, Jashimuddin M, Nath B, Nath TK. Land use classification and change detection by using multi-temporal remotely sensed imagery: the case of Chunati wildlife sanctuary, Bangladesh. *Egypt J Remote Sens Space Sci.* 2018;21(1):37–47.
15. Nalina P, Meenambal T, Sathyanarayan Sridhar R. Land-use/land cover dynamics of Nilgiris district, India inferred from satellite images. *Am J Appl Sci.* 2014;11(3):455–461.
16. Longley PA. Geographical information systems: will developments in urban remote sensing and GIS lead to better urban geography? *Prog Hum Geogr.* 2002;26(2):231–239.
17. Tso B, Mather PM. *Classification Methods for Remotely Sensed Data.* 2nd ed. America: Taylor and Francis Group; 2009. Chapter 2–3.
18. Challinor AJ, Watson J, Lobell DB, Howden SM, Smith DR, Chhetri N. A meta-analysis of crop yield under climate change and adaptation. *J Nat Clim Chang.* 2014;4:287–291.
19. Fuchs R, Herold M, Verburg PH, Clevers JGPW. A high-resolution and harmonized model approach for reconstructing and analyzing historic land changes in Europe. *J Biogeosci.* 2013;10:1543–1559.

20. Karami KD. *Model of Land Use Optimization in Khorramabad Region Using Remote Sensing and Geographic Information System* [Master Thesis in Remote Sensing]. University of Sari; 2000.
21. Borri D, Caprioli M, Tarantino E. *Spayial Informattion Extraction from VHR Satellite Data to Detect Land Cover Transformations.* Italy: Polytechnic University of Bari; 2005.
22. Zhou W, Austin T, Morgan GR. *Measuring Urban arcel Lawn Greenness by Using an Object oriented Classification Approach.* Rubenstein School of Environment and Natural Resources, University of Vermont, George D. Aiken Center, 81 Carrigan Drive; 2005.
23. Davoudi Z, Hajinezhad A, Abbas Nia M, Pourhashemi S. Monitoring land use change using a remote sensing technique (case study: Shahriar City). In: 2014:245–261. The Application of Remote Sensing and GIS in Natural Resources Science; vol. 5.
24. Khani F, Mousavi S, Arooji H, Alizadeh M. Evaluation of the effects of the free construction of the North Tehran road in the district of ken-Sulaghan districts. *Q J Hum Geogr.* 2016;92:421–434.

Further reading

Loucks DP, Van Beek E. *Water Resource Systems Planning and Management: An Introduction to Methods, Models, and Applications.* vol. 4. Springer; 2017:63–78.

Index

Note: Page numbers followed by *f* indicate figures and *t* indicate tables.

Printed in the United States
by Baker & Taylor Publisher Services